The Growth of Biological Thought

The Growth of Biological Thought

Diversity, Evolution, and Inheritance

ERNST MAYR

The Belknap Press of
Harvard University Press
Cambridge, Massachusetts   London, England

10 9 8 7 6

Library of Congress Cataloging in Publication Data
Mayr, Ernst, 1904–
    The growth of biological thought.

    Bibliography: p.
    Includes index.
    1. Biology—History.   2. Biology—Philosophy—History.
I. Title.
QH305.M26   574′.09   81-13204
ISBN 0-674-36445-7 (cloth)   AACR2
ISBN 0-674-36446-5 (paper)

For Gretel

Preface

MUCH OF MODERN BIOLOGY, particularly the various controversies between different schools of thought, cannot be fully understood without a knowledge of the historical background of the problems. Whenever I made this point to my students, they would ask me in what book they could read up on these matters. To my embarrassment, I had to admit that none of the published volumes filled this need. To be sure, there is much literature on the lives of biologists and their discoveries, but these writings are invariably inadequate as far as an analysis of the major problems of biology are concerned or as a history of concepts and ideas in biology. While some of the histories of individual biological disciplines, like genetics and physiology, are indeed histories of ideas, there is nothing available that covers biology as a whole. To fill this gap in the literature is the object of this work. This volume is not, and this must be stressed, a history of biology, and it is not intended to displace existing histories of biology, such as that of Nordenskiöld. The emphasis is on the background and the development of the ideas dominating modern biology; in other words, it is a developmental, not a purely descriptive, history. Such a treatment justifies, indeed necessitates, the neglect of certain temporary developments in biology that left no impact on the subsequent history of ideas.

When I first conceived the plan to write a history of ideas in biology, the goal seemed impossibly remote. The first years (1970–1975) were devoted to reading, notetaking, and the preparation of a first draft. Soon it became obvious that the subject was too vast for a single volume, and I decided to prepare first a volume on the biology of "ultimate" (evolutionary) causations. But even this limited objective is a hopelessly vast undertaking. If I have been successful at all, it is because I have myself done a considerable amount of research in most areas covered by this volume. This means that I was already reasonably familiar with the problems and some of the literature of the areas involved. I hope to deal with the biology of "proximate" (functional) causations in a later volume that will cover physiology in all of its aspects, developmental biology, and neurobiology. When a biological discipline, for instance genetics, deals both with ultimate and proximate causations, only the ultimate causations are treated in the present volume. There are two areas of biology that might have been (at least in part) but were not included in this volume: the conceptual history of ecology and that of behavioral biology (particularly ethology). Fortunately, this omission will not be quite as painful as it might otherwise be, because several volumes by other authors dealing with the history of ecology and ethology are now in active preparation.

viii

The professional historian is not likely to learn much from chapters 1 and 3; in fact he may consider them somewhat amateurish. I have added these two chapters for the benefit of non-historians, believing that it will help them to see the purely scientific developments of the other chapters with a deepened perception.

I owe an immense debt of gratitude to numerous individuals and institutions. Peter Ashlock, F. J. Ayala, John Beatty, Walter Bock, Robert Brandon, Arthur Cain, Fred Churchill, Bill Coleman, Lindley Darden, Max Delbrück, Michael Ghiselin, John Greene, Carl Gustav Hempel, Sandra Herbert, Jon Hodge, David Hull, David Layzer, E. B. Lewis, Robert Merton, J. A. Moore, Ron Munson, Edward Reed, Phillip Sloan, Frank Sulloway, Mary Williams, and others have read drafts of various chapters, have pointed out errors and omissions, and have made numerous constructive suggestions. I did not always follow their advice and am thus solely responsible for remaining errors and deficiencies. To P. Ax, Muriel Blaisdell, and B. Werner I am indebted for useful factual information.

Gillian Brown, Cheryl Burgdorf, Sally Loth, Agnes I. Martin, Maureen Sepkoski, and Charlotte Ward have typed innumerable drafts and helped with the bibliography. Walter Borawski not only typed preliminary versions but also the entire final copy of the manuscript and of the bibliography and prepared the manuscript of the index. Randy Bird contributed to filling gaps in the references. Susan Wallace edited the entire manuscript and in the process eliminated numerous inconsistencies, redundancies, and stylistic infelicities. All of these people materially contributed to the quality of the final product. It is obvious how great a debt of gratitude I owe to them.

The Museum of Comparative Zoology, through the courtesy of its Director, Professor A. W. Crompton, has provided office space, secretarial help, and library facilities, even after my retirement. Research periods at the Institute for Advanced Study (Princeton, spring 1970), at the library of the Max Planck Institute of Biology (Tübingen, 1970), a senior fellowship of the Alexander von Humboldt Foundation (Würzburg, 1977), a fellowship awarded by the Rockefeller Foundation (Villa Serbelloni, Bellagio, 1977), and a grant (No. GS 32176) by the National Science Foundation have greatly facilitated my work.

Whenever secretarial help was not available, my wife took over, transcribed dictations, excerpted literature, and aided the work on the manuscript in countless ways. It is impossible to acknowledge appropriately her inestimable contributions to this volume.

Ernst Mayr
Museum of Comparative Zoology
Harvard University

ix

Contents

Part II   Evolution

The Growth of Biological Thought

1 ✍ Introduction: How to write history of biology

ANYTHING THAT changes in time has, by definition, a history—the universe, countries, dynasties, art and philosophy, and ideas. Science also, ever since its emergence from myths and early philosophies, has experienced a steady historical change and is thus a legitimate subject for the historian. Because the essence of science is the continuing process of problem solving in the quest for an understanding of the world in which we live, a history of science is first a history of the problems of science and their solution or attempted solutions. But it is also a history of the development of the principles that form the conceptual framework of science. Because the great controversies of the past often reach into modern science, many current arguments cannot be fully understood unless one understands their history.

Written histories, like science itself, are constantly in need of revision. Erroneous interpretations of an earlier author eventually become myths, accepted without question and carried forward from generation to generation. A particular endeavor of mine has been to expose and eliminate as many of these myths as possible—without, I hope, creating too many new ones. The main reason, however, why histories are in constant need of revision is that at any given time they merely reflect the present state of understanding; they depend on how the author interpreted the current zeitgeist of biology and on his own conceptual framework and background. Thus, by necessity the writing of history is subjective and ephemeral.[1]

When we compare published histories of science, it becomes at once apparent that different historians have quite different concepts of science and also of history writing. Ultimately all of them attempt to portray the increase in scientific knowledge and the changes in interpretive concepts. But not all historians of science have attempted to answer the six principal questions that must be addressed by anyone who wants to describe the progress of science critically and comprehensively: Who? When? Where?

What? How? and Why? On the basis of the author's selection from among these questions, most of the histories known to me can be classified as follows (cf. Passmore, 1965: 857–861), though it must be recognized that nearly all histories are a combination of the various approaches or strategies:

Lexicographic Histories

These are more or less descriptive histories with a strong emphasis on the questions What? When? and Where? What were the principal scientific activities at any given past period? What were the centers of science where the leading scientists were working, and how did they shift in the course of time? No one will argue about the value of such histories. A correct presentation of the true facts is indispensable because much of the traditional history of science (and its standard texts) is encrusted with myths and spurious anecdotes. Yet, a purely descriptive history provides only part of the story.

Chronological Histories

A consideration of time sequences is crucial to any kind of history writing. Indeed, one can even make chronology the primary organizing criterion, and some authors have done so. They have asked, for instance, what happened in biology between 1749 and 1789, or between 1789 and 1830? Chronological histories present a sequence of cross sections through the entirety of developments in all branches of biology. This is not only a legitimate but indeed a most revealing approach. It creates a feeling for the zeitgeist and the totality of contemporary influences. It permits one to investigate how developments in other branches of science have influenced biology, and how even within biology advances made by the experimentalists have affected the thinking of the naturalists, and vice versa. The understanding of many problems in the development of biology is greatly facilitated by this chronological approach. However, it suffers from the drawback that each major scientific problem is atomized.

Biographical Histories

The endeavor in these volumes is to portray the progress of science through the lives of leading scientists. This approach is also legitimate, since science is made by people and the impact of individual scientists like Newton, Darwin, and Mendel has often been of quasi-revolutionary nature. However, this approach shares with the purely chronological approach one very serious weakness: it atomizes each major scientific problem. The species problem, for example, will have to be discussed under Plato, Aristotle, Cesalpino and the herbalists, Buffon, Linnaeus, Cuvier, Darwin, Weismann, Nägeli, de Vries, Jordan, Morgan, Huxley, Mayr, Simpson, and so on. All of these discussions of the same problem are separated from each other by many pages, if not chapters.

Cultural and Sociological Histories 3

This approach stresses the point that science is a form of human endeavor and is therefore inseparable from the intellectual and institutional milieu of the period. This view is particularly fascinating to those who come to the history of science from the field of general history. They might ask such questions as why was British science from 1700 to 1850 so strongly experimental and mechanical while contemporary French science tended to be mathematical and rationalistic? Why did natural theology dominate science for 75 years longer in Britain than on the continent? To what extent was Darwin's theory of natural selection a child of the industrial revolution?

Even if the historian of biology chooses not to adopt this approach, he must carefully study the cultural and intellectual milieu of a scientist if he wants to determine the causes for the rise of new concepts. This is of evident importance in the present work, since one of the major objectives of my treatment is to investigate the reasons for the changes in biological theories. What enabled one investigator to make a discovery that has escaped his contemporaries? Why did he reject the traditional interpretations and advance a new one? From where did he get the inspiration for his new approach? These are the kind of questions that need to be asked.

Most early histories of science, particularly those of special scientific disciplines, were written by working scientists, who took it for granted that the intellectual impetus for scientific change came from within the field itself ("internal" influences). Later on, when the history of science became more professionalized and historians and sociologists began to analyze the progress of scientific thought, they tended to stress the influence of the general intellectual, cultural, and social milieu of the period ("external" influences). No one would want to doubt that both kinds of influences exist, but there is a great deal of disagreement on their relative importance, particularly with reference to specific developments, such as Darwin's theory of natural selection.

Often it is rather difficult even to distinguish external from internal factors. The Great Chain of Being (*scala naturae*) was a philosophical concept which clearly had an impact on concept formation in the case of Lamarck and other early evolutionists. Yet, Aristotle had developed this concept on the basis of empirical observations of organisms. On the other hand, universally adopted ideologies are among the most uncontroversial of external factors. The Christian dogma of creationism and the argument from design coming from natural theology dominated biological thinking for centuries. Essentialism (from Plato) is another all-powerful ideology. Interestingly, its displacement by Darwin was largely due to the observations of animal breeders and taxonomists—that is, to internal factors.

External factors do not necessarily originate in religion, philosophy, cultural life or politics, but—as far as biology is concerned—they may originate in a different science. The extreme physicalism (including determinism and extreme reductionism) that was prevalent in Western thinking after the scientific revolution strongly influenced theory formation in biology for several centuries, often quite adversely as is now evident. Scholastic logic, to cite another example, dominated taxonomic method from Cesalpino to Linnaeus. These examples, to which many others could be added, document without doubt the importance of external influences on theory formation in biology. They will be analyzed in full detail in the relevant chapters.

It is important to realize that external factors influence science in two entirely different ways: They may either affect the overall level of scientific activity at a given place at a given time, or they may affect or even give rise to a particular scientific theory. All too often in the past these two aspects have been lumped together, resulting in much controversy over the relative importance of external versus internal factors.

The effect of environmental conditions on the *level* of scientific activities has been appreciated as long as there has been a history of science. It has been speculated endlessly as to why the Greeks had such an interest in scientific questions and why there was a revival of science during the Renaissance. What was the effect of Protestantism on science (Merton, 1938)? Why did science during the nineteenth century flourish to such an extent in Germany? Sometimes important external factors can be specified, for instance (as Merz, 1896–1914, has pointed out), the replacement in 1694 of Latin by German at Halle University, and the founding in 1737 of a University at Göttingen in which "Wissenschaft" played an important role. Institutional changes of all sorts, including the founding of the Royal Society, political events such as wars and the launching of Sputnik, as well as technological needs have had either a stimulating or a depressing effect on the level of scientific activity. Yet, this still leaves open the highly controversial question of to what extent such external factors have favored or inhibited *specific* scientific theories.

5

In recent years Marxist historiographers in particular have voiced the thesis that social ideologies influence the ideas of a scientist, and that the history of science as practiced until now has totally neglected the social context. The result, they believe, has been a bourgeois history of science, which is quite different from what a proletarian history of science would be. What is needed instead, they say, is "radical" history. This demand ultimately goes back to Marx's claim that ruling ideas cannot be separated from ruling classes. Therefore, bourgeois history of science will be quite different from proletarian history of science.

However, the thesis that there is a proletarian way to write the history of science is in conflict with three sets of facts: First, the masses do not establish scientific theories that are different from those of the scientific class. If there is any difference, it is that the "common man" often retains ideas long after they have been discarded by scientists. Second, there is high social mobility among scientists, with from one quarter to one third of each new crop of scientists coming from the lower socioeconomic classes. Third, birth order within a social class tends to be far more important in determining those who originate rebellious new ideas than does membership in a particular class (Sulloway, MS). All of this is in conflict with the thesis that the socioeconomic environment has a dominant impact on the birth of particular new scientific ideas and concepts. The burden of proof is clearly on those who make such claims, and so far they have failed to supply any concrete evidence whatsoever (see Chapter 11).

Of course no one lives in a vacuum, and anyone who reads voraciously, as for example Darwin did after his return from the voyage of the *Beagle,* is bound to be influenced by his reading (Schweber, 1977). Darwin's notebooks are ample evidence for the correctness of this inference. But, as Hodge (1974) points out, this by itself does not prove the thesis of the Marxists that "Darwin and Wallace were extending the laissez-faire capitalist ethos from society to all nature." Up to now it appears that the influence of social factors on the development of specific biological advances has been negligible. The reverse, of course, is not true. But the study of the impact of science on social theory, social institutions, and politics belongs to the domains of history, sociology, and political science, and not to that of the history of science. I agree with Alexander Koyré (1965: 856) that it is futile to "deduce the existence" of certain scientists and sciences from their environment. "Athens does not explain Plato anymore than Syracuse explains Archimedes or Florence Galileo. To look for explanations along these lines is an entirely futile enterprise, as futile as trying to predict the future evolution of science or of the sciences as a function of the structure of the social context." Thomas Kuhn (1971: 280) has likewise observed that the historian seems invariably to give "excessive emphasis to the role of the surrounding climate of extra-scientific ideas" (see also Passmore, 1965).

Problematic Histories

More than one hundred years ago Lord Acton advised historians, "Study problems, not periods." This advice is particularly appropriate for the history of biology, which is characterized by the longevity of its scientific problems. Most of the great controversies of the nineteenth and early twentieth centuries relate to problems already known to Aristotle. Such controversies endure from generation to generation, and from century to century. They are processes, not events, and can be fully understood only through a historical treatment. As R. G. Collingwood said of history (1939: 98), it "is concerned not with events but with processes. Processes are things which do not begin and end but turn into one another." This must be stressed particularly in the face of the static views of the logical positivists who thought that logical structure was the real problem of science: "The philosophy of science is conceived [by them] primarily as a careful and detailed analysis of the logical structure and the conceptual problems of contemporary science" (Laudan, 1968). Actually most scientific problems are far better understood by studying their history than their logic. However, it must be remembered that problematic history does not replace chronological history. The two approaches are complementary.

In the problematic approach the chief emphasis is placed on the history of attempts to solve problems—for instance, the nature of fertilization or the direction-giving factor in evolution. The history not only of the successful but also of the unsuccessful attempts to solve these problems is presented. In the treatment of the major controversies in the field, an endeavor is made to analyze the ideologies (or dogmas) as well as the particular evidence by which the adversaries supported their opposing theories. In problematic history the emphasis is on the working scientist and his conceptual world. What were the scientific problems of his time? What were the conceptual and technical tools available to him in his quest for a solution? What were the methods he could employ? What prevalent ideas of his period directed his research and influenced his decisions? Questions of this nature dominate the approach in problematic history.

7

I have chosen this approach for the present book. The reader should be aware of the fact that this is not a traditional history of science. Owing to its concentration on the history of scientific problems and concepts, it slights by necessity the biographical and sociological aspects of the history of biology. It should therefore be used in conjunction with a general history of biology (like Nordenskiöld, 1926), with the *Dictionary of Scientific Biography*, and with available histories of special areas of biology. Since I am a biologist, I am better qualified to write a history of the problems and concepts of biology than a biographical or sociological history.

It is the essence of problematic history to ask why. Why was it in England that the theory of natural selection was developed, in fact independently four times? Why did genuine population genetics arise in Russia? Why were Bateson's explanatory attempts in genetics almost uniformly wrong? Why did Correns get distracted into all sorts of peripheral problems and therefore contributed so little to major advances in genetics after 1900? Why did the Morgan school devote their efforts for so many years to reinforcing the already well-established chromosome theory of inheritance, instead of opening up new frontiers? Why were de Vries and Johannsen so much less successful in the evolutionary application of their findings than in their straight genetic work? Attempts to answer such questions require the collecting and scrutiny of much evidence, and this almost invariably leads to new insights even if the respective question turns out to have been invalid. Answers to why-questions are inevitably somewhat speculative and subjective, but they force one into the ordering of observations and into the constant testing of one's conclusions consistent with the hypothetico-deductive method. Now that the legitimacy of why-questions has been established even for scientific research, particularly in evolutionary biology, there should be even less question about the legitimacy of such questions in the writing of history. At the worst, the detailed analysis necessitated by such a question may establish that the assumptions underlying the question were wrong. Even this would advance our understanding.

8 Throughout this volume I have endeavored to carry the analysis of each problem as far as possible and to dissect heterogeneous theories and concepts into their individual components. Not all historians have been aware how complex many biological concepts are—in fact, how complex the structure of biology as a whole is. As a consequence, some exceedingly confused accounts of the history of biology have been published by authors who did not understand that there are two biologies, that of functional and that of evolutionary causations. Similarly, anyone who writes about "Darwin's theory of evolution" in the singular, without segregating the theories of gradual evolution, common descent, speciation, and the mechanism of natural selection, will be quite unable to discuss the subject competently. Most major theories of biology were, when first proposed, such composites. Their history and their impact cannot be understood unless the various components are separated and studied independently. They often belong to very different conceptual lineages.

It is my conviction that one cannot understand the growth of biological thought unless one understands the thought-structure of biology. For this reason I have attempted to present the insights and concepts of biology in considerable detail. This was particularly necessary in the treatment of diversity (Part I) because no other adequate treatment or conceptual framework of the science of diversity is available. I am aware of the danger that some critic might exclaim, "But this is a textbook of biology, historically arranged!" Perhaps this is what a problematical history of biology ought to be. Perhaps the greatest difficulty any conceptual history of biology must cope with is the longevity of the controversies. Many of the current controversies had their origin generations or even centuries ago, some indeed going all the way back to the Greeks. A more or less "timeless" presentation of the issues is more constructive in such cases than a chronological one.

I have tried to make each of the major sections of this volume (Diversity, Evolution, Inheritance) a self-contained unit. A similar separation is attempted for each separate problem within these three major areas. This leads to a certain amount of overlap and redundancy because there are numerous cross-connections between different topics and each topical strand will pass through the same sequence of time-dependent intellectual milieus. I have made a special effort to strike a balance between a certain amount of unavoidable duplication and convenient cross references to other chapters.

Subjectivity and Bias

A well-known Soviet theoretician of Marxism has once referred to my writings as "pure dialectical materialism." I am not a Marxist and I do not know the latest definition of dialectical materialism, but I do admit that I share some of Engels' antireductionist views, as stated in his *Anti-Dühring,* and that I am greatly attracted by Hegel's scheme of thesis-antithesis-synthesis. Furthermore, I believe that an antithesis is most easily provoked by a categorical statement of a thesis, and that the issue is most readily resolved by such a confrontation of an uncompromising thesis and antithesis and that the ultimate synthesis is thus most quickly achieved. Many examples for this can be found in the history of biology.

This view has dominated my presentation. Whenever possible, I have attempted a synthesis of opposing viewpoints (unless one of them is clearly in error). Where the situation is quite unresolved, I have described the opposing viewpoints in categorical, sometimes almost one-sided, terms in order to provoke a rejoinder, if such is justified. Because I hate beating around the bush, I have sometimes been called dogmatic. I think this is the wrong epithet for my attitude. A dogmatic person insists on being right, regardless of opposing evidence. This has never been my attitude and, indeed, I pride myself on having changed my mind on frequent occasions. However, it is true that my tactic is to make sweeping categorical statements. Whether or not this is a fault, in the free world of the interchange of scientific ideas, is debatable. My own feeling is that it leads more quickly to the ultimate solution of scientific problems than a cautious sitting on the fence. Indeed, I agree with Passmore (1965) that histories should even be polemical. Such histories will arouse contradiction and they will challenge the reader to come up with a refutation. By a dialectical process this will speed up a synthesis of perspective. The unambiguous adoption of a definite viewpoint should not be confused with subjectivity.

The traditional admonition to historians has always been to be strictly objective. This ideal was well expressed by the great historian Leopold von Ranke when he said the historian should "show how it really was." History was envisioned by him as the accurate reconstruction of a series of past events. Such objectivity is entirely appropriate when one attempts to answer who, what, when, and where, although it must be pointed out that even in presenting facts the historian is subjective because he uses value judgments when sorting the facts and is selective when deciding which ones to accept and how to relate them to one another.

Subjectivity enters at every stage of history writing, especially when one is seeking explanations and asks why, as is necessary in problematic history. One cannot arrive at explanations without using one's own personal judgment, and this is inevitably subjective. A subjective treatment is usually far more stimulating than a coldly objective one because it has a greater heuristic value.

To what extent is subjectivity permissible and where does it become bias? Radl (1907–08), for instance, had such a strong anti-Darwinian bias that he was not even able to present the Darwinian theory adequately. This clearly went too far. Subjectivity is apt to become bias whenever the evaluation of scientists of former periods is involved. Here historians tend to go either to one or to the other extreme. Either they adopt a strictly retrospective approach in which the past is evaluated entirely in the light of present knowledge and understanding, or else they suppress hindsight completely and describe past events strictly in terms of the thinking at that period. It seems to me that neither approach is entirely satisfactory.

A better procedure would be to combine the best aspects of both approaches. This would first attempt to reconstruct the intellectual milieu of the period as faithfully as possible. But it would not be satisfactory to treat past controversies strictly in terms of the information available at the time. This would leave such controversies as unresolved and opaque as they were when they took place. Instead, modern knowledge should be used whenever this helps in the understanding of past difficulties. Only such an approach will enable us to determine the reasons for the controversy and for the failure to resolve it. Was it a semantic difficulty (for example, the use of the same word in different meanings), or a conceptual disagreement (such as essentialist vs. population thinking), or an outright error (like the confusion of ultimate and proximate causes)? A study of past controversies is particularly illuminating when the arguments and opposing viewpoints are analyzed in terms of our present knowledge.

Semantic problems are particularly bothersome because they are so often undiscovered. The Greeks, for instance, had a very limited technical vocabulary and often used the same term for rather different things or concepts. Both Plato and Aristotle used the term *eidos* (and Aristotle at least, used it in several senses!), but the major meaning of the term is totally different in the two authors. Plato was an essentialist, but Aristotle was essentialist only to a very limited extent (Balme, 1980). Aristotle used the term *genos* occasionally as a collective noun (corresponding to the taxonomists' genus) but far more often in the sense of species. When Aristotle was rediscovered in the late Middle Ages and translated into Latin and western European languages, his terms were translated in "equivalent" terms available in medieval dictionaries. These misleading translations have had an unfortunate influence on our understanding of Aristotle's thought. Some modern authors have had the courage to use modern terms to reveal Aristotle's thought, terms he would have quite likely used if he were living today. I am thinking of Delbrück's use of "genetic program" to make clear Aristotle's intention when using *eidos* in the description of individual development. Likewise, one should use "teleonomy" (instead of "teleology") when Aristotle discusses goal-directedness controlled by an *eidos* (program). This is not anachronistic but simply a way of making clear what an ancient author thought, by using a terminology that is unambiguous for a modern reader.

It would, however, be quite inappropriate to use modern hindsight for value judgments. Lamarck, for instance, was not nearly as wrong as he seems to those familiar with selectionism and Mendelian genetics, when judged in terms of the facts known to him and of the ideas prevailing at his period. The phrase "Whig interpretation of history" was introduced by the historian Herbert Butterfield (1931) to characterize the habit of some English constitutional historians to see their subject as a progressive broadening of human rights, in which good "forwardlooking" liberals were continuously struggling with the "backwardlooking" conservatives. Butterfield later (1957) applied the term *whiggish* to that kind of history of science in which every scientist is judged by the extent of his contribution toward the establishment of our current interpretation of science. Instead of evaluating a scientist in terms

11

of the intellectual milieu in which he was active, he is evaluated strictly in terms of current concepts. The complete context of problems and concepts in which the earlier scientists had worked is ignored in this approach. The history of biology is rich in such biased whig interpretations.

Whenever there is a scientific controversy, the views of the losing side are almost invariably later misrepresented by the victors. Examples are the treatment of Buffon by the Linnaeans, of Lamarck by the Cuvierians, of Linnaeus by the Darwinians, of the biometricians by the Mendelians, and so forth. The historian of biology must attempt to present a better balanced account. Many, now rejected theories, like the inheritance of acquired characters espoused by Lamarck, seemed formerly so consistent with the known facts that authors should not be criticized for having adopted such prevailing theories even if they have since been shown to be wrong. Almost always those who held an erroneous theory had seemingly valid reasons for doing so. They were trying to emphasize something that was neglected by their opponents. The preformationists, for instance, attempted to stress something which was later resurrected as the genetic program. The biometricians upheld Darwin's views of gradual evolution against the saltationism of the Mendelians. In both instances correct ideas were lumped together with erroneous ones and went down together with the errors. In my case I tend to pay special attention to underdogs (both persons and theories) because in the past they have often been treated unfairly or at least inadequately.

The path of science is never straight. There are always competing theories and most of the attention of a period may be directed toward a side issue which eventually turns out to be a dead end. These developments often illuminate the zeitgeist of a period more successfully than the straightforward advances of science. Regrettably, lack of space precludes an adequate treatment of many of these developments. No history can afford to deal with every lost cause and every deviation. There are, however, exceptions. Some of the failures or errors of the past very suitably reveal aspects of contemporary thinking which we might otherwise miss. Macleay's and Swainson's quinarianism, for instance, which was totally eclipsed by the *Origin of Species,* represented a sincere endeavor to reconcile the seeming chaotic diversity of nature with the then prevailing conviction that there had to be some "higher" order in nature. It also reveals the still powerful hold of the old myth that all order in the world is ultimately numerical. As ill-conceived and ephemeral as the theory of quinarianism was, it nevertheless contributes to our understanding of the thinking of its period. The same can be said of almost any theory or school of the past that is no longer considered valid. The interests of a historian necessarily influence his decision as to which subjects to treat in detail and which others in a more cursory fashion. I tend to agree with Schuster, who said in *The Progress of Physics* (1911), "I prefer to be frankly subjective, and warn you beforehand that my account will be fragmentary, and to a great extent reminiscent of those aspects which have come under my own personal view."   *13*

Historians versus Scientists

Two groups of scholars with entirely different viewpoints and backgrounds—historians and scientists—have claimed the history of science as their own. Their respective contributions are somewhat different, dictated by differences in their interests and competence. A scientist tends to select for analysis and discussion rather different problems from a historian or sociologist. For instance, in recent accounts of evolution by various evolutionists, H. Spencer has hardly received any attention. There are good reasons for this neglect. Not only was Spencer vague and confused, but the ideas he championed were those of others and already obsolete when taken up by Spencer. That Spencer's borrowed ideas were quite popular and influential, as far as the general public was concerned, is without question true, but it is not the task of the scientist-historian to trespass in the domain of the sociologist. Biologists usually lack the competence to deal with social history. On the other hand, it would be just as ridiculous to demand that a social historian present a competent analysis of the scientific issues. The history of science derives inspiration, information, and methodological assistance both from science and from history and, in turn, contributes by its findings to both fields.

There are valid reasons for the interest of both historians and scientists in the history of science. The Greeks had no science, as we now define it, and whatever science they had was practiced by philosophers and physicians. After the Middle Ages there was a continuing trend of emancipation of science from philosophy and from the general zeitgeist. In the Renaissance period and during the eighteenth century scientific beliefs were strongly influenced by the scientists' attitude toward religion and philosophy. A Cartesian, an orthodox Christian, or a Deist would inevitably have different ideas on cosmology, generation, and all aspects of the interpretation of life, matter, and origins. Nothing signaled the emancipation of science from religion and philosophy more definitely than the Darwinian revolution. Since that time it has become quite impossible to say by looking at an author's scientific publications whether he was a devout Christian or an atheist. Except for a few fundamentalists, this is true even for the writings of biologists on the subject of evolution.

14 This trend toward the emancipation of science had a considerable effect on the writing of the history of science. The farther back we go in time, the less important becomes the store of scientific knowledge of the period and the more important the general intellectual atmosphere. As far as biology is concerned, it is not until after about 1740 that the scientific problems begin to separate themselves from the general intellectual controversies of the period. There is no question that historians are particularly well qualified to deal with the earlier time span in the history of biology. However, the nineteenth and twentieth centuries' history of special biological disciplines was entirely dominated by scientists, until its rather recent professionalisation. This is well illustrated by such recent histories of special areas in biology, as those of Dunn, Stubbe, and Sturtevant in genetics, of Fruton, Edsall, and Leicester in biochemistry, of Needham and Oppenheimer in embryology, of Baker and Hughes in cytology, of Stresemann in ornithology, to mention only a few names in the extensive literature. They demonstrate the qualification of scientists for historical research.

The Bias of the Physical Scientists

Most general histories of "science" have been written by historians of physics who have never quite gotten over the parochial attitude that anything that is not applicable to physics is not science. Physical scientists tend to rate biologists on a scale of values depending on the extent to which each biologist has used "laws," measurements, experiments, and other aspects of scientific research that are rated highly in the physical sciences. As a result, the judgments on fields of biology made by certain historians of the physical sciences that one may find in that literature are so ludicrous that one can only smile. For example, knowing that Darwin developed his theory of evolution largely on the basis of his observations as a naturalist, one can only marvel at this statement, made by a well-known historian of Newton: "The naturalist is indeed a trained observer, but his observations differ from those of a gamekeeper only in degree, not in kind; his sole esoteric qualification is familiarity with systematic nomenclature." This kind of biased physicalist thinking is entirely out of place in the study of evolutionary biology, as we shall see in Chapter 2. Theory formation and its history in evolutionary and systematic biology require a radically different approach, an approach which is in some ways more similar to that adopted by a historian of archeology or by an interpreter of modern world history.

Other Biases

Not just the physicist but every specialist, quite naturally, considers his particular field of research to be the most interesting and its methods to be the most productive. As a result, often an invidious kind of chauvinism exists among fields, and even within a field such as biology. It is chauvinism, for instance, when Hartmann (1947) allotted 98 percent of his large *General Biology* to physiological biology and only 2 percent to evolutionary biology. It is chauvinism when certain historians ascribe the occurrence of the evolutionary synthesis entirely to the findings of genetics, completely ignoring the contribution made by systematics, paleontology, and other branches of evolutionary biology (Mayr and Provine 1980).

There is sometimes also a national chauvinism within a field which tends to exaggerate or even to misrepresent the importance of the scientists from the writer's own country and to belittle or ignore scientists of other nations. This is not necessarily due to misplaced patriotism but is often the result of an inability to read the languages in which important contributions by scientists of other countries have been published. In my own work I am keenly aware of the probability of bias introduced by my inability to read Slavic languages and Japanese.

Pitfalls and Difficulties

The greatest difficulty in the endeavor to identify the vast number of problems of biology and to reconstruct the development of its conceptual framework is the vast amount of material to be studied. This consists, in principle, of the entire store of knowledge of biology, including all books and periodical articles published by biologists, their letters and biographies, information on the institutions with which they were associated, contemporary social history, and much else. Not even the most conscientious historian would be able to cover even one tenth of one percent of all this material. The situation is aggravated by the exponential acceleration in the rate of current scientific output. In an amazingly short span of years more papers (and pages!) are now published than in the whole preceding history of science. Even specialists complain that they can no longer keep up with the avalanche of research output in their own field. Curiously, exactly the same is true for history writing. In the United States there are now perhaps five times as many historians of biology as there were only twenty-five years ago.

16.

Even though I have valiantly attempted to read the most important publications, I know that every specialist will discover numerous omissions in my treatment and presumably not infrequent errors. The first draft of most of the manuscript was written from 1970 to 1976 and the more recently published literature is not always as adequately incorporated as would have been desirable. My task would have been altogether impossible if it had not been for the richness and excellence of the modern secondary literature. The older literature was often rather superficial, and author after author would copy the same myths or errors, as one discovered when one consulted the original publications. Obviously, in a volume such as this one, which may contain more than 20,000 individual items of information, it is impossible to verify each item in the original source. Since my work is not a lexicographic history, an occasional factual error is not fatal. My major objective has been to synthesize an enormous literature with a consistent emphasis on interpretation and the analysis of causation.

Timeliness

A criticism often raised against historians of science, and not without justification, is that they are preoccupied too exclusively with the "prehistory" of science, that is, with periods the events of which are largely irrelevant for modern science. To avoid this complaint I have tried to bring the story as close to the present as is possible for a nonspecialist. In some cases, for instance the discovery in the last five to ten years in molecular biology of numerous families of DNA, the conceptual consequences (for instance on evolution) are still too uncertain to be dealt with.

I disagree with the statement of a recent historian that "the object of the history of science is investigation and disputes that have been concluded rather than issues that are presently alive." This is quite in error. Most scientific controversies extend over far longer periods of time than is generally conceived. Even today's controversies have a root that usually goes far back in time. It is precisely the historical study of such controversies that often contributes materially to a conceptual clarification and thus makes the ultimate solution possible. Analogous to the field of world history, where "current history" is recognized as a legitimate field, there is "current history" in the history of science. Nothing would be more misleading than to assume that the history of science deals only with dead issues. On the contrary, one might even go so far as to consider as prehistory the accounts of long-dead issues of earlier centuries and millennia.

17

Simplification

A historian who covers such a vast area as is dealt with in this volume is forced to present a highly streamlined account. The reader is warned that the seeming simplicity of many of the developments is quite deceiving. Detailed accounts that concentrate on special developments or short periods must be consulted if one is to appreciate the full flavor of the many cross currents, false starts, and unsuccessful hypotheses that prevailed at any given period. Developments virtually never were as straightforward and logical as they appear to be in a simplified retrospective account. It is particularly difficult to emphasize adequately the often quite paralyzing power of entrenched conceptualizations when confronted by new discoveries or new concepts.

Inaccuracy is also introduced by labeling certain authors as vitalists, preformationists, teleologists, saltationists, or neo-Darwinians, as if these labels would refer to homogeneous types. Actually, these categories consist of individuals no two of whom had exactly the same views. This is particularly true for the epithets "Lamarckians" and "neo-Lamarckians," some of whom had nothing in common with each other except a belief in an inheritance of acquired characters.

Silent Assumptions

A further difficulty for the historian is posed by most scientists' unawareness of their own framework of ideas. They rarely articulate—if they think about it at all—what truths or concepts they accept without question and what others they totally reject. In many cases the historian can piece this together only by reconstructing the total intellectual milieu of the period. And yet an understanding of these silent assumptions may be necessary in order to answer previously puzzling questions. In science one always deals with priorities and value systems; they determine the direction of new research after a previous piece of research has been completed; they determine which theories the investigator is most anxious either to confirm or to refute; they also determine whether or not he considers an area of research exhausted. And yet a study of the factors that determine such priorities or value systems has been greatly neglected until now. The historian must attempt to find out what went on in the mind of a worker when *18* he gave a new interpretation to a long familiar set of facts. It is perhaps legitimate to say that the truly crucial events in the history of science always take place in the mind of a scientist. One must, so to speak, attempt to think as the scientist thought when he performed the work one is trying to analyze.

Most scientists tend to concentrate in their publications on new facts or rather on new discoveries, and in particular on anything that is spectacular. At the same time they usually fail to record important ongoing changes of concepts or emphasis. They may even fail to recognize such changes or may consider them as negligible even when they are aware of them. When the modern historian attempts to reconstruct such changes in past centuries, he cannot help but project into history the interests and scale of values of the present. This danger of interpretation can be minimized only if the historian is fully aware of what he is doing.

Why Study the History of Biology?

My own interest in the history of science was aroused by reading A. O. Lovejoy's *The Great Chain of Being,* where the attempt is made—and it was eminently successful—to trace the life history, so to speak, of a single idea (or a cohesive complex of ideas) from the ancients to the end of the eighteenth century. I have learned more from this one volume than from almost anything else I have read. Others who have attempted a similar approach are Ernst Cassirer and Alexander Koyré. They have provided entirely new standards for scientific historiography.

In the case of the history of science, the focal points are problems rather than ideas, but the approach of the historian of science is not much different from that of a historian of ideas such as Lovejoy. Like Lovejoy, he attempts to trace the problem back to its beginning and to follow up its fate and its ramifications from such a beginning either to its solution or to the present time.

It is the principal objective of this volume to discover for each branch of biology and for each period what the open problems were and what proposals were made to solve them; the nature of the dominant concepts, their changes, and the causes for their modification and for the development of new concepts; and finally, what effect prevailing or newly arising concepts had in delaying or accelerating the solution of the open problems of the period. At its best this approach would portray the complete life history of each problem of biology.

Preoccupation with this sort of conceptual history of science *19* is sometimes belittled as a hobby of retired scientists. Such an attitude ignores the manifold contributions which this branch of scholarship makes. The history of science, as has been pointed out frequently, is particularly suitable as a first introduction to science. It helps to bridge the gap between "general beliefs" and the actual findings of science, since it shows in what manner and for what reasons science has advanced beyond the beliefs of folklore. To illustrate this for a single branch of biology, in the history of genetics it can be shown by what discoveries and arguments rather widely held erroneous beliefs were refuted, as for instance that there is an inheritance of acquired characters; that the genetic materials of the parents "blend"; that the "blood" of a female is tainted so that she can never again produce "pure" offspring once she has been inseminated, even if only a single time; that a single egg is simultaneously fertilized by the sperm of several males; or that accidents of a pregnant mother can lead to the production of heritable characters. Similar erroneous beliefs, derived from folklore, myths, religious documents, or from early philosophies, had originally been held in many fields of biology. The historical demonstration of the gradual replacement of these prescientific or early scientific beliefs by better based scientific theories and concepts greatly assists in explaining the current framework of biological theories.

The layperson often excuses his ignorance of science with the comment that he finds science too technical or too mathematical. Let me assure the prospective reader of this volume that he will hardly find any mathematics in its pages and that it is not technical to the extent that a layperson would have difficulty with the exposition. It is a major advantage of the history of ideas in biology that one can study it without a background knowledge of the name of a single species of animal or plant or of the major taxonomic groups and their classification. However, a student of the history of ideas must acquire some knowledge of the dominant concepts in biology, like inheritance, program, population, variation, emergence, or organismic. It is the objective of Chapter 2 to provide an introduction into the world of major biological concepts. Many of these concepts (and the terms that go with them) have now also been incorporated into various branches of the humanities, and it has simply become a matter of education to be acquainted with them. All of these concepts are indispensable for an understanding of man and the world in which he lives. Any endeavor to elucidate the origin and nature of man must be based on a thorough understanding of the concepts and theories of biology. Finally, it is helpful to become familiar with a very small repertory of technical terms like gamete, zygote, species, gene, chromosome, and so on, terms that are defined in the Glossary. Yet, the total vocabulary of such technical terms is far smaller than what a student in any field of the humanities has to learn, whether it be music, literature, or current history.

It is not only the layperson whose horizon will be greatly extended by the study of the history of ideas in biology. Advances in many areas of biology are so precipitous at the present time that specialists can no longer keep up with developments in areas of biology outside their own. The broad survey of biology and its dominant concepts that is presented in this volume will help in filling some of the gaps. My survey is also directed toward those who have entered biology in recent years from the outside, that is, from chemistry, physics, mathematics, or other adjacent fields. The technical sophistication of these "neo-biologists" is, unfortunately, rarely matched by an equivalent conceptual sophistication. Indeed, those who know organisms in nature and understand the ways of evolution are often appalled by the naiveté of some of the generalizations made in some papers in molecular biology. Admittedly, there is no quick and easy way to compensate for this deficiency. Like Conant, I feel that the study of the history of a field is the best way of acquiring an understanding of its concepts. Only by going over the hard way by which these concepts were worked out—by learning all the earlier wrong assumptions that had to be refuted one by one, in other words by learning all past mistakes—can one hope to acquire a really thorough and sound understanding. In science one learns not only by one's own mistakes but by the history of the mistakes of others.

2 ⁄ The place of biology in the sciences and its conceptual structure

IT IS QUITE impossible to try to understand the development
of any particular concept or problem in the history of biology
unless one has first answered for oneself these questions: What is
science? What is the place of biology among the sciences? And
what is the conceptual structure of biology? Entirely misleading
answers have been given to all three of these questions, particularly by philosophers and other nonbiologists, and this has greatly
impeded an understanding of the growth of biological thought.
To try to answer these basic questions correctly, then, is the first
task of my analysis. It will provide a secure basis for the study of
the history of specific concepts.

THE NATURE OF SCIENCE

From the earliest times on man has asked questions about the
origin and the meaning of the world and frequently about its purpose. His tentative answers can be found in the myths characteristic of every culture, even the most primitive ones. He has advanced beyond these simple beginnings in two rather different
directions. In one his ideas became formalized in religions, which
proclaimed a set of dogmas, usually based on revelation. The
Western world, for instance, at the end of the Middle Ages was
completely dominated by an implicit trust in the teachings of the
Bible, and beyond that, by a universal belief in the supernatural.

Philosophy, and later science, is the alternative way of dealing
with the mysteries of the world, although science was not strictly
separated from religion in its early history. Science confronts these
mysteries with questions, with doubts, with curiosity, and with explanatory endeavors, thus with a rather different attitude from
religion. The pre-Socratic (Ionian) philosophers initiated this different approach by searching for "natural" explanations, in terms
of observable forces of nature, like fire, water, and air (see Chap-

ter 3). This endeavor to understand the causation of natural phenomena was the beginning of science. For many centuries after the fall of Rome this tradition was virtually forgotten, but it was revived again in the late Middle Ages and during the scientific revolution. The belief grew that the divine truth was revealed to us not only in Scripture but also in God's creation.

Galileo's statement of this opinion is well known: "I think that in the discussion of natural problems we ought not to begin at the authority of places of Scripture, but at sensible experiments and necessary demonstrations. For, from the Divine Word the sacred Scripture and Nature did both alike proceed." He continued that "God equally admirably reveals Himself to us in Nature's actions as in the Scriptures' sacred dictions." He thought that a god who governs the world with the help of eternal laws inspires trust and faith at least as much as one who forever intervenes in the course of events. It was this thought which gave rise to the birth of science as we now understand it. Science for Galileo was not an alternative to religion but an inseparable part of it. Likewise, many great philosophers from the seventeenth to the nineteenth centuries—for instance, Kant—included God in their explanatory schemes. So-called natural theology was, in spite of its name, as much science as it was theology. The conflict between science and theology developed only later when science explained more and more processes and phenomena of nature by "natural laws" which previously had been considered inexplicable except by the intervention of the Creator or by special laws ordained by Him.[1]

A fundamental difference between religion and science, then, is that religion usually consists of a set of dogmas, often "revealed" dogmas, to which there is no alternative nor much leeway in interpretation. In science, by contrast, there is virtually a premium on alternative explanations and a readiness to replace one theory by another. The discovery of an alternate explanatory scheme is often the source of great elation. The goodness of a scientific idea is judged only to a minor degree by criteria extrinsic to science because it is on the whole judged entirely by its efficacy in explanation and, sometimes, prediction.

Scientists, curiously, have been rather inarticulate about stating what science is all about. In the heyday of empiricism and inductionism the objective of science was most often described as the accumulation of new knowledge. By contrast, when one reads the writings of philosophers of science, one gets the impression that for them science is a methodology. Even though no one will want to question the indispensability of method, the almost exclusive preoccupation of some philosophers of science with method has deflected attention from the more basic purpose of science, which is to increase our understanding of the world in which we live, and of ourselves.

Science has a number of objectives. Ayala (1968) describes them this way: (1) Science seeks to organize knowledge in a systematic way, endeavoring to discover patterns of relationship among phenomena and processes. (2) Science strives to provide *23* explanations for the occurrence of events. (3) Science proposes explanatory hypotheses that must be testable, that is, accessible to the possibility of rejection. More broadly, science attempts to subsume the vast diversity of the phenomena and processes of nature under a much smaller number of explanatory principles.

Discovery of New Facts or Development of New Concepts?

Discoveries are the symbol of science in the public mind. The discovery of a new fact is usually easily reportable, and thus the news media also see science in terms of new discoveries. When Alfred Nobel wrote out the conditions for Nobel prizes, he thought entirely in terms of new discoveries, particularly those useful to mankind. Yet to think of science merely as an accumulation of facts is very misleading. In biological science, and this is perhaps rather more true for evolutionary than for functional biology, most major progress was made by the introduction of new concepts, or the improvement of existing concepts. Our understanding of the world is achieved more effectively by conceptual improvements than by the discovery of new facts, even though the two are not mutually exclusive.

Let me illustrate this by one or two examples. Ratios of 3:1 had been discovered by plant breeders many times before Mendel. Even Darwin had obtained a number of such ratios in his plant-breeding work. Nevertheless, all this was meaningless until Mendel introduced the appropriate concepts and until Weismann introduced additional concepts that made Mendelian segregation more meaningful. Similarly, the phenomena that are now explained by natural selection were widely known long before Darwin but made no sense until the concept of populations consisting of unique individuals was introduced. Then natural selection acquired great explanatory power. The concepts of population thinking and geographic variation, together with that of isolation, were, in turn, the prerequisites for the development of the theory of geographic speciation. That the acquisition of reproductive isolation is a crucial component in the process of speciation was not fully realized until the concept of isolating mechanisms had been clarified. The true role of isolating mechanisms was not understood as long as geographical barriers were included among the isolating mechanisms, as was still done by Dobzhansky (1937).

24

One can take almost any advance, either in evolutionary biology or in systematics, and show that it did not depend as much on discoveries as on the introduction of improved concepts. Historians of science have long known this, but this fact is unfortunately far too little understood among nonscientists. To be sure, discoveries are an essential component of scientific advance, and some current bottlenecks in biology, such as the problem of the origin of life and of the organization of the central nervous system, are primarily due to a deficiency of certain basic facts. Nevertheless, the contribution made by new concepts or by the more or less radical transformation of old concepts is equally and often more important than facts and their discovery. In evolutionary biology, concepts like evolution, common descent, geographic speciation, isolating mechanisms, or natural selection have led to a drastic reorientation in a previously confusing area of biology and to new theory formation and countless new investigations. Those are not far wrong who insist that the progress of science consists principally in the progress of scientific concepts.

The use of concepts is, of course, not confined to science, since there are concepts in art, in history (and other areas of the humanities), in philosophy, and indeed in any activity of the human mind. What criteria, then, beyond the use of concepts, can be used for the demarcation between science and these other human endeavors? The answer to this question is not as simple as one might expect, as is documented by asking to what extent the social sciences are sciences. Tentatively one might suggest that what characterizes science is the rigor of its methodology, the possibility of testing or falsifying its conclusions, and of establishing noncontradictory "paradigms" (systems of theories). Method, even if it is not all of science, is one of its important aspects, particularly since it differs somewhat in different scientific disciplines.

Method in Science

The Greeks always looked for rational explanations in the world of phenomena. The school of Hippocrates, for example, when trying to determine the cause of a disease, did not look for it in a divine influence but attributed it to natural causes such as climate or nutrition. The Ionian philosophers, likewise, attempted to give rational explanations for the phenomena of the inanimate and living world. Aristotle, unquestionably the father of scientific methodology, gives in his *Posterior analytics* such a remarkable account of how one ought to go about a scientific explanation (McKeon, 1947; Foley, 1953; Vogel, 1952) that almost up to the nineteenth century, says Laudan (1977: 13) in a somewhat extreme statement, "philosophers of science were still working largely within the confines of the methodological problems discussed by Aristotle and his commentators." The Greek philosophers, including Aristotle, were primarily rationalists. They thought—Empedocles being a typical example—that they could solve scientific problems simply by sharp reasoning, involving ordinarily what we would now call deduction. The undoubted success which these ancient physicians and philosophers had in their explanations led to an overrating of a purely rational approach, which reached its climax in Descartes. Even though he did some empirical research (dissections, for example), many statements of this philosopher read as though he had believed that everything could be solved simply by concentrated thinking.

The ensuing attacks on Cartesianism by inductivists and experimentalists made it very clear that method was considered of great importance in science. This is as true today as it was in the seventeenth century. All too many philosophers, unfortunately, continued to believe until far into the nineteenth century that they could solve the riddles of the universe simply by reasoning or philosophizing. When their conclusions were in conflict with the findings of science, some of them insisted that they were right and science was wrong. It was this attitude which induced Helmholtz to complain bitterly about the arrogance of the philosophers. The reaction of philosophers to natural selection, relativity, and quantum mechanics shows that this attitude has by no means been entirely overcome.

Descartes endeavored to present only such conclusions and theories as had the certainty of a mathematical proof. Although there have always been some dissenters, the belief that a scientist had to supply absolute proof for all of his findings and theories prevailed until modern times. It dominated not only the physical sciences, where proof of the nature of a mathematical proof is often possible, but also the biological sciences. Even here, inferences are often so conclusive that they can be accepted as proof, as for instance the claim that the blood circulates or that a particular kind of caterpillar is the larval stage of a particular species of butterfly. The fact that the most minute exploration of the face of the earth has failed to reveal the presence of dinosaurs can be accepted as proof that they have become extinct. So far I have referred to facts, and to prove whether or not an assertion corresponds to a fact can usually be done. In many cases, however, and perhaps in the majority of the conclusions of the biologists, it is impossible to supply proof of such certainty (Hume, 1738). How are we to "prove" that natural selection is the directing agent guiding the evolution of organisms?

Eventually the physicists also realized that they could not always give absolute proof (Lakatos, 1976), and the new theory of science no longer demands it. Instead, scientists are satisfied to consider as true either that which appears most probable on the basis of the available evidence, or that which is consistent with more, or more compelling, facts than competing hypotheses. Realizing the impossibility of supplying absolute proof for many scientific conclusions, the philosopher Karl Popper has proposed that falsifiability be made the test of their validity instead. The burden of the argument thus is shifted to the opponent of a scientific theory. Accordingly, that theory is accepted which has withstood successfully the greatest number and variety of attempts to refute it. Popper's claim also allows one rather neatly to delimit science from nonscience: any claim which in principle cannot be falsified is outside the realm of science. Thus, the assertion that there are men on the Andromeda nebula is not a scientific hypothesis.

Falsification, however, is sometimes as difficult to provide as positive proof. It is therefore not considered the only measure for obtaining scientific acceptability. As the history of science demonstrates, when scientific theories were rejected, it was often not because they had been clearly refuted but rather because an alternative new theory seemed more probable, simpler, or more elegant. Furthermore, rejected theories are often tenaciously adhered to by a minority of followers, in spite of a series of seemingly successful refutations.

The new theory of science, based on a probabilistic interpretation of scientific conclusions, makes it inappropriate to speak of truth or proof as something absolute. This is of greater consequence for some branches of biology than for others. Every evolutionist who has had a discussion with lay people has been asked: "Has evolution been proven?" or "How do you prove that man descended from apes?" He is then obliged first to discuss the nature of scientific proof.

The working scientist, by contrast, has always been pragmatic. He was always reasonably happy with a theory until a better one was proposed. Factors that were inaccessible to explanation were treated as a black box, as was done by Darwin with the source of genetic variability, one of the chief components of his theory of natural selection. It did not and does not bother a scientist unduly that many of his generalizations are only probabilistic and that there is a remarkably high stochastic component in many, if not most, natural processes. Accepting great flexibility as one of the attributes of scientific theories, the scientist is willing to test numerous theories, to combine elements of different theories, and sometimes even to consider several alternate theories (multiple working hypotheses) simultaneously, while in search for evidence that would permit him to adopt one in preference to the others (Chamberlin, 1890). It should not be concealed, however, that the open-mindedness of scientists is not without limitations. When theories are "strange" or alien to the current intellectual milieu, they tend to be ignored or silenced. As we shall see, this is true, for instance, for the concepts of emergentism and of level-specific properties of hierarchies.

Interestingly, Darwin's approach was completely in line with modern theory. He realized he could never demonstrate evolutionary conclusions with the certainty of a mathematical proof. Instead, in about twenty different parts of the *Origin* he asks: "Is this particular finding—whether a pattern of distribution or an anatomical structure—more easily explained by special creation or by evolutionary opportunism?" Invariably he insists that the second alternative is the more probable one. Darwin anticipated many of the most important tenets of the current philosophy of science. Although scientists have now universally adopted the probabilistic interpretation of scientific truth—and indeed the complete impossibility of supplying demonstrations of the certainty of mathematical proof for most of their conclusions—this new insight is still not yet appreciated by many nonscientists. It would be desirable if this new concept of scientific truth were made part of scientific instruction to a much greater extent.

There are indications, however, that the importance attributed to the choice of method has been exaggerated. In this I agree with Koyré (1965), who opined that "abstract methodology is of relatively little importance for the concrete development of scientific thought." Goodfield (1974) could discover no difference in scientific success and theory formation between reductionists and antireductionists among physiologists. Kuhn and others have likewise minimized the importance of choice of method. Scientists in their actual research often go back and forth between a phase in which they collect material or conduct purely descriptive or classificatory research and another phase of concept formation or testing of theories.

Induction

For centuries there have been arguments as to the respective merits of the inductive versus the deductive method (Medawar, 1967). It is now clear that this is a relatively irrelevant argument. Inductivism claims that a scientist can arrive at objective, unbiased conclusions only by simply recording, measuring, and describing what he encounters without having any prior hypotheses or preconceived expectations. Francis Bacon (1561–1626) was the chief promoter of inductivism, though he never applied this method consistently in his own work. Darwin, who boasted that he was following "the true Baconian method," was anything but an inductivist. Indeed, he made fun of this method, saying that if one did believe in this method, "one might as well go into a gravel pit and count the pebbles and describe the colours." Yet, in the philosophical literature Darwin has often been classified as an inductivist. Inductivism had a great vogue in the eighteenth and early nineteenth centuries, but it is now clear that a purely inductive approach is quite sterile. This is illustrated by the plant breeder Gaertner, who patiently made and recorded tens of thousands of crosses without arriving at any generalizations. Liebig (1863) was the first prominent scientist to repudiate Baconian inductivism, arguing convincingly that no scientist ever had or ever could follow the methods described in the *Novum Organum*. Liebig's incisive critique brought the reign of inductivism to an end (Laudan, 1968).

Hypothetico-Deductive Method

Inductivism was replaced more and more consciously by the so-called hypothetico-deductive method.[2] According to this method the first step is to "speculate," as Darwin called it, that is, to generate a hypothesis. The second step is to conduct experiments or gather observations permitting the testing of this hypothesis. Darwin's employment of this method has been excellently described by Ghiselin (1969), Hull (1973a), and Ruse (1975b). There is a strong commonsense element in this method, and one can argue that it is already implicit in the Aristotelian method and certainly in much of the so-called deductivism of Descartes and his followers. Although temporarily eclipsed during the eighteenth-century vogue of inductivism, it became the prevailing method in the nineteenth century.

The reason why the hypothetico-deductive method has been so widely adopted is that it has two great advantages. First, it fits right in with the growing conviction that there is no absolute truth and that our conclusions and theories should continually be tested. And second, connected with this new relativism, it encourages the continuous establishment of new theories and the search for new observations and new experiments that either confirm or refute the new hypotheses. It makes science more flexible and more enterprising, and has made some scientific controversies less acrimonious, since victory in the battle for ultimate truth is no longer involved.

To what extent scientists actually employ the hypothetico-deductive approach is arguable. Collingwood (1939) has stated quite correctly that a hypothesis is always a tentative answer to a question and that the posing of a question is really the first step on the path toward a theory. The history of science knows scores of instances where an investigator was in the possession of all the important facts for a new theory but simply failed to ask the right question. Accepting the importance of questions, however, immediately leads to new doubts: Why, in the first place, was the question raised? The answer must be because a scientist observed something which he did not understand or something the origin of which puzzled him, or because he encountered some seemingly contradictory phenomena from which he wanted to remove the contradiction. In other words, observations of facts gave rise to the questions.

The anti-inductivists are, of course, entirely right in claiming that such facts by themselves never lead to a theory. They acquire significance only when an inquiring mind asks an important question. The creative mind is able, as Schopenhauer has stated it, "to think something that nobody has thought yet, while looking at something that everybody sees." Imagination, thus, is ultimately the most important prerequisite of scientific progress.

The hypothetico-deductive method is, in essence, the modern scientific method of discovery, although the establishment of a tentative hypothesis is invariably preceded by observations and the posing of questions.

30

Experiment versus Comparison

The difference between physical and biological research is not, as is often claimed, a difference of methodology. Experimentation is not restricted to the physical sciences but is a major method of biology, particularly of functional biology (see below). Observation and classification are clearly more important in the biological than in the physical sciences, yet it is evident that these are the dominant methods in such physical sciences as geology, meteorology, and astronomy. Analysis is of equal importance in the physical and in the biological sciences, as we will see.

In philosophies of science written by physical scientists, the experiment is often referred to as *the* method of science.[3] This is not true, because other strictly scientific methods are of major importance in sciences like evolutionary biology and oceanography. Each science demands its own appropriate methods. For Galileo, the student of mechanics, measurement and quantification were of paramount importance. For Aristotle, the student of living systems and of organic diversity, the analysis of what we now call teleonomic processes and the establishment of categories were favored approaches. In physiology and other functional sciences the experimental method is not only appropriate but almost the only approach leading to results.

Most historians of the physical sciences display an extraordinary ignorance when discussing methods other than the experimental one. Morgan (1926) well documents the arrogance of the experimentalist. He denied to the paleontologist any competence in theory formation: "My good friend the paleontologist [he was, no doubt, referring to H. F. Osborn] is in greater danger than he realizes, when he leaves descriptions and attempts explanation. He has no way to check up his speculations . . . [and when looking at the gaps in the fossil record] the geneticist says to the paleontologist, since you do not know, and from the nature of your case can never know, whether your differences are due to one change [a single mutation] or to a thousand, you cannot with certainty tell us anything about the hereditary units which have made the process of evolution." As if the paleontologist could not draw quite valid inferences from his material, inferences that can be tested in numerous ways. To imply that experimental work is never descriptive is also misleading. When reporting the results of their experiments the practitioners of the experimental method are as descriptive as the naturalists when reporting their observations. *31* The alternate to experiment, clearly, is observation. Progress in many branches of science depends on observations made in order to answer carefully posed questions. Modern evolutionary, behavioral, and ecological biology have demonstrated conclusively that these largely observational sciences are anything but descriptive. Actually, many papers based on experiments made without adequate *Fragestellung* (alas, there are far too many!) are more descriptive than most nonexperimental publications in evolutionary biology.

Mere observation, however, is not sufficient. It was not until late in the eighteenth century that a method was first seriously employed which is peculiarly suitable for the study of diversity; it is the comparative method. Although he had some forerunners, Cuvier without question was the first great champion of the comparative method (see Chapter 4). It is frequently overlooked that the application of the comparative method must be preceded by a classification of the items to be compared. Indeed, the success of the comparative analysis depends to a large extent on the goodness of the preceding classification. At the same time, discrepancies revealed by the comparison often lead to an improved classification of the phenomena. Such a back-and-forth between two methods characterizes many branches of science and is not in the least circular (Hull, 1967).

The difference between the experimental and the comparative method is not as great as it may appear at first. In both of them data are gathered and in both of them observation plays a crucial role (though the experimentalist usually does not mention the fact that his results are due to the observation of the conducted experiments). In the so-called observational sciences the observer studies experiments of nature. The principal difference between the two sets of observations is that in the artificial experiment one can choose the conditions and is thus able to test the factors that determine the outcome of the experiment. In an experiment of nature, whether it is an earthquake or the production of an insular fauna, it is our main task to infer or reconstruct the conditions under which this experiment of nature had taken place. By searching for the right constellation of factors, it is sometimes almost possible in a "controlled" observation to reach the reliability of a controlled experiment. As Pantin (1968: 17) has stated, "In astronomy, in geology, and in biology observation of natural events at chosen times and places can sometimes provide information as wholly sufficient for a conclusion to be drawn as that which can be obtained by experiment."

32

It is important to emphasize the scientific legitimacy of the observational-comparative method because the experimental method is inapplicable to many scientific problems. Yet, contrary to the claims of some physicists, the branches of science which depend on the comparative method are not inferior. As a wise scientist, E. B. Wilson, expressed a long time ago, "The experiments performed in our laboratories but supplement those that have taken place and are always taking place in nature, and their results must be wrought into the same fabric." Wilson opposed consistently those who claimed that progress in biology could be achieved "by experiment alone." Observation led to the discovery of foreign faunas and floras and became the basis of biogeography; observation revealed the diversity of organic nature and led to the establishment of the Linnaean hierarchy and to the theory of common descent; observation led to the foundations of ethology and ecology. Observation in biology has probably produced more insights than all experiments combined.

THE POSITION OF BIOLOGY WITHIN THE SCIENCES

When confronted by mythology or religion, science offers a unified front. All sciences, in spite of manifold differences, have in common that they are devoted to the endeavor to understand the world. Science wants to explain, it wants to generalize, and it wants to determine the causation of things, events, and processes. To that extent, at least, there is a unity of science (Causey, 1977).

From this observation the conclusion is often drawn that what is true for one science, let us say physics, must be equally true for all sciences. To give illustration, I must have some six or seven volumes on my shelves which claim to deal with the "philosophy of science" but all of them actually deal only with the philosophy of the physical sciences. Philosophers of science, most of them with a background in physics, have unfortunately based their treatment of the philosophy and methodology of science almost exclusively on the physical sciences. Such treatments are very incomplete because they fail to cover the rich domain of phenomena and processes found in the world of living organisms. Philosophers and humanists, when they describe or criticize "science," almost invariably have only the physical sciences (or even technology) in mind. When historians speak of the scientific revolution, which was primarily a revolution of the mechanical sciences, they silently imply all too often that this was a revolution pertaining equally to the biological sciences.

That there are important differences between biology and the physical sciences is often entirely ignored. Most physicists seem to take it for granted that physics is the paradigm of science and that once one understands physics, one can understand any other science, including biology. The "arrogance of the physicists" (Hull, 1973) has become proverbial among scientists. The physicist Ernest Rutherford, for example, referred to biology as "postage stamp collecting." Even V. Weisskopf, normally quite free of the usual hubris of the physicists, forgot himself recently sufficiently to claim that "the scientific world view is based upon the great discoveries of the nineteenth century concerning the nature of electricity and heat and the existence of atoms and molecules" (1977: 405), as if Darwin, Bernard, Mendel, and Freud (not to mention hundreds of other biologists) had not made a tremendous contribution to our scientific world view, indeed, perhaps a greater one than the physicists.

To counterbalance this attitude, it is sometimes beneficial or even necessary to stress the plurality of science. Too often Newton and the natural laws are considered as co-extensive with science. Yet, if one looks at the intellectual scene during the sixteenth, seventeenth, and eighteenth centuries, one finds that simultaneously there were several other traditions which had virtually nothing to do with each other or with mechanics. The botany of the herbalists, the magnificent plates in the anatomy of Vesalius, the ubiquitous natural-history cabinets, the scientific voyages, the *jardins des plantes*, the menageries—what did all of this have to do with Newton? And yet this other science is what inspired Rousseau's romanticism and the dogma of the noble savage.

It has become apparent only in recent years how naive and misleading the assumption of the sameness of physical and biological sciences is. The physicist C. F. von Weizsaecker (1971) admits that the conventional physical explanation "and the abstract mathematical form in which it is dressed, does not . . . satisfy our need for a real understanding of nature. Moreover, a common world view no longer unites the great groups of sciences . . . the physicist finds an autonomous biology."

34   A study of biological phenomena, thus, leads to the legitimate question: To what extent are methodology and conceptual framework of the physical sciences appropriate models for the biological sciences? This question pertains not merely to such somewhat exceptional problems as that of "consciousness" or "mind" but to any biological phenomenon or concept, such as population, species, adaptation, digestion, selection, competition, and the like. Are these biological phenomena and concepts without equivalent in the physical sciences?

Nowhere is the difference between different sciences more conspicuous than in their philosophical applications. Many philosophers have pointed out that there is no conceivable connection between the physical sciences and ethics. Yet, it is equally evident that there is a seeming potential for a connection between biological sciences and ethics: Social Spencerism is one example; eugenics is another one. There is some validity to the physicist's claim that there is no connection between the physical sciences and ethics (but think of nuclear physics!). However, if he proclaims, as so many physicists have done, that there is no connection between "science" and ethics, he displays parochial narrowness. Political ideologies have always shown far more interest in the biological than in the physical sciences. Lysenkoism and the *tabula rasa* teaching of behaviorism (and its Marxist followers) are just some examples. For all these reasons it is wrong to speak of philosophy of science when one means the philosophy of the physical sciences.

The conviction of many physical scientists that all the insights of biology can be reduced to the laws of physics has led many biologists, in self-defense, to assert the autonomy of biology. Although this emancipation movement of the biologists has, quite naturally, encountered considerable resistance, not only among physical scientists but also among essentialism-committed philosophers, it has continued to gain in strength in recent decades. A dispassionate discussion of the question whether the principles, theories, and laws of the physical sciences explain everything in the biological sciences, or whether biology is, at least in part, an autonomous science has been made very difficult by a conspicuous rivalry—even mutual hostility—among the sciences, both within physical and biological sciences and between these two camps. Numerous have been the attempts (for example, that of Comte) to rank the sciences, with mathematics (or geometry in particular) appointed the queen of the sciences. The rivalry becomes apparent in the competition for honors such as Nobel prizes, budgets within universities and governments, positions, and general prestige *vis à vis* nonscientists.[4]

The preceding discussion might convey the impression that I too am pleading for a complete autonomy of the biological sciences—in other words, that I want to abandon radically the concept of the unity of science and replace it by the concept of two separate sciences, the physical and the biological sciences. This is not my position. All I wish to assert is that the physical sciences are not an appropriate yardstick of science. Physics is quite unsuited for this role because, as the physicist Eugene Wigner has stated very correctly, "present day physics deals with a limiting case." To use an analogy, physics corresponds to Euclidean geometry, which is the limiting case of all geometries (including non-Euclidean geometry). No one has described this situation better than G. G. Simpson (1964b: 106–107): "Insistence that the study of organisms requires principles additional to those of the physical sciences does not imply a dualistic or vitalistic view of nature. Life . . . is not thereby necessarily considered as nonphysical or nonmaterial. It is just that living things have been affected for . . . billions of years by historical processes . . . The results of those processes are systems different in kind from any nonliving systems and almost incomparably more complicated. They are not for that reason necessarily any less material or less physical in nature. The point is that *all* known material processes and explanatory principles apply to organisms, while only a limited number of them apply to nonliving systems . . . Biology, then, is the science that stands at the center of all science . . . And it is here, in the field where all the principles of all the sciences are embodied, that science can truly become unified."

The recognition that in the biological sciences we deal with phenomena unknown for inanimate objects is by no means new. The history of science, from Aristotle on, has been a history of endeavors to assert the autonomy of biology, and of attempts to stem the tide of facile mechanistic-quantitative explanations. However, when naturalists and other biologists as well as some philosophers stressed the importance of quality, uniqueness, and history in biology, their efforts were often ridiculed and simply brushed aside as "bad science." This fate happened even to Kant, who argued rather convincingly in his *Kritik der Urteilskraft* (1790) that biology is different from the physical sciences and that living organisms are different from inanimate objects. Regrettably, often such endeavors were labeled as vitalism and hence as outside the boundary of science. Claims for the autonomy of biology have begun to be taken seriously only within the last generation or so, that is, after the final extinction of any genuine vitalism.

It is becoming quite evident that one will never be able to make universally valid statements about science in general until one has first compared the various sciences with each other, and has determined what they have in common and what distinguishes them.

How and Why Is Biology Different?

The word "biology" is a child of the nineteenth century. Prior to that date, there was no such science. When Bacon, Descartes, Leibniz, and Kant wrote about science and its methodology, biology as such did not exist, only medicine (including anatomy and physiology), natural history, and botany (somewhat of a mixture). Anatomy, the dissection of the human body, was until far into the eighteenth century a branch of medicine, and botany likewise was practiced primarily by physicians interested in medicinal herbs. The natural history of animals was studied mainly as part of natural theology, in order to bolster the argument from design. The scientific revolution in the physical sciences had left the biological sciences virtually untouched. The major innovations in biological thinking did not take place until the nineteenth and twentieth centuries. It is not surprising, therefore, that the philosophy of science, when it developed in the seventeenth and eighteenth centuries, was based exclusively on the physical sciences and that it has been very difficult, subsequently, to revise it in such a way as to encompass also the biological sciences. It is only in recent decades that several philosophers (such as Scriven, Beckner, Hull, and Campbell) have attempted to characterize the differences between biology and the physical sciences (Ayala 1968). Thought about this problem is still so new that one can only make tentative statements. The object of the following discussion is more to delineate the nature of the problems than to provide definite solutions.

Laws in the Physical and Biological Sciences

Laws play an important explanatory role in the physical sciences. A particular happening is considered to be explained when it can be shown to be due to particular causal factors that are consistent with general laws. Some philosophers have designated the establishment of laws as the diagnostic criterion of science. *37* Such laws are believed to be strictly deterministic and thus to permit precise predictions.

The question has been raised in recent years whether or not laws are as important in biology as they seem to be in the physical sciences. Some philosophers, like Smart (1963; 1968), have gone so far as to deny that there are any universal laws in biology such as characterize physics. Other philosophers, like Ruse (1973) and to a lesser extent Hull (1974), have emphatically defended the existence of biological laws. Biologists have paid virtually no attention to the argument, implying that this question is of little relevance for the working biologist.

If one looks back over the history of biology one finds that nineteenth-century authors like Lamarck, Agassiz, Darwin, Haeckel, Cope, and most of their contemporaries referred frequently to laws. If one looks at a modern textbook of almost any branch of biology, one may not encounter the term "law" even a single time. This does not mean that regularities do not occur in biology; it simply means that they are either too obvious to be mentioned, or too trivial. This is well illustrated by the one hundred evolutionary "laws" listed by Rensch (1968: 109–114). They all refer to adaptive trends effected by natural selection. Most of them have occasional or frequent exceptions and are only "rules," not universal laws. They are explanatory as far as past events are concerned, but not predictive, except in a statistical (probabilistic) sense. When I say: "A territory-holding male songbird has a 98.7% (or whatever the correct figure may be) chance to be victorious over an intruder," I can hardly claim to have established a law. When they say that proteins do not translate information back into the nucleic acids, molecular biologists consider this a fact rather than a law.

Generalizations in biology are almost invariably of a probabilistic nature. As one wit has formulated it, there is only one universal law in biology: "All biological laws have exceptions." This probabilistic conceptualization contrasts strikingly with the view during the early period of the scientific revolution that causation in nature is regulated by laws that can be stated in mathematical terms. Actually, this idea occurred apparently first to Pythagoras. It has remained a dominant idea, particularly in the physical sciences, up to the present day. Again and again it was made the basis of some comprehensive philosophy, but taking very different forms in the hands of various authors. With Plato it gave rise to essentialism, with Galileo to a mechanistic world picture, and with Descartes to the deductive method. All three philosophies had a fundamental impact on biology.

38

Plato's thinking was that of a student of geometry: A triangle, no matter what combination of angles it has, always has the *form* of a triangle, and is thus discontinuously different from a quadrangle or any other polygon. For Plato, the variable world of phenomena in an analogous manner was nothing but the reflection of a limited number of fixed and unchanging forms, *eide* (as Plato called them) or *essences* as they were called by the Thomists in the Middle Ages. These essences are what is real and important in this world. As ideas they can exist independent of any objects. Constancy and discontinuity are the points of special emphasis for the essentialists. Variation is attributed to the imperfect manifestation of the underlying essences. This conceptualization was the basis not only of the realism of the Thomists but also of so-called idealism or of the positivism of later philosophers, up to the twentieth century. Whitehead, who was a peculiar mixture of a mathematician and a mystic (perhaps one should call him a Pythagorean), once stated: "The safest general characterization of the European philosophical tradition is that it consists in a series of footnotes to Plato." No doubt, this was meant as praise, but it really was a condemnation, so far as it was true at all. What it really says is that European philosophy through all the centuries was unable to free itself from the strait jacket of Plato's essentialism. Essentialism, with its emphasis on discontinuity, constancy, and typical values ("typology"), dominated the thinking of the western world to a degree that is still not yet fully appreciated by the historians of ideas. Darwin, one of the first thinkers to reject essentialism (at least in part), was not at all understood by the contemporary philosophers (all of whom were essentialists), and his concept of evolution through natural selection was therefore found unacceptable. Genuine change, according to essentialism, is possible only through the saltational origin of new essences. Because evolution as explained by Darwin is by necessity gradual, it is quite incompatible with essentialism. However, the philosophy of essentialism fitted well with the thinking of the physical scientists, whose "classes" consist of identical entities, be they sodium atoms, protons, or pi-mesons.

For Galileo, likewise, geometry was the key to the laws of nature. But he applied it in a far more mathematical manner than Plato: "Philosophy is written in this grand book, the universe, which stands continually open to our gaze. But the book cannot be understood unless one first learns to comprehend the language and read the letters in which it is composed. It is written in the language of mathematics, and its characters are triangles, circles, and other geometric figures without which it is humanly impossible to understand a single word of it; without these, one wanders about in a dark labyrinth" (*The Assayer,* 1623, as quoted by Kearney, 1964). It was, however, not only geometry that was considered basic by him, but all aspects of mathematics and particularly any kind of quantification of measurement.

The "mechanization of the world picture"—the belief in a highly orderly world such as one would expect if the world was designed by the creator to obey a limited set of eternal laws (Maier, 1938; Dijksterhuis, 1961)—made rapid progress in the ensuing centuries and achieved its greatest triumph in Newton's unification of terrestrial and celestial mechanics. These splendid successes led to an almost unlimited prestige of mathematics. It resulted in Kant's famous—or infamous—dictum "that only so much genuine science can be found in any branch of the natural sciences as it contains mathematics." If this were true where would the *Origin of Species* stand as a work of science? Not surprisingly, Darwin had a rather low opinion of mathematics (Hull, 1973: 12).

The blind faith in the magic of numbers and quantities perhaps reached its climax in the middle of the nineteenth century. Even so perceptive a thinker as Merz (1896: 30) stated, "Modern science defines the method, not the aim of its work. It is based upon numbering and calculating—in short, upon mathematical processes; and the progress of science depends as much upon introducing mathematical notions into subjects which are apparently not mathematical, as upon the extention of mathematical methods and conceptions themselves."

In spite of rather devastating subsequent refutations (Ghiselin, 1969: 21), philosophers with a background in mathematics or

39

physics continue to cling to the myth of mathematics as the queen of the sciences. For instance, the mathematician Jacob Bronowski (1960, p. 218) stated that "to this day, our confidence in any science is roughly proportional to the amount of mathematics it employs . . . We feel that physics is truly a science, but that there somehow clings to chemistry the less formal odor (and odium) of the cookbook. And as we proceed further to biology, then to economics, and last to social studies, we know that we are fast slipping down a slope away from science." These misconceptions concerning qualitative and historical sciences, or sciences dealing with systems too complex to be expressed in mathematical formulas, culminated in the arrogant claim that biology is an inferior science. This has led to facile and thoroughly misleading mathematical explanations in various areas of biology.

40

No one was more impressed by the importance of mathematics than Descartes, but the consequences of this admiration on his thinking were quite different from that on Galileo or Newton. Descartes was impressed by the rigor of mathematical proofs and by the certainty by which conclusions issued from definite statements, going so far as to claim that the laws of mathematics had been legislated by God in the same manner as a king ordains laws in his realm. Descartes developed a logic in which the methods of mathematics were used, in a strictly deductive manner, to acquire rational knowledge. It was the structure of thought that was mathematical rather than a language of equations or mathematical formulas. Nevertheless, it favored strictly deterministic explanations and essentialistic thinking. Leibniz, who followed Descartes' methodology, was the founder of mathematical logic.

As overwhelming as the dominance of mathematics over the sciences was for several centuries, there were voices of dissent almost from the beginning. Pierre Bayle (1647–1706) was apparently the first to deny the claim that mathematical knowledge was the only kind of knowledge attainable by the scientific method. He asserted, for instance, that historical certainty was not inferior but simply different from mathematical certainty. Facts of history, such as that the Roman empire once existed, were as certain as anything in mathematics. A biologist would likewise insist that the former existence of dinosaurs and trilobites was as certain as any mathematical theorem. Another author to launch a devastating attack on the mathematical-geometrical interpretations of the world by Descartes was Giambattista Vico. He asserted that the methods of observation, classification, and hypothesis were indeed able to give one a genuine though modest "outside" knowledge of the material world.

Natural history was a second source of rebellion against Galileo's mathematical ideal of science. It was particularly promoted by Buffon, who asserted emphatically (*Oeuvr. Phil.:* 26) that some subjects are far too complicated for a useful employment of mathematics, among these subjects being all parts of natural history. Here observation and comparison were the appropriate methods. Buffon's *Histoire naturelle* in turn decisively influenced Herder and through him the Romantics and *Naturphilosophie*. Even Kant, by 1790, had abandoned his subservience to mathematics. If the invalidity of the mathematical ideal of science had not been obvious before, it certainly became so with the publication of the *Origin of Species*.

41

It might be mentioned, incidentally, how misleading it is to refer to mathematics as the "queen of the sciences." Mathematics, of course, is as little a science as grammar is a language (comparable to Latin or Russian); mathematics is a language relating to all sciences or to none, although to highly varying degrees. There are some sciences, like the physical sciences and much of functional biology, in which quantification and other mathematical approaches have a high explanatory or heuristic value. There are other sciences, like systematics and much of evolutionary biology, in which the contributions of mathematics are very minor.

In fact, an ill-advised application of mathematics in these branches of biology has sometimes led to typological thinking, and thus to misconceptions. The geneticist Johannsen, for instance, fell prey to this temptation and "simplified" genetically variable populations to "pure lines," thereby losing the very meaning of "population" and drawing erroneous conclusions concerning the importance of natural selection. The founders of mathematical population genetics, likewise, for the sake of mathematical tractability, oversimplified the factors entering their formulas. This led to a stress on absolute fitness values of genes, to an overvaluation of additive gene effects, and to the assumption that genes, rather than individuals, are the target of natural selection. This invariably led to unrealistic results.

When Darwin calculated that the earth must be more than a thousand million years old in order to account for the phenomena of geology and phylogeny, Lord Kelvin pronounced him emphatically wrong on the basis of calculations of the heat loss of a globe of the size of the sun: 24 million years was the maximum he could allow (Burchfield, 1975). It is rather amusing with what assurance Kelvin assumed the correctness of his own and the error of the naturalists' age determinations. Since biology was an inferior science, there could be no question where the error was. Kelvin never allowed for the possibility of the existence of an unknown physical factor that would eventually validate the calculations of the biologists. In this intellectual climate biologists went out of their way to interpret their findings in terms of simple physics. Weismann (in his early work) ascribed inheritance as due to "molecular movements" and Bateson as due to "the movement of vortices," explanations that simply retarded scientific progress.

42

The situation has changed rather dramatically during the last fifty years. The indeterminacy of most strictly biological processes no longer stands in sharp contrast to a strict determinacy in physical processes. The study of the effects of turbulence in galaxies and nebulae, as well as in the oceans and weather systems, has shown how frequent and powerful are stochastic processes in inanimate nature. This conclusion has not been acceptable to some physicists, leading Einstein, for example, to exclaim, "God does not play dice!" Yet, stochastic processes occur at every hierarchical level from the atomic nucleus up to the systems produced by the big bang. And stochastic processes, even though making predictions probabilistic (or impossible) rather than absolute, are just as causal as deterministic processes. Only absolute predictions are impossible owing to the complexity of the hierarchical systems, the high number of possible options at each step, and the numerous interactions of simultaneously occurring processes. Weather systems and cosmic nebulae are, in that respect, not in principle different from living systems. The number of potentially possible interactions in such highly complex systems is far too large to permit prediction as to which one will actually take place. The students of natural selection and of other evolutionary processes, of quantum mechanics, and of astrophysics reached this conclusion at different times and more or less independently.

For all of these reasons, physics is no longer the yardstick of science. Particularly where the study of man is concerned, it is biology that provides methodology and conceptualization. The President of France has recently formulated this conviction in these words: "There is no doubt that mathematics, physics, and other sciences rather ill-advisedly referred to as 'exact' . . . will continue to afford surprising discoveries—yet, I cannot help feeling that the real scientific revolution of the future must come from biology."

Concepts in the Biological Sciences

Instead of formulating laws, biologists usually organize their generalizations into a framework of concepts. The claim has been made that laws versus concepts is only a formal difference, since every concept can be translated into one or several laws. Even if this were formally true, of which I am not at all sure, such a translation would not be helpful in the actual performance of biological research. Laws lack the flexibility and heuristic usefulness of concepts.

43

Progress in biological science is, perhaps, largely a matter of the development of these concepts or principles. The progress of systematics was characterized by the crystallization and refinement of such concepts as classification, species, category, taxon, and so on; evolutionary science by such concepts as descent, selection, and fitness. Similar key concepts could be listed for every branch of biology.[5]

Scientific progress consists in the development of new concepts, like selection or biological species, and the repeated refinement of definitions by which these concepts are articulated. Particularly important is the occasional recognition that a more or less technical term, previously believed to characterize or designate a certain concept, was in reality used for a mixture of two or more concepts, like "isolation" for geographical and reproductive isolation, or "variety" (as used, for instance, by Darwin) for individuals and populations, or "teleological" for four different phenomena.

It is strange how little attention the philosophy of science has paid to the overwhelming importance of concepts. For this reason, it is not yet possible to describe in detail the processes of discovery and maturation of concepts. This much is evident, however, that the major contribution of the leaders of biological thought has been the development and refinement of concepts and occasionally the elimination of erroneous ones. Evolutionary biology owes a remarkably large portion of its concepts to Charles Darwin, and ethology to Konrad Lorenz.

The history of concepts, so neglected up to now, is full of surprises. "Affinity" or "relationship," as used in pre-evolutionary systematics to designate not much more than simple similarity, was transferred after 1859 to "proximity of descent" without causing any confusion or difficulty. On the other hand, the attempt by Hennig to transfer the term "monophyletic" from a characterization of a taxonomic group to a characterization of a pathway of descent caused painful upheavals in taxonomy. The study of concepts sometimes reveals serious terminological deficiencies in certain languages. The term "resource," for instance, so important for ecology (partitioning of, competition for, and so on) had no equivalent in German, until the English word was Germanicized to "Ressourcen."

44   There are concepts of the most diverse kind. Biology, for instance, has greatly benefitted from a refinement of quasi-philosophical or methodological concepts, like proximate and evolutionary causation, or the clear demarcation of the comparative from the experimental method. The recognition of the existence of a comparative approach constituted the introduction of a new concept into biology.

The difficulties within a science are particularly great when a truly new concept is introduced. This occurred, for instance, through the introduction of population thinking as the replacement of Plato's essence concept, or the introduction of concepts like selection or like genetic (closed) or open programs. This is what Kuhn, in part, was talking about when he spoke of scientific revolutions.

Sometimes simply the introduction of a new term, like "isolating mechanisms," "taxon," or "teleonomic," has greatly helped to clarify a previously confused conceptual situation. More often the conceptual morass had to be eliminated first, before the introduction of a new terminology could do any good. This was true for Johannsen's terms "genotype" and "phenotype" (even though Johannsen himself was somewhat confused; see Roll-Hansen, 1978a).

One further difficulty is posed by the fact that the same term may be used for different concepts in different sciences, or even disciplines of the same science. The term "evolution" meant something very different for embryologists from the eighteenth century (Bonnet) or for Louis Agassiz (1874) than it meant for the Darwinians; likewise, it meant something very different for most anthropologists (at least those directly or indirectly influenced by Herbert Spencer) than for selectionists. Many celebrated controversies in the history of science were caused almost entirely because the opponents referred to very different concepts by the same term.

In the history of biology the phrasing of definitions has often proven rather difficult, and most definitions have been modified repeatedly. This is not surprising since definitions are temporary verbalizations of concepts, and concepts—particularly difficult concepts—are usually revised repeatedly as our knowledge and understanding grows. This is well illustrated by the definitions of such concepts as species, mutation, territory, gene, individual, adaptation, and fitness.

One very important methodological aspect of science is frequently misunderstood and has been a major cause of controversy over such concepts as homology or classification. It is the relation between a definition and the evidence that the definition is met in a particular instance (Simpson, 1961: 68–70). This is best illustrated by an example: The term "homologous" existed already prior to 1859, but it acquired its currently accepted meaning only when Darwin established the theory of common descent. Under this theory the biologically most meaningful definition of "homologous" is: "A feature in two or more taxa is homologous when it is derived from the same (or a corresponding) feature of their common ancestor." What is the nature of the evidence that can be used to demonstrate probable homology in a given case? There is a whole set of such criteria (like the position of a structure in relation to others), but it is completely misleading to include such evidence in the definition of "homologous," as has been done by some authors. The same relation between definition and the evidence that the definition is met exists in the definition of virtually all terms used in biology. For instance, an author may attempt a "phylogenetic classification" but may rely entirely on morphological evidence to infer relationship. This does not make it a morphological classification. The currently most widely accepted species definition includes the criterion of the reproductive community ("interbreeding"). A paleontologist cannot test interbreeding in his fossil material but he can usually bring together various other kinds of evidence (association, similarity, and so on) to strengthen the probability of conspecificity. A definition articulates a concept, but it does not need to include the evidence that the definition is met.

Let me now discuss some of the concepts of particular importance in biology.

Population Thinking versus Essentialism

Western thinking for more than two thousand years after Plato was dominated by essentialism. It was not until the nineteenth

century that a new and different way of thinking about nature
began to spread, so-called population thinking. What is population thinking and how does it differ from essentialism? Population
thinkers stress the uniqueness of everything in the organic world.
What is important for them is the individual, not the type. They
emphasize that every individual in sexually reproducing species is
uniquely different from all others, with much individuality even
existing in uniparentally reproducing ones. There is no "typical"
individual, and mean values are abstractions. Much of what in the
past has been designated in biology as "classes" are populations
consisting of unique individuals (Ghiselin, 1974b; Hull, 1976).

46

There was a potential for population thinking in Leibniz's
theory of monads, for Leibniz postulated that each monad was
individualistically different from every other monad, a major departure from essentialism. But essentialism had such a strong hold
in Germany that Leibniz's suggestion did not result in any population thinking. When it finally developed elsewhere, it had two
roots; one consisted of the British animal breeders (Bakewell, Sebright, and many others) who had come to realize that every individual in their herds had different heritable characteristics, on
the basis of which they selected the sires and dams of the next
generation. The other root was systematics. All practicing naturalists were struck by the observation that when collecting a "series" of specimens of a single species they found that no two specimens were ever completely alike. Not only did Darwin stress this
in his barnacle work, but even Darwin's critics concurred on this
point. Wollaston (1860), for instance, wrote "amongst the millions
of people who have been born into the world, we are certain that
no two have ever been precisely alike in every respect; and in a
similar manner it is not too much to affirm the same of all living
creatures (however alike some of them may seem to our uneducated eyes) that have ever existed." Similar statements were made
by many mid-nineteenth-century taxonomists. This uniqueness is
true not only for individuals but even for stages in the life cycle
of any individual, and for aggregations of individuals whether they
be demes, species, or plant and animal associations. Considering
the large number of genes that are either turned on or turned off
in a given cell, it is quite possible that not even any two cells in
the body are completely identical. This uniqueness of biological
individuals means that we must approach groups of biological entities in a very different spirit from the way we deal with groups
of identical inorganic entities. This is the basic meaning of popu-

lation thinking. The differences between biological individuals are real, while the mean values which we may calculate in the comparison of groups of individuals (species, for example) are man-made inferences. This fundamental difference between the classes of the physical scientists and the populations of the biologists has various consequences. For instance, he who does not understand the uniqueness of individuals is unable to understand the working of natural selection.[6]

The statistics of the essentialist are quite different from those of the populationist. When we measure a physical constant—for instance, the speed of light—we know that under equivalent circumstances it is constant and that any variation in the observational results is due to inaccuracy of measurement, the statistics simply indicating the degree of reliability of our results. The early statistics from Petty and Graunt to Quetelet (Hilts, 1973) was essentialistic statistics, attempting to arrive at true values in order to overcome the confusing effects of variation. Quetelet, a follower of Laplace, was interested in deterministic laws. He hoped by his method to be able to calculate the characteristics of the "average man," that is, to discover the "essence" of man. Variation was nothing but "errors" around the mean values.

Francis Galton was perhaps the first to realize fully that the mean value of variable biological populations is a construct. Differences in height among a group of people are real and not the result of inaccuracies of measurement. The most interesting parameter in the statistics of natural populations is the actual variation, its amount, and its nature. The amount of variation is different from character to character and from species to species. Darwin could not have arrived at a theory of natural selection if he had not adopted populational thinking. The sweeping statements in the racist literature, on the other hand, are almost invariably based on essentialistic (typological) thinking.

As important as the introduction of new concepts like population thinking was the elimination or revision of erroneous concepts. This is well illustrated by the concept of teleology.

The Problem of Teleology

From Plato, Aristotle, and the Stoics on, a belief was prevalent (opposed by the Epicureans) that there is a purpose, a predetermined end, in nature and its processes. Those with this view in the seventeenth and eighteenth centuries—the teleologists—saw the clear expression of a purpose not only in the *scala naturae,* culminating in man, but also in the total unity and harmony of nature and its manifold adaptations. Opposed to the teleologists were the strict mechanists, who viewed the universe as a mechanism functioning according to natural laws. The seeming purposefulness of the universe, of the goal-directed processes in the development of individuals, and of the adaptations of organs was too conspicuous for the mechanist to ignore, however. How could a mechanism have all these properties purely as the result of laws, without recourse to final causes? No one perhaps was more keenly aware of this dilemma than Kant. To be for or against teleology remained a battle cry throughout the nineteenth century and right up to modern times.

48

Only within the last twenty-five years or so has the solution become evident. It is now clear that seemingly goal-directed processes exist in nature which are not in any way in conflict with a strictly physico-chemical explanation. The solution, as so often in the history of science, was achieved by dissecting a complex problem and partitioning it into its components. It became obvious (Mayr, 1974d) that the term "teleological" had been applied to four different concepts or processes.

(1) Teleonomic activities. The discovery of the existence of genetic programs has provided a mechanistic explanation of one class of teleological phenomena. A physiological process or a behavior that owes its goal-directedness to the operation of a program can be designated as "teleonomic" (Pittendrigh, 1958). All the processes of individual development (ontogeny) as well as all seemingly goal-directed behaviors of individuals fall in this category, and are characterized by two components: they are guided by a program, and they depend on the existence of some endpoint or goal which is foreseen in the program regulating the behavior. The endpoint might be a structure, a physiological function or steady state, the attainment of a new geographical position, or a consummatory behavioral act. Each particular program is the result of natural selection and is constantly adjusted by the selective value of the achieved endpoint (Mayr, 1974d). Aristotle called these causes "the *for-the-sake-of-which* causes" (Gotthelf, 1976). From the point of view of causation it is important to state that the program as well as the stimuli which elicit the goal-seeking behavior precede in time the seemingly purposive behavior. There are usually manifold feedback devices which improve the precision of the teleonomic process, but the truly characteristic aspect of teleonomic behavior is that mechanisms exist which initiate or "cause" this

goal-seeking behavior. Teleonomic processes are particularly important in ontogeny, physiology, and behavior. They belong to the area of proximate causation, even though the programs were acquired during evolutionary history. It is the endpoints that produce the selection pressure which cause the historical construction of the genetic program.

(2) *Teleomatic processes.* Any process, particularly one relating to inanimate objects, in which a definite end is reached strictly as a consequence of physical laws may be designated as "teleomatic" (Mayr, 1974d). When a falling rock reaches its endpoint, the ground, no goal-seeking or intentional or programmed behavior is involved, but simply conformance to the law of gravitation. So it is with a river inexorably flowing toward the ocean. When a red-hot piece of iron reaches an end state where its temperature and that of its environment are equal, the reaching of this endpoint is, again, due to strict compliance with a physical law, the first law of thermodynamics. The entire process of cosmic evolution, from the first big bang to the present time, is strictly due to a sequence of teleomatic processes on which stochastic perturbations are superimposed. The laws of gravitation and of thermodynamics are among the natural laws which most frequently govern teleomatic processes. Already Aristotle was aware of the separate existence of this class of processes, referring to them as caused by "necessity."

(3) *Adapted systems.* The natural theologians were particularly impressed by the design of all the structures responsible for physiological functions: the heart that is built to pump the blood through the body, the kidneys that are built to eliminate the by-products of protein metabolism, the intestinal tract that performs digestion and makes nutritional material available to the body, and so on. It was one of the most decisive achievements of Darwin to have shown that the origin and gradual improvements of these organs could be explained through natural selection. It is therefore advisable not to use the term teleological ("end-directed") to designate organs which owe their adaptedness to a *past* selectionist process. Adaptational or selectionist language is here more appropriate (Munson, 1971; Wimsatt, 1972) than teleological language which may imply the existence of orthogenetic forces as responsible for the origin of these organs.

One studies adapted systems by asking why-questions. Why are there valves in the veins? Sherrington (1906: 235) stressed this very appropriately for the reflex: "We cannot . . . obtain due

49

profit from the study of any particular type-reflex unless we can discuss its immediate purpose as an adapted act . . . The purpose of a reflex seems as legitimate and urgent an object for natural inquiry as the purpose of the coloring of an insect or blossom. And the importance to physiology is, that the reflex can not be really intelligible to the physiologist until he knows its aim."

(4) Cosmic teleology. Although Aristotle had developed his concept of teleology on the basis of the study of individual development, where it is entirely legitimate, he eventually applied it also to the universe as a whole. This being two thousand years before the proposal of the theory of natural selection, Aristotle could think of only two alternatives when encountering instances of adaptation: coincidence (chance) or purpose. Since it cannot be coincidence that the grinding molars are always flat and the cutting teeth (incisors) sharp-edged, the difference must be ascribed to purpose. "There is purpose, then, in what is, and in what happens in Nature." Indeed, so much in the universe reflects seeming purpose that final causation must be postulated.[7]

In due time this concept of cosmic teleology, particularly when combined with Christian dogma, became the prevailing concept of teleology. It is this teleology which modern science rejects without reservation. There is and never was any program on the basis of which either cosmic or biological evolution has occurred. If there is a seeming aspect of progression in biological evolution, from the prokaryotes of two or three billion years ago to the higher animals and plants, this can be explained entirely as the result of selection forces generated by competition among individuals and species and by the colonization of new adaptive zones.

Until natural selection was fully understood, many evolutionists, from Lamarck to H. F. Osborn and Teilhard de Chardin, postulated the existence of a nonphysical (perhaps even nonmaterial) force which drove the living world upward toward ever greater perfection (orthogenesis). It was not too difficult for materialistic biologists to show that there is no evidence for such a force, that evolution rarely produces perfection, and that the seeming progress toward greater perfection can be explained quite well by natural selection. The rectilinearity of many evolutionary trends is due to the many constraints which the genotype and the epigenetic system impose on the response to selection pressures.

The theory of orthogenesis has recently been revived by hard-nosed physical scientists. Eigen (1971) in his theory of hypercycles is convinced "that the evolution of life . . . must be considered

50

an inevitable process despite its indeterminate course." Monod (1974a: 22) refers to Eigen (and Prigogine) as "animists," owing to their efforts to "show first, that life could not have failed to start on earth, and secondly that evolution could not have failed to occur." Biologists, of course, would reject the deterministic aspects of Eigen's theory, but rather similar conclusions can be based on stochastic processes constantly "ordered" by natural selection. Monod, in his theorizing, curiously failed to give due weight to natural selection.

The partitioning, under four headings, of the aggregate of concepts lumped under the term "teleological" should eliminate teleology as a source of argument. One would wish, however, that these recent advances were more widely known among nonbiologists. For instance, many psychologists in their discussion of goal-directed behavior still operate with such undefinable terms as "intentions" and "consciousness" which make an objective analysis impossible. Since we have no way of determining which of the animals (and plants) have intentions or consciousness, the use of these terms adds nothing to the analysis; indeed, it only obfuscates it. Progress in the solution of these psychological problems depends on a reconceptualization of intention or consciousness in terms of our new evolutionary insights.

Special Characteristics of Living Organisms

The question why some objects of nature are inanimate while others are alive, and what the special characteristics of living organisms are, already occupied the thoughts of the ancients. From the days of the Epicureans and Aristotle up to the early decades of this century there have always been two opposing interpretations of the phenomena of life. According to one school, the mechanists, organisms are nothing but machines, the workings of which can be explained by the laws of mechanics, physics, and chemistry. Many of the mechanists of the seventeenth and eighteenth centuries could not see that there was any significant difference between a rock and a living organism. Did not both share the same characteristics—gravity, inertia, temperature, and so on— and obey the same physical forces? When Newton proposed his law of gravitation in purely mathematical terms, many of his followers postulated an invisible but strictly materialistic gravitational force to explain the planetary motions as well as terrestrial gravity. By assumed analogy, some biologists invoked an equally materialistic and equally invisible force (*vis viva*) to explain living processes.

Later authors, however, believed that such a vital force was outside the realm of the chemico-physical laws. They thus continued a tradition started by Aristotle and other ancient philosophers. This vitalistic school opposed the mechanists, believing that there are processes in living organisms which do not obey the laws of physics and chemistry. Vitalism had representatives well into the twentieth century, one of the last being the embryologist Hans Driesch. However, by the 1920s or 1930s biologists had almost universally rejected vitalism, primarily for two reasons. First, because it virtually leaves the realm of science by falling back on an unknown and presumably unknowable factor, and second, because it became eventually possible to explain in physico-chemical terms all the phenomena which according to the vitalists "demanded" a vitalistic explanation. It is fair to say that for biologists vitalism has been a dead issue for more than fifty years. Curiously, during that period it has still been defended by a number of physicists and philosophers.

This rejection of vitalism was made possible by the simultaneous rejection of a crude "animals are nothing but machines" conceptualization. Like Kant in his later years, most biologists realized that organisms are different from inanimate matter and that the difference had to be explained not by postulating a vital force but by modifying rather drastically the mechanistic theory. Such a theory begins by granting that there is nothing in the processes, functions, and activities of living organisms that is in conflict with or outside of any of the laws of physics and chemistry. All biologists are thorough-going "materialists" in the sense that they recognize no supernatural or immaterial forces, but only such that are physico-chemical. But they do not accept the naive mechanistic explanation of the seventeenth century and disagree with the statement that animals are "nothing but" machines. Organismic biologists stress the fact that organisms have many characteristics that are without parallel in the world of inanimate objects. The explanatory equipment of the physical sciences is insufficient to explain complex living systems and, in particular, the interplay between historically acquired information and the responses of these genetic programs to the physical world. The phenomena of life have a much broader scope than the relatively simple phenomena dealt with by physics and chemistry. This is why it is just as impossible to include biology in physics as it is to include physics in geometry.

Attempts have been made again and again to define "life." These endeavors are rather futile since it is now quite clear that there is no special substance, object, or force that can be identified with life. The process of living, however, can be defined. There is no doubt that living organisms possess certain attributes that are not or not in the same manner found in inanimate objects. Different authors have stressed different characteristics, but I have been unable to find in the literature an adequate listing of such features. The list which I herewith present is presumably both incomplete and somewhat redundant. It may serve, however, for the want of a better tabulation, to illustrate the kinds of characteristics by which living organisms differ from inanimate matter.

53

Complexity and Organization

Complexity per se is not a fundamental difference between organic and inorganic systems. There are some highly complex inanimate systems (the air masses of the world weather system or any galaxy), and there are quite a few relatively simple organic systems like many macromolecules. Systems may have any degree of complexity, but, on the average, systems in the world of organisms are infinitely more complex than those of inanimate objects. Simon (1962) has defined complex systems as those in which "the whole is more than the sum of the parts, not in an ultimate, metaphysical sense but in the important pragmatic sense that, given the properties of the parts and the laws of their interaction, it is not a trivial matter to infer the properties of the whole." I accept this definition, except that we may continue to consider some relatively simple systems, such as the solar system, as complex even after we have succeeded in explaining their complexity. Complexity in living systems exists at every level from the nucleus (with its DNA program), to the cell, to any organ system (like kidney, liver, or brain), to the individual, the ecosystem, or the society. Living systems are invariably characterized by elaborate feedback mechanisms unknown in their precision and complexity in any inanimate system. They have the capacity to respond to external stimuli, the capacity for metabolism (binding or release of energy), and the capacity to grow and to differentiate.

Living systems do not have a random complexity but are highly organized. Most structures of an organism are meaningless without the rest of the organism; wings, legs, heads, kidneys cannot live by themselves but only as parts of the ensemble. Consequently, all parts have an adaptive significance and may be able to perform teleonomic activities. Such mutual adaptation of parts is unknown in the inanimate world. This co-adapted function of parts was already recognized by Aristotle when he said, "As every instrument and every bodily member subserves some partial end, that is to say, some specialization, so the whole body must be destined to minister to some plenary sphere of action" (*De Partibus* 1.5.645a10–15).

Chemical Uniqueness

Living organisms are composed of macromolecules with the most extraordinary characteristics. For instance, there are nucleic acids that can be translated into polypeptides, enzymes that serve as catalysts in metabolic processes, phosphates that permit energy transfer, and lipids that can build membranes. Many of these molecules are so specific and so uniquely capable of carrying out one particular function, like rhodopsine in the process of photoreception, that they occur in the animal and plant kingdoms whenever there is a need for that particular function. These organic macromolecules do not differ in principle from other molecules. They are, however, far more complex than the low molecular weight molecules that are the regular constituents of inanimate nature. The larger organic macromolecules are not normally found in inanimate matter.

Quality

The physical world is a world of quantification (Newton's movements and forces) and of mass actions. By contrast, the world of life can be designated as a world of qualities. Individual differences, communication systems, stored information, properties of the macromolecules, interactions in ecosystems, and many other aspects of living organisms are prevailingly qualitative in nature. One can translate these qualitative aspects into quantitative ones, but one loses thereby the real significance of the respective biological phenomena, exactly as if one would describe a painting of Rembrandt in terms of the wave lengths of the prevailing color reflected by each square millimeter of the painting.

In a like manner, many times in the history of biology, brave efforts to translate qualitative biological phenomena into mathematical terms have ultimately proved complete failures because they lost touch with reality. Early efforts to emphasize the importance of quality, like those of Galen, Paracelsus, and van Helmont, were likewise failures owing to the choice of wrong parameters, but they were the first steps in the right direction. The champions of quantification tend to consider the recognition of quality as something unscientific or at best as something purely descriptive and classificatory. They reveal by this bias how little they understand the nature of biological phenomena. Quantification is important in many fields of biology, but not to the exclusion of all qualitative aspects.

These are particularly important in relational phenomena, which are precisely the phenomena that dominate living nature. Species, classification, ecosystems, communicatory behavior, regulation, and just about every other biological process deals with relational properties. These can be expressed, in most cases, only qualitatively, not quantitatively.

Uniqueness and Variability

In biology, one rarely deals with classes of identical entities, but nearly always studies populations consisting of unique individuals. This is true for every level of the hierarchy, from cells to ecosystems. Many biological phenomena, particularly population phenomena, are characterized by extremely high variances. Rates of evolution or rates of speciation may differ from each other by three to five orders of magnitude, a degree of variability rarely if ever recorded for physical phenomena.

While entities in the physical sciences, let us say atoms or elementary particles, have constant characteristics, biological entities are characterized by their changeability. Cells, for instance, continuously change their properties and so do individuals. Every individual undergoes a drastic change from birth to death, that is, from the original zygote, through adolescence, adulthood, senescence, to death. Again, there is nothing like it in inanimate nature, except for radioactive decay, the behavior of highly complex systems (such as the Gulf Stream and weather systems), and some vague analogies in astrophysics.

Possession of a Genetic Program

All organisms possess a historically evolved genetic program, coded in the DNA of the nucleus of the zygote (or in RNA in some viruses). Nothing comparable to it exists in the inanimate world, except for manmade computers. The presence of this program gives organisms a peculiar duality, consisting of a phenotype and a genotype (see Chapter 16). Two aspects of this program must be specially emphasized: the first is that it is the result of a history that goes back to the origin of life and thus incorporates the "experiences" of all ancestors (Delbrück, 1949). The second is that it endows organisms with the capacity for teleonomic processes and activities, a capacity totally absent in the inanimate world. Except for the twilight zone of the origin of life, the possession of a genetic program provides for an absolute difference between organisms and inanimate matter.

56

One of the properties of the genetic program is that it can supervise its own precise replication and that of other living systems such as organelles, cells, and whole organisms. There is nothing exactly equivalent in inorganic nature. An occasional error may occur during replication (let us say one error in 10,000 or 100,000 replications). Once such a mutation has taken place, it becomes a constant feature of the genetic program. Mutation is the primary source of all genetic variation.

The full understanding of the nature of the genetic program was achieved by molecular biology only in the 1950s after the elucidation of the structure of DNA. Yet, it was already felt by the ancients that there had to be something that ordered the raw material into the patterned systems of living beings. As Delbrück (1971) pointed out quite correctly, Aristotle's *eidos* (even though considered immaterial, because invisible) was conceptually virtually identical with the ontogenetic program of the developmental physiologist. Buffon's *moule intérieure* was a similar ordering device. Yet, it required the rise of computer science before the concept of such a program became reputable. What is particularly important is that the genetic program itself remains unchanged while it sends out its instructions to the body. The whole concept of program is so novel that it is still resisted by many philosophers.

Historical Nature

One result of having an inherited genetic program is that classes of living organisms are not primarily assembled or recognized by similarity but by common descent, that is, by a set of joint properties due to a common history. Hence, many of the attributes of classes recognized by logicians are not at all appropriate characteristics of species or higher taxa. This is even true for cell lineages in ontogeny. In other words, the "classes" of the biologist often are not equivalent to the "classes" of the logician. This must be remembered in many definitional arguments, but in none more than the one whether species taxa are "individuals" or classes.

Natural Selection

Natural selection, the differential reproduction of individuals that differ uniquely in their adaptive superiority, is a process without exact equivalent among processes of change in the inanimate world. Considering how often natural selection is still being misunderstood, it is worth quoting Sewall Wright's perceptive remark (1967a): "The Darwinian process of continuous interplay of a random and a selective process is not intermediate between pure chance and pure determinism, but in its consequences qualitatively utterly different from either."

This process, at least in sexually reproducing species, is further characterized by the fact that, by recombination, a new gene pool is organized in each generation, and hence a new and unpredictable start is made for the selection procedure of the next generation.

Indeterminacy

There has long been argument among biologists and philosophers whether physical and biological processes differ in determinacy and predictability. Unfortunately, epistemological and ontological aspects have been consistently confused and this has beclouded the issue.

The word *prediction* is being used in two entirely different senses. When the philosopher of science speaks of prediction, he means *logical prediction,* that is, conformance of individual observations with a theory or a scientific law. Darwin's theory of common descent, for instance, permitted Haeckel the prediction that "missing links" between apes and humans would be found in the fossil record. Theories are tested by the predictions which they permit. Since the physical sciences are a system of theories to a far greater extent than biology, prediction plays in them a much greater role than in biology.

Prediction, in daily usage, is an inference from the present to the future, it deals with a sequence of events, it is *temporal prediction.* In strictly deterministic physical laws absolute temporal predictions are often possible, such as predictions of the occurrence of eclipses. Temporal predictions are much more rarely possible in the biological sciences. The sex of the next child in a family cannot be predicted. No one would have predicted at the beginning of the Cretaceous that the flourishing group of the dinosaurs would become extinct by the end of this era. Predictions in biology are, on the average, far more probabilistic than in the physical sciences.

The existence of the two kinds of predictions must be kept in mind in the discussion of causation and explanation. G. Bergmann defines a causal explanation as one "which owing to a law of nature permits to make predictions about future states of a system when its state at the present time is known." Essentially, this is only a rewording of the notorious boast of Laplace. Such claims have been rejected by Scriven (1959: 477) who asserts that (temporal) prediction is not part of causality and "that one cannot regard explanations as unsatisfactory when they . . . are not such as to enable the event in question to have been predicted."

In biology, and particularly in evolutionary biology, explanations ordinarily concern historical narratives. As far back as 1909 Baldwin specified two reasons why biological events are so often unpredictable: the great complexity of biological systems and the frequency at which unexpected novelties emerge at higher hierarchical levels. I can think of several others. Some of these might be considered ontological, others epistemological, indeterminacies. These factors do not weaken the principle of causality, conceived in a "postdictive" sense.[8]

Randomness of an event with respect to its significance. Spontaneous mutation caused by an error in DNA replication illustrates this cause for indeterminacy very well. There is no connection between the molecular event and its potential significance. The same is true for such events as crossing over, chromosomal segregation, gametic selection, mate selection, and for much in survival. Neither the underlying molecular phenomena nor the mechanical motions involved in some of these processes are related to their biological effects.

Uniqueness. The properties of a unique event or newly produced unique entity cannot be predicted (see above).

Magnitude of stochastic perturbations. Let me illustrate the effect of this factor by an example. Let us say a species consists of one million uniquely different individuals. Each individual has a chance to be killed by an enemy, to succumb to a pathogen, to encounter a weather catastrophe, to suffer from malnutrition, to fail to find a mate, or to lose its offspring before they can reproduce. These are some of the numerous factors determining reproductive success. Which of these factors will become active depends on highly variable environmental constellations which are unique and unpredictable. We have, therefore, two highly variable systems (unique individuals and unique environmental constellations) interacting with each other. Chance determines to a large extent how they mesh together.

Complexity. Each organic system is so rich in feedbacks, homeostatic devices, and potential multiple pathways that a complete description is quite impossible. Hence a prediction of its productions is also impossible. Furthermore, the analysis of such a system would require its destruction and thus preclude the completion of the analysis.

Emergence of new and unpredictable qualities at hierarchical levels (this will be discussed in more detail below).

The stated eight characteristics, together with additional ones to be mentioned below under the discussion of reductionism, make it clear that a living system is something quite different from any inanimate object. At the same time, not one of these characteristics is in conflict with a strictly mechanistic interpretation of the world.

Reduction and Biology

The claim of an autonomy of the science of living organisms, as manifested in the eight unique or special characters of life listed above, has been rather unpopular with many physical scientists and philosophers of the physical sciences. They have reacted by asserting that the seeming autonomy of the world of life does not really exist, but that all theories of biology can, at least in principle, be reduced to theories of physics. This, they claim, restores the unity of science.

The claim that reductionism is the only justifiable approach is often reinforced by the additional claim that its alternative is vitalism. This is not true. Even though some antireductionists have indeed been vitalists, virtually all recent antireductionists have emphatically rejected vitalism.

Actually, it would be difficult to find a more ambiguous word than the word "reduce." When one studies the reductionist literature, one finds that the term "reduction" has been used in at least three different meanings (Dobzhansky and Ayala, 1974; Hull, 1973b; Schaffner, 1969; Nagel, 1961).

Constitutive Reductionism

It asserts that the material composition of organisms is exactly the same as that found in the inorganic world. Furthermore, it posits that none of the events and processes encountered in the world of living organisms is in any conflict with the physico-chemical phenomena at the level of atoms and molecules. These claims are accepted by modern biologists. The difference between inorganic matter and living organisms does not consist in the substance of which they are composed but in the organization of biological systems. Constitutive reductionism is therefore not controversial. Virtually all biologists accept the assertations of constitutive reductionism, and have done so (except the vitalists) for the last two hundred years or more. Authors who accept constitutive reduction but reject other forms of reduction are *not* vitalists, the claims of some philosophers notwithstanding.

Explanatory Reductionism

This type of reductionism claims that one cannot understand a whole until one has dissected it into its components, and again these components into theirs, down to the lowest hierarchical level of integration. In biological phenomena it would mean reducing the study of all phenomena to the molecular level, that is, "Molecular biology is all of biology." Indeed, it is true that such explanatory reduction is sometimes illuminating. The functioning of genes was not understood until Watson and Crick had figured out the structure of DNA. In physiology, likewise, the functioning of an organ is usually not fully understood until the molecular processes at the cellular level are clarified.

There are, however, a number of severe limitations to such explanatory reduction. One is that the processes at the higher hierarchical level are often largely independent of those of the lower levels. The units of the lower levels may be integrated so completely that they operate as units at the higher levels. The functioning of an articulation, for example, can be explained without a knowledge of the chemical composition of cartilage. Furthermore, replacing the articulating surface by a plastic, as is done in modern surgery, may completely restore the normal functioning of an articulation. There are probably as many cases where a dissection of a functional system into its components is unhelpful or at least irrelevant as there are cases where this is of explanatory value. A facile application of explanatory reduction in the history of biology has often done more harm than good. Examples are the early cell theory, which interpreted the organism as "an aggregate of cells," or early population genetics, which considered the genotype an aggregate of independent genes with constant fitness values.

Extreme analytical reductionism is a failure because it cannot give proper weight to the interaction of components of a complex system. An isolated component almost invariably has characteristics that are different from those of the same component when it is part of its ensemble, and does not reveal, when isolated, its contribution to the interactions. René Dubos (1965: 337) has well stated the reasons why the atomized approach is singularly unproductive when applied to complex systems: "In the most common and probably the most important phenomena of life, the constituent parts are so interdependent that they lose their character, their meaning, and indeed their very existence, when dissected from the functional whole. In order to deal with problems of organized complexity, it is therefore essential to investigate situations in which several interrelated systems function in an integrated manner."

61

The most important conclusions one can draw from a critical study of explanatory reductionism is that the lower levels in hierarchies or systems supply only a limited amount of information on the characteristics and processes of the higher levels. As the physicist P. W. Anderson said (1972: 393–396), "The more the elementary particle physicists tell us about the nature of the fundamental laws, the less relevance they seem to have to the very real problems of the rest of science, much less of society." Furthermore, it is rather misleading to apply the term "reduction" to an analytical method.

There are numerous other ways in which analysis of complex biological systems can be facilitated. The genetics of animals, for instance, was originally studied in horses, cattle, dogs, and other large mammals. Geneticists later shifted to fowl and various species of rodents. To get more numerous generations per year and perhaps simpler genetic systems, rodents were replaced after 1910 in most genetic labs by *Drosophila melanogaster* and other species of *Drosophila*. This was followed by a shift in the 1930s to *Neurospora* and other species of fungi (yeasts). Finally, most molecular genetics was done with bacteria (for example, *Escherichia coli*) and various viruses. In addition to a more rapid sequence of generations, the endeavor was to find ever simpler genetic systems and extrapolate from them to the more complex systems. By and large this hope was fulfilled, except that it was shown eventually that the genetic system of prokaryotes (bacteria) and of viruses is not quite comparable to that of eukaryotes, where the genetic material is organized in complex chromosomes. Simplification, thus, must be undertaken with care. There is always the danger that one shifts to a system that, in its simplicity, is sufficiently different to be no longer comparable.

Theory Reductionism

62   This type of reductionism postulates that the theories and laws formulated in one field of science (usually a more complex field or one higher in the hierarchy) can be shown to be special cases of theories and laws formulated in some other branch of science. If this is done successfully, one branch of science has been "reduced" to the other one, in the quaint language of certain philosophers of science. To take a specific case, biology is considered to be reduced to physics when the terms of biology are defined in terms of physics and the laws of biology are deduced from the laws of physics.

Such theory reduction was repeatedly attempted within the physical sciences, but never, according to Popper (1974), with complete success. I am not aware of any biological theory that has ever been reduced to a physico-chemical theory. The claim that genetics has been reduced to chemistry after the discovery of the structure of DNA, RNA, and certain enzymes cannot be justified. To be sure, the chemical nature of a number of black boxes in the classical genetic theory was filled in, but this did not affect in any way the nature of the theory of transmission genetics. As gratifying as it is to be able to supplement the classical genetic theory by a chemical analysis, this does not in the least reduce genetics to chemistry. The essential concepts of genetics, like gene, genotype, mutation, diploidy, heterozygosity, segregation, recombination, and so on, are not chemical concepts at all and one would look for them in vain in a textbook on chemistry.

Theory reductionism is a fallacy because it confuses *processes* and *concepts*. As Beckner (1974) has pointed out, such biological processes as meiosis, gastrulation, and predation are also chemical and physical processes, but they are only biological concepts and cannot be reduced to physico-chemical concepts. Furthermore, any adapted structure is the result of selection but this again is a concept which cannot be expressed in strictly physico-chemical terms.

It is a fallacy because it fails to consider the fact that the same event may have entirely different meanings in several different conceptual schemes. The courtship of a male, for instance, can be described entirely in the language and conceptual framework of the physical sciences (locomotion, energy turnover, metabolic processes, and so on), but it can also be described in the framework of behavioral or reproductive biology. The same is true for many other events, properties, relations, and processes relating to living organisms. Species, competition, territory, migration, and hibernation are examples of organismic phenomena for which a purely physical description is at best incomplete and usually biologically irrelevant.

63

This discussion of reductionism can be summarized by saying that the analysis of systems is a valuable method, but that attempts at a "reduction" of purely biological phenomena or concepts to laws of the physical sciences has rarely, if ever, led to any advance in our understanding. Reduction is at best a vacuous, but more often a thoroughly misleading and futile, approach. This futility is particularly well illustrated by the phenomenon of emergence.

Emergence

Systems almost always have the peculiarity that the characteristics of the whole cannot (not even in theory) be deduced from the most complete knowledge of the components, taken separately or in other partial combinations. This appearance of new characteristics in wholes has been designated as *emergence*.[9] Emergence has often been invoked in attempts to explain such difficult phenomena as life, mind, and consciousness. Actually, emergence is equally characteristic of inorganic systems. As far back as 1868, T. H. Huxley asserted that the peculiar properties of water, its "aquosity," could not be deduced from our understanding of the properties of hydrogen and oxygen. The person, however, who was more responsible than anyone else for the recognition of the importance of emergence was Lloyd Morgan (1894). There is no question, he said, "that at various grades of organization, material configurations display new and unexpected phenomena and that these include the most striking features of adaptive machinery." Such emergence is quite universal and, as Popper said, "We live in a universe of emergent novelty" (1974: 281). Emergence is a descriptive notion which, particularly in more complex systems, seems to resist analysis. Simply to say, as has been done, that emergence is due to complexity is, of course, not an explanation.

Perhaps the two most interesting characteristics of new wholes are (1) that they, in turn, can become parts of still higher-level systems, and (2) that wholes can affect properties of components at lower levels. The latter phenomenon is sometimes referred to as "downward causation" (Campbell, 1974: 182). Emergentism is a thoroughly materialistic philosophy. Those who deny it, like Rensch (1971; 1974), are forced to adopt pan-psychic or hylozoic theories of matter.

Two false claims against emergentism must be rejected. The first is that emergentists are vitalists. This claim, indeed, was valid for some of the nineteenth-century and early twentieth-century emergentists, but it is not valid for modern emergentists, who accept constitutive reduction without reservation and are therefore, by definition, nonvitalists. The second is the assertion that it is part of emergentism to believe that organisms can only be studied as wholes, and that any further analysis is to be rejected. Perhaps there have been some holists who have made such a claim, but this view is certainly alien to 99 percent of all emergentists. All they claim is that explanatory reduction is incomplete, since new and previously unpredictable characters emerge at higher levels of complexity in hierarchical systems. Hence, complex systems must be studied at every level, because each level has properties not shown at lower levels.

Some recent authors have rejected the term "emergence" as tainted with an unwanted metaphysical flavor. Simpson (1964b) has referred to it as the "compositional" method, Lorenz (1973) as fulguration. So many authors, however, have now adopted the term "emergence"—and, like the term "selection," it has been "purified" by this frequent usage to such an extent (through the elimination of vitalistic and finalistic connotations)—that I see no reason for not adopting it.

The Hierarchical Structure of Living Systems

Complex systems very often have a hierarchical structure (Simon, 1962), the entities of one level being compounded into new entities at the next higher level, like cells into tissues, tissues into organs, and organs into functional systems. Hierarchical organization is also present in the inanimate world, such as, for instance, elementary particles, atoms, molecules, crystals, and so on, but it is in living systems that hierarchical structure is of special significance. Pattee (1973) asserts that all problems of biology, particu-

64

larly those relating to emergence (see below), are ultimately problems of hierarchical organization.

In spite of the widespread interest in hierarchies, we are still
rather uncertain about the classification of hierarchies and about
the special attributes of different kinds of hierarchies. In biology
one deals apparently with two kinds of hierarchies. One is represented by *constitutive hierarchies,* like the series macromolecule, cellular organelle, cell, tissue, organ, and so forth. In such a hierarchy the members of a lower level, let us say tissues, are combined
into new units (organs) that have unitary functions and emergent
properties. The formation of constitutive hierarchies is one of the
most characteristic properties of living organisms. At each level
there are different problems, different questions to be asked, and
different theories to be formulated. Each of these levels has given
rise to a separate branch of biology: molecules to molecular biology, cells to cytology, tissues to histology, and so forth, up to biogeography and the study of ecosystems. Traditionally, the recognition of these hierarchical levels has been one of the ways of
subdividing biology into fields. To which particular level an investigator will turn, depends on his interests. A molecular biologist is
simply not interested in the problems studied by the functional
morphologist or the zoogeographer and vice versa. The problems
and findings at other levels are usually largely irrelevant for those
working at a given hierarchical level. For a full understanding of
living phenomena every level must be studied but, as was pointed
out above, the findings made at lower levels usually add very little
toward solving the problems posed at the higher levels. When a
well-known Nobel laureate in biochemistry said, "There is only
one biology, and it is molecular biology," he simply revealed his
ignorance and lack of understanding of biology.

With so many components contributing to the functioning of
a biological system, it is for the working scientist a matter of strategy and interest to decide the study of which level would make
the greatest contribution toward the full understanding of the system under present circumstances. This includes the decision to
leave certain black boxes unopened.

An altogether different kind of hierarchy may be designated
as an *aggregational hierarchy.* The best known paradigm for it is
the Linnaean hierarchy of taxonomic categories, from the species
through genus and family up to phylum and kingdom. It is strictly
an arrangement of convenience. The units at the lower level—for
example, the species of a genus, or the genera of a family—are

not compounded by any interaction into emerging new higher level units as a whole. Instead, groups of taxa are ranked in ever higher categories by the taxonomist. The validity of this statement is not weakened by the fact that the members of a (natural) higher taxon are descendants of a common ancestor. Such hierarchies produced by rank assignments to categories are essentially only classificatory devices. It is unknown to me to what extent there might be still other kinds of hierarchies.

Holism–Organicism

66

Perceptive biologists, all the way back to Aristotle, have been dissatisfied with a purely atomistic-reductionist approach to the problems of biology. Most biologists simply stressed wholeness, that is, the integration of systems. Others sidestepped scientific explanation by invoking metaphysical forces. Vitalism was the favorite explanation right into the twentieth century. When Smuts (1926) introduced the convenient term "holism" to express that the whole is more than the sum of its parts, he combined it with vitalistic ideas which unfortunately tainted the otherwise suitable term "holism" from its very beginning. The terms "organismal" and "organicism" were apparently introduced by Ritter (1919) and are now rather widely used, for instance by Beckner (1974: 163). Bertalanffy (1952) listed some thirty authors who had declared their sympathy for a holistic-organismic approach. This list is, however, very incomplete, not even including the names of Lloyd Morgan, Jan Smuts, and J. S. Haldane. Francois Jacob's (1970) concept of the integron is a particularly well-argued endorsement of organismic thinking.

In contrast to the earlier holistic proposals which usually were more or less vitalistic, the newer ones are strictly materialistic. They stress that the units at higher hierarchical levels are more than the sums of their parts and, hence, that a dissection into parts always leaves an unresolved residue—in other words, that explanatory reduction is unsuccessful. More importantly, they stress the autonomous problems and theories of each level and ultimately the autonomy of biology as a whole. The philosophy of science can no longer afford to ignore the organismic concept of biology as being vitalistic and hence belonging to metaphysics. A philosophy of science restricted to that which can be observed in inanimate objects is deplorably incomplete.

There are many scientists who concentrate on the study of isolated objects and processes. They deal with them as if they existed in a vacuum. Perhaps the most important aspect of holism is that it emphasizes relationships. I myself have always felt that relationships are not given sufficient weight. This is why I have called the species concept a relational concept, and why my work on genetic revolution (1954) and on the cohesion of the genotype (1975) both deal with relational phenomena. My attack on bean-bag genetics (1959d) springs from the same source (see Chapter 13).

Others have felt the same way. The painter Georges Braque (1882–1963) declared, "I do not believe in things, I believe only in their relationships." Einstein, of course, based his entire relativity theory on the consideration of relationship. And when discussing changing selective values of genes in different genetic environments, I have called this concept, somewhat jokingly, the relativity theory of genes.

THE CONCEPTUAL STRUCTURE OF BIOLOGY

When comparing biology with the physical sciences, I have, up to now, treated biology as if it were a homogeneous science. This is not correct. Actually, biology is diversified and heterogeneous in several important ways. For thousands of years biological phenomena were subsumed under two labels, medicine (physiology) and natural history. This was actually a remarkably perceptive division, far more penetrating than such later labels of convenience as zoology, botany, mycology, cytology, or genetics. For biology can be divided into the study of proximate causes, the subject of the physiological sciences (broadly conceived), and into the study of ultimate (evolutionary) causes, the subject matter of natural history (Mayr, 1961).

What proximate and evolutionary causes are is best made clear by a concrete example. Why does a certain warbler individual in temperate North America start its southward migration in the night of August 25? The proximate causes are that the bird, belonging to a migratory species responding to photoperiodicity, had become physiologically ready to migrate at that date since the numbers of hours of daylight had dropped to a certain threshold and since weather conditions (wind, temperature, barometric pressure) favored departure during that night. Yet a screech owl and a nuthatch living in the same piece of woodlands and exposed to the same decrease in daylight and to the same weather conditions did not depart southward; in fact, these other species remained in the same area throughout the year because they lacked the migratory urge. Obviously, then, there has to be an entirely different second set of causal factors to account for the differences between the migratory and the sedentary species. It consists in a genotype acquired through natural selection during thousands and millions of years of evolution which determines whether or not a population or species is migratory. An insect-eating warbler or flycatcher will have been selected to migrate because it otherwise would starve to death during the winter. Other species which can find their food throughout the winter will have been selected to avoid the perilous and, for them, unnecessary migration.

To give another example, the proximate causation of sexual dimorphism might be hormonal or some genetic growth factors, while sexual selection or a selective advantage of differential utilization of the food niche might be the ultimate causation. Any biological phenomenon is due to these two independent kinds of causations.

There is considerable uncertainty concerning the origin of the terminology proximate-ultimate. Herbert Spencer and George Romanes have used these terms rather vaguely, but John Baker was apparently the first author to distinguish clearly between ultimate causes responsible for the evolution of a given genetic program (selection) and proximate causes responsible, so to speak, for the release of the stored genetic information in response to current environmental stimuli: "Thus abundance of insect food for the young [at certain months] might be the ultimate, and length of day the proximate cause of a breeding season" (Baker, 1938: 162).

The two biologies that are concerned with the two kinds of causations are remarkably self-contained. Proximate causes relate to the functions of an organism and its parts as well as its development, from functional morphology down to biochemistry. Evolutionary, historical, or ultimate causes, on the other hand, attempt to explain why an organism is the way it is. Organisms, in contrast to inanimate objects, have two different sets of causes because organisms have a genetic program. Proximate causes have to do with the decoding of the program of a given individual; evolutionary causes have to do with the changes of genetic programs through time, and with the reasons for these changes.

The functional biologist is vitally concerned with the operation and interaction of structural elements, from molecules up to organs and whole individuals. His ever-repeated question is "How?" How does something operate, how does it function? The functional anatomist who studies an articulation shares this method and approach with the molecular biologist who studies the function of DNA molecules in the transfer of genetic information. The functional biologist attempts to isolate the particular component he studies, and in any given study he usually deals with a single individual, a single organ, a single cell, or a single part of a cell. He attempts to eliminate and control all variables, and he repeats his experiments under constant or varying conditions until he believes he has clarified the function of the element he studies. The chief technique of the functional biologist is the experiment, and his approach is essentially the same as that of the physicist and the chemist. Indeed, by isolating the studied phenomenon sufficiently from the complexities of the organism, he may achieve the ideal of a purely physical or chemical experiment. In spite of certain limitations of this method, one must agree with the functional biologist that such a simplified approach is an absolute necessity for achieving his particular objectives. The spectacular success of biochemical and biophysical research justifies this direct, although distinctly simplistic, approach (Mayr, 1961). There is little argument about the methodology and the achievements of functional biology from William Harvey to Claude Bernard and to molecular biology.

Every organism, whether an individual or a species, is the product of a long history, a history that dates back more than 3000 million years. As Max Delbrück (1949: 173) has said, "A mature physicist, acquainting himself for the first time with the problems of biology, is puzzled by the circumstance that there are no 'absolute phenomena' in biology. Everything is time-bound and space-bound. The animal or plant or micro-organism he is working with is but a link in an evolutionary chain of changing forms, none of which has any permanent validity."

There is hardly any structure or function in an organism that can be fully understood unless it is studied against this historical background. To find the causes for the existing characteristics, and particularly adaptations, of organisms is the main preoccupation of the evolutionary biologist. He is impressed by the enormous diversity as well as the pathway by which it has been achieved. He studies the forces that bring about changes in faunas and floras (as in part documented by paleontology), and he studies the steps by which have evolved the miraculous adaptations so characteristic of every aspect of the organic world.

In evolutionary biology almost all phenomena and processes are explained through inferences based on comparative studies. These, in turn, are made possible by very careful and detailed descriptive studies. It is sometimes overlooked how essential a component in the methodology of evolutionary biology the underlying descriptive work is. The conceptual breakthroughs of Darwin, Weismann, Jordan, Rensch, Simpson, and Whitman would have been quite impossible without the sound foundation of descriptive research on which they could erect their conceptual framework (Lorenz, 1973). Natural history, by necessity, was strictly descriptive in its early history and so was early anatomy. The endeavors of the systematists of the eighteenth and early nineteenth centuries to classify the diversity of nature rose more and more above simple description. After 1859 the autonomy of evolutionary biology as a legitimate biological discipline was no longer in question.

Functional biology has often been designated as quantitative; by contrast it is in many cases quite legitimate to refer to evolutionary biology as qualitative. The term "qualitative" was a pejorative designation during the anti-Aristotelian period of the scientific revolution. In spite of the efforts of Leibniz and other perceptive authors, it remained that way until the Darwinian revolution. Under its liberating impact a change in the intellectual climate occurred that made the development of evolutionary biology possible.

This revolution was not successful instantaneously. Many physical scientists and functional biologists have consistently failed to understand the special nature of evolutionary biology. Driesch in his autobiography, written in the 1930s, comments with considerable satisfaction that biological professorships are nowadays awarded "only to experimentalists. Systematic problems have receded entirely to the background." He totally ignored the existence of evolutionary biology. This attitude was widespread among experimental biologists.

Haeckel (1877) was perhaps the first biologist to object vigorously to the notion that all science had to be like the physical sciences or to be based on mathematics. Evolutionary biology, he insisted, is a historical science. Particularly the studies of embryology, paleontology, and phylogeny are historical, he said. Instead, of "historical," we would perhaps now say "regulated by historically acquired genetic programs and their changes in historical time." Alas, this viewpoint made only very slow progress. When Baldwin in 1909 pointed out how much the acceptance of Darwinism had changed the thinking of biologists, he concluded that "the reign of physical science and of mechanical law over the scientific and philosophical mind is over now, at the opening of the 20th century." In this optimism he was mistaken; there are still many philosophers who write as if Darwin had never existed and as if evolutionary biology were not part of science.

71

Historical Narratives and Evolutionary Biology

Philosophy of science, when first developed, was firmly grounded in physics, and more specifically in mechanics, where processes and events can be explained as the consequences of specific laws, with prediction symmetrical to causation. History-bound scientific phenomena, by contrast, do not fit well into this conceptualization. As the physicist Hermann Bondi (1977: 6) stated correctly, "Any theory of the origin of the solar system, of the origin of life on earth, of the origin of the universe is of an exceptional nature [as compared to conventional theories of physics] in that it tries to describe an event in some sense unique." Uniqueness, indeed, is the outstanding characteristic of any event in evolutionary history.

Several philosophers of science have, therefore, argued that explanations in evolutionary biology are not provided by theories but by "historical narratives." As T. A. Goudge (1961: 65–79) has stated: "Narrative explanations enter into evolutionary theory at points where singular events of major importance for the history of life are being discussed . . . Narrative explanations are constructed without mentioning any general laws . . . Whenever a narrative explanation of an event in evolution is called for, the event is not an instance of a kind [class], but is a singular occurrence, something which has happened just once and which can not recur [in the same way] . . . Historical explanations form an essential part of evolutionary theory." Morton White (1963) has developed these ideas further. The notion of central subjects is crucial to the logical structure of historical narratives. Any phyletic line, any fauna (in zoogeography), or any higher taxon is a central subject in terms of the historical narrative theory and has continuity through time. Sciences in which historical narratives play an important role include cosmogony, geology, paleontology (phylogeny), and biogeography.

Historical narratives have explanatory value because earlier events in a historical sequence usually make a causal contribution to later events. For instance, the extinction of the dinosaurs at the end of the Cretaceous vacated a large number of ecological niches and thus set the stage for the spectacular radiation of mammals during the Paleocene and Eocene. One of the objects of a historical narrative, thus, is to discover causes responsible for ensuing events.

Philosophers trained in the axioms of essentialistic logic seem to have great difficulty in understanding the peculiar nature of uniqueness and of historical sequences of events. Their attempts to deny the importance of historical narratives or to axiomatize them in terms of covering laws fail to convince.

The most characteristic aspect of evolutionary biology are the questions it asks. Instead of concentrating on what? and how? as does the biology of proximate causes, it asks why? Why are certain organisms very similar to each other, while others are utterly different? Why are there two sexes in most species of organisms? Why is there such a diversity of animal and plant life? Why are the faunas of some areas rich in species and those of others poor?

If an organism has certain characteristics, they must have been derived from those of an ancestor or have been acquired because they were of selective advantage. The question "why?" in the sense of "what for?" is meaningless in the world of inanimate objects. One can ask, "Why is the sun hot?" but only in the sense of "how come?" By contrast, in the living world the question "what for?" has a powerful heuristic value. The question, "Why are there valves in the veins?" contributed to Harvey's discovery of the circulation of blood. By asking, "Why do the nuclei in cells undergo the complex process of reorganization during mitosis, instead of simply dividing in half?," Roux (1883) gave the first correct interpretation of cell division. He fully understood that "the question concerning the significance of a biological process can be asked in two ways. First with reference to its function for the biological structure in which it occurs, but secondly one can . . . also [ask for] the causes responsible for the origin and progress of this process." Therefore the evolutionary biologist must always ask why questions when he attempts to analyze evolutionary causations.

All biological processes have both a proximate cause and an evolutionary cause. Much confusion in the history of biology has

72

resulted when authors concentrated exclusively either on proximate or on evolutionary causation. For instance, let us consider the question, "What is the reason for sexual dimorphism?" T. H. Morgan (1932) castigated the evolutionists for speculating about this question when, as he said, the answer is so simple: male and female tissues during ontogeny respond to different hormonal influences. He never considered the evolutionary question why the hormonal systems of males and females are different. The role of sexual dimorphism in courtship and other behavioral and ecological contexts was of no interest to him.

73

Or to take another example: What is the meaning of fertilization? Several functional biologists, when considering this question, were impressed by the fact that the unfertilized egg is quiescent while development (indicated by the first cleavage division) begins almost immediately after the spermatozoon had entered the egg. Fertilization, therefore, it was stated by some functional biologists, has as its objective the initiation of development. The evolutionary biologist, by contrast, pointed out that fertilization was not needed in parthenogenetic species in order to initiate development, and he concluded therefore that the true objective of fertilization is the achievement of a recombination of paternal and maternal genes, such recombination producing the genetic variability required as material for natural selection (Weismann, 1886).

It is evident from these case histories that no biological problem is fully solved until both the proximate and the evolutionary causation has been elucidated. Furthermore, the study of evolutionary causes is as legitimate a part of biology as is the study of the usually physico-chemical proximate causes. The biology of the origin of genetic programs and their changes through evolutionary history is as important as the biology of the translating (decoding) of genetic programs, that is, the study of proximate causes. The assumption of Julius von Sachs, Jacques Loeb, and other naive mechanists that biology consists entirely of a study of proximate causations is demonstrably wrong.

A New Philosophy of Biology

It is now clear that a new philosophy of biology is needed. This will include and combine the cybernetic-functional-organizational ideas of functional biology with the populational-historical program-uniqueness-adaptedness concepts of evolutionary biology. Although obvious in its essential outlines, this new philosophy of biology is, at the present time, more of a manifesto of something to be achieved than the statement of a mature conceptual system. It is most explicit in its criticism of logical positivism, essentialism, physicalism, and reductionism but is still rather hesitant and inchoate in its major theses. The various authors who have written on the subject in recent years, like Simpson, Rensch, Mainx, the contributors to the Ayala and Dobzhansky volume (1974), and the authors of philosophies of biology (Beckner, Campbell, Hull, Munson, and so on), still differ widely from one another not only in emphasis but even in some basic tenets (for instance, acceptance or rejection of emergentism). There is, however, one very encouraging development. All of the more perceptive writers on the subject reject the extreme views of the past: Not a single one of them accepts vitalism in any form whatsoever. Nor does any of them endorse any kind of atomistic or explanatory reductionism. With the borderlines of a new philosophy of biology clearly staked out, there is every hope for a true synthesis in the not-too-distant future.

Philosophers of science, when dealing with the subject of biology, have devoted a good deal of time and attention to the problems of mind, consciousness, and life. I feel that they have created for themselves unnecessary difficulties. As far as consciousness is concerned, it is impossible to define it. Various criteria indicate that even lower invertebrates have consciousness, possibly even protozoans in their avoidance reactions. Whether one wants to pursue this down to the prokaryotes (for example, magnetic bacteria) is a matter of taste. At any rate, the concept of consciousness cannot even approximately be defined and therefore detailed discussion is impossible.

As far as the words "life" and "mind" are concerned, they merely refer to reifications of activities and have no separate existence as entities. "Mind" refers not to an object but to mental activity and since mental activities occur throughout much of the animal kingdom (depending on how we define "mental"), one can say that mind occurs whenever organisms are found that can be shown to have mental processes. Life, likewise, is simply the reification of the processes of living. Criteria for living can be stated and adopted, but there is no such thing as an independent "life" in a living organism. The danger is too great that a separate existence is assigned to such "life" analogous to that of a soul (Blandino, 1969). The avoidance of nouns that are nothing but reifications of processes greatly facilitates the analysis of the phenomena that are characteristic for biology.

The gradual emergence of an autonomous philosophy of biology has been a long, drawn-out, and painful process. Early attempts were doomed to failure owing to a lack of knowledge of the facts of biology and the prevalence of unsuitable or erroneous concepts. This is well illustrated by Kant's philosophy of biology. What Kant did not realize was that the subject matter of biology first had to be straightened out by the biologists themselves (by science!)—for instance, that it was the task of the systematists to explain causally the Linnaean hierarchy (done by Darwin in his theory of common descent), or that it was the task of the evolutionist to explain the origin of adaptation without invoking supernatural forces (done by Darwin and Wallace through their theory of natural selection). Once these explanations had become available, the philosophers again could have joined the enterprise. They did so, but alas—on the whole—by fighting Darwin and by endorsing biologically unsound theories. This has continued into modern times, to wit, the publications of authors like Marjorie Grene, Hans Jonas, and so on.

I think it is fair to state that biologists like Rensch, Waddington, Simpson, Bertalanffy, Medawar, Ayala, Mayr, and Ghiselin have made a far greater contribution to a philosophy of biology than the whole older generation of philosophers, including Cassirer, Popper, Russell, Bloch, Bunge, Hempel, and Nagel. It is only the generation of the youngest philosophers (Beckner, Hull, Munson, Wimsatt, Beatty, Brandon) who are finally able to get away from the obsolete biological theories of vitalism, orthogenesis, macrogenesis, and dualism or the positivist-reductionist theories of the older philosophers.[10] One needs only to read what such an otherwise so brilliant philosopher as Ernst Cassirer says about Kant's *Critique of Judgment* to realize how difficult it is for a traditional philosopher to understand the problems of biology. To their excuse, it must be stated that the guilt is shared by the biologists, who have failed to present a clear analysis of the conceptual problems of biology. They were unable to see the forest for the trees.

What principles or concepts would form a good basis on which to found a philosophy of biology? I shall not even attempt to be exhaustive, but from the preceding discussion it should be evident

(1) that a full understanding of organisms cannot be secured through the theories of physics and chemistry alone;

(2) that the historical nature of organisms must be fully considered, in particular their possession of an historically acquired genetic program;

(3) that individuals at most hierarchical levels, from the cell up, are unique and form populations, the variance of which is one of their major characteristics;

(4) that there are two biologies, functional biology, which asks proximate questions, and evolutionary biology, which asks ultimate questions;

(5) that the history of biology has been dominated by the establishment of concepts and by their maturation, modification, and—occasionally—their rejection;

(6) that the patterned complexity of living systems is hierarchically organized and that higher levels in the hierarchy are characterized by the emergence of novelties;

(7) that observation and comparison are methods in biological research that are fully as scientific and heuristic as the experiment;

(8) that an insistence on the autonomy of biology does not mean an endorsement of vitalism, orthogenesis, or any other theory that is in conflict with the laws of chemistry or physics.

A philosophy of biology must include a consideration of all major specifically biological concepts, not only those of molecular biology, physiology, and development but also those of evolutionary biology (such as natural selection, inclusive fitness, adaptation, progress, descent), systematics (species, category, classification), behavioral biology and ecology (competition, resource utilization, ecosystem).

I might even add a few "don'ts." For instance, a philosophy of biology should not waste any time on a futile attempt at theory reduction. It should not take one of the existing philosophies of physics as a starting point. (It is depressing to discover how little some prestigious volumes in this field have to do with the actual practices in scientific research, at least in biology.) It should not focus most of its attention on laws, considering what small role laws actually play in much of biological theory. In other words, what is needed is an uncommitted philosophy of biology which stays equally far away from vitalism and other unscientific ideologies and from a physicalist reductionism that is unable to do justice to specifically biological phenomena and systems.

Biology and Human Thought

C. P. Snow in a well-known essay (1959) claimed that there is an unbridgeable gap between the cultures of science and the humanities. He is right about the communication gap between physicists and humanists, but there is almost as great a gap between, let us say, physicists and naturalists. Also, there is even a rather pronounced failure of communication between representatives of functional biology and evolutionary biology. Furthermore, functional biology shares with the physical sciences an interest in laws, prediction, all aspects of quantity and quantification and the functional aspects of processes, while in evolutionary biology questions like quality, historicity, information, and selective value are of special interest, questions that are also of concern in the behavioral and social sciences but not in physics. It is not at all unreasonable, therefore, to consider evolutionary biology as something of a bridge between the physical sciences on one hand and the social sciences and the humanities on the other.

In a comparison of history and the sciences Carr (1961: 62) states that history is believed to differ from all the sciences in five respects: (1) history deals exclusively with the unique, science with the general; (2) history teaches no lessons; (3) history is unable to predict; (4) history is necessarily subjective; and (5) history (unlike science) involves issues of religion and morality. These differences are valid only for the physical sciences. Statements 1, 3, 4, and 5 are largely true also for evolutionary biology and, as Carr admits, some of these claims (statement 2, for instance) are not even strictly true for history. In other words, the sharp break between science and the nonsciences does not exist.

The nature of the impact which science has had on man and his thinking has been controversial. That Copernicus, Darwin, and Freud have profoundly altered man's thinking can hardly be questioned. The impact of the physical sciences in the last few hundred years has been primarily via technology. Kuhn (1971) claims that for a scientist to exert a real influence on human thinking, he must be read by lay people. No matter how distinguished certain mathematical physicists (including Einstein and Bohr) were, "none of them, as far as I can see, has had more than the most tenuous and indirect impact upon the development of extrascientific thought." Whether or not Kuhn is right, one can certainly claim that some scientists have more influence on the thinking of the intelligent layman than others. Quite likely it depends on the degree with which the scientist's subject matter is of immediate concern to the layperson. Hence, biology, psychology, anthropology, and related sciences naturally have a much greater impact on human thought than the physical sciences.

Prior to the rise of science it was the philosophers who were, so to speak, charged with the task of furthering the understanding of this world. Since the nineteenth century, philosophy has more and more retreated into the study of the logic and methodology of science, largely abandoning vast areas such as metaphysics, ontology, epistemology that used to be the major concern of philosophy. Much of this area, unfortunately, has been left a virtual no man's land because most scientists are entirely satisfied to pursue their special researches, not at all concerned how the general conclusions derivable from these studies might affect basic issues of human concern and of general epistemology. Philosophers, on the other hand, find it difficult if not impossible to keep up with the rapid advances of science, and as a result turn to trivial or esoteric problems. Opportunities of joint approaches by philosophers and scientists, as profitable as they would be, are all too rarely taken.

Biology and Human Values

It is sometimes stated that, in contrast to religious interpretations, science has the great advantage of being impersonal, detached, unemotional, and thus completely objective. This may well be true for most explanations in the physical sciences but it is not at all true for much of explanation in the biological sciences. The findings and theories of the biologist are quite often in conflict with traditional values of our society. For instance, Darwin's teacher Adam Sedgwick vigorously rejected the theory of natural selection because it implied the refutation of the argument from design and would thus permit a materialistic explanation of the world, that is, as he saw it, an elimination of God from the explanation of order and adaptation in the world. Biological theory very often is, indeed, very value-laden. As examples one might mention Darwin's theory of common descent which deprived man of his unique place in the universe. More recently, the argument as to whether and to what extent IQ is genetically determined, particularly when joined to the race problem, and the arguments about sociobiology are apt illustrations. In all these cases conflicts arose between certain scientific findings or interpretations and certain traditional value systems. No matter how objective scientific research might be, its findings often lead to conclusions that are value-laden.

Literary critics have long been aware of the impact which the writings of some scientists have had on novelists and essayists and through them on the public at large. The reports of the happiness and innocence of primitive natives of exotic countries brought home by eighteenth-century explorers, as erroneous as they were, greatly affected eighteenth- and nineteenth-century writers and, ultimately, political ideologies.

It was a tragedy both for biology and for mankind that the currently prevailing framework of our social and political ideals developed and was adopted when the thinking of western man was largely dominated by the ideas of the scientific revolution, that is, by a set of ideas based on the principles of the physical sciences. This included essentialistic thinking and, as a correlate, a belief in the essential identity of members of a class. Even though the ideological revolution of the eighteenth century was, to a large extent, a rebellion against feudalism and class privileges, it cannot be denied that the ideals of democracy were in part derived from the stated principles of physicalism. As a consequence, democracy can be interpreted to assert not only equality before the law but also essentialistic identity in all respects. This is expressed in the claim, "All men are *created* equal," which is something very different from the statement, "All men have equal rights and are equal before the law." Anyone who believes in the genetic uniqueness of every individual thereby believes in the conclusion, "No two individuals are *created* equal."

When evolutionary biology developed in the nineteenth century, it demonstrated the inapplicability of these physical principles to unique biological individuals, to heterogeneous populations, and to evolutionary systems. Nevertheless, the fused ideology of physicalism and antifeudalism, usually called democracy (no two people have exactly the same concept of democracy), has taken over in the western world to such an extent that even the slightest implied criticism (as in these lines) is usually rejected with complete intolerance. Democratic ideology and evolutionary thinking share a high regard for the individual but differ on many other aspects of our value system. The recent controversy over sociobiology is a sad illustration of the intolerance displayed by a segment of our society when statements of a scientist come into conflict with political doctrines. Orwell (1972) has well described this: "At any given moment there is an orthodoxy, a body of ideas

79

which it is assumed that all right-thinking people will accept without question. It is not exactly forbidden to state this or that or the other, but it is 'not done' so to say . . . Anyone who challenges the prevailing orthodoxy finds himself silenced with surprising effectiveness. A genuinely unfashionable opinion is almost never given a fair hearing, either in the popular press or in the highbrow periodicals." Scientists, I am afraid, are not entirely innocent of such intolerance.

80 All social reformers from Helvetius, Rousseau, and Robert Owen to certain Marxists (but not Karl Marx himself) have accepted Locke's claim that man at birth is a *tabula rasa* on which any characteristics can be stamped. Hence, by providing the proper environment and education, one can make anything out of any individual, considering that all of them are potentially identical. This led Robert Owen (1813) to the claim that "by judicious training, the infants of any one class in the world may be readily formed into men of any other class." Since classes were defined socioeconomically (at least by implication), Owen's statement had considerable validity. When extended to individuals and stated in a somewhat more extreme form, as was done by the behaviorist John B. Watson in 1924, it becomes very questionable. No wonder that those who hold such optimistic views are dismayed by the claims of those who have investigated the genetics of human characteristics in twin and adoption studies.

The demonstration by systematics, physical anthropology, genetics, and behavioral biology that no two individuals in any species (including the human one) are identical has created deep concern among all those who are sincere believers in the principle of human equality. As Haldane and Dobzhansky have pointed out, the dilemma can be escaped by giving a definition of equality which is consonant with modern biological findings. All individuals should be equal before the law and be entitled to equal opportunities. However, considering their biological inequalities, they must be given a diverse environment (for example, diverse educational opportunities) in order to assure *equal* opportunities. Paradoxically, identicism that ignores biological nonidentity is democracy's worst enemy when it comes to implementing the ideals of equal opportunity.

Biology has an awesome responsibility. It can hardly be denied that it has helped to undermine traditional beliefs and value systems. Many of the most optimistic ideas of the Enlightenment, including equality and the possibility of a perfect society, were ultimately (although very subconsciously) part of physico-theology. It was God who had made this near-perfect world. A belief in such a world was bound to collapse when the belief in God as designer was undermined. Hence Sedgwick's justified anguish. Losing a belief in God led to an existential vacuum and an unswered question as to the meaning of life. Leading thinkers, from the Enlightenment on, felt strongly that biology should not be merely a destroyer of traditional values but also the creator of new value systems. Virtually all biologists are religious, in the deeper sense of this word, even though it may be a religion without revelation, as it was called by Julian Huxley. The unknown and maybe unknowable instills in us a sense of humility and awe, but most of those who tried to replace a belief in God by a belief in man took the wrong path. They defined man as the self, the personal ego, and promoted an ideology of self-concern and egotism which not only fails to bring happiness but is conspicuously destructive in the long run. *81*

It would, of course, be equally simple-minded and dangerous to treat man simply as a biological creature, that is, as if he were nothing but an animal. Man, owing to his many unique features, has the capability to develop culture and to transmit acquired information, as well as value systems and ethical norms, to later generations. One would, therefore, get a very one-sided and indeed misleading concept of man, if one were to base one's evaluation of man entirely on the study of subhuman creatures. And yet, the study of animals has given us some of the most significant insights on the nature of man, even where these studies have revealed nothing more but how different man is in some characteristics from his nearest simian relatives.

If, instead of defining man as the personal ego or merely a biological creature, one defines man as mankind, an entirely different ethics and ideology is possible. It would be an ideology that is quite compatible with the traditional social value of wanting to "better mankind" and yet which is compatible with any of the new findings of biology. If this approach is chosen, there will be no conflict between science and the most profound human values (Campbell, 1974: 183–185; Rensch, 1971).

Such an approach, at first sight, would seem to be in conflict with the principle of inclusive fitness. This, however, is not necessarily so, for two reasons. First, in the anonymous mass society of modern mankind, it might well contribute to one's own inclusive fitness to work for the improvement of society as a whole.

Second, man is a unique species, in that a large amount of cultural "inheritance" has been added to biological inheritance, and that the nature of this cultural inheritance can affect Darwinian fitness. This interaction has, so far, not been sufficiently considered by those who were interested in the effect of Darwinism on human evolution. It is my personal conviction that the seeming conflicts between inclusive fitness, cultural inheritance, and a sound ethics can be resolved.

3 ✑ The changing intellectual milieu of biology

WRITING a history of ideas requires that the science of a given historical period be divided into its major problems and that the development of each problem be traced in time. Such a strictly topical treatment has its advantages but it isolates each problem from its connections with other contemporary problems in science as well as from the total cultural and intellectual milieu of the period. In order to compensate for this grave deficiency, I am providing in this chapter a brief history of biology as a whole and an attempt to relate it to the intellectual milieu of the time. The more specialized treatment of individual biological problems presented in the later chapters should be read against this overview. This introductory chapter will also establish a few connections with areas of functional biology (anatomy, physiology, embryology, behavior) which are not covered elsewhere in this volume.[1]

Each age has its own "mood" or conceptual framework which, though far from being uniform, somehow affects most thought and action. The Athenian culture of the fifth and fourth centuries B.C., the other-worldliness of much of the Middle Ages, or the scientific revolution of the seventeenth century are examples of strikingly differing intellectual milieus. However, it is wrong to assume that any era is always dominated by one mode of thinking, that is, by one explanatory framework or ideology, to be replaced eventually by a new and often very different conceptual framework. In the eighteenth century, for example, the intellectual framework of Linnaeus differed in just about every respect from that of his contemporary Buffon. Two very different research traditions may co-exist, with the adherents of each working in intellectual isolation. For instance, the positivism of the physicists in the second half of the nineteenth century, resting on an essentialist foundation, co-existed with the Darwinism of the naturalists, which was based on population thinking and asked adaptational questions that were quite meaningless to a positivist physicist.[2]

ANTIQUITY

All primitive men are keen naturalists, not surprisingly since their survival depends on a knowledge of nature. They must know potential enemies and sources of livelihood; they are interested in life and death, in illness and birth, in the "mind" and the differences between man and other living beings. Almost universal among primitive peoples of the world is a belief that all of nature is "alive," that even rocks, mountains, and sky are inhabited by spirits, souls, or gods. The powers of the gods are part of nature, and nature herself is active and creative. All religions prior to Judaism were more or less animistic, and their attitude toward the divine was altogether different from the monotheism of Judaism. The interpretation of the world by early man was a direct consequence of his animistic beliefs (Sarton, Thorndike).

There are reasons to believe that the science of early civilizations advanced considerably beyond this primitive state, but except for some medical lore, we have next to no information about the biological knowledge of the Sumerians, Babylonians, Egyptians, and other civilizations preceding that of the Greeks. There is no evidence that attempts were made to devise explanatory schemes for whatever facts had been accumulated.

The great Greek epics of Homer and Hesiod vividly portray the polytheism of the early Greeks, which contrasted so strikingly with the monotheism of Judaism, Christianity, and Islam. It seems that this polytheism permitted the development of philosophy and early science. For the Greeks there was no powerful single God with a "revealed" book that would make it a sacrilege to think about natural causes. Nor was there a powerful priesthood, as in Babylon, Egypt, and Israel, to claim a monopoly on thought about the natural and the supernatural. Nothing, therefore, prevented different thinkers in Greece from coming to different conclusions.

As far as Greek biology is concerned, we can distinguish three great traditions. The first is a natural-history tradition, based on the knowledge of local plants and animals, a tradition going back to our pre-human ancestors. This knowledge was handed by word of mouth from generation to generation, and it is rather certain that the little we know of it through Aristotle's *Historia animalium* and the writings of Theophrastus on plants provides only a glimpse at a vastly greater store of knowledge. The information about wild animals was valuably supplemented in many cultures by experience with domestic animals. Individual behavior, birth, growth, nutrition, sickness, death, and many other phenomena of biological significance are far more easily observed in domestic than in wild animals. Since most of these manifestations of life in animals are the same as in man, they encouraged comparative studies. In due time this made an important contribution to the development of research in anatomy and medical science.

The second Greek tradition, that of philosophy, originated with the Ionian philosophers Thales, Anaximander, Anaximenes, and their followers, who started a radically new approach.[3] They related natural phenomena to natural causes and natural origins, not to spirits, gods, or other supernatural agents. In their search for a unifying concept that would account for many different phenomena, they often postulated an ultimate cause or element from which all else originated, such as water, air, earth, or nondescript matter. Apparently, these Ionian philosophers had considerable knowledge of the achievements of the Babylonian and other near Eastern cultures and adopted some of their interpretations, primarily those relating to inanimate nature. The speculations of the Ionians on the origin of living beings had no lasting influence. Of a little more significance were their thoughts about human physiology. The real importance of the Ionian school is that it signifies the beginnings of science; that is, they sought natural causes for natural phenomena.

The center of philosophical thinking shifted later, in the sixth and in the fifth century B.C., to the Greek colonies in Sicily and southern Italy, where the key figures were Pythagoras, Xenophanes, Parmenides, and Empedocles. Pythagoras, with his emphasis on numbers and quantities, started a powerful tradition affecting not only the physical sciences but also biology. Empedocles seems to have thought more about biological matters than any of his predecessors, but little of his teaching is preserved. He is now best known for his postulation of the existence of four elements: fire, air, water, and earth. The entire material world, according to him, is composed of varying combinations of these four elements, either leading to greater homogeneity or else to greater mixing. A belief in these four elements continued for more than two thousand years. A concern with heterogeneity versus homogeneity appears again in the writings of the nineteenth-century zoologist K. E. von Baer and in those of the philosopher Herbert Spencer.

The ensuing decades saw the establishment of two great philosophical traditions, that of Heraclitus, who stressed change ("Everything is in flux") and that of Democritus, the founder of atomism, who by contrast stressed the unchanging constancy of the atoms, the ultimate components of everything. Democritus seems to have written a great deal about biological matters, though little survives, and some of Aristotle's ideas are believed to have been derived from him. Apparently he was the first to have posed a problem that has split philosophers ever since: Does organization of phenomena, particularly in the world of life, result purely from chance or is it necessary, owing to the structure of the elementary components, the atoms? Chance or necessity has ever since been the theme of controversies among philosophers.[4] It provided Monod (1970) with the title of his well-known book. It was Darwin, more than 2,200 years later, who showed that chance and necessity are not the only two options, and that the two-step process of natural selection avoids Democritus's dilemma.

These early Greek philosophers recognized that such familiar physiological phenomena as locomotion, nutrition, perception, and reproduction require explanation. What strikes the modern student as strange is the fact that they thought they could provide such explanation merely by concentrated thinking about the respective problem. Admittedly, at the time in which they lived, this was perhaps the only conceivable approach to these problems. The situation slowly changed, particularly when experimental science began to emancipate itself from philosophy during the late Middle Ages and in the Renaissance.

The lingering tradition of providing scientific explanation by mere philosophizing had an increasingly deleterious effect on scientific research in the eighteenth and nineteenth centuries, leading to Helmholtz's bitter complaint about the arrogance of the philosophers who rejected his experimental findings because these were in conflict with their deductions. The objections of essentialistic philosophers to Darwinism is another illustration of this attitude. In ancient Greece, however, the deductive philosophical approach helped to raise questions which no one had asked before, it led to an ever-more precise formulation of these questions, and it thereby set the stage for a purely scientific approach which ultimately replaced philosophizing.

The third great ancient tradition, co-existing with natural history and the philosophical tradition, was the biomedical tradition of the school of Hippocrates (about 450–377 B.C.), which developed an extensive corpus of anatomical and physiological knowledge and theory. This body of knowledge, further developed by the Alexandrians (Herophilus and Erasistratus) and by Galen and his school, formed the basis of the revival of anatomy and physiology in the Renaissance, particularly in the Italian schools. Research in human anatomy and physiology was the main interest of biology from the post-Aristotelian period until the eighteenth century. For science as a whole, however—indeed for the entire western thinking—developments in philosophy were far more important than concrete discoveries in anatomy and physiology.

Two Greek philosophers, Plato and Aristotle, had a greater influence on the subsequent development of science than any others. Plato (ca. 427–347 B.C.) had a special interest in geometry which powerfully affected his thinking. His observation that a triangle, no matter what combination of angles it has, is always a triangle, discontinuously different from a quadrangle or any other polygon, became the basis for his essentialism,[5] a philosophy quite unsuitable for biology. It took more than two thousand years for biology, under the influence of Darwin, to escape the paralyzing grip of essentialism. Plato's influence was equally unfortunate in matters more strictly biological. With the roots of his thinking in geometry, it is not surprising that he had little use for natural-history observations. Indeed, in the *Timaeus* he expressly states that no true knowledge can be acquired through the observations of the senses, but only a pleasure to the eye. His emphasis on the soul and on the architect (demiurg) of the cosmos permitted, through the neo-Platonists, a connection with Christian dogma which dominated the thinking of western man up to the seventeenth century. Without questioning the importance of Plato for the history of philosophy, I must say that for biology he was a disaster. His inappropriate concepts influenced biology adversely for centuries. The rise of modern biological thought is, in part, the emancipation from Platonic thinking.

With Aristotle, the story is different.

Aristotle

No one prior to Darwin has made a greater contribution to our understanding of the living world than Aristotle (384–322 B.C.).[6] His knowledge of biological matters was vast and had diverse sources. In his youth he was educated by Asclepiadic physicians; later he spent three years of his life on the island of Lesbos, where he evidently devoted much time to the study of marine organisms. In almost any portion of the history of biology one has to start with Aristotle. He was the first to distinguish various of the disciplines of biology and to devote to them monographic treatments (*De partibus animalium, De generatione animalium,* and so forth). He was the first to discover the great heuristic value of comparison and is rightly celebrated as the founder of the comparative method. He was the first to give detailed life histories of a large number of species of animals. He devoted an entire book to reproductive biology and life histories (Egerton, 1975). He was intensely interested in the phenomenon of organic diversity, as well as in the meaning of the differences among animals and plants. Although he did not propose a formal classification, he classified animals according to certain criteria and his arrangement of the invertebrates was superior to that of Linnaeus two thousand years later. Perhaps the least distinguished part of his biological corpus is his physiology, where he largely adopted traditional ideas. Far more than his predecessors, he was an empiricist. His speculations always go back to observations he had made. At one time (*De generatione animalium* 760b28) he states rather clearly that the information one receives from one's senses has primacy over what reason tells one. In that respect he was a world apart from the so-called Aristotelians among the scholastics, who reasoned out all problems.

Aristotle's outstanding characteristic was that he searched for causes. He was not satisfied merely to ask how-questions, but was amazingly modern by asking also why-questions. Why does an organism grow from a fertilized egg to the perfect adult form? Why is the world of living organisms so rich in end-directed activities and behaviors? He clearly saw that raw matter lacks the capacity to develop the complex form of an organism. Something additional had to be present, for which he used the word *eidos,* a term which he defined entirely differently from Plato. Aristotle's *eidos* is a teleonomic principle which performed in Aristotle's thinking precisely what the genetic program of the modern biologist performs. In contrast to Plato, who posited an outside force to explain the regularity of nature and especially its tendency toward reaching complexity and goals, Aristotle taught that natural substances act according to their own properties, and that all phenomena of nature are processes or the manifestations of processes. And since all processes have an end, he considered the study of ends as an essential component of the study of nature.

Consequently, for Aristotle all structures and biological activities have a biological meaning or, as we would now say, an adaptive significance. One of Aristotle's major aims was to elucidate such meanings. Aristotelian why-questions have played an important heuristic role in the history of biology. "Why?" is the most important question the evolutionary biologist asks in all of his researches.

There are four ways of conceiving the origin and the nature of the world: (1) a static world of short duration (the Judeo-Christian created world), (2) a static world of unlimited duration (Aristotle's world view), (3) a cyclical change in the state of the world in which periods of golden ages alternate with periods of decay and rebirth, and (4) a gradually evolving world (Lamarck, Darwin). Aristotle's belief in an essentially perfect world precluded any belief in evolution.

Aristotle has received full recognition for his pioneering thoughts only within recent decades. The bad reputation he had in past centuries had several causes. One is that the Thomists adopted him as their authoritative philosopher and when scholasticism came into disrepute, Aristotle automatically shared in the eclipse. Even more important is the fact that during the scientific revolution of the sixteenth and seventeenth centuries, almost the entire emphasis was on the physical sciences. Aristotle, who developed a remarkable philosophy of biology, unfortunately simultaneously believed that one could deal with macrocosm and microcosm in the same way and applied his biological thinking to physics and cosmology. The results were rather unfortunate, as Francis Bacon, Descartes, and many other authors of the sixteenth, seventeenth, and eighteenth centuries never tired of pointing out. The scorn heaped on Aristotle by these authors is difficult to understand considering the excellence and originality of most of his work.

The renewed appreciation in modern times of Aristotle's importance grew to the degree that the biological sciences emancipated themselves from the physical sciences. Only when the dual nature of living organisms was fully understood in our time was it realized that the blueprint of development and activity—the genetic program—represents the formative principle which Aristotle had postulated. As a result, we are beginning to be more tolerant of Aristotle. The world of philosophers and physicists was for hundreds of years completely deaf to the assertion of naturalists such as Aristotle that something more than the laws of physics was needed to produce a frog from a frog egg and a chicken from a chicken egg (Mayr, 1976). Nor does this require any *élan vital, nisus formativus, Entelechie,* or living spirit. All it requires is the recognition that complex biological systems are the product of genetic programs that have a history of more than three thousand million years. Nothing has caused more time-consuming and adrenalin-producing controversies than the myth that macrocosm and microcosm obey exactly the same laws. There is little indication that this insight has yet reached most philosophers, but it is beginning to be understood among biologists.

90

After Aristotle there was a continuation of the three Greek biological traditions. Natural history, particularly the description and classification of plants, reached a height in the writings of Theophrastus and Dioscorides, while Pliny (A.D. 23–79), whose interests were zoological, was an encyclopedic compiler. The biomedical tradition reached its height in Galen (A.D.131–200), whose influence endured until the nineteenth century.

In post-Aristotelian philosophy, a polarity developed between the Epicureans and the Stoics. Epicurus (342 to 271 B.C.), building on the foundation laid by Democritus, believed everything is made of unchanging atoms which whirl about and collide at random. He established a well thought-out materialistic explanation of the inanimate and living world, all things happening through natural causes. Life was viewed by him as due to the motions of lifeless matter. His explanation as to how manifestations of life originate through the assembly of appropriate configurations of atoms was remarkably modern. His follower Lucretius (99–55 B.C.) was an equally uncompromising atomistic materialist. Both of them rejected Aristotle's teleological ideas, Lucretius presenting a well-reasoned argument against the concept of design. He states many of the arguments that were raised again in the eighteenth and nineteenth centuries. Yet, Aristotle was entirely justified in his criticism of those atomists who, through a purely accidental interaction of water and fire, produced lions and oak trees. Galen agreed with him.

The argument of the Epicureans was mainly directed against the Stoics, who supported pantheistic ideas and believed in a designed world created for the benefit of man. According to them, it is the object of philosophy to understand the order of the world; later, natural theology was derived from Stoic thought. The Stoics rejected chance as a factor in the world; everything is teleological and deterministic. Their attitude was strictly anthropocentric, stressing the differences between intelligent man and instinct-guided animals (Pohlenz, 1948).

Nothing of any real consequence happened in biology after Lucretius and Galen until the Renaissance. The Arabs, so far as I can determine, made no important contributions to biology. This is even true for two Arab scholars, Avicenna (980–1037) and Aberrhos (Ibn Rosh, 1120–1198), who showed a particular interest in biological matters. It was, however, through Arab translations that Aristotle again became known to the western world. This was perhaps the greatest contribution the Arabs made to the history of biology. Another of their contributions was more indirect. The Greeks were great thinkers but experimented only to a limited extent (Regenbogen, 1931). By contrast, the Arabs were great experimenters, and one can go so far as saying that they laid the foundation on which experimental science later arose. The pathway to this ultimate goal was quite tortuous, with alchemy being the most important intermediate step.

THE CHRISTIAN WORLD PICTURE

When Christianity conquered the West, the Greek concept of an eternal, essentially static world was replaced by an entirely new one. Christian theology is dominated by the concept of creation. The world according to the Bible is a recently created one, all knowledge of which is contained in the revealed word. This dogma precluded the need, indeed the possibility, of asking any why-questions or of harboring any thought of evolution. The world, having been created by God, was, as later expressed by Leibniz, "the best of all possible worlds." Man's attitude toward nature was governed by God's command to "be fruitful and multiply, and fill the earth and subdue it; and have dominion over the fish of the sea, and over the birds of the air, and over every living thing that moves upon the earth" (Genesis 1:28). Nature was subservient to man; there was nothing in the Hebrew or Christian dogma of that oneness with nature felt by the animists or reflected in many Buddhist beliefs. The recent reverence for the environment was alien to the great monotheistic religions of the Near East (White, 1967).

No other development in Christianity was as important for biology as the world view known as natural theology. In the writings of the Church fathers, nature is sometimes compared to a book, a natural analogue to the revealed book of the Christian religion, the Bible. The equivalence of the two books suggests that a study of the book of nature, God's creation, would allow the development of a natural theology, supplementing the revealed theology of the Bible.

Christian natural theology was not a new concept. The harmony of the world, and the seeming perfection of the adaptations of the world of life, had impressed observers again and again, long before the rise of Christianity. As far back as the Old Kingdom of Egypt (Memphis), two thousand years before the Greeks and Hebrews, a creative intelligence had been postulated that had designed the phenomena of nature. More definite teleological statements can be found in Herodotus and Xenophon. Plato saw the world as the creation of an intelligent, good, reasoning, and divine artisan. The idea that the earth is a designed and fit environment for life was further cultivated and enriched by the Stoics. Galen strongly endorsed the concept of design, the work of a wise and powerful creator. But no one was more important for the development of natural theology than Saint Thomas Aquinas. A teleological world view became dominant in western thinking through his writings. In his *Summa theologiae* the fifth argument proving the existence of God is based on the order and harmony of the world which requires that there must be an intelligent being directing all natural things to their end.

Despite the teachings of natural theology, the scholastic age was not favorable for the development of the natural sciences. The scholastics were rationalists; it was their endeavor to determine the truth by logic, not by observation or experiment. Hence their interminable disputations. Teaching and the search for truth, as they exercised it, was the privilege of the clerics. The study of natural things, indeed any empirical approach, was on the whole despised. The dominant philosophy of scholasticism was the Thomistic one, believed by Aquinas to have been mainly derived from Aristotle. This philosophy is known under the strangely misleading name of realism. Its most characteristic aspect, as it appears to a modern biologist, is its total support of essentialism. Nominalism, the only other powerful school of scholastic philosophy, stressed that only individuals exist, bracketed together into classes by names. Nominalism had no influence on biology during the Middle Ages, and it is still not at all clear whether and to what extent nominalism contributed to the eventual rise of empiricism and population thinking.

The concept of the Christian Church that the "revealed word" had overriding authority was, curiously, extended in medieval times to other writings, particularly to Aristotle's work and even to the writings of Arab scholars like Avicenna. When an argument arose as to how many teeth the horse has, one looked it up in Aristotle rather than in the mouth of a horse. The inner-directed world of medieval Christianity paid little attention to nature. This began to change somewhat in the twelfth and thirteenth centuries. Hildegard of Bingen (1098–1179) and Albertus Magnus (1193–1280) wrote on natural history but their work is not in the same class with that of the splendid observer Frederick II (1194–1250), whose superb *Art of Falconry* (*De arte venandi*) was many centuries ahead of its time in its interest in the morphology and biology of birds. Frederick's genuine understanding of the living animal, so obviously based on personal experience, loomed large above the level of other contemporary natural-history writings, exemplified by the uncritical compilations of a Cantimpré or Beauvais (Stresemann, 1975). Frederick's influence was manifold. He had some of Aristotle's writings translated into Latin and was a patron of the medical school of Salerno (founded 1150), where human bodies were dissected for the first time in more than one thousand years.

93

Beginning with Salerno, universities were founded in various parts of Europe, particularly in Italy (Bologna, Padova), France (Paris, Montpellier), and England (Oxford and Cambridge). They had exceedingly different backgrounds, some originating as medical schools or law schools, others, like the Sorbonne (founded around 1200) as schools of theology. Most of them soon became centers of scholasticism, and it has been argued whether their existence had a beneficial or detrimental effect on western learning. In some areas (anatomy, for instance), they eventually became the centers of progressive scholarship. As far as biology as a whole is concerned, it was not until the late eighteenth and early nineteenth century that the universities became centers of biological research.

Logic, cosmology, and physics (Crombie, 1952) had a remarkable renaissance in the later Middle Ages, the high intellectual level of which has been appreciated only in the last generation. By comparison, biology continued to be dormant. The only aspects of living nature that received attention were problems of medicine and human biology. One looks in vain for any attempt to grapple with the deeper problems of life so fascinating to later centuries and to the modern mind. One has the feeling that this lack of interest was somehow connected with the extreme piety of the period, which did not allow questions about God's creation, but then one wonders why this taboo did not extend to physics and cosmology. Was it that the prestige of mathematics and its theological neutrality led automatically to physics and cosmology, while there was no such entering wedge for biology? Natural theology eventually provided such an opening, but effectively not until the seventeenth century. Was it the discovery of exotic countries, where the same heavenly bodies occurred and the same laws of physics held as in Europe, but where entirely different faunas and floras were found? Was it that the phenomena of life require asking far more sophisticated questions than those raised by the study of falling bodies? Who knows? We still lack an adequate analysis for the lag between the awakening in the mechanical sciences and the post-medieval revival of biology.

THE RENAISSANCE

During the Renaissance a new interest developed in natural history and anatomy. Both, in a way, were parts of medicine, and the most active investigators in these fields were usually professors of medicine or practicing physicians.

The study of medicinal plants was popular throughout the later Middle Ages, as reflected in the number of herbals, particularly after the works of Theophrastus and Dioscorides had again become available. But it was the plant books of Brunfels, Bock, and Fuchs which heralded a new "back-to-nature" movement in the study of plants (see Chapter 4). The liberating influence of travel eventually made itself felt as well. It began with the crusades, continued with the travels of the Venetian merchants (such as Marco Polo's visit to China) and the voyages of the Portuguese seafarers, and culminated in the discovery of the New World by Columbus (1492). One of the decisive consequences of these travels was the sudden recognition of the immense diversity of animal and plant life in all parts of the globe. This realization led to the publication of several encyclopedic natural histories by Wotton, Gesner, and Aldrovandi, and to more specialized ones by Belon on birds and by Rondelet on marine organisms.

Anatomy was taught in medieval medical schools, particularly in Italy and France, but in a peculiarly literary way. The professor of medicine would recite Galen, while an assistant ("surgeon") dissected the corresponding parts of the body. This was poorly done, and the oratory and the disputations of the professors, all of them merely interpreting Galen, were considered to be far more important than the dissection. It was Andreas Vesalius (1514–1564), more than anyone else, who changed all this. He himself actively participated in the dissections, invented new dissecting instruments, and finally published an anatomical work with magnificent illustrations: *De Humani Corporis Fabrica* (1543). In this he corrected a number of errors of Galen, but he himself made only a rather limited number of discoveries and retained the Aristotelian framework of physiological explanation. Nevertheless, a new era in anatomy started with Vesalius, in which the scholastic reliance on the traditional texts was replaced by personal observations. Vesalius' successors, including Fallopio, Fabricius ab Aquapendente, Eustacchi, Cesalpino, and Severino, not only made important discoveries in human anatomy but several of them also made important contributions to comparative anatomy and to embryology. The particular importance of this development is that it provided the basis for the new start in physiology.

95

Applied science, that is, technology and the engineering arts, prepared the ground during the Renaissance for an entirely new way of looking at things. The mechanization of the world picture which resulted from this movement reached a first culmination in the thought of Galileo Galilei (1564–1642) and in that of his students and associates. Nature for them was a law-bound system of matter in motion. Motion was the gist of everything and everything had to have a mechanical cause. His stress on quantification was expressed in the admonition "to measure what can be measured and to make measurable what cannot be measured." This led to the development and use of instruments in order to determine quantities, to the calculation of regularities which led to the establishment of general laws, and to a dependence on observation and experiment rather than on the word of authority. This meant in particular a rejection of certain aspects of Aristotelianism which had become so authoritative through the influence of the Thomists.

The attacks on Aristotle came not only from the physicists but also from philosophy. Francis Bacon, who was particularly scathing in his anti-Aristotelianism, became the prophet of the method of induction, though his own biological theories were entirely deductive constructions. Bacon's great merit, however, was his unceasing questioning of authority and his emphasis on the incompleteness of our knowledge, in contrast to the medieval belief that knowledge was complete.

The most positive contributions of the scientific revolution, as far as biology is concerned, was the development of a new attitude toward investigation. It consisted in a complete rejection of sterile scholasticism, which endeavored to find the truth merely by logic. There was greater emphasis on experiment and observation, that is, on the collecting of facts. This favored the explanation of regularities in the phenomena of the world by natural laws, which it became the task of the scientist to discover. The actual number of concrete contributions to biology made by a mechanistic approach is very small. It includes Harvey's measurements of blood volume, which were an important link in the chain of his argument in favor of a circulation of blood, and the studies of some anatomists, particularly Giovanni Alfonso Borelli (1608–1679), on locomotion. Indeed, no other branch of physiology is as suitable for a mechanical analysis as the movement of extremities, articulations, and muscles.

The publication of Newton's *Principia* in 1687, which proposed a mechanization of the entire inanimate world on a mathematical basis, greatly reinforced the mechanistic approach to physiology. More than ever, it now became fashionable to explain everything in physical terms of forces and motion, as inappropriate as such an explanation was for most biological phenomena. The explanation of warm-bloodedness in mammals and birds as due to friction of the blood in the blood vessels, for instance, was accepted for about 150 years even though this would have been refuted by a few simple experiments or by the observation of blood circulation in cold-blooded amphibians and fish of the same body size as mice or birds. Facile physicalist explanations were a great impediment of biological research during the seventeenth and eighteenth centuries, and sometimes even later.

As Radl (1913: viii) pointed out long ago, the triumph of the physical sciences during the scientific revolution was, in many ways, a defeat for biology, and for all those specifically biological modes of thinking that did not receive recognition again until the nineteenth and twentieth centuries: teleonomy (maligned as the search for final causes), systems thinking, the study of qualitative and emergent properties and of historical developments. All this was neglected, if not actively opposed and ridiculed. The response of the life scientists to the attacks of the physical scientists was either a futile attempt to express biological processes in the quite unsuitable terms of the physicists ("movements and forces") or an equally futile escape into vitalism or supernatural explanations. It is embarrassingly recent that biologists have had the intellectual strength to develop an explanatory paradigm that fully takes into consideration the unique properties of the world of life and yet is fully consistent with the laws of chemistry and physics (see Chapter 1).

Descartes

No one perhaps contributed more to the spread of the mechanistic world picture than the philosopher René Descartes (1596–1650). As with Plato, his thought was greatly influenced by mathematics, his most brilliant contribution probably being the invention of analytical geometry. His attacks on Aristotelian cosmology were legitimate and constructive even though ultimately his own proposals did not prevail either. However, his proposal to reduce organisms to a class of automata had the unfortunate consequence of offending every biologist who had even the slightest understanding of organisms. Descartes's crass mechanism encountered, therefore, violent opposition. This expressed itself usually in an equally absurd teleological vitalism. It is presumably no coincidence that France, the country with the most extreme mechanists from Descartes to La Mettrie and Holbach, was perhaps also the most active center of vitalism. Descartes's claims that organisms are merely automata, that the human species differs from them by having a soul, that all science must be based on mathematics, and many other of his sweepingly dogmatic statements, since proven to be quite erroneous, created a millstone around the neck of biology, the effects of which (in the mechanism-vitalism controversy) have carried through to the end of the nineteenth century. One of the weakest components of Descartes's thinking concerned origins. He thought that organisms were formed by the fortuitous coming together of particles. Ultimately this meant explaining nature as the result of blind accident. This thesis, however, was clearly contradicted by the order of nature and the remarkable adaptations of all creatures, as demonstrated by the naturalists.

What is most astounding about Descartes is that, his own protestations notwithstanding, much of the framework of his reasoning is Thomistic. His way of thinking is well illustrated by his conclusions concerning his own existence: "I concluded that I was a substance whose whole essence or nature consists in thinking, and whose existence depends neither on its location in space nor on any material thing. Thus the self, or rather the soul, by which I am what I am, is entirely distinct from the body, is indeed easier to know than the body, and would not cease to be what it is, even if there were no body" (*Discourse on Method*, p. 4). Most of his physiological conclusions were arrived at by deduction rather than by observation or experiment. Like Plato before him, Descartes demonstrated by the failure of his method that one cannot solve biological problems through mathematical reasoning. Much still remains to be done in the investigation of Descartes's influence on the subsequent development of biology, particularly in France. This includes the question how far Cartesianism was responsible for the cool reception of evolutionary thinking (for example, of Lamarck) in France in later centuries. What is particularly remarkable, in hindsight, is the naiveté with which purely physical explanations of the most simplistic form were accepted by Descartes and some of his followers. Buffon, for instance, concluded that "a single force," namely gravitational attraction, "is the cause of all phenomena of brute matter and this force, combined with that of heat, produces the living molecules on which depend all the effects of organized bodies" (*Oeuvr. Phil.*: 41).

Perhaps biology had to go through a phase in which the sterile physicalism of Descartes was adopted. Aristotle's perfectly sound demonstration that biological form could not be understood in terms of mere inanimate matter had unfortunately been vulgarized by the scholastics, who replaced Aristotle's psyche by the soul of Christian dogma. Aristotelian-Galenic physiology had indeed become scientifically unacceptable when interpreted in terms of the Christian soul. Under these circumstances Descartes had two options. He could either go back to the Aristotelian "form" and redefine it, as does the modern biologist in his genetic program, or he could entirely throw out the Christian soul, as far as animals are concerned, without replacing it by anything, leaving the organism a piece of inanimate matter like all other inanimate things. The latter was the option he adopted, an option obviously unacceptable to any biologist who knew that an organism is more than just inanimate matter. Not being much of a biologist, Descartes did not think so. It was only when he contemplated man that Descartes realized that his thesis would not do. He then adopted the dualism between body and soul, a dualism (not new with Descartes) which has plagued us ever since.

The dominance of the mechanical world view was not complete. Indeed, the claims of the Galilean mechanists and of the Cartesians were so extreme that they elicited a number of countermovements almost at once. Two among these are of particular interest in the history of biology: the rise of a qualitative-chemical tradition and the study of diversity. Both movements were, in part, rooted in the scientific revolution.

A novel movement in sixteenth-century physiology concentrated on quality and chemical constituents instead of on movement and forces. This approach was by no means antiphysicalist in principle, because it utilized concepts, laws, and mechanisms in the explanation of living processes that had first been developed to explain processes in the inanimate world. I refer to Paracelsus (1493–1541) and his followers, to the alchemists, and to the school usually referred to as the iatrochemists. As unpromising as this new movement was at first, and as wrong in particulars, in the long run it had a far longer lasting impact on the explanation of biological processes than the strictly mechanistic ones. Paracelsus, part genius and part charlatan, a believer in magic and occult forces, rejected the importance of the traditional four elements of the Greeks, replacing them by actual chemicals, particularly sulphur, mercury, and salt. His concept of life processes as chemical processes started an entirely new tradition which, through J. H. van Helmont (1577–1644), was the beginning of a new phase in the history of physiology. In van Helmont's writings we find a peculiar mixture of superstition, vitalism, and extraordinarily perceptive observations. He coined the term "gas" and did significant research on CO_2. He recognized the acidity of the stomach and the alkalinity of the small intestines, thereby initiating a whole new field of research in nutritional biology. The chemicalization of physiology continued through his followers such as Stahl.

THE DISCOVERY OF DIVERSITY

One of the objectives of attempts to provide a mechanistic explanation of all phenomena was to further the unity of science. It became the ambition of physical scientists to reduce the phenomena of the universe to a minimal number of laws. Owing to the discovery of the almost unlimited diversity of animals and plants, an almost diametrically opposite tendency developed in the study of living organisms. The herbalists and encyclopedists had revived the tradition of Theophrastus and Aristotle, discovering and describing with loving detail diverse kinds of organisms. More and more naturalists devoted themselves to the study of nature's diversity and discovered that the world of creation is far richer than anyone had imagined. And the glory of God could be studied in any of his creatures, from the lowliest ones up to the rhinoceroses and elephants admired by Dürer or Gesner.

The scientific revolution coincidentally made a major contribution to the interest in diversity. The development of all sorts of new instruments was one of the products of the spirit of mechanization, the most important among them for the biologist being the microscope. Microscopy opened up a new world for the biologist. Even though the earliest microscopes permitted only a tenfold magnification, this was sufficient to reveal the existence of an entirely unexpected living microcosmos, particularly of aquatic organisms invisible to the naked eye.

The two outstanding early practitioners of microscopy were Anton van Leeuwenhoek (1632–1723) and Marcello Malpighi (1628–1694). They provided descriptions of animal and plant tissues (the birth of histology), and they discovered freshwater plankton, blood cells, and even the spermatozoon. The work of these early microscopists is characterized by the pleasure of discovery. Without a goal, they looked at almost any magnifiable object and described what they saw. One finds very little biological theory in their writings. Incidentally, the earliest applications of electron microscopy, three hundred years later, were characterized by a similar attitude.

It was also in this period that the insects were discovered to be a proper subject for scientific study. Francesco Redi in 1668 showed that insects are not the results of spontaneous generation but develop from eggs laid by fertilized females. Jan Swammerdam (1637–1680) did superb anatomical work on the honey bee and other insects. Pierre Lyonnet, Ferchault de Réaumur, de Serres, Leonhard Frisch, and Roesel von Rosenhof were other naturalists of the seventeenth and eighteenth centuries who made major contributions to the knowledge of insects. Most of them were motivated by the sheer joy of describing what they discovered, even if it was nothing more than the 4,041 muscles of a caterpillar (Lyonnet, 1762; see Chapter 4).

The enthusiasm over the extraordinary diversity of the living world was fired still further by the success of voyages and individual explorers bringing back exotic plants and animals from all continents. Captain Cook had the Forsters, father and son, as naturalists on one of his voyages. The younger Forster inspired Alexander von Humboldt, who in turn inspired the young Charles Darwin. The era of overseas travel and explorations resulted in a veritable obsession with exotic organisms and led to the establishment of vast collections, as illustrated by those of the patrons of Linnaeus in the Netherlands, of Banks and his competitors in London, and of the Jardin du Roi in Paris, which was directed by Buffon.

The exponential growth of the collections produced the *101* foremost need of the period: classification. Beginning with Cesalpino (1583), Tournefort, and John Ray (whose work is analyzed in Chapter 4), the age of classification reached its climax with Carl Linnaeus (1707–1778). His importance was exalted in his lifetime beyond that of any naturalist since Aristotle. Yet one hundred years later Linnaeus was denigrated as a pedantic throwback to the scholastic period. We now see him as a child of his time, outstanding in some ways and blind in others. As a local naturalist, like John Ray before him, he observed the clear-cut discontinuity between species and assumed the impossibility of one species changing into another one. His insistence on the constancy and sharp delimitation of species, at least in his earlier writings, set the stage for the subsequent development of an evolutionary theory. It is only in recent years that one has remembered again Linnaeus's contribution to phytogeography and ecology. Most of his followers unfortunately lacked Linnaeus's flair and found ample satisfaction in describing new species.

Not all naturalists of the period succumbed to the craze of species description. J. G. Kölreuter (1733–1806), for instance, although starting out from a rather traditional interest in the nature of species, made pioneering contributions to genetics, fertilization, and flower biology in plants. These studies were extended by C. K. Sprengel (1750–1816) through copious experiments on fertilization in plants. The work of these two investigators, although virtually ignored in their lifetime, was part of the foundation on which Darwin later based his experimental research on fertilization (and fertility) in plants.

A tradition in natural history very different from that of Linnaeus was initiated by Buffon, whose *Histoire naturelle* (1749ff) was read by practically every educated European. With its emphasis on living animals and their life history, this work had a tremendous impact on natural-history studies, an impact that did not come to full fruition until the modern age of ethology and ecology. The study of natural history in the eighteenth and early nineteenth centuries was almost completely in the hands of amateurs, particularly country parsons such as Zorn, White (vicar of Selborne), and C. L. Brehm. Buffon, as brilliant as he was as a popularizer, exerted perhaps his greatest influence through his stimulating, often daringly novel, ideas. He had an enormously liberating influence on contemporary thinking, in such divergent fields as cosmology, embryonic development, the species, the natural system, and the history of the earth. He never quite advanced to a theory of evolution, but unquestionably he prepared the ground for Lamarck (see Chapter 7). I entirely agree with Nordenskiöld's evaluation of Buffon (1928: 229): "In the purely theoretical sphere he was the foremost biologist of the eighteenth century, the one who possessed the greatest wealth of ideas, of real benefit to subsequent ages and exerting an influence stretching far into the future."

Diversity, of course, is a phenomenon that does not seem to fit at all the Newtonian paradigm of physical laws. Yet, since laws were evidence for the existence of a law-giving creator, the discovery of laws regulating diversity became a challenge for students of diversity from Kielmayer to the quinarians and to Louis Agassiz. These endeavors, largely against the intention of their authors, provided much evidence for evolution.

Linnaeus, for all intents and purposes, founded the science of systematics and Buffon made the study of natural history everybody's pastime. Physiology reached new heights with Haller and embryology with Bonnet and Wolff. As a consequence, biology, so greatly eclipsed by the physical sciences in the seventeenth century, began to come into its own in the middle of the eighteenth century.

The dominant interest of the century, clearly, was the description, comparison, and classification of organisms. Anatomy, which had been primarily a method of physiological research since its very beginning, now became increasingly comparative. It developed into another method of studying diversity. The comparative method, one of the two great methods of science (experimentation being the other one) had its real beginning in the second half of the eighteenth century. To be sure, comparative studies had been conducted since the sixteenth century by Belon, Fabrizio, and Serverino, but they became a systematic research method only in the hands of Camper, Hunter, Pallas, Daubenton, and

102

particularly Vicq-d'Azyr. The new tradition, thus established, reached a first peak in the work of Cuvier who in a series of methodical studies, with particular emphasis on the invertebrates, demonstrated the absence of any intermediates between the major phyla of animals, thus completely refuting the existence of a *scala naturae*. After 1859 comparative anatomical studies, of course, provided some of the most convincing evidence in favor of Darwin's theory of common descent.

Natural Theology

It is difficult for a modern person to appreciate the unity of science and Christian religion that existed at the time of the Renaissance and far into the eighteenth century. The reason why there was no conflict between science and theology was that the two had been synthesized as natural theology (physico-theology), the science of the day. The natural theologian studied the works of the creator for the sake of theology. Nature for him was convincing proof for the existence of a supreme being, for how else could one explain the harmony and purposiveness of the creation? This justified the study of nature, an activity about which many of the devout were rather a little self-conscious, particularly in the seventeenth century. The spirit of natural theology still dominated authors as late as Leibniz, Linnaeus, and Herder, and British science up to the middle of the nineteenth century. The total domination of all scientific activities and thinking by the concepts of natural theology has long been understood by the historians of science and we have a large number of perceptive treatments.

The mechanization of the world picture caused a serious dilemma for the devout. If he followed the claims of the physical scientist, he had to assume that the world had been created at a single time and that at the same time natural laws ("secondary causes") had been established which required only a minimal amount of divine intervention in subsequent periods. The task of the "natural philosopher" was to study the proximate causes by which these divine laws manifested themselves. This interpretation fitted the phenomena of the physical world reasonably well but was completely contradicted by the phenomena of the living world. Here such a diversity of individual actions and interactions is observed that it becomes inconceivable to explain it by a limited number of basic laws. Everything in the living world seemed to be so unpredictable, so special, and so unique that the observing naturalist found it necessary to invoke the creator, his thought, and his activity in every detail of the life of every individual of every kind of organism. This, however, seemed likewise unthinkable because, as one of the commentators said, a ruler supervises his workers but does not himself perform all the tasks of a working man. Thus neither alternative seemed acceptable. The next two hundred years were filled with endeavors to escape this dilemma, but there was no escape within the framework of creationist dogma. Consequently, the two schools of thought continued: The physical scientists saw in God the power who had instituted, at the time of creation, the laws governing the processes of this world. By contrast, the devout naturalists who studied living nature concluded that the basic laws of Galileo and Newton were meaningless as far as the diversity and adaptation of the living world is concerned. Rather, they saw the hand of God even in the smallest aspect of adaptation or diversity. John Ray's *The Wisdom of God Manifested in the Works of the Creation* (1691) is not only a powerful argument from design but also very sound natural history, indeed, one might say one of the earliest works of ecology. The excellence of observations on which the writings of the naturalist-theologians were based gave them a wide circulation and greatly contributed to the spread of the study of natural history. Natural theology was a necessary development because design was really the only possible explanation for adaptation in a static "created" world. Any new finding in this early age of natural history was grist on the mill of natural theology. The supposedly idyllic life of the inhabitants of the tropics, in particular, was seen as evidence for the providential design by the creator. The discovery of infusorians and zoophytes seemed to confirm the Great Chain of Being, leading up to man. But the hour of triumph of natural theology was short. It was implicitly questioned in much of Buffon's writings and quite explicitly criticized in Hume's *Dialogues* (1779) concerning natural religion and in Kant's *Critique of Judgment* (1790).

Evolutionary biology greatly benefited from natural theology. This sounds like a rather paradoxical claim, considering that evolution received hardly any attention prior to 1859, and yet it is true, although in an indirect sense. What natural theology did was to ask questions concerning the wisdom of the creator and the ingenious way in which he had adapted all organisms to each other and to their environment. This led to the seminal studies of Rei-

104

marus and Kirby on animal instincts, and to C. K. Sprengel's discovery of the adaptations of flowers for insect pollination and the corresponding adaptations of the pollinators. From Ray and Derham, to Paley, to the authors of the Bridgewater treatises, and to numerous of their contemporaries, all natural theologians described what we would now call adaptations. When "the hand of the creator" was replaced in the explanatory scheme by "natural selection," it permitted incorporating most of the natural theology literature on living organisms almost unchanged into evolutionary biology. No one can question that natural theology laid a remarkably rich and solid foundation for evolutionary biology and that it was not until far into the Darwinian period that studies in adaptation were again pursued as eagerly as it had been done by natural theology.

Natural theology represents an intensely optimistic world view. Yet much happened in the second half of the eighteenth century to destroy this unbounded optimism, beginning with the earthquake of Lisbon, the horrors of the French Revolution, and the realization of the intensity of the struggle for existence. The hold of natural theology on the thinking of western man ended in France and Germany before the end of the eighteenth century. Curiously, it had a new flowering in England in the first half of the nineteenth century. Paley's *Natural Theology* (1803) and the *Bridgewater Treatises* (1832–1840) once more advanced quite emphatically the "argument from design." The leading English paleontologists and biologists of the day were natural theologians, including Charles Lyell and other of Darwin's friends. This fact explains much of the intellectual structure of the *Origin of Species* (see Chapter 9).

Life and Generation

Except for natural history, the study of living organisms from the Renaissance to the nineteenth century was largely in the hands of the medical profession. Even the great botanists (with the exception of Ray) had been educated as medical doctors. Their main interest was, of course, the functioning of the healthy or sick body, and second, the problem of "generation," that is, the origin of new organisms. By the beginning of the eighteenth century it became the task of physiology to find a compromise between the ever more radical mechanistic and opposing vitalistic extremes. It was Albrecht von Haller (1707–1777) who gave physiology a new direction. He returned to the empirical tradition of Harvey and of the vivisectionists and attempted to determine the function of various organs by innumerable animal experiments. Even though he found no evidence for a "soul" directing physiological activities, his experiments convinced him that the structures of the living body have certain properties (like irritability) which are not found in inanimate nature.

In spite of Haller's balanced conclusions, the pendulum continued to swing back and forth right up to the first quarter of the twentieth century. Vitalism and mechanism continued to battle each other. Vitalism, for instance, was defended by the Montpellier school (Bordeu, Barthez), by the German *Naturphilosophen,* by Bichat and Claude Bernard, and by Driesch, while an uncompromising mechanism was preached by Ludwig, duBois-Reymond, Julius Sachs, and Jacques Loeb. It is perhaps legitimate to say that this controversy was not totally eliminated until it was recognized that all manifestations of development and life are controlled by genetic programs.

To go back to the seventeenth and eighteenth centuries, the second great controversy concerned the nature of development. The question to be answered was how can the "amorphous" egg of a frog develop into an adult frog and a fish egg into a fish? The defenders of *preformation* thought that there was something preformed in the egg which was responsible for turning the egg of a grassfrog into a grassfrog and that of a trout into a trout. Unfortunately the extreme representatives of the preformationist school postulated pre-existence, that is, that a miniaturized adult (homunculus) was somehow encapsuled in the egg (or in the spermatozoon), an assumption the absurdity of which was rather easily demonstrated. Their opponents who upheld the thesis of *epigenesis,* that is the gradual differentiation of an entirely amorphous egg into the organs of the adult, were hardly more convincing since they were quite unable to account for the species specificity of this process and thus they had to invoke vital forces. They were the leaders of vitalism. As so often in the history of biology, neither of the opposing theories prevailed in the end but rather their eclective fusion. The epigenesists were correct in stating that the egg at its beginning is essentially undifferentiated, and the preformationists were correct that its development is controlled by something preformed, now recognized as the genetic program. Among the participants in this controversy, in addition to Haller one might mention Bonnet, Spallanzani, and C. F. Wolff (Roe, 1981).

BIOLOGY IN THE ENLIGHTENMENT

As indicated by the term "Enlightenment," the eighteenth century, from Buffon, Voltaire, and Rousseau to Diderot, Condillac, Helvetius, and Condorcet, was an intellectually liberating period. The predominant form of religion was deism. Even though admitting the existence of God, the enlightened deist could find no evidence that God had created the world for the benefit of man. His God was the supreme intelligence, the creator of the world and its universal order, the promulgator of general and immutable laws. His was a God remote from man, with whom he is little concerned. It was not a very large step from deism through agnosticism to outright atheism. Many thinkers took the step.

107

The Enlightenment was a time when any previously held dogma, whether theological, philosophical or scientific, was critically questioned. The persecutions of the philosophers by the French government ("the King") should warn us, however, that many of the teachings of the philosophers were considered as political as they were philosophical.

Condorcet's egalitarianism, for instance, was a rebellion against class privileges (feudalism), completely ignoring any possible biological aspects. He recognized only three kinds of inequality, relating to wealth, social status, and education, but not allowing for any differences in native endowment. Total equality could be established, he thought, as soon as wealth, status, and education were equalized. A concept like natural selection, or even evolution, would make no sense to one committed to such uncompromising egalitarianism.

One must remember, however, that the Enlightenment was not a homogeneous movement. Almost as many views were represented as there were different philosophers.

Paris from Buffon to Cuvier

The history of biology is rich in episodes of a meteoric rise of some center of research. The north Italian universities in the sixteenth and seventeenth century are one example, the German universities in the second half of the nineteenth century are another one, and Paris from Buffon (1749) to Cuvier (1832) is a third one. The specific contributions of the principal figures in the galaxy of Parisian stars will be discussed in the respective chapters, but at this time it is important to single out one name, that of Lamarck (1744–1829), because his proposal of evolutionism (first articulated in the *Discours* of 1800) was such a radical departure from tradition.

It is often said that only young people have revolutionary new inspirations, but Lamarck was more than fifty years old when he developed his heterodox ideas. His geological studies had convinced him that the earth was very old and that conditions on it were constantly changing. Fully understanding that organisms are adapted to their environment, he was forced to the conclusion that they must change in order to maintain their adaptation to the ever changing world. His comparison of fossil mollusks through the Tertiary strata to the present confirmed this conclusion. This led Lamarck to propose a theory of transformation (1809), which postulated an intrinsic tendency of organisms to strive toward perfection and an ability to adjust to the demands of the environment. Virtually all of his explanatory endeavors were unsuccessful largely because he depended on conventional beliefs such as an inheritance of acquired characters. Even though Lamarck was viciously attacked by Cuvier, he nevertheless impressed many of his readers, including Chambers, the author of *Vestiges* (1844). Opposing protestations notwithstanding, he clearly prepared the ground for Darwin. Even without his evolutionary theory Lamarck would be an honored name in the history of science owing to his manifold contributions to botany and to the knowledge of the biology and classification of the invertebrates.

Lamarck is sometimes credited for having ushered in a new era of biology by his theory of evolution (1800; 1809) and by his coining the word "biology" in 1802 (independently proposed also by Burdach in 1800 and by Treviranus in 1802). A broad look at the biological sciences does not substantiate this claim. Lamarck's evolutionary theory had exceedingly little impact, and the coining of the *word* "biology" did not create a *science* of biology. In the early 1800s there was really no biology yet, regardless of Lamarck's grandiose scheme (Grassé, 1940) and the work of some of the *Naturphilosophen* in Germany. These were only prospectuses of a to-be-created biology. What existed was natural history and medical physiology. The unification of biology had to wait for the establishment of evolutionary biology and for the development of such disciplines as cytology.

Lamarck's great opponent was Cuvier (1769–1832), whose contributions to science are almost too extensive to be listed. He clearly established the science of paleontology, and his analysis of the vertebrate faunas of the Paris basin were as important a contribution to stratigraphy as was the work of William Smith in England. I have already mentioned Cuvier's important studies in comparative anatomy and his destruction of the *scala naturae*. When E. Geoffroy Saint-Hilaire attempted once more to revive the concept of unity of plan in the entire animal world, Cuvier refuted these claims devastatingly. The so-called Academy dispute (1831) with Geoffroy Saint-Hilaire was not a debate about evolution, as is sometimes claimed, but about whether or not the structural plans of all animals can be reduced to a single archetype.

Cuvier had an enormous impact on his age, an impact that was both good and bad. He inspired research in comparative anatomy, perhaps more in Germany than in France, and in paleontology, but he also impressed his conservative frame of mind on generations of French biologists. As a result, evolution, in spite of the priority of Lamarck, had a tougher time getting accepted in France than in any other scientifically active European nation. Cuvier played a remarkably paradoxical role in the history of evolution. He opposed it in its foremost representative, Lamarck, with all the power of his knowledge and logic, and yet his own researches in comparative anatomy, systematics, and paleontology supplied some of the best evidence for those who subsequently embraced evolutionism.

THE RISE OF SCIENCE FROM THE SEVENTEENTH TO THE NINETEENTH CENTURY

Much happened in these three centuries, but it is often impossible to say what is cause and what is effect. The movement of Latin-speaking scholars from country to country, so characteristic of the late Middle Ages and the Renaissance, declined strikingly and with it the popularity of the Latin language. As a result, nationalism in science grew, aided and abetted by the use of national languages in the scholarly literature. Work published in the foreign literature was ever more rarely referred to. This parochialism reached its height in the nineteenth century, with the result that each country had its own intellectual and spiritual milieu.

There was perhaps no other era in western history when national moods were as different as in the period from 1790 to 1860. In Britain empiricism was dominant. It rested on a (nominalistic) tradition going back to William of Ockham, was primarily developed by John Locke, and was adopted by the eighteenth-century chemists Hales, Black, Cavendish, and Priestley. In France there was first the ferocity of rebellion and then an extraordinary reaction after the restoration of the monarchy. Even though neither natural theology nor the church played a role, there was a spirit of great conservativism through Cuvier. The mood was entirely different in Germany. Here a country was finding itself after the extreme trials and deprivations of the seventeenth and eighteenth centuries and the new spirit expressed itself in great enthusiasms, first for classical antiquity, then for various romantic movements, culminating in *Naturphilosophie* (as developed by Schelling, Oken, Carus). As in France, physico-theology played no role after about 1780. England represented a complete contrast. Here natural theology was completely dominant. Science, particularly biological science, was rather neglected, being almost entirely in the hand of amateurs, if not dilettantes. This was the background against which the rise of Darwinism must be considered.

Professionalization in science developed in France after the revolution of 1789 and at about the same time in Germany (I know of no thorough analysis; see Mendelsohn, 1964), but in Britain it was delayed until about the middle of the nineteenth century. The now generally accepted concept of science and its pursuit largely developed at the German universities. It is here that the first teaching laboratories were established in the 1830s (those of Purkinje, Liebig, Leuckart). The German universities of the nineteenth century were devoted to research and scholarship to a higher degree than those of any other country. No one saw a conflict between pure science and useful knowledge. There was a remarkable similarity in Germany between the university system and the apprentice system in the crafts. It strongly encouraged excellence and achievement.

When science began to prosper in the United States and graduate schools were established at the universities, it was the system of the German university that was most widely adopted. Again, a massive movement of scholars between countries began to develop in the later decades of the nineteenth century, a movement in which the marine biological station at Naples played an important role. Science once more became truly international, a fact which has strongly affected the development of experimental biology in the United States (Allen, 1960).

One further word on the geographic situation. Nearly all the major contributions to biological advance from the fifteenth to the end of the nineteenth century were made by only six or seven countries. The center was first Italy, but then it shifted to Switzerland, France and the Netherlands, later to Sweden, and finally to Germany and England. There was always a free movement of scholars and it was primarily for economic or sociological reasons that either one or the other country temporarily held the hegemony. One of the reasons, for instance, for the primacy of Germany in biology in the nineteenth century was the early establishment of chairs of zoology, botany, and physiology at the German universities. At a time when Richard Owen was about the only professional biologist in Britain (all teaching was done either by divines or by M.D.s), zoology and botany had already been professionalized in Germany.

The Nature of Scientific Publication

Up to well into the nineteenth century, science was progressing at quite a leisurely pace. In many disciplines and subdisciplines there was only a single specialist at any one time. So few people were working in the different branches of biology that Darwin thought he could afford to wait twenty years before publishing his theory of natural selection. He was quite thunderstruck when somebody else (A. R. Wallace) had the same idea. When the professionalization of biology began with the establishment of chairs for diverse branches of biology at many universities and when each professor began educating numerous young specialists, an exponential acceleration in the rate of scientific production occurred.

The numerical increase of specialists brought about an important change in the nature of biological publication. This, as Julius Sachs pointed out in his history of botany, took place in the first half of the nineteenth century. The great books which had characterized the eighteenth century, like Buffon's *Histoire naturelle* or Linnaeus' *Systema Naturae* began to be supplemented not only by shorter monographs, but—more significantly—by short journal papers. This caused a need for many new journals. Up to 1830 there had only been the publications of the Royal Society, of the French and other academies, as well as such general publications as the *Göttinger Wissenschaftliche Nachrichten*. Now, special societies like the Zoological Society, the Linnean Society, and the Geological Society of London began publishing. Independent journals, such as *Annals and Magazine, American Journal of Science, Zeitschrift für wissenschaftliche Zoologie,* and *Jahrbücher für wissenschaftliche Botanik,* made their appearance. We still lack a history of the biological journals, but that they had an important impact on the development of biology is without question.

112

As biology became more and more specialized in modern times, *Chromosoma, Evolution, Ecology,* and the *Zeitschrift für Tierpsychologie* (again only a few random samples) became the rallying points of newly developing subdisciplines. It has now reached the point where more papers (and pages) are published in the course of a few decades than in the whole antecedent history of biology. This broadens and deepens biology immensely, but if we should attempt to list the ten most basic problems of biology, we would probably discover that most of these problems had already been posed at least fifty or one hundred years ago. Even if the historian cannot follow every problem or controversy into the 1980s, he can certainly lay a foundation that facilitates the understanding of the current activities.

DIVISIVE DEVELOPMENTS IN THE NINETEENTH CENTURY

The development of the comparative approach around the turn of the nineteenth century provided for the first time an excellent opportunity for the unification of biology, that is, the formation of a bridge between the naturalists and the anatomists-physiologists. Cuvier's emphasis on function strengthened this bond. But only few biologists took advantage of this opportunity, no one more so than Johannes Müller (1801–1858) who switched in the 1830s from pure physiology to comparative embryology and invertebrate morphology. But Müller's own students aggravated the split in biology by their aggressive promotion of a physicalist-reductionist approach toward the study of life, quite unsuited for the study of the phenomena the naturalists were interested in. From the 1840s on, more than ever, there was a lack of communication between the naturalists and the physiologists, or as it was possible to frame it after 1859, between the students of evolutionary (ultimate) and those of physiological (proximate) causations. In some ways this polarity was nothing but a continuation of the old contrast between the herbalist-naturalists and the physician-physiologists of the sixteenth century, but the conflicts and differences of interests were now much more precisely defined, particularly after 1859. Two well defined biologies—evolutionary and functional—were now coexisting side by side. They competed with each other for talent and resources. They engaged in one controversy after the other, engendered by the difficulty of understanding the opposing viewpoints.

Some historians of science like to distinguish different periods, each with a single dominant paradigm (Kuhn), episteme (Foucault), or research tradition. This interpretation does not fit the situation in biology. Ever since the later seventeenth century, one finds more and more often that even within a given biological discipline or specialization, two seemingly incompatible paradigms may exist side by side, like preformation and epigenesis, mechanism and vitalism, iatrophysics and iatrochemistry, deism and natural theology, or catastrophism and uniformitarianism, to mention only a few of the numerous polarities. This creates formidable difficulties of interpretation. How can it be explained, on the basis of the total intellectual, cultural, and spiritual context, that is, the zeitgeist of the period, that two diametrically opposed interpretations could have originated and been maintained?

Two additional problems exist for the historiographer. The various controversies, a sample of which I have just listed, do not coincide with each other, and their terminations (by whatever means) fall in separate periods. Worse than that, as I have already described, the sequence of events in different countries is often very different: *Naturphilosophie* was largely confined to Germany (exceptions E. Geoffroy Saint-Hilaire, quinarianism, Richard Owen); natural theology dominated British science in the first half of the nineteenth century but had played out in the eighteenth century in France and Germany. Foucault's ideal, to paint the progress of science (and its milieu) as a series of consecutive *epistemes,* is clearly not encountered in the real world.

What we find instead are two sets of phenomena. First, a gradual change of the structure, institutionalization, and normative aspects of that which we now call science, and secondly, some definite periods in individual branches of science. Therefore, the best I can do is to present a regrettably disjunct set of thumbnail sketches of advances in various biological disciplines. Further study will unquestionably succeed in establishing whether and to what extent connections between the events in the various branches of biology have existed and also what the connections (if any) are between the scientific advances and the general intellectual and social milieu. The establishment of such connections is regrettably slighted in my own account. The two branches of biology that were best defined in the middle of the nineteenth century were physiological biology and evolutionary biology. I shall deal with them first, before taking up later developments.

114

Physiology Comes of Age

In no other area of biology has the pendulum between opposing viewpoints swung back and forth as frequently and as violently as in physiology. Extreme mechanistic interpretations, considering organisms as nothing but machines to be explained in terms of movements and forces, and extreme vitalism, considering organisms as being completely controlled by a sensitive if not thinking soul, were opposing each other from the time of Descartes and Galileo virtually to the end of the nineteenth century.

The physicalist movement was greatly strengthened by the popular philosophical publications of three natural scientists, Karl Vogt, Jacob Moleschott, and Ludwig Büchner, generally referred to as the German scientific materialists (Gregory, 1977). In spite of their name, they were sincere idealists, but equally sincere atheists. By their unswerving opposition to vitalism, supernaturalism, and any other kind of nonmaterialistic explanation, they served as watchdogs, so-to-speak, of physiology, attacking relentlessly any interpretation that was not physico-chemical.

There were two reasons for the rise of an almost rampant reductionist physicalism in mid-nineteenth-century physiology. One was that the still widespread power of vitalism evoked a justifiable opposition. The other reason was the enormous current prestige of the physical sciences which physiologists were able to extend to themselves by adopting uncompromising physicalism and "mechanical" explanations. Helmholtz was one of the leaders in this endeavor and proposed in 1869 at the Innsbruck meeting of the German naturalists the motto: "Endziel der Naturwissenschaften ist, die allen anderen Veränderungen zugrundeliegenden Bewegungen und deren Triebkräfte zu finden, also sie in Mechanik aufzulösen" ("The ultimate objective of the natural sciences is to reduce all processes in nature to the movements that underlie them and to find their driving forces, that is, to reduce them to mechanics").

Such a reduction is, indeed, often possible in those areas of biology that deal with proximate causes, and attempts of such an analysis are usually heuristic even where they are unsuccessful. The high prestige of this reduction, however, resulted in its application to many biological problems, particularly in evolutionary biology, where this approach is entirely unsuitable. Helmholtz, for example, went back and forth between the physical and biological sciences, a movement which was facilitated by the fact that all physiological processes are, indeed, ultimately chemical or physical processes. But his fashionable concept was readily applied also to branches of biology where it is inappropriate. Haeckel (1866), in the preface of his *Generelle Morphologie,* gives himself the task of lifting the science of organisms "durch mechanisch-kausale Begründung"—to the level of the inorganic sciences. Nägeli calls his great treatise of evolution the *Mechanisch-Physiologische Theorie der Abstammungslehre* (1884) and, at about the same time, Roux recasts embryology into "Entwicklungsmechanik."

There were two great weaknesses in these endeavors. First, "mechanistic" or "mechanical" was rarely clearly defined, sometimes meaning mechanical quite literally, as in studies of functional morphology, sometimes, however, meaning simply the opposite of supernatural. The second weakness is that the prophets of mechanism never made any distinction between proximate and ultimate causations, failing to see that the mechanistic approach, although quite indispensable in the study of proximate causations, is usually quite meaningless in the analysis of evolutionary causations.

The methodology of physiology underwent drastic changes in the nineteenth century, including a much more refined application of physical methods, particularly by Helmholtz and Ludwig, and, even more so, an increasing application of chemical methods. Each bodily process and the function of each organ and each gland were studied separately by a large army of medical, zoological, and chemical physiologists. Human physiology was, on the whole, conducted in separate laboratories from animal or plant physiology, although the human physiologists made extensive use of animal experimentation (including vivisection). The publication in 1859 of the *Origin of Species* caused hardly a ripple since explanation in physiology was explanation of proximate causation.

Darwinism

Evolutionism did not perish with the death of Lamarck in 1829. It remained a popular thought in Germany with the *Naturphilosophen* and with a few other zoologists and botanists such as Schaaffhausen and Unger. In England it was revived by Chambers in his *Vestiges* (1844), a plea for evolutionism that was highly popular in spite of the violent criticism of the professionals. Yet, natural theology and the argument from design continued to be dominant, supported by virtually all the leading scientists of the era, including Charles Lyell. This was the background against which Darwin in 1859 proposed his new theory.

Evolution consists in changes of adaptation and in diversity. Lamarck in his theory had virtually ignored diversity, assuming that new kinds of organisms originate continuously by spontaneous generation. As a result of reading Lyell's *Principles* and of his studies of the fauna of the Galapagos Islands and of South America, Darwin's attention centered on the origin of diversity, that is, on the origin of new species. His evolutionary theory was one of "common descent," ultimately deriving all organisms from a few original ancestors or possibly from a single first life. Man thereby was inexorably made part of the total evolutionary stream and was demoted from the exalted position given to him by the stoics, by the Christian dogma, and by the philosophy of Descartes. This theory of common descent can be designated as the first Darwinian revolution.

Darwin was equally revolutionary in his theory of the causation of evolution. First of all, he rejected the saltational theories of the essentialists and insisted on completely gradual evolution. He likewise rejected the Lamarckian notion of evolution by an automatic intrinsic drive toward perfection, proposing instead a strict and separate causation of every single evolutionary change. This causation for Darwin was a two-step phenomenon, the first step being the continuous production of an inexhaustible supply of genetic variation. Here, Darwin was not embarrassed to admit that he did not understand at all how such variation was produced. He treated it as a "black box." The second step was the differential survival and reproduction ("selection") among the

116

oversupply of individuals produced in every generation. This natural selection was not a "chance phenomenon," as Darwin was so often accused of having adopted, but strictly caused (although in a probabilistic sense) by the interaction between genetic endowment and environmental circumstances. This theory of evolutionary causation was Darwin's second revolution. He explained design (the harmony of the living world) in a strictly materialistic fashion and so, according to his opponents, he had "dethroned God."

The first Darwinian revolution, that is, the theory of common descent, was soon adopted by nearly all knowledgeable biologists (though some of his original opponents, such as Sedgwick and Agassiz, resisted it to their death). The second Darwinian revolution, the acceptance by biologists of natural selection as the only direction-giving factor in evolution, was not completed until the period of the "evolutionary synthesis," about 1936–1947.

117

Darwin's theory of common descent was one of the most heuristic theories ever proposed. It put a whole army of zoologists, anatomists, and embryologists to work to determine relationships and the probable characteristics of the inferred common ancestors. This was an almost endless task, not at all completed even at the present day, since there is still great uncertainty about the nearest relatives and the presumptive common ancestor of many of the major groups of plants and animals. Curiously, comparative anatomy limited itself almost entirely to the application of Darwin's theory of common descent and was, it cannot be denied, an unconscious continuation of the tradition of idealistic morphology. Hardly anyone asked direct questions concerning the causes of the structural changes in phylogeny. It was not until the 1950s that comparative morphology consciously became evolutionary morphology, by establishing contact with ecology and behavioral biology and consistently asking why-questions.

Haeckel's theory of recapitulation, that is, the theory that an organism during its ontogeny passes through the morphological stages of its ancestors, resulted in an enormous stimulation of comparative embryology. Kovalevsky's discovery that the ascidians are near relatives of the vertebrates, both belonging to the phylum Chordata, was a typical achievement of this type of research.

Comparative embryology asked almost exclusively the questions of evolutionary biology and was thus quite unsatisfactory to the representatives of functional biology. Goette, His, and Roux eventually rebelled against this one-sidedness and sought to establish an embryology devoted to the study of proximate causations, a purely mechanistic embryology, not one solely of speculation and history. This new embryology, characteristically designated by Roux as "Entwicklungsmechanik," dominated embryology from the 1880s to the 1930s. However, it soon ran into difficulties when it was found that two perfect embryos could develop from an egg separated into two after the first cleavage division. What machine, if cut in half, could function normally? This unexpected amount of self-regulation induced Driesch, who had performed this experiment, to embrace a rather extreme form of vitalism and to postulate a nonmechanical "entelechy." Even those embryologists who did not follow Driesch tended to adopt interpretations that were tinted with vitalism, such as Spemann's "organizer." Interestingly, though not antievolutionists, embryologists were virtually unanimously opposed to Darwinism. But then, so were most biologists of the time.

A minor reorientation in European biology occurred around 1870. This is when the students of J. Müller's successors took over, when the impact of Darwin's *Origin* gathered momentum, when microscopy truly came of age, when the gradual professionalization of British science began to be noticeable (Thistleton-Dyer, Michael Foster), and when France began to emancipate itself from Cuvier's influence. Progress, however, was very uneven in the different areas of biology. As a result of rapid technological advances in microscopes and in fixing and staining methods, no area was more successful in the 1870s, 80s, and 90s than the study of cells and their nuclei. In this period the process of fertilization was finally understood. Weismann, Hertwig, Strasburger, and Kölliker concluded in 1884 that the nucleus contained the genetic material. Darwin's theory of pangenesis was proposed before this information on cells was available. The subsequent research on cells led to the establishment of rather elaborate genetic theories, culminating in the detailed analysis and synthesis by Weismann (1892). With exception of Nägeli (1884) and O. Hertwig, all these authors postulated particulate inheritance, and with the exception of de Vries (1889), they all concentrated on the developmental aspect of inheritance. With the benefit of modern hindsight, we can see that they made two major assumptions that proved to be incorrect. First they assumed—in order to explain differentiation and quantitative inheritance—that the determinants for a character could be represented in a nucleus by multiple identical particles that could be distributed unequally during cell division, and second, they thought that these determinants converted directly into the structures of the developing organism. The first of these assumptions was refuted by Mendel, the second by Avery and molecular biology.

In the year 1900 de Vries and Correns rediscovered Mendel's rules, demonstrating that each parent contributes to each segregating character only one genetic unit, later called a gene (see Chapters 16 and 17). Within two decades most of the principles of transmission genetics were worked out by a whole army of geneticists, under the leadership of Bateson, Punnett, Cuénot, Correns, Johannsen, Castle, East, Baur, and T. H. Morgan. All the evidence they brought together indicated that the genetic material was unchanging, that is, inheritance is "hard." Changes in the genetic material are discontinuous and were designated "mutations." Unfortunately, de Vries and Bateson used the discovery of Mendelian inheritance as the basis for a new saltational theory of evolution, rejecting Darwin's concept of gradual evolution and more or less ignoring his theory of natural selection.

This interpretation of evolution was altogether unacceptable to the naturalists, whose understanding of the nature of species and of geographic variation had made immense progress during the preceding fifty years. Most importantly they had begun to understand the nature of populations and had developed "population thinking," according to which each individual is unique in its characteristics. Their evidence completely confirmed Darwin's conclusion that evolution is gradual (except in cases of polyploidy) and that speciation is normally geographic speciation. The literature of the taxonomists, ultimately culminating in the "new systematics," was as regrettably ignored by the experimental biologists as was much of the post-1910 genetic literature by the naturalists. The result was a deplorable communication gap between these two camps of biologists.

The difficulties and misunderstandings were finally resolved during the period between 1936–1947, resulting in a unified evolutionary theory often referred to as the "evolutionary synthesis" (Mayr and Provine, 1980). Dobzhansky, Rensch, Mayr, Huxley, Simpson, and Stebbins, among others, showed that the major evolutionary phenomena such as speciation, evolutionary trends, the origin of evolutionary novelties, and the entire systematic hierarchy could be explained in terms of the genetic theory as matured during the 1920s and 30s. Except for shifts in emphasis and for a far more precise analysis of all the various mechanisms, the synthetic theory of evolution is the paradigm of today.

BIOLOGY IN THE TWENTIETH CENTURY

120

In the same period during which the evolutionary theory was refined, whole new fields of biology emerged. Of particular importance have been the fields of ethology (the comparative study of animal behavior), of ecology, and of molecular biology.

Ethology and Ecology

After the pioneering (but largely ignored) work of Darwin (1872), Whitmann (1898), and O. Heinroth (1910), the field of ethology owes its real development to Konrad Lorenz (1927ff) and subsequently to Niko Tinbergen. While the previous schools of animal psychologists had devoted most of their attention to the study of proximate causes of behavior, generally working with a single species and concentrating on learning processes, the ethologists focused on the interaction between the genetic program and subsequent experience. They were most successful in the study of species-specific behaviors, particularly courtship behavior, behavior that is largely controlled by closed programs. The arguments between Lorenz and von Holst on one side and authors like Schneirla and Lehrman on the other side about the size of the genetic contribution to behavior were in some ways a replay of similar arguments that had gone on all the way back to the eighteenth century (Reimarus vs. Condillac) and nineteenth century (Altum vs. A. Brehm). The controversies of the 1940s and 50s in this field are now a matter of the past. There is little difference of principle among students of animal behavior, and what differences remain are largely a matter of emphasis.

The study of behavior is now expanding primarily in two directions. On one hand it is merging with neurophysiology and sensory physiology, and on the other hand with ecology: species-specific behavior is studied from the point of view of its selective significance within the niche of the respective species. Finally, much of behavior consists of the exchange of signals, most frequently among conspecific individuals. The science of signals and messages (semiotics) and the role that communication plays in the social structure of species are now particularly active areas of research.

The twentieth century is also usually credited with having given birth to *ecology*. It is true that the importance of the study of the environment had never before been appreciated quite so keenly as since the 1960s, but ecological thought goes back to the ancients (Glacken, 1970). It is prominent in the writings of Buffon and Linnaeus, and played an important role in the travelogues of the great explorers of the eighteenth and nineteenth centuries (the Forsters and Humboldt, for example). For these travelers the ultimate goal was no longer collecting and describing species but understanding the interaction of organisms and their environment. Alexander von Humboldt became the father of ecological plant geography, but his interests later turned almost entirely to geophysics. Many of Darwin's discussions and considerations would be quite appropriate for a textbook of ecology. The term "ecology" was proposed in 1866 by Haeckel as the science dealing with "the household of nature." Semper (1880) provided a first general text on the subject. In the ensuing years, there was little contact among various groups that studied "the conditions of life" or "associations" of various kinds of organisms. Möbius (1877) published his classical study of an oysterbank, Hensen and others concentrated on marine ecology, Warming on plant ecology, and still others founded limnology, the science (mostly ecological) of freshwaters.

Ecology long remained rather static and descriptive, literally thousands of papers dealing with the number of species and individuals within a certain measured area. Various authors competed with each other in proposing fancy nomenclatures for any and all terms being used in the field; even the spade with which plants were dug out was renamed "geotome."

A revitalization of the field was due to three developments. One was the calculations of Lotka-Volterra dealing with the cyclic changes of populations owing to predator-prey relations, and more broadly with various other aspects of the growth, decline, and cycling of populations. The second development was a greater emphasis on competition. It led to the establishment of the principle of competitive exclusion and its experimental testing by Gause. In due time the study of the competitive relationships of species became one of the major subdivisions of ecology under the leadership of David Lack and Robert MacArthur. The subject occupies the border area between ecology and evolutionary biology since these competitive relationships not only determine the presence and absence of species, their relative frequency, and the total species diversity but also the adaptive changes of these species in the course of evolution. A third development leading to the vitalization of ecology was attention to energy turnover problems, particularly in freshwater and in the ocean. The question of how much computer-based modeling is contributing to the understanding of interactions in ecosystems is still controversial.

Since many ecological factors are ultimately behavioral characteristics, such as predator thwarting, feeding strategies, niche selection, niche recognition, all evaluations of aspects of the environment, and many others, one can perhaps even go so far as saying that, at least in animals, the greater part of ecological research is now concerned with behavioral problems. Furthermore, all work in plant as well as animal ecology ultimately deals with natural selection.

The Emergence of Molecular Biology

As the analysis of physiological and developmental processes became more detailed and more sophisticated, it became evident that ultimately many of these processes can be reduced to the action of biological molecules. The study of such molecules at first was strictly the domain of chemistry and biochemistry. The early roots of biochemistry go far back into the nineteenth century but originally there was no clear delimitation from organic chemistry, and biochemical research was usually conducted in chemical institutions. Indeed, much early biochemistry had little to do with biology, being merely the chemistry of compounds extracted from organisms or, at best, compounds of importance in biological processes. To the present day, some biochemistry is still of this complexion. A second pathway leading to molecular biology branched off from physiology (Florkin, 1972ff; Fruton, 1972; Leicester, 1974).

Some of the achievements of biochemistry are of particular significance for the biologist. One is the elucidation, step by step, of certain metabolic pathways, for instance the citric acid cycle, as well as the eventual demonstration that each step is normally controlled by a specific gene. Such work is no longer simply biochemistry and it became customary and quite legitimate to refer to it as molecular biology. Indeed, one deals here with the biology of molecules, their modifications, interactions, and even their evolutionary history.

Another important development was the realization that the assumptions of colloid chemistry were unrealistic and that many biologically important materials consisted of polymers of high molecular weight. This development, particularly connected with the name of Staudinger in the 1920s, greatly facilitated the eventual understanding of collagen, muscle protein, and most importantly, DNA and RNA. Polymerized organic molecules have some of the properties of crystals and it was discovered that their complex three-dimensional structure could be elucidated with X-ray crystallography (as demonstrated by Bragg, Perutz, Wilkins, and others). Through these studies it became clear that the three-dimensional structure of macromolecules, that is, their morphology, is the basis of their functioning. Although most biological macromolecules are ultimately aggregates of the same limited number of atoms, mainly carbon, hydrogen, oxygen, sulphur, phosphorus, and nitrogen, they all have extraordinarily specific and often totally unique properties. The study of the three-dimensional configuration of these macromolecules greatly added to our understanding of these properties.

Molecular biologists have worked out the structure of literally thousands of biological compounds and clarified the pathways in which they are involved, but few of their researches created as much excitement as those clarifying the chemical nature of the genetic material. As far back as 1869, Miescher had discovered that a high proportion of the nuclear material consisted of nucleic acids. For a while (1880s and 90s) it was postulated that nuclein (nucleic acid) was the genetic material, but this hypothesis eventually lost in popularity (Chapter 19). It was not until Avery and his associates demonstrated in 1944 that the transforming substance of the pneumococcus was DNA that a reorientation occurred. Although several biologists were at once fully aware of the importance of Avery's discovery, they did not have the technical know-how for a detailed study of this fascinating molecule. The problem was quite obvious. How could this seemingly simple molecule (believed at that time to be simple compared to most proteins) carry the entire information in the nucleus of the fertilized egg to control the species-specific development of the resulting organism? One needed to know the exact structure of DNA before one could begin to speculate how it could perform its unique function. A hot race to achieve this goal developed among a number of laboratories, with Watson and Crick of the Cavendish Laboratory in Cambridge, England, emerging in 1953 as victors. If they had not been successful, somebody else would have been a few months or years later.

Everybody has heard of the story of the double helix, but not everybody fully understands the significance of this discovery. It turned out that the DNA does not participate directly in development or in the physiological functions of the body but simply supplies a set of instructions (a genetic program) which is translated into the appropriate proteins. DNA is a blueprint, identical in every cell of the body, which through fertilization is handed on from generation to generation. The crucial component of the DNA molecules are four base pairs (always one purine and one pyrimidine). A sequence of three base pairs (a triplet) functions like a letter in a code and controls the translation into a specific amino acid. The sequence of such triplets determines the particular peptide that is formed. The discovery that it is triplets of DNA that are translated into amino acids was made by M. Nirenberg in 1961. The sequence of the bases in the triplet is the code.

The discovery of the double helix of DNA and of its code was a breakthrough of the first order. It clarified once and for all some of the most confused areas of biology and led to the posing of clear-cut new questions, some of which are now among the current frontiers of biology. It showed why organisms are fundamentally different from any kind of nonliving material. There is nothing in the inanimate world that has a genetic program which stores information with a history of three thousand million years! At the same time, this purely materialistic explanation elucidates many of the phenomena which the vitalists had claimed could not be explained chemically or physically. To be sure, it still is a physicalist explanation, but one infinitely more sophisticated than the gross mechanistic explanations of earlier centuries.

Simultaneously with the purely chemical developments of molecular biology occurred others of a different nature. The invention of the electron microscope in the 1930s, for instance, permitted an entirely new understanding of cell structure. What the nineteenth-century investigators had called protoplasm and had considered to be the basic substance of life turned out to be a highly complex system of intracellular organelles with various functions. Most of them are membrane systems which serve as the "habitat" of specific macromolecules. Molecular biology is advancing at far more frontiers than can here be mentioned, many of them of considerable medical importance.

MAJOR PERIODS IN THE HISTORY OF BIOLOGY

It is traditional in historiography to distinguish periods. Western world history, for instance, has been divided into three periods: ancient, medieval, and modern. The break between medieval and modern is usually placed around 1500, or to be more precise between 1447 and 1517. Within this period, it was said, all those decisive events occurred—or movements started—that gave the new West its characteristic tone: the invention of printing with movable type (1447), the Renaissance (supposed to have started with the fall of Constantinople in 1453), the discovery of the New World (1492), and the Reformation (1517). These events signify rather drastic changes, even though one may question the legitimacy of postulating a clear-cut division between medieval and modern. After all, there were also numerous notable developments in the two hundred years before 1447.

125

Historians of science have attempted similarly to distinguish well-defined eras in science. Much has been made of the fact that the chief works of Copernicus and Vesalius were both published in 1543. More importantly, the events of the period from Galileo (1564–1642) to Newton (1642–1727) have been designated as "the Scientific Revolution" (Hall, 1954). As significant as the advances were that occurred in the physical sciences during this period and also in philosophy (with Bacon and Descartes), no world-shaking changes happened simultaneously in biology. And for one who is critical, the *Fabrica* of Vesalius is, except for the artistic superiority of its illustrations, hardly a revolutionary treatise. It cannot in any way be compared in importance with *De Revolutionibus* of Copernicus (see also Radl, 1913: 99–107).

The sixteenth century was a difficult and often contradictory period, a period of rapid changes of mood. It saw the height of humanism (as exemplified in the work of Erasmus of Rotterdam), Luther's reformation (1517), but also a vigorous beginning of the counterreformation (with the founding of the Jesuit order), and the beginning of the scientific revolution. The rediscovery of the real Aristotle (as distinct from that of the scholastics) had a clear impact on biology in the work of Cesalpino and Harvey. Although in no way comparable with the flowering of the mechanical sciences, both physiology and natural history showed definite signs of increasing activity toward the end of the sixteenth and early seventeenth century.

All the indications are that there is little congruence between what happened in the physical and in the life sciences. Nor can one delimit well-defined ideological periods in biology, as has been pointed out quite rightly by John Greene (1967) in a thoughtful review of Foucault's *Les mots et les choses*. Jacob's *Logic of Life* (1970) is written in the Foucault tradition even though Jacob likewise does not accept Foucault's periods. Holmes (1977), in turn, has questioned whether Jacob's delimitation of periods is any better.

126

None of these authors has quite faced up to the question why the partitioning of the history of biology leads to such different results when done by different authors. Could it be that perhaps such periods are purely imaginary and hence can be established only by arbitrary divisions that will be made in different ways by different authors? It is rather unlikely that this suggestion is correct. Many of the periods recognized by certain historians are far too real. It seems to me that the answer is a different one. It is, that these periods are not universal. They differ to some extent in different countries and they differ quite decidedly in different sciences and in different parts of biology and in particular between functional biology and evolutionary biology. The correlation between changes in these two branches of biology is not at all close.

The biological sciences lack the unity of the physical sciences and, as I have mentioned, each of the various disciplines has had its own chronology of birth and flowering. Until the seventeenth century or thereabouts, biological science as we would now call it, consisted of two fields, connected only in the most tenuous manner: natural history and medicine. Then, during the 17th and 18th centuries natural history rather definitely separated into zoology and botany, although many of the practitioners, up to Linnaeus and Lamarck, moved freely from one to the other. In medicine, at the same time, anatomy, physiology, surgery, and clinical medicine became increasingly segregated. Fields that became dominant in the twentieth century, like genetics, biochemistry, ecology, and evolutionary biology, simply did not exist prior to 1800. The rise—and occasional setbacks—of each of these fields is a fascinating story that will be one of the major themes of the ensuing chapters of this volume.

A taxonomist, a geneticist, or a physiologist may recognize different periods and so might a German, a Frenchman, or an Englishman. It is regrettable that history is not more tidy, but this is the way it is. Unfortunately, this makes the task of the historian rather difficult because he may have to study simultaneously five or six different contemporary "research traditions" (as Larry Laudan calls it). As provocative as the problem of intellectual periods is, their recognition is sufficiently new so that no good analysis so far is available for all of biology and for all the world.

Each of the numerous biological disciplines, such as embryology, cytology, physiology, or neurology, had periods of stagnation and periods of greatly accelerated advance. The question is sometimes asked whether there was ever any period during which the biological sciences experienced as drastic a reorientation as did the physical sciences during the scientific revolution. The answer must be no. To be sure there have been particular years during which a new beginning was made in one or the other branch of biology: 1828 for embryology, 1839 for cytology, 1859 for evolutionary biology, and 1900 for genetics. However, each branch of biology had its own cycle and there was no broad-based general revolution. Even the publication of the *Origin of Species* in 1859 was virtually without impact on the experimental branches of biology. The replacement of essentialistic by populational thinking, so fundamental in evolutionary biology, hardly touched functional biology until almost a hundred years later. The elucidation of DNA (1953) had a powerful impact on cellular and molecular biology but was irrelevant for much of organismic biology.

The closest equivalent to a revolution in the biological sciences can be found from about 1830 to 1860, one of the most exciting periods in the history of biology (see Jacob, 1973: 178). It was then that embryology received a quantum boost through the work of K. E. von Baer, when cytology got its start with the discovery of the nucleus by Brown and the work of Schwann, Schleiden, and Virchow, when the new physiology took form under Helmholtz, duBois-Reymond, Ludwig, Bernard, when the foundations were laid for organic chemistry by Wöhler, Liebig, and others, when invertebrate zoology was placed on a new foundation by Johannes Müller, Leuckart, Siebold, and Sars, and finally and most importantly when the new theory of evolution was conceived by Darwin and Wallace. These manifold activities were by no means part of a single unified movement, in fact, they were largely independent. Much of the activity was due to the increasing professionalization of science, the improvement of the microscope, and the rapid development of chemistry. Some of it however, was the direct result of the inexplicable rise of a particular genius.

BIOLOGY AND PHILOSOPHY

Among the Greeks there was no separation of science and philosophy. Philosophy was the science of the day, as was particularly true for the Ionian philosophers from Thales on. Some mathematicians-engineers, like Archimedes, and some physicians-physiologists, like Hippocrates and later Galen, came closest to being genuine scientists but the outstanding philosophers of the period, like Aristotle, were as much scientists as they were philosophers.

The two disciplines began to separate after the end of scholasticism. Anatomists like Vesalius, physicist-astronomers like Galileo, botanist-anatomists like Cesalpino, and physiologists like Harvey were primarily scientists even though some of them had a very strong philosophical Aristotelian or anti-Aristotelian commitment. The philosophers, in turn, became increasingly "pure" philosophers. Descartes was one of the few who was both a scientist and a philosopher, while Berkeley, Hobbes, Locke, and Hume are already pure philosophers. Kant was perhaps the last philosopher also to make distinguished theoretical contributions to science (namely to anthropology and cosmology), contributions that one will find quoted in strictly scientific historiographies. After him it was the scientists and mathematicians who contributed to philosophy (Herschel, Darwin, Helmholtz, Mach, Russell, Einstein, Heisenberg, K. Lorenz) rather than the reverse.

Philosophy flowered during the eighteenth and nineteenth centuries. The hold of Aristotle had been broken by Descartes, and the hold of Descartes was broken by Locke, Hume, and Kant. Curiously, no matter how different they were in most of their other views, all the philosophers of this period asked most of their questions in the framework of essentialism. The nineteenth century witnessed various new departures, among which Comte's positivism, a philosophy of science, was most important. A strongly reductionist materialism, represented in Germany by Vogt, Büchner, and Moleschott (Gregory, 1977), was quite influential, if for no other reason than that its exaggerations led to the development of holist, emergentist, or even vitalist movements. Yet by its consistent and never falsified denials of any dualism and supernaturalism, it had a lasting effect.

Within biology these philosophical movements had their greatest impact on physiology and psychobiology, that is, on disciplines of biology dealing with proximate causes. What has not yet been properly analyzed is the exact nature of the relation between these philosophies and physiological research. Some opposing claims notwithstanding, philosophy seems to have played only a minor, if not negligible, role in the process of discovery, but philosophical dogmas or principles played a large role in the framing of explanatory hypotheses.

Among the philosophers, Gottfried Wilhelm Leibniz (1646–1717), in contrast to the physicalist philosophers of his day, had a real concern for understanding nature as a whole. He showed how unsatisfactory it is to explain the workings of the living world strictly with the help of secondary, physical causes. Even though his own answers (a pre-established harmony and a law of sufficient reason) were not the sought-for solutions, he posed problems that deeply puzzled the following generations of philosophers, including Kant. In spite of his mathematical brilliancy, Leibniz saw clearly that there was more to nature than mere quantity and became one of the first to appreciate the importance of quality. In an age dominated by the discontinuity concept of essentialism, he stressed continuity. His interest in the *scala naturae*, static though he conceived it, helped to prepare the ground for evolutionary thinking. He profoundly affected the thinking of Buffon, Maupertuis, Diderot, and others of the philosophers of the Enlightenment, and through them Lamarck. He was perhaps the most important counterinfluence to the essentialistic, mechanistic thinking of the Galilei-Newton tradition.

The philosophical foundations of evolutionary biology are far less clear than those of functional biology. The concept of a directionality in life ("higher and lower") goes back to Aristotle and to the *scala naturae* (Lovejoy, 1936), but population thinking apparently had only very tenuous roots in philosophy (late nominalism). The crucial insight of the importance of history (in contrast with the timelessness of physical laws) did receive a considerable input from philosophy (Vico, Leibniz, Herder). An acceptance of the importance of history almost inevitably led to a recognition of the process of development. Development was important for Schelling (and the *Naturphilosophen*), Hegel, Comte, Marx, and Spencer. The importance of these thinkers is well stated in Mandelbaum's (1971: 42) definition of historicism: "Historicism is the belief that an adequate understanding of the nature of any phenomenon and an adequate assessment of its value are to be gained through con-

129

sidering it in terms of the place which it occupied and the role it played within a process of development."

It would be tempting to suggest that the theory of evolution originated in this type of thinking, but there is little evidence that this was the case, except for Spencer's evolutionism, which, however, was not seminal for the thinking of Darwin, Wallace, Huxley, or Haeckel. Indeed, and rather unexpectedly, historicism never seems to have had close relations to evolutionary biology, except perhaps in anthropology. However, historicism and logical positivism were two completely incompatible developments. It is only in relatively recent times that the concept of "historical narratives" has been accepted by some philosophers of science. And yet, it might have been evident soon after 1859 that the concept of law is far less helpful in evolutionary biology (and for that matter in any science dealing with time-dominated processes such as cosmology, meteorology, paleontology, paleoclimatology, or oceanography) than the concept of historical narratives.

The opponents of Cartesianism asked questions that never occurred to the mechanists. These questions made it embarrassingly evident how incomplete the explanations of the mechanists were. Not only did they ask questions about time and history, but also why-questions were asked increasingly often, that is, "ultimate causations" were searched for. It was in Germany, toward the end of the eighteenth and early in the nineteenth century, that the most determined resistance developed against the mechanistic approach of the followers of Newton, who were satisfied in asking simple questions concerning proximate causations. Even authors outside of biology, like Herder, exerted a powerful influence on this dissent. Unfortunately, no constructive new paradigm emerged from these efforts (in which Goethe and Kant were also involved); instead, this movement fell into the hands of an Oken, Schelling, and Carus, authors whose fantasies the experts could only meet with ridicule and whose silly constructions the modern reader can only read with embarrassment. Yet some of their basic interests were to be very much the same as those of Darwin. Revolted by the excesses of the *Naturphilosophen*, the antimechanistic naturalists retreated into nonproblematic description, a field in which the scope was unlimited but, as the best minds soon pointed out, also intellectually unrewarding.

It is still subject to controversy whether or not philosophy has made any contribution to science after 1800. Not surprisingly, philosophers generally tend to answer this question in the affirm-ative, scientists often in the negative. There is little doubt, however, that the formulation of Darwin's research program was influenced by philosophy (Ruse, 1979; Hodge, 1982). In recent generations philosophy rather clearly has retreated into metascience, that is, an analysis of scientific methodology, semantics, linguistics, semiotics, and other subjects at the periphery of science.

BIOLOGY TODAY

If one wanted to characterize modern biology in a few words, what would one say? Perhaps the most impressive aspect of current biology is its unification. Virtually all the great controversies of former centuries have been resolved. Vitalism in all of its forms has been totally refuted and has had no serious adherency for several generations. The numerous competing evolutionary theories have been abandoned one by one and replaced by a synthetic one, which rejects essentialism, an inheritance of acquired characters, orthogenetic trends, and saltationism.

More and more biologists have learned that functional and evolutionary biology are not an "either-or" situation, but that no biological problem is solved until both proximate *and* ultimate (= evolutionary) causations are determined. As a result, many molecular biologists now study evolutionary problems, and many evolutionary biologists deal with molecular problems. There is far more mutual understanding than prevailed even twenty-five years ago.

The last twenty-five years have also seen the final emancipation of biology from the physical sciences. It is now widely admitted not only that the complexity of biological systems is of a different order of magnitude, but also that the existence of historically evolved programs is unknown in the inanimate world. Teleonomic processes and adapted systems, made possible by these programs, are unknown in physical systems.

The process of emergence—the origination of previously unsuspected new qualities or properties at higher levels of integration in complex hierarchical systems—is vastly more important in living than in inanimate systems. This also contributes to the differences between the physical and biological sciences, and to differences in the strategies and explanatory models in these fields.

The question of what the major current problems of biology are cannot be answered, for I do not know of a single biological discipline that does not have major unsolved problems, this being true even for such classical fields as systematics, biogeography, and comparative anatomy. Still, the most burning and as yet most intractable problems are those that involve complex systems. The simplest of these, currently in the center of interest in molecular biology, is the structure and function of the eukaryote chromosome. To understand this we must also know what is the specific function and the interaction of the various kinds of DNA (coding for soluable or non-soluable proteins, silent DNA, middle repetitive, high repetitive, and so forth). Although chemically all these DNAs are in principle the same, some produce building materials, some have a regulatory function, and still others are believed by some molecular biologists to have no function at all (to be "parasitic"). This may be true, but is not very convincing for a dyed-in-the-wool Darwinian like myself. However, I have no doubt that the whole complex DNA system will be understood within a few years.

I am less sanguine about the rate of progress in our understanding of the more complex physiological systems such as those that control differentiation and the working of the central nervous system. One cannot solve these problems without dissecting the systems into their components and yet the destruction of the systems during the analysis makes it very difficult to understand the nature of all the interactions and control mechanisms within the systems. It will require much time and patience before we will fully understand complex biological systems. And it will come only through a combination of reductionist and emergentist approaches.

Biology has now become so large and so diversified that it can no longer be dominated entirely by one particular fashion, such as species describing in the age of Linnaeus, the construction of phylogenies in the post-Darwinian period, or *Entwicklungsmechanik* in the 1920s. To be sure, molecular biology is particularly active at present, but neurobiology is also vigorous and flourishing, and so are ecology and behavioral biology. And even the less active branches of biology have their own journals (including many new ones), organize symposia, and pose new questions all the time. What is most important is that in spite of the seeming fractionation, there is now more of a spirit of unity than in several hundred years.

I

Diversity of Life

$$\text{※※※※※※※※※※※※※※※※※※※※※※※※※※※}$$

Hardly ANY aspect of life is more characteristic than its almost unlimited diversity. No two individuals in sexually reproducing populations are the same, no two populations of the same species, no two species, no two higher taxa, nor any associations, and so *ad infinitum*. Wherever we look, we find uniqueness, and uniqueness spells diversity.

Diversity in the living world exists at every hierarchical level. There are at least 10,000 different kinds of macromolecules in a higher organism (some estimates going even much higher than that). Taking all the different states of repression and derepression of all the genes in a nucleus into consideration, there are millions, if not billions, of different cells in a higher organism. There are thousands of different organs, glands, muscles, neurocenters, tissues, and so on. Any two individuals in sexually reproducing species are different not only because they are genetically unique but also because they may differ by age and sex and by having accumulated different types of information in their open memory programs and in their immune systems. This diversity is the basis of ecosystems and the cause of competition and symbiosis; it also makes natural selection possible. Every organism depends for its survival on a knowledge of the diversity of its environment, or at least on an ability to cope with it. Indeed, there is hardly any biological process or phenomenon where diversity is not involved.

What is particularly significant is that one can ask very similar questions concerning diversity at each hierarchical level, such as the extent or variance of the diversity, its mean value, its origin, its functional role, and its selective significance. As is characteristic for so much in the biological sciences, the answers to most of these questions are of a qualitative rather than a quantitative nature. Whatever level of diversity one is dealing with, the first step in its study is obviously that of inventory taking. It is the discovery and description of the different "kinds" of which a particular class consists, whether they be different tissues and organs in anatomy, different normal and abnormal cells and cellular organelles in cytology, different kinds of associations and biota in ecology and biogeography, or different kinds of species and higher taxa in taxonomy. The foundation which description and inventory taking lays forms the basis on which all further progress in the relevant sciences depends. In the ensuing chapters I shall concentrate on one particular component of the diversity of life, the diversity of kinds or organisms.[1]

134

The Discovery of the Extent of Diversity

Diversity has occupied man's mind ever since there has been man. No matter how ignorant a native tribe might be in other matters biological, invariably it has a considerable vocabulary of names for the various kinds of animals and plants occurring locally. The first creatures to be named are, of course, those of immediate concern to man, whether predators (bears, wolves), sources of food (hares, deer, fish, clams, vegetables, fruits, and so on), sources of clothing (skins, fur, feathers), or possessors of magical qualities. These are still today the predominant "kinds" in folklore.

That this preoccupation with the diversity of nature is worldwide became apparent when European naturalists returned from expeditions and collecting trips. They would invariably report on an amazing knowledge of birds, plants, fishes, or shallow-water sea life which they had encountered in every tribe of native peoples they had visited. Such knowledge is handed by word of mouth from generation to generation. Each tribe, not surprisingly, concentrates on the natural history of special interest to its daily life. A coastal tribe may know all about the shellfish in the intertidal zone but hardly anything about the bird life of the adjacent forest. Since the number of kinds of birds in a district is usually quite limited, a tribe may have a separate name for each species (Diamond, 1966). In the case of rich local floras, the emphasis may be on generic names, a tradition continued by the botanist Linnaeus. There is usually a rich vocabulary for cultivated plants and domesticated animals, but members of tribes with a hunting tradition may also have a superb knowledge of wild animals and native plants. It is a great pity that this knowledge has been neglected by anthropologists for such a long time. Since these traditions are quickly lost under the impact of civilization, it is too late in many areas to study folk taxonomy. Fortunately, some excellent studies have been published in recent years.[2] What is particularly interesting is how often not only species and varieties were recognized but also higher taxa.

The early naturalists knew only the limited fauna or flora of their home country. Even Aristotle mentions only about 550 kinds of animals, and the early Renaissance herbals contained between 250 and 600 kinds of plants. That not all parts of the world have the same biota was, however, already known to the ancients from the accounts of travelers as recorded by Herodotus, Pliny, and other authors. They mentioned elephants, giraffes, tigers, and many other animals not found in the Mediterranean, or at least not on the European shores.

135

The existence of such strange creatures greatly excited the European imagination, because of the universal fascination of civilized man with the unknown, whether it be exotic countries, strange people, or bizarre animals and plants. To discover and describe all the marvelous creatures in this wonderful world of ours was the great passion of travelers and compilers from Pliny to Gesner and the disciples of Linnaeus. The ancients, of course, had not the slightest inkling of the extent of the geographical localization of faunas and floras which we now know to exist. This did not happen until travelers penetrated far into Asia, like Marco Polo (1254–1323), or into Africa. When the Portuguese began their voyages in the fifteenth century and Columbus discovered the New World (1492), a whole new dimension was added to the appreciation of the world's biotic diversity. Cook's voyages, which opened up the exploration of Australia and the Pacific Islands, were the capstone in the erection of this edifice. Yet, this was only the beginning, for the early travelers and collectors obtained only small samples of the distant faunas and floras. Even in Europe new species of mammals and butterflies were described as late as the 1940s and 1950s. And as far as less conspicuous groups and less accessible areas are concerned, the treasure chest of undescribed species seems to be inexhaustible. In the tropics even at this date we may not know more than a fifth or tenth of the existing species.

The increase in knowledge was accompanied by a noticeable change in attitude. The early travelers were interested in the spectacular. They liked nothing better than to come home with tales of monsters and fabulous creatures of all sorts. Soon this was replaced by a genuine interest in the purely exotic. Private collectors in Britain, France, Holland, and Germany established natural-history cabinets, with an attitude that was hardly better than that of stamp or coin collectors. However, true naturalists like Linnaeus and Artedi benefited from the enthusiasm of such collector-patrons. Marcgrave in Brazil and Rumphius in the East Indies were among visitors to the colonies who made important contributions to the natural history of previously almost unknown areas (see Stresemann, 1975).

136

The eighteenth century was the beginning of the era of great voyages. Bougainville and other French expeditions, as well as Cook and other British ones, brought home wonderful treasures. This activity accelerated in the nineteenth century, with Russia (Kotzebue) and the United States joining. Travelers went to the corners of the globe, collecting natural-history specimens of all descriptions, filling the private museums to the point of bursting, and forcing the building of great national and state museums and herbaria.[3] There never were too many specimens because every collection produced more novelties. In a group as well known as the birds, a single expedition (Whitney South Sea Expedition), visiting nearly all islands of the south seas, discovered more than thirty new species as recently as the 1920s and 30s.

The work of Humboldt and Bonpland in South America, of Darwin on the *Beagle* (1831–1836), of A. R. Wallace in the East Indies (1854–1862) and of Bates and Spruce in Amazonia is well known, but it is usually forgotten that there were quite literally thousands of other collectors. Linnaeus sent his students to bring back exotic plants, but some of the best succumbed to tropical diseases: Bartsch (d. 1738), Ternström (d. 1746), Hasselquist (d. 1752), Loefling (d. 1756), and Forskål (d. 1763). The tragedy was even greater in the East Indies, where the flower of European zoology succumbed to tropical diseases or murder during a thirty-year period: Kuhl (d. 1821), van Hasselt (d. 1823), Boie (d. 1827), Macklot (d. 1832), van Oort (d. 1834), Horner (d. 1838), Forsten (d. 1843), and Schwaner (d. 1851). This included the most enthusiastic and gifted naturalists of the period, their dream being to contribute to the knowledge of the animal life of the tropics. Kuhl and Boie had been the most brilliant young naturalists of Germany (Stresemann, 1975). The gap caused by their death contributed to the ensuing decline in the quality of German natural history, for there is always only a limited number of first-rate minds at a given period.

Unexplored or poorly known countries were only one of the many frontiers rolled back by the students of diversity. Other life forms and exotic environments were studied as well. Parasites, for example, became fit subjects for serious study. Human intestinal parasites were already mentioned in the papyrus Ebers (1500 B.C.) and were discussed by the ancient Greek physicians; when their ubiquity in man and animals was established, it led to the belief that they originated by spontaneous generation. Not until the nineteenth century was it realized that many, if not most, parasites are restricted to a single host and that a host species may be infested simultaneously by several different kinds of parasites: tapeworms (cestodes), flukes (trematodes), threadworms (nematodes), blood parasites, and cell parasites. Beginning with the work of zoologists like Rudolphi, von Siebold, Küchenmeister, and Leuckart, an ever larger army of parasitologists specialized in this brand of diversity.[4] Owing to the complex life cycles of most parasites, their study requires particular perseverance and ingenuity. Since parasites are among the most serious causes of human diseases (malaria, sleeping sickness, schistosomiasis, rickettsias, and so on), their study was rightly given special attention. Plants likewise are extensively parasitized—by gall insects, mites, and a vast array of fungi and viruses. It probably would be no exaggeration to claim that there are more species of plant parasites than of higher plants. Their gradual discovery led to an enormous expansion of the realm of organic diversity.

Another frontier of diversity was found in fresh water and the oceans. Aristotle, during his stay on Lesbos, had been fascinated with marine life. Yet, as late as 1758 Linnaeus mentioned ridiculously few marine organisms in the *Systema Naturae*, except for a few fish, mollusks, and corals. This changed rapidly, owing to the interests of Pallas, St. Müller, and a series of Scandinavian investigators. Soon discovery followed on discovery. But here again, the end of the quest is not yet in sight. Sars was the first to open the gate to the deep-sea fauna, which received the special attention of the British Challenger Expedition (1872–1876). The Scandinavians, Dutch, French, and Germans followed with oceanographic expeditions, and specialists are still busy describing their discoveries. Marine life itself provided still another frontier: marine parasites. Marine organisms are in part parasitized by the same higher taxa as terrestrial organisms (such as cestodes and trematodes), but other parasites are restricted to the oceans (mesozoans, parasitic copepodes, Rhizocephala), where they have experienced a rich radiation.

The microscope opened up still another frontier of diversity: the world of organisms that could not be seen with the naked eye or at least not well (Nordenskiöld, 1928). The use of simple lenses to magnify small objects may go back to the ancients. A combination of lenses—that is, a microscope—was apparently first constructed by some Dutch lensmakers early in the seventeenth century. A study of the structure of the bee (based on a five-fold magnification) by the Italian Francisco Stelluti, published in Rome in 1625, was the first work in biological microscopy. All of the microscopical work in the next two hundred years was done with exceedingly simple instruments. Much of it was devoted to the study of plant tissues (Hooke, Grew, Malpighi) or of the fine structure of animals, particularly insects (Malpighi, Swammerdam). Swammerdam discovered *Daphnia* in 1669 but neither described it in detail nor followed it up by studying other plankton organisms (Schierbeck, 1967; Nordenskiöld, 1928).

As important as the role of these investigators was for the history of cytology and of animal and plant morphology, it was van Leeuwenhoek who deserves the principal credit for having employed the microscope to expand the frontiers of diversity (Dobell, 1960). With quite an amazing instrument, a single-lens microscope, he was able to achieve magnifications up to, it is said, 270 times. It was he who discovered in 1674, 1675, 1676, and later years, the rich life of protists (protozoans and one-celled algae) and other plankton organisms (rotifers, small crustaceans, and so on) in water and thus laid the foundation for several of the subsequently most flourishing branches of biology. Indeed, he even discovered and described bacteria. His discovery of the Infusoria (one-celled animals and plants) had an enormous impact on the thinking of his period and on the problem of spontaneous generation. But most importantly van Leeuwenhoek was the first to make biologists aware of the vast realm of microscopic life, which raised entirely new problems for the students of classification.

It was not until 1838 that Ehrenberg gave the first comprehensive treatment of the protozoans but, this being prior to the cell theory, he considered them "vollkommene Organismen," endowed with the same organs as higher organisms (nerves, muscles, intestines, gonads, and so forth). C. T. von Siebold established the phylum Protozoa in 1848 and demonstrated their single-cell nature.[5] Rapid progress was made in the first half of the nine-teenth century also in the knowledge of all kinds of plankton an-imals and algae. Every improvement of the microscope added to our knowledge, the invention of the electron microscope in the 1930s even permitting the study of the morphology of viruses.

The focus in my story so far has been the opening of the world of animal diversity. This coincided with a similar activity in plant exploration. Here also one can speak of a series of frontiers. Even before the flowering plants (angiosperms) had been reason-ably well described, certain botanists had begun to specialize in the cryptogams (ferns, mosses, lichens, algae) and the rich world of fungi (Mägdefrau, 1973).

Fossils

But this is not the end! The diversity of the living world is more than matched by the life of bygone ages, represented in the fossil state. The highest estimates of the number of living animals and plants is about 10 million species. Considering that life on earth began about 3.5 billion years ago, also that a rather rich biota has existed for at least 500 million years, and allowing for a reasonable turnover in the species composition of the biota, an estimate of one billion extinct species is presumably rather on the low side. In paleontology, the time of the great discoveries, like *Archaeopteryx,* a link between reptiles and birds, and *Ichthyostega,* a link between fishes and amphibians, is perhaps nearing an end, but even today an occasional new phylum of fossil invertebrates is still described and there seems to be no end to the new orders, families, and genera.

The discovery of fossil faunas and floras has a long history, going back to the ancients (see also Part II).[6] Fossil marine mol-lusca were mentioned by Herodotus, Strabo, Plutarch, and partic-ularly Xenophanes and were recognized as due to marine regres-sions. Fossil mammals, reptiles, and amphibians, however, were beginning to get attention only in the seventeenth century, with an ever increasing number of discoveries being made in the eigh-teenth and nineteenth centuries. Who is not familiar with the dis-covery of mastodons, dinosaurs, ichthyosaurs, pterodactyls, moas, and other, often giant, fossil vertebrates?

A parallel expansion of our knowledge was experienced by paleobotany (Mägdefrau, 1973: 231–251). The problems in this field are great owing to the difficulty of matching stems, leaves, flowers, pollen, and fruits (seeds), but the number of known fossils has steadily grown, and with it our understanding of their distribution in space and time. The study of fossil pollen has made a particularly important contribution. But there are still many great puzzles, including that of the origin of the angiosperms (Doyle, 1978).

Until the 1950s the oldest known fossils (latest Precambrian) were about 625 million years old. Barghoorn, Cloud, and Schopf have, since then, pushed this frontier back by an order of five with their discovery of fossil prokaryotes in rocks that are about 3.5 billion years old (Schopf, 1978).

The prokaryotes, living or fossil, are now the most challenging frontier of descriptive systematics. A careful study of the biochemistry and physiology of the bacteria has revealed that they are far more diversified than had been previously realized. Woese and collaborators, indeed, have proposed to recognize a separate kingdom (Archaebacteria) for the methanobacteria and relatives and still another one for those prokaryotes that are believed to have given rise to the symbiotic organelles of the eukaryote cells (mitochondria, plastids, and so on). The study of ribosomal RNA and of other molecules has finally brought considerable light into the previously rather controversial classification of the bacteria (Fox et al., 1980). Amazingly enough, there is always something new, often startlingly new, being discovered in taxonomy, the oldest of the branches of biology, for instance, the rediscovery of *Trichoplax,* seemingly the most primitive of the metazoans (Grell, 1972).

Systematics, the Science of Diversity

When one reviews the history of the exploration of organic diversity, one cannot help being awed by the overwhelming diversity of nature in space (all continents), in time (from 3.5 billion years ago to the present), in size (from viruses to whales), in habitat (air, land, fresh water, ocean), and in life-style (free-living versus parasitic). Not surprisingly, man did not ignore the incredible richness of organic life all around him; in fact, he had diverse reasons for studying it. There was, first of all, his ever present curiosity about his environment, and his wish to know it and to understand it. There was also the purely practical need to know which animals and plants might be useful to him, particularly as food and, in the case of plants, as medicine. When Linnaeus was asked what good the study of diversity was, he, as a pious creationist, answered in his dissertation, "Cui bono?": All created things have to serve a purpose. Some plants are for medicine, some organisms are meant for human food, and so on. The all-wise creator did not do anything in vain but created everything for a special purpose or for the benefit of somebody or something. It is our task to discover these intended uses, and this is the purpose of natural history.

The passion of some of seventeenth- and eighteenth-century authors for the study of nature had, however, still another reason. Already the Greeks had extolled the harmony of nature: The whole world forms a *Kosmos,* a word implying for the Greeks beauty and order. Whether nature was considered the perfect product of the creator or, as interpreted by Seneca and the pantheists, as being one and the same with god, many devout scientists, such as John Ray, Isaac Newton, and Carl Linnaeus were convinced of the existence of a deep-seated hidden order and harmony in nature which it was their task to unravel and explain.

The laws of physics stress universality and uniformity. If only chance and the blind action of physical laws were active in the universe, thus argued the natural scientists of the seventeenth and early eighteenth centuries, one should find either a homogeneous or a totally chaotic world of things. Consequently, only the existence of a creator can account for the well-integrated diversity of living beings one actually finds. As Newton phrased it, "We know him only by his most wise and excellent contrivances of things, and final causes; we admire him for his perfections; but we reverence and adore him on account of his dominion; for we adore him as his servants; and a God without dominion, providence, and final causes, is nothing else but Fate and Nature. Blind metaphysical necessity, which is certainly the same always and everywhere, could produce no variety of things. All the diversity of natural things which we find, suited to different times and places, could arise from nothing but the ideas and will of a Being necessarily existing."

The study of the perfect harmony of nature, and of its diversity, was, thus, the best way to know God. And it became part of the great vogue of natural theology in the eighteenth and early nineteenth centuries. Not only adaptation, as evidence for design, was the subject matter of natural theology but diversity itself. No one felt this more keenly than Louis Agassiz, who considered the natural system (as he described it in his *Essay on Classification*) as the most decisive proof of the existence of God.[7]

The well-nigh inconceivable richness of kinds of organisms

posed a serious challenge to the human mind, however. The western world was preoccupied with a search for laws ever since the scientific revolution in mechanics and physics. Yet, no aspect of nature was as unyielding to the discovery of laws as was organic diversity. The only way such laws could be discovered, it was felt, was to order diversity by classifying it. This explains why the naturalists in the seventeenth, eighteenth, and nineteenth centuries were so obsessed with classification. It permitted them to put the bewildering diversity at least into some sort of order. As it happens, classification eventually did lead to the searched for law: descent (by modification) from a common ancestor. So important seemed this ordering procedure to the zoologists and botanists in the eighteenth century that classification was treated almost as synonymous with science.

142

As all other branches of science, taxonomy had gifted as well as rather dull witted practitioners. Some specialists did nothing throughout their professional lives but describe new species. This seemed acceptable in the era of Linnaeus when taxonomy had enormous prestige. At that period the dominance of systematics resulted in the neglect of all other contemporary biological researches, for instance, those of Kölreuter. But eventually, quite rightly, the question was asked: Does such purely descriptive activity qualify as science when it does not include a search for laws nor any endeavor to reach generalizations? The splendid successes of von Baer, Magendie, Claude Bernard, Schleiden, von Helmholtz, and Virchow (from the 1830s to 1850s) in other branches of biology resulted in a rapid decline of the prestige of systematics. However, it took a new lease on life after 1859, when Darwin's theory of the origin of taxa by common descent provided the first nonsupernatural explanation for the existence of higher taxa. This new intellectual impulse, however, was soon exhausted, and the exciting advances in functional biology in the last third of the nineteenth century led to a renewed decline of systematics. Physiologists and experimental embryologists considered it a purely descriptive activity and quite unworthy of the attention of a "true scientist." Physical scientists as well as experimental biologists agreed in considering natural history a form of stamp collecting. A leading zoologist, when visiting Cambridge University late in the nineteenth century, observed: "Natural history is discouraged as much as possible, and regarded as idle trifling, by the thousands and one mathematicians of that venerated university."

A well-known historian of the physical sciences stated as recently as 1960: "Taxonomy little tempts the historian of scientific ideas."

What these critics failed to see was to what extent the study of diversity was the basis for research in major parts of biology (Mayr, 1974b). They failed to see what it had become in the hands of Aristotle, Cuvier, Weismann, or Lorenz. Natural history is one of the most fertile and original branches of biology. Is it not true that Darwin's *Origin of Species* was essentially based on natural-history research and that the sciences of ethology and ecology developed out of natural history? Biology would be an exceedingly narrow discipline if restricted to experimental laboratory researches, deprived of the contact with the continuing, invigorating input from natural history.

143

Unfortunately, no one has yet written a history of the influence which natural history has exerted on the development of biology, although D. E. Allen's *The Naturalist in Britain* (1976) does an excellent job for nineteenth-century England. Stresemann's *Ornithology* (1975) covers the same subject as far as birds are concerned. In every group of naturalists there have been keen and inquisitive minds who asked deeper questions. They contributed the most valuable of the writings of natural theology (for example, Ray, Zorn, and Kirby), they founded natural-history journals and societies, and they outlined the basic problems which eventually became the subject matter of evolutionary biology, biogeography, ecology, and behavior. Interestingly, all the great pioneers in these areas were amateurs—dedicated and enthusiastic amateurs. Natural history was the last branch of biology to become professionalized. Only in our age is it being appreciated what a great conceptual contribution natural history has made to biology.

There is no dearth of so-called histories of taxonomy, but almost without exception they are merely histories of classifications. They record the gradual improvements (as well as occasional backward steps) in the concrete classification of groups of animals or plants proposed by authors from Aristotle, Theophrastus, and Dioscorides to Adanson, Linnaeus, Pallas, Cuvier, Lamarck, de Jussieu, Lindley, Hooker, Engler, Ehrenberg, Leuckart, Haeckel, Huxley, and many others. These historians show that the incessant endeavors to regroup genera, families, and orders have succeeded in establishing more homogeneous groupings, groupings which reflect common descent and degree of evolutionary divergence. It is a fascinating history of trial and error.[8]

The focus on classification in this genre of literature fails, however, to come to grips with the history of the changing ideas and concepts of the field. The most important aspect of the history of systematics is that it is, like the history of evolutionary biology, a history of concepts rather than of facts, and that certain competing, if not totally antagonistic, concepts and interpretations have continued to co-exist from the beginning of the eighteenth century to the present time, a period of some 250 years. The heterogeneity of conceptualization in taxonomy is in part due to the fact that a different tradition prevails in the taxonomy of each group of organisms. This is true not only for bacteria, plants, and animals but even for different groups of plants, insects, or vertebrates. The entry of such new concepts as multiple character classifications, polytypic species, sibling species (vs. biological races), or biological species into the taxonomy of different higher taxa occurred at very different times.

One's first impression is that the history of systematics was a never-ending struggle with the same old problems, such as: What is a species? What is relationship? How are higher taxa best delimited? How does one assemble species into higher taxa? What are the most reliable characters? What principles have to be applied when ranking taxa in higher categories? What is the function of a classification? and so forth.

Evidently, the history of systematics does not conform at all to the concept of the progress of science described by Thomas Kuhn in his theory of scientific revolutions. Not even the Darwinian revolution in 1859 produced as decisive a change in systematics as one might have expected. What the reasons for this state of affairs are will become apparent from the ensuing presentation. Yet, it will also show that there was by no means a total conceptual stasis in the last three hundred years. Concepts have changed and have been clarified, as is best demonstrated by the changes in the use and meaning of some frequently employed terms at different periods and in the writings of different authors.[9]

How could a truly unified theory of systematics develop as long as the term "affinity" was used both for mere similarity and for genetic relationship, when the term "variety" was used for geographically circumscribed populations and for intrapopulation variants (individuals), when the term "species" was used for morphologically different individuals and for reproductively isolated populations, and the term "classification" for identification schemes and for true classifications? The term "natural system" meant very different things at different periods, and certain terms, like "category," were often used by the same author for very different concepts. Most of the authors who employed the same term (for example, "category" or "variety") in very different ways were quite unaware of the ambiguity of their own usage. It is probably legitimate to say that more progress was made in clarifying the concepts of taxonomy during the last forty years than during the preceding two hundred years.

The Structure of Systematics

Uniqueness is the first overwhelming impression one gets when one looks at the elephant, the giraffe, the emperor penguin, the swallowtail butterfly, the oak tree, and the mushroom. If this diversity were really chaotic, it could not be studied. But there are regularities, and more than that, as Darwin and so many others have shown, these regularities can be explained. In addition to a chance element, there are determinable causes in the production of diversity. It is therefore legitimate to recognize a science called systematics, which has diversity as its subject matter. As Simpson (1961) defined it, "Systematics is the scientific study of the kinds and diversity of organisms and of any and all relationships among them." As Simpson has furthermore said, systematics "is at the same time the most elementary and the most inclusive part of [biology], most elementary because [organisms] can not be discussed or treated in a scientific way until some taxonomy has been achieved, and most inclusive because [systematics] in its various branches gathers together, utilizes, summarizes, and implements everything that is known about [organisms], whether morphological, physiological, psychological, or ecological."

Because systematics covers such an enormous field, endeavors have been made to subdivide it. For instance, it has been stated that the classification of a taxon goes through various stages of maturation. "These have sometimes been informally referred to as alpha, beta, and gamma taxonomy. Alpha taxonomy refers to the level at which species are characterized and named; beta taxonomy to the arranging of these species into a natural system of lesser and higher categories; and gamma taxonomy . . . to evolutionary studies" based on taxonomic research (Mayr, Linsley, and Usinger, 1953: 19). Actually, work at the alpha and beta level proceeds simultaneously—they are not stages—and gamma taxonomy is not strictly taxonomy. The history of the field is best understood, if two subfields of taxonomy are recognized: (1) *microtaxonomy,* which deals with the methods and principles by which kinds ("species") of organisms are recognized and delimited, and (2) *macrotaxonomy,* which deals with the methods and principles by which kinds of organisms are classified, that is, arranged in the form of classifications. *Taxonomy* as a whole, then, is defined (somewhat more narrowly than systematics) as "the theory and practice of delimiting kinds of organisms and of classifying them" (Simpson, 1961; Mayr, 1969).

146

4 ✐ Macrotaxonomy, the science of classifying

CLASSIFICATIONS are necessary wherever one has to deal with diversity. Thus one has classifications of languages, of goods in any manufacturing or marketing system, of books in a library, or of animals and plants in nature. In all these cases the process of classifying consists in the grouping of individual objects into categories or classes. There is no argument about this basic procedure, but what has been controversial for centuries is how this is best done, what classifying criteria should be used, and what the ultimate purpose of a classification should be. It is the task of the history of macrotaxonomy to recount and discuss the diverse and frequently changing answers to these questions.

Before treating the subject historically, it is necessary to deal critically with a number of concepts which have often, one is tempted to say consistently, been confused in the history of taxonomy.

Identification versus Classification

Identification schemes are *not* classifications. The procedure of identification is based on deductive reasoning. Its purpose is to place an investigated individual into one of the classes of an already existing classification. If one succeeds, one has "identified" the specimen. Identification deals with just a few characters which assign the specimen into one or the other couplet of an identification key (Mayr, 1969: 4, 66, 112–115). By contrast, classification, as now conceived, assembles populations and taxa into groups, and these, in turn, into even larger groups, this process making use of large numbers of characters.

An understanding of the difference between classification and identification schemes is crucial in the evaluation of so-called "special purpose classifications," such as "classifications" of medicinal plants on the basis of specific curative properties. Such "classifications" are actually nothing but identification schemes, or so it seems to the modern taxonomist. When the Greek physician Dioscorides ordered plants according to their curative properties, he wanted to safeguard the use of the right species for its specific medicinal purpose. Since most medicines were derived from plants almost up to modern times, pharmacopeias served simultaneously as manuals of plant identification.

Some special-purpose classifications, however, are not identification keys but actually serve the purpose implied by their name. This is true, for instance, when plants are classified in the ecological literature according to growth form or habitat. The usefulness of such classifications is very limited. Yet, prior to the sixteenth century virtually all attempts at "classification" were of this utilitarian type. In the consideration of classifications, it is therefore very important to have a clear understanding of the various possible objectives of a classification.

The Functions of Classifications

Philosophers as well as taxonomists have realized almost from the beginning that classifications serve a dual purpose, a practical and a general (that is, scientific or metaphysical) one. But there has been considerable disagreement as to the nature of these two objectives. The practical purpose that was stressed particularly by the early authors was that of serving as an identification key. In more recent times the practical purpose that is most often stressed is that a classification should serve as an index to an information storage and retrieval system. In order to accomplish this purpose best, a classification should consist of classes of objects that have the greatest number of attributes in common. Such a classification is automatically the key to the information stored in it. Ease of information retrieval is generally the principal or exclusive objective of the classification of items like books in a library and most other inanimate objects that are ordered according to more or less arbitrary criteria. By contrast, the classification of items that are connected by causation (as in a classification of diseases) or by origin (as in biological classification) are subject to considerable constraints but have the valuable capacity to serve as the basis of far-reaching generalizations.

As far as the general meaning of biological classification is concerned, there has been much change over time. For Aristotle it reflected the harmony of nature, in particular, as far as it was expressed in the *scala naturae*. For the natural theologians, as was clearly expressed by Louis Agassiz (1857), the classification demonstrated the plan of creation of the designer of this world. The natural system is the expression of this plan. After the proposal of the theory of common descent by Darwin, the metaphysical interpretation of classification was replaced by a scientific one. Since the observations in all comparative branches of biology were organized with the help of "the natural system" (now evolutionarily defined), it became the primary function of classification to delimit taxa and construct a hierarchy of higher taxa which permitted the greatest number of valid generalizations. This was based on the assumption that members of a taxon, sharing a common heritage as descendants from a common ancestor, will have more *149* characters in common with each other than with species not so related. Evolutionary classifications thus have considerable heuristic value in all comparative studies. They are open to testing either with additional characters or by comparison with other taxa (Warburton, 1967).

The existence of these two kinds (practical and general) of objectives of a biological classification has led to controversy. For instance, it has been questioned whether the object of information retrieval is compatible with that of generalization. What is the nature of the embodied generalizations? Can they be considered a theory?

This short statement of the problems connected with the various roles of classification may sharpen the attention of the reader when following the historical changes in the attitude toward these problems.

ARISTOTLE

The history of taxonomy starts with Aristotle (384–322 B.C.). Though much seems to have been known about animals and plants before him, the few earlier writings that have come down to us do not contain any classifications. As far as factual information is concerned, Aristotle evidently incorporated in his work anything he was able to obtain from his predecessors, presumably mostly from the school of Hippocrates. Yet, the loving detail with which Aristotle describes various marine animals indicates that much of his information is original, or at least was obtained by him personally from fishermen and similar "folklore" sources. It is believed that he devoted himself largely to natural-history studies during his years at Lesbos. His main work of descriptive zoology is *Historia animalium*, but there are numerous statements of systematic relevance also in *De partibus,* in *De generatione,* and in other writings.

Aristotle is traditionally celebrated as the father of the science of classification, and yet from the Renaissance to the present there has been great uncertainty and much disagreement as to what really his principles of classification were.[1] In part this seems to be due to the fact that Aristotle advanced different methods in his earlier writings (in which he developed his principles of logic) from those in his later biological works, in part owing to his conviction that the method of definition by logical division was not capable of providing a reasonably comprehensive description and characterization of groups of animals.

Aristotle's method of logic is best explained by reference to a well-known parlor game. A person is asked to guess an object which the other members of the party have chosen during his absence. "Is it alive?" might be his first question. This separates all conceivable objects into two classes: organic beings and inanimate objects. If the answer is yes, he may ask, "Does it belong to the animal kingdom?" thus dividing the class of living beings into two classes, animals and nonanimals. A continuation of this procedure of always splitting the remaining class of objects into two parts (dichotomous division) will sooner or later lead to the correct choice.

Expressed in terms of Aristotelian logic, the largest observed class, the *summum genus* (for instance, plants), is divided by a deductive process into two (or more) subordinated subclasses, which are called "species." Each "species" in turn becomes at the next lower level of division a "genus" which is again subdivided into "species." This process is repeated until the lowest set of species can no longer be divided. The "species" of the logician need, of course, have nothing to do with the biological species, although the products of the last step in the division of a class of organisms may, in fact, be biological species. Classification by logical division is *downward classification.* It is as applicable to inanimate objects (furniture can be divided into chairs, tables, beds, and so on) as it is to organisms.

What has confused later writers is that Aristotle when describing his method of logic actually used as examples discriminating criteria that applied to animals, such as "hairy or not," "with blood or bloodless," "quadrupedal or not." However, logical division was *not* the method by which Aristotle classified animals, as is evident from the fact that Aristotle's system of animals is not an elaborate hierarchy,[2] and even more so from the fact that he quite specifically ridicules dichotomous division as a classifying principle (*De partibus animalium* 642b5–644a11) and goes to considerable lengths to show why it would not work. But in spite of its rejection by Aristotle, logical division in biological classification was the preferred method from the Renaissance (Cesalpino) to Linnaeus (see below). Contrary to what is found in much of the historical literature, it is not legitimate to designate this method of classifying as Aristotelian.

How then did Aristotle proceed when classifying the diversity of animals? He did so in a very modern way: he formed groups by inspection: "The proper course is to endeavor to take the animals according to their groups, following the lead of the bulk of mankind, who have marked off by many differentiae, not by means of dichotomy" (643b9–14). "It is by resemblance of the shapes of their parts, or of their whole body, that the groups are marked off from each other" (644b7–9). Only after he had established the groups did Aristotle select some convenient differentiating characters. On this commonsense phenetic approach Aristotle superimposed a system of evaluating the attributes that serve to characterize these groups and rank them in some sort of sequence. This ranking is the aspect of Aristotle's taxonomy that is hardest to understand for the modern. As is well known, Aristotle was greatly impressed by the importance of the four elements—fire, water, earth, and air—and thus the attributes hot versus cold, or moist versus dry, were of crucial importance to him. Hot ranked above cold, and moist above dry. Blood, being both hot and moist, became thus a particularly important characteristic. Aristotle, consequently, had a scale of values for different physiological functions, as they seemed to be characteristic for different kinds of animals. Hotter, moister creatures were supposed to be rational, whereas colder, dryer creatures had less vital heat and lacked the higher type of "soul." This type of speculation particularly appealed to the Renaissance Aristotelians and induced them to propose concrete rank scales of taxonomic characters, based on their assumed physiological importance.

This must be remembered, if one wants to understand why Aristotle's classifications were not meant to be identification schemes or purely phenetic schemes. Aristotle recognized certain groups primarily to illustrate his physiological theories and to be able to organize information on reproduction, life cycle (degrees of perfection of the offspring), and habitat (air, land, water). For his purposes it was therefore quite legitimate to separate the aquatic cetaceans from the terrestrial mammals, and the soft, free-swimming cephalopods from the marine and terrestrial hard-shelled mollusks. On the whole, in spite of some incongruous combinations and unclassified residues, Aristotle's higher taxa of animals were distinctly superior to those of Linnaeus, whose primary interest was in plants.

When we study Aristotle's zoological works, we are impressed by three findings. First, Aristotle was enormously interested in the diversity of the world. Second, there is no evidence that Aristotle had a particular interest in animal classification per se; he nowhere tabulated the nine higher taxa recognized by him. Finally, to repeat, whatever classification he had was not the result of logical division. It is remarkable how little of Aristotle's system of logic is reflected in the *Historia animalium.* In this work one has far more the impression of an empirical, almost pragmatic approach rather than that of deductive logic.

Aristotle simply wanted to tell in the most efficient way what he knew about animals in order "that we may first get a clear notion of distinctive characters and common properties" (491a8). The quickest way to achieve this objective was comparison. Indeed, the entire book is built on comparisons: of structure (comparative anatomy), of reproductive biology, and of behavior (animal psychology). To facilitate comparison he grouped the 580 kinds of animals mentioned by him into assemblages like birds and fishes, often making use of groupings that are as old as the Greek language.

His classification of all animals into "blooded" and "bloodless" was accepted until the groups were renamed by Lamarck "vertebrates" and "invertebrates." Within the blooded animals Aristotle recognized the birds and fishes as separate genera, but ran into some trouble with the remainder. Accepting vivipary versus ovipary as an important characteristic, he separated the hairy ones (now called mammals) from the cold-blooded oviparous ones (reptiles and amphibians). He clearly separated the cetaceans both from the fishes and from the terrestrial mammals. Different kinds of flying animals were separated by him quite widely from each other, the birds having feathered wings, bats leathery wings, and insects membranaceous wings. However, among the invertebrates his Testacea (hard-shelled mollusks) included such heterogeneous elements as barnacles, sea urchins, snails, and mussels.

Aristotle provided a wealth of observations on structural differences of various animal groups with particular reference to the digestive and reproductive systems. Yet he seems to have been at least equally interested in the ecology of animals (their habitat and mode of living), their reproductive biology, and their temperament. "Animals differ from one another in their manner of life, in their activities, in their habits, and in their parts," particularly with relation to the elements water, air, and earth (487a11–12). It is now clear that it was not Aristotle's objective to provide a classification of animals that would be useful in identifying them.

What then was Aristotle's importance in the history of systematics? Perhaps his most important contribution was that he, an outstanding philosopher, took such an interest in animals and their properties. This greatly facilitated the revival of zoology in the late Middle Ages and in the Renaissance. Whether in relation to structure, food habits, behavior, or reproduction, everywhere he asks the kind of significant questions that made the study of animals a science. He also laid the groundwork for the eventual organization of biology into morphology, systematics, physiology, embryology, and ethology and gave instructions on how a researcher should proceed. His formalization of individual kinds (species) and of collective groups (genera) was the point of departure for the more perceptive and elaborate classifications of the later period.

Now that Aristotle is no longer considered merely one of the fathers of scholasticism but also a philosophizing biologist, many aspects of his work appear in an entirely new light. A modern analysis of Aristotle's conceptual framework of taxonomy is still wanting, however.[3]

As a broad generalization one can probably say that the level of natural history went steadily downhill after the death of Aristotle. Pliny and Aelian were busy compilers who quite uncritically placed good natural history next to fabulous creatures of various mythologies.[4] In the ensuing period animals were written about not for the sake of providing knowledge about them but for the sake of moralizing; they became symbols. If one wanted to moralize about diligence, one wrote about the ant; about courage, the lion. With the rise of Christianity, stories about animals were often made part of religious tracts. Animals became symbols for certain ideas in Christian dogma and were introduced in paintings and other art as symbols of these ideas. One might say that the study of animals became a purely spiritual or aesthetic endeavor, almost totally divorced from natural history as such. This was true,

153

broadly speaking, at least for the more than 1,000-year period from Pliny (d. A.D. 79) to the fifteenth century (Stannard, 1979). Frederick II's *De arte venandi* (1250) and the writings of Albertus Magnus (ca. 1200–1280) were quite exceptional for their period.

154

A rapid change occurred in the ensuing centuries, furthered by a number of developments. One was the rediscovery of the biological writings of Aristotle and their being made available in new translations. Another was a general improvement in the standard of living, with more emphasis on medical art and a corresponding emphasis on medical herbs. Finally, a kind of back-to-nature movement, away from the exclusive emphasis on spiritual life, developed in the late Middle Ages. From Hildegard von Bingen (1098–1179) and Albertus Magnus on, more and more people apparently looked at living plants and animals in nature, and what is more, they wrote about them, and what is even more, when the art of printing came up, they printed books about them. Yet, it was a slow and gradual process. The encyclopedic tradition of that great and uncritical compiler Pliny continued unto the days of Gesner and Aldrovandi. But by then, in the sixteenth century, the authors of all nature books were physicians.

THE CLASSIFICATION OF PLANTS BY THE ANCIENTS AND THE HERBALISTS

Aristotle also wrote on plants, but his writings are lost. Thus the history of botany starts with the *Inquiry into Plants* of his pupil, Theophrastus (371–287 B.C.). As important as his contributions are to plant morphology and plant biology, Theophrastus adopted no formal system of classification, the form of growth (trees, shrubs, undershrubs, and herbs) constituting the primary criterion of division; other criteria are the presence or absence of spines, cultivation by man or not, and so on. Theophrastus apparently took many of his groupings from folklore, with the result that some of them are quite natural (oaks, willows), while others are, taxonomically speaking, quite artificial, like "daphne," a conglomerate of plants with evergreen leaves.

Far more important for the immediate history of botany was Dioscorides (ca. A.D. 60). As a Greek physician attached to the Roman army, he had traveled very widely and acquired an enormous treasure of information on plants that were of use to man. His *Materia medica* contains descriptions of between five hundred and six hundred plants that are either of medicinal use or provide spices, oils, resin, or fruit. The arrangement of the plants in his five books is based principally on their practical use (medicinal roots, herbs used as condiments, perfumes, and so on). Nevertheless, he often lists related plants in sequence—for instance most of his 22 species of *Labiatae* or 36 species of *Umbelliferae*. Indeed, he criticized the alphabetical arrangement adopted by some previous authors on the ground that it separated related plants possessing similar properties. The principal importance of Dioscorides is that his *Materia medica* was the chief botany text for one *155* and a half millenia (Mägdefrau, 1973: 4–11). Dioscorides was considered the supreme authority in all matters relating to plants, particularly their medicinal properties. Yet, as in the case of Galen's anatomy, the tradition became more and more book knowledge, more and more divorced from nature and from actual organisms.

Beginning with the thirteenth century, however, a number of herbals were published in which one notices a return to actual observation of nature, a trend which greatly accelerated after the invention of the art of printing. A Latin translation of Dioscorides was published in 1478, and one of Theophrastus in 1483, and many of the handwritten herbals of the preceding several centuries were printed for the first time at that period.[5] The increasing interest in the identification of plants, the discovery of rich floras of local species unknown to Dioscorides, as well as the search for new medical properties in newly discovered plants led to the founding of chairs of botany at European medical schools, the first at Padua, in 1533.

A new era started with the work of the "German fathers of botany," Brunfels (1488–1534), Bock (1489–1554), and Fuchs (1501–1566). These naturalists represent a return to nature and to personal observation. Their accounts are not a set of compilations and endlessly copied myths and allegories but are descriptions, based on real, living plants, observed in nature. They also represent attempts to describe and illustrate local floras; the illustrations produced by the excellent draftsmen and woodcarvers they employed reached a level of accuracy and artisanship not surpassed for generations. These played the same role in botany that the illustrations in Vesalius had played in anatomy. The title of Brunfels' work, *Herbarum Vivae Eicones* (1530) stresses the fact that the plants had been drawn from nature (by Hans Weiditz). All three herbals describe and illustrate many central European species that had been entirely unknown to the ancient botanists.

Brunfels illustrated 260 plants, Fuchs in his *Historia Stirpium* (1542) no less than 500 plants.

Hieronymus Bock was the most original of the three. All of his descriptions, written in precise and picturesque colloquial German, were clearly based on his own observations. Furthermore, he rejected expressly the alphabetical arrangements of other herbals and gave as his method "to place together, yet to keep distinct, all plants which are related and connected, or otherwise resemble one another." Not only did he write good descriptions, but he noted the localities and habitats (including properties of soil) of the plants he describes and their flowering seasons and other aspects of their life history. In this way Bock's work was the prototype of all future local floras, and along with other herbals printed in France and England, was among the most popular books of the era.[6]

156

Classification among the Herbalists

Perhaps the most conspicuous aspect of the "classifications" of the herbalists is the absence of any consistent system, their interest being not at all in classification but in the properties of individual species. In the case of Brunfels (1530), the sequence seems to be quite arbitrary, at least as far as genera are concerned. However, closely related species, for instance *Plantago major, P. minor,* and *P. rubea,* are placed next to each other. Fuchs (1542) arranges his plants largely alphabetically, the contents of the first four chapters being *Absinthium, Abrotonum, Asarum,* and *Acorum.* This sequence is retained in the German edition (1543), even though the German names for these four genera, *Wermut, Taubwurtz, Haselwurtz,* and *Drachenwurtz,* are now in the reverse sequence of the alphabet. Amusingly, Fuchs comments that he has left out of the greatly abbreviated German edition that which the "common man" does not need to know.

Three classificatory aspects of the herbals are worth singling out. First, there is a vague recognition of kinds (species) and groups (genera). Second, many recognized groups, like grasses, are quite natural but are often expanded by the addition of superficially similar forms. For instance, among the nettles would be classified the true nettles (*Urtica*) as well as labiates with similar leaves, the false nettles. Next to wheat (a grass) one finds buckwheat (a dicotyledon), merely because the word "wheat" is part of its vernacular name. Such juxtaposition was of considerable value for identification but was no basis for a sound classification. Finally, there were only limited attempts to establish higher taxa. In his *Herball* (1597, based on Dodoens and Lobel), for instance, Gerard devotes his first chapter to "grasses, rushes, corne, flags, bulbose or onion-rooted plants," that is, to largely monocotyledonous plants. His second chapter, however, contains "all sorts of herbs for meate, medicine, or sweet smelling use"—botanically speaking, a complete hodgepodge.

The tradition of the herbals reached its climax with the publication of Caspar Bauhin's *Pinax* (1623). It shows the remarkable progress made in the ninety years since the publication of the *Eicones* of Brunfels. Some 6,000 kinds of plants are described in 12 books divided into 72 sections. All kinds are assigned to a genus and a species, although no generic diagnoses are given. Related plants are frequently placed together on the basis of their overall similarity or owing to common properties. The groups thus formed are not given taxa names and no diagnoses of higher taxa are provided. Nevertheless, there is an implicit recognition of the monocotyledons, and the species and genera of some nine or ten families and subfamilies of dicotyledons are brought together. Although Bauhin nowhere explains his method, it is evident that he simultaneously considered a large number of different characters and grouped those genera together which share a large number of characters. Considering that it was the main purpose of the *Pinax* to provide a convenient catalogue of plant names, Bauhin's ability to find related genera and to group them together is altogether astonishing.

The beginnings of almost every later development of systematic botany can be found in the writings of the herbalists: attempts to group plants on the basis of similarity or shared characteristics, beginnings of a binary nomenclature and even of dichotomous keys, a search for new characteristics, and an endeavor to provide more accurate and more detailed descriptions. Perhaps the most valuable contribution of the herbalists was their empirical attitude. No longer were they satisfied merely to copy the writings of Dioscorides and Theophrastus; they actually studied the plant in nature and described "wie eyn yedes seiner Art und Geschlecht nach auffwachs/wie es blüe/und besame/zu welcher zeit im jar/ und in welcherley erdtrich eyn yedes am besten zu finden seie" ("how each grows according to its species and genus, how it flowers and sets seed in what season of the year, and on what soil each is best found"; Bock, 1539). But each of the herbalists had his own way of doing things and all of them were utterly inconsistent in whatever methods they used.

Since at that time relatively few plants were known, one could find a species by simply thumbing through an herbal until one encountered something reasonably similar, and only then did one carefully read the description and study the illustration, in order to ensure the identification. This simple method became insufficient, however, when the number of known plants increased during the sixteenth and seventeenth centuries at an almost exponential rate. Whereas Fuchs (1542) knew some 500 species and Bauhin (1623) 6,000 species, John Ray in 1682 listed already 18,000. An alphabetical or otherwise arbitrary arrangement was no longer sufficient. To cope with this avalanche of new "kinds" of plants, a far more careful discrimination of species within the broader "kinds" (genera) became necessary, and a more serious endeavor was made to recognize groupings of related genera, that is, higher taxa. Some system or method by which one could rather quickly identify a given specimen was also required.

DOWNWARD CLASSIFICATION BY LOGICAL DIVISION

The theory of classification would seem deceptively simple: one orders the to-be-classified objects on the basis of their similarity. When one deals with organisms, this at once leads to the question, How does one determine, indeed preferably measure, similarity? The answer is by a careful analysis of characters. The choice and evaluation of characters has therefore occupied the center of attention in recent discussions. However, we shall never understand the contrast between the earlier theories of classification, dominant from Cesalpino to Linnaeus, with those prevalent since Darwin if we start with a consideration of characters. Rather one must begin by asking what kinds of classifications are possible.

Actually, ordering by similarity was not the way classifications were constructed during the Renaissance. The need for identification was preeminent and the first comprehensive methodology of plant taxonomy was developed to satisfy this demand. The Italian anatomist-physiologist Andrea Cesalpino (1519–1603) is generally and rightly credited with having been the first to have done this consistently in his great work *De Plantis* (1583). He considered himself a follower of Theophrastus and, like him, divided plants into trees, shrubs, undershrubs (perennials), and herbaceous plants. But for a system of easy identification he turned to Aristotle and borrowed from him the method of logical division, with which everybody was familiar who had attended school between the Middle Ages and the end of the eighteenth century (see discussion of Aristotle, above).

The principles of downward classification by logical division are basically very simple. Yet, in the writings of the classifiers from Cesalpino to Linnaeus they were embedded in such a complex matrix of scholastic dogma and jargon (as indicated by such terms as "essence," "universals," "accidentals," "differentiae," "characters," and so on) that a specialized study is required for their understanding.[7]

159

The method of logical division did not originate with Aristotle. Already Plato was interested in distinguishing general groups from subordinated items (*Messon* 72; *Parmenides* 129C; *Politicus* 261B), but it reached its full importance only in Aristotle's followers—for example, the tree of Porphyry, also called the Ramean tree (Jevons, 1877: 702). The most characteristic feature of this method is the division of a "genus" into two "species" ("tertium non dat"). This is called dichotomous division. This method is ideally suited for the construction of identification keys, but it often leads to highly artificial and unbalanced classifications. Aristotle himself, as pointed out above, ridiculed the idea of basing a classification on dichotomy, but his use of zoological examples in his logical exercises misled his followers.

There were several reasons for the great popularity of downward classification between Cesalpino and the nineteenth century. Its most important practical advantage was that it started with a number of easily recognizable classes—let us say with trees, shrubs, herbs, or, in the case of animals, with birds, butterflies, or beetles—and divided them into subordinate sets of subclasses with the help of appropriate differentiating characters ("differentiae"). No prior knowledge of species was required, only an ability to carry out the procedure of logical division. Any layperson could do this. It would be a mistake, however, to think that the only reason for the popularity of logical division was its practicality. Its popularity was greatest in the centuries when everyone looked for order and logic in the created world. Hence, if the world represents an orderly system, what better way could there be to study and analyze it than with the tools and procedures of logic? A classification could reflect the order of nature adequately only if it was based on the true essences of the organisms. It was the method of logical division which would help in the discovery and definition of these essences. The method thus was a perfect reflection of the dominant essentialistic philosophy of the period.

Nothing in the whole method of logical division is more important than the choice of the differentiating characters. The dependence on single characters implicit in this method necessitates a careful *weighting* of characters.[8] Cesalpino, fully aware of this, devoted a great deal of care and attention to the study of plant morphology. He discovered many useful characters and was one of the first (Gesner preceded him) to recognize the taxonomic value of fructification.

Cesalpino was altogether on the wrong track in his theory of weighting characters, however. A true follower of Aristotle, he selected characters on the basis of their physiological importance. The two kinds of attributes he considered to be most important for a plant were those relating to nutrition and to generation. The nutritive aspect (growth) was considered the most important, and this is why his first division was into trees, herbs, and so on. The importance of generation was reflected in the emphasis on fructification, seed, and seedlings (in analogy with Aristotle's emphasis on the animal embryo). Comparison was an important element in his method, but he carried it to ludicrous extremes when he attempted to match the functionally most important structures of animals and plants. Thus, he equated the roots of plants with stomach and intestinal tract in animals, and he included the stem and stalk of plants in their reproductive system because they bear the seeds and fruits.

Considering the frequency of convergence, parallelism, the loss of characters, and other irregularities of character evolution, one would expect that the method of logical division based on single characters would lead to absolute chaos. A study of Cesalpino's plant classification shows, however, that the 32 groups of plants recognized by him are, on the whole, remarkably "natural."[9] It is quite evident that Cesalpino could not have arrived at his groupings merely by applying logical division. As Stafleu (1969: 23) correctly remarked, Cesalpino evidently "started out with certain natural groups known to him intuitively or by tradition, and added [the] rather irrelevant and certainly unimportant superstructure" of logical division. Cesalpino, thus, followed a two-step procedure. He first sorted his plants by inspection into more or less natural groups and then searched for suitable key characters that would permit him to arrange these groups in accordance with the principles of logical division. Only thus was he able to achieve simultaneously two objectives; to provide a convenient identification key and to group his plants into classes according to their "affinity" (see below). He was not always successful in compromising between the conflicting demands of the two methods, as when the principles of logical division forced him to separate the herbaceous and the tree-like legumes into two different families.

In spite of the evident shortcomings of his system, Cesalpino exerted a powerful influence on botany during the ensuing two hundred years. Up to and including Linnaeus, all systems of plant classifications were variations and improvements upon the approach first taken by Cesalpino. They all based their classifications on the method of logical division and on some a priori weighting of characters. Cesalpino had such an impact not because his classification was particularly useful but simply because he was the first author to provide a more or less consistent method of classification. It would have to do until someone came up with a better one.

The choice of the characters during the early steps of the division results by necessity in entirely different classifications. This is why the systems of the great botanists of the seventeenth and eighteenth centuries, who followed in Cesalpino's footsteps, differed so drastically from each other. Only the specialist is interested in the details by which the classifications of Magnol, Tournefort, Rivinus, Bauhin, Ray, and the various lesser figures disagreed with each other. All these botanists differed in their knowledge of plants and this influenced their choice of the characters for the first division. In the classification of animals, likewise, it led to entirely different classifications whether one chose as the first *differentia* with blood or without, hairy or hairless, two-footed or four-footed.

Another consequence of the method of downward classification is that it cannot be improved gradually and piecemeal. The replacement of one character by a different one will result in an entirely new classification. Under this system the possible number of different classifications is virtually unlimited. And yet these botanists somehow succeeded in adjusting the choice and sequence of their characters in such a way as not to break up certain well-known natural groups of plants. How clearly the "naturalness" of certain groups was appreciated is demonstrated by the fact that, as Larson (1971: 41) points out, "Many plant families—

161

Coniferae, Cruciferae, Graminaceae and *Umbelliferae,* for example—
were established in the 16th century, and remained intact
throughout the vicissitutes of the struggle over systems." More
and more such groupings were recognized, particularly when certain seemingly isolated European genera were discovered to belong to rich tropical families.

162

The botanists of the seventeenth century differed from each
other, however, not only in the weight they gave to different characters but also in whether their major interest was in the genus or
the species, and how sancrosanct they held the principle of logical
division and the reputedly Aristotelian weighting system of characters. It is in these two points in which the two greatest botanists
of the late seventeenth century most strongly differed from each
other.

Ray and Tournefort

John Ray (1627–1705), of course, was far more than a botanist.[10] He co-authored the most important zoological treatises of
the period and wrote one of the great books in natural theology.
But he was also a practical Englishman, whose primary aim was
to produce a plant book that would permit the unequivocal identification of plants. Consequently, he was particularly concerned
with the nature of the species. In his *Historia Plantarum* he deals
with no less than 18,655 plant "species" and gives a definition of
the species category (see Glossary) which was largely adopted for
the next 150 years. Almost unique among the early botanists, he
was not educated as a physician, and was less touched by the scholastic tradition than his contemporaries, including even Tournefort, who had been educated at a Jesuit college. Consequently, it
is not surprising that John Ray from his earliest botanical publications on was far less consistent in his application of dichotomous
division than Cesalpino or Tournefort. He not only used different
sets of subordinate characters in different of his classes, but he
did not even hesitate to shift from the fructification to vegetative
characters (presence of a stem or of bulbous roots) when this
seemed convenient. Tournefort and Rivinus attacked him vigorously for these deviations, but Ray answered the criticism with the
pragmatic advice, "An acceptable classification is one . . . that joins
together those plants which are similar, and agree in primary parts,
or in total external aspect, and that separates those which differ
in these aspects" (*Synopsis,* 1690: 33). He repeats this guiding prin-

ciple in all of his subsequent publications. For instance, "The first condition of a natural method must be that it neither splits plant groups between which apparent natural similarities exist nor lumps such with natural distinctions" (*Sylloge,* 1694: 17). Cesalpino and other defenders of logical division had, of course, claimed that this is precisely what their method would do. Ray is thus forced to go further. In his *De Variis* (1696) he points out that one really has no objective method of determining which characters reflect the essence and which are accidental. In other words, he implicitly rejects the method of a priori weighting. (It is important to note that he does not reject the concept of an essence or the difference between essential and accidental characters.) From this he concludes that not only flower and fruit but other parts of the plant as well may reflect the essence. He even goes so far as saying that species may differ from each other by their sets of accidentals (*Ornithology,* 1678).

163

Sloan (1972) advocated the thesis that it was Ray's study of the writings of Locke which led him to these heretical views. There is much to indicate, however, that Ray had arrived at the unorthodox evaluation of characters by a purely pragmatic approach and had taken up "philosophical studies" in order to gather ammunition for his reply to Tournefort (April 29, 1696, letter to Robinson). Since it is very doubtful that a single character can reflect the essence of a genus, Ray recommends in his *Methodus Plantarum* (1703: 6–7): "The best arrangement of plants is that in which all genera, from the highest through the subordinate and lowest, have several common attributes or agree in several parts or accidents." He goes so far as to employ even ecological criteria for groupings, a set of characters strictly "forbidden" since Cesalpino. Actually Magnol (*Prodromus,* 1689) had already recommended combinations of characters.

Ray's contribution to the true classification of plants was rather minor. Like Albertus Magnus, Pena, Lobel, and Bauhin, he distinguished monocotyledons and dicotyledons without recognizing the fundamental nature of the difference. He still retains Theophrastus' division of plants into trees, shrubs, herbs, and so on, and his classifications of the Caryophyllaceae and Solanaceae, for instance, are quite inferior to those of Bauhin and other predecessors. The history of botanical classifications indicates that Ray's influence was limited. Nevertheless, it can hardly be doubted that he contributed to a weakening of the hold of the method of logical division.

Ray's illustrious contemporary in France, Joseph Pitton de Tournefort (1656–1708), was perhaps the first botanist fully to realize the wealth of exotic floras (Sloan, 1972: 39–52; Mägdefrau, 1963: 46–48). Purely practical considerations were therefore more important for him than the development of a universal or natural method. His aim was to provide a convenient key to the diversity of plants: "To know the plants is to know the exact names one has given to them on the basis of the structure of some of their parts" (Tournefort, *Institutiones,* 1694: 1). Since the number

of genera was still manageable in his day, he concentrated on this rank. In contrast to most of his predecessors, he used a single word as the generic name. Tournefort's greatest merit consists in the first clear formulation of the genus concept and the judicious delimitation and clear description of 698 genera of plants, most of which (sometimes under different names) were adopted by Linnaeus. As a result some of the best-known generic names in plants go back to Tournefort. Since flowers and fruits offer the greatest number of easily visible characters, these were the parts of the plant on which he based most of his descriptions, but he occasionally referred to other structures when this seemed helpful. Tournefort was far more willing to make concessions to practical needs than Linnaeus. In the case of plants without fruits and flowers or those in which these structures are too small to be seen with the naked eye, he recommended that "for the correct determination of [such] genera not only all the rest of the parts of the plants should be utilized, but also their accidental characters, means of propagation, and overall character and external appearance" (*Institutiones:* 61).

In spite of his careful analysis of characters, his classification of higher taxa was more or less artificial. Of the 22 classes he had established, only 6 correspond to natural groups. For purposes of identification, however, Tournefort's *Methode* was more successful than the systems of his contemporaries Ray, Morison, or Rivinus. It was widely adopted not only in France but also in the Netherlands and eventually in England and Germany. The systems of Boerhaave (1710), Magnol (1729), and Siegesbeck (1737) were variants of Tournefort's. They differed primarily in the choice of the character which they considered most important. The primary objective of all these systems was identification with the help of logical division. None of them succeeded in achieving a consistent delimitation of natural groups, as is indeed impossible with the method of logical division.

Downward classification was not a bad strategy at the time of Cesalpino, for everything about classification was uncertain at that period. A realistic species concept had not yet been developed, and the number of newly discovered kinds of organisms was increasing at an exponential rate. At a time when few people knew anything about natural history, correct identification was the greatest need, and divisional classification was very well suited for this objective. In retrospect it is obvious that it was a suitable, if not inevitable, first step toward a superior method of classification.

The botanists of this period were frequently maligned as "Aristotelians," implying a deductive approach and a blind reliance on tradition and authority. This is quite unjustified. To be sure, they used the methods of logical division as being the system best suited for successful identification, but their work was not at all based on authority but rather on study in nature, extensive traveling, and careful analysis of specimens. They laid a sound empirical foundation for the improved systems of the post-Linnaean period.

Attention must be called here to striking difference in the historical development of natural history and of the physical sciences. The sixteenth and seventeenth centuries witnessed the so-called scientific revolution, which, however, was essentially limited to the physical sciences and, to a lesser extent, to some parts of functional biology. Natural history and systematics, however, were almost totally untouched by these sweeping changes in the neighboring sciences. From Cesalpino through Tournefort and Ray (not to mention Jungius or Rivinus) to Linnaeus there continued an unbroken tradition of essentialism and the method of logical division. It has been claimed, not without justification, that natural history, almost up to the time of Darwin, continued to be dominated by the metaphysics of Plato and Aristotle. What should be added, however, is that it was also dominated by a different strand in the skein of Aristotle's thinking: the spirit of the naturalist, the joy of observing nature and the fascination with diversity. That part of the Aristotelian heritage has continued to the present day, while in systematics the Aristotelian metaphysics, already greatly weakened during the transition period between Adanson and 1859, was completely routed by Darwin.

The rapid accumulation of knowledge on plant classification between the early 1500s and Linnaeus would have been impossible without an important technological advance, the invention of the herbarium (Lanjouw and Stafleu, 1956). The idea to press and dry plants seems to have originated with Luca Ghini (1490–1556), among whose pupils were Cibo (whose herbarium from 1532 is still extant), Turner, Aldrovandi, and Cesalpino, all of whom made herbaria. Herbaria were quite indispensable for the collecting of exotic plants. Most of Linnaeus' descriptions of non-Swedish plants, for instance, were made from herbarium specimens. Each of the great herbaria of the world has now three to six million specimens, to which botanists actively refer for description and identification. There are good reasons to believe that the great advances made in the classification of plants during the second half of the sixteenth century were considerably facilitated by the new technology of herbaria, which permitted a referring back to specimens at all seasons of the year. The second important technological advance, of course, was woodcut making.

166

Luca Ghini was a great innovator in another way. He established in 1543 (or 1544) the first university botanic garden at Pisa. A second one was established in Padova in 1545. At a time when herbaria were few and illustrations poor, the value of botanic gardens for teaching purposes cannot be exaggerated. By the end of the sixteenth century public botanic gardens had been established at Florence, Bologna, Paris, and Montpellier.

PRE-LINNAEAN ZOOLOGISTS

Animal classification, by comparison to that of plants, had a considerable head start when science reawakened during the Renaissance. While flowering plants are rather uniform in their structure, there are conspicuous differences between a vertebrate, an insect, or a jellyfish, and even within the vertebrates, between a mammal, a bird, a frog, or a fish. Not surprisingly, therefore, the major groups of animals had been distinguished since before the time of Aristotle. No elaborate theory was needed to recognize them. As a result of such striking differences among well-defined animal taxa, zoologists tended to specialize and to concentrate on one particular group, like mammals, birds (Turner, Belon), or fishes (Rondelet).

But there is an even more important difference between the treatment of plants and animals. Plants are very numerous, but in spite of their seeming similarity, certain species were believed to have very specific curative properties. Correct identification was thus the overriding need. Although identification played also some role in the animal books, everybody knew the lion, the fox, the hare, the crow, and it did not seem particularly interesting or important how they were classified. However, there was the tradition of the moralizing animal books like the *physiologus* or Konrad von Megenberg's *Puch der natur* which dwelled on the habits of animals. As a result, from the very beginning the emphasis of the new zoologies was on what we would now call behavior and ecology. To be sure, there was still a tradition of quoting faithfully from the classical authors and of indulging in learned philological analyses of the meaning of animal names; also there was still considerable credulity in travelers' tales and in the existence of monsters. But the authors showed a genuine interest in the living animal and clear evidence that they had studied their subject matter in nature. Yet they had little interest in classification, and animal taxonomy soon lagged behind that of plants. *167*

At the beginning of the sixteenth century five naturalists born within 22 years of each other were responsible for the revival of zoology after the Middle Ages.[11] William Turner (1508–1568), although an Englishman, spent much of his adult life on the continent, where he published in Cologne in 1544 an *Avium . . . Historia* containing life histories of individual birds, clearly based on his own observations. Turner is also known for botanical publications, but they were not as pioneering as his ornithology. A far more weighty tome is Pierre Belon's (1517–1564) *L'histoire de la nature des oyseaux* (1555). Belon had acquired considerable fame owing to his travels through the eastern Mediterranean and the countries of the Near East. Using ecological and morphological characters, he classified birds into raptors, waterfowl with webbed feet, marsh birds with web-less feet, terrestrial birds, and large and small arboreal birds. Thus, adaptation to the habitat was his major classifying criterion. Yet some of Belon's groupings survived, particularly in the French ornithological literature, until the latter part of the nineteenth century. Belon also published on fishes and other aquatic animals (1551; 1553), but these works were almost immediately eclipsed by Guillaume Rondelet's (1507–1566) *De Piscibus Libri 18* (1554), which included the description of about two hundred species of real fishes as well as cetaceans, cephalopods, crustaceans, hard-shelled mollusks, annelids, echinoderms, coelenterates, and sponges. A number of monsters were also included, as if they were regular inhabitants of the Mediterranean.

In 1551 began the publication of Konrad Gesner's *Historia*

Animalium. This was an immense encyclopedia comprising more than 4,000 pages, where Gesner (1516–1565) had compiled everything he could find in the literature on the various species of animals. Pliny rather than Aristotle was evidently his ideal. Gesner was far too busy to contribute many personal observations on animals, but his numerous correspondents supplied him with original material. In spite of his intense interest in anything and everything he could learn about animals, Gesner was obviously not interested in classification. The species are listed alphabetically in each volume, "in order to facilitate the use of the work." In two other works, *Icones* (1553) and *Nomenclator* (1560), Gesner grouped the species systematically but did not demonstrate any progress beyond the earlier attempts of Aristotle and Rondelet. His superior botanical work, unfortunately, was not published until long after his death (1751–1771) and thus exerted little influence.

Gesner's single volume on birds was expanded by Ulisse Aldrovandi (1522–1605) into three large volumes, apparently without adding any original observations except the anatomical findings of some of his friends and their students. It was nothing but a vast compilation, of which Buffon said, "By eliminating everything useless or irrelevant to the subject, one could reduce it to a tenth of the original." Aldrovandi's *Ornithologia* (1599; 1600; 1603) differed in one respect from Gesner's *Historia:* the species were not listed alphabetically but grouped into such totally artificial categories as birds with hard beaks, birds that bathe in dust or in dust and water, those that sing well, that are water birds, and so on—quite a caricature of a classification, not even following the principles of logical division.

The one hundred years after Gesner which saw great advances in plant classification saw none in zoology. Progress was not made until function and habitat as classifying criteria were replaced by structure. This was first done in Francis Willughby's (1635–1672) posthumously published *Ornithologiae libri tres* (1676), where birds were classified on the basis of structural characters, such as the form of the bill and feet and body size. Even though the principles of logical division were used, Willughby had obviously known birds very well, and most of the groups recognized by him are still considered natural by current standards (Stresemann, 1975). We shall never know how much of this classification was contributed by his friend John Ray, the editor of Willughby's manuscripts. At any rate, Ray himself soon published small synopses of mammals and reptiles (1693) and insects (1705), and his synopses of birds (1713) and fishes (1713) were published post-humously. As artificial as much of Ray's method was, the resulting classifications were not only the best up to that time but in certain details superior even to the later ones of Linnaeus.

Animal classification became a burning problem when the world of insects was "discovered" in the seventeenth century. It was soon realized that the number of insect species far exceeded that of plants, and various naturalists (Swammerdam, Merian, Réaumur, de Geer, and Roesel) began to devote much or all of their attention to insects and their classification. Among these René Antoine Ferchault de Réaumur (1683–1757) was the greatest. His famous six-volume natural history of insects, even though in part patterned after Jan Swammerdam's (1637–1680) work, was pioneering in various ways. His superb observations of the living insect set an example for Buffon's *Histoire naturelle,* and his emphasis on the higher taxa (rather than the tedious description of species) was followed by Cuvier in his 1795 *Mémoir.* Although Réaumur was not particularly interested in classification, he made numerous perceptive observations such as that female glow-worms (fireflies), even though lacking the diagnostic character of hard elytra, were nevertheless beetles. He realized that the delimitation of natural groups does not depend on single diagnostic characters. Réaumur's views were clearly indicative of the growing resistance to the method of logical division, and together with the writings of Adanson, pioneered the principles of upward classification (see below). Réaumur's work was continued by C. de Geer (1720–1778), who made major contributions to insect classification which apparently influenced Linnaeus' system of insects considerably (Tuxen, 1973; Winsor, 1976a).

Natural history from the sixteenth through the eighteenth centuries was not as strictly divided into zoology and botany as it was in the nineteenth century. Authors like Turner, Gesner, Ray, Linnaeus, Adanson, Lamarck, and others wrote books on animals and on plants. But even in these centuries most authors specialized either on animals (Belon, Rondelet, Swammerdam, Réaumur, Buffon) or on plants (Cesalpino, Bauhin, Morison, Tournefort). After 1800 no taxonomist was any longer able to cover both the animal and plant kingdoms. Owing to this increasing separation, it is not surprising that rather different traditions developed gradually in the taxonomy of the two kingdoms, in spite of the partial transfer of the botanical methodology from plants to animals by Ray and Linnaeus.

While specialization among the zoologists was pronounced from the beginning, the botanists, owing to the structural uniformity of the flowering plants (angiosperms), could readily switch from the study of one family to that of any other, without having to learn any new techniques or terminologies. It was not until rather late in the nineteenth century that some botanists began to become specialists of certain families, be they orchids, grasses, or palms—a trend which has become pronounced in the last fifty years. Specialization among zoologists increased further when they began to study insects and aquatic animals, (though an occasional zoologist has specialized simultaneously in very different taxa, like the French arachnologist Eugène Simon (1848–1924), who was also a hummingbird specialist). Such specialization on a single higher taxon distracted from a concern with the methods and principles of higher classification. One can hardly deny that up to the end of the eighteenth century animal taxonomy was trailing behind plant taxonomy.

There is another reason for the lag of zoology: plants are much easier to preserve than animals. While herbaria had been popular since the middle of the sixteenth century, it was not until the end of the eighteenth century that appropriate methods were invented for the protection of animal collections against the ravages of moths and dermestid beetles. Preservation in alcohol had long been used, but who can study a collection of birds preserved in alcohol? This method is suitable for fishes and certain marine organisms and for specimens to be used for dissection, but not for birds whose color is important. Salt and alum were used for a time to preserve bird and mammal skins, but not until Becoeur had invented arsenical soap in the 1750s did it become possible to preserve bird skins permanently (Farber, 1977). This single technical advance is responsible for the existence of the vast modern bird and mammal collections.

Insects likewise were extremely vulnerable to destruction by dermestid beetles, and it was not possible to have permanent insect collections until naphthalene and tightly fitting collection cases and occasional fumigation were introduced. There also had to be constant supervision by a curatorial staff. Even though the same problems existed in principle for plants, the actual danger was far smaller and the ease of making and maintaining collections correspondingly greater. The rapid ascendancy of animal systematics after 1800 is to be explained in part as a result of new technologies in preservation of animal collections.

170

Two other major differences between animals and plants must be pointed out. When the extraordinary diversity of the internal anatomy of invertebrate animals was discovered by Cuvier and Lamarck (see below), it led to a great flourishing of comparative anatomy, and this in turn to a great interest among zoologists in the classification of classes and phyla. The far greater internal uniformity of plants, or perhaps more correctly the greater difficulty of interpreting plant anatomy, precluded a similar development in botany. Finally the species is a far more complex phenomenon in plants than in animals (at least higher animals) and this has resulted in the fact that zoologists have held a rather different species concept from that of the botanists (see Chapter 6).

171

When writing a history of systematics it leads to misinterpretations if one lumps indiscriminately the statements of botanists and zoologists. The views of these two kinds of biologists will have to be presented and interpreted in the context not only of their material but also of their conceptual development. But even within one of these branches of taxonomy different conceptual worlds may co-exist. For instance, in plant taxonomy the Linnaean school was for a long time so dominant that nonconformists were systematically ignored, if not suppressed. This is, in part, the reason for the neglect of botanists like Magnol and Adanson who, in some ways, were greater scientists than Linnaeus. Even today the theory of classification and the species concept are usually vastly different, when the views of a specialist of a well-known group of animals (for example, birds) are compared to those of a poorly known one of insects or other invertebrates.

CARL LINNAEUS

No other naturalist has had as great a fame in his own lifetime as Carl Linnaeus (1707–1778),[12] sometimes called the "father of taxonomy." Yet one hundred years after his death he was widely considered nothing but a narrow-minded pedant. Now, through the researches of Cain, von Hofsten, Stearn, Larson, Stafleu, and other Linnaeus scholars, we can paint a more balanced picture.[13] This is not an easy task since Linnaeus was a very complex person, having seemingly incompatible traits. In his methodology he was, indeed, of a pedantic pragmatism, yet he also had great literary powers. He was a numerologist (with a fondness for the numbers 5, 12, 365) and, particularly in later years, quite a bit of a mystic;

and yet he was the model of a painstakingly descriptive taxonomist. He spent years in Holland and visited Germany, France, and England, yet he spoke only Swedish and Latin and knew little of any foreign language. By the time he reached Holland (1735), his method and conceptual framework had already matured to a remarkable degree, but even though his method changed subsequently only little (his later invention of binomialism was not considered by Linnaeus an important modification of his system), his philosophical ideas changed rather decisively. He took a deeper

172   interest in only one aspect of the biology of the individual species, their sexual biology (Ritterbush, 1964: 109–122), but, as his essays show (*Amoenitates Academicae*), Linnaeus was interested in a wide variety of biogeographical and ecological subjects (Linné, 1972).[14] Classifying, however, was his primary interest, indeed his obsession to classify anything he came in contact with went so far that he proposed an elaborate classification of botanists into phytologists, botanophiles, collectores, methodici, Adonides, oratores, eristici, and so on (*Philosophia Botanica:* para. 6–52).

   In 1753 Linnaeus knew about 6,000 species of plants and believed the total might be about 10,000 and the number of species of animals (he listed 4,000 in 1758) to be about the same. (His contemporary Zimmermann (1778) made the remarkably more realistic estimates of 150,000 species of plants and seven million species of animals, to be discovered eventually.) His entire method (for instance, a botanist must remember the diagnosis of every genus!) was based on his assumption of a limited number of taxa; and yet we now know more than 200,000 species of phanerogams alone. Linnaeus knew 236 Swedish species of algae, lichens, and fungi, as compared to about 13,000 species known from Sweden today. He assumed that the tropics of all parts of the world had a rather uniform plant life. But such insufficiencies in his knowledge were not nearly as deleterious for the development of his methodology as his conceptual conflicts. On one hand, as we shall presently see, Linnaeus was a practitioner of scholastic logic and a strict essentialist; on the other hand, he also accepted the principle of plenitude which stresses continuity. The major aim of his method was the eminently practical one of assuring the correct identification of plants and animals, yet the procedure by which he tried to achieve this was the highly artificial one of logical division. No wonder his critics were able to discover so many inconsistencies in his writings.

Still, Linnaeus deserves all the praise he got. His technical innovations (including the invention of binominal nomenclature), his introduction of a rigorous system of telegram-style diagnoses, his development of an elaborate terminology of plant morphology (Bremekamp, 1953a), his standardization of synonymies, and of every other conceivable aspect of taxonomic research brought consensus and simplicity back into taxonomy and nomenclature where there had been a threat of total chaos. This was the secret of his popularity and success. By his authority Linnaeus had been able to impose his methods on the world of systematics and this was largely responsible for the unprecedented flowering of taxonomic research on animals and plants during the eighteenth and early nineteenth centuries.   *173*

Nevertheless, various post-Linnaean authors, both botanists and zoologists, deplored the fact that Linnaeus' work had resulted in such an emphasis on classifying and name-giving that it had led to a near obliteration of all other aspects of natural history, "in particular the study of the living animal has been completely suppressed . . . as a further result not only varieties, but also juvenals and larvae of known species were described as separate species" (Siebold, 1854). The neglect of Kölreuter and the struggles, in both botany and zoology, to attract talent to physiology and embryology deplored by Nägeli and Sachs support this evaluation.

The reason why modern writers have had so much difficulty in understanding Linnaeus is that many of the terms used by him, like "genus," "species," "name," to "know," and "natural system," have the very special meanings which these terms have in the system of scholastic logic. In school Linnaeus had excelled in logic, and he was evidently deeply impressed by the precision of this method. From Cesalpino on, every botanist had applied logical division with more or less consistency, and it still dominated Linnaeus (Cain, 1958).[15]

There is one respect in which Linnaeus differed significantly from his forerunners. In their downward classifications they had applied dichotomy as often as was necessary to reach the looked for "genus" or "species." By contrast, Linnaeus applied the full rigor of the method only at the level of the genus. He was less interested in categories higher than the genus and was vague and inconsistent with respect to intraspecific variation.

Linnaeus and the Higher Categories

Instead of a system of consistent downward dichotomy, he adopted a system dominated within a kingdom, by a hierarchy of only four categorical levels: class, order, genus, and species. Classifying the entire diversity of nature into taxa at these four levels gave a clarity and consistency to his system that was altogether absent from the cumbersome dichotomies of most of his forerunners.

174   The modern taxonomist recognizes an elaborate hierarchy of higher categories. The complete series from the species to the kingdom is often called the *Linnaean hierarchy* (Simpson, 1961; Mayr, 1969), though Linnaeus was not the first author to recognize categories above the genus. As we have seen, Aristotle vaguely indicated a hierarchy in his arrangement of animals. He divided all animals into bloodless animals and those with blood. The latter have subcategories like four-footed, many-footed, or footless ones, and so forth. Most of these groupings are made with the help of single diagnostic characters and Aristotle's successors have usually interpreted his arrangements as diagnostic keys. Yet, as I mentioned earlier, Aristotle himself made fun of the artificial method of dichotomous keys. He realized that his criterion "live-bearing" (versus "egg-laying"), for example, does not produce a natural group. At no time did Aristotle make a terminological distinction between various levels of higher categories.

When the Aristotelian tradition was revived during the Renaissance, it included a lack of interest in the higher categories. The herbalists as well as the encyclopedists either recognized no higher categories at all above the genus or they designated the groups which resulted from their logical division as "books," "chapters," or some other nontaxonomic term. Ray's suprageneric groups, likewise, were entirely informal designations. Tournefort was apparently the first botanist to develop a formal classification of categories higher than the genus. He divided plants into 22 classes; these classes in turn he subdivided into 122 sections.

The terminology of these higher categories varied at first from author to author. What Tournefort had called "sections" Magnol and Adanson called "families" and Linnaeus "orders." As the number of genera and species of plants grew, and with it the need for a more elaborate hierarchy, all these alternative terms were incorporated into a single terminology. The family category was rather consistently used by 1800 to designate a level between genus and order. Cuvier is still inconsistent in the application of these terms in succeeding publications. They became completely formalized only in the writings of the entomologist Latreille.

Linnaeus was singularly uncommunicative when it came to defining his categories order, class, and kingdom. One has the feeling that he introduced these higher categories not for theoretical but for purely practical reasons. Indeed, he states quite frankly that class and order are less "natural" than the genus. As he writes in the *Philosophia Botanica* (para. 160): "A class is an agreement of several genera in the parts of their fructifications in accordance with the principles of nature and art." In other words, classes are to some extent artificial, but Linnaeus intimates that they will be replaced by natural classes when all the genera of plants have been discovered and described. The order was for Linnaeus even more of a choice of convenience: "An order is a subdivision of classes needed to avoid placing together more genera than the mind can easily follow" (para. 161). It is evident that the higher categories were for Linnaeus primarily convenient information-retrieval devices. His lack of interest in the higher categories is documented by the fact that the higher taxa of animals recognized by Linnaeus are decidedly inferior (that is, more heterogeneous) than those that had been delimited by Aristotle more than two thousand years earlier.

There are various inconsistencies in Linnaeus' attitude toward the higher categories. The genus represents his essentialist thinking par excellence, and all genera are separated by sharp discontinuities. However, he has a rather nominalist attitude toward classes and orders. For them he adopts Leibniz's motto that nature makes no jumps. The more plants we know, the more gaps between the higher taxa will be filled until the boundaries between orders and classes may finally disappear. His adherence to the principle of plenitude is documented by his statement that all taxa of plants have relationships on all sides as do neighboring countries on a map of the world (par. 77; for map, see Greene, 1959: 135).

The Genus

The genus is for the modern taxonomist the lowest collective category, an aggregate of species sharing certain joint properties. This was not the concept of the genus among the practitioners of logical division. For them the genus was a class with a definable essence which can be divided into species with the help of differ-

175

entiae. A genus did not designate a fixed rank in a hierarchy of categories and the generic "name" was often a polynomial, particularly at the lower levels of division. Originally there was little uniformity in usage, and Aristotle sometimes used the term *genos* even where we would now speak of species (Balme, Grene). The gradual narrowing down of the term "genus" to the categorical level which we now call genus was a slow process. It began among the herbalists and encyclopedists, among whom Cordus (1541) and Gesner (1551) already uséd the generic name in a remarkably 176 modern way, although the German herbalists used the word *Geschlecht* (genus) more often in the sense of species than of genus. The rather vague usage of the words "genus" and "species" began to acquire biological meaning in the writings of the great seventeenth-century taxonomists Ray and Tournefort.

The genus was for Linnaeus the cornerstone of classification.[16] When establishing order in one's environment, one does not classify things but their "essences." It was an axiom for Linnaeus that natural genera exist, that these—that is, the "essences"—had been created as such, and that they could be recognized on the basis of their characters of fructification. He admitted as many genera as there are different groups of species which agree in the structure of their fructification. It is not the taxonomist who "makes" the genus; he only discovers the genera that had been created in the beginning. There is a very close association in Linnaeus' theory of classification between creationist dogma and essentialist logic.

In his earliest writings Linnaeus still adhered to the strict code of logic, so that he called the whole plant kingdom the *summum genus,* the species of which were the classes of plants. He abandoned this usage after 1735, and restricted the term "genus" to the hierarchical level immediately above the species. In 1764 he listed 1,239 genera of plants. Linnaeus is quite emphatic about his method, which he describes in great detail in his *Philosophia Botanica* (para. 186–209). The definition of a genus is a statement of its essence. "The 'character' is the definition of the genus, it is threefold: the *factitious,* the *essential,* and the *natural.* The *generic character* is the same as the *definition of the genus* [para. 186] . . . The *essential definition* attributes to the genus to which it applies a characteristic which is very particularly restricted to it, and which is special. The essential definition [character] distinguishes, by means of a unique idea, each genus from its neighbors in the same natural order" (para. 187).

One would need only the essential definition if there were a way to determine what the essential characters of a genus are. However, Linnaeus, by implication, concedes to Ray that no such method is known. For this reason one must also present a *factitious definition* which "distinguishes a genus from the other genera in an artificial order" (para. 188). Finally, "The *natural definition* lists all possible generic characters; and so it includes both essential and factitious definitions" (para. 189).[17]

Although Linnaeus greatly changed his ideas on the sharp delimitation and fixity of species in the course of his scientific life (see Chapter 6), he never wavered on the genus. One has the impression that he perceived genera intuitively (by inspection), which prompted his famous motto: "It is not the character (diagnosis) which makes the genus, but the genus which gives the character." Indeed, he often ignored deviations in somewhat aberrant species as long as these still "obviously" belonged to a given genus. The genus, to him, was the most convenient information-retrieval package, because in the limited representation of the animal and plant kingdoms known to him, the genera, on the whole, were separated from each other by clear-cut discontinuities. But more importantly, for reasons of his essentialist philosophy, the genus (with its essence) was the God-given real unit of diversity.

In some respects, at least conceptually, the genus of Linnaeus, in its essentialistic, monolithic, independent existence, was a backward step from the genus of Tournefort, which was an aggregate of species, hence a collective category. The genus is a device, Tournefort said, "to bring together as in bouquets plants which resemble one another, and to separate them from those which they do not resemble" (*Elemens de botanique,* 1694: 13). The modern genus concept thus traces back to Tournefort rather than to Linnaeus. Stafleu (1971: 74) is quite correct in pointing out "that it was not really Linnaeus who produced for the first time consistently composed definitions (diagnoses) and therefore comparable descriptions of genera. The honor for this goes to the pragmatic empiricist Tournefort."

The Sexual System of Linnaeus

A classification, for Linnaeus, was a system that permitted the botanist to "know" the plants, that is, to name them quickly and with certainty. One could devise such a system only by using well-defined, stable characters. The vegetative portions of the plant show many adaptations to special conditions and are thus subject to convergent trends (such as between cacti and euphorbs) which misled the early plant taxonomists. The flower, which Linnaeus chose as the major source of his characters, had the great advantage that the numerical differences in stamens and pistils (and several other flower characters) were not ad hoc adaptations but, as we would now say, either an incidental by-product of the underlying genotype or else adaptations to facilitate pollination, independently of the habitat.

178   Linnaeus, rather misleadingly, called his method the "sexual system." This terminology reflected Linnaeus' evaluation of the preeminent importance of reproduction. Reproduction, for him, indicated the secret working plan of the creator. Actually, of course, differences in the number of stamens and pistils, as practical as they may be for identification, are of little functional significance, if any. But Linnaeus would have considered it bad taste to admit this frankly, and in order to give his system philosophical justification, he called it the sexual system. It was first set out in a key in the *Systema Naturae* (1st ed.) of 1735. Four basic criteria were used: number, shape, proportion, and situation. Absolute number, thus, was only one of Linnaeus' character sets. Whether the flowers were visible (later called phanerogams) or not, how many stamens and pistils there are, whether or not they are fused, and whether or not male and female elements occur in the same flower were among the characters which Linnaeus used to distinguish 24 classes (Monandria, Diandria, and so on). The classes, in turn, were divided into orders with the help of additional characters.

As artificial as this system was, it was remarkably useful for the practical purposes of identification and for information storage and retrieval. Any botanist using the sexual system would come to the same result as Linnaeus. All he had to do was to learn a rather limited number of names of the parts of the flower and fruit and then he could identify any plant. No wonder nearly everybody adopted the Linnaean system. When, already in 1739, Bernard de Jussieu, the leader of French botany, pronounced the Linnaean method preferable to that of his compatriot Tournefort, because it was more exact, the victory was complete.

In a classification based on common descent any species (or higher taxon) can be encountered only a single time. It has a unique position in the hierarchy. This constraint does not exist for an artificial identification key. A variable taxon may be entered repeatedly in different couplets. This must be remembered in connection with Linnaeus' classification of hard-shelled invertebrates. He placed types with a shell (mollusks, cirripedia, certain polychaetes) in the order Testacea, but soft animals, that is, mollusks without a shell (such as slugs and cephalopods), coelenterates, and most polychaetes in the order Mollusca. But when he listed the genera of Testacea, he gives in each case also a molluscan generic name for the soft animal. For instance: *Chiton* (animal *Doris*), *Cypraea* (animal *Limax*), *Nautilus* (animal *Sepia*), *Lepas* (animal *Triton*), and so on. The genera *Doris, Limax, Sepia,* and *Triton* are again listed as valid genera in the order Mollusca. The overriding concern for Linnaeus was the practical one of identification, and this is what his double-entry system attempted to facilitate (von Hofsten, 1963). It was clearly a compromise, with the shell serving for identification, while the animal indicated the real position in the system. One might also interpret it as the simultaneous attempt at an artificial and a natural classification.

179

Considering the seeming artificiality of the method of logical division, surprisingly many of the genera recognized by Linnaeus consist of well characterized groups of species, many still accepted as genera or families today. A close examination of these classifications solves the riddle. It is rather obvious that Linnaeus, like Cesalpino, first recognized such groups by visual inspection and elaborated the definition (essence) only subsequently. This Linnaeus confessed openly in his *Philosophia Botanica* (para. 168), where he says: "One must consult the habit secretly, under the table (so to speak), in order to avoid the formation of incorrect genera." When Linnaeus' son was asked what his father's secret was to be able to create so many natural genera, in spite of the artificiality of his method, he answered: "It was no other than his experience in knowing plants from their external appearance. Therefore he often departed from his own method by not being disturbed by variation in the number of parts if only the character of the genus could be preserved." As a result, Linnaeus sometimes went even so far as placing in a single genus species which differed in their number of stamens and should have gone into different classes of his sexual system! Also, he often transferred the diagnosis of a genus untouched into later editions of his works, even though subsequently added species had attributes that were in conflict with the old generic diagnosis. He was equally inconsistent in his animal classifications. The sheep-tick, a wingless fly, was without hesitation classified among the "two-winged insects" (Diptera). There are scores of similar cases in Linnaeus' zoological works where pragmatic considerations won out over philosophical principles (See also Winsor, 1976a).

BUFFON

The eighteenth century was the great age of natural history. It saw the heroic voyages of Captain Cook, of Bougainville, and of Commerson (Stresemann, 1975), and a new interest in nature was reflected not only in the writings of Rousseau but in those of most of the "philosophes" of the Enlightenment. It was the century of natural-history cabinets and herbaria owned not only by kings and princes but also by wealthy citizens such as George Clifford (1685–1760) in Holland, Sir Hans Sloane (1660–1753), and Sir Joseph Banks (1743–1820) in England, and others in France and other countries on the continent.[18] One of the ambitions of these patrons of natural history was the publication of a scientific catalogue of their collections.

Books on nature became increasingly popular, but none had the spectacular success of Buffon's *Histoire naturelle.* Though dealing, like Linnaeus' taxonomic treatises, with the diversity of nature, Buffon's approach was fundamentally different. Identification was the least of his concerns; rather he wanted to paint vivid pictures of different kinds of animals. He rejected the pedantry of the scholastics and humanists and wanted to have nothing to do with their emphasis on logical categories, essences, and discontinuities. He was rather inclined to favor ideas promoted by Leibniz in which plenitude and continuity were emphasized as well as the Aristotelian concept of the scale of perfection. To Buffon this seemed a far superior view of nature than the dry-as-dust compartmentalization of the "nomenclateurs," by which term he contemptuously referred to Linnaeus and his disciples. Buffon's studies of Newton led him in the same direction. Did not the law of gravitation and the other laws of physics show that there was a unity in nature effected by general laws? Why dissect and destroy this unity by cutting it up into species, genera, and classes? Nature knows no species, genera, and other categories; it knows only individuals, he declared in the first volume of his *Histoire naturelle* in 1749, all is continuity (but already in 1749 he excluded species from this sweeping claim). Buffon's first love had been physics and mathematics and, although he had had some previous familiarity with natural history, it was not until he had been appointed director of the Jardin du Roi (now Jardin des Plantes) in 1739 at the age of 32 that he became vitally interested in the diversity of the organic world.

Buffon and Linnaeus were both born in 1707, and yet the contrast between the two men could not have been greater, and this was, at first, also true for their followers. The Linnaeans emphasized all those aspects of the taxonomic procedure that would facilitate identification, while Buffon and the French school concentrated on the understanding of natural diversity. The Linnaeans stressed discontinuity, Buffon continuity. Linnaeus adhered to Platonic philosophy and Thomistic logic while Buffon was influenced by Newton, Leibniz, and nominalism. Linnaeus concentrated on "essential" characters, quite often a single diagnostic one because, as he claimed, attention to descriptive detail would prevent recognition of the essential characters. Buffon, by contrast, insisted that we "must make use of all parts of the object which we have under consideration," including internal anatomy, behavior, and distribution.

181

Buffon's approach was well suited for the treatment of mammals and merely a continuation of the tradition of earlier classifiers (for example, Gesner). The number of mammal species was quite limited, and identification was rarely a problem. Only botanists like Ray and Linnaeus had applied the principles of logical division to the classification of animals. When Buffon classified mammals into domestic and wild animals, he justified the division as being "the most natural one." For him "natural" meant practical, not "reflecting the essence," as for Linnaeus.[19]

Around 1749 Buffon's views began to change, eventually becoming considerably modified under the impact of his growing knowledge of organisms (Roger, 1963: 566). While in 1749 he expressed a radical skepticism of the possibility of any classification of living organisms, by 1755 he admitted that there are related species. In 1758 he still ridiculed the idea of genera, but in 1761 he accepted them to facilitate the difficult enumeration of the "smallest objects of nature," and by 1770 the genus is made the basis of his classification of birds, presumably still with the mental reservation of its arbitrary nature. Even though he admits a common descent of "genera" of domestic animals, they are, of course, only biological species. Also, from 1761 on, he adopted the concept of the family. Yet, it must be remembered that Buffon never attempted to classify the entire animal and plant kingdoms. Indeed, much of his *Histoire naturelle* is a series of monographs of individual mammalian species. These are superbly done, from the literary as well as the scientific point of view, and had an enormous impact on the education of young zoologists. Yet, they were not the material with which to develop a general theory of systematics, something Buffon simply was not interested in.

Although starting at opposite poles, Linnaeus and Buffon got increasingly closer in their ideas as their work progressed. Linnaeus liberalized his ideas on the fixity of species, and Buffon admitted (contrary to the nominalists' views) that species could be nonarbitrarily defined as reproductive communities (*Hist. nat.,* 1753, IV: 384–386). Buffon, however, never accepted Linnaeus' views on the nature of the genus, that is, the belief that it is the most objective of all categories. Furthermore, his criteria for the recognition of higher taxa were entirely different from those which Linnaeus professed to use (total habitus vs. single characters revealing the essence).

182

Toward the end of their lives, let us say in the 1770s, the contrast between the taxonomic methods of Linnaeus and Buffon had been reduced to such an extent that their respective traditions merged in their pupils. Lamarck, a protégé of Buffon, still proclaimed loudly that categories do not exist, only individuals, but once he had recorded this article of faith, he no longer paid much attention to it in his taxonomic works. The same is true for Lacépède. In Cuvier, finally, it is no longer possible to discern the Buffonian nominalist tradition.

A NEW START IN ANIMAL CLASSIFICATION

Little progress was made in animal classification during the seventeenth and eighteenth centuries. Indeed, the Linnaean classification of the invertebrates was a backward step from that of Aristotle. All this changed overnight with the publication in 1795 of Georges Cuvier's (1769–1832) *Memoir on the Classification of the Animals Named Worms.*[20] The catch-all taxon recognized by Linnaeus under the name Vermes was divided by Cuvier into six new classes of equal rank: mollusks, crustaceans, insects, worms, echinoderms, and zoophytes. Seventeen years later he deprived the vertebrates of their favored position by raising some of the invertebrates to an equal rank, and classified all animals in four phyla ("embranchements"): vertebrates, mollusks, articulates, and radiates (Cuvier, 1812). Within these highest taxa, a number of new classes, orders, and families were recognized which up to then had been confounded with each other or totally overlooked. He consolidated the Linnaean Mollusca and Testacea into the class Mollusca and removed the jellyfish (Medusa) and sea anemone (Actinia) from the mollusks to the zoophytes.

What was most important in Cuvier's contribution to animal classification was his discovery of the great information content of the internal anatomy of the invertebrates. When dissecting numerous marine animals, he found an abundance of new characters and types of organization. This initiated the great tradition of the comparative zoology of the invertebrates. His findings permitted Cuvier to recognize for the first time a number of taxa that are still accepted today.

183

What is rather remarkable about Cuvier's enormous contribution is that even though it is based on a carefully worked out system of concepts and laws, conceptually it represents no advance over the principles of Aristotelian logic. Again the emphasis is on classification from the top down by a process of division and there still is a search for the essence, the true nature of each group, and characters are still weighted on the basis of functional importance. Nevertheless, he introduced some innovations.

Cuvier and the Correlation of Characters

Cuvier thought that certain physiological systems were of such importance that they would control the conformation of all other characters. This was a new conceptual departure. Taxonomists prior to Cuvier had acted on the whole as if each character was independent of every other character, and as if an organism with a different character had a different essence. Buffon had been the first to disagree with this atomistic approach. An organism was not an arbitrary jumble of characters, as it appeared from the writings of the Linnaeans; rather the composition of the characters was dictated by their "correlation." Cuvier expanded Buffon's rather general ideas into a concrete principle, that of the *correlation of parts* (see Chapter 8). The various parts of an organism are interdependent to such a degree that if given the tooth of an artiodactyle ungulate, the anatomist can at once make numerous statements about the probable structure of other parts of the anatomy of this animal. All the functions of an organism are mutually dependent on each other to such a degree that they cannot vary independently:

> It is in this mutual dependence of the functions and the aid which they reciprocally lend one another that are founded the laws which determine the relations of their organs and which possess a necessity equal to that of metaphysical or mathematical laws, since it is evident that the seemly harmony between organs which interact is a necessary condition of existence of the creature to which they belong and that if one of these functions were modified in a manner incompatible with the modification of the others, the creature could no longer continue to exist. (*Leçons d'anatomie comparée*, 1800, I: 51)

184

Ever since Cuvier's proclamation, experienced taxonomists have utilized correlated variation as one of the most important clues in the evaluation of characters. It may reveal both ad hoc specializations in connection with the occupation of special adaptive zones as well as deep-seated genetic integration, as expressed in character constancy in higher taxa. Lamarck was apparently the first to call attention to the importance of such constancy in his *Flore Françoise* (1778), soon followed by de Jussieu. But Cuvier went further than merely calling attention to a correlation of parts; he also had an elaborate system of weighting characters, embodied in his principle of the *subordination of characters* (see below).

Where Cuvier differs in his approach from an author like Linnaeus is that he is genuinely interested in classification and its principles, rather than in an identification scheme. In his 1795 *Memoir,* like Réaumur before him, he does not even bother to describe genera or species. What his real aim is, he expressed in these words: "In conclusion, I have not presented this essay on division that it may serve as the beginnings of the determination of the name of species; an artificial system would be easier for this, and this is only proper. My aim has been to make known more exactly the nature and true relationships of the *animaux à sang blanc* [invertebrates], by reducing to general principles what is known of their structure and general properties."

Lamarck

As different as Jean Baptiste Lamarck (1744–1829) and Cuvier were from each other philosophically, their contributions to classification were remarkably similar (Burkhardt, 1977). Lamarck also made numerous valuable innovations in the classification of the invertebrates, dealing with such problems as the position of the cirripedes and tunicates and the recognition of the arachnids and annelids as distinct taxa. Indeed, from the protozoans to the mollusks, Lamarck made numerous taxonomic contributions. But when it comes to the theory of classification, his concepts were as conventional as those of Cuvier. Lamarck started out with a belief in a single series of animals beginning with the simplest Infusoria and culminating with man. Consequently he tried to rank each higher taxon according to its "degree of perfection." Later on, in part under the impact of Cuvier's replacement of the single series by four *embranchements,* but also in part as a result of his own comparative studies, Lamarck increasingly abandoned the concept of a single series. At first he merely admitted that certain species and genera diverged from the straight line owing to the "force of circumstances," but eventually he admitted the branching of "masses" (higher taxa), and his final presentation of animal relationships (1815) does not differ in principle from a phylogenetic tree such as one would expect to find in the literature at the end of the nineteenth century. Lamarck frequently emphasized how important he considered the activity of classifying, since "the study of affinities . . . should now be regarded as the chief instrument for the progress of natural science."

185

TAXONOMIC CHARACTERS

Classification is the ordering of organisms into taxa on the basis of their similarity and relationship as determined by or inferred from their taxonomic characters. This definition indicates the decisive importance of taxonomic characters for the construction of classifications. Yet, from the beginnings of the history of taxonomy to the present day there has been much disagreement as to what characters are most useful and, indeed, legitimate in taxonomic analysis. Much of the history of classification is a history of controversy on this point. The Greeks were fully aware of the fact that utilitarian characters, like medicinal properties or the presence of spines, have little to do with other, seemingly more deep-seated, properties of a plant. The essentialists, who classified by logical division, also felt that some characters were more fundamental than others. Even though their terminology of essential and accidental characters was burdened by scholastic dogma, they did sense a truth which was not understood until centuries later. From Cesalpino on, it was recognized that such nonmorphological characters as relation to man (cultivated versus wild), seasonality (deciduous or evergreen), or habitat were less apt to produce useful classifications than structural characteristics. Consequently, use of structural characters has dominated taxonomy since the sixteenth century.

Taking the entire period from Cesalpino to the present, taxonomic characters have provoked three major controversies: (1) Should one use only a single key character (*fundamentum divisionis*) or multiple ("all possible") characters? (2) Should only morphological characters be allowed, or also ecological, physiological, and behavioral characters? (3) Should characters be "weighted" or not—and if weighted, by what criteria?

Already Aristotle stated that some characters are more useful in delimiting groups of animals than others, and very few authors throughout the whole history of taxonomy have disagreed with this conclusion. (The numerical pheneticists [see below], in some of their earlier publications [Sokal and Sneath, 1963], are among the few exceptions by promoting equal weighting of all characters.) The problem thus was not to weight or not but two other problems: What principles should one use to determine the weight of a character? and How should one translate a scale of weights into a classification? It must be remembered that rejecting a particular criterion of weighting of one author by another one does not mean a rejection of the method of weighting as such. Authors like Buffon and Adanson, who favored the use of "as many characters as possible," did not in the least propose that they be weighted equally.

As long as classifications were basically identification schemes, they required, by necessity, a reliance on single characters. It did not matter if the groups produced by this method were heterogeneous, as long as the goal of identification could be reached. In the case of plants, experienced botanists knew that no other part of a plant provides more and better diagnostic features than the "fructification" (flowers, fruits, and seeds). A particular advantage of this structural system is that it includes a great number of quantifiable characters, such as the number of flower petals, stamens, and pistils. Flowers had the additional advantage of being comparatively invariant within a species (compared to most aspects of the leafy vegetation) and yet composed of a rich assortment of variable parts displaying species-specific differences. No one was more diligent and successful in pointing out such differ-

186

ences than Linnaeus, even though some contemporaries cursed him for using characters which one could only see with the help of a hand lens.

None of the essentialists would have admitted that he used fructification because of its practical advantages. Instead, they constructed an elaborate myth, in which evidently they believed themselves, that certain aspects of a plant in some way were more important than others, and therefore reflected the essence better. Cesalpino rated nutrition and the results of nutrition (growth) highest, while he ranked reproduction, reflected in fructification, next highest: since providing for the continuation of a plant into the next generation is the next most important clue to the essence, all aspects of fructification (flowers and seeds) are the next most important characters. Linnaeus differed from him by ranking fructification above growth, and stated quite simply (*Phil. Bot.*, para. 88) that "the essence of the plant consists in its fructification." Perhaps the best evidence for the fact that flowers were chosen for their usefulness rather than for philosophical reasons is that even today they are most prominently used in identification keys in spite of the fact that the argument of their "functional importance" was abandoned two hundred years ago.

187

Even though botanists from Gesner (1567) and Cesalpino (1583) to Linnaeus all agreed on the importance of fructification, this still left a great deal of choice owing to the multitude of characters available, all relating to fructification. Different botanists chose different characters on which to base their first division: Tournefort and Rivinus the corolla, Magnol the calyx, Boerhave the fruit, Siegesbeck the seeds, and Linnaeus the stamens and pistils. It would have been difficult to decide how these components of fructification should be ranked according to their functional significance. As a result, pre-Linnaean botanists largely split on national lines. The British followed Ray, the Germans Rivinus (Bachmann), and the French Tournefort. Since identification was the principal objective, Tournefort's system, which was simpler, more concise, and more easily memorized than the other two, was more and more widely adopted by botanists, until it was replaced by the even more practical sexual system of Linnaeus.

When the number of known animals increased rapidly during the seventeenth and eighteenth centuries, morphological characters were used ever more frequently, but among the zoologists there was none of that intense interest in methodology that characterized contemporary botanists. Ecological criteria were still often given preference, particularly for groups other than the vertebrates. Vallisnieri (1713), for instance, divided the insects into four major groups, those that inhabit plants, those that live in the water (including crustaceans), those that dwell in rocks and in the soil, and those that live in or on animals. Even when morphological characters were used, they were often poorly chosen, such as when Linnaeus classed the fish-shaped whales with the fish or when he combined the majority of the invertebrates into the worms (Vermes).

Cuvier's principle of the subordination of characters, according to which various parts in an organism differ in taxonomic value, was a system of weighting. In his earlier work (prior to about 1805) the organs of nutrition, and particularly of circulation, are most prominent as the diagnostic characters of Cuvier's higher taxa. By 1807, however, the nervous system had been clearly raised to the top rank, and it now played the most important role in the delimitation and ranking of his four *embranchements* (Coleman, 1964). At the level of the lower categories Cuvier often assigned the same character different weight in different animal groups. For instance, tooth characters define orders among the mammals, genera among the reptiles, but only species among the fishes. The structure of the feet, to give another example, has the value of an ordinal character for mammals, being their principal means of locomotion, while in birds where the wings are preeminent, the feet have a much lower value as a taxonomic character. Nevertheless, Cuvier thought that certain characters were associated with a certain rank in the hierarchy of categories. Evidently his subordination of characters is nothing more or less than the a priori system of weighting of the botanists except that in animals, in traditionally Aristotelian manner, "sensitivity" is ranked highest, hence the primary characters are derived from the nervous system. Even though Cuvier revolutionized the classification of the invertebrates, it was not through the introduction of new concepts but rather by making available a whole new set of characters, those derived from the internal anatomy.

A second revolution in the use of animal characters, again not involving any new concepts, was produced by a technological advance, the invention of the microscope. Van Leeuwenhoek's introduction of optical instruments into the study of natural history (about 1673) was an innovation the full impact of which has not yet run its course (as indicated by recent discoveries made with the help of the scanning electron microscope). Even stamens and pistils, the key characters in Linnaeus' system, are best seen with the help of magnification. The study of the sculpture of beetle elytra or of antennae, wing veins, and genital armatures of all insects require at least a hand lens. Most aquatic invertebrates, and of course algae, protozoans, and other protists, require a microscope for their study.

The extent of microscopical studies greatly accelerated after the 1820s. The careful histological examination of organisms of all sizes led to the discovery of taxonomically important sense organs, glands, accessory structures of the reproductive and digestive systems, and previously unknown details of the nervous system. Entirely novel characters (for example, chromosomal and biochemical differences) were added in due time, also made possible by technological advances. Even though the number of characters available to the taxonomist seemed to mushroom, the new information was not enough to settle important controversies on relationship.

189

The dogma that one particular type of characters is best suited as the basis of classification came under heavy attack already in Linnaeus' lifetime. However, it was not the principle of weighting as such that was attacked but the reasoning on the basis of which weighting was to be done. At first, as was described above, the only weighting criterion that was admitted as determining taxonomic usefulness was that of functional importance. In due time, however, entirely new criteria of weighting were proposed. Lamarck, Cuvier, and de Jussieu stressed the importance of "constant" characters. De Candolle emphasized growth symmetries which indeed in plants often characterize genera and entire families. Such symmetries can be found in the flowers, in the insertion of the leaves, and in other characteristics of plants.

Polythetic Taxa

The genus (at any level) is represented for the essentialist by the totality of all "species" (meaning subordinate taxa) sharing in the same essence, or, as it was later expressed by taxonomists, by all those that had certain "characters" in common. It was a source of considerable distress, from the earliest period of classification on, that certain individuals or species were found which lacked one or the other character "typical" (that is, essential) for the taxon. Pedants would separate such species generically; the more experienced taxonomists, for instance Linnaeus, would simply ignore the discrepancy. Indeed, higher taxa were found that could be reliably defined only by a combination of characters, each of which might occur also outside the given taxon or which might be occasionally absent in a member of the taxon. In such cases no single feature is either necessary for membership in such a taxon or sufficient.

190

Adanson seems to have been the first to have recognized this clearly, although it is already implied in some statements made by Ray. Vicq-d'Azyr (1786) stated that "a group may be perfectly natural, and yet have not a single character common to all the species that compose it." Heincke (1898) showed that two species of fish, the herring and the sprat, differ from each other in eight structural characters, but only 10 percent of the individuals differ from each other in all of these characters. Beckner (1959) was the first to give formal recognition to this principle by designating taxa based on character combinations as "polytypic." However, since the term "polytypic" was already employed in taxonomy in a different sense, Sneath (1962) introduced the replacement term *polythetic*.

To allow the characterization of higher taxa by polythetic character combinations signaled the final demise of an essentialistic definition. However, long before that the entire concept of relying on particularly important characters, necessitated by the method of logical division, had come under attack and had led in due time to an entirely new concept of classifying.

UPWARD CLASSIFICATION BY EMPIRICAL GROUPING

The dominant method of classifying from Cesalpino to Linnaeus—downward classification by logical division—became more and more unsatisfactory as European botanists and zoologists were inundated by the avalanche of new genera and families from the tropics. The method of classification according to the principle of logical division was expected to achieve two objectives: to reveal the order of nature (the plan of creation) and to provide a convenient identification scheme. When the method was practiced, however, it became apparent that the two objectives were incompatible and that a consistent application of the principles of logical division usually led to absurd results. A retrospective analysis of this theory of classification shows that it had at least three basic weaknesses:

(1) When only a small fauna or flora needs to be classified, an identification scheme is sufficient, such as logical division is able to provide. The method is, however, unable to assemble "natural" groups of species and genera, such as one needs for classifications when one deals with very large faunas and floras.

(2) Only a single character can be used at each step. The choice of this character was dictated by its supposed ability to reveal the essence of the "genus." However, the claim that certain characters, for instance those of greater functional importance, are better qualified to reveal the essence of a taxon than others is neither theoretically nor practically substantiated. Hence, the whole system of weighting characters according to their presumed functional importance lacks validity.

191

(3) The whole philosophy of essentialism, on which the method of logical division was based, is invalid and therefore unsuitable as a basis for a theory of classification.

The drastic revolution in philosophical thinking that had been taking place in the seventeenth and eighteenth centuries could not fail to have an effect on the thinking of the classifying naturalists. It has been a fascinating challenge to various historians to infer the relative influence of the scientific revolution and the Enlightenment, of the philosophy of Locke with its emphasis on nominalism and empiricism, of Kant, and of the ideas of Newton and Leibniz with their stress on continuity, on the thinking of Buffon, Linnaeus, and their schools. Buffon's ridiculing of the "nomenclateurs" (he meant the Linnaeans) was one of the manifestations of these philosophical influences.

Yet, when one closely studies the taxonomic work of the eighteenth century, it is quite evident that purely practical considerations played a major, if not the dominant, role in shaping taxonomic concepts. The practical difficulties encountered by downward classification daily became more apparent. How valid was a method which forced even the great Linnaeus to "cheat" and to sort his species, so to speak, "under the table," because divisional logic was unable to accomplish this? What should his less experienced followers do to avoid coming up with altogether absurd classifications? One will not understand the nature of the fundamental changes in taxonomic theory between 1750 and 1850 unless he pays equal attention to the new demands made by taxonomic practice and to the realization that the philosophical foundations of downward classification were gradually being eroded away.

Eventually it became clear that it was futile to attempt to salvage downward, divisional classification by modifying it, and that the only way out was to replace it by a completely different method: *upward or compositional classification.* In this method one starts at the bottom, sorts species into groups of similar ones, and combines these groups into a hierarchy of higher taxa. The method is, at least in principle, a strictly empirical one. In spite of various controversies (see below), it is by and large the method employed by every modern taxonomist, at least in the initial stages of the classifying procedure.

The adoption of classification by inspection and by grouping, instead of by division, signified a total methodological revolution. Not only was the direction of the classificatory steps reversed, but reliance on a single character (*fundamentum divisionis*) was replaced by the utilization and simultaneous consideration of numerous characters, or of "all characters," as some of the proponents of upward classification have insisted.

In spite of the drastic conceptual difference of the two methods, the replacement of divisional by compositional classification occurred so gradually during the period from the end of the seventeenth to the nineteenth century that apparently no one was fully aware that it was taking place.

There were several reasons for the gradual nature of the change. First of all, the method of classifying items by "inspection" was, of course, not at all a new invention. Already Aristotle had delimited his higher taxa by a combination of characters. If one were asked to sort a basket of mixed fruit, one would have little trouble sorting them by "inspection" into apples, pears, and oranges. Such a preliminary sorting was apparently undertaken by all the early botanists, even by those professing to be practitioners of logical division. It was done quite openly by Bock and Bauhin, and clandestinely by Cesalpino, Tournefort, and Linnaeus. Evidently a certain amount of compositional classification had been incorporated into the divisional method from the very beginning. (Conversely, after logical division had been rejected in principle, some elements of it were retained owing to their usefulness in identification.)

There were several prerequisites for the occurrence of the changeover, but no thorough analysis of its history has so far been undertaken. First an upward classification is possible only if one understands what one is grouping—that is, species. Thus, a pre-

192

requisite of the compositional approach was a knowledge of species, even if essentialistically defined. The early herbalists and other pre-Linnaean authors, who sometimes lumped all species of a genus or treated variants as full species, would have had considerable difficulty with the compositional method. The development of a natural-history tradition in the seventeenth and eighteenth centuries made here a crucial contribution (see Chapter 6). The second prerequisite was the weakening of the hold of essentialism, as described above. Finally, there developed at this period, in part as a result of the decline of essentialism, an empirical attitude in which a greater interest was displayed in results than in underlying principles.

193

Three botanists in the 1680s pioneered in the method of sorting species on the basis of groups of characters. The British botanist Morison used a variety of characters, and Ray stated that in order to draw the right inferences on the essence of a genus, "there can be no more certain sign or evidence than the possession of several common attributes" (*De Variis,* 1696: 13). He repeated this in 1703 by saying, "The best arrangement of plants is that in which all genera from the highest, through the subordinate and lowest, have several common attributes or agree in several parts or accidents" (*Meth. Plant.:* 6–7).

In France, at about the same time, Magnol (1689) rejected the reliance on the method of delimiting higher taxa by division. In order to draw inferences on relationship, he used characters from all parts of the plant, not only the fructification. More significantly, he quite specifically stressed the importance of a holistic approach, that is, the grouping of species "by inspection": "There is even in numerous plants a certain likeness, an affinity that does not consist of the parts as considered separately, but as a whole; an important affinity, but which cannot be expressed" (*Prodromus,* 1689). The special importance of Magnol was that he had a large impact on Adanson, whose ideas he helped to shape. His refusal to classify characters into essential and accidental ones (as demanded by the essentialists), although ignored by Linnaeus, was adopted by Adanson and by the entire empirical school.

Buffon (*Oeuvr. Phil.,* 1749: 13) was quite emphatic in backing classification by inspection: "It would seem to me that the only way to design an instructive and natural method is to group together things that resemble each other and to separate things that differ from each other." He also stressed that one should take all characters into consideration, and this advice was adopted by

Merrem, Blumenbach, Pallas, Illiger, Meckel, and other zoologists (Stresemann, 1975: 107).

The first author who had the intellectual courage to question openly the validity of the method of logical division was Michel Adanson (1727–1806). In his *Les familles naturelles des plantes* (1763), he suggested replacing it by an empirical inductive approach "because the botanical methods that consider only one part, or only a small number of parts of plants, are arbitrary, hypothetical, and abstract. They cannot be natural . . . the only natural method in botany is one that takes into consideration all parts of plants . . . [and this is how] we find the affinity which brings plants together and which separates them in classes and families." Adanson went further than that and developed an elaborate method of testing taxonomic characters.

Adanson and the Use of Multiple Characters

The rejection of the single-character divisional method posed new problems. If the delimitation of groups is to be based on several characters, how many characters should be used, and should there be any preference for certain characters? Adanson was the first botanist who investigated these questions quite systematically. In order to find out what effect on the classification the choice of characters would have, Adanson experimentally prepared 65 artificial groupings of plants, each based on a particular character, such as the shape of the corolla, the position of the seeds, or the presence of thorns. These arrangements showed him that it is impossible to arrive at a satisfactory system based either on a single character or a combination of only two. Since Adanson calculated the proportion of natural groupings produced by each of these arrangements, he has sometimes been called a numerical taxonomist, first by Adrien de Jussieu in 1848. This claim is quite misleading, since Adanson did not use this arithmetic approach in the actual delimiting of genera and families. This he did, following Magnol's example, by the visual perception of groups. Although he first worked out the differences between genera and species, "It was by the overall view [*ensemble*] of these comparative descriptions, that I perceived that plants sort themselves naturally into classes or families" (*Fam. pl.*, 1763: clviii).

Adanson clearly saw that different characters differed in their taxonomic significance. "Giving equal weight to all attributes would have been in logical contradiction to Adanson's inductive method. Such an arbitrary procedure would have meant an a priori eval-

194

uation of the characters" (Stafleu, 1963: 201; see also Burtt, 1966). What Adanson promoted was the potential consideration of all parts of the plant, not merely of the fructification. He particularly stressed two points: (1) that certain characters contribute nothing to the improvement of a' classification and should be ignored, and (2) that the characters with the greatest information content are different from family to family. Each family has its own "génie."

Some of Adanson's opponents criticized him for the peculiar reason that his method required too great a knowledge of plants. This would have been a legitimate criticism if identification were the only object of classification; but, as the history of systematics has proven again and again, satisfactory classifications—classifications based upon a critical evaluation of all the evidence—can be constructed only by those who do have a thorough knowledge of the group concerned. One can summarize Adanson's attitude toward characters by saying that he did indeed favor the weighting of characters, but not on the basis of any preconceived notion or a priori principles (like physiological importance) but rather by an a posteriori method based on a comparison of groups that had been previously established by inspection.

Almost every principle promoted by Adanson has now become part of the taxonomic methodology. Yet, in an age still dominated by Thomistic logic and by a virtually dictatorial Linnaean authority, Adanson was almost totally ignored. It is hard to say how much of an impact his *Familles des plantes* actually had. It was praised by Lamarck, but others who were clearly influenced by it, like A. L. de Jussieu, rather ungenerously failed to mention the source of their ideas. When numerous practicing taxonomists arrived at the same principles in later years and generations, it was independently and empirically rather than by studying Adanson's largely forgotten writings. It was not until almost a century later that Adanson's greatness was rediscovered (Stafleu, 1963).

TRANSITION PERIOD (1758–1859)

The century after the publication of the tenth edition of Linnaeus' *Systema Naturae* (1758) was an era of unprecedented taxonomic activity, much of it engendered by the enormous prestige which Linnaeus had given to the study of diversity. As more and more organisms were discovered, more and more young people became zoologists and botanists. The search for new species and their classification threatened to smother all other interests in biology. For instance, the exciting researches in flower biology of a Kölreuter or Sprengel were ignored, because they did not produce new species. Nägeli (1865), who was not a taxonomist, deplored the fact that all other branches of botany were being drowned by the "stream of systematics."

The enormous accumulation of animal and plant specimens in private and public collections resulted in profound changes in the taxonomic profession. Taxonomists became more professional and more specialized. New journals were founded to accommodate the descriptions of the numerous novelties, and amateurs discovered that they could achieve a high level of competence by specializing in a single family. The annual output of taxonomic research climbed steadily.

The frontiers of taxonomy expanded into entirely new areas. Up to now zoology had been dominated by an interest in the vertebrates and botany by the study of the flowering plants. Now in zoology an interest developed in the invertebrates, particularly the marine ones, and eventually (beginning with Sars) even in deep-sea organisms. The botanists, in turn, paid ever greater attention to the cryptogams.

It was in the period between the publication of Adanson's *Familles des plantes* (1763) and Darwin's *Origin of Species* (1859) that downward classification was gradually replaced by upward classification. France, among the European countries perhaps the one least dominated by essentialism, clearly led in introducing the new methods of taxonomy. This is evident not only from the early pioneering efforts of Magnol, Buffon, and Adanson but also from the writings of Lamarck and Cuvier. Lamarck (1809; 1815), even though still adhering to much obsolete philosophy, classified by grouping rather than by logical division, and Cuvier's principle of the correlation of parts strongly reinforced the trend toward a multiple-character conceptualization of taxa and the search for new characters. This ushered in a new pragmatic tradition in zoological classification, in which characters were evaluated by the ability to make a contribution to the formation of seemingly "natural" groupings; that is, they were evaluated a posteriori. Furthermore, it was recognized that the relative importance of a character (its weight) could change from one higher taxon to another; that is, the taxonomic value of characters is not absolute.

This also led to a reconceptualization of the taxonomic categories. No longer were they considered steps in logical division (from *summum genus* down to the lowest species) but rather as ranks in a hierarchy. The genus now became a collective higher category, thus something entirely different, ontologically and epistemologically, from the essentialistic genus of logical division. This change in the meaning and role of the genus was often ignored by taxonomists and philosophers, resulting in misunderstandings and conceptual confusion.

There was also a subtle shift in the relative importance of different categories. For Linnaeus the genus was the center of the universe. As the genera became larger and larger, owing to the continuing discovery of new species, most genera had to be split again and again, and the emphasis shifted to the next higher rank, the family. In many, but by no means all, groups of organisms, the family became the most stable unit of classification.

197

The shift from downward to upward classification (together with the associated methodological and conceptual changes) was slow, gradual, and irregular, as has been true for nearly all scientific "revolutions." As I have mentioned, the emphasis on families started with Magnol (1689); the use of several characters often derived from different organ systems had already been adopted, more or less hesitantly, by Bauhin, Morison, Ray, Magnol, and Tournefort; Bauhin (1623) was perhaps the first to classify by grouping plants "according to their natural affinities." Yet, all of these authors were inconsistent, particularly since they always wanted their classifications also to serve, to a lesser or greater extent, as identification systems.

One factor to which Stafleu (1963: 126) calls attention quite correctly is that not only practical considerations but also Descartes and his principles helped to undermine the authority of Aristotelian logical division. Adanson, for instance, modeled his own method very much according to the four basic Cartesian methodical rules: doubt, analysis, synthesis, and enumeration. The Cartesian influence, as well as the Newtonian and Leibnizian influences (through Buffon), were among the reasons why Linnaeus had less impact in France than in any other taxonomically active country. His many practical innovations (binominalism, rules of nomenclature, and so on) were, of course, eventually adopted, but his Aristotelianism was accepted only as a convenient method of identification, and not as a sound philosophical basis for classification. Perhaps the most conspicuous development in post-Linnaean taxonomy was that classifications became more and more clearly hierarchical (see below).

Plant taxonomy, which had had such a magnificent flowering in the two hundred years between Cesalpino (1583) and Linnaeus, continued to have steady but unspectacular growth in the post-Linnaean period. Three developments characterized this era. The most important was the endeavor (not yet fully achieved even in our day) to develop a "natural system" of plants. De Jussieu, de Candolle, Endlicher, Lindley, Bentham, and Hooker all contributed more or less successfully to this end. Furthermore, increasing attention was paid among the cryptogams, not only to ferns and mosses but also to fungi, algae, and one-celled aquatic plants (protists). Finally—even though botanists rarely specialized to the extent zoologists did—an age of specialization also arrived in botany, monographs being published dealing with special groups of plants and thus leading to very intensive researches of selected portions of the plant kingdom.

One development that is far too little appreciated was the fact that animal taxonomy during this period became a major branch of academic zoology. Naturalists like Siebold, Leuckart, Ehrenberg, Sars, Dujardin, and many others (one might even include Darwin in this category) started out as taxonomists but, by becoming interested in the living animal as a whole, made major contributions to general zoology, such as the clarification of the life cycle of parasites, the alternation of generations, the sequence of larval stages of marine invertebrates, the structure of internal organs and their function, and almost any other aspect of the living animal. Quite often it can be demonstrated unequivocally that such studies arose directly from taxonomic researches and yet taxonomy has rarely been given credit for its role in initiating new approaches in biology. For instance, it is only within recent years (Ghiselin, 1969) that it has been fully realized how important Darwin's monographic work on the cirripedes was for the maturation of his evolutionary theory.

The Search for a Natural System

Most proponents of downward classification were fully aware of the fact that the classifications produced by their methods were "artificial." Linnaeus deplored in several of his works that the time had not yet come for a truly "natural" classification (as he interpreted it). On several occasions he published fragments of a "natural" classification of plants (Stafleu, 1971) and, no matter how artificial some of his major divisions were, within them he grouped the majority of the genera very much as would a modern evolutionary taxonomist. However, simply replacing downward by upward classification was not sufficient to produce a natural classification. There had to be some organizing principle, some basic concept that would serve as a guideline for the taxonomist.

Ever since the Greeks a conviction had prevailed that the diversity of nature was the reflection of some deeper order or harmony. No other group of philosophers, as far as we can judge from what little of their writing has survived, devoted so much thought to this harmony than the Pythagoreans. Natural theology revived the concept of a harmonious balance of nature and saw signs of it everywhere in the evident "design" of all adaptations. But diversity at first seemed quite chaotic and did not seem to fit too well with this philosophy. The situation became particularly unsatisfactory in the post-Linnaean period when the number of known species and higher taxa of animals and plants was growing at an almost exponential rate. When viewing the almost chaotic mountains of new species, how could one avoid asking, "Where is that harmony of nature of which every naturalist is dreaming? What are the laws that control diversity? What plan did the father of all things have when he designed little creatures and big ones?"

It was simply inconceivable, in a period so strongly dominated by natural theology, that organic diversity could be totally without rhyme or reason, that it could be simply the result of "accident." Consequently, it was the task of the taxonomist to find the laws that regulate diversity or, as others would put it, to discover the plan of creation.

The classification which would reflect this divine plan most perfectly would be "the natural system," and to achieve it was the ideal of every naturalist. Yet, when one studies what various authors had in mind who used the term "natural," one encounters great diversity. A discussion of some of the usages of the term will help us in understanding the thinking of this period. The several meanings are best made clear by giving their respective antonyms.

(1) "Natural" is that which reflects the true "nature" (that is, essence), as opposed to that which is due to "accident." Essentialist classifiers from Cesalpino to Linnaeus attempted to provide classifications that were natural in this sense (Cain, 1958). In principle this was the ideal of Linnaeus and this is what he seems to have had in mind when he expressed his dissatisfaction with his artificial sexual system. Natural did not at all signify to Linnaeus what it means to us, because for him the "nature" of a species, of a

genus, or of a higher taxon was its essence. All Linnaeus scholars agree in this point (Stafleu, 1971; Larson, 1971).

One must never forget Linnaeus' belief that genera and higher taxa, as God's creations, represented unchangeable essences, and that one would truly know them only after one had fully recognized these essences. As stated by Cain (1958: 155), "It is likely that for [Linnaeus] a 'natural' system is one that showed the *natures* of things, and natures meant in practice essences." To realize this helps us to understand his essays on the "natural method" (meaning natural system).

His theory of the origin of classes and genera (in an appendix to the *Genera Plantarum*, 1764) is a strictly creationist one. From all this, it is rather obvious, what Linnaeus really had in mind, when talking about the "natural system": a system in which the intuitive definition of higher taxa (based on overall resemblance) is replaced by a determination of the true essence of these taxa. Among Linnaeus' successors, of course, the term "natural system" gradually acquired an entirely different meaning.

(2) As the power of essentialist philosophy weakened, the term "natural" began to mean that which is rational rather than capricious. This interpretation reflected the widespread eighteenth-century attitude that the order of nature was rational and could be perceived and understood by reasoning. Everything in nature obeys God-given laws and the order of nature conforms to God's plan. The "natural system," if it could be found, would reflect the blue print of creation (Agassiz, 1857).

(3) For still others the term "natural" meant "empirical" as contrasting with "artificial" (that is, purely utilitarian). A natural classification under this conception would satisfy John Stuart Mill's demands: "The ends of scientific classification are best answered when the objects are formed into groups respecting which a greater number of general propositions can be made, and those propositions more important, than could be made respecting any other groups into which the same things could be distributed." Basically, considerations of this sort had been underlying Adanson's endeavors. It is a tradition, initiated by Bauhin, half-heartedly supported by Morison and Ray, but rather definitely advocated by Magnol.

(4) Finally, after 1859, the term "natural" when used to describe a classification system meant "of common descent." A natural classification after Darwin is one in which all taxa consist of the descendants of a common ancestor.

A listing of these various meanings of the term "natural" does not describe exhaustively the conceptual foundations of the classifications proposed during this period, however. The search for a harmony or plan in nature was influenced by some other concepts which we have already, in part, encountered in other contexts. Three concepts, in particular, were popular in successive periods.

Scala Naturae

For centuries the scale of perfection seemed to be the only conceivable scheme to bring order into diversity.[21] Blumenbach (1782: 8–9) was one of many authors who saw in the *scala naturae* the sound basis of a natural system which would permit man "to order the natural bodies according to their greatest and most manifold affinity, to bring together the similar ones, and to remove from each other the dissimilar ones." Lamarck, particularly in his earlier writings, expressed similar sentiments. Among botanists the idea of the *scala naturae* was less popular, since little of a trend toward perfection was discernible among plants, except for the advance from the algae and other cryptogams to the phanerogams. As a result Linnaeus preferred to compare a classification to a map, where each country touches several others.

The claim that there is a continuous sequence from the least perfect atom of matter up to the most perfect organism, man, was challenged increasingly often as more was learned about diversity. Lamarck no longer defended any continuity between inorganic and organic, even though he postulated frequent spontaneous generation. The so-called "zoophytes" (corals, polyps, and so on) were scrutinized with particular care. Were they truly intermediate between plants and animals, and, if not, were they plants or animals? It produced great excitement and no little consternation when Trembley[22] discovered in 1740 that the green hydra (*Chlorohydra viridissima*) was definitely an animal and yet it had chlorophyll and amazing powers of regeneration, a capacity one had hitherto thought was typically restricted to plants. Soon Trembley demonstrated that corals and bryozoans were also animals, rather than intermediates between plants and animals. The great deal of branching which Lamarck admitted in different lines of animalian affinity was also quite incompatible with a single scale of perfection.

It received its final death blow when Cuvier (1812) asserted emphatically that there are four distinct phyla of animals, no more and no fewer, and that there was absolutely no connection among them. It was still sometimes possible, after Cuvier, to recognize affinity among the members of smaller groupings, but the ordering principle of "growing perfection" was no longer applicable. Furthermore, it became ever less convincing to postulate connections between dissimilar groups. The unity of the organic world, at one time symbolized by the *scala naturae,* seemed to disintegrate the more, the better one knew the world of life. When it was realized that a one-dimensional line or guiding principle was inadequate, the search began for multidimensional schemes.

202

Affinity and Analogy

The placing of a group of organisms in the scale of perfection was determined by its *affinity* to less perfect or more perfect neighbors. A modern biologist has a good deal of difficulty in visualizing what "affinity" meant in pre-evolutionary discussions; perhaps it simply meant similarity. Yet, there was a conviction that this similarity reflected some kind of causal relationship, such as represented in the *scala naturae* or as seen by Louis Agassiz in the blueprint of creation.

Some of the difficulties of the *scala naturae* seemed to be due to existence of two kinds of similarity, true affinity and another one which was designated by Schelling, Oken, and their followers as *analogy.* Penguins are related to the ducks or auks by true affinity, but to the aquatic mammals (such as whales) by analogy. Hawks show affinity to parrots and pigeons, but are analogous to the carnivores among the mammals. As bizarre as some of the thinking of the *Naturphilosophen* was, their partitioning of "relationship" into affinity and analogy proved to be of great importance in the ensuing history of biology. It was on this basis that Richard Owen developed his concepts of homology and analogy which henceforth dominated comparative anatomy, particularly after the terms had received an evolutionary redefinition.

But prior to the proposal of evolution, how could one convert this idea of affinity and analogy into a system? Here the *Naturphilosophen* borrowed their thoughts from the Pythagoreans. How can law and harmony in nature be better expressed than in numbers? The entomologist W. S. MacLeay (1819) chose the number 5, and although others subsequently experimented with 3 and 4, 5 remained the most popular number, referred to as the *quinarian* system.[23] MacLeay thought that all taxa were arranged on circles, five per circle, and that adjacent circles were touching each other ("osculating"). Taxa on the same circle showed affinity, relationship to other circles illustrated analogy. One of the tasks of the taxonomist, thus, was to search for such cross connections.

Even though these often quite bizarre schemes were scathingly criticized by the more sober naturalists, one can sympathize with the adherents of quinarianism. They looked for laws that would account for the diversity of nature, and numerical schemes seemed the best available, prior to evolutionism. Even T. H. Huxley was, for a while, enamored with quinarianism and made numerous attempts to arrange the higher taxa of invertebrates in corresponding circles or parallel rows (Winsor, 1976b). Quinarianism was so popular in England in the 1840s and 1850s that even Darwin gave it some thought. For, if organisms were truly arranged in multiples of five, then this would indicate that they had been designed by a superior author and, thus, diversification by natural selection would be refuted. However, even a superficial study showed to Darwin that the facts of taxonomic diversity were incompatible with any of the numerological schemes. In particular his barnacle research provided no evidence for quinarianism.

203

Even most of those who rejected quinarianism could not help recognizing that there were several kinds of similarity. Affinity and analogy were always acknowledged, but mere incidental "resemblance" and other kinds of similarity were sometimes also recognized. Affinity was the most puzzling of these similarities, but there was broad agreement that it was "the direct result of those Laws of Organic Life which the Creator has enacted for his own guidance in the act of Creation" (Strickland, 1846: 356). This is why affinity became for Agassiz one of the most powerful proofs for the existence of the creator.

Pragmatism and Hierarchism

The failure of the *scala naturae,* of the grand schemes of the *Naturphilosophen,* and of the Pythagorean endeavors of the numerologists had a rather sobering effect on taxonomy. In the fifty years before the *Origin* most taxonomists shunned theorizing altogether and, when adopting the principles of upward classification, they were satisfied with the simple pragmatic activity of grouping seemingly similar species and genera together.

There was little conceptual advance in this period. Cuvier, even in his last publication, merely restated the principles he had enunciated twenty years earlier. The situation in botany was no better. A.-P. de Candolle's *Théorie élémentaire* (1813), opposing claims notwithstanding, still adheres to the classical essentialist a priori methods.[24] Yet, virtually unnoticed by the practitioners themselves, the use of single key characters for the establishment of higher taxa was being replaced by the grouping of species (or other lower taxa) into higher taxa on the basis of character combinations. Upward classification was becoming a matter of course. To begin "at the bottom" greatly stimulated the development of specialists (Lindroth, 1973).

204   The result of the new approach was the discovery that many previously recognized taxa were highly heterogeneous. For instance, Meckel (1821) and Leuckart (1848) were able to show that Cuvier's Radiata, distinguished on the basis of the key character "radial versus bilateral symmetry," was an unnatural assemblage of two highly distinct phyla, the echinoderms and the coelenterates. At every level, from the phylum down to the genus, previously recognized higher taxa were reexamined and separated into more homogenous components if found to be unnatural aggregates. By 1859 a large part of the animal taxa had been redefined and restricted to groups of species that largely agreed in their structural and biological characteristics.

The enthusiasm for this theory-free purely pragmatic approach was somewhat dampened when certain phenomena were discovered which undermined too great a trust in phenetic resemblance. Of course, it had long been known that caterpiller and butterfly are the same animal, but with the increased interest in classifying one could no longer avoid asking whether a classification based on caterpillars would be the same as that based on the butterflies into which they metamorphosed. In the first half of the nineteenth century it was discovered that such a metamorphosis occurs in many invertebrate groups; indeed it is normal for most groups of sessile marine groups. From the beginning of any zoological system the barnacles had been classified among the Mollusca or their subdivision, the Testacea. It created a sensation when John Vaughan Thompson observed on May 8, 1826, a crustacean larva attached to the bottom of a glass vessel change into a young barnacle (Winsor, 1969). Further studies left no doubt that the barnacles were sessile crustaceans. Thompson and other students of marine life found, furthermore, that many plankton organisms are nothing but the larval stages of well-known invertebrates, and that even the free-living crustaceans may metamorphose through several larval stages (nauplius, zoea, cypris).

The comforting idea of types that could be arranged either according to the primacy of function (Cuvier) or by the determination of a discrete *Bauplan* (von Baer and the *Naturphilosophen*) was thrown into even more complete disarray by two further discoveries made in the first half of the nineteenth century. One consisted in the complicated life cycles of cestodes, trematodes, and other internal parasites. The stages in the alternation of generations—for example, cysticercus and tapeworm, and cercaria and fluke—are so utterly different although produced by the same genotypes as to cast a great deal of doubt on the validity of a purely phenetic approach to classification. Even more startling was the discovery of a regular alternation of generations in salps (tunicates) by Adelbert von Chamisso (1819), and in coelenterates by Michael Sars (1838–1846) and J. J. Steenstrup (1842). The free-swimming and the sessile generations of a single species are so different that up to this discovery they had been placed into entirely different taxa (Winsor, 1976b; Churchill, 1979). The problem was not unknown to the botanist where, among various groups of cryptogams, the gametophyte and the sporophyte are usually entirely different.

205

Fortunately, these somewhat unsettling discoveries did not lead to another bout of metaphysical speculations but simply inspired the taxonomists to redouble their efforts to bring together "natural" groups of "related" organisms. These efforts, almost automatically, resulted in a classification of subordinated categories, nowadays usually referred to as the Linnaean hierarchy. What is the meaning of the term "hierarchy" in taxonomic theory?

HIERARCHICAL CLASSIFICATIONS

Most classifications, whether of inanimate objects or of organisms, are hierarchical. There are "higher" and "lower" categories, there are higher and lower ranks. What is usually overlooked is that the use of the term "hierarchy" is ambiguous, and that two fundamentally different kinds of arrangements have been designated as hierarchical. A hierarchy can be either *exclusive* or *inclusive*. Military ranks from private, corporal, sergeant, lieutenant, captain, up to general are a typical example of an exclusive hierarchy. A lower rank is not a subdivision of a higher rank; thus, lieutenants are not a subdivision of captains. The *scala naturae*, which so strongly dominated thinking from the sixteenth to the eighteenth century, is another good illustration of an exclusive hierarchy. Each level of perfection was considered an advance (or degradation) from the next lower (or higher) level in the hierarchy, but did *not* include it. The hierarchy of functions, as defended from Cesalpino to Cuvier, is another example of an exclusive hierarchy. That growth has the highest rank and reproduction the next highest in this hierarchy does not mean that reproduction is a subdivision of growth, such as genera are subdivisions of families.

206

The modern hierarchy of taxonomic categories is a typical example of an inclusive hierarchy. To illustrate this with a concrete example, dog-like species, such as wolf, coyote, and jackal, are grouped in the genus *Canis* (dogs); the various dog-like and fox-like genera are combined in the family Canidae. These, together with the cats, bears, weasels, and other related families, are combined in the order Carnivora. Class, subphylum, phylum, and kingdom are successive higher ranks in this hierarchy. Each higher taxon contains the taxa of the lower, subordinate, ranks.

In theory, inclusive classificatory hierarchies can be produced both by logical division and by compositional classification. However, historically, logical division never led to a well-defined taxonomic hierarchy, because each level was treated separately, since each "species" (as defined in divisional logic) became a "genus" at the next lower hierarchical level. And even when both Tournefort and Linnaeus stabilized the genus to a large extent, they recognized only two higher categories and showed little interest in them (see above).

Linnaeus was the first author who used the higher categories rationally and, on the whole, consistently. Nevertheless, his thinking was still too much dominated by the principle of logical division to permit him to propose a fully consistent, inclusive hierarchy of all organisms. This development did not truly take place until the two decades between 1795 to 1815 (Winsor, 1976b: 2–3). The remarkable group of taxonomists at the Paris Museum d'Histoire Naturelle was largely responsible for this conceptual change. However, different authors adopted the new way of thinking to different degrees. For instance, Lamarck's arrangement of the higher taxa (masses) still represented a strictly exclusive hierarchy even though there were compositional, hence inclusive, elements at the level of the lower categories. Cuvier's four phyla (*embranchements*) had no hierarchical connection, or if so, a strictly exclusive one. However, within these phyla, at the level of the lower categories, some inclusive hierarchical tendencies can be observed.

The method of constructing inclusive hierarchies by a compositional procedure is important for a number of reasons. Obviously, a theory of common descent, such as Darwin proposed in 1859, would not occur to anyone except for the existence of the inclusive hierarchy of taxonomic categories. Of more immediate concern is that the new approach raises both practical and conceptual questions concerning the principles on which the construction of an inclusive hierarchy should be based. This was a particularly burning problem, because the thinking of most taxonomists was still influenced if not dominated by an adherence to the *scala naturae,* an exclusive hierarchy.

207

The Reality of the Higher Categories and Taxa

As far back as the seventeenth century, and with more or less vigor ever since, there has been an argument as to the "reality" of the higher categories. Essentialists like Linnaeus insisted dogmatically that at least the genus, characterized by its essence, was a "real" phenomenon. Taxonomists with nominalist leanings, led by Buffon (1749), insisted with equal force that only individuals exist and that at least the higher categories, like genus, family, and class, are only arbitrary conventions of the human mind. The fact that no two botanists of the seventeenth century arrived at the same classification certainly seemed to support the nominalist argument. By making a distinction between the abstract-ideal and the concrete-real, Buffon laid the foundation for a solution, but the controversy continued for another two hundred years.

The reason why it lasted so long was primarily a terminological confusion: The term "category" was used in two entirely different senses. It was not until a new term, the word "taxon," had been introduced for one of the two former meanings of the word "category" that this confusion was eliminated.[25]

A *taxon* is a "group of organisms of any taxonomic rank that is sufficiently distinct to be worthy of being named and assigned to a definite category." In terms of logic, a taxon is an individual, and the individual animals or plants of which the taxon is composed are *parts* of the taxon (Ghiselin, 1975; Hull, 1976). A *category,* in its restricted modern meaning, designates rank or level in a hierarchic classification; it is a class, the members of which are all the taxa to which a given rank is assigned. The difference between taxon and category is best made clear by an example: robins, thrushes, songbirds, Passerine birds, birds, vertebrates, chordates, and animals are groups of real organisms; they are taxa. The rank which is given to the mentioned taxa in the hierarchical classification is indicated by the categories in which they are placed: species, family, suborder, order, class, subphylum, phylum, and kingdom.

208   The question, Are the higher categories real? must thus be dissolved into two separate questions: (1) Are (most of) the groups (taxa) which we rank in the higher categories well delimited? and (2) Is it possible to give an objective (nonarbitrary) definition of such higher categories as genus, family, or order? The answer to the first question is clearly yes, but to the second one it is clearly no. Such taxa as the hummingbirds, anthropoid apes, or penguins are exceedingly "natural" or "real" (that is, well delimited), and yet the categorical rank given to them is subjective, at least for taxa above the species level. A taxon might be placed in the family category by one author, in a lower category (tribe) by a second author, and in a higher one (superfamily) by a third author. In other words, categorical rank is largely an arbitrary decision. Those who so heatedly argued about the reality or not of categories were simply talking about different things. This was clearly understood by some earlier authors (for instance, Plate, 1914), but the distinction continued to be ignored for lack of a terminological handle.

5 ✐ Grouping according to common ancestry

THE EMPIRICAL taxonomists had no causal explanation for the fact that one is able to group species according to "relationship" or "affinity." When Strickland (1840) defined affinity as "the relation which subsists between two or more members of a natural group, or in other words, an agreement in essential characters," he left the key words, "natural and essential," undefined. It was Darwin who filled the explanatory gap and showed why there are natural groups and why they share "essential" characters. It was he who provided the basic theory of biological classification. No one prior to him had stated so unequivocally that the members of a taxon are similar because they are descendants of a common ancestor. To be sure, the idea was not entirely new, and Buffon had played with the possibility that similar species, like horse and ass, or all cats, might have been derived from an ancestral species, and so had Erasmus Darwin and some of the German evolutionists. Linnaeus, in his later years, had suggested that the members of a higher taxon might be the products of hybridization. Yet, neither Buffon nor Linnaeus converted these speculations into either a theory of classification or of evolution. Nor did it have any influence on hierarchical classification when Pallas in 1766 and Lamarck in 1809 and 1815 proposed tree-like diagrams of relationship (Simpson, 1961: 52).

That Darwin was the founder of the whole field of evolutionary taxonomy is realized by only few. As Simpson has rightly said, "Evolutionary taxonomy stems explicitly and almost exclusively from Darwin." By this is meant not only that the theory of common descent accounts automatically for most of the degrees of similarity among organisms (as indeed it does), but also that Darwin developed a well thought out theory with a detailed statement of methods and difficulties.[1] The entire thirteenth chapter of the *Origin* (pp. 411–458 of the 1st ed.) is devoted by him to the development of his theory of classification. It starts with the often quoted sentences: "From the first dawn of life, all organic beings are found to resemble each other in descending degrees, so that they can be classed in groups under groups. This classification is evidently not arbitrary like the grouping of the stars and constellations" (p. 411). Darwin here implicitly rejects the frequently made claim that the classifications, which by 1859 had reached a considerable level of refinement and elaboration, were an arbitrary and artificial product of the taxonomist. He continues:

210

> Naturalists try to arrange the species, genera and families in each class on what is called the Natural System. But what is meant by this system? Some authors look at it merely as a scheme for arranging together those living objects which are most alike, and for separating those which are most unlike . . . But many naturalists think that something more is meant by the Natural System; they believe that it reveals the plan of the Creator; but unless it be specified whether in time or space, or what else is meant by the plan of the Creator, it seems to me that nothing is thus added to our knowledge . . . I believe that something more is included [in our classification], than mere resemblance; and that propinquity of descent—the only known cause of the similarity of organic beings—is the bond, hidden as it is by various degrees of modification which is partially revealed to us by our classifications. (p. 413)

In the *Origin,* and in his correspondence, Darwin stresses again and again that "all true classification is genealogical" (p. 420), but "genealogy by itself does not give classification" (*L.L.D.,* II: 247). To be sure, Darwin believed "that the arrangement of the groups within each class, in due subordination and relation to the other groups, must be strictly genealogical in order to be natural"; but he also realized that this was not the whole story, "but that the amount of difference in the several branches or groups, though allied in the same degree in blood to their common progenitor, may differ greatly, being due to the different degrees of modification which they have undergone; and this is expressed by the forms being ranked under different genera, families, sections, or orders" (*Origin:* 420). This is a very important statement, because it focuses on the principal difference between two modern schools of taxonomy, the cladists and the evolutionary taxonomists, as will be discussed below.

At this point Darwin refers back to his famous phylogenetic diagram (*Origin:* 116) in which each of three congeneric Silurian species (A, F, and I) has modern descendants of very different rank. The line derived from species F changed so little that it is still classified in the Silurian genus, while its two sister groups A and I now constitute different families or even orders (p. 125). In his classification of the Cirripedia Darwin usually applied the principle of determining rank by degree of divergence rather than by proximity to the branching point (Ghiselin and Jaffe, 1973; Mayr, 1974c).

Darwin had spent something like eight years on the classification of the Cirripedia (barnacles) and this had given him great insight into classification, both theoretically and practically (Ghiselin, 1969). This allowed him to draw up a set of recommendations to help the taxonomist in finding what resemblances will be most useful in order to determine "propinquity of descent." In particular, he emphasized again and again the importance of weighting the taxonomic value (information content) of all characters: *211*

> It might have been thought (and was in ancient times thought) that those parts of the structure which determined the habits of life, and the general place of being in the economy of nature, would be of very high importance in classification. Nothing can be more false . . . it may even be given as a general rule, that the less any part of the organization is concerned with special habits, the more important it becomes for classification. (p. 414; see also p. 425)

In particular, Darwin rejects the idea so widespread among seventeenth- and eighteenth-century botanists, and since Cuvier also among zoologists, that the more important a structure is for the survival and perpetuation of an organism, the more important it will also be for its classification. He quotes case after case (pp. 415–416) refuting such an assumption: "That the merely physiological importance of an organ does not determine its classificatory value is almost shown by the one fact, that in allied groups, in which the same organ . . . has nearly the same physiological value, its classificatory value is widely different" (p. 415). He illustrates this by the highly unequal value of the antennae as a taxonomic character in different families of insects.

Darwin's advice does not constitute a denial of the importance of natural selection. What Darwin felt is rather that special adaptations may involve only a limited portion of the genetic endowment of a group and thus be less informative than the general habitus. Furthermore, special adaptations may be acquired independently in several unrelated evolutionary lines; in other words they are subject to convergence. An awareness of these potential shortcomings of special adaptations may help to protect the taxonomist from interpreting convergence as evidence for common descent. Other types of characters, sometimes seemingly very slight ones, are more informative: "The importance for classification, of trifling characters, mainly depends on their being correlated with several other characters, of more or less importance. The value indeed of an aggregate of characters is very evident in natural history" (p. 417). Darwin, thus, was fully aware of the importance of the concordant variation of several characters. After discussing the particular properties of other characters such as embryological, rudimentary, and distributional characters Darwin reaches the following conclusion:

> All the foregoing rules and aids and difficulties in classification are explained . . . on the view that the natural system is founded on descent with modification; that the characters which naturalists consider as showing true affinity between any two or more species, are those which have been inherited from a common parent, and, insofar all true classification is genealogical; that community of descent is the hidden bond which naturalists have been unconsciously seeking. (p. 420)

In the evaluation of characters Darwin proposes certain rules, some of which we have already mentioned. Like Ray, Lamarck, de Jussieu, Cuvier, de Candolle, and most classifiers of the preceding centuries, he emphasizes the high taxonomic weight of characters that are *constant* over large groups. Furthermore, he emphasizes the importance of correlated character complexes, provided they are not merely the result of a similar mode of life. He devotes quite a section to spurious similarities caused by *convergent* evolution (p. 427) and warns the taxonomist not to be misled by such "analogical or adaptive resemblances" (p. 427).[2]

Theoretical discussions of evolutionary classification in the ensuing century have consisted in little more than footnotes to Darwin. None of Darwin's rules or principles has been refuted and none of any special consequence has since been added. Two of Darwin's recommendations are of particular importance. One is to separate similarities due to descent from spurious similarities due to convergence. For instance, a character like the chorda has high taxonomic weight because it is part of a complex character system which could have hardly originated twice independently. On the other hand, metamerism (segmentation) is not nearly as basic a character because there is a great deal of evidence that it has originated at least twice in the animal kingdom. It is highly unlikely that the metamerism of the vertebrates has any phylogenetic connection with that of the arthropods.

Darwin's other recommendation is to "weight" characters. Such an evaluation is important because some characters have much greater information content than others. Phylogenetic weighting as practiced by Darwin is a process of a posteriori weighting. The weight of a character is given by its correlation with the most securely established (by various methods of testing) parts of classifications. Some taxonomists have found it difficult to distinguish it from a priori weighting (as practiced by Cesalpino and Cuvier). However, this can be done by appropriate analysis, and ever since phylogenetic a posteriori weighting has been reemphasized (Mayr, 1959a; Cain, 1959b), it has been found a useful method (Mayr, 1969) and is now merging with computerized weighting methods.

The reason for the highly unequal information content of so-called taxonomic characters has not yet been determined unequivocally, but is believed to be due to the fact that some components of the phenotype are built far more tightly into the genotype than others. The more deep-seated genetically a character or character complex is, the more likely it will be useful in revealing relationship. The work of Schmalhausen, Waddington, and Lerner has shown that its architecture provides the genotype with such a stable integration that certain components of the phenotype may remain unchanged during phyletic divergence. The underlying canalizations and regulatory mechanisms seem occasionally to remain virtually untouched during evolution, and this accounts for the sometimes quite unexpected stability of seemingly trifling components of the phenotype.

As far as the methodology of classification is concerned, the Darwinian revolution had only minor impact. It is evident that the real turning point in the history of taxonomy was the abandonment of essentialism and of "downward classification," and this had been largely completed well before 1859. The decisive contribution which Darwin made to taxonomy was twofold: by his theory of common descent he provided an explanatory theory for the existence of the Linnaean hierarchy and for the homogeneity of taxa in a "natural" classification, and he restored, at least in principle, the concept of continuity among groups of organisms, which had been rejected by Cuvier and by the *Naturphilosophen* in their theory of archetypes. Let us look at some aspects of these contributions more closely.

The Meaning of Affinity

As we saw, the quinarians and various other zoologists and botanists between the 1820s and 1840s had clearly recognized that there were two kinds of similarity among organisms. The similarity of a whale and a terrestrial mammal are due to affinity, the similarity of a whale and a fish are due to analogy. The more perceptive of these students, such as Strickland and Owen, appreciated the fact that analogies were due to similarity of function, but they were quite at a loss to explain affinity, except by invoking the "plan of creation." Darwin solved this problem by simply stating that affinity is proximity of descent. This led to the postulate that all taxa should be composed of descendants from the nearest common ancestor, or to use Haeckel's terminology, that they should be monophyletic. In order to delimit such taxa, it was necessary to scrutinize carefully all similarities and differences in order to discriminate between characters due to common descent (the only ones useful for classification) and analogical (convergent) characters, like hooked bills in hawks and owls or webbed feet in aquatic birds, that were acquired independently owing to similarity of function.

The Restoration of Continuity

The rejection of the scale of perfection by the comparative anatomists early in the nineteenth century had resulted in the recognition of as many unconnected units as there were archetypes (*Baupläne*). To be sure, there was still a search for what was higher or lower as reflected in the advice which Louis Agassiz gave to his students: "Any fact that you may bring to show that one Order is higher than any other is true scientific research." Darwin's interpretation of the Linnaean hierarchy as being due to common descent not only restored the principle of continuity but represented a powerful research program. No one perceived this more clearly than Haeckel, whose ambition it was to connect all taxa of animals and plants on the basis of their descent, and to represent this in the form of phylogenetic trees, which have graced textbooks of systematics ever since. Haeckel was an artist and presented his phylogenies quite literally as picturesque trees. They were increasingly replaced by tree-like diagrams, so-called dendrograms, an early example of which can be found in the *Origin* (Voss, 1952).

The relation between postulated phylogeny and classification has remained controversial from 1859 on to the present. As early as 1863, T. H. Huxley rejected all phylogenetic considerations and demanded that all classifications be based "upon purely structural considerations . . . such classification acquires its highest importance as a statement of the empirical laws of the correlation of structures." Huxley here clearly differed from Darwin, whose principle it was that one cannot make observations without having a theory. The modern trend has been to apply Darwin's principle by asking about each taxon whether the characteristics of the included species indicate that the taxon is monophyletic—that is, by consistently making phylogenetic postulates and then testing whether they are supported by the taxonomic evidence.

For Haeckel there was no doubt that classification had to be based on relationship and that relationship was known as soon as phylogeny was understood. The principal task in classification thus was to develop methods that would reveal phylogeny. Among these methods the one that excited Haeckel and his contemporaries the most was the theory of *recapitulation* (Gould, 1977). It stated, in its most classical form, that the ontogenetic stages recapitulate the adult stages of the ancestors (see Chapter 10). The theory is now known to be invalid, but it was nevertheless a most heuristic theory, having given rise to comparative embryology and having generated many spectacular discoveries. Its greatest triumph was Kowalewsky's demonstration of a chorda in the larvae of the sea squirts (tunicates), thus proving that they are chordates and not mollusks, as had been thought previously. The fact that mammalian embryos have gill arches as their fish ancestors (discovered by H. Rathke, 1825), and many similar discoveries of comparative embryology showed that a modified theory of recapitulation is acceptable, which states that embryos frequently recapitulate embryonic stages of their ancestors. The heuristic value of comparative embryology had already been abundantly demonstrated between 1820 and 1859 by the discovery of the typically crustacean larvae of barnacles (Winsor, 1969), by the elucidation of the life cycle of some parasitic crustaceans, and finally by the demonstration that the brittle starfish *Comatula* is nothing but the head of the crinoid *Pentacrinus*. Indeed, Louis Agassiz's 1848–49 Lowell lectures had as their theme the value of comparative embryology as supplementing comparative anatomy. Since 1836 Agassiz had had the idea that there is a threefold parallelism between fossil history, embryological development, and rank in classification (Winsor, 1976b: 108).

The ultimate result of researches in comparative anatomy and embryology was that one unnatural class or phylum of animals after another was made natural by the removal of unrelated elements, such as the removal of barnacles and tunicates from the mollusks and, most importantly, the breaking up of Cuvier's phylum Radiata (Winsor, 1976b). This process of understanding the relationship of the major groups of invertebrates, which owed its original impetus to Cuvier and Lamarck, probably made as much (or more) progress in the fifty years before the *Origin* as in the fifty years afterwards. Careful morphological analysis contributed more to the recognition and delimitation of natural taxa than phylogenetic theory. Yet, the endeavor to construct phylogenetic trees was the most active preoccupation of zoologists in the second half of the nineteenth century.[3]

The classification of the major taxa of plants was perhaps even less affected by evolutionary theory than that of animals. Unconsciously the principles of the *scala naturae,* a progression from simple (primitive) to complex, continued for a long time the guiding principle of botanists. The classification of the flowering plants (angiosperms) was handicapped in two different ways: First, it was based almost entirely on flower structure, and it is only in the last thirty or forty years that wood anatomy and chemical constituents have been seriously added to the repertory of useful characters. The second handicap was a misconception as to what the most primitive flowers were. It was long assumed that the earliest angiosperms had been wind pollinated and had no petals, and therefore among living families, those like Betulaceae, Fagaceae, and related families (Amentiferae), that are wind pollinated were considered to be most primitive. It is now realized that wind pollination and correlated floral reduction is secondary, and that an entirely different group of families, related to the Magnoliaceae and Ranunculaceae (Ranales), are most primitive. The connecting links between them and the seed ferns, the inferred ancestors of the angiosperms, have not yet been found in the fossil record. The treatment in the leading textbooks indicates not only the level of activity in this field in recent decades but also the remarkable progress that has been made in delimiting reasonably homogeneous, natural taxa.

Perhaps nothing in the nineteenth century influenced the development of botany as profoundly as Hofmeister's researches on the life cycle and reproduction of the cryptogams and the homologies of their reproductive structures. This not only gave the first clear view of relationships within the cryptogams but it broke down the previously insurmountable barrier between cryptogams and flowering plants. Hofmeister's investigations clearly established that a more or less uniform plan of organization ran through the entire vegetable kingdom. His *Comparative Studies of the Cryptogams* (1851) laid a sound foundation for an establishment of a phylogeny of the cryptogams after 1859. Once the characteristics of the various groups had been clearly worked out, it was a relatively easy matter to order them with the help of the principle of common descent.

Soon the main attention was directed toward the study of the variation in reproduction in various groups of cryptogams and to relationships within these groups. None of them needed clarification as much as the highly heterogeneous fungi. The great pioneer in these studies was Anton de Bary (1831–1888), who summarized his numerous detailed analyses of the life cycle of various groups of fungi in 1866 and 1888, laying a sound foundation for the active work of his successors. The importance and uniqueness of the fungi is recognized by various recent investigators by separating them from the plants in a separate kingdom.

THE DECLINE OF MACROTAXONOMIC RESEARCH

After the 1880s there was a gradual but noticeable decline of interest in macrotaxonomy and in phylogenetic studies. This had numerous reasons, some internal to the field and some external. Most important perhaps was a feeling of disappointment over the difficulty of getting clear-cut results. Similarity is usually a reasonably accurate indicator of relationship where the classification of taxa below the rank of orders is involved. In the classification of the higher taxa (orders, classes, and phyla) similarity is no longer a reliable guide and disappointingly little progress was therefore made. It comes as rather a surprise to most nontaxonomists how uncertain our understanding of degrees of relationship among organisms still is today. For instance, it is still unknown for most orders of birds which other order is a given order's nearest relative. The same is true for many mammalian families and genera, for instance the Lagomorpha, Tubulidentata, Xenarthra, and *Tupaia*.

Yet these uncertainties in the classification of higher vertebrates are very minor compared to those of the invertebrates, the lower plants, and most of all, the prokaryotes and viruses. When one reads recent discussions on the classification of the lower invertebrates one is struck by the fact that some of the same questions are still controversial that were argued about in the 1870s, 80s, and 90s. There are usually majority opinions, but the mere fact that unorthodox alternatives have vigorous proponents indicates the degree of still-prevailing uncertainty. To provide a flavor of the kind of problems that are controversial I might pose some questions: From what group of protozoans did the metazoans evolve? Do all metazoans have a single protozoan ancestor, or did the sponges evolve separately? Are the mesozoans, the coelenterates, or the turbellarians the most primitive metazoans above the sponges? Is the division of the higher invertebrates into Protostomia and Deuterostomia a natural one? To which of these two groups (if they are recognized) do the Tentaculata (lophophorates) belong? How sound is the archicoelomate theory?

Many problems concerning the relationship of the taxa of arthropods are also still unsolved, and likewise the derivation of the arthropods from the annelids. Kerkut (1960) quite rightly has called attention to these uncertainties, of which of course no one is better aware than the specialists in the field. This being a history of ideas, it is impossible even to begin giving a history of the sequence of classifications for the various higher taxa of animals and plants that have been proposed in the last two hundred years. Yet it is a fascinating story.[4] In each generation new hopes were raised by new principles (such as recapitulation) or newly discovered characters, but progress has been slow.

The futile attempts to establish the relationship of the major phyla of animals induced at least one competent zoologist at the turn of the century to deny common descent. Fleischmann (1901) called the theory a beautiful myth not substantiated by any factual foundation. Kerkut, fifty years later, does not draw such an extreme conclusion but he is almost equally pessimistic about ever achieving an understanding of the relationship of the higher animal taxa. Honesty compels us to admit that our ignorance concerning these relationships is still great, not to say overwhelming. This is a depressing state of affairs considering that more than one hundred years have passed since the great post-*Origin* period of phylogeny construction. The morphological and embryological clues are simply not sufficient for the task.

A second reason for the post-Darwinian disenchantment with macrotaxonomy was conceptual confusion. When Haeckel and his followers insisted that only those classifications were natural that were based on the phylogeny of the groups concerned, his opponents objected by saying, How do we know the phylogeny? Is it not true that phylogenies are deduced from the findings made during the establishment of classifications? Hence, how can we base classifications on phylogeny without getting involved in a hopelessly circular argument? It is only in relatively recent years that the argument was resolved. Neither is phylogeny based on classification, nor classification based on phylogeny. Both are based on a study of "natural groups" found in nature, groups having character combinations one would expect in the descendants of a common ancestor. Both classification and phylogeny are based on the same comparison of organisms and their characteristics and on a careful evaluation of their similarities and differences (Mayr, 1969). Evolutionary taxonomists are now in agreement that biological classifications must be consistent with the inferred phylogeny. This conceptual clarification has opened the way for a renewed interest in the classification of higher taxa.

219

The other causes for the decline of interest in macrotaxonomy after 1900 were external. Owing to the claim of the Mendelians that mutations make new species, most of the excitement in taxonomy shifted to microtaxonomy (the "species problem"), eventually culminating in the new systematics. With subspecies widely considered to be incipient species, many specialists, particularly students of birds, mammals, butterflies, and snails, devoted all of their attention to the description of new subspecies. Concentration on the species level also revealed an endless supply of still undescribed species, all of it contributing to a neglect of macrotaxonomy.

Perhaps the most important factor in the decline of macrotaxonomy was growing competition from other branches of biology. With the exciting discoveries made in experimental biology (*Entwicklungsmechanik*, cytology, Mendelian genetics, physiology, biochemistry) most of the brightest young biologists moved into these areas. This resulted in a manpower shortage in taxonomy and in reduced institutional support for this branch of biology.

Among 29 papers given at the symposium "Systematics of Today" held at Uppsala in commemoration of the 250th anniversary of the birth of Linnaeus in 1957 (Hedberg, 1958), only four dealt with problems of macrotaxonomy. This illustrates well the dominant interest in the species level, characteristic for most taxonomists in the first half of the twentieth century. Nevertheless, a low-level interest in the classification of higher taxa continued throughout the period, and a number of significant papers dealing with the problems and concepts of classification were published, such as those of Bather (1927), Simpson (1945), Rensch (1947), and Huxley (1958). By the 1960s the task of the new systematics in microtaxonomy was largely completed (at least as far as the development of concepts is concerned) and the time had come for a new preoccupation with macrotaxonomy.

The Need for a New Methodology

The splendid start which Darwin had made in developing a theory and methodology of macrotaxonomy was largely ignored in the post-Darwinian period. The standards by which genera, families, and orders were recognized and combined into still higher taxa were highly uneven in different groups of organisms. In the more poorly known groups, single-character "classifications"—or, more correctly, identification schemes—were still very much in vogue. Since different authors might choose different key characters, the same kind of controversies developed as had characterized botany in the seventeenth century. Taxonomists frequently proposed new classifications without adequate justification, except the claim that they were "better." For Linnaeus the names of the higher taxa were meant to serve as aids to the memory, but this objective was completely lost sight of by those zoologists and botanists who split genera and families into ever smaller pieces. This went so far that in birds, for instance, certain authors in the 1920s and 30s recognized a separate genus for nearly every species. There were no standards in the application of the higher categories. One distinguished ornithologist placed the families of birds in 25 orders, another equally distinguished author placed them in 48 orders. To anyone looking at macrotaxonomy from the outside, let us say from applied sciences such as medicine, agriculture or ecology, the situation in taxonomy looked rather chaotic, as, in fact, it was.

However, the situation was not all black. There were at least some useful textbooks of the theory and practice of zoological systematics (Ferris, 1928; Rensch, 1934; Mayr, Linsley, and Usinger, 1953). Scattered through the literature were occasionally incisive discussions of certain aspects of the theory of classification, for instance, by Mayr (1942: 280–291) on the meaning of the genus, and, more importantly, by Simpson (1945) on the theory of macrotaxonomy. Perhaps the outstanding constructive development of the period was to give an ecological meaning to the higher taxa. It was recognized that higher taxa are composed of species which, in the aggregate, occupy a specific niche or adaptive zone. In other words, the primacy shifted from the morphological character by which a higher taxon is recognized, to its biological meaning in the household of nature. Nevertheless, as far as the average biologist was concerned, classification offered serious problems, to put it mildly. The new systematics, by largely concentrating on the species level, had provided no solution for the needs of macrotaxonomy. Help had to come from somewhere else.

221

Independently, two drastically different solutions were proposed: numerical phenetics and cladistics. Both new methodologies were not proposed as reforms but rather as revolutionary replacements of existing procedures.

NUMERICAL PHENETICS

Part of almost any method of classification is the ordering of items into groups on the basis of similarity. However, the construction of biological classifications by the method of a posteriori weighting as exercised by empirical taxonomists from Adanson on, and theoretically justified by Darwin, requires considerable knowledge and experience. Quite legitimately, the question was therefore occasionally raised whether it should not be possible to develop a method by which even a totally inexperienced person, in fact even a nonbiologist, could group species into "natural" genera and higher taxa. Indeed, the availability of an automatic and objective method would be advantageous even for the experienced taxonomist in cases where two or more authors disagree on the optimal classification. The essential element in such an approach would be the development of methods that permit one to quantitize degrees of similarity and to convert a qualitative or subjective taxonomy into an objective, numerical taxonomy.

The history of numerical taxonomy has not yet been written. Pioneering endeavors go back to the middle of the nineteenth century, although most of them deal with intraspecific variation, particularly geographic variation. Attempts to use numerical methods for the classification of species, genera, and still higher taxa are usually buried in the taxonomic literature and are known only to a few specialists. Some references to this literature can be found in Simpson, Roe, and Lewontin's *Quantitative Zoology* (1960).

One of the almost forgotten pioneers was the geneticist A. H. Sturtevant (1939; 1942). He took considerable precautions to escape bias and excluded from his calculations any characters known to be adaptationally or developmentally related. In an analysis of 39 characters in *Drosophila* he was able to order the 58 species into a number of groups of relatives and, more importantly, to arrive at several rather broader generalizations that have since been confirmed repeatedly. The first of these is that strictly phenetic methods are most reliable when applied to close relatives but tend to give contradictory results when applied to distantly related forms. Furthermore he was able to arrive at a definition of *good characters* by constructing a table showing the correlation between the different characters and discovering that those were the "best" which could serve as useful indicators of the probable nature of other characters, in other words, characters that were co-variant with other characters.

After the electronic computer had been invented it occurred independently to three groups of taxonomists—C.D. Michener, and R. R. Sokal (1957) in Kansas, P. H. A. Sneath (1957), a bacteriologist in London, and A. J. Cain and G. A. Harrison (1958) in Oxford, England—to propose computer methods for quantifying similarity and for the grouping of species and higher taxa with the help of such quantitative methods. The most important aspect of their proposals was to replace the integrating capacity of the human brain, which had achieved in classical taxonomy the grouping of taxa simply by inspection or by a tabulation of similarities, by the mechanical operations of the computer. This, they believed, would replace the arbitrary and subjective evaluation customary in the past with an objective and invariably repeatable method. At first all three groups agreed that all characters be given equal weight, but Cain and Harrison (1960) soon recognized the different information content of different characters and therefore proposed "phyletic weighting." Michener soon also distanced himself from his early proposal, but the two remaining pioneers, Sokal and Sneath, joined forces and presented their methodology and philosophy in 1963 in a classical treatise, *Principles of Numerical Taxonomy*. The label "numerical taxonomy" was misleading for, as Simpson and others have pointed out, numerical methods had been used in taxonomy for generations, and by the most diverging schools of taxonomy. It has therefore become customary to refer to the taxonomic methodology of Sokal and Sneath as *numerical phenetics*. Unfortunately, the new methods were promoted by making sweeping claims that could not be subsequently substantiated. For instance, it was claimed that any two scientists, given the same set of characters but working entirely independently, would arrive at identical estimates of the resemblance of two organisms when using the new phenetic methods. The evident invalidity of these claims evoked considerable hostility among experienced traditional taxonomists.[5] A drastically revised second edition (1973) of Sokal and Sneath includes many improvements. Other texts in numerical taxonomy are Jardine and Sibson (1971) and Clifford and Stephenson (1975). For a somewhat different approach see also Throckmorton (1968).

223

As Darwin had already pointed out, different characters have different information content, and rather different classifications will result when different mixes of characters are chosen. Different parts of the body, different stages in the life cycle, and morphological versus biochemical characters all lead to different similarity estimates. In order to demonstrate their objectivity, the numerical pheneticists proposed to reject species as the unit of classification and replace them by "operational taxonomic units" (OTUs), as if this were an improvement. Actually this led them into the same practical difficulties which had caused the abandonment of the typological species concept. Either the pheneticists had to treat the different sexes, age classes, and morphs as different OTUs and then separate females and males and other strongly varying phenotypes into different taxa, or else they had to analyze biological variants (phena) very carefully and combine variants into OTUs which coincide with biological species. Such an evaluation of variation, although far more realistic, required precisely the kind of subjective judgments that were claimed to be excluded by the "objective" phenetic method.

By far the most important difference between traditional taxonomists and numerical pheneticists consists in their attitude toward weighting. There are three and only three possible attitudes toward weighting. According to the first, all characters are equivalent, that is, they have the same importance in classification. Although the pheneticists have referred to this as a "nonweighting" method, it is of course a method of a priori weighting to give each character equal weight. This is as misleading as the a priori weighting methods of Aristotle, Cesalpino, and Cuvier. Whether or not a marine invertebrate has a chorda or not is of greater taxonomic value than a hundred other characters. That some characters contain a great deal of information concerning relationship while others are merely "noise" was pointed out already by Adanson more than two hundred years ago (*Fam. pl.*, 1763, I: clxvii).

The second option concerning weighting is that there exists a set of fixed criteria, for instance physiological importance, by which one can determine the relative taxonomic importance of different characters. This was the a priori method of Aristotle and Cuvier. The third option is a posteriori weighting, in which organisms are first ordered into seemingly natural groups (through a consideration of numerous characters and character combinations) and then those characters are given the greatest weight which seem to be correlated with the most natural groups. This was Darwin's approach; he concluded, "The importance for classification of trifling characters mainly depends on their being correlated with several other characters of more or less importance" (*Origin:* 417).

Throughout the history of taxonomy virtually all experienced taxonomists have known and frequently emphasized how different the taxonomic value of different characters can be. A classification of anthropoid apes and hominoids based primarily on the structure of the cortex of the brain would be appreciably different from a classification based on the major physiological macromolecules (hemoglobin, and so forth). The younger generation of numerical taxonomists appreciates the striking differences in information content of various characters and is now concentrating on replacing intuition and subjective evaluation by objective methods of a posteriori weighting (through correlation analysis, for example).

Pheneticists convert the totality of the similarity values of individual characters into a single "phenetic distance" or overall similarity value. But, as Simpson (1964a) has said correctly, "A single measure of similarity involves an enormous loss of information, mainly on the character direction and origin of differences." It is fundamentally unsound to quantify similarity in a comparison of entities as highly heterogeneous as the character complexes of different taxa. This is why phenetics has been called a typological method, and why Simpson concluded that phenetics has led "to retrogression in taxonomic principle . . . a conscious revival of pre-evolutionary, 18th century, principles."

One might ignore the conceptual weaknesses of the method if it was able to produce practical results. However, to compensate at least in part for mosaic evolution and for the noise introduced by characters without information content, pheneticists need to program very large numbers of characters (preferably more than a hundred). It is usually possible to find such large numbers of characters in groups as morphologically complex as the arthropods (insects, arachnids, and so on). But there is a serious scarcity of taxonomically useful characters in most other groups of organisms. This alone precludes the use of a method based on the unweighted analysis of characters. Furthermore, even in insects the method is very laborious, and the programming of a hundred or more characters of a large number of taxa is exceedingly time-consuming. One of the pioneers of phenetics, C. D. Michener, for this reason went back to traditional methods when classifying a large collection of Australian bees (containing numerous new species).

225

Now, some twenty-five years after the phenetic philosophy was first pronounced, it is possible to give a tentative verdict on the validity and usefulness of its methods. Clearly, the basic approach in all classification is phenetic, endeavoring to establish groups of "similar" entities. The success of this endeavor depends on the methods and principles by which the similarity is established. Numerical phenetics has clearly failed in this respect by its insistence on equal weighting and by its decision to ignore all phyletic information.

However, the failure of the basic philosophy of numerical phenetics is no reason not to acknowledge the usefulness of many of the numerical and particularly multivariate methods developed and employed by the pheneticists. The methods pioneered by the numerical taxonomists are now widely used in many areas of science and in other fields where the sorting and classifying of data is important. To have advocated and introduced into taxonomy the use of such methods would seem to be the greatest contribution of numerical phenetics. Another one is the emphasis on the principle, traditionally endorsed by the best taxonomists, of using as many different characters and character systems as feasible and as yield new information.

Phenetic methods are most useful in the grouping of species in large genera and in the classification of previously confused groups. On the other hand, I do not know of a single substantial contribution made by numerical phenetics to the classification of any mature groups, or to classification at the level of orders, classes, or phyla.

The most promising future development of numerical phenetics is presumably the further development of weighting procedures. These are either based on the correlated variation (co-variation) of characters or on various empirical guidelines. The inferred descendants of a common ancestor are almost invariably recognizable by the joint possession of certain characters, and it is nothing but common sense to give greater weight to some characters than to others. Any approach to classification that does not make use of the weighting of characters is clearly inefficient.

In the endeavor to be "strictly objective" the phenetic schools refrain from taking any evidence from descent into consideration. To do so is the chief characteristic of an opposing school of classification, cladistics.

CLADISTICS

The proposal of the taxonomic school known by the name "cladistics," was motivated by the same consideration as phenetics, to eliminate subjectivity and arbitrariness from classifying by developing a virtually automatic method. Its first comprehensive statement was published in 1950 by Willi Hennig in his *Theorie der phylogenetischen Systematik*. According to him, classifications ought to be based exclusively on genealogy, that is, on the branching pattern of phylogeny. Phylogeny, he asserts, consists of a sequence of dichotomies, each representing the splitting of a parental species into two sister (that is, daughter) species; the ancestral species is assumed to cease to exist at the time of the dichotomy. Sister groups must be given the same categorical rank, and the ancestral species, together with all of its descendants, must be included in a single holophyletic taxon.

Hennig's work is written in rather difficult German, some sentences being virtually unintelligible. It nowhere refers to the writings of Huxley, Mayr, Rensch, Simpson, and other authors who had covered in part the same ground in the preceding decades. New terms and definitions are casually introduced, but there is no index that would guide one to the relevant pages. Not surprisingly, the volume was at first rather universally ignored, except by a few German authors. It did not become more widely known until in 1965 and 1966, when English versions of Hennig's methodology were published. By the 1970s a virtual Hennig cult had developed, even though some of his so-called followers have moved rather far from Hennig's original principles.

Although Hennig had designated his method as *phylogenetic systematics*, it was based only on a single component in phylogeny, the branching of lineages. It was therefore renamed by other authors as cladistics (or cladism), under which name it is now generally known.

The crucial aspect of cladistics is the careful analysis of all characters in the comparison of related taxa and in the partitioning of these characters into ancestral (plesiomorph) and uniquely derived (apomorph) characters. Branching points in the phylogeny are determined by the backwards tracing of uniquely derived characters (synapomorphies) because such apomorph characters are believed to be found only among the descendants of the ancestor in which the character first occurred. The method is said to permit the reconstruction of phylogeny without fossil evidence, and this claim has indeed been substantiated to some extent.

227

Ever since Darwin it has been the objective of the evolutionary taxonomist to recognize only monophyletic taxa, that is, taxa exclusively composed of descendants of a common ancestor. Groups presumed to be monophyletic are constantly tested against ever new characters to determine whether or not the postulate of monophyly is substantiated. This method is noncircular, as was shown by Hull (1967). The careful comparison of species and genera included in a higher taxon and the analysis of all similarities in order to determine whether they are truly homologous had indeed established by 1950 that the majority of the recognized taxa of animals (this was less true for plants) were monophyletic. However, Willi Hennig was the first author to articulate explicitly the principle that branching points in the genealogy should be based exclusively on synapomorph characters. It is the joint possession of uniquely derived characters which proves the common ancestry of a given set of species, he said.

Cladistic Analysis

In principle the method of cladistic analysis for the delimitation of monophyletic groups is a superb procedure. It spells out objective criteria for the establishment of the community of descent. It forces the careful analysis of all characters, and introduces a new principle of character weighting, that of joint possession of synapomorph characters. Groups sharing the same synapomorphies are sister groups. However, a number of objections have been raised against cladistic analysis.

The first is a problem of terminology. Hennig introduced a considerable number of new technical terms, most of which were unnecessary (though "plesiomorph" and "apomorph" are still commonly in use). Also confusing was Hennig's attempt to transfer well-established terms to entirely different concepts—for instance, to restrict the term "phylogeny" to the branching component of phylogeny, to define "relationship" strictly in terms of proximity to the nearest branching point, and, worst of all, to shift the term "monophyletic" from its customary use as a designation of a taxon to the process of descent. From Haeckel to 1950 the sequence of operations had always been first to delimit a taxon on the basis of phenetic considerations and then to test whether it was monophyletic. The cladists simply combine all inferred descendants of a given species into a "monophyletic" taxon, even if they are as different as birds and crocodiles.

A second problem is the difficulty of determining synapomorphy. The joint possession by two taxa of a derived character can have either of two causations. Either the character was derived from the nearest common ancestor (genuine or homologous synapomorphy) or it was acquired by *convergence* (nonhomologous or pseudoapomorphy). The reliability of the determination of monophyly of a group depends to a large extent on the care that is taken to discriminate between these two classes of similarities. The frequency of nonhomologous "apomorphies" has been underestimated by many cladists. How often a given, even seemingly rather improbable, adaptation can be acquired independently is well illustrated by the evolution of eyes. Photoreceptors originated at least forty times independently in the animal kingdom, and in another twenty cases it cannot be determined whether the eyes found in related taxa were patristic or convergent developments (Plawen and Mayr, 1977). This and many other cases (see Gingerich, 1979) illustrate how difficult it often is to partition synapomorphies into those that are homologous and those that are not. The independent loss of a character in separate lineages is a particularly frequent form of convergence.

Another formidable difficulty in the determination of synapomorphies is the establishment of the direction of evolutionary change, that is, the determination which character state is ancestral and which is derived. For instance, the placement of apetalous genera and families of angiosperms depends on the decision whether the absence of flower petals is considered a primary (ancestral) or derived condition; or, to take a case from the animal kingdom, the tunicates can either be considered as primitive, and the Acrania (*Amphioxus*) and the vertebrates as being due to neoteny (reproduction by larvae) or else *Amphioxus* can be considered as the ancestral condition and the tunicates (Ascidians) as a specialized, secondarily sessile side branch. The systems of animals and plants are riddled with situations where the arrangement of higher taxa depends entirely on the reading of the direction in   *229* which evolution is believed to have occurred. Cases where the evolutionary direction was reversed are particularly troublesome, but far more common than is generally admitted.

There are various methods, including a study of the fossil record, by which either one or the other polarity can be made probable, but the fact remains that an unequivocal determination of the ancestral condition is at present often impossible.

The results of the cladistic analysis are recorded in a branching diagram designated as a *cladogram*. It consists of a series of dichotomies depicting the successive splits of the phyletic lines. Two assumptions in the construction of cladograms are strictly arbitrary. The first one is that every existing species is eliminated when a new species originates, and the second one is that every splitting event is a dichotomy. With the realization that most speciation events occur in small, isolated founder populations, it becomes obvious that such speciation is of no influence whatsoever on the genetics and morphology of the parent species, which may continue essentially unchanged for millions of years and may continue to bud off new daughter species at frequent intervals. Strict dichotomy is likewise an unrealistic assumption. A large taxon, having attained a grade-like stage, may produce simultaneously several specialized daughter lines, which although technically sister groups, may go their own separate ways, having no more in common than the derivation from the same parental taxon. In some recently constructed cladograms this is recognized, and certain dichotomies have been replaced by polytomies (Ashlock, 1981). For all these reasons Hull (1979) has correctly stressed that the claim of many cladists that their method is completely objective and not arbitrary is not substantiated by the facts. This is important to remember in connection with the criticism by the cladists of the weaknesses of competing taxonomic methodologies.

Cladistic Classification

The various difficulties of the cladistic analysis are also encountered by the traditional taxonomists and are not the major reason for their opposition to cladistics. Rather, it is the relation between cladistic analysis and cladistic classification. For a cladist his task is completed with the cladistic analysis. The reconstructed genealogy, represented in the branching diagram (cladogram), gives him the classification directly. The cladistic classification reflects the branching pattern accurately and permits reading off the phylogeny of a group quite directly. If the only information one wants to retrieve from a classification is the sequence of branching points of the phylogeny, then a cladistic classification is the answer. If one wants more of the history of a group reflected in the classification, one will look for a method which does not completely disregard evolutionary divergence and autapomorph characters.[6] The cladist disagrees with Darwin's insight that genealogy by itself does not give classification. He delimits his taxa not by similarity but by the principle of holophyly, that is, by uniting all the descendants of a common ancestor into a single taxon. This results in such incongruent combinations as a joint taxon for crocodiles and birds or for chimpanzee and man. Classifications are based entirely on synapomorphies, even in cases, like the evolution of birds from reptiles, where the autapomorph characters vastly outnumber the synapomorphies with their nearest reptilian relatives.

In other words, the method ignores the fact that phylogeny has two components, the splitting of evolutionary lines and the subsequent evolutionary changes of the split lines. The reason why this latter component is so important for classification is that the evolutionary history of sister groups is often strikingly different. Among two related groups derived from the nearest common ancestor, one may hardly differ from the ancestral group, while the other one may have entered a new adaptive zone and have evolved there into a drastically novel type. Even though they are "sister groups" in the terminology of cladistics, taxonomists traditionally have ranked such groups at different categorical levels. Nothing illustrates better the difference between cladistics and traditional taxonomy than Hennig's decree that sister groups must be assigned the same rank regardless of how greatly they differ in their divergence since their separation.

For Hennig, cladistics was "phylogenetic classification" and it was his endeavor (even though the methodology was unsuited for this objective) to represent phylogenetic evolution in his classifications. This interest is not shared by some of his followers, who either refrain from any reference to evolution and phylogeny or even deliberately deny that evolution should be reflected in a classification. For more extensive critiques of cladistic classification I must refer to the recent literature.[7]

Let me attempt to do justice to the cladistic method in a few final words. The greatest virtue of cladistic analysis is that it is an effective method for testing the "naturalness" (that is, the monophyly) of groups originally delimited by phenetic methods. Since similarity of species and genera may have various reasons, monophyly can be confirmed only by a rigorous analysis of the homology of the characters on which the similarity is based.

231

In order to appreciate how fundamental the impact of Hennig's methodology has been, one needs only to look at recent taxonomic revisions, particularly those relating to fishes and certain groups of insects. Even many authors, like Michener, who reject the claim that the cladogram can be translated directly into a classification carefully attempt to use the principle of synapomorphy in the delimitation of taxa. The cladistic analysis has been particularly successful where numerous characters are available and where the existing classifications have been rather immature. The new cladograms it has produced have shown that many previously accepted taxa were polyphyletic. The translation of the detailed analysis into a strictly corresponding classification, as for instance by Rosen (1973) for the higher teleosteans, leads, however, to a proliferation of new usages of previously existing taxa names and to the coining of many new names, and, even more disturbingly, to the introduction of many new categorical levels. The objection that such a procedure is clearly incompatible with the ideal of a convenient classification is rejected by Bonde (1974: 567) as not being a "valid argument against Hennig's theory." It certainly is not against cladistic analysis, but very much so against cladistic classification.

Perhaps the most important contribution of Hennig's method is that it has helped to clarify the relation between phylogeny and classification. Simpson, Mayr, and other taxonomists had equivocated when discussing the relationship between phylogeny and classification. Angiosperm taxonomists, discouraged by the difficulties of reconstructing the phylogeny of the flowering plants, disagreed with the conclusions of the zoologists that taxa should be consistent with the findings of phylogeny and that the higher taxa must be monophyletic units (Mayr, 1942: 277–280). Davis and Heywood write in their textbook: "Classification, taxonomists claim, should be based on, or reflect, phylogeny. This aim, we believe, is unrealistic in a group with an extremely inadequate fossil record . . . Indeed, the whole conception of phylogenetic classification is, we believe, a mistake" (1963: xviii). What these authors overlooked is that the fossil record is equally inadequate for most groups of animals and that phylogeny must be inferred in all these groups. It is Hennig's great merit to have articulated a methodology that permits such inferences and permits their repeated testing. Thus, the absence of fossils does not preclude the establishment of phylogenies. So far as I know, the entire accepted phylogeny of the orders of mammals had originally been based on comparative anatomical research (via homology) and in no case was the previously established phylogeny refuted by later fossil finds.

The seemingly eternal argument whether classification should express phylogeny, or should be based on phylogeny, or should be consistent with phylogeny, or should have nothing whatsoever to do with phylogeny is now beginning to become clarified. It is evident that in both classification and phylogeny one proceeds according to the hypothetico-deductive method. This means that one must test a series of propositions: (1) that the members of each taxon are each other's nearest "relatives" (that is, most similar to each other); (2) that all members of a taxon are descendants of the nearest common ancestor (monophyly sensu stricto); (3) that the Linnaean hierarchy of taxa is consistent with the inferred phylogeny.

There are numerous ways of testing each of these propositions, all of them ultimately going back to an analysis of homology. In the study of homology it is most important "to distinguish between definition and the evidence available and used to determine whether the definition can be taken as applicable" (Simpson, 1975: 17; also 1961: 68–70). After 1859 there has been only one definition of homologous that makes biological sense: "A feature [character, structure, and so on] is homologous in two or more taxa if it can be traced back to [derived from] the same [a corresponding] feature in the presumptive common ancestor of these taxa."

Numerous authors have contributed to the assembling of criteria, helping us to decide whether or not the definition is met in a particular case. As far as morphological characters are concerned, the best such enumeration of criteria is that of Remane (1952). Some of his criteria (for example, position in relation to other structures) are not applicable to behavioral or biochemical homologies; in fact, it may be necessary to develop a different set of criteria as evidence for homology for each type of characters. It was therefore unfortunate, and quite inappropriate, that Remane raised the criteria that served as *evidence* for homology to the *definition* for homology.

233

THE TRADITIONAL OR EVOLUTIONARY METHODOLOGY

Both phenetics and cladistics have recruited numerous followers. Nevertheless, the majority of taxonomists, even though adopting one or the other methodological advance from the two new schools, have retained the traditional methodology of classifying. It consists in attempting to represent in the classification not only the branching of phyletic lines but also their subsequent divergence. This can be done by indicating in the ranking of the various taxa whether or not they have become drastically different from their sister groups owing to invasion of a new niche or adaptive zone. This results in the conversion of the cladogram into a phylogram (Mayr, 1969). This school is sometimes referred to as evolutionary taxonomy, because in its philosophy it follows Darwin almost to the letter. It is also sometimes referred to as eclectic taxonomy, because in its methodology it makes use of newly developed methods such as certain numerical methods of phenetics and the ancestral-derived partitioning of characters of cladistics. The methods and principles of evolutionary taxonomy are well described in the textbooks of Simpson (1961) and Mayr (1969), as well as in a number of essays by Bock, Ghiselin, Michener, and Ashlock.[8]

The main difference between this method and cladistics is in the considerable weight given to autapomorph characters. These are derived characters acquired by one sister group but not by the other. Since the number of characters which birds have acquired since branching off from the archosaurian branch is many times as large as the number of characters which distinguish archosaurians from the other reptiles, birds are recognized as a separate class of vertebrates rather than combined with the crocodilians (only surviving archosaurians) into the same class or order. Likewise, fleas are recognized as a separate order or suborder of insects, even though they are clearly derived from one particular subdivision of the dipterans; lice (anoplurans) are recognized as a separate higher taxon even though clearly derived from one group of mallophagans, which in turn are derived from one group of psocopterans. In these cases, and all others where one sideline ("ex-group") has acquired numerous autapomorphies by becoming drastically modified through special adaptations, a purely cladistic treatment leads to a misleading picture of relationships, as this term is traditionally understood (Kim and Ludwig, 1978). The ranking of a taxon in the evolutionary method, thus, is based on the relative weight of the autapomorphies as compared to the synapomorphies of the sister groups.

234

As Rensch, Huxley, and others have emphasized, the anagenetic component of evolution often leads to the development of definite "grades," or levels of evolutionary change, which must receive recognition in classification. The objection raised by the cladists that this would introduce subjectivity into classification has been rejected by the evolutionary taxonomist with two arguments. The first is that the cladistic method likewise is replete with subjectivity owing to decision making as to the polarity of evolutionary change, owing to mosaic evolution, and owing to decisions concerning evolutionary parallelism (Hull, 1979). The second counterargument is that in most cases it is not too difficult to calculate an approximate ratio between autapomorphies and synapomorphies of two sister groups. Whenever a clade (a phyletic lineage) has entered a new adaptive zone, resulting in a drastic reorganization, the transformation may have to be given greater taxonomic weight than the proximity of joint ancestry. The particular importance of the autapomorphies is that they reflect the occupation of new niches and new adaptive zones, which often are of far greater biological significance than the cladistic synapomorphies.

The concept and term "grade" have a long history. Ray Lankester (1909) spoke of Protozoa and Metazoa as successive grades, and, after separating the sponges as Parazoa, he divided the remaining Metazoa into two grades, Enterocoela (Coelenterata) and Coelomata. Bather (1927) made extensive use of the grade concept, and attempted to illustrate how certain phyletic lines passed through several grades in successive geological periods. More recently, Huxley (1958) showed how useful the grade concept is in illustrating evolutionary developments and serving as the basis for the ranking of taxa. Rensch and Simpson have also called attention to the existence of such evolutionary plateaus on which a good deal of speciation (cladogenesis) may occur without any significant anagenesis.

The cladist ignores the existence of grades because this approach condones the recognition of "paraphyletic" taxa. A monophyletic group is "paraphyletic," in the terminology of the cladists, if it is not "holophyletic," that is, if it does not include all of the descendants of the common ancestor. The class Reptilia, for instance, as traditionally recognized, is a paraphyletic group, because it does not include the Aves and the Mammalia, two groups that were separated as having reached a grade level differing from that of the remaining Reptilia. The recognition of paraphyletic groups prevents the automatic translation of a classification into a branching pattern but is able to express degrees of divergence, something the cladogram cannot do.

235

NEW TAXONOMIC CHARACTERS

The backbone of all three methods of classification—numerical phenetics, cladistics, and evolutionary taxonomy—is the analysis and evaluation of taxonomic characters. An insufficiency of informative characters is the usual reason why disagreements between competing classifications cannot be resolved. Not surprisingly, the most frequent complaint made by a taxonomist is that the group of animals or plants on which he is working does not supply sufficient characters to allow an unequivocal decision on relationship. Two phenomena in particular contribute to this difficulty. One is the well-known fact that the phenotype in certain groups of organisms is remarkably "standardized," as in the hundreds of species of *Rana* or the thousands of species of *Drosophila,* and thus provides only few morphological clues to relationship. The other is that any deviation from this standard type usually affects only a single functional complex, correlated with some special ad hoc adaptation. A shift to a new source of food or the adoption of a new set of courtship signals may perhaps result in a noticeable morphological reconstruction that can be divided into a considerable number of individual characters. To count these as separate characters would, however, be quite misleading, since, phylogenetically speaking, they are merely reflections of a single functional shift. Darwin, already, warned against too much trust in ad hoc specializations (*Origin:* 414).

A difficulty encountered by the taxonomist even more frequently consists in a conflict between the conclusions based on different structures. The study of the extremities, for instance, may indicate that taxon *b* is most clearly related to taxon *a,* while features of the intestinal tract suggest that taxon *b* is the nearest relative of taxon *c.* In such a case an evaluation of additional features of the extremities or the intestinal tract rarely produces a satisfying solution.

236

One encounters numerous such cases in every higher taxon, and taxonomists have, therefore, in recent decades given much attention to the search for new taxonomic characters. Even though a careful morphological analysis continuously reveals new characters, nonmorphological characters play an increasingly important role in the establishment of classifications. This includes components of behavior, life history, and annual cycle (compare Aristotle's live-bearing vs. egg-laying), physiology, ecology (for example, niche utilization), parasites, and any other conceivable attribute of an organism. Many of these characters are most useful in the discrimination of species, but some are indicative of the relationship of groups of species.

Geographical distribution often provides unexpected clues, as already Darwin intimated. An aberrant Australian genus is far more likely to be related to an indigenous Australian family than to have its nearest relatives in Africa or South America. This principle of searching for the nearest relative in a geographically easily accessible area does not work in the case of some relicts and groups with unusually great dispersal facilities, but it does work in remarkably many instances, as has been documented by Simpson for animals and by Thorne for plants. A combination of a cladistic and a biogeographical analysis, as Hennig and his followers have shown, is sometimes particularly revealing.

What only a few decades ago was the most recent frontier of taxonomy, the study of biochemical characters, has now become one of its most active and most useful branches.[9] It began, shortly after 1900, with immunological studies (Nuttall, 1904). Immunological methods are still in use today (Leone, 1964) but a large battery of new methods has since been added. More specifically, what is studied is the distribution, variation, and evolution of molecules. Relatively small molecules, like alkaloids and saponins in plants, have often a rather restrictive taxonomic distribution and thus may be indicative of relationship (Hegnauer, 1962; Hawkes, 1968). In the case of larger molecules one can study their evolution by various methods, particularly by the determination of changes in the amino acid sequence. Such differences can often be quantitized and used to construct dendrograms of phenetic distance. The study of individual macromolecules, such as hemoglobin, lysozyme, or cytochrome-c, are expensive in time and equipment, however, and depend, for broader use, on methods of automation. Biochemical methods are most useful where a morphological analysis has failed or at least has only been able to produce ambiguous results. The analysis of alleles of enzyme genes (isozymes) by the method of electrophoresis has been particularly productive (Ayala, 1976). Not only have numerous sibling species been revealed by this method, but the amount of difference between two species has been shown to be roughly correlated with the time at which the evolutionary lines leading to the two species separated. The electrophoresis analysis, if based on a sufficiently large number of gene loci, is therefore most valuable in providing an independent check on the results of morphological analysis. The method of DNA hybridization progresses directly to the genotype. In this method the compatibilities of a large part of the genomes of two species are measured, and the degree of matching indicates quite directly the nearness of relationship in those cases in which certain technical difficulties have been overcome. Single molecular characters are, of course, as susceptible to convergence as are morphological characters. It is therefore as dangerous to construct molecular single-character classifications as it is to construct morphological single-character classifications.

238 *237*

Molecular methods are needed desperately to discover the connecting points of the higher taxa, let us say the orders of birds or the phyla of invertebrates. Here morphological analysis has so far failed, because it has been impossible to find a sufficient number of clearly homologous characters and because the polarity of evolutionary trends is often uncertain.

The results of the morphological and the molecular analysis are not always congruent, as a comparison of man and chimpanzee shows. Some authors have, therefore, suggested that it might be necessary to have two classifications, one based on morphological and one on molecular characters. Such a suggestion seems to be ill-advised for a number of reasons: not only is there the probability that different molecular characters might require different molecular classifications, but the suggestion also implies that we have several phylogenies, which clearly is wrong. Classifications are not classifications of separate characters but classifications of whole organisms. It will be the task of future synthesis to integrate the findings made on morphological, behavioral, and various kinds of molecular characters into a single optimal classification.

THE EPISTEMOLOGY OF CLASSIFICATION

Philosophers have traditionally shown a considerable interest in the principles of classification. Indeed, classification (though not the classification of organisms per se) was one of Aristotle's central concerns (see Chapter 4). The replacement of downward classification (logical division) by grouping (upward classification) in the post-Linnaean period was a major philosophical advance, and philosophers in the nineteenth century continued to be very much interested in classification—for instance Mill, Jevons, and the Thomists. Yet somehow philosophers failed to draw the necessary conclusions concerning biological classification which the Darwinian revolution had made possible. Almost unanimously they continued to support essentialism and various other concepts made obsolete by evolutionary thinking. For instance, they tended to confuse identification and classification, and referred to classification as a process involving individual specimens, while actually classification deals with populations (species), while individual organisms are merely assigned to species (that is, are identified). Even today some philosophers (Hempel, 1965) seem to think "that classification consists in the dividing of large classes into sub-sets" (downward classification), though in fact evolutionary classifications operate through the grouping of related taxa into higher taxa.

By far the most serious deficiency in the approach of most philosophers has been the assumption that "the classification of animals and plants . . . is essentially similar in principle to the classification of inanimate objects" (Gilmour, 1940: 465). The methodology of phenetics is based on the same assumption. Unfortunately, it is not valid. Artificial or arbitrary classifications are legitimate for objects that are classified strictly on the basis of some quality or characteristic, like books in a library. Definite constraints, however, exist for the classification of items about which explanatory theories exist (Mayr, 1981). This is true, for instance, for causal classifications of diseases as well as for the classification of organisms that is based on the theory that the relationship of organisms is due to common descent. Indeed, it is impossible to arrive at meaningful classifications of items that are the product of a developmental history unless the historical processes responsible for their origin are duly taken into consideration. To classify kinds of stars, geological phenomena, components of human culture, or biological diversity on the basis of Gilmour's advice produces similarity classes which in most cases reflect the actual relatedness of the items only incompletely. As a consequence, ever since Darwin it has been agreed among evolutionary taxonomists that natural taxa must be monophyletic, in the classical usage of this word, that is, they must consist of descendants from a common ancestor. This theoretical basis of all biological classification is a powerful constraint and completely refutes the claim that theories of classification are equally applicable to inanimate objects and organisms. Members of the younger generation of philosophers (such as Beckner, Hull) are cognizant of these developments and are in the process of developing a philosophy of biological classification in collaboration with those biologists who have given most thought to the relation between evolutionary theory and classification, such as Simpson, Mayr, and Bock.[10]

FACILITATION OF INFORMATION RETRIEVAL

The conclusions on relationship at which the evolutionary taxonomist arrives represented by a phylogram, places equal weight on the exact positioning of branching points and on the degree of divergence (that is, the number of autapomorphies) of each phyletic lineage. It is this phylogram which is used by the evolutionists as the basis of their generalizations.

A classification, however, has a second function: to serve as a key to the information stored in the system. In order for the classification to serve as a maximally useful information retrieval system, a number of aspects of a classification must be considered when translating the phylogram into a classification, aspects known under the terms "rank," "size of taxa," "symmetry," and "sequence." Some arbitrariness is involved in the determination of each of these aspects and they will therefore presumably be controversial forever.

Ranking

Rank in the Linnaean hierarchy is indicated by the category in which a given taxon is placed. Rank determination is one of the most difficult and arbitrary decision processes in classification. For the cladist, rank is automatically given by the nearest branching point of the phylogenetic tree, since sister taxa must have the same rank. The evolutionary taxonomist, by contrast, must decide what number and weight of autapomorphous characters justify a difference in rank between two sister groups; such a decision becomes particularly difficult when the evidence from different kinds of characters is conflicting. A molecular taxonomist, for instance, might place *Pan* (chimpanzee) and *Homo* in the same family, owing to the similarity of their macromolecules, while Julian Huxley proposed to raise man to the rank of a separate kingdom (Psychozoa), owing to the unique characteristics of man with respect to his central nervous system and its capacities. There are no firm rules as to how to resolve such conflicts, except to say that one should look for overall balance in the system, and adopt a scale of ranking that will permit the most useful generalizations.

Size

There is even more discordance among taxonomists concerning the optimal size of taxa. Some authors consider even relatively minor differences as justifying the recognition of new genera, families, and higher taxa. They are referred to as "splitters" in the taxonomic jargon. The majority of taxonomists prefer rather large, comprehensive taxa, as being better able to express relationship and reducing the burden on the memory. They are referred to as "lumpers." The conflict between lumpers and splitters has been with us since the days of Linnaeus, who himself was a lumper. He was able to cope with organic diversity by using, in addition to the kingdom, only four levels in the hierarchy of categories (species, genus, order, and phylum). Today even relatively conservative taxonomists recognize 21 categorical ranks (Simpson, 1945). Where Linnaeus recognized only 312 genera for all animals, the modern zoologist knows more than 100,000, including 2,045 for birds alone. As a general rule one can say that most taxonomic groups pass through a phase of rather intensive splitting when they are studied more actively, but that the splitting phase is reversed when the knowledge of the group reaches greater maturity. There is broad agreement that the function of a classification to serve as an index to an information-retrieval system

240

imposes constraints on the size of the taxa and the number of ranks in the hierarchy.

The pheneticists are the only recent taxonomists who have made a serious effort to introduce some uniformity and nonarbitrariness into the ranking of taxa. By using various distance measures, either based on morphological characters (Sokal and Sneath, 1973) or on genetic distance (Nei, 1975), they have proposed absolute degrees of difference which qualify (or do not) a species group for generic separation. When the distance measure has a sufficiently broad basis (such as DNA matching or the isozymes of at least 30 or 40, preferably far more, gene loci), it may reflect rather well the amount of evolutionary divergence of the various species groups. There are indications that standards of generic recognition, based on degree of molecular difference, should be different in different higher taxa, if degrees of morphological divergence are in strong conflict with the molecular divergence. Morphologically exceedingly similar species of frogs and toads may display remarkable molecular divergence, while in groups like birds and mammals pronounced divergence in morphology and color pattern may not reflect any major molecular divergence. If a uniform molecular yardstick were adopted, many long-recognized genera of warmblooded vertebrates would have to be synonymized, while in the anurans and in the gastropods new genera would have to be introduced for morphologically very similar species groups. It is doubtful whether this would be desirable considering the primary function of classifications.

Symmetry

The problem of symmetry is one created by evolution rather than by taxonomists. Ideal symmetry would exist if all taxa at every categorical level were of equal size. For the quinarians, the ideal number was five. The thought that all taxa should have approximately the same number of species first arose when natural theology still dominated the thought of naturalists. The problem was first treated by A. von Humboldt, later by von Buch, and in 1835 by an anonymous entomologist (*Ent. Mag.* 2: 44–54, 280–286) whose article drew Darwin's attention. To have taxa of highly uneven size seemed too capricious to be worthy of the planning of the creator. Unfortunately, evolution (together with extinction) is indeed so capricious. There are whole orders of animals with only a single species, and numerous genera, particularly among insects, with over 1,000 species. It is now evident that rates of speciation, as well as of survival, are highly unequal in different areas of the natural system.

Sequence

Perhaps the most intractable problem in classification is that of converting the phylogenetic tree into a linear sequence. As long as one thought that there was only a single scale of perfection, the task was simple in principle. As Lamarck stated it, one starts with the least perfect organisms and ends with the most perfect. When Cuvier destroyed the *scala naturae,* he found in the subordination of characters a new sequencing criterion. He rejected any continuity between the four animal phyla recognized by him; nevertheless, their ranking according to the development of the central nervous system clearly dictated a sequence. The basic idea of the scale of perfection was thus still maintained. The acceptance of evolution had singularly little influence on the theory of taxonomic sequencing. The language of the *scala naturae* was simply given an evolutionary ring. The "more perfect" organisms became the "more highly evolved" or simply "the higher" organisms. Virtually all classifications of animals and plants are based on the explicit or implicit principle that the more primitive or lower organisms are placed first, and the higher ones later. In due time, however, considerable soul-searching began, concerning the meaning of the word "higher." Why should fish be higher than the honeybee? Why should mammals be higher than birds? Is a parasite higher or lower than the free-living form from which it was derived?

As the study of animal and plant relationships matured, it became ever clearer that neither a scale of perfection nor even a simple phylogenetic tree correctly describes organic diversity. Rather, most groups of organisms are best visualized as highly complex phylogenetic bushes, with numerous equivalent branches, each of them beginning with rather simple, primitive ancestors and terminating with rather complex and specialized descendants. The fact of adaptive radiation makes the establishment of a truly logical theory of taxonomic sequencing impossible. In large parts of the natural system it is impossible to demonstrate that one particular taxonomic sequence is superior to other alternatives. As a result, there has been an increasing tendency to adopt purely practical criteria that would aid in information retrieval (Mayr, 1969).[11] The most important principle is to retain any widely accepted sequence unless it is clearly demonstrated that it had brought unrelated taxa together. The continuing arguments in the taxonomic literature concerning the "best" sequence of the orders of angiosperms or of the families of song birds show that even these minimal constraints are not sufficient to ensure stability; and yet a linear sequence is a practical necessity. Specimens in collections are arranged in a linear sequence, and so is the printed word in all revisions, catalogues, and reviews.

CURRENT STATE AND FUTURE OF SYSTEMATICS

243

Considering the fact that taxonomy is the oldest branch of biology, its current vigor is quite remarkable. It manifests itself in the founding of new journals explicitly devoted to taxonomy (*Taxon, Systematic Zoology, Systematic Botany,* and so on), to a whole series of major texts, to numerous international symposia, and to an ever-growing annual bibliography. There is activity on many fronts, not merely on the methodology of taxonomy. The mere description of new species is an unending business. What is more surprising is the number of major new types discovered or at least recognized in recent decades. For instance, the new phylum Pogonophora was described as late as 1937 and Gnathostomulida even more recently (1956). The only surviving coelacanth, *Latimeria,* was discovered in 1938, the primitive mollusk *Neopilina* in 1956, the ancient crustacean group Cephalocarida in 1955. Just about all we know about the rich interstitial fauna of marine sands and muds was discovered in the last fifty years. That *Trichoplax* is the most primitive metazoan was realized only in the 1970s.

Perhaps the most astounding discovery is represented by the pre-Cambrian fossils described by Barghoorn, Cloud, and Schopf. They pushed the history of life back from about 650 million years to about 3.5 billion years ago. But discoveries are sometimes made simply by a more careful study of existing fossils, as shown by the recent description of the Agmata, an extinct phylum of invertebrates from the early Cambrian.

Indicative of the vigor of recent taxonomic research are the improvements in the classification of higher taxa in all groups of organisms from the bacteria, fungi, and protozoa up to the vertebrates, including the primates. The old controversy, whether the polyp or the medusa is the ancestral form of the coelenterates, has been clarified by many recent researches with the polyp now having far more supporters than the medusa. The Scyphozoa seem to have more ancestral characters than any other class of coelenterates, and the newly recognized class of Cubozoa (Werner, 1975) connects them very nicely with the Hydrozoa. In the plants the work of Thorne, Carlquist, Cronquist, Stebbins, and Takhtajan have led to a complete reclassification of the angiosperms. The number of higher taxa of unknown or at least uncertain relationship is, however, still very large, and even more progress is to be expected in the two or three next decades than in the preceding ones.

244

As far back as Linnaeus, and even earlier (Aristotle and Theophrastus), two kingdoms of organisms were recognized, Plantae and Animalia. Fungi and bacteria were called plants. The more the study of one-celled organisms and microbes has advanced in recent times, the more the artificiality of this classification has been recognized. First of all, it was realized that the blue-green algae (better called cyanobacteria) and bacteria differ radically from all other organisms, and have therefore been segregated as prokaryotes (Stanier and van Niel, 1942). They lack an organized cell nucleus and complex chromosomes and differ from the remaining organisms (eukaryotes) in most of their macromolecules. There is great diversity (metabolically and otherwise) among the bacteria, but even the most divergent (and apparently most primitive) group of the bacteria, the Methanobacteria, have so many characteristics in common with the other bacteria that they are best combined with them in the kingdom Monera.

The fungi are now also rather generally separated as a separate kingdom from the plants, from whom they differ not only in metabolism (no photosynthesis) but also in cellular structure (always haploid) and in other ways. Whether or not to recognize still another kingdom (Protista) for the one-celled animals and plants, as is advocated by some authors, is a matter of taste. Since the literature on protozoans and one-celled algae is rather separate from that of metazoans and metaphytes, such a separation might facilitate information retrieval. Such questions of the best structuring of the classification of all organisms have been discussed by Margulis (1981).

Among the many reasons for the steady advances in the classification of organisms, the improvements in methodology are foremost. It is now realized that classification is not a one-step procedure, and that therefore simplistic methods rarely lead to satisfactory results. Classifying consists of a whole series of steps

(Mayr, 1981) and at each of these steps different procedures are needed and are most useful. For instance, phenetic methods are most useful in the first tentative delimitation of taxa, and again in the ranking of taxa on the basis of degrees of patristic and cladistic differences. Cladistic methods are most useful in the testing of inferred branching patterns (cladistic analysis).

The question, to what extent numerical methods are useful and indeed superior to the human computer is not yet settled. Most morphological characters are so riddled with convergences, polyphyly, and mosaic evolution that they are very vulnerable as raw material for numerical analysis. Convergence and polyphyly occurs also in the evolution of macromolecules, and presumably in that of DNA, but there are indications that certain changes in macromolecules pose such strong constraints on the subsequent evolution of these molecules as to suggest that molecular similarities, if sufficiently large numbers of molecules are evaluated simultaneously, are more reliable than an indiscriminate morpho-phenetic analysis, as originally proposed by numerical phenetics.

245

THE STUDY OF DIVERSITY

The terms "taxonomy" and "systematics" were generally considered as synonymous during the first half of this century. If asked what the tasks of systematics are, the taxonomist would have answered, "To describe the diversity of nature (meaning: to describe the species of which the diversity exists) and to classify it." And yet as far back as the days of Leeuwenhoek and Swammerdam in the seventeenth century it had become evident that the study of organic diversity comprised more than the description and classification of species. Already then (and as a matter of fact as far back as Aristotle) it was apparent that the study of diversity did not exhaust itself with these elementary preoccupations of the taxonomist. From the very beginning the study of diversity included the analysis of stages of the life cycle and of sexual dimorphism. When living animals were studied in nature, it was also found that different species occurred in different habitats, preferred different foods, and had different behaviors. But it was not until the middle of this century that the great importance of the study of diversity was fully realized, in the wake of the new systematics and the evolutionary synthesis. It then became apparent that the traditional definition of the function of systematics was far too limited, and failed to reflect the true situation.

As a consequence Simpson (1961) made a clear terminological distinction between taxonomy and systematics. He retained the term "taxonomy" in its traditional meaning, but gave to "systematics" a much broader scope, defining it as "the scientific study of the kinds and diversity of organisms, and of any and all relationships among them." Systematics thus was conceived as the science of diversity and this new broadened concept of systematics has been widely adopted. The new definition raised at once the question as to what should be included among the functions of this broadly conceived science of diversity, and what role it should play in contemporary biology.

246

Taxonomy, narrowly defined, remains the backbone and foundation of the entire science of systematics. To make a complete inventory of the existing species of animals and plants and to order them in a classification seems to be a never-ending task. A specialist of the taxonomy of mites (Acarina), nematodes, spiders, or of some neglected group of insects or marine invertebrates can still productively spend his entire life doing nothing but describing new species and assigning them to appropriate genera. The diversity of organic nature seems to be virtually unlimited. At the present time about 10,000 new species of animals are described annually, and even if we accept the lowest of the estimates of undescribed species, it would take another two hundred years to complete the task of simply describing and naming all existing species.

A curious aspect of taxonomy is the far-reaching autonomy of its different branches. Depending on the degree of maturity in the knowledge of a group, the methods and concepts applied in each group show different degrees of sophistication. Indeed, one can find, in various specialized areas of contemporary taxonomy, each of the different stages in the conceptual struggles of taxonomy from Linnaeus and Cuvier right up to the new systematics. Even today, for instance, there are some authors for whom the word "classification" simply means an identification system. Polytypic species taxa are universally accepted in ornithology but are unheard of in many other areas of animal taxonomy. The independence of the various groups is well illustrated by the fact that zoologists, botanists, and bacteriologists each have their own different codes of nomenclature.

Diversity is one of the two great aspects of organic nature, the life processes being the other one. Yet this importance of the study of diversity was not always recognized. As a result, systematics has had its ups and downs in the history of biology. At the time of Linnaeus it virtually monopolized the field, and it received another boost during the post-Darwinian period of phylogeny construction (from Haeckel and his followers). But, in part as a reaction to the excesses of the preceding period, there followed a period of neglect, if not suppression of the study of diversity. When one looks at the writings of Max Hartmann (general biology), Hans Driesch, T. H. Morgan, Jacques Loeb, and other experimental biologists, one would never guess that the study of diversity was also an important and flourishing field of biology. Part of this neglect was deserved, since the work of those who studied diversity at this period was usually excessively descriptive (as in synecology and most of taxonomy) or one-sidedly concentrated on phylogenetic problems (comparative anatomy, the ethology of Heinroth and Whitman). When students of diversity were interested at all in more general problems, the reconstruction of the common ancestor often seemed to be their ultimate objective.

At the present time we lack any useful historical analysis that would tell us when and how this situation changed. It is evident, however, that in the 1920s, 30s, and 40s, new developments took place. There are many indications that population systematics was the entering wedge. It led, in Russia, to the origin of population genetics in the work of Chetverikov (1926; see Adams, 1972; Mayr and Provine, 1980). Population systematics culminated in the new systematics (Rensch, Huxley, Mayr), which in turn contributed decisively to the evolutionary synthesis (Mayr, 1963). The spread of evolutionary thinking, and, more particularly, of population thinking, led to a new conceptualization in paleontology (Simpson, 1944; 1953), in evolutionary morphology (Davis, 1960; Bock, 1959), in ecology (Lack, MacArthur) and in ethology (Lorenz, Tinbergen). Questions relating to diversity as well as a comparative approach based upon diversity played a dominant role in all these developments.

The new emphasis on diversity drastically affected the conceptual climate of entire branches of biology. For decades, for instance, evolution was described as the change in gene frequencies in populations. This reductionist definition limited evolutionary biology to the modification of existing species, that is, to the adaptational component of evolution. The origin of diversity was neglected as if it were not even part of evolution. The same attitude was displayed in the writings of most paleontologists up to the 1950s and 60s. Simpson (1944; 1953) and other contemporary paleontologists confined themselves almost entirely to the treatment of the vertical component of evolution and omitted the problems of the *origin* of diversity even in the discussion of adaptive radiation. Not until 1972 (Eldredge and Gould) was the origin of diversity given proper attention by paleontologists. In evolutionary morphology the study of diversity also led to new concepts. The single-minded interest in the common ancestor (through the study of homologous similarities) is now being replaced by an interest in the origin of differences among the descendants, that is, by an interest in diversity. It seems that the same is happening in ethology, although there the development is still very much at its beginnings.

248

Seeing the extent to which the influence of a modern attitude toward diversity has pervaded all of organismic biology, it may be worthwhile to study in more detail the specific contributions made by systematics. This is also necessary to refute the widespread impression of outsiders that systematics is nothing but a kind of glorified stamp collecting. There has been a tendency to credit some of the most important contributions of systematics to some neighboring field such as population genetics, ecology, or ethology, even when the advances were actually made by practicing taxonomists and had been made possible only through the experience they had gained as taxonomists. It is quite misleading to limit the labels "taxonomy" or "systematics" to purely clerical, descriptive operations, and to give a different label to the broader findings and concepts that emerge from the more elementary descriptive operations.

It must be remembered that in the beginning (seventeenth and eighteenth centuries) systematics and natural history were very much a single field. Most now-recognized branches of organismic biology developed out of systematics. A large part of ecology deals with the interactions of different species, whether they be competition, symbiosis, or predator-prey relationships. The nature of these interactions cannot be understood except through a close study of the interacting species. Almost all the work of the MacArthur school in ecology deals with diversity, and so does the study of ecosystems. Since much of animal behavior is species specific, and since most of the evolution of behavior comes from the comparison of different species, it is again obvious how closely the study of diversity is integrated with this branch of biology. There are many branches of biology that depend entirely on systematics. This includes biogeography, cytogenetics, biological oceanography, and stratigraphy. I shall not again dwell on the indispensability of systematics for applied sciences such as public health, agriculture, and conservation.

As important as systematics is to the foundation of the mentioned branches of organismic biology, it is perhaps even more important through its contribution to the conceptual broadening of modern biology. The greatest unifying theory in biology, the theory of evolution, was largely a contribution made by systematics. It is no coincidence that Darwin wrote his *Origin of Species* after encountering taxonomic problems during the voyage of the *Beagle* and after eight years of concentrated work on barnacle taxonomy. Taxonomists also supplied the major clues to the solution of many individual evolutionary problems, including the role of isolation, the mechanisms of speciation, the nature of isolating mechanisms, rates of evolution, trends of evolution, and the problem of the emergence of evolutionary novelties. Taxonomists (including paleotaxonomists) more than any other kind of biologists have made significant contributions to all these subjects.

Taxonomists were active participants in the evolutionary synthesis (Mayr and Provine, 1980). Most of the authors who most successfully integrated genetics with the major problems of evolution, like Chetverikov, Rensch, Dobzhansky, Mayr, and Simpson, were taxonomists by background.

Environmental physiology owes much to systematics. Zoological systematists, like C. L. Gloger, J. A. Allen, and Bernhard Rensch, have made major contributions to the discovery of adaptive geographic variation and to the establishment of climatic rules. It was zoologists with taxonomic competence who demonstrated the genetic basis for adaptive differences among geographic races.

Perhaps the greatest contribution made by the study of diversity is that it helped to develop new approaches to philosophy. It was the study of diversity more than anything else which undermined essentialism, the most insidious of all philosophies. By emphasizing that each individual is uniquely different from every other one, the students of diversity focused attention on the role of the individual; this in turn led to population thinking, a type of thinking that is of the utmost importance for the interaction of human subgroups, human societies, and human races. By showing that each species is unique and thus irreplaceable, the student of diversity has taught us a reverence for every single product of evolution, one of the most important components of conservation thinking. By stressing the importance of the individual, by developing and applying population thinking, and by giving us a reverence for the diversity of nature, systematics has supplied a dimension to man's conceptualization that had been largely ignored if not denied by the physical sciences; and yet it is a component which is crucial to the well-being of human society, and for the planning of the future of mankind.

6 ✒ Microtaxonomy, the science of species

THE ENTITIES which the taxonomist assembles into genera and still higher taxa are the species. They are the basic kinds of living beings that make up the diversity of nature. They represent the lowest level of genuine discontinuity above the level of the individual. The song sparrow and the fox sparrow are different species and so are the red oak and the pin oak. The entity designated by the term "species" would, at first sight, seem obvious, simple, and easily defined. But this is not the case. There is probably no other concept in biology that has remained so consistently controversial as the species concept.[1] One should have thought that the animated debate of the post-Darwinian period would have produced clarity and unanimity or, at least, that the new systematics of the 1930s and 40s would have brought final clarity, but this was not the case. Even today several papers on the species problem are published each year and they reveal almost as much difference of opinion as existed one hundred years ago. The advance that has been made is that the nature of the disagreement is much more clearly formulated than in earlier periods. What is particularly interesting for the student of ideas is that the history of the species problem is, to a large degree, quite independent of the history of the problem of classification. The branch of systematics dealing with the species problem can be designated as microtaxonomy, and its history will be treated in this chapter.

When one speaks of species, one ordinarily has species of plants and animals in mind. Actually the term is often applied to all sorts of objects, in the sense of "kinds of." The chemist may speak of species of molecules and the mineralogist of species of minerals (Niggli, 1949; Hooykaas, 1952). Yet, the species concept in chemistry and mineralogy is fundamentally different from that of contemporary biological systematics. For a species name in mineralogy is on the whole a class name, defined in terms of a set of properties essential for membership in the class. Species of inanimate objects, thus, correspond more or less to the Linnaean or pre-Linnaean species but not in the slightest to the modern biological species.

But even if we limit our attention to species of organisms we encounter a great diversity of viewpoint, in part because the species category performs different functions in different branches of biology. For the working taxonomist the species taxon is the elementary "kind" that needs to be identified and classified; for the laboratory biologist it is the organism that has definite species-specific characters with respect to physiological, biochemical, or behavioral attributes; for the evolutionist it is the unit of evolution (Monod, 1974b) and for the paleontologist it is a section of a phyletic lineage. Different specialists at best will emphasize different aspects; at worst they will arrive at widely diverging conclusions. The result is continuing controversy.

It seems to be one of the most elementary urges of man to want to know what the different kinds of things are of which his environment is composed. Even primitive people have names for kinds of birds, fishes, flowers or trees, and the species recognized by them are usually exactly the same ones recognized by the modern taxonomist (Gould, 1979). Such naming of kinds is made possible because the diversity of nature is not continuous but consists of discrete entities, separated from each other by discontinuities. One finds in nature not merely individuals but also "species," that is, groups of individuals that share certain characteristics with each other.

The concept of species is needed because the term "kind of" is not sufficiently precise. The problem of delimiting species taxa against groupings of higher and lower categorical rank is one of demarcation. The discrimination of true biological species within genera, thus, is the problem of demarcations against more comprehensive groupings. But each biological species contains many phena[2] which are often so different from each other that they were first described as different species. If the term "species" is equated with "different kind," there is no discriminating criterion that permits an unequivocal attribution of different "kinds" to the three categories phenon, true species, and genus. It is the task of the species concept to serve as a yardstick in the proper classification of "kinds."

This at once poses a problem. What are the characteristics that permit the assignment of individuals to species? This question is easily answered, when the difference between two species is as clear-cut as that between the lion and the tiger. In many other cases, variation among the individuals of a species seems, at first sight, to be of the same order of magnitude as that between species. For there is an enormous amount of variation within species of animals and plants, reflected in sexual dimorphism, the existence of different stages in the life cycle (such as caterpillar and butterfly), an alternation of generations, and many other forms of individual variation. These cause great difficulties in the delimitation of species. If one wants to solve these problems, it is necessary to have not only sufficient biological information but also a clear concept as to what is meant by the term "species."   *253*

Species Taxon and Species Category

In retrospect it has become evident that great confusion was caused by the application of the term "species" to two fundamentally different logical categories, both of them referred to as the species. The introduction of the new term *taxon* [3] now permits a clear distinction between the two concepts. A taxon is a concrete zoological or botanical object. Groups of individuals like wolves, bluebirds, or houseflies are species taxa (see Chapter 4).

When a taxonomist first encounters specimens or individuals in nature that he wants to assign to a species, he deals with a strictly zoological or botanical problem. Are the individuals from a given district members of the same population? He is concerned not with problems of rank, as in the case of the problem of the species category, but with problems of delimitation. He deals with a given zoological object, let us say snow geese (*Anser caerulescens*) and attempts to determine whether white and blue birds are products of the same gene pool. But he also deals with an ontological problem. Are the animals that belong to a species members of a class or are they not? Ghiselin (1974b) has come out quite emphatically in favor of the interpretation (see also Dobzhansky, 1951) to consider all the products of the gene pool of a species as *parts* of the species (not as members of a class!) and to consider the species as a whole as an individual, ontologically speaking. That species taxa are not classes but have a different status had been said by perceptive zoologists for some time. Species taxa are individuals in the sense that each species has spatiotemporal unity and historical continuity (Hull, 1976; 1978). Each species has reasonably discrete boundaries, internal coherence at any one time, and, with limits, continuity through time. Any aggregate of populations that satisfies the definition of the species category is a species taxon.

The practical problems relating to the species taxon are two-fold: (1) the assignment of individual variants ("phena"; see Mayr, 1969) to the appropriate species taxon, and (2) the delimitation of taxa against each other, particularly the decision which populations of a single variable aggregate of populations in time and space to include in a single species.

The *species taxon* must be sharply distinguished from the *species category*. The species category is the class, the members of which are species taxa. The particular definition of the species category which an author adopts determines which taxa he must rank as species. The problem of the species category is simply one of definition. How do we define the term "species"? The changing definitions represent the history of the species concept.

The determination of species status is thus a two-step procedure. The first step consists of the delimitation of the presumptive species taxon against others and the second is the ranking of the given taxon into the appropriate category, for example, "population," "subspecies," or "species." This clear recognition of the fundamental difference between the species taxon and the species category is a development of only the last few decades, and has finally eliminated a major source of confusion, at least in principle. Many controversies, supposedly about the species concept, actually concerned the recognition of species taxa, and the assignment of individual variants (or other phena) to species taxa. Polytypic species, for instance, are not a separate category of species but only a special kind of species taxa. Most taxonomists, including myself, were confused about this until a few years ago.

EARLY SPECIES CONCEPTS

The ancients did not recognize the biological integrity of each species. Aristotle, for instance, accepted the frequent occurrence of hybridization among species, as between fox and dog or between tiger and dog. Both Aristotle and Theophrastus accepted the folklore belief that seeds of one species of plant could germinate into plants of another species (heterogony). Most herbalists and early botanists also accepted this as true or at least made no effort to refute it.[4] Albertus Magnus described five ways of transforming one plant into another.

In view of these uncertainties about the nature of species, it is not surprising that no consistent terminology existed. According to our dictionaries, the Greek word for species is *eidos* and for genus *genos,* yet Plato used the two words completely interchangeably. He never used *eidos* in the sense of "species," subordinated to a category "genus." Aristotle did make a distinction between the two words, but primarily in his treatises on logic. In his biological writings the word *genos* is used 413 times, but in 354 cases it refers to a kind of animal and only in the remaining cases to a generic category. In the 96 instances where *eidos* is used, only 24 denote kinds of animals. Thus the term *eidos* is used in only 6 percent of the 378 cases where reference is made to a kind of animal, in all others the term *genos.* "The traditional assumption that Aristotle did actually classify animals into genera and species . . . is not supported by the evidence" (Balme, 1962).

255

The principal use of the terms "genus" and "species" in Greek philosophy was in discussions on logic. In logical division the genus was divided into species, regardless of the rank of the genus. *Canis* would be a species in the genus of carnivores, but the poodle would be a species in the genus dog. The terms "genus" and "species" regulated the inclusion of members in larger classes. This usage, which emphasized relative rank, continued from the ancients to the time of Linnaeus, who in one of his earlier publications wrote: "Vegetabilium species sunt: Lithophyta, Algae, Fungi . . ." and so on (*Fundamenta,* 1735).

The adoption of Christianity and acceptance of the dogma of creation changed the situation remarkably little at first. Saint Augustine declared that plants on the third day of creation had been brought forth *causaliter,* that is, the earth had then received the power to produce them. This allowed for spontaneous generation, heterogony, and all sorts of other changes in the subsequent history of the earth. His species definition ("similia atque ad unam originem pertinentia": "What is similar and of a single origin") foreshadows that of Ray.

The attitude toward species changed drastically after the Reformation. The fixity and complete constancy of species now became a firm dogma. A literal interpretation of Genesis required the belief in the individual creation of every species of plants and animals on the days prior to Adam's creation. The species, thus, was the unit of creation. The rapid progress of natural history at that period favored this development. Most of the herbalists, in their studies of wild plants, likewise arrived at the idea that species were well-defined units of nature and that they were constant and sharply separated from each other.

THE ESSENTIALIST SPECIES CONCEPT

The creationist interpretation of species by the Christian fundamentalists agreed well with the essentialist species concept, according to which each species is characterized by its unchanging essence (*eidos*) and separated from all other species by a sharp discontinuity. Essentialism assumes that the diversity of inanimate as well as of organic nature is the reflection of a limited number of unchanging universals (Hull, 1975). This concept ultimately goes back to Plato's concept of the *eidos,* and this is what later authors had in mind when they spoke of the essence, or "nature," of some object or organism. All those objects belong to the same species that share the same essence.

The presence of the same essence is inferred on the basis of similarity. Species, thus, were simply defined as groups of *similar* individuals that are *different* from individuals belonging to other species. Species, thus conceived, represent different "types" of organisms. Individuals, according to this concept, do not stand in any special relation to each other; they are merely expressions of the same *eidos.* Variation is the result of imperfect manifestations of the *eidos.*

The criterion of similarity worked reasonably well in the sorting of "species" of minerals and other inanimate objects. Similarity, however, is a rather unreliable yardstick when one has to classify highly variable organisms. How can one know whether or not two individuals share the same essence? This can be assumed for those that are very similar, those "sharing the same characters." But what are we to do with individuals that are as different as are males and females in sexually dimorphic animals, or as are larvae and adults in invertebrates, or as are any of the other strikingly different variants so often found in a species? The method of inference from similarity completely broke down in all cases of pronounced sexual and age variation or any kind of polymorphism. One was forced to ask, Is there any other criterion by which one can determine "shared essence"?

John Ray (*Hist. Plant.,* 1686; tr. E. Silk in Beddall, 1957) was the first to provide a biological answer to this question:

> In order that an inventory of plants may be begun and a classification of them correctly established, we must try to discover criteria of some sort for distinguishing what are called "species." After a long and considerable investigation, no surer

criterion for determining species has occurred to me than the distinguishing features that perpetuate themselves in propagation from seed. Thus, no matter what variations occur in the individuals or the species, if they spring from the seed of one and the same plant, they are accidental variations and not such as to distinguish a species . . . Animals likewise that differ specifically preserve their distinct species permanently; one species never springs from the seed of another nor vice versa.

Here was a splendid compromise between the practical experience of the naturalist, who can observe in nature what belongs to a species, and the essentialist definition, which demands an underlying shared essence. Quite obviously the entire range of amplitude of variation that any given pair of conspecific parents can produce in their own offspring is contained within the potential of the essence of a single species. The importance of reproduction for the species concept is that it permits inferences on the amount of variation compatible with the existence of a single essence.

257

Ray's definition was enthusiastically adopted by generations of naturalists. It had the additional advantage that it fitted so well with the creationist dogma. This is what Cuvier had in mind when he defined species as "individus descendants des parents communs."[5] He explained this in a letter to his friend Pfaff. "We imagine that a species is the total descendence of the first couple created by God, almost as all men are represented as the children of Adam and Eve. What means have we, at this time, to rediscover the path of this genealogy? It is assuredly not in structural resemblance. There remains in reality only reproduction and I maintain that this is the sole certain and even infallible character for the recognition of the species" (Coleman, 1964: 145). Actually this was nothing but Ray's criterion, and later Cuvier himself admitted that in practice similarity was the primary criterion in the delimitation of species taxa. Clearly, there are no evolutionary overtones in Cuvier's species definition.

Numerous species definitions from Ray to the end of the nineteenth century affirmed on one hand the fixity, permanence, and bridgeless discontinuity of species, and used at the same time biological criteria to reconcile the seeming contradiction between conspicuous variation and the presence of a single essence. The words "common descent" so frequently used by writers of that period had the purely operational meaning of blood relationship, rather than any belief in evolution. When such an emphatically antievolutionary author as von Baer (1828) defines the species as "the sum of the individuals that are united by common descent," it is quite evident that he does not refer to evolution, nor does Kant when he says, "The natural classification deals with lines of descent, grouping animals according to blood kinship" (Lovejoy, 1959d: 180). To a creationist it simply meant descent from the pair that had been originally created. Such "descent" was reaffirmed by Linnaeus.

Linnaeus

258   Carl Linnaeus, the great Swedish botanist, is always described as a champion of the essentialist species. That he was, but this characterization by no means describes adequately the versatility of his species concept, for Linnaeus combined the experiences of a local naturalist, a pious creationist, and a disciple of logical division.[6] Even though all three components of his thinking stressed the constancy and sharp delimitation of species, it is necessary, for a complete understanding of Linnaeus' thinking, to keep all three sources of his concept in mind. He first (1736) articulated his species concept in the famous aphorism, "We count as many species as different forms were created in the beginning." In 1751 in the *Philosophia Botanica* (para. 157) he expanded it in the statement: "There are as many species as the infinite being created diverse forms in the beginning, which, following the laws of generation, produced as many others but always similar to them: Therefore there are as many species as we have different structures before us today."

When Linnaeus said "created," he meant this quite literally. In an essay he stated his belief "that at the beginning of the world, there was created only a single sexual pair of every species of living things . . . by a sexual pair I mean one male and one female in every species where the individuals differ in sex: But there are certain classes of animals natural Hermaphrodites, and of these only a single individual was originally formed in each kind." He came to this conclusion not only on the basis of his religious convictions but also because this expressed the then "modern" scientific findings. Spallanzani and Redi had refuted the occurrence of spontaneous generation, and Ray as well as Linnaeus had convinced themselves that the conversion of seeds of one species into such of another species (heterogony) was likewise impossible. The assumptions of Saint Augustine were not confirmed.

The species never played as important a role in Linnaeus' thinking as the genus. As a result he was often rather casual in his treatment of particular species in his taxonomic catalogues, those on plants (*Species Plantarum*) and on animals (*Systema Naturae*), two works in which his compilations of species are rich in errors. This necessitated frequent revisions of these writings.

The observations of the naturalists, the requirements of Christian faith, and the dogma of essentialism all led to the conclusion of the existence of well-defined and completely constant species, a concept which had an enormous influence for the ensuing one hundred years. As long as it was believed that species readily changed into others (heterogony) or were equally readily produced by spontaneous generation, the whole problem of evolution could not arise. Poulton (1903), Mayr (1957), and Zirkle (1959) have called attention to the fact that Linnaeus' insistence on the reality, sharp delimitation, and constancy of species did perhaps more to encourage subsequent evolutionary studies than if he had endorsed the traditional belief in a great plasticity of species. It was his species concept which generated a contradiction between the many indications of evolution in nature and the seeming constancy of species, a contradiction which had to be resolved.

Curiously, Linnaeus later in life reneged on the typological species concept of constant species, so well known under his name. He removed the statement "nullae species novae" (no new species) from the twelfth edition of the *Systema Naturae* (1766), and crossed out the words "Natura non facit saltus" in his own copy of the *Philosophia Botanica* (Hofsten, 1958). A number of botanical discoveries were responsible for this change of mind (Zimmermann, 1953: 201–210). First he encountered a striking mutation of flower structure (*Peloria*) in the plant *Linaria* which he thought was a newly arisen good species and genus, and later he encountered a number of putative species hybrids. It led him to a curious belief that perhaps only genera had been created in the beginning and that species were the product of hybridization among these genera. This hypothesis was, of course, not only inconsistent with everything he had said and believed before but was in fact irreconcilable with essentialism. Not surprisingly, Linnaeus was at once bitterly attacked from all sides, because the production of new essences by hybridization was unthinkable for any consistent essentialist. No one made this point more strongly than Kölreuter, who in a series of experiments (1761–1766) showed that newly produced hybrids between species are not constant new species but highly variable and could be returned to the parental species by continuous back-crossing (Olby, 1966).[7] These later ideas of Linnaeus were almost entirely forgotten in the ensuing period and had apparently no influence whatsoever on any later evolutionary thinking.

His contemporary Michel Adanson, so revolutionary in some of his thinking, had an entirely orthodox species concept. He made a careful analysis of the species problem and then concluded "that the transmutation of species does not happen among plants, no more than among animals, and there is not even direct proof of it among minerals, following the accepted principle that constancy is essential in the determination of a species" (1769: 418). This quotation illustrates particularly well how formalistic and nonbiological a species concept was held even by perceptive and otherwise enlightened biologists.

The essentialist species concept was accepted by taxonomists in the post-Linnaean period almost unanimously. This concept postulated four species characteristics: (1) species consist of similar individuals sharing in the same essence; (2) each species is separated from all others by a sharp discontinuity; (3) each species is constant through time; and (4) there are severe limitations to the possible variation of any one species. This, for instance, was Lyell's species concept.

Buffon

In his thought on species, Georges Louis Buffon, although earlier in time, was nearer to current thinking than Linnaeus or Cuvier. It is rather difficult to present a concise summary of Buffon's ideas on species, not only because they are scattered through numerous volumes of his *Histoire naturelle* but also because his thinking changed through time, from his first statement in 1749 to his last one in 1766. Different students of Buffon have therefore presented different interpretations.[8]

Buffon's earliest pronouncements on the species had a strongly nominalistic flavor, and seem to emphasize the existence of individuals rather than species, and of continuity between them: "Nature progresses by unknown gradations and consequently does not submit to our absolute divisions when passing by imperceptible nuances from one species to another and often from one genus to another. Inevitably there are a great number of doubtful species and intermediate specimens which one does not know where to place" (*Oeuvr. Phil.:* 10, trans. Farber, 1972).

Actually, this statement in vol. 1 of *Histoire naturelle* was part of an attack on the Linnaean system, and in two other volumes of the *Histoire naturelle* (all three published simultaneously in 1749) Buffon supported the concept of constant, well-delimited species. Though it has been denied by various specialists, Buffon's species were essentialistically conceived. Each species was characterized by a species-specific *moule intérieur* which, although differently derived, shared many attributes with Plato's *eidos*. Furthermore, each species was clearly separated from all other species:

> There exists in nature a general prototype in each species upon which all individuals are moulded. The individuals, however, are altered or improved, depending on the circumstances, in the process of realization. Relative to certain characteristics, then, there is an irregular appearance in the succession of individuals, yet at the same time there is a striking constancy in the species considered as a whole. The first animal, the first horse for example, was the exterior model and the internal mould from which all past, present, and future horses have been formed. (*Hist. nat.*, IV: 215–216, from Farber, 1972: 266)

261

It was this succession of individuals which Buffon considered the most important characteristic of species, because each succession of individuals is sharply separated from all others:

> It is then in the characteristic diversity of species that the intervals between the nuances of nature are the most noticeable and best marked. One could even say that these intervals between species are the most uniform and the least variable of all, since one can always draw a line between two species, that is, between two successions of individuals that cannot reproduce with each other. This distinction is the strongest that we have in natural history . . . Each species—each succession of individuals that can successfully reproduce with each other—will be considered as a unit and treated separately . . . The species then is only a constant succession of similar individuals that can reproduce together. (*Hist. nat.*, IV: 384–385)

Supplementing Ray's criterion, which had demonstrated that exceedingly different-appearing organisms could belong to the same species if sharing a common descent, Buffon discovered a criterion by which one could decide whether or not two very similar "kinds" were different species. For instance, are donkey and horse one species? His solution was that individuals that cannot produce fertile offspring belong to different species. "We should regard two animals as belonging to the same species if, by means of copulation, they can perpetuate themselves and preserve the likeness of the species; and we should regard them as belonging to different species if they are incapable of producing progeny by the same means" (*Hist. nat.*, II: 10). "A species is a constant succession of similar individuals that can reproduce together" (p. 385). The conspicuous novelty in Buffon's species concept is that the criterion of conspecificity is no longer, as in Ray, the range of morphological variation in the descendants of one set of parents but rather their ability to produce fertile offspring. By introducing this entirely new criterion, Buffon had gone a long way toward the biological species concept. Yet, by considering species as constant and invariable, Buffon still adhered to the essentialistic species concept.

There is a second aspect by which Buffon differs from Linnaeus and other orthodox taxonomists: namely, his down-grading of morphological characters and his emphasis on habits, temperament, and instinct, which he considered far more important characteristics of species than purely structural features. It is not sufficient, he says, to identify a species by a few key characters; if one wants to know an animal, one must know all of its characteristics. No one took this admonition more seriously than the field naturalists; the great flowering of the natural history of living animals, particularly birds, in the ensuing generations owes a great deal to Buffon's concepts. One can always recognize a species by characteristics of its life history. A species, therefore, is something natural and real, in contrast to the Linnaean genus, which is a purely arbitrary construct.

Later in life (after 1765) Buffon somewhat modified his species concept, by defining the word "species" in a narrower and rather restricted sense (Roger, 1963: 576). When he realized, particularly by studying birds, that there were closely related groups of species, some of them apparently producing fertile hybrids, he assigned to such "families" of species the attributes he had previously assigned only to species. Yet, he maintained at the same time his concept of well-defined species at a lower level. This groping for a new species concept, foreshadowing the idea that groups of species might have a unity owing to common descent, had apparently no lasting impact on his readers and played no further role in the later history of the species concept.

On the other hand, Buffon's rather "biological" way of looking at species exerted an important influence. Zimmermann (1778, I: 130) states that he is following Buffon, Blumenbach, and Spallanzani in adopting cross-fertility as the species criterion, and that he is including all dogs in a single species, "first because they all mate with each other and, what is most important, produce fertile young; secondly because all the races of dogs have the same instincts, the same attachment to man, and the same capacity to be tamed." Such a biological concept of species was widespread between 1750 and 1860, as reflected in the writings of Pallas, Gloger, Faber, Altum, and the best naturalists of the nineteenth century. Yet, simultaneously a strictly essentialistic concept survived, particularly among the collector types, who described every variant as a new species. Pastor C. L. Brehm named no less than 14 "species" of house sparrows from his little village in Thuringia; a French specialist of freshwater clams based more than 250 species names on variants of a single species. For these authors, species were types and any consideration of species as populations was alien to their thinking. It is this way of thinking of species which is often referred to in the systematic literature as the typological species concept. There is hardly any higher taxon of animals and plants in which there have not been one or two such "species mongers" active, accounting for hundreds and thousands of synonyms (Mayr, 1969: 144–162).

263

In botany, perhaps even more so than in zoology, variation was made the excuse for the description of innumerable new species, particularly in so-called "difficult" genera like *Rubus* or *Crataegus*. The situation was aggravated by the almost universal failure of botanists to distinguish terminologically between individual and geographical varieties. The first beginning of an improvement came when the International Botanical Congress of 1867 adopted proposals by Alphonse de Candolle to recognize subspecies, varieties, and other subdivisions of the species. In the ensuing years the publications of Kerner (1866; 1869) and Wettstein (1898) helped to clarify the situation. But even after the rise of the new systematics, all too many botanists still used the term "variety" indiscriminately for geographical populations and for intrapopulational variants.

THE NOMINALISTIC SPECIES CONCEPT

Opposition to the essentialistic species concept developed on two fronts, among naturalists and among philosophers. The two philosophers who exercised the greatest influence in the early and middle eighteenth century, Leibniz and Locke, were both uncomfortable with the concept of well-defined, sharply separated species. Locke did not necessarily deny the existence of species, but said: "I think it nevertheless true that the boundaries of species, whereby men sort them, are made by men." He exclaimed that he was unable to see why two breeds of dogs "are not as distinct species as a spaniel and an elephant . . . so uncertain are the boundaries of species of animals to us."

When the concepts of plenitude and continuity began to dominate western thinking in the post-Leibnizian period (Lovejoy, 1936: 229–241), the concept of discontinuous systematic categories, including that of the species, became a liability, and philosophers fell back on a nominalistic definition of species. For the nominalist, only individuals exist while species or any other "classes" are man-made constructs.

Nominalism, a medieval school of philosophy, rejected the notion of essentialism that similar things share the same substance (essence) and claimed instead that all that classes of similar things share is a name. This interpretation was also applied to the species by several eighteenth-century authors (Crombie, 1950). Thus, Robinet claimed, "There are only individuals, and no kingdoms or classes or genera or species" (*De la nature*, IV: 1–2). Similar statements can be found in the writings of several French naturalists beginning with the first volume of Buffon (1749) and continuing in Lamarck (Burkhardt, 1977) and Lacépède (1800).

Buffon quickly abandoned this concept (if he ever truly believed in it) and the other naturalists like Lamarck and Lacépède dealt with species in a thoroughly orthodox manner in their actual taxonomic treatises. In his later years (1817), Lamarck became more and more convinced of the importance of species. He emphasized that the species of inanimate objects was something entirely different from the species of organisms. Species of organisms are complex systems of heterogeneous molecules, which explains their capacity for variation and change. Finally, he raised questions concerning their evolutionary change and whether "they do not multiply and thus become diversified." This prophetic view of species was a long way from Lamarck's earlier nominalistic claim that only individuals exist.[9]

The nominalistic concept of species remained popular among botanists throughout the nineteenth century. Schleiden and Nägeli were among its chief proponents. "Messy" genera like *Rubus* and *Hieracium* were the evidence, most frequently cited, to defend this view. Among paleontologists it was likewise popular, particularly among authors working on such "artifacts" as conodonts where a delimitation of species taxa is indeed often difficult. Spirited de-

264

fenses in favor of considering species a purely arbitrary convention were published by prominent botanists such as Cronquist and paleontologists such as A. B. Shaw within the last decade. The botanist Bessey (1908) stated this viewpoint particularly well: "Nature produces individuals and nothing more . . . species have no actual existence in nature. They are mental concepts and nothing more . . . species have been invented in order that we may refer to great numbers of individuals collectively."[10] Some recent opponents of the biological concept (for example, Sokal and Crovello, 1970) also endorse ideas that are basically nominalistic, though they are very much in the minority. The evidence for intrinsically maintained discontinuities between sympatric natural populations is so overwhelmingly conclusive that most students of local faunas and floras have adopted the biological species concept.

265

The reason why eighteenth- and nineteenth-century authors who were dissatisfied with the essentialist species concept adopted the nominalist concept was not necessarily because they were impressed by its superiority but simply because they could not think of any other alternative. It lost this advantage with the coming of the biological species concept and is now no longer in vogue, at least not among biologists.

DARWIN'S SPECIES CONCEPT

No author reflects the struggle with the species concept more vividly than Darwin. The species which he encountered as a youthful collector and naturalist at Shrewsbury, Edinburgh, and Cambridge was the typological, "nondimensional" species of the local fauna. This was also the species of his beetle-collecting friends, and of Henslow and Lyell (Mayr, 1972b). It was still Darwin's species concept when he landed in the Galapagos on September 16, 1835. The *Beagle* visited four islands (Chatham, Charles, Albemarle, and James), all within less than one hundred miles of each other. Never before having been exposed to geographic variation, Darwin took it for granted that the fauna of all these neighboring islands was the same, and apparently labeled all of his specimens simply as coming from the "Galapagos Islands" (Sulloway ms.). The fact that the local Spaniards could distinguish each island race of giant tortoise apparently made at first little impression on Darwin, whose mind at that time was very much preoccupied with geology. When Darwin subsequently sorted his collections of birds,

he was confronted by the problem of how to classify the populations on different islands. For instance, there is a mockingbird
(*Mimus*) on every island of the Galapagos, but the birds of a given
island are somewhat different from the birds on most of the other
islands. Are the inhabitants of the various islands different species
or are they varieties? was the question Darwin asked. There was
no doubt that they were different taxa, because the differences
could be seen and described. The problem was one of ranking,
that is, of placing them in the appropriate category. One has to

266 keep this in mind when one analyzes Darwin's statements on species. It is even more important to realize that Darwin's species
concept underwent a considerable change in the 1840s and 50s
(Kottler, 1978; Sulloway, 1979). In the 1830s Darwin's concepts
of species and speciation were determined almost exclusively by
the zoological evidence. Indeed, he conceived of species as being
maintained by reproductive isolation. That this is the way Darwin
thought about species at that period was unknown to Darwin
scholars until his notebooks were rediscovered. Here he wrote, for
instance, "My definition of species has nothing to do with hybrid-
ity, is simply an instinctive impulse to keep separate, which will no
doubt be overcome [or else no hybrids would be produced], but
until it is these animals are distinct species" (NBT, C: 161).[11] Here
we have a clear description of reproductive isolation, maintained
by ethological isolating mechanisms. There are repeated references in the notebooks to mutual "repugnance" of species to in-
tercrossing. "The dislike of two species to each other is evidently
an instinct; and this prevents breeding" (B: 197). "Definition of
species: one that remains at large with constant characters, together with other beings of very near structure" (B: 213). In these
notebooks Darwin emphasized repeatedly that species status had
little if anything to do with degree of difference. "Hence species
may be good ones and differ scarcely in any external character"
(B: 213). Here he refers to the two sibling species of the leaf warbler *Phylloscopus trochilus* (*collybita* and *sibilatrix*), discovered in England by Gilbert White in 1768, which were so similar that they
were not formally recognized by taxonomists until 1817. It is no
exaggeration to state that in the 1830s Darwin had what was very
close to the modern biological species concept.

When one goes to the *Origin* of 1859 and reads what it says
about species, one cannot help but feel that one is dealing with an
altogether different author (Mayr, 1959b). Since until the notebooks were rediscovered this is the Darwin known to the biologi-

cal world from 1859 on, it is of historical importance to quote what Darwin said in the *Origin:*

No one definition has as yet satisfied all naturalists; yet every naturalist knows vaguely what he means when he speaks of a species. (p. 44)

In determining whether a form should be ranked as a species or a variety, the opinion of naturalists having sound judgment and wide experience seems the only guide to follow. (p. 47)

From these remarks it will be seen that I look at the term species, as one arbitrarily given for the sake of convenience to a set of individuals closely resembling each other, and that it does not essentially differ from the term variety, which is given to less distinct and more fluctuating forms. (p. 52; see also p. 469)

267

Hence the amount of difference is one very important criterion in settling whether two forms should be ranked as species or varieties. (pp. 56–57)

Varieties have the same general characters as species, for they cannot be distinguished from species. (p. 58; and a similar statement, p. 175)

It can thus be shown that neither sterility nor fertility affords any clear distinction between species and varieties. (p. 248)

In short, we shall have to treat species in the same manner as those naturalists treat genera, who admit that genera are merely artificial combinations made for convenience. (p. 485)

And in a letter to Hooker (December 24, 1856) Darwin wrote: "I have just been comparing definitions of species . . . It is really laughable to see what different ideas are prominent in various naturalists' minds, when they speak of 'species'; in some, resemblance is everything and descent of little weight—in some, resemblance seems to go for nothing, and creation the reigning idea—in some descent is the key—in some, sterility an unfailing test, with others it is not worth a farthing. It all comes, I believe, from trying to define the undefinable" (*L.L.D.,* II: 88).

What could have brought about this complete turn around in Darwin's species concept? His reading as well as his correspondence indicate that after 1840, and particularly from the 1850s on, Darwin was increasingly influenced by the botanical literature and by correspondence with his botanist friends. As he himself said: "All my notions about *how* species change are derived from a long continued study of the works of (and converse with) agriculturists and horticulturalists" (*L.L.D.*, II: 79). Perhaps no other botanist influenced Darwin's thinking more than William Herbert, who, among other things said: "There is no real or natural line of difference between species and permanent or discernible variety . . . nor do there exist any features on which reliance can be placed to pronounce whether two plants are distinguishable as species or varieties" (1837: 341). Statements almost exactly like this one can be found in the botanical literature from that time until the present. Only rarely is an attempt made to discriminate between sympatric and allopatric situations. Herbert did not give cross fertility primacy over degree of morphological similarity since he believed "that the fertility of the hybrid or mixed offspring depends more upon the constitutional [whatever that means!] than the closer botanical affinities of the parents" (1837: 342). Not reproductive isolation but degree of difference had now become the yardstick of species status. For Herbert the genus was the only "natural" category.

Many of Darwin's statements are perfectly legitimate if one translates the word "variety" into "geographical isolate." It is as true today as it was in the days of Darwin that the ranking of geographical isolates, particularly those that are strongly marked, is arbitrary. There are literally hundreds, perhaps even thousands, of geographical isolates among birds which as recently as the 1970s were ranked as species by some ornithologists and as subspecies by others.

If all Darwin had wanted to say is that it is difficult, and often impossible, to rank isolated populations, no one could have found fault with him. Geographical isolates indeed are incipient species. Unfortunately, Darwin used strictly typological language, and by using terms like "forms" and "varieties," instead of "individuals" or "populations," he introduced confusing ambiguity. Furthermore, instead of using the term "variety" consistently for geographic races, he frequently employed it, particularly in his later writings, as a designation for a variant or aberrant individual. By this extension of the meaning of the term "variety," Darwin confounded two rather different modes of speciation, geographic and sympatric speciation.

When one glances over the statements about species made by Darwin in the *Origin,* one might get the impression that he considered species as something purely arbitrary and invented merely for the convenience of taxonomists. Some of his comments remind one of Lamarck's statement that species do not exist, only individuals. And yet, in their taxonomic work both men treated species in a perfectly orthodox manner (Lamarck with mollusks, Darwin with barnacles), as if they were so many independent creations. And this, let me add, was quite legitimate because in these taxonomic monographs they listed and described species taxa, and the definition of the species category was, except in borderline cases, an irrelevant consideration.

Somehow, Darwin was very pleased with himself for having "solved" the species problem: Since species continue to evolve, they cannot be defined, they are purely arbitrary designations. The taxonomist no longer will have to worry what a species is: "When the views entertained in this volume . . . are generally admitted . . . systematists . . . will not be incessantly haunted by the shadowy doubt whether this or that form be in essence a species. This, I feel sure, and I speak after experience, will be of no slight relief" (*Origin:* 484). This explains why Darwin no longer attempted to define what a species is. He treated it purely typologically as characterized by "degree of difference." Ghiselin (1969: 101) has stated quite correctly that "there is no solid evidence that [Darwin] conceived of species as reproductively isolated populations." This is surely true for the period when he wrote the *Origin*.

One must further remember that in the *Origin* Darwin dealt with species in the context of the problem of the gradual origin of species. There was a strong, even though perhaps unconscious, motivation for Darwin to demonstrate that species lack the constancy and distinctiveness claimed for them by the creationists. For how could they be the result of gradual change through natural selection if it were true, as Darwin's opponents continued to claim for the next hundred years, that species are sharply delimited and separated by "bridgeless gaps"? Hence, it was good strategy to deny the distinctness of species. Considerable support for this claim could be marshaled, provided one defined species simply by degree of difference rather than by reproductive isolation, and provided one made no distinction between geographical and intrapopulational "varieties." When species are thus conceived, the origin of new species is not an unsurmountable problem. But the switch from Darwin's species concept of the 1830s to that of the 1850s laid the foundation for controversies that lasted for a century.

269

THE RISE OF THE BIOLOGICAL SPECIES CONCEPT

The publication of the *Origin* created a formidable dilemma for the students of species. Species evidently had descended from common ancestors and, as Darwin claimed, by a slow, gradual process. Yet, the local naturalists found species to be separated in nature by bridgeless gaps and not at all consisting of arbitrary aggregates of specimens, as Darwin seemed to claim in the *Origin*. As a result, species continued to be dealt with as if no one had established a theory of evolution. Among the museum taxonomists, the essentialist interpretation of species continued to dominate (Stresemann, 1975). It was called the *morphological species concept* because degree of morphological difference was used as the criterion by which it was decided whether certain individuals belonged to the same or to different species. As late as 1900, a group of leading British biologists and taxonomists, including Ray Lankester, W. F. R. Weldon, William Bateson, and A. R. Wallace, unanimously endorsed a strictly morphological species definition (Cock, 1977). The definition provided by Wallace—"A species is a group of individuals which reproduce their like within definite limits of variation, and which are not connected with their nearest allied species by insensible variations"—would raise every isolated geographical race to the rank of a separate species. Whenever variation was encountered, Ray's prescription was applied, that is, to consider as conspecific whatever any given pair of conspecific parents could produce in their offspring. This species concept was not only adopted by the majority of taxonomists, but it was also the dominant concept of the experimental biologists. De Vries' species of *Oenothera* were based on such a morphological definition and as recently as 1957 Sonneborn refused to designate the "varieties" of *Paramecium* as species even though, on the basis of their biological characteristics and reproductive behavior, it was abundantly evident that this is what they were, as Sonneborn himself eventually admitted.[12]

Far superior to the morphological concept was a species concept found all along in the writings of field naturalists. Such authors as F. A. Pernau (1660–1731) and Johann Heinrich Zorn (1698–1748) studied every aspect of the biology of the birds in their surroundings and never questioned that all of them belonged to well-defined species, clearly separated from all others by biological characteristics (song, nest, migratory pattern, and soon) and by reproductive isolation. Zorn, like Ray, was in the tradition of natural theology, and in the ensuing 150 years the finest work on species in nature was done by natural theologians. Indeed, the foremost students of birds during this period, Gilbert White, C. L. Brehm, and Bernard Altum, were priests or ministers (Stresemann, 1975). In the study of the species of insects in nature, natural theologians like William Kirby were also at the forefront. It was this tradition of the field naturalists which, when it became self-conscious and scientific, led to the development of the biological species concept.

The old species concept, based on the metaphysical concept of an essence, is so fundamentally different from the biological concept of a reproductively isolated population that a gradual changeover from one into the other was not possible. What was required was a conscious rejection of the essentialist concept. This was facilitated by the clear recognition of a number of difficulties encountered by the students of species when trying to apply the criterion of "degree of difference" (Mayr, 1969: 24–25). The first was that no evidence could be found for the existence of an underlying essence or "form" responsible for the sharply defined discontinuities in nature. In other words, there is no way of determining the essence of a species, hence no way of using the essence as a yardstick in doubtful cases. The second difficulty was posed by conspicuous polymorphism, that is, the occurrence of strikingly different individuals in nature which nevertheless, by their breeding habits or life histories, could be shown to belong to a single reproductive community. The third difficulty was the reverse of the second one, that is, the occurrence in nature of "forms" which clearly differed in their biology (behavior, ecology) and were reproductively isolated from each other yet could not be distinguished morphologically (sibling species; see below).

When one looks at many of the historical species discussions, one is impressed with how tantalizingly close to a biological species concept some of the earlier authors had come. To a modern biologist, it would seem only a small step from Ray's modified essentialistic definition—"A species is an assemblage of all variants that are potentially the offspring of the same parents"—to a species definition based on the concept of reproductive communities alone. Even closer was Buffon's definition, "A species is a constant succession of similar individuals that can reproduce together" and whose hybrids are sterile. Yet, Buffon still considered species essentialistically constant. Girtanner (Sloan, 1978) and Illiger (Mayr, 1968) in some of their statements likewise came close to a statement of the biological species, but were also unable to shed the essentialistic framework of their thinking. The same is true for many other authors of the nineteenth century. None of them took the seemingly small step of defining the species in terms of a reproductively isolated assemblage of populations. Why was there such a long delay?

272

There are three aspects of the biological species that required the adoption of new concepts. The first is to envision species not as types but as populations (or groups of populations), that is, to shift from essentialism to population thinking. The second is to define species not in terms of degree of difference but by distinctness, that is, by the reproductive gap. And third, to define species not by intrinsic properties but by their relation to other co-existing species, a relation expressed both behaviorally (noninterbreeding) and ecologically (not fatally competing). When these three conceptual changes are adopted, it becomes obvious that the species concept is meaningful only in the nondimensional situation: multidimensional considerations are important in the delimitation of species taxa but not in the development of the conceptual yardstick. It also becomes evident that the concept is called biological not because it deals with biological taxa but because the definition is biological, being quite inapplicable to species of inanimate objects; and that one must not confuse matters relating to the species taxon with matters relating to the concept of the species category.

The clear pronouncement and explicit analysis of these characteristics of the biological species was not achieved until the 1940s and 1950s.[13] However, the essential points were grasped by a series of pioneers. The first two authors who clearly described and defined the biological species were the entomologists K. Jordan (1896; 1905) and Poulton (1903; see Mayr, 1955). Poulton defined the species "as an interbreeding community, as syngamic," and Jordan stated, "Individuals connected by blood relationship form a single faunistic unit in an area . . . The units, of which the fauna of an area is composed, are separated from each other by gaps which at this point are not bridged by anything" (1905: 157).

The Properties of the Biological Species

Leading ornithologists like Stresemann and Rensch applied the biological species concept consistently in the 1920s and 1930s. In 1919 (p. 64), Stresemann emphasized that it is not degree of difference that characterizes species, but "that forms when they have risen [during geographic isolation] to species rank have become so different from each other physiologically that they . . . can come together again without interbreeding." Dobzhansky's definition of species as forms "which are physiologically incapable of interbreeding" (1937: 312) is virtually the same. The history of the numerous attempts to achieve a satisfactory biological species definition has been told repeatedly (for example, Mayr, 1957; 1963). Mayr's 1942 definition—"Species are groups of actually or potentially interbreeding natural populations which are reproductively isolated from other such groups" (p. 120)—still had some weaknesses. The "actual vs. potential" distinction is unnecessary since "reproductively isolated" refers to the possession of isolating mechanisms, and it is irrelevant for species status whether or not they are challenged at a given moment. A more descriptive definition is: *A species is a reproductive community of populations (reproductively isolated from others) that occupies a specific niche in nature.*

273

This definition does not instruct one how to delimit species taxa. What it does do is to permit one to determine the categorical rank of taxa. By contrast, degree of morphological distinctness is not a suitable criterion, as proven by sibling species and conspicuous morphs. The biological species concept, expressing a relation among populations, is meaningful and truly applicable only in the nondimensional situation. It can be extended to multidimensional situations only by inference.

The words "reproductively isolated" are the key words of the biological species definition. They pose at once the problem as to the cause of this isolation, a problem that was solved by the development of the concept of *isolating mechanisms.* The crude beginning of this concept goes back all the way to Buffon's sterility criterion, a criterion popular among botanists far into the twentieth century. Zoologists, and particularly ornithologists and students of butterflies, however, observed that in nature the sterility barrier is rarely tested in animals and that conspecificity is usually determined by behavioral compatibility. In the course of time, more and more devices that prevent species from interbreeding were discovered, for instance differences in breeding or flowering season and occupation of different habitats. The Swedish botanist Du Rietz (1930) was apparently the first to provide a detailed listing and classification of such barriers to the interbreeding of species. Their study was clearly handicapped by the absence of a technical term. Dobzhansky provided the term "isolating mechanism" for "any agent that hinders the interbreeding of groups of individuals . . . The isolating mechanisms may be divided into two large categories, the geographical and the physiological" (1937: 230). Even though Dobzhansky realized that geographical isolation was "on a different plane from any kind of physiological one," he did not appreciate that only the latter are genuine properties of species. For this reason, Mayr restricted the term "isolating mechanisms" to biological properties of species, expressly excluding geographical barriers (1942: 247). There still remained one difficulty: an occasional individual in an otherwise perfectly good species may hybridize. In other words, isolating mechanisms can only provide the integrity of populations, but not of every last single individual. This recognition led Mayr to an improved definition: "Isolating mechanisms are biological properties of individuals which prevent the interbreeding of populations that are actually or potentially sympatric" (1963: 91). In the last forty years the study of isolating mechanisms has become one of the most active fields of biology.[14]

Reproductive isolation, however, is only one of the two major characteristics of the species. Even the earliest naturalists had observed that species are restricted to certain habitats, and that each species fits into a particular niche. These ideas were prominent in the writings of Buffon and of all the eighteenth- and nineteenth-century writers who spoke of the economy of nature. Darwin was convinced that the geographic range of a species was largely determined by the species borders of its competitors.[15] Yet, during the development of the modern species concept, the emphasis was at first almost exclusively on reproductive isolation. The person who more than anyone else deserves credit for reviving an interest in the ecological significance of species was David Lack (1944; 1949). It is historically interesting to compare his evolutionary interpretation of bill size in different species of Galapagos finches. In an earlier paper (1945, but actually written prior to 1940) he had interpreted bill size as a species recognition signal, thus as an isolating mechanism, while in his later book (1947) he interpreted it as adaptation to a species-specific food niche, an interpretation that has since been abundantly confirmed.

It is now quite clear that the process of speciation is not completed by the acquisition of isolating mechanisms but requires also the acquisition of adaptations that permit co-existence with potential competitors. How difficult it is for a species to invade the range of a potential competitor is documented by the great frequency of parapatric distribution patterns of closely related species. (Populations or species are *parapatric* when in contact geographically but not overlapping and rarely or never inter-breeding.) In such cases one species is superior at one side of the line of division, the other species at the other side. Parapatry can also be caused by cross-sterility but in absence of premating isolating mechanisms.

An attempt was made by Van Valen (1976: 233) to base a species definition on niche occupation: "A species is . . . a lineage . . . which occupies an adaptive zone minimally different from that of any other lineage in its range." This reflects the principle of competitive exclusion, but is not very practical as a species definition because it is often exceedingly difficult to discover the "minimal" niche difference between two species, as demonstrated by much ecological research. Furthermore, many species (for example, the caterpillar-butterfly) occupy very different niches at different stages of their life cycle and in different portions of their geographical range. Is each of these therefore a different lineage and species? Such cases show graphically that the reproductive community is the real core of the species concept. As a matter of fact, niche occupation and reproductive isolation are two aspects of the species that are not mutually exclusive (except in parapatric ones), as shown by Lack (1947), Dobzhansky (1951), Mayr (1963: 66–88), and others. Indeed, the major biological meaning of reproductive isolation is that it provides protection for a genotype adapted for the utilization of a specific niche. Reproductive isolation and niche specialization (competitive exclusion) are, thus, simply two sides of the same coin. It is only where the criterion of reproductive isolation breaks down, as in the case of asexual clones, that one makes use of the criterion of niche occupation (Mayr, 1969: 31).

THE NEW SYSTEMATICS

The replacement of the essentialist species concept by a populationally conceived biological species has been an exceedingly slow process. Prerequisites for this replacement were the development of the theory of polytypic species taxa, a refinement in the terminology of infraspecific categories, and, most importantly, a growing realization of the immense variability of natural populations. Taxonomists, biometricians, population geneticists, and most recently biochemists (through enzyme analysis) have contributed to the increasing weakening of the typological species concept. Experimental physiologists and embryologists have perhaps been the last to become converted to population thinking. Findings made by the new technique of enzyme electrophoresis have contributed to their gradual conversion.

The rate at which the populational species concept has been applied to different groups of animals and plants has been highly uneven. When species can be readily studied in nature, the conversion to the biological species concept was virtually completed more than thirty years ago. Where only preserved material is studied, as is true for many groups of insects and other invertebrates, the prevailing species concept is rather typological even today.

Particularly perceptive students of birds, mammals, fishes, snails, and butterflies arrived independently at very similar conclusions. Yet the opinions of these progressive leaders of systematics was a minority opinion until the 1930s. Most other taxonomists dealt with species and their variation in a manner not drastically different from that of Linnaeus nearly two hundred years earlier. By 1940, however, the new movement was sufficiently visible so that a nontaxonomist, Julian Huxley, referred to it as *the new systematics* in a volume thus entitled, even though, curiously, there was little new systematics in that volume.

What was the new systematics? It was not a specific technique and is, perhaps, best described as a viewpoint, an attitude, or a general philosophy. It started primarily as a rebellion against the nominalistic, typological, and thoroughly nonbiological approach of certain (alas, all too many) taxonomists of the preceding period. The new systematist appreciates that all organisms occur in nature as members of populations. He studies the biological properties of organisms rather than the static characters of dead specimens. He utilizes the greatest number of kinds of characters—physiological, biochemical, and behavioral characters as well as morphological ones. He uses new techniques not only to measure specimens but also to record their sounds, to perform chemical analysis, and to make statistical and correlational computations. The earliest systematic treatments in zoology of the ideas of the new systematics were by Rensch (1929; 1933; 1934) and Mayr (1942).

The situation was more complex in botany. Here, a formidable chasm had developed between the herbarium taxonomists, who continued to cultivate the tradition of Linnaeus, and the field naturalists and experimentalists, who became increasingly dissatisfied with the typological-morphological approach of the herbarium workers, particularly since the efforts of a few imaginative pioneers seemed to have had no permanent impact.[16] The Swedish plant ecologist Turesson (1922) finally rebelled against this tradition, asserting that the conventional terminology of species and variety was quite unsuited to describe the dynamics of variation in natural populations. To cope with this situation Turesson introduced the new terms *ecospecies* for the "Linnaean species from an ecological point of view" and *ecotype* for "the product arising as a result of the genotypical response of an ecospecies to a particular habitat." More than that, he asserted that the study of the genetic and ecological variation of natural populations had nothing to do with taxonomy and should be made the subject matter of a separate biological science, *genecology* (by others called *biosystematics*).

Turesson himself was rather typological in his thinking. One gets the impression from his writings that he considered plant species as a mosaic of ecotypes rather than as an aggregate of variable populations. To some extent the same tendency toward typological thinking can be found in the writings of other Scandinavian authors. Nevertheless, Turesson's revolutionary concepts and his experimental analysis of samples of wild plant populations have had an impact on plant taxonomy that can hardly be exaggerated. He inspired numerous studies of the adaptive characteristics of local populations, which greatly advanced our understanding of the population structure of plant species and their ability to respond to local selection pressures. It was a liberating rebellion against the Linnaean tradition in the herbaria with its commitment to identification and typological thinking. Turesson's cry for a new botany, whether called genecology or biosystematics,

277

was taken up by Anderson, Turrill, Stebbins, Epling, Camp, Gregor, Fassett, and other students of plant populations.[17]

Botanists were clearly ahead of zoologists in two respects. They introduced experimental methods earlier and employed them far more extensively, aided by the fact that it is far easier to transplant and cultivate plants and to breed them in cultivation than to breed most animals. Botanists also utilized chromosome studies earlier and more intensively, in part by necessity owing to the frequency of polyploidy in plants. On the other hand, the introduction of the polytypic species greatly lagged in plant taxonomy, and the findings of the chromosomal studies were often interpreted in a strictly typological manner. For a few decades there seemed to be a complete split between the Linnaeanism of the herbaria and the experimental approach of the population botanists. However, in due time the new thinking, thus introduced into botany, spread also into the herbaria, and the gap that had existed in 1922 became increasingly smaller and finally disappeared. Most herbaria now have plant breeding facilities and supplement their understanding of the variation of natural species by studying genetic and karyological variation as well as sometimes the variation of enzymes and other molecules (Mayr, 1963: 351–354; Ehrendorfer, 1970; Grant, 1971; Solbrig, 1979; 1980).

When reviewing the species concept in botany, one must realize that species are a far more complex phenomenon in many groups of plants than in most groups of animals, particularly birds. It is not only the immobility of plant individuals, favoring the formation of ecotypes, that leads to complications but also the widespread occurrence of polyploidy, hybridization, and various forms of asexuality and self-fertilization. Certain botanists, not without justification, have raised the question whether the wide spectrum of breeding systems that can be found in plants can all be subsumed under the single concept (and term) "species." In addition to the term "ecospecies," Turesson introduced the term "coenospecies" for the totality of populations (and species) able to exchange genes with each other by hybridization. The most ambitious scheme to distinguish terminologically between different breeding systems in plants was proposed by Camp and Gilly (1943), who distinguished by special technical names twelve different kinds of species. Actually there is so much overlap in the criteria used and so little correlation between genetic mechanisms and visible morphological variation that this elaborate scheme was not adopted by any other author. Yet, the diversity of plant breeding systems

278

may help to explain why there has been so much resistance among botanists to the adoption of the biological species concept.

Attempts to recognize different kinds of species have not been altogether absent in zoology either. Certain authors (such as Cain, 1954) have attempted to distinguish morphospecies, biospecies, paleospecies, ecospecies, ethospecies, and so on, but one does not have the impression that these terminological endeavors led to any new insights. The term that is perhaps best justified is "agamospecies" for species of asexually reproducing organisms (see below).

279

THE VALIDITY OF THE BIOLOGICAL SPECIES CONCEPT

The biological species concept has not remained unchallenged. The early attacks, from the 1920s to 40s, questioned primarily its practicality: "How can a paleontologist test the reproductive isolation of fossils?" or, "The items I arrange in my collections are discrete and distinct types, and they are best called species." No questions of biological meaning were asked by these opponents, only those of administrative, curatorial convenience. The proponents of the biological species concept had relatively little trouble demonstrating that the opponents confused species taxon and species category, that they did not know the difference between evidence and inference (as Simpson has pointed out perceptively), and that going back to the morphological species concept takes one right back to the arbitrariness of having to decide how different an individual or a population has to be to deserve species status.

Another set of criticism made at this period (also largely owing to a confusion of species taxon and species category) was based on the wish to define species "quantitatively" or "experimentally." Since the biological species concept is neither based on quantitative nor experimental criteria, so it was claimed, it is to be rejected. This rejection rests on the fallacious claim that the methodologies and theories of the physical sciences are, without adjustment, applicable to evolutionary biology. Any naturalist can observe the genetically programmed reproductive and ecological discontinuities that exist in nature without applying a sophisticated computer analysis.

In the 1950s to 1970s a new set of arguments against the

biological species concept came to the fore. Various authors claimed that in the particular organisms which they were studying, they were unable to find the clear-cut gaps between sympatric populations described by the adherents of the biological species concept. In other words, it was claimed that there is no valid observational basis for the biological species concept and that the biological species is a special situation in a few groups that cannot be generalized and extended to all organisms. To cope with the diversity of nature one must therefore adopt either a different, more comprehensive concept, or else one must adopt several species concepts to cope with different types of organisms.

These are serious objections and they have a certain amount of validity. This leads to the question whether the cases that do not seem to fit are exceptions or whether perhaps it is the biological species concept which is based on an exceptional situation? It is sometimes claimed that the biological species concept was "invented" by ornithologists and is valid only for birds. The historical facts refute this assertion. To be sure, a number of ornithologists (Hartert, Stresemann, Rensch, Mayr) were very active in promoting the concept, but Poulton and K. Jordan, the two great pioneers of the concept, were entomologists, and the *Drosophila* workers from Timofeeff-Ressovsky, Dobzhansky, and J. T. Patterson to Spieth and Carson were staunch supporters of the biological species. As unorthodox as some of M. J. D. White's ideas on speciation may seem, he vigorously affirms his acceptance of the biological species, based on his thorough knowledge of orthopterans and other insects (White, 1978). It is evident, then, that the concept does not describe an exceptional situation.

The frequency at which the biological species does not work can be determined only by a careful statistical analysis of all the species of a higher taxon. The first author who undertook such an analysis was Verne Grant (1957). He took eleven genera of Californian plants and determined the percentage of "good" species, that is, well-delimited species that cannot be confused with other species nor interbreed with others. In contrast to the situation in birds, less than half of the species were "good." Only in the milkweed genus *Asclepias* were all 108 species "good." In an analysis of all the species of North American birds, Mayr and Short (1970) showed that 46 of the 607 species had strongly differentiated, peripherally isolated populations which some ornithologists considered full species, others subspecies. In only about four other cases were there any questions concerning species status. The biological species concept was of great help in deciding on species status in sibling species, polymorphic species, and in cases of hybridization. In only a single case (two species in the genus *Pipilo*) did the concept break down completely. In *Drosophila,* where species, on the whole, are quite orthodox, a few situations have been found (for example, in the South American *D. willistoni* complex) that are rather exceptional. The validity of the frequently made claim that the biological species concept cannot be applied to certain higher taxa of animals or plants can be judged only after thorough quantitative analyses of such taxa have been made, such as those described above.

281

The biological factors which create most difficulties for the biological species concept are the following:

Morphological Similarity (or Identity)

The morphological species concept was so well entrenched when the biological concept was introduced that many biologists balked at recognizing morphologically identical populations as sibling species when it was discovered that they were reproductively isolated. The discrimination in 1768 between three species of leaf-warblers (*Phylloscopus*) by Gilbert White and, in the 1820s, between two species of brown creepers (*Certhia*) and between two species of black-capped titmice (*Parus*) by C. L. Brehm, were perhaps the first cases of a recognition of cryptic or sibling species, as such exceedingly similar species have since been called (Mayr, 1942; 1948; 1963). Soon sibling species were also recognized in insects (Walsh, 1864; 1865), even though the majority of entomologists, firmly wedded to a morphological species concept, generally referred to them as "biological races" (Thorpe, 1930; 1940). It was not until the 1930s and 1940s that the enormous importance of sibling species in agriculture and public health was recognized. In particular the discovery by various students of the malaria mosquitos that so-called *Anopheles maculipennis* was actually a complex of six sibling species permitted a dramatic advance in malaria control. Yet, the resistance against the concept of morphologically very similar species, even by outstanding biologists, continued until the 1940s and 50s. When Dobzhansky and Epling (1944) described *Drosophila persimilis,* Sturtevant (1944) objected and continued to call this species *D. pseudoobscura* B. After it had become quite evident that the so-called "varieties" of *Paramecium* were reproductively isolated species, Sonneborn (1957) refused to accept this conclusion and referred to them as syngens. It was not until 1975 that he finally conceded their species status. From the protozoans to the mammals there is no group of animals in which numerous sibling species have not been described in recent years.[18]

The recognition of sibling species faces quite legitimate objections in three areas. (1) Among largely featureless protists or prokaryotes very special techniques (such as nuclear transplants, biochemical analyses) are needed to establish specific distinctness. (2) Among fossils, when all the evidence is missing that would be needed to discriminate among sibling species. (3) In autopolyploids among plants, individuals with different chromosome numbers may be reproductively isolated but morphologically indistinguishable. None of these special situations refutes the biological species concept, even though the practicing taxonomist may be forced occasionally to use morphological criteria to delimit species taxa and thus treat groups of sibling species under a single binomen.

Two interpretations of sibling species were argued about in the 1950s and 1960s. According to Mayr (1948; 1963), sibling species provide evidence that the correlation between morphological divergence and the acquisition of isolating mechanisms is not very strong. Sibling species are biological species that have acquired reproductive isolation but not yet morphological difference. If a genus includes both sibling species and morphologically distinct species, the latter are usually more different genetically, but such a relation does not necessarily hold in intergeneric comparisons. According to another group of students, sibling species are incipient species representing a stage in the speciation process. Subsequent researches have revealed rather convincingly that, from the point of view of reproductive isolation, sibling species do not differ from morphologically distinct species. Furthermore, morphologically distinct species, like *Drosophila silvestris* and *D. heteroneura* on Hawaii, are sometimes far more similar genetically than sibling species. It is now evident that sibling species are *not* incipient species.

Borderline Cases or Incipient Species

Since most species originate as geographical isolates, one should expect that a certain percentage of such isolated populations are on the borderline between subspecies and species status. The decision whether or not to call such populations species is by necessity somewhat arbitrary. The existence of such borderline cases is what is to be expected if one believes in evolution. Many of these cases are likewise an embarrassment to the morphological species concept, since they are as intermediate morphologically as they are reproductively. Of the 607 species of North American birds, for example, 46 have populations belonging to the class of incipient species.

Uniparental Reproduction (Asexuality)

The biological species concept is based on the reproductive isolation of populations. The concept, therefore, cannot be applied in groups of animals and plants that have abandoned bisexual reproduction. For these organisms, populations in the conventional meaning of biology do not exist. In an asexually reproducing species every individual and every clone is reproductively isolated. It would be absurd to call each of them a separate species. But how can such individuals and clones be combined into species? This has long been a source of argument among biologists. Whatever solution is adopted, it is at best a compromise. The one that seems to fit most situations better than any other is based on the knowledge that a species is characterized not only by reproductive isolation but also by the fact that it occupies a species-specific ecological niche. This second characteristic of the species can usually be applied to asexual organisms. It is thus customary to combine into species those asexual individuals and clones that fill the same ecological niche or that play the same role in the ecosystem. The ecological landscape is highly diversified, and as a result ecological niches are discontinuous and so are the occupants of such ecological niches. These discontinuities can often be used to discriminate species taxa among uniparentally reproducing organisms. In most cases, the different niche occupations are correlated with certain morphological, physiological, or biochemical differences and thus one can use these kinds of differences in order to infer ecological differences. Usually whole clusters of more or less correlated differences are involved, and such clusters are called species (Stanier et al., 1970: 525). It is not so, as is sometimes said, that species in asexual organisms are morphologically defined, but rather that morphological differences permit an inference on their niche occupation and thus on their species status.

Evidently, the last word has not yet been said about the species concept in uniparentally reproducing organisms, particularly prokaryotes. In these organisms, many puzzling phenomena are encountered, not matched by anything known in the higher eukaryotes. This includes, for instance. the extreme constancy of

283

"species" of Cyanophyceae (blue-green algae) and the evidence for a good deal of gene exchange in certain genera of bacteria. Far more factual knowledge on these situations is required before one can speculate about what species concept is best applied to these organisms.

Leakage of Isolating Mechanisms (Hybridization)

In mobile animals with well-developed behavioral isolating mechanisms, hybridization is rare, in fact in most species exceptional. A complete breakdown of reproductive isolation among certain populations of sympatric species of animals has, however, been described in a number of cases (Mayr, 1963: 114–125). Yet, neither in birds nor in *Drosophila* is it a serious threat to the biological species concept. Most animal groups are not yet sufficiently well-known to determine the frequency of hybridization. The impression one gets from the literature is that it is as rare in most other groups of animals as it is in birds and in *Drosophila*. Where it is more common, as in freshwater fishes, it still does not lead to a serious breakdown of the integrity of species, since the hybrids are usually sterile. Hybridization, of course, creates just as great a difficulty for the morphological as for the biological species concept.

The situation clearly is different in plants. Edgar Anderson (1949) introduced the useful concept of introgression to designate the incorporation of genes of one species into the gene pool of another species as a result of successful hybridization and back crossing. Botanists are in agreement that such a leakage of genes from one species into another is frequent, although there is still considerable argument as to how common it is (Grant, 1971: 163–184). Sometimes the leakage leads to a complete breakdown of the species border, corresponding to a similar breakdown between the bird species *Passer domesticus* and *P. hispaniolensis*, or *Pipilo erythrophthalmus* and *P. ocai*, but far more frequently the two parental species continue to exist side by side in spite of continuous introgression. Stebbins has recorded a case of two species of California oaks (*Quercus*), hybrids between which are known from the Pliocene to the present, and yet where the two species have retained their essential integrity. Additional similar cases, particularly among oaks, have been described in the recent botanical literature. The genetics of such situations is not understood at all, for it seems as if some part of the genotype of the two species is

284

not affected by the hybridization. The two species, in such a case, seem to remain "reproductively isolated," in the sense that they do not fuse into a single population, in spite of the leakage of certain of their genes. Whether or not in such cases of introgression the biological species concept must be abandoned is still controversial (Van Valen, 1976). I do not think so. Introducing a new technical term for such occurrences may help to call attention to their existence but is of no explanatory value. However, there may be plant genera, like *Rubus, Crataegus,* or *Taraxacum,* without discrete species.

285

The so-called species problem in biology can be reduced to a simple choice between two alternatives: Are species realities of nature or are they simply theoretical constructs of the human mind? Attacks on the species concept had evoked a spirited outburst from Karl Jordan (1905). Wherever he went on his field excursions, he exclaimed, he found well-defined natural discontinuities, each of them characterized not only by visible characteristics but also by a wide array of biological characteristics (vocalization, seasonality, ecological preferences, and so on). For him, an entomologist, species were a simple fact of nature. The ornithologist feels exactly the same way. Indeed, he has trouble understanding why anyone should worry about the species problem.

Then, where do the attacks on the biological species concept come from? They come either from mathematicians, like Sokal, who have only a limited acquaintance with species in nature, or from botanists. I am not a botanist myself, but I have collected and identified plants since my earliest youth and on three continents. To be sure there are "messy" situations, as mentioned above, but I am far more impressed by the clear distinction of most "kinds" of plants I encounter in nature than by the occasional messes. A myopic preoccupation with the "messy" situation has prevented many botanists (but by no means all of them, perhaps not even the majority) from seeing that the concept species describes the situation of natural diversity in plants quite adequately in most cases.[19] Amusingly, some of those who have been most active in belittling the species concept seem to put on a different hat when writing taxonomic revisions and monographs, because there they are thoroughly conventional, thus displaying the same conceptual inconsistency as Lamarck did when applying the nominalist species concept.

APPLYING THE BIOLOGICAL SPECIES CONCEPT TO MULTIDIMENSIONAL SPECIES TAXA

In the course of the eighteenth and early nineteenth centuries the species lost more and more the burden imposed upon it by essentialistic dogma and gradually became the unit of observation of the local naturalist. He knew that what he encountered in his study area was neither simply an assortment of individuals, as claimed by the nominalists, nor the reflections of an essence. Local populations had a unity maintained by the interbreeding of the individuals of which these populations were composed. Differences of sex and age, or other kinds of individual variation, rarely confused a naturalist for long. As encountered by him, species were "real" objects, constant and separated from each other by clearcut gaps. They were the "nondimensional" species known to John Ray and Gilbert White in England and to Linnaeus in Sweden.

286

Mayr, in a series of analyses from 1946 on, pointed out that the species concept has its full meaning only where populations belonging to different species come into contact. This takes place in local situations without the dimensions of space (geography) and time. The word "species" in such a nondimensional situation, designates a relational concept, like the word "brother." To be a brother is not an intrinsic property of an individual, since it depends entirely on the existence of a sibling. A population, likewise, is a species in relation to other sympatric populations. The function of the species concept is to determine the status of coexisting individuals and populations. To know whether or not such individuals belong to the same species is of fundamental importance for the ecologist and the student of behavior. These biologists deal almost exclusively with nondimensional situations. Whether or not two populations that are not in contact with each other either in space or time are conspecific is in most cases biologically uninteresting, if not altogether irrelevant.

There are, however, three groups of biologists whose research problems force them to go beyond the nondimensional situations: taxonomists, paleontologists, and evolutionists. They are forced to categorize populations which represent each other in space and time and which show, as it is called, geographical or temporal variation. How do these workers cope with the problems posed by multidimensional species?

Variation in the Space Dimension

Let me begin with the problem of *geographic variation*. When populations of different countries were compared, all sorts of difficulties in ranking them in the appropriate category were found.[20] The first author to encounter this problem was apparently Buffon, when studying the animals of North America. There he found not only mammals that were very much like European species, as the beaver, the moose, and the red deer (elk) but also many birds seemingly related to well-known species of Europe. These observations induced Buffon to state: "Without reversing the order of nature, it could possibly be that all the animals of the New World are basically the same as those of the Old World from whom they originated. One could further suggest that having been separated from the remaining animals by immense seas or by impassable land, and with time having received all the impressions and suffered all the effects of the climate, which itself has been changed by the very causes that produced the separation, these animals have shrunk, have become distorted, etc. This, however, should not hinder us from regarding them presently as animals of different species" (*Oeuvr. Phil.:* 382). For the sake of convenience most authors, when classifying vicariants, also ranked them as species.

287

Some naturalists preferred a different solution. Pallas during his travels to Siberia, and other Russian explorers into eastern Asia (whose collections Pallas studied), likewise discovered numerous vicariants of European species. Even though he recognized them as new, Pallas usually ranked them as "varieties." These two solutions to the problem of how to rank geographical representatives of previously known species continued to compete with each other for the next 150 years. Most taxonomists called every geographical variant a different species, but the ornithologist Gloger (1833), emphasizing the fact that many, if not most, species of birds vary geographically, recommended calling all such geographical variants races or varieties.

Two developments were necessary in order to terminate this controversy. The first was the rejection of essentialism with its insistence of the constancy of the species essence. This permitted the recognition of the fact that there are geographically variable species taxa and that one had to develop an infraspecific terminology in order to cope with this variability. The second was the recognition that the delimitation of these taxa is a quite separate problem from that of developing unambiguous criteria for ranking them, either in the species category or in an infraspecific category. In order to be able to rank a taxon, one must have a concept of the category in which to place it. This is why it was necessary to develop a clear-cut species concept.

Infraspecific Categories

Essentialists did not know how to deal with variation since, by definition, all members of a species have the same essence. When individuals were found that differed strongly from the norm of the species, they were considered a different species; when they differed only slightly, a "variety." The variety (*varietas*) was the only subdivision of the species recognized by Linnaeus and the early taxonomists, a variety being anything that deviated from the ideal type of the species. In his *Philosophia Botanica* (1751, para. 158) Linnaeus characterized the variety as follows: "There are as many varieties as there are different plants produced from the seed of the same species. A variety is a plant changed by an accidental cause: climate, soil, temperature, winds, etc. A variety consequently reverts to its original condition when the soil is changed." Here Linnaeus defined the variety as a nongenetic modification of the phenotype. Yet, in his discussion of varieties in the animal kingdom (para. 259), Linnaeus included also genetic variants such as races of domestic animals and various kinds of intrapopulation variants. As examples he lists "white and black cows, small and big ones, fat and lean ones, smooth and woolly ones; likewise the races of domestic dogs." It is evident that the category "variety" in the writings of Linnaeus consisted of a highly heterogeneous lot of deviations from the species type. He did not distinguish between inheritable and noninheritable varieties, nor between those that refer to individuals and those that represent different populations (such as domestic and geographic races). This confusion continued for two hundred years, and some residues of it can be found even in the contemporary literature. The application of the term "variety" to such different phenomena as intrapopulation variants and distinguishable populations was one of the reasons why Darwin did not see more clearly the fundamental problem of speciation (Mayr, 1959b).

Geographical varieties became particularly important in the history of systematics and evolutionism. For instance, Pallas and Esper (Mayr, 1963: 335) recognized as early as the eighteenth century that geographical races are something rather different from ordinary varieties and attempted to express this terminologically. In due time, such varieties were designated as "subspecies" but were still treated quite typologically. The subspecies was considered, almost to the end of the nineteenth century, as a taxonomic unit like the morphological species but of a lower categorical rank.[21] This typological interpretation of the subspecies was only very slowly replaced by a populational one. A subspecies is now defined as "an aggregate of phenotypically similar populations of a species, inhabiting a geographic subdivision of the range of a species, and differing taxonomically from other populations of the species" (Mayr, 1969: 41). It is a unit of convenience for the taxonomist, but not a unit of evolution.

289

By the 1850s the most progressive zoologists, particularly students of birds, fishes, butterflies, and snails, began to realize not only that no two individuals of a population are entirely identical, but also that most populations differ from each other in the mean values of many characters. The consequences of this new insight on evolutionary theory will be discussed in a later chapter but it also had an impact on the classification of species.

When a population differed "taxonomically" (which usually meant morphologically) from previously named ones, it was described as a new subspecies. A name was added to the species name, and thus the name of the subspecies became a trinominal. The British race of the white wagtail *Motacilla alba,* for instance, became *Motacilla alba* var. *lugubris.* In due time the designation "var." was omitted and subspecies were referred to simply by trinominals, such as *Motacilla alba lugubris.* The first author to employ trinomials routinely was Schlegel (1844).

At the same time a tendency already present in the writings of Esper, became a tradition: to restrict the term "variety" to individual (intrapopulational) variants and the term "subspecies" to geographic races (Mayr, 1942: 108–113).

The consistency with which the term "subspecies" is applied to geographic races varies from one taxonomic group to the other. Many botanists even today call geographic races varieties. In certain areas of zoology the term "variety" is used only for individual variants, geographic races being either ignored (if slight) or raised to the rank of full species. We are still far from consistency in plant and animal taxonomy.

Polytypic Species Taxa

It is only in some species that subspecies are recognized. Certain authors felt that such species taxa should be terminologically distinguished and various names for them were suggested. Rensch (1929) proposed the term *Rassenkreise,* while Mayr (1942) employed the internationally more suitable term *polytypic species,* originally introduced by Julian Huxley. This is the term now in universal use to describe species that are composed of a number of subspecies (Mayr, 1969: 37–52).

At first it was thought that the adoption of polytypic species would demand the establishment of a new species concept. However, as soon as the terminological distinction between "category" and "taxon" was made, it became evident that the polytypic species is merely a special kind of species taxon but does not require any change in the concept of the biological species category.

The recognition of polytypic species taxa by the new systematics led to a great clarification and simplification of classification at the species level. In ornithology, for instance, the consistent application of polytypic species permitted a reduction in the number of recognized species taxa from over 20,000 in 1920 to around 9,000 at present.

Progress in the modernization of species taxonomy was highly uneven in different areas of biology. For instance, more than 95 percent of all species of birds had been described by 1930, and only about three to four new species per year were discovered subsequently. As a result, most taxonomic effort was devoted to the evaluation of the validity of subspecies and to the delimitation of polytypic species. In other groups, so many new species are still discovered that the use of the polytypic species concept has hardly begun.

To the present day there is not yet complete unanimity among ornithologists in the treatment of subspecies. The situation was almost chaotic in the nineteenth century (Stresemann, 1975: 243–268), however. Some authors ignored geographically isolated populations altogether unless strikingly different, others described them as subspecies, still others called each such population a full species. By 1890 the disagreement had narrowed down. It had been agreed that distinguishable populations should be recognized, but there was still disagreement as to which of these populations to call subspecies and which species. Under the influence of the leading American ornithologists Baird, Coues, and Ridgway, the principle was adopted of treating as subspecies all those populations the variation of which overlapped that of the parent population. This principle was expressed in the slogan, "Intergradation is the touchstone of trinomialism." True to the morphol-

290

ogical species concept, to which these authors adhered, any isolate was called a species that showed a clear-cut morphological or color difference. This criterion for the recognition of species taxa was widely adopted internationally, for instance in the Lankester-Wallace species definition referred to above.

The German ornithologist Ernst Hartert took exception to this narrow conceptualization of subspecies and replaced it by a new criterion, that of geographical representation. Even if a geographically representative population differed, and "even though one does not have intermediate forms," by calling it a subspecies one "shows the closeness of the relationship." His definition of subspecies, thus, was by inference based on the biological species concept. Although vigorously opposed by the ornithological establishment, in America as well as in Europe, Hartert's principle soon found followers in Germany (Meyer, Erlanger, Schalow) and in Austria (Tschusi, Hellmayr) and became completely victorious in the 1920s under the leadership of Erwin Stresemann.

291

However, Stresemann and some of his followers often went too far in the application of the principle of geographical vicariance. They tended to reduce every allopatric species to the rank of a subspecies, in part influenced by the *Formenkreislehre* of Otto Kleinschmidt (see Stresemann, 1975). It was Rensch (1929) who called for a halt to this excessive lumping. He proposed to recognize not only groups of geographically representative subspecies, that is, polytypic species, but also groups of geographically representative species which he called *Artenkreise*, renamed "superspecies" by Mayr (1931). Superspecies are groups of geographically vicariant populations (previously considered polytypic species), the members of which (called "allospecies" by Amadon) have been isolated sufficiently long to have reached species level. Much of the activity of avian taxonomy in recent decades has consisted in the careful scrutiny of polytypic species, particularly in insular regions, to determine which isolated and pronounced subspecies should be raised to the rank of allospecies. The main value in the recognition of superspecies is in zoogeographic research.

Superspecies are common also in many other groups of organisms, but this does not become apparent unless a specialist depicts the distribution of the most nearly related species on the same map. Quite often the borders of such species are in contact with each other or overlap slightly (parapatric distribution), with or without a slight amount of hybridization. The *Rana pipiens* group of frogs in North America, in the 1940s considered a widespread polytypic species, has since been shown to be such a super-species, consisting of at least six component (allo-)species.

The taxonomy of no other group of animals is as mature as that of birds. For this reason birds have been particularly valuable not only in evolutionary studies but also in ecology. In most other groups of animals the application of polytypic species, biologically defined, has progressed to a much lesser degree. There are indications that the situation in birds is a particularly simple one since many of the difficulties encountered by students of other taxa of animals and of plants seem to be absent from birds. Chromosomal variation, for instance, seems very slight and polyploidy absent. Interspecific hybridization is sufficiently rare not to cause any problems, nor is there any ecological specialization or adaptation to specific hosts that would create difficulties. Incipient speciation seems to proceed exclusively by way of geographical isolates. This gives the bird species a simplicity that is rarely found in other taxa (Mayr and Short, 1970). It will require a great deal of additional research to determine whether or not the prevailing species concept has to be modified or additional kinds of species taxa recognized in order to cope with complexities in other groups of animals and plants.

Variation in the Time Dimension

Paleontologists face particularly formidable difficulties in the delimitation of species taxa. At different fossil localities they have to deal both with the time dimension and with the geographical dimension. The student of recent biota deals with an instantaneous cross-section of the evolutionary continuity, while the paleontologist deals with diversity as a continuum. In view of these difficulties paleontologists, until rather recent times, have on the whole sidestepped the species problem. In his *Tempo and Mode in Evolution* (1944), Simpson hardly mentions the species at all. When he does, it is only in connection with speciation, as for instance, "The phyletic lines . . . are composed of successive species, but successive species are quite different things from the contemporaneous species that are involved in speciation as that word is usually used" (p. 202).

Like the neontologist, the paleontologist must attempt to solve his problem by beginning with the nondimensional situation. This approach is possible, since a sample taken at a fossil locality (at a restricted horizon) normally represents a nondimensional situation. Here the paleontologist is able to make an unequivocal decision. The variation here encountered is either that of a single population, or it represents several species. Subspecies in space and time are excluded by the nature of the situation. The analysis of such monotopic samples provides the yardstick to apply in the comparison of samples differing in space and time. To split the material collected at a single exposure into numerous "varieties," as was done by some invertebrate paleontologists, may be helpful in stratigraphic researches but is indefensible biologically. It is even less meaningful biologically when paleobotanists recognize separate "species" for leaves, stems, inflorescences, and seeds collected in the same deposits. Admittedly even such nondimensional samples pose difficult problems. It is not always easy to determine whether certain phena represent different species or whether they represent age or sex differences. In groups where sibling species occur these can presumably never be separated in fossils. However, these are technical difficulties rather than conceptual problems.

Such conceptual difficulties arise, however, when the paleontologist is forced to expand the local species of a single locality into the multidimensional space of the history of life. By what criteria should he delimit his species taxa? Each phyletic lineage is an open system. Where should one set the beginnings and the end of a species in such continuity? Hennig (1950), of the cladistic school, attempts to escape this dilemma by defining the species simply as the segment of a phyletic lineage between two branching points. This omits any reference to reproductive isolation and is strongly typological in its exclusive reliance on a limited number of ancestral or derived characters. Furthermore, it is strictly formalistic, since in this scheme species a automatically becomes species b when another species, c, branches off, even when there is no evidence for any difference between species a and b. The cladist E. O. Wiley (1978) has recently stated that "no presumed separate, single, evolutionary lineage may be subdivided into a series of ancestral and descendant species." The claims of Hennig and his followers ignore the fact that speciation events that take place in peripherally isolated populations have no effect on the main body of the species, which continues its evolutionary life without changing its species status, since it is unaffected by the budding off of a peripheral daughter species.

The formalistic "solution" of the problem of species in time, thus, is no solution. As Simpson (1961: 165) rightly points out, all evolutionary lineages (except in cases of saltational evolution) have a complete evolutionary continuity, and if one would not divide such a line into ancestral and descendant species one "could start with man and run back to a protist still in the species *Homo sapiens.*" But how does one divide such a lineage into a sequence of species?

Simpson has attempted to solve the problem by introducing a new species concept, that of the *evolutionary species:* "An evolutionary species is a lineage (an ancestral-descendant sequence of populations) evolving separately from others and with its own unitary evolutionary role and tendencies" (1961: 153). The vulnerability of this definition is, of course, that it applies equally to most incipient species, such as geographically isolated subspecies. These also evolve separately and have their own unitary evolutionary role but are not species until they have acquired reproductive isolation. Also, what exactly is meant by "unitary evolutionary role"? Simpson's definition is that of a phyletic lineage, but not of a species.

Furthermore, this definition does not tell us at all how to delimit a sequence of species taxa in time. Are the sequences of temporal species taxa described by Gingerich (1977) in *Plesiadapis* and other Paleocene-Eocene mammals good species or only subspecies? The answer, they are species if they have different "unitary evolutionary roles," is no answer, for how are we ever to determine this? Simpson's definition is basically a typological description which quite ignores the fact that the evolutionarily most interesting species taxa are polytypic species. Many of the populations and races of such complexes differ significantly in niche utilization; they do not have a unitary evolutionary role. The paleontological species definition is trying to replace the nondimensional by a one-dimensional (time-dimension) definition, but becomes involved in contradictions by ignoring the "horizontal" dimensions (longitude and latitude). The principal weakness of so-called evolutionary species definitions is that they minimize (if not ignore) the crucial species problem, the causation and maintenance of discontinuities between species, and concentrate instead on the problem of how to delimit multidimensional species taxa. Yet, they do not even meet the limited objective of how to delimit such open-ended systems.[22] Incidentally, there have been earlier attempts to incorporate the criterion of "evolving" in a species definition (for example, that of Alfred Emerson), but they were not adopted when the irrelevance and impracticality of this criterion was recognized.

The situation, however, is not altogether hopeless. Many phyletic lineages become extinct and this sets a natural end at the last species of the lineage. Likewise, many species originate quite rapidly in a peripherally isolated founder population or in a temporary refuge. Here, the beginning of the first species of the new lineage is given. It is only in the cases of a sequence of ancestor-descendant species which transform gradually into each other in a single phyletic lineage that a sharp delimitation between temporal species taxa is impossible. Here biological evolution fails to accommodate the wishes of the taxonomist. Fortunately, the fossil record is more accommodating. Its deficiencies usually provide sufficient gaps in the lineages to permit a delimitation of vertical species taxa, as artificial as this may be. It seems that we will have to accept this compromise solution since the evidence does not seem to support the claim of some proponents of the theory of "punctuated equilibria" (see Chapter 13) that there is never any phyletic speciation and that all new species originate in founder (or refuge) populations or even by saltations.

THE SIGNIFICANCE OF SPECIES IN BIOLOGY

Owing to the never-ending arguments about the definition of the species category, those who do not work with species but with cells or molecules may think that the species is an arbitrary and insignificant concept in biology. Actually, the recognition of the importance of species is steadily growing (Mayr, 1969). At one time the species category was largely the concern of the taxonomists. At a later stage the geneticists began to emphasize the genetic unity of species, expressing this either by referring to the species as a single large gene pool or by stressing the coadaptation of the genes with each other. This in turn affected the evaluation of the process of speciation. This is no longer seen, as it was by Darwin, as a casual, ever present change but as a rather drastic event. We have again become conscious that it requires a concrete step to get from one species to the next. It is now seen that the genetic changes from population to population within a species are often of a different kind and definitely of a lower order of magnitude than the genetic changes from one species to another.

With one of the important aspects of the species being its reproductive isolation from other species, it is evident that the acquisition of the isolating mechanisms is of crucial importance in the history of species. One may find a great deal of genetic change in adjacent and interconnected populations of a species, but they all belong to the same species when they are not separated by reproductive barriers. As long as such barriers are absent, the diversity of populations may be reversed by gene flow, by drastic hybridization, or by converging selection pressures. The point of irreversibility is reached when an evolving population has acquired isolating mechanisms from its parental population. From this "point of no return" on, the new species can invade new niches and new adaptive zones. The origin of new higher taxa and of all evolutionary novelties ultimately goes back to a founder species. The species, therefore, is the basic unit of evolutionary biology (Mayr, 1970: 373–374).

296

The role of the species in evolution has often been underrated. Huxley (1942) considered most of species formation "a biological luxury, without bearing upon the major and continuing trends of evolutionary process." Mayr (1963: 620–621) disagreed with this interpretation: "A prodigious multiplication of species is a prerequisite for evolutionary progress. Each species is a biological experiment . . . There is no way to predict, as far as the incipient species is concerned, whether the new niche it enters is a dead end or the entrance into a large new adaptive zone . . . Although the evolutionist may speak of broad phenomena, such as trends, adaptations, specializations and regressions, they are not separable from the progression of the entities that display these trends, the species. The species are the real units of evolution."

The species also to a large extent is the basic unit of ecology. Since ecosystems are composed of species, no ecosystem can be fully understood until it has been dissected into its component species and until the mutual interactions of these species are understood. A species, regardless of the individuals composing it, interacts as a unit with other species with which it shares the environment (cf. Cody and Diamond, 1974). This interaction of species is the principal subject of ecology.

In view of the fact that isolating mechanisms make a species a reproductive community, the animal species is also an important unit in behavioral science. Individuals that belong to the same species share the same signaling system as far as all components of the courtship behavior is concerned. Likewise, members of a species share many other behavioral patterns, particularly all those that have to do with social behavior.

Because the species is one of, if not the most significant of, the units of evolution, of systematics, of ecology, and of ethology, the species is as important a unit of biology as is the cell at a lower level of integration. It is an immensely useful ordering device for many significant biological phenomena. Even though there is no special name for the "science of species," as is the name "cytology" for the science of cells, there is no doubt that such a science exists and that it has become one of the most active areas in biology.

The species is also of great practical importance. Much confusion in various branches of biology, including physiology, has been caused by the imprecise, if not erroneous, identification of the species with which the investigator worked. Applied biologists, whether dealing with disease vectors, pathogens, agricultural or forestry pests, or wildlife or fisheries problems, constantly deal with species. In spite of the variability caused by the genetic uniqueness of every individual, there is a species-specific unity to the genetic program (DNA) of nearly every species. This pervading presence of species poses a multitude of problems of origin and meaning, problems which occupy a major portion of the current research in biology.

II

Evolution

THERE IS probably not a primitive tribe in the world that does not have some myth about the origin of man, of trees, of the sun, and perhaps even of the world as a whole. A big snake or a giant bird, a fish or a lion or some other organism with supernatural powers or capacities of generation was the kind of acting force involved in these origins. When religions developed, with concrete deities, it was these gods who created things and life. Among the Greeks it was Zeus, Athena, Poseidon, and other gods who filled this role. The story of creation in Genesis is the prototype of this concept of origin. Most of these early stories of origin have in common that the creation was a once-and-for-all-time event. It resulted in a static, timeless world, the only change in which is the coming and going of seasons and of human generations. A process of evolution was an altogether alien concept to the early creationists. Genuine evolutionary thinking rose remarkably late in history, opposing claims notwithstanding.[1]

7 ✐ Origins without evolution

THE SEVENTH, sixth, and fifth centuries B.C. were periods of unprecedented flourishing of commerce and trade in the eastern Mediterranean and the Near East. Greeks, particularly the Ionian colonists in Asia Minor, traveled to Egypt and Mesopotamia and became acquainted with Egyptian geometry and Babylonian astronomy. It gradually became apparent that one could explain many phenomena "naturally" which had previously been ascribed to the activities of gods. Why not, then, also ask questions about the origin of matter, of the earth, of life?

And who should ask such questions? Not scientists, because until the late Middle Ages and the Renaissance there was no science as we now understand this word. The place of scientists was taken by philosophers, who searched for the truth and attempted to understand the world in which they lived. Various schools of philosophy were described in Chapter 3; here will be mentioned only what may have a bearing on evolution (Guthrie, 1962). Unfortunately, only fragments are left of the writings of the pre-Socratic philosophers, but there is enough to make us suspect that much of what they taught was apparently part of the Babylonian or Egyptian tradition.

The first philosopher we know was Thales of Milet, who lived ca. 625–547 B.C. He was primarily an astronomer, geometer, and meteorologist and evidently not interested in biological phenomena. He considered water the first principle, and Aristotle speculated later that Thales might have done so because water plays such an important role in animal and plant life, even semen being moist. Also, so many animals have some connection with water in their life cycle.

His student Anaximander (ca. 610–546 B.C.), although best known as a geographer and astronomer, was more interested in the world of life. He had a complete but quite fanciful cosmogony in which fire, earth, water, and air played important roles. Anaximander imagined the first generation of organisms as coming into being through a metamorphosis, like that through which an insect emerges from the chrysalis stage:

> The first animals were generated in the moisture, and were enclosed within spiny barks. As they grew older, they migrated onto the drier land; and, once their outer bark was split and shed, they survived for a short time in the new mode of existence.

> Man, to begin with, was generated from living things of another kind, since, whereas others can quickly hunt for their own food, men alone require prolonged nursing. If he had been like that in the beginning, he would never have survived. Thus men were formed within these [fish-like creatures] and remained within them like embryos until they had reached maturity. Then at last the creatures burst open, and out of them came men and women who were able to fend for themselves. (from Toulmin and Goodfield, 1965: 36)

This is not an anticipation of evolution, as has sometimes been claimed, but rather refers to the ontogeny of spontaneous generations. The subsequent generations of philosophers—Anaximenes (ca. 555 B.C.), Diogenes of Apollonia (ca. 435 B.C.), Xenophanes (ca. 500 B.C.), and Parmenides (ca. 475 B.C.)—accepted spontaneous generation from slime or moist earth.

Empedocles (ca. 492–432 B.C.) proposed a preposterous theory of the origin of living beings. At first, only body parts originated: heads or limbs without bodies, heads without eyes or mouths, and so on. While floating, these parts were attracted to each other until perfect combinations were achieved; imperfect ones perished. It is quite ridiculous to call this a forerunner of Darwin's theory of natural selection, since no selection is involved in bringing together complementary parts, nor is the elimination of imperfect pieces a process of selection. Empedocles was perhaps originally inspired to propose his theory by the existence of monstrosities such as two-headed calves.

In the writings of Anaxagoras (ca. 500–428 B.C.) and Democritus (ca. 500–404 B.C.) we encounter the first hints of adaptation. For Anaxagoras a nonmaterial NOUS provided the impetus to get the world going but without directing the future course of the origin of things. This was not a theory of creation by design, as has sometimes been claimed. Democritus, who apparently admired organic adaptations, carefully refrained from postulating any directive agency. Rather, he thought, the building up of organization—of systems—was a necessary consequence of the property of the atoms. Democritus was thus the first to pose the problem of chance mechanisms versus immanent goal-directed tendencies. He also believed in the orderliness of the world, posing problems which Aristotle later attempted to solve through the concept of teleology.

Two aspects, in particular, characterize the concepts of world origins of the early Greek philosophers. (1) The acts of "creation" are de-deified, that is, the world, or life, or specific organisms, are not the products of the action of a god, as was universally believed in the prephilosophical period, but are the result of the generative power of nature. (2) The origins were nonteleological, that is, without an underlying design or goal; rather, what took place was the result of chance or of an irrational necessity.

Thus, these philosophers were the first to provide a "natural" explanation of the world of phenomena, that is, a rational explanation that invokes only known forces and materials, like the sun's heat or water and earth. As naive and primitive as these speculations appear to the modern mind, they constitute the first scientific revolution, so to speak, a rejection of supernatural in favor of materialistic explanations.

There is another seemingly fundamental difference between the world concept of the Greek philosophers and the priest-authors of the Bible. The world of the Bible is new, the creation having taken place only about 4,000 years B.C., as Bishop Ussher later calculated. Furthermore, this world was soon going to come again to an end, on the day of reckoning. Thus, time was a negligible aspect of the world view. The treatment of time by the Greek philosophers, on the other hand, appears to us inconsistent. Time for us moderns means change, but the predominant concept among the pre-Socratics was an eternal world, without significant change, or at best with cyclical changes that sooner or later resulted in a return to the original condition—a steady-state world. This apparently was true even for Heraclitus of the *panta rhei* motto ("all is in flux"). Hence, even though time was unlimited, it was of little consequence for the Greek world view; it did not require a replacement of a world of origins by an evolving world. And origins were indeed of enormous interest to them: the origin of the universe, of the earth, of life, of animals, of man, and of language. But little thought, if any, was given to subsequent change.

The approach of the school of Hippocrates (ca. 460–370 B.C.) was drastically different. These physicians placed greater weight on observation and on an empirical approach than on reasoning. They believed unquestioningly in an inheritance of acquired char-

303

acters and in the principle of use and disuse. Climate and other regional factors were responsible for differences among people living in different places.

Plato

Scattered through the teachings of the Ionian philosophers were promising points for the development of evolutionary thinking, such as unlimited.time, spontaneous generation, changes in the environment, and an emphasis on ontogenetic change in the individual. But it went no further. Indeed, Greek philosophy soon changed its direction rather drastically. Owing to the influence of Parmenides, and even more so of the Pythagoreans of southern Italy, the thinking of Greek philosophy moved more and more toward abstract metaphysics and was increasingly influenced by mathematics, particularly geometry. This was the first of countless episodes in the history of biology where mathematics or the physical sciences exerted a harmful influence on the development of biology. Preoccupation with geometry led to the search for "unchanging realities," *Idealgestalten*, underlying the fleeting flux of appearances. In other words it led to the development of essentialism (see Chapter 2), and this philosophy is, of course, totally incompatible with evolutionary thinking.

> Once the axiom was accepted that all temporal changes observed by the senses were merely permutations and combinations of "eternal principles", the historical sequence of events (which formed a part of the 'flux') [individual variation being another part] lost all fundamental significance. It became interesting only to the extent that it offered clues to the nature of the enduring realities . . . philosophers concerned themselves instead with matters of *general principle*—the geometric layout of the heavens, the mathematical forms associated with the different material elements . . . More and more they became obsessed with the idea of a changeless universal order, or "cosmos": the eternal and unending scheme of Nature—society included—whose basic principles it was their particular task to discover. (Toulmin and Goodfield, 1965: 40)

These new concepts found their most brilliant spokesman in Plato, the great antihero of evolutionism. Plato's thinking was that of a geometrician, and he evidently understood very little of biological phenomena. Four of Plato's dogmas had a particularly deleterious impact on biology through the ensuing two thousand years. One, as stated, was essentialism, the belief in constant *eide*,

304

fixed ideas, separate from and independent of the phenomena of appearance. The second was the concept of an animate cosmos, a living, harmonious whole (Hall, 1969: 93), which made it so difficult in later periods to explain how evolution could have taken place, because any change would disturb the harmony. Third, he replaced spontaneous generation by a creative power, a demiurge. Since Plato was a polytheist and pagan, his demiurge was something of a less concrete person than the creator-god of the great monotheistic religions. Yet, it was in terms of monotheism that the demiurge, the craftsman who made the world, was later interpreted. And it was this interpretation which led to the later Christian tradition that "it is the task of the philosopher to reveal the blueprint of the creator," a tradition still powerful up to the middle of the nineteenth century (natural theology, Louis Agassiz). The fourth of Plato's influential dogmas was his great stress on "soul." References to noncorporeal principles can be found also in pre-Socratic philosophers but nowhere as specific, detailed, and all-pervasive as in Plato. When later this merged with Christian concepts, the belief in the soul made it exceedingly difficult for the devout to accept evolution, or at least to include man and his soul in the evolutionary scheme. It has often been pointed out what a disaster Plato's writings have been for biology, but nowhere else as much as for evolutionary thinking.[1]

Aristotle

The first great naturalist of whom we know, Aristotle, would seem to have been the ideal person to become the first to develop a theory of evolution.[2] He was an excellent observer and was the first to see a gradation in living nature. Indeed, he thought that "nature passes from inanimate objects through plants to animals in an unbroken sequence." Many marine animals, he said, like sponges, ascidians, and sea anemones, resemble plants more than animals. Later writers converted this into the grand concept of the *scala naturae* or Great Chain of Being (Lovejoy, 1936), which in the eighteenth century facilitated the emergence of evolutionary thinking among the followers of Leibniz.

Not so with Aristotle. He held too many other concepts irreconcilable with evolution. Movement in the organic world, from conception to birth to death, does not lead to permanent change, only to a steady-state continuity. Constancy and perpetuity are thus reconcilable with movement and with the evanescence of individuals and individual phenomena.

As a naturalist, he found everywhere well-defined species, fixed and unchanging, and in spite of all of his stress on continuity in nature, this fixity of species and their forms (*eide*) had to be eternal. Not only was Aristotle not an evolutionist, in fact he had great trouble in imagining beginnings of any kind. For him the natural order was eternal and unchanging, and he would have gladly endorsed Hutton's proclamation, "No Vestige of a Beginning—no Prospect of an End!"

It must be pointed out that the unilinear gradation which Aristotle saw in the world was a strictly static concept. He repeatedly rejected the "evolution" theory of Empedocles. There is order in nature, and everything in nature has its purpose. He stated clearly (Gen. An. 2.1.731b35) that man and the genera of animals and plants are eternal; they can neither vanish nor have they been created. The idea that the universe could have evolved from an original chaos, or that higher organisms could have evolved from lower ones, was totally alien to Aristotle's thought. To repeat, Aristotle was opposed to evolution of any kind. Biologists, including Charles Darwin, always have had great admiration for Aristotle, but they have had to admit regretfully that they could not count him among the evolutionists. This antievolutionary position of Aristotle was of decisive importance for the developments of the next two thousand years, considering Aristotle's enormous influence during that period.

Among the post-Aristotelian thinkers, the Epicureans are sometimes mentioned as potential evolutionists. This is a misinterpretation. Admittedly, in contrast to Aristotle, they were interested in origins. In Lucretius' poem "On the Nature of Things" the spontaneous origin in a bygone golden age of all kinds of creatures is postulated, even that of man (Bailey, 1928; De Witt, 1954). Yet, he resolutely rejects evolutionary change:

> But each thing has its own process of growth;
> All must preserve their mutual differences,
> Governed by Nature's irreversible law.

He imagined that the earth was so prodigious in its creativity that it produced not only viable creatures but also monsters and weaklings which could not survive and were weeded out. This process of elimination has sometimes been designated as an early theory of natural selection, an interpretation which is, of course, quite misleading, as we shall see.

Thus, at the end of the classical period, the thinkers had not yet been able to emancipate themselves from a concept of either an entirely static or a steady-state world. At best they concerned themselves with origins. A historical change in the organic world— organic evolution—was quite outside the conceptual framework of the era.

Many historians have speculated as to why the Greeks were so unsuccessful in founding evolutionism. We have already touched upon all these reasons: There was the absence of a concept of time, and if there was an idea of time, it was that of an unchanging eternity, or of an ever-continuing cyclical change always returning to the same beginning. There was the concept of a perfect *kosmos*. There was essentialism, which is completely incompatible with a concept of variation or change. All this had to be weakened or broken down before evolution could be thought of. And yet, in one way, the Greeks did lay a foundation for evolutionary biology, and it was Aristotle more than anyone else who was responsible for this. Evolution, as we now realize, can be inferred only by indirect evidence, supplied by natural history, and it was Aristotle who founded natural history.

307

THE IMPACT OF CHRISTIANITY

During the fall of the Roman Empire a new ideology, Christianity, took over western thought. Its influence and that of the all-powerful Church hierarchy cannot be exaggerated. They introduced a drastically different type of conceptualization and abolished freedom of thought. No longer was man able to think and to speculate as he pleased. Now, the word of God was the measure of all things, and this word was revealed in the holy scriptures; thus the Bible became the measure of all things. Instead of a timeless eternity, the Christian and Jewish religions believed in an almighty author of all things who had created the world out of nothing and who would someday, on the day of judgment, bring it abruptly to an end. His making of the earth took six days, enough for all sorts of origins but not for any evolution. Nor had there been time for evolution since creation, since on the basis of the genealogies recorded in the Bible, it was calculated that the world had been created as recently as 4000 B.C. And yet, in the Old Testament there are many allusions to linear sequences (as in the six days of creation) which were more suitable as a base for evolutionary thought than either the constant or cyclical world of the Greeks.

The early Church fathers could afford to be rather liberal in their interpretation of the Bible because with everybody a convinced Christian, heresies posed no danger. Nor was there any science in existence that might have necessitated a more rigid position. Saint Augustine (Gilson, 1960), for instance, even though recognizing the Holy Scriptures as the only authority, adopted a somewhat allegorical interpretation of creation. Not only finished products were created at the beginning but much of God's creation, according to him, consisted in giving nature the potential to produce organisms. Their essences, their "naturae," were indeed created in the beginning, but they emerged or were activated often only much later. All parts of nature, land or water, have the capacity to generate something new, inanimate or alive. Thus, spontaneous generation would be no problem to the faithful; it can happen any day.

The thousand years after the founding of Christianity were a period of depressing intellectual stagnation. Characteristic for the scholastic universities were their attempts to establish truth by legalistic, deductive arguments, but the long, drawn-out controversies resulting from this approach were doomed to failure. A reawakening came from an entirely different direction, from a revival of an interest in nature, a rebirth of natural history, as is evident in the activities of Frederick II and Albertus Magnus (see Chapter 4). Whether Aristotelians or not, the Catholic scholars of the Middle Ages, in spite of frequent references to a chain of being or a grand hierarchy of the phenomena of the world, maintained a belief in the strict fixity of all species.

Perhaps the most important event during the scholastic period was a revolt in the camp of the scholastic philosophers themselves. A faction developed, later called the *nominalists,* who rejected the basic tenets of essentialism.[3] There are no essences, they said, and all we actually have are names that are attached to groups of objects. Once we have the name "chair," we can bracket together all objects that fall under the definition of chair, whether they are diningroom chairs, lawn chairs, or upholstered chairs. The attacks on essentialism (called *realism*) made by the nominalists signaled the first weakening of essentialism. Some of the thinking of the inductive philosophers and empiricists of England, from Bacon on, had nominalist overtones, and there is the possibility of an ideological continuity. Indeed, nominalism was perhaps an anticipation of population thinking (see Chapter 2).

The Reformation represented a definite set-back for evolutionism, because the coming of Protestantism reinforced the authority of the Bible. Indeed, it led to a completely literal interpretation of "the Word," that is, to fundamentalism. Liberal interpretations, such as that of Saint Augustine, were now totally rejected.

Curiously, the so-called scientific revolution of the sixteenth and seventeenth centuries, a movement largely confined to the physical sciences, caused no change at all in this attitude toward creationism. All the leading physical scientists and mathematicians—Descartes, Huyghens, Boyle, and Newton—were believers in a personal god and strict creationists. The mechanization of the world picture (Dijksterhuis, 1961), the dominant conceptual revolution of the time, did not require, indeed could not even tolerate, evolution. A stable, uniquely created world maintained by general laws made complete sense to one who was steeped in essentialism and who believed in a perfect universe.

Philosophy was equally reactionary. One finds no trace of genuine evolutionary thinking in the writings of Bacon, Descartes, or Spinoza. Descartes emphasized that, considering God's omnipotence, he can have created only perfection, and nothing can evolve that has been perfect from the beginning.[4] Curiously, it was theology, in the form of natural theology, which prepared the way for evolutionary thinking to a far greater degree than philosophy.

THE COMING OF EVOLUTIONISM

Evolution, in a way, contradicts common sense. The offspring of any organism always develops again into the parental type. A cat always produces only cats. To be sure, prior to the acceptance of evolution there have been theories of sudden change. For instance, there was the belief in spontaneous generation, as well as that of *heterogony*, a belief that seeds of one species of plants, let us say wheat, could occasionally produce plants of a different species, such as rye.[5] But both of these were theories of origins and neither of them had anything to do with evolution. It required a veritable intellectual revolution before one could even conceive of evolution.

The greatest obstacle to the establishment of the theory of evolution was the fact that evolution cannot be observed directly like the phenomena of physics, such as a falling stone or boiling water or any other process that takes place in seconds, minutes, or hours during which ongoing changes can be carefully recorded. Instead, evolution can only be inferred. But in order to draw inferences one must first have an appropriate conceptual framework. Fossils, facts of variation and of inheritance, and the existence of a natural hierarchy of organisms can serve as evidence only after someone has postulated the occurrence of evolution. However, the consecutive *Weltanschauungen* prevailing from the Greeks to the eighteenth century were incompatible with evolutionary thinking, or at least inimical to it. An indispensable prerequisite for the proposal of a theory of evolution was, therefore, an erosion of the world view prevailing in the western world prior to the adoption of evolutionary thinking. It had two major theses. The first was the belief that the universe in every detail was designed by an intelligent creator. This, together with the other one, the concept of a static, unchanging world of short duration, were so firmly entrenched in the western mind by the end of the Middle Ages that it seemed quite inconceivable that they could ever be dislodged. And yet this is what happened gradually during the seventeenth, eighteenth, and early nineteenth centuries. What were the causes of this astonishing intellectual revolution? Was it the result of scientific research or of an all-pervasive change in the cultural and intellectual milieu? The correct answer to this is apparently that both were important.

Beginning with the fourteenth century, a new spirit seems to have awakened in the west. The age of the voyages, the rediscovery of the thought of the ancients, the Reformation, the new philosophies of Bacon and Descartes, the development of secular literature, and finally the scientific revolution all weakened previously held beliefs. The more the scientific revolution in the physical sciences stressed the need for a rational treatment of natural phenomena, the less acceptable became supernatural explanations.

The changes were not restricted to science. Ferment was found everywhere. A concept of history began to develop in the late seventeenth and eighteenth centuries, no doubt stimulated by the revival of the Greek tradition, the study of the Greek classics, and the new interest in classical Greek architecture and culture. The voyages acquainted the western world with the existence of primitive man, and the question suddenly was asked, How did civilized man develop from an earlier primitive state? This led for the first time to the asking of questions that now occupy the social sciences. The Italian Giambattista Vico wrote the great pioneering text *Scienzia Nuova* (1725), dealing with the philosophy of history (Croce, 1913; Berlin, 1960). For him, the various periods of human history were not different aspects of essentially the same story; rather, they were successive stages of a continuing process, a process of necessary evolution.

The gradual emancipation from the spiritual and intellectual straitjacket of the Church was accompanied by the development of profane literature. Forbidden thoughts were introduced in works of fiction, and new theories concerning the origin of the earth, of man, or of human society were tried out in various utopian works, many of which were published in the sixteenth, seventeenth, and eighteenth centuries.

311

Two of these works of fiction are particularly important as indicating the new thinking. One is Bernard de Fontenelle's *Conversations on the Plurality of the Worlds* (1686), in which Descartes' theory of vortices is consistently and in a radical manner used in the development of a theory of the origin of the world. It postulates the existence of living beings on other planets and on the moon, and infers their probable characteristics on the assumed temperature and atmospheric conditions of these heavenly bodies. In addition to our solar system, an infinity of other solar systems is postulated, and an infinity of space. And—although this was not made explicit—if space is infinite, why not also time?

Fontenelle's *Plurality of the Worlds* was strictly a work of fiction, with strong metaphysical overtones. By contrast, *Telliamed,* a work of fiction by de Maillet (1748), had a solid foundation in its author's long-continued geological studies. As stated in the subtitle, this work purports to record "Conversations between an Indian Philosopher and a French Missionary on the Diminution of the Sea." It is an extraordinarily imaginative piece of work, in which the most daring, the most heterodox ideas are ascribed to (and placed in the mouth of) the Indian philosopher. The work is divided into three conversations, the first two of which deal almost exclusively with geological matters, being in many respects remarkably advanced for the period and, perhaps, far too much neglected in the history of geology. The third conversation, the longest of the three, has a great deal to say about the origin of life and the metamorphosis of living beings.

De Maillet's major geological thesis is that the earth had once been entirely covered by the sea, from which it is now gradually emerging. This process occupied millions of years. Originally there were only aquatic plants and animals, some of which, when they came out on land, became transformed into their terrestrial equivalents. The earth, as we now see it, is not the product of an instantaneous creation but was formed gradually, by natural processes. The air is always full of "seeds" of all kinds of organisms, which will come to life whenever environmental conditions are favorable. Existing species become transformed whenever a transformation is required by newly arising conditions. For instance, flying fishes can become converted into birds, and human beings previously existed in the seas in the form of mermaids and mermen. Indeed, all terrestrial organisms are merely transformed aquatic organisms. Since there is always merely a transformation of a previously existing organism into a new form, a concept of genuine evolution did not exist for de Maillet. Nevertheless, *Telliamed* is important by showing to what extent the thinkers of the eighteenth century had emancipated themselves from the restrictions of earlier centuries.

312

Although *Telliamed* was not published until 1748, it had actually been written around 1715, some thirty years after Fontenelle's work (1686). Both works reflect the deep impression made on the intellectuals of the period by the writings of Descartes, Newton, and Leibniz and by such scientific discoveries as those of Leeuwenhoek and other naturalists. The massive impact of science on the thinking of the period is evident.

Let us now attempt to study in more detail the scientific advances that were responsible for this profound change in the western mind. There were three rather independent streams of scientific advance in the sixteenth, seventeenth, and eighteenth centuries, helping to prepare the ground for evolutionary theory but in very different ways: natural philosophy (physical sciences), geology, and natural history (broadly conceived).

The Role of Cosmology

The scientific revolution of the physical sciences (from Copernicus and Galileo to Newton and Laplace) laid particular stress on basic, general laws, like the law of gravity, which govern all physical phenomena. They explain not only the movements of bodies, including suns and planets, but also functional phenomena in living organisms. As it was stated by Boyle:

This philosophy . . . teaches that God, indeed, gave motion to matter. But that in the beginning, he so guided the various motions of the parts of it, as to contrive them into the world he designed they should compose, and established those rules of motion, and that order amongst things corporeal, which we call the laws of nature. Thus, the universe being once formed by God, and the laws of motion settled, and all upheld by his perpetual concourse, the general providence. The same philosophy teaches that the phenomena of the world are physically produced by the mechanical properties of the parts of matter; and that they operate upon one another according to mechanical laws. (Boyle, 1738: 187)

313

The widespread Greek concept of the universe as an organism (with a soul) was being replaced by that of the universe as a machine kept going by a set of laws. The gradual acceptance of the new way of thinking referred to as the mechanization of the world picture (Dijksterhuis, 1961), pervaded not only the physical sciences but also physiology and other branches of biology. The new interpretation demanded a mechanical explanation for all natural phenomena. If the planets, for instance, moved in their orbits, as dictated by the laws of planetary motion, then there was no longer a need for the creator to intervene incessantly. He still was the first cause for everything that exists, but subsequent to creation all natural processes were regulated by "secondary causes," exemplified by general laws of the physical sciences. The explanation of all natural phenomena by such laws and the search for previously undiscovered laws became the goal of science.

This new way of thinking was particularly successful in cosmology. The universe of the Bible and even of the Greek astronomers like Ptolemy was of very limited size. The discovery of the telescope put an end to this. The more powerful the telescopes became, the further the world seemed to expand, no limit ever being discovered. The concept of the infinity of the universe became increasingly accepted, and this process has continued right up to modern astronomy. The more man became accustomed to the concept of the infinity of space, the more frequently the question will have occurred to him whether there was not also an infinity of time.

Not only did the concept that the universe was infinite in space and time emerge, but eventually the idea arose that it was not even constant but forever changing. Yet everything that happened had to be consistent with the accounts in the Bible. Indeed, any new findings in natural philosophy had to be reconciled with the Mosaic accounts. In Britain, the first to publish a revolutionary geology was the Reverend Thomas Burnet with his *Sacred Theory of the Earth* (1681), explaining the entire history of the earth from creation to the present. The great event was the Flood, caused by a bursting of the outer crust and an eruption of the subterranean waters to the surface. All of these events, including the final conflagration on the Day of Judgment, were due to natural causes that God had set in train at the creation. John Woodward, in his *Essay towards a Natural History of the Earth* (1695), was far more traditional. The Deluge was due to direct intervention by the Lord, but since then the world has again been more or less stationary. All fossils are the product of the Flood and prove its actual occurrence, thus confirming the account of the Bible. It was a very comforting interpretation. A third writer on the earth's history, William Whiston, attempted to interpret the story of the Bible in terms of Newtonian physics. The most interesting speculation in his *New Theory of the Earth* (1696) was that Noah's Flood had resulted from a near approach of a comet.

314

What was important in all three explanations was the urge to find a "natural" explanation for various events of the earth's history without deviating too far from the literal accounts of the Bible (Greene, 1959: 15, 39). This was the foot in the door, and henceforth philosophers and cosmologists speculated ever more freely and boldly about the history of the earth, the sun, and the stars. However, the idea that the universe as a whole had evolved came surprisingly late. It was for the first time logically and consistently developed by the famous German philosopher Immanuel Kant (1724–1804) in an early publication, entitled *A General History of Nature and a Theory of the Heavens* (1755). Here Kant systematically developed the now familiar idea that the world had its beginning with a chaotic universal nebula that started to rotate and eventually formed the galaxies, suns, and planets. What is particularly striking about Kant's interpretation is the gradualness of the entire process: "The future succession of time, by which eternity is unexhausted, will entirely animate the whole range of Space to which God is present, and will gradually put it into that regular order which is conformable to the excellence of His plan . . . The Creation is never finished or complete. It did indeed once have a beginning, but it will never cease." Ever new stars and galaxies will evolve.

This was no longer a static world but a dynamic, continuously evolving one, only remotely governed by secondary causes—a truly heretical thought. By this revolutionary new idea, "Kant deliberately set on one side Newton's careful distinction between the creation of the present Order of Nature and its maintenance: the only creation we need demand was the progressive victory of order over chaos through an infinity of time" (Toulmin and Goodfield, 1965: 133).

Kant far outdid Buffon's estimates that the world might be 168,000 or even half a million years old. He clearly thought in terms of infinity, and thereby contributed to a change in thinking of the period that was later reflected in the writings of Hutton and Lamarck, even though neither of them was presumably directly acquainted with Kant's publication.

315

The Role of Geology

Changes in thinking even more fundamental than those in cosmology took place in geology.[6] It was in the eighteenth century that the students of nature for the first time became fully aware of the constant changes which the surface of the earth is and has been undergoing. A new science began to develop, geology, whose foremost task was a historical one, the reconstruction of the sequence of events which had taken place on the earth through time. The evidence that led to the discovery that the surface of the earth had not always been as it is now—indeed that the earth has a history—came from several sources.

One of them was the discovery of extinct volcanoes in central France (Puy de Dôme district). This contributed to the realization that basalt, a rock of widespread occurrence, is nothing but ancient lava, a remnant of ancient volcanic extrusions, that layers of such lava are widespread, and that the deeper ones must be very old.

At about the same period it was understood for the first time that many, indeed most, geological strata are sedimentary deposits.[7] Furthermore, when these sedimentary deposits were studied carefully, it was discovered that they occupy an enormous column, often as much as 10,000 feet, sometimes more than 100,000 feet, deep. The shock which this discovery caused was so profound because it made the great age of the earth inevitable, since it must have taken an immense amount of time for sedimentary strata of such thickness to be deposited. It was further discovered that neither the volcanic nor the sedimentary deposits had remained inviolate after they had been laid down. They were subsequently eroded by water, cutting valleys, sometimes deep ones, through them. More than that, many sedimentary layers had been folded subsequently, often quite violently; in some cases layers were completely turned over. All this is now so much taken for granted that it is difficult to appreciate how revolutionary these interpretations were in the seventeenth and eighteenth centuries and how widely they were at first resisted.

For a while a bitter controversy raged among different schools of geologists whether the forces of water (neptunism) or of fire (volcanism) had made a greater contribution to the present configuration of the earth's crust. In due time the respective roles of volcanism, erosion (and sedimentation), and mountain building were properly apportioned. But the understanding of the forces acting on the crust of the earth continued to grow, with highly important contributions (such as those of plate tectonics) being made as recently as the 1960s. Whatever the various geological discoveries were, they all had one thing in common: they mutually reenforced the realization of the immense age of the earth (Albritton, 1980). Inevitably this had to lead to a clash with anyone who accepted the literal interpretation of the Bible.

The Church, which had more or less officially adopted 4000 B.C. as the date of creation, considered any substantial departure from this date to be heresy. Nevertheless, Buffon in his *Les époques de la nature* (1779) had the courage to calculate the age of the earth as at least 168,000 years (Roger, 1962). (His unpublished private estimate was half a million years, a good deal larger.) Buffon gave much thought to these problems and seems to have been the first thinker who had a rational and internally consistent concept of the history of the earth. In his 1779 publication, which was a greatly expanded version of an essay he had published about 25 years earlier, Buffon recognized seven "epochs," as he called them: the first, when the earth and planets were formed; the second, when the great mountain ranges originated; the third, when water covered the mainland; the fourth, when the water subsided and the volcanoes began their activity; the fifth (a very interesting one), when elephants and other tropical animals inhabited the north (their fossils had been found in the north, and Buffon could not imagine that these tropical animals could have lived in any other climatic zone than a tropical one); the sixth, when the continents were separated from each other (Buffon postulated this because he recognized clearly the similarity of the fauna of North America and that of Europe and Asia; since these continents are now separated by water, he concluded that they must have been formerly connected); finally, the seventh epoch, when man appeared. This was the last period, so recent, indeed, that man does not occur in the fossil record. Biological facts played a large role in Buffon's reconstruction of the history of the earth, and we must now turn to the biological discoveries that prepared the way for evolutionary thinking.

317

The Role of Natural History

Those who concentrate on the discoveries in the physical sciences tend to ascribe the intellectual changes in the seventeenth and eighteenth centuries entirely to the mechanization of the world picture. This ignores the important role played by developments in the various areas of natural history. These produced a wealth of new facts which eventually turned out to be incompatible with the story of a single creation. Thus, anything that contributed to a flowering of natural history is part of the history of evolutionary biology.

Most important is perhaps the simple fact that nature was being rediscovered after the Middle Ages. Increasingly, authors expressed a delight in birds and flowers. Beginning around 1520, a series of beautifully illustrated works on the native plants of south Germany and other parts of Europe began to appear (see Chapter 4). This stimulated a desire to go outdoors and look for these plants, and even to discover new ones not previously described and illustrated. A similar interest in birds, fishes, and other productions of nature also developed. This led to the discovery that most of these western European species had not been mentioned in the Bible at all nor in the writings of Theophrastus, Aristotle, or Pliny. People began to wonder, What do we really know about the world we live in?

The Bible knew only the fauna and flora of the Near East, and the salvage of this very limited fauna in Noah's Ark was conceivable. However, when the great voyages began in the fourteenth and fifteenth centuries, leading to ever more astonishing discoveries in the sixteenth, seventeenth, and eighteenth centuries, the credibility of the biblical story was fatally undermined by the description of entirely new faunas in Africa, the East Indies, the Americas, and Australia. How could all these rich faunas have been accommodated in the Ark? If all animals had spread out from Mount Ararat (Armenia), the reputed landing place of the Ark, why is not the fauna of the entire world uniform? How were the isolated continents of America and Australia colonized?[8] Facts of biogeography posed some of the most insoluble dilemmas for the creationists and were eventually used by Darwin as his most convincing evidence in favor of evolution (see Chapter 10).

318   New doubts about the credibility of the biblical story were raised by the ever-increasing knowledge of fossils. To be sure, fossils had already been known to the ancients. Xenophanes of Colophon (about 500 B.C.) found fossil fishes in Tertiary quarries at Syracuse in Sicily and fossils of marine mollusks on the island of Malta. Rather remarkably, he did not interpret them as the record of past catastrophes but rather as the result of the gradual shifting of sea levels, somewhat along the line of ideas expressed by Anaximander. Aristotle stated similar views in his *Meteorologia*. Being strongly anticatastrophic, he also explained fossils as the result of slow shifts in the sea level. Nevertheless two erroneous interpretations of fossils dominated, up to the eighteenth century (in part also going back to Aristotle).

At first, it was widely believed that fossils "grew out of rocks," as do crystals or metal ores, and were nothing but a "lusus naturae," an accident of nature. Either nature was credited with a *vis plastica* capable of forming all sorts of shapes in rocks, or fossils were ascribed to a universal occurrence of "germs" in nature, manifesting themselves either in spontaneous generations or as fossils in rocks. Numerous distinguished authors, like Albertus Magnus, Mattioli, Falloppio, Agricola, Kircher, Gesner, Camerarius, and Tournefort, not to mention scores of lesser writers, held such opinions.

At the time when, finally, the idea gained general recognition that fossils are the remains of formerly living organisms, the literal interpretation of the Bible was the dominant practice, and accordingly fossils were interpreted as the remnants of the creatures that had perished in Noah's flood (particularly by Steno, Woodward, and Scheuchzer). Even though Leonardo da Vinci, Fracastoro, and other pioneers advanced much evidence against the simultaneity of all fossils, the dogma of the young age of the earth was for a long time too powerful to permit adoption of the theory of a sequence of distinct fossil faunas.

Two developments eventually undermined the simple-minded explanation of fossils as remnants of Noah's flood. One was the discovery of unknown, hence presumably extinct, animals and plants among the fossils, and the other was the development of stratigraphy. The discovery of extinct animals was not so much directly in conflict with the Bible as with the somewhat peculiar concept of God held in the seventeenth and eighteenth centuries. According to the principle of plenitude, adhered to by most of the leading thinkers of the period but particularly by Leibniz, God in the ampleness of his mind had surely created any creature that was possible. But God in his benevolence could not possibly permit any of his own creatures to become extinct. The fossil remains of seemingly extinct organisms thus posed a real dilemma for which various solutions were proposed (see Chapter 8 under Lamarck).

The second of the difficulties was caused by the discovery that fossil beds are stratified and that each of the strata has a distinct fauna and flora. The full appreciation of this situation came remarkably late, considering that fossils had been known for over two thousand years. Xenophanes already had noted that one might find different fossils in different quarries, that is, that different rocks might contain different fossils. Other authors made similar observations. Yet this evidence was ignored as long as fossils were considered artifacts of nature or remnants of Noah's Flood. The rapid progress of geological research in the eighteenth century made it impossible to ignore this any longer. Numerous authors, partly working independently, partly being stimulated by each other, began to understand that rocks occurred in a definite sequence, that most of them were stratified, and that certain strata had a wide distribution. At first the strata were identified primarily by petrographic characteristics (schists, slates, limestones, chalks, and so on), but a few perceptive pioneers discovered that certain fossils were associated with certain strata. Various histories of geology have attempted to give due credit to the work of such authors as Steno, Lister, Woodward, Hooke, Holloway, Strachey, Arduino, Lehmann, Füchsel, Werner, Michell, Bergmann, Soulavie, Walch, and others.[9] Unfortunately, no good comparative history of this early phase of stratigraphy is available. The observations published by these authors are piecemeal and unsystematic. Nevertheless, it is now agreed that there were two men who converted the scattered information on fossils and their occurrence into a science of stratigraphy, the English surveyor William Smith and the French zoologist Georges Cuvier.

Smith, by profession a surveyor and engineer, discovered, while helping to build canals and attempting to trace seams of coal- and ore-bearing strata in mines, that one can identify geological strata by the fossils they contain. It was sometimes possible to trace such strata for hundreds of miles, even when the lithology (rock formation) was changing. Smith developed these principles between 1791 and 1799, but he did not publish his famous map of the strata of England and Wales until 1815 (Eyles, 1969). In the meantime French naturalists had actively collected fossils in the limestone quarries of the Paris basin, and Cuvier and his collaborators worked out the exact stratigraphy of these fossils (primarily mammals) and characterized each of the faunas in admirable detail.[10] Schlotheim in Germany (1804; 1813) had come to similar conclusions.

The findings, both in France and in England, left no doubt—as distasteful as this conclusion was to many geologists—that a time sequence was involved and that the lowest strata were the oldest. Eventually the additional discovery was made that it was often possible to correlate the strata not just through England and continental Europe but through large parts of the world, if one makes allowances for the same kinds of regional differences that are found even today between the faunas of, let us say, Europe and Australia or the marine faunas of the Atlantic and the Pacific oceans. Still, differences between contemporary faunas in different parts of the world are not nearly as great as differences between the faunas of different geological periods, let us say between recent organisms and those of the mid-Tertiary, not to mention those of the Mesozoic or the Paleozoic.

Neither Cuvier nor the great British geologists (including Lyell) of the first half of the nineteenth century drew the conclusion from this evidence, which to us seems so inescapable, that there was a continuing evolutionary change of these faunas. Instead, they maintained for another fifty years either that each fossil fauna was wiped out by a catastrophe, to be replaced through special creation by an entirely new fauna, or else that the extinction was more piecemeal, the replacement nevertheless being due to individual special creations (see Chapter 8). Origins rather than evolution remained the explanatory concept.

Further Developments in Biology

One discovery after the other in natural history shook the foundations of previous beliefs. The invention of the microscope, for instance, led to the discovery by van Leeuwenhoek of a previously unknown set of organisms.[11] It added an entirely new dimension to the diversity of the living world and seemed to supply the long looked-for bridge between visible organisms and inanimate nature. Most importantly, it seemed to give powerful support to the hypothesis of spontaneous generation (Farley, 1977). In spite of the demonstration of Redi and Spallanzani that maggots would not develop in meat if egg-laying by flies was prevented, the belief became widespread that microscopic organisms, particularly infusorians, could be generated from inanimate matter. Soon everybody knew the recipe to produce such organisms: Place some dry hay in water, and after a couple of days this water will be full of microscopic organisms. This demonstration of "spontaneous generation" was, of course, totally in conflict with the dogma of a single creation at the beginning of the world. Spontaneous generation subsequently became a key component of Lamarck's theory of evolution.

321

Finally, there was one further development in biology which in due time significantly affected evolutionary thinking: the rise of systematics. From Cesalpino and Gesner on, there was a steady advance in the inventory taking of animals and plants (see Chapter 4). For a long time, it had seemed possible to arrange these organisms in a single *scala naturae,* from the simplest to the most perfect, and such a scale of perfection seemed to conform well with the eighteenth-century concept of the creator. Yet, the more the knowledge of plants and animals advanced, the more difficult it became to arrange them in a single file. Rather, they fell into well-defined and frequently rather isolated groups like mammals, birds, and reptiles and their subdivisions, and these were much more conveniently arranged in an inclusive hierarchy of categories. Cuvier claimed that all animals belong to no more and no fewer than four groups ("embranchements"): Vertebrata, Mollusca, Articulata, and Radiata. He insisted that these four phyla were quite unrelated to each other, yet he admitted a rather elaborate system of relationships within each of these four branches. In the denial of any connection between inanimate matter and organisms, and between plants and animals, Cuvier agreed with Lamarck, but he went further by denying the existence of a single lineage of animals. His decisive destruction of the *scala naturae* led to the posing of entirely new questions, and opened the way for the construction of evolutionary classifications (see Chapter 4), even though Cuvier himself failed to take this step.

THE FRENCH ENLIGHTENMENT

The gradual emancipation from traditional roles in religion, philosophy, and politics became a veritable revolutionary movement during the Enlightenment.[12] Although this movement had started in Britain (particularly Scotland) at the end of the seventeenth and the beginning of the eighteenth centuries, France led in developing new concepts in thinking about the world of life. It was therefore no accident that eventually a Frenchman was the first to develop a genuine theory of evolution.

The eighteenth century was an era of particularly strong and unresolved intellectual tensions. In philosophy it was characterized by attempts to reconcile the opposing thoughts of Descartes, Newton, and Leibniz. A belief in revelation became more and more unfashionable, as more and more contradictions were found in the Bible and as the mechanization of the world picture left ever less room for supernatural phenomena. *Theism,* a belief in a personal god of revelation who forever intervened in natural processes and performed miracles, became unacceptable to most philosophers and scientists. Yet, even *deism,* a belief in a god who had created the world in the beginning and with it the laws ("secondary causes") that have governed the world ever since, ran into formidable difficulties. Could his "design" have been so detailed that it included every particular structure and functioning of the countless species of animals and plants and their equally countless mutual interactions? And how could such an original design be reconciled with the changes which seemed so evident everywhere on earth? More specifically, as we shall presently see, how could either design or general laws explain such biological phenomena as extinction or vestigial organs? Throughout the eighteenth and the first half of the nineteenth centuries, one naturalist and philosopher after the other tried to compromise between creationist and deist interpretations of the living world. Still other authors quite openly became atheists, denying not only design but even the existence of a creator. The world for them simply was a large machine. But how could that explain the attributes of man and the harmonious adaptations of all organisms to the environment in which they live? There were seemingly unanswerable questions whether one was a theist, a deist, or an atheist. The intellectual turmoil which the clash between these ideologies produced, together with the continuing increase in our knowledge of the living world, culminated in Darwin's thought.

The century from about 1740 to 1840 is crucial for the story of evolution because this was the period when the concept of evolution made its breakthrough in the minds of the most advanced thinkers. It was a period of change not only in geology and natural history but also in political and social thinking. The erosion in the natural sciences of the belief in a steady-state world was matched in the political sciences and, indeed, in the practical world of government and society, by a questioning of the belief in god-given dynasties and feudal hierarchies, with their emphasis on the status quo. This was challenged by the concept of "progress," perhaps the dominant theme in the writings of the philosophers of the Enlightenment. A connection between the two themes, evolution in the natural world and progress in the social world, is obvious. What is far less obvious is where the ideas in the two fields originated, and what the relative contribution of the natural and the social sciences were to this trend of thought.

323

An answer to this question is important in connection with the controversy between externalism and internalism in science. Did the concept of progress originate in the political area and, as the externalists would claim, reappear in the natural sciences as the concept of evolution? An analysis of the concept of evolution is indispensable for answering this question.

The Ideas of Progress and Evolution

Progress always means growth and development, even when it involves only immanent potentialities. As far as mankind is concerned, it was predicted by Fontenelle (1688): "There will be no end to the growth and development of human wisdom." In a way this was a new concept, but in another way it was a very old one because all the components of the concept of progress such as growth and development (from Aristotle), continuity, necessity, unfolding purpose, final end, and so on were widespread not only among the ancients but also in the world view of Saint Augustine. A short time before Fontenelle, Pascal (1647) had likewise compared the development of mankind with the growth of the individual.

Development was equally important in the thought of Leibniz, with his emphasis on continuity and plenitude. In many ways, this was in contrast to the teaching of Descartes, with his stress on uniformity and mathematical constancy. No one stated the importance of potentiality more emphatically than Leibniz:

324

> Although many substances have already attained a great perfection, yet on account of the infinite divisibility of the continuous, there always remain in the abyss of things slumbering parts which have yet to be awakened, to grow in size and worth, and in a word, to advance to a more perfect state . . . there is a perpetual and a most free progress of the whole universe in fulfillment of the universal beauty and perfection of the works of God, so that it is always advancing towards a greater development. (Nisbet, 1969: 115)[13]

The eighteenth century, in its optimism, forever stressed progress in the continuing improvement of man. Herder, Kant, and other leading thinkers of the period expressed this belief and joined what one might call the search for a law of progress (Nisbet, 1969: 104–136). Such progress characterizes not only nature but also all human institutions, and it was of course this emphasis which was so important in the shaping of the United States Constitution and the French Revolution.

The apogee of this trend of thought was reached in Condorcet's *Progress of the Human Mind* (1795), where he states that "Nature has set no term to the perfection of human faculties, that the perfectibility of man is truly indefinite; and that the progress of this perfectibility, from now onwards independent of any power that might wish to halt it, has no other limit than the duration of the globe upon which nature has cast us."

If there were a necessary pathway from a concept of continuous and limitless progress to a theory of evolution, the great naturalists of the eighteenth century should have speedily taken it. But this did not happen. Neither Buffon nor Needham, Robinet, Diderot, Bonnet, or Haller converted the philosophical-political concept of progress into a scientific theory of evolution. Indeed, it was not until after the reaction to the Enlightenment had set in, with Napoleon's usurpation of power in France, that Lamarck developed his theory of evolution.

There are many reasons for questioning the concept of an inevitable conversion of the political theory of progress into a biological theory of evolution. For instance, the naturalists considered the idea of progress rather incompatible with the countless facts indicating regressive evolution (including parasitism and vestigial organs). Perhaps the most restraining factor was the power of essentialism. Was not all progress merely an unfolding of already existing potentialities without any change in the underlying essence, that is, without any actual evolution? Fontenelle, for instance, rejected any idea of change beyond growth, for has it not been shown by Descartes and others, he said, that nature is uniform in her workings; that she does not vary her prescription from generation to generation? All he could accept was an unfolding of an existing potential. There is a considerable difference between growth and history. Growth is merely an unfolding of an immanent potential, while history is actual change.

325

Leibniz, by contrast, goes beyond such essentialistic unfolding. For him the potentiality of nature was unlimited, "and hence no end of progress is ever to be reached." This optimism was a logical consequence of the principles of plenitude, immanence, and continuity, though Voltaire ridiculed it. In spite of this criticism, Leibniz's thoughts were taken up by most social philosophers of the nineteenth century such as Marx, Comte, and Spencer, who said that progress is "not an accident but a beneficient necessity."

There were two elements in Leibniz's thought which affected the future history of evolutionary biology. The concept of continuity and gradualness ("Everything goes by degrees in nature, and nothing leaps, and this rule controlling changes is part of my law of continuity"; 1712: 376), with its explicit rejection of Platonism, was an important positive contribution and indispensable prerequisite for modern evolutionary thinking. It was one of the foundation stones of Darwin's explanation of evolution. The other concept, that of an inner drive toward progress, if not perfection, was nothing but a handicap. It forced those, like Spencer, who came to evolution through a belief in progress to adopt entirely erroneous theories for the mechanism of evolution (see Chapter 11). Those who rejected inevitable progress, which includes the school of Scottish philosophers, actually were closer to Darwin's thinking than the French apostles of progress. It is now widely held that a belief in inevitable and continuing progress is disastrous for any ideology incorporating such a belief (Monod, 1970).

The idea of progress is closely allied to and in part originated from the concept of the *scala naturae* or great chain of being (Lovejoy, 1936), a concept that goes back to Plato but took new forms among the scholastics of the Middle Ages and then again in the seventeenth and eighteenth centuries. It is based on the belief in a linear continuity (but also rank order) from the world of inanimate objects through plants to lower animals to higher animals to man (and ideally through the angels to God). The additional concept of *plenitude*, postulating that everything that is possible actually exists, was usually associated with that of the *scala naturae*. There could be no voids, and the gaps between neighboring links of the chain were so infinitesimally small that the chain was practically continuous. In Leibniz, who particularly stressed this continuity, the influence of his mathematical interests is very evident. Indeed, he expresses his thoughts on this subject often in mathematical terms. Prior to Leibniz, the great chain of being was a strictly static concept, since the chain of being was perfect when it was created and, since, with a movement toward greater perfection being impossible, any change could only be a deterioration, a degradation.

The increasing perfection on which the *scala naturae* was based could be expressed in various ways, by "more soul" (in the sense of Aristotle), more consciousness, more reason, greater advance toward God, or what not. Basically it was largely a postulated ideal because observation by no means confirmed the existence of such a perfect, continuous, strictly linear chain. On the contrary, one saw everywhere conspicuous gaps, such as between mammals and birds, fishes and invertebrates, ferns and mosses. Hence the delight when corals and other organisms (such as zoophytes) were discovered that seemed to connect plants and animals most felicitously. It was bravely postulated that the other gaps would likewise be filled by future discoveries. Among Leibniz's numerous followers none was as consistent as Charles Bonnet (1720–1793), who established a most elaborate "échelle des êtres naturels," in which flying squirrel, bat, and ostrich represented the continuity between mammals and birds.[14] For him, "organization" was the criterion that determined rank in the chain. Where his statements imply evolution, it is clearly merely an unfolding of a pre-existing potential.

The existence of fossils and other evidence of seeming extinction was in apparent conflict with the principle of plenitude and required explanation. Leibniz, in his *Protogaea* (1693), admits that many kinds of organisms that existed in earlier geological periods have since disappeared, and that many of those that now live were then apparently nonexistent. This leads him to postulate that "even the species of animals have many times been transformed" in the course of the vast changes which have taken place in the condition of the earth's crust. The number of monads, however, stayed constant, and no descent was involved as we now understand it but merely an evoking of pre-existing potentials. Thus the outward appearance of the chain of being underwent a change in time without the underlying essences changing at all. *327* This new version of the chain of being was referred to by Lovejoy as the "temporalizing" of the chain of being. Contrary claims notwithstanding, this was not the proposal of a theory of evolution.

Since the principle of plenitude could not permit the occurrence of extinction, extinct animals had to be interpreted as earlier stages of organisms still existing. This, for instance, was clearly the interpretation of Bonnet (*Palingénèse*), while Robinet had a more fanciful idea of an origin of new types by a combination of earlier prototypes. Nothing new was created, however, since the potential for everything was pre-existing. For Robinet it is "the first axiom of natural philosophy" that "the Scale of Being constitutes an infinitely graduated whole, with no real lines of separation; that there are only individuals, and no kingdoms or classes or genera or species" (Guyénot, 1941: 386). The chain, for him, was due to successive acts of creation by nature, but there was no evolution and no genetic continuity. Curiously, one finds rather similar ideas, although phrased in somewhat more creationist terms, as late as 1857, in the writings of Louis Agassiz.

The concept of evolution was, so to speak, "in the air" throughout the second half of the eighteenth century, and certain historians of science have designated three Frenchmen—Maupertuis, Buffon, and Diderot—as evolutionists. The same honor has been bestowed by German historians upon Rodig, Herder, Goethe, and Kant. Subsequent researches have failed to substantiate any of these claims. All these "forerunners" were essentialists and either postulated new origins (rather than an evolution of existing types) or else merely believed in an unfolding ("evolution" sensu stricto) of immanent potentialities. Nevertheless, the writings of these authors are highly interesting, not only because they illuminate the steady approach toward evolutionary thought, but also because they illustrate the intellectual milieu in which evolutionary thinking had to attempt to establish itself. In some sense all of these writers were indeed forerunners of Lamarck; in another sense none of them was, for Lamarck was the first to make a complete break with the essentialistic impediments against evolutionism.

Maupertuis

Pierre Louis Moreau de Maupertuis (1698–1759) was one of the most advanced thinkers of his time.[15] It was he who first brought Newtonian thought to France, where it was eagerly taken up by Voltaire and others. Yet, Maupertuis was also the first person in France to recognize that the simple Newtonian paradigm of "forces and motion" was insufficient for biology, indeed even for chemistry, and this is why he incorporated Leibnizian thought into his conceptual framework. Through him and Madame Chatelet, Buffon became acquainted with Leibnizianism and, as a consequence, Leibnizian ideas were strong in the writings of most of the French *philosophes* and eighteenth-century scientists, including Lamarck.

Even though Maupertuis' major concern was mathematics and astronomy, he had a keen interest in biological phenomena and was one of the pioneers of genetics (see Chapter 14). Yet, contrary claims notwithstanding, he was neither an evolutionist nor one of the founders of the theory of natural selection, and his explanations were more those of a cosmologist than of a biologist. His real importance was that he opposed the strongly deterministic and creationist component in Newtonianism and went back to Lucretius and the Epicureans in ascribing origins largely to accident. There was far too much diversity and heterogeneity in nature for the world to have been produced by design. He strongly criticized the natural theologians, with arguments such as that an existence of poisonous plants and venomous animals is incompatible with a concept of "wisdom and the goodness of the Creator."

Materialists (atheists) who denied the existence of a creator had to account somehow for the existence of organisms. They went back to the ideas of Lucretius: Organisms can originate by "spontaneous generation." But this *deus ex machina* process allowed for various versions. One could believe in the existence of ever-present living germs or molecules which by fortuitous coming together could produce even the highest organism. Ideas like this were held as late as the middle of the eighteenth century, by not only Maupertuis but La Mettrie, Diderot, and others. Another possibility would be to combine the concepts of spontaneous generation and the *scala naturae*. There are no living germs pervading all of nature; hence spontaneous generation must be able to derive life from inanimate matter. However, this process can generate only the simplest organisms from lifeless material, and these gradually become transformed into more and more complex creatures, by moving up a "temporalized" *scala naturae*. This, in essence, as we shall see, was Lamarck's theory of evolution.

Maupertuis' explanation of the origin of the world of organisms involved the massive spontaneous generation of new kinds of animals and plants and an equally massive elimination of those that are deficient. This is, of course, a theory of origins, as they were so widespread among the Greeks, but not a theory of evolution. It must be emphasized, as correctly pointed out by Roger (1963), that this theory of elimination of deleterious new variants has nothing whatsoever to do with natural selection.

329

Not having any background in natural history, Maupertuis found nothing ludicrous in the idea that any kind of organism, even an elephant, could be the product of a fortuitous combination of material. "Chance, one might say, turned out a vast number of individuals; a small proportion of these were organized in such a manner that the animals' organs could satisfy their needs. A much greater number showed neither adaptation nor order; these last have all perished. Thus the species which we see today are but a small part of all those that a blind destiny has produced" (*Essaie de cosmologie,* 1750).

Yet, Maupertuis did not rely entirely on spontaneous generation for the origin of new kinds. His genetic studies led him to a theory of what we would now call speciation by mutation. A new species for Maupertuis was nothing but a mutant individual, and in this thought he was a forerunner of de Vries. Races, for him, began with fortuitous individuals. Maupertuis was clearly an essentialist, and even though he could think of the production of new essences, he was unable to conceive of a gradual and continuous improvement of a population by the selection (this is, reproduction) of the best-adapted individuals. Nevertheless, his world was not a static one but one in which time played an important role.

Buffon

The two greatest naturalists of the eighteenth century, Buffon and Linnaeus, were born in the same year, 1707. Yet, except for the identity of the year of their birth and their great interest in natural history, the two men were about as different as any two men can be. Buffon (1707–1788) was wealthy, a member of an aristocratic family, and able to take up the study of science as an avocation.[16] Linnaeus was poor and had a hard struggle to obtain a position and make a living. In most of their concepts of science, likewise, they held opposing views (see Chapter 4).

As a young man, Buffon spent a year in England during which he studied mathematics, physics, and plant physiology. After he had returned to France, he published a translation of Newton's *Fluxions* and of Stephen Hales' *Vegetable Statics*. Owing to the special patronage of the minister Maurepas, Buffon was appointed in 1739 the Intendant (director) of the Jardin du Roi, even though he was not particularly well qualified for this position. However, he threw himself with great enthusiasm into the new job and developed the plan of writing a universal natural history from minerals to man. Thirty-five large quarto volumes of this work were published between 1749 and Buffon's death in 1788, and nine further volumes were added to the series subsequently. In this monumental and fascinating *Histoire naturelle*, Buffon dealt in a stimulating manner with almost all the problems that would subsequently be raised by evolutionists. Written in a brilliant style, this work was read in French or in one of the numerous translations by every educated person in Europe. It is no exaggeration to claim that virtually all the well-known writers of the Enlightenment, and even of later generations, in France as well as in other European countries were Buffonians, either directly or indirectly. Truly, Buffon was the father of all thought in natural history in the second half of the eighteenth century.[17] Even though, as we shall see presently, he himself was not an evolutionist, it is nevertheless true that he was the father of evolutionism. And it was surely he who was responsible for the immense interest in France in natural history (Burkhardt, 1977: 14–17).

There are few thinkers who are as difficult to interpret correctly as Buffon. This has many reasons. For instance, Buffon's great work is a literary encyclopedia of natural history, and references to a given general topic—let us say evolution, species, or variation—are scattered through many different volumes. Furthermore, Buffon's ideas quite clearly evolved during his long and active life, but all attempts to classify his thought into well-defined periods have been rather unsatisfactory. With his versatile, indeed almost mercurial mind, Buffon looked at many subjects from so many different sides that he not infrequently contradicted himself. It requires a study of the totality of his work before one can state with conviction which of Buffon's views should be considered as most typical. Finally, there is the probability that in his earlier publications Buffon was not able to write with complete frankness. In the 1740s, the theologians at the Sorbonne still had a great deal of power, and at one time (1751) Buffon indeed had to recant some statements he had made on the history and age of the earth. It is probable that at least some of Buffon's observations were phrased in such a manner as to placate the theologians.

When Buffon published in 1749 the first three volumes of his natural history, he was a rather strict Newtonian. As a result, he was impressed by concepts of movement and continuity, and a preoccupation with large numbers of static, discontinuous entities such as species, genera, and families appeared quite meaningless to him. When he was appointed director of the Jardin du Roi (now Jardin des Plantes), his acquaintance with systematics was quite limited, but he turned this deficiency into a virtue by attacking the "nomenclateurs" (Linnaeans) as dry-as-dust pedants and by preaching instead a study of living animals and their characteristics in life. In the introductory discourse, he states that it is quite impossible to distribute kinds of organisms among distinct categories because there are always intermediates between one genus and another. Furthermore, if one were to adopt any classification, it would have to be based on the totality of all characters and not, as was done by Linnaeus, on an arbitrary selection of a few. In spite of a stress on continuity, Buffon offers no suggestion of evolution in the first three volumes. He did not propose a temporalization of the chain of being nor suggest that one species had originated or developed from another. Indeed, in the first volume, the idea is defended that the only real entities in nature are individuals.

The sequence of species which Buffon adopts for his natural history is a purely utilitarian one. He begins with the ones that are most important, most useful, or most familiar to man. Hence, domestic species like horse, dog, and cow are treated before the wild animals, and those of the temperate zone, in turn, before the exotic animals. This arbitrary classification was clearly as unsuited as can be to serve as a basis for evolutionary considerations. As far as man is concerned, he is the most superior living being: "Everything, even his outward appearance, demonstrates man's superiority over all other living beings." Very much in the spirit of Descartes, Buffon considers man's ability to think as his outstanding characteristic: "To be and to think are for us the same thing."

331

Since he is convinced that animals cannot think, there is for him a tremendous gap between animals and man. This made it quite impossible to consider man as having evolved from animals.

The wording of the first three volumes of the *Histoire naturelle* (1749) indicates that Buffon at that time might have been an atheist. In 1764, he definitely uses the language of a deist. When, in 1774, Buffon writes, "The deeper I penetrate into the depths of nature, the more I admire and profoundly respect her author," he seems to express genuine feelings. When Buffon comes to believe in an eternal order and in laws of nature, he needs a lawgiver who is responsible for the observed secondary causes. Science would be meaningless if the world were not governed by an immutable and universal order. In this concept, Buffon is remarkably close to Aristotle who, on the basis of the same concept of an eternal order of the universe, also came to reject evolution.

Buffon was fully aware of the possibility of "common descent" and was perhaps the first author ever to articulate it clearly:

> Not only the ass and the horse, but also man, the apes, the quadrupeds, and all the animals might be regarded as constituting but a single family . . . If it were admitted that the ass is of the family of the horse, and different from the horse only because it has varied from the original form, one could equally well say that the ape is of the family of man, that he is a degenerate man, that man and ape have a common origin; that, in fact, all the families, among plants as well as animals, have come from a single stock, and that all animals are descended from a single animal, from which have sprung in the course of time, as a result of progress or of degeneration, all the other races of animals. For if it were once shown that we are justified in establishing these families; if it were granted that among animals and plants there has been (I do not say several species) but even a single one, which has been produced in the course of direct descent from another species; if, for example, it were true that the ass is but a degeneration from the horse—then there would no longer be any limit to the power of nature, and we should not be wrong in supposing that, with sufficient time, she has been able from a single being to derive all the other organized beings. But this is by no means a proper representation of nature. We are assured by the authority of revelation that all animals have participated equally in the grace of direct Creation and that the first pair of every species issued fully formed from the hands of the Creator. (Buffon, 1766)

This statement could be, and indeed it has sometimes been, interpreted as a mere pro forma refutation (for the benefit of the theologians) of a genuine belief in evolution by Buffon. All recent interpreters of Buffon (Lovejoy, Wilkie, Roger) agree, however, that the statement, when studied in the context of the essay in which it is included, is indeed a serious rejection of the possibility of common descent. The quoted statement is directly followed by a number of different arguments against the possibility of the descent of a genuine species from another. In particular, Buffon lists three arguments. First, within recorded history no new species have been known to appear. Second, the infertility of hybrids establishes an impassable barrier between species. And third, if one species originated from another one, "for example the ass species came from the horse," the result could have been brought about only slowly and by gradations. There would therefore be between the horse and the ass a large number of intermediate animals. Why, then, do we not today see the representatives, the descendants, of those intermediate species? Why is it that only the two extremes remain? These three arguments lead Buffon to this conclusion: "Though it cannot be demonstrated that the production of a species by degeneration from another species is an impossibility for nature, the number of probabilities against it is so enormous that even on philosophical grounds one can scarcely have any doubt upon the point."

Then, how do species originate? Living matter (organic molecules) is continuously formed as the result of spontaneous chemical combination. Organic molecules, in turn, combine spontaneously to form the first individual of all fundamental species. This primary being, thus formed, becomes the prototype of a species. It becomes the *moule intérieur* (epigenetic inner form) for its descendants and thus guarantees the permanence of the species. This permanence is incessantly challenged by the "circumstances" which induce the production of varieties. However, the permanence of the *moule intérieur* prevents variations from transgressing certain limits. In this respect, the *moule intérieur* plays a role similar to Aristotle's *eidos* (form). Many lower organisms are continuously produced from organic molecules by spontaneous generation. There are as many kinds of animals and plants as there are viable combinations of organic molecules. Inviable combinations perish.

There is quite a contrast between the first three volumes of the *Histoire naturelle* (published in 1749) and the fourth (1753) and

333

subsequent volumes. One reason is that Buffon in the early 1750s had become acquainted with the work of Leibniz, with his emphasis on the chain of being, plenitude, the perfection of the universe, and its hints of evolution. Buffon's writings from this point on contain a mixture of Newtonian and Leibnizian thoughts. On one hand he continued to believe in plenitude and stated: "It seems that everything exists which can be." On the other hand, he rejected final causes, and his attitude throughout is antiteleological. The world was created perfect in the beginning, and there was nothing that would necessitate its movement toward greater perfection. Occasionally he clearly rejects Plato's essentialism, as when stating that we have to abstract from the diversity of phenomena, but these abstractions are the product of our own intelligence, and not real. Yet most of his interpretations are typological, as is evident from his treatment of the species.

334

In the first volume of the *Histoire naturelle,* Buffon denied the existence of species, stating that only individuals exist. This viewpoint is completely abandoned in the second volume, where he defines species as follows:

> We should regard two animals as belonging to the same species if by means of copulation they perpetuate themselves and preserve the likeness of the species, and we should regard them as belonging to different species if they are incapable of producing progeny by the same means. Thus, the fox will be known to be a different species from the dog if it proves to be a fact that from the mating of a male and female of these two kinds of animals no offspring is born. And even if there should result a hybrid offspring, a sort of mule, this would suffice to prove that fox and dog are not of the same species inasmuch as this mule would be sterile.

The production of a sterile hybrid proves that different species are involved because, for the maintenance of a species, "a continuous, perpetual and unvarying reproduction is necessary." As Lovejoy rightly says, this language implies not only that species are real but also that they are constant and invariable entities. Species, for Buffon, are types and not populations. Having such a rigid species concept, together with the phenomenon of sterility in the hybrids, precluded the concept of an evolutionary derivation of one species from another one. Buffon's species definition has the further disadvantage that it does not truly define a concept but only provides a method for testing whether two individuals belong to the same or to different species. It is a discriminant function.

Buffon's most important discussion of the kind of questions we would now consider to fall under the heading of "evolutionary biology" is in his essay on the degeneration of animals (1766). Here he strongly expresses his view that most variation is nongenetic and caused by the environment. This is indicated by the fact that the domestic breeds are the most variable of all animals because man has transported them into all climates and has exposed them to a great variety of foods, an opinion later endorsed by Darwin.

335

Buffon's background in the physical sciences is particularly evident in his discussion of variation. Since he believes that the same causes should have the same effects, he contends that animals that live in the same places must resemble each other, because the same climate produces the same animals and the same plants. Owing to his belief in the primacy of physical causes, he is quite convinced that there is life on other planets, and calculates, on the basis of estimates of their rate of cooling, at what time life would have originated on these planets. Buffon's concept of organisms being the "product" of a place where they live played an important role in the thinking of biogeographers of the ensuing one hundred years.[18]

These discussions must have made it clear why there is no contradiction between the statements that Buffon was not an evolutionist, and yet that he was the father of evolutionism. He was the first person to discuss a large number of evolutionary problems, problems that prior to Buffon had not been raised by anybody. Even though he often arrived at wrong conclusions, it was he who added these topics to the repertory of scientific problems. Even though Buffon himself rejected evolutionary explanations, he brought them to the attention of the scientific world. We owe to him extensive discussions of the origin of the earth in general, and of sedimentary rocks in particular; he established the importance of the problem of the extinction of species of animals; he raised the question whether closely related species, like horse and ass, could have descended from a common ancestor; and, finally, he was the first to focus full attention on the problem of problems, the establishment of reproductive isolation (as we would now call it) between two incipient species.

What, on balance, was the net effect of Buffon's thought on

the future development of evolutionary thinking? He clearly occupies the ambiguous position of having both hindered and furthered evolutionism. He hindered it by his frequent endorsements of the doctrine of the immutability of species. He hindered it, furthermore, by providing a species criterion—fertility among members of a species—which his contemporaries considered quite impregnable to evolutionary change. Indeed, the problem how a species could give rise to another species from which it is separated by the sterility barrier bothered some geneticists as recently as the first half of this century (Bateson, 1922; Goldschmidt, 1940). These reservations of Buffon's, shared by all the more knowledgeable of his contemporaries, were the reason why a mere demonstration of evolutionary change was not sufficient to establish evolutionism. What was needed, and this was subsequently provided by the proponents of geographic speciation, was a demonstration of how the gap between species could be bridged.

Far more important are Buffon's positive contributions to evolutionism.

(1) Through his detailed analysis, he brought the idea of evolution into the realm of science, to be treated henceforth as a proper subject of investigation.

(2) He generalized the results of his dissections (with his collaborator Daubenton) by developing the concept of the "unity of the type." This gave rise at first to the school of idealistic morphology and, subsequently, to comparative anatomy, which produced so much evidence in favor of evolution.

(3) He, more than anyone else, was responsible for a new chronology of the earth, that is, for the acceptance of a vast time scale.

(4) He was the founder of biogeography. When, out of opposition to Linnaeus, he arranged species according to their country of origin, he grouped them into faunas. The compilation of faunal lists by Buffon and his followers served as the basis of far-reaching generalizations. Darwin, indeed, derived more evidence for evolution from the fact of distribution than from any other biological phenomenon (see Chapter 10).

Natural history, prior to Buffon, had all the earmarks of an avocation, a hobby. Buffon is the one who raised it to the status of a science. Much of the *Histoire naturelle* consisted of what today we would call "ecology"; other parts were devoted to the study of behavior. Here was a splendid reaffirmation of the value of study-

336

ing whole animals as a counterweight against the atomizing influences of contemporary physiology, particularly since Buffon himself was equally interested in physiology, development, and organic molecules. It makes no difference which of the authors in the second half of the eighteenth century one reads, their discussions are, in the last analysis, merely commentaries on Buffon's work. Except for Aristotle and Darwin, there has been no other student of organisms who has had as far-reaching an influence.

Diderot

Among the leaders of the Enlightenment, none was more interested in the living organism than Denis Diderot (1713–1784). In various articles in the *Encyclopédie,* and particularly in a series of imaginative essays, he grappled again and again with the origin and nature of life, accident or determinism, the interaction of molecules, spontaneous generation, the role of the environment, and similar problems.[19] Diderot, evidently, was an avid reader, and his speculations borrow freely from Buffon, Leibniz, Maupertuis, Condillac, Bordeu, Haller, and other contemporaries. He produced few, if any, original thoughts, but the brilliant manner in which he fused current speculations into an explanatory fabric had a great impact on French intellectuals. Perhaps his most daring essay was *Le rêve de D'Alembert* (D'Alembert's Dream). Although written in 1769, it was not officially published until 1830. Yet, a clandestine version circulated through Paris soon after it was written. Hence, the contents were evidently well known in the Paris salons, and it is almost certain that Lamarck was familiar with it. The flavor of the work is well expressed in the ravings of the fever-ridden D'Alembert:

> All creatures are involved in the life of all others . . . All nature is in a perpetual state of flux. Every animal is more or less a human being, every mineral more or less a plant, every plant more or less an animal . . . There is nothing clear defined in nature . . . Is there in nature any one atom exactly similar to any other? No . . . Don't you agree that in nature everything is bound up with everything else and that there cannot be a gap in the chain? . . . There is but one great individual, and that is the whole . . . You poor philosophers, and you talk about essences! Drop your idea of essences . . . and what of species? Species are merely tendencies toward a common end peculiar to them . . . And life? A series of actions and reactions . . . The living molecule is the origin of

all, there's not a speck in the whole of nature that does not feel pain or pleasure.

This short monologue contains a catalogue of virtually all the ideas on life and matter that had been held from the ancients to the then-modern philosophers like Leibniz and Buffon. Although some of the elements of Diderot's later thought played a role in evolutionary theories, Diderot himself was not in any way an evolutionist. There is no implication in his writings that life on earth changes with time. By the time Diderot wrote the *Dream* he had become an uncompromising atheist. His world was not "created"; it had none of the "designed" properties of the world of the natural theologians. It was a thoroughly materialistic world of molecules. Perhaps the most memorable sentence of the *Dream* is: "The organs produce the needs, and conversely the needs produce the organs." This thought, apparently originating with Condillac, eventually became one of the cornerstones of Lamarck's theory of evolution.

DEVELOPMENTS ELSEWHERE IN EUROPE

Most of the writers so far mentioned were French, and France, indeed, occupied the intellectual leadership of Europe in the eighteenth century. But there was much ferment also in Britain (particularly Scotland), Germany, the Netherlands, and Scandinavia. In fact, Britain and Germany took over the field after the deaths of Lamarck and Cuvier. In Germany things had been rather quiet after Leibniz, with his extraordinary originality. Indications that rigid theism was weakening were widespread, however. A liberal deism, that is, a rejection of all revelation including the Bible, reached its finest flowering in the writings of Reimarus.[20] In biology, his main impact was on the interpretation of animal behavior. The most influential thinker of the period, however, was the historian Johann Gottfried Herder,[21] whose major contribution was his emphasis on historical thinking and on diversity. In his four-volume work, *Ideas towards a Philosophy of the History of Man* (1784–1791), he deals not only with the "rise of man" but also, at length, with the universe and the world of animals and plants. Herder exerted great influence on the thinking of Goethe, Kant, and the *Naturphilosophen* by his consistently historical approach to all questions. But, like all other Germans, he was an essentialist, for whom a transformation of one species into another one was quite unthinkable. Herder's basic concept of the living world was that of a temporalized *scala naturae*, but he never came to grips with the problem of how one could get from plants to animals or from simple animals to higher animals. Yet, he insisted that "we see the form of organization ascend, and with it the powers and propensities of the creatures become more various until, finally, they all, so far as possible, unite in the form of man." Many of Herder's thoughts are derived from Buffon, even though he often expands them considerably, as in his treatment of the struggle for existence.

Kant has often been called a forerunner of Darwin, but without justification, as has been clearly demonstrated by several writers, particularly well by Lovejoy (1959d). Even though Kant saw the problems rather clearly, as is evident from his discussion of adaptation in the *Critique of Judgment* (1790), being a thoroughgoing essentialist, he simply could not conceive of evolution. He was much impressed by Buffon's argument that the sterility barrier maintained the sharp delimitation of species against each other and he took this to be clear-cut proof for the impossibility of getting from one species to another one by anything like evolution. He never resolved the conflict between the discontinuity of species and the continuity in the universe which he expressed in his cosmology and in his adherence to the Great Chain of Being. The seeming conflict between the purely mechanical laws of physics and chemistry and the perfect adaptation of all organisms which seemed to demand ad hoc creation posed a dilemma for Kant which he was unable to solve (Mayr, 1974d: 383–404; Lovejoy, 1959d: 173–206).

No one reflects late eighteenth-century thinking in Germany better than J. F. Blumenbach, who in his influential natural history wrote extensively on mutability, extinction, spontaneous generation, degeneration, final causes, creation, catastrophes, and *Bildungstrieb*. Blumenbach was remarkably knowledgeable but quite unable to emancipate himself from the dominant ideas of his time.

England, which in the seventeenth and early eighteenth centuries had displayed such leadership in philosophy (Locke, Berkeley, Hume), in physics, and in physiology, made virtually no contribution to evolutionary thinking in the eighteenth century. The only exception is Erasmus Darwin,[22] Charles Darwin's grandfather, who in his *Zoonomia* (1794) indulged in some casual evolutionary speculations. He never amplified them further, and thus they had remarkably little impact on subsequent developments.

There is no justification for a detailed presentation of his thought beyond emphasizing that three assumptions concerning the ideas of Erasmus Darwin are erroneous:

(1) That he anticipated Lamarck, or even that Lamarck got his ideas from him. The belief in an inheritance of acquired characters and other ideas found in both authors were widespread at that period. Lamarck clearly was not acquainted with Erasmus Darwin. (2) That he greatly influenced his grandson. There is hardly a trace of Erasmus Darwin's ideas in the *Origin,* and Charles Darwin explicitly denied such an influence, even though Darwin's notebooks reflect the reading of *Zoonomia* (Hodge, 1981). (3) That he was a highly original thinker. He was primarily a synthesizer and popularizer; virtually all of his individual ideas can be traced to earlier authors with whom Erasmus Darwin was familiar owing to his extensive reading. His so-called evolutionary ideas were widely held by natural theologians and British animal breeders.

The lack of interest in evolution in England during the eighteenth century had several reasons. The great flowering of empiricism which occurred at that time resulted in an overemphasis on the physical and experimental sciences. The pursuit of natural history was entirely in the hands of ordained ministers and inevitably led to a belief in the perfect design of a created world, a belief totally incompatible with the concept of evolution.

Linnaeus

At first sight it might seem altogether inappropriate to mention Carl Linnaeus (1707–1778), often considered the arch foe of evolutionism, in a discussion of the history of evolutionary thinking. Yet he played an important role (see Chapter 4). Although starting out with a theory of classification based on the scholastic theory of logical division, Linnaeus laid the groundwork for the development of a natural, hierarchical classification, which in time virtually forced the acceptance of the concept of common descent. He had an inkling of the relationship of orders and classes as indicated by his well-known statement: "All groups of plants show relationships on all sides, like countries on a map of the world" (*Philosophia Botanica,* 1750: 77). And yet, by recognizing genera, orders, and classes, Linnaeus destroyed the "continuity of life" and replaced it by a hierarchy of discontinuities. This conformed perfectly with essentialistic thinking, but it created a conflict with the continuism of evolutionary thinking. To reconcile continuity and discontinuity, thus, became one of the great challenges of evolutionary biology.

By his insistence on the constancy and fixity of species, in contrast to the vagueness of the rather nominalistic French school, Linnaeus made the origin of species a scientific problem (Poulton, 1903; Mayr, 1957). The problem was compounded by his theory of an origin of species by hybridization, proposed late in his life. Like Ray, Linnaeus decisively rejected heterogony. In fact he denied, at least in his major writings, any or all transmutation of one *341* species into another one.

His keen interest in the balance of nature and in the struggle for existence was important in the development of the thought of later natural theologians, and of de Candolle and other pre-Darwinians (Hofsten, 1958; Limoges, 1970). It formed an important part of the conceptual background of the theory of natural selection. In fact, a good deal of Darwin's argument can be traced back to Linnaeus, even where it consists of a refutation of Linnaean ideas. In short, Linnaeus made a major contribution to the conceptual milieu that gave rise to later evolutionary theories.[23]

THE HERITAGE OF THE PRE-LAMARCKIAN PERIOD

The seventeenth and eighteenth centuries, as we have seen, experienced an almost total revolution in man's concept of nature. In an "age of reason," revelation could no longer be accepted as the final authority concerning the explanation of natural phenomena. Theism was widely replaced by deism or even atheism. Discoveries in all fields discredited the Bible as a source of scientific explanation. The God of interventions and miracles was replaced by an image of God as the author of general laws which function as secondary causes in the production of all concrete phenomena. This interpretation was consistent with the discovery of great physical laws which keep suns and planets moving automatically without divine intervention. Infinity of time, infinity of space, and cosmological evolution (Kant, Laplace) were being accepted. Discoveries in the biological sciences posed particularly serious challenges to the creationist-interventionist interpretation. These included the heterogeneity of faunas and floras, increasing difference of fossils in lower strata, increasing evidence for frequency of extinction, the inclusive Linnaean hierarchy, the discovery of morphological types, the discovery of microscopic organisms, the recognition of the incredible adaptedness of organisms, the beginnings of a replacement of typological by populational thinking.

By the end of the century, it had become evident that two major problems demanded an explanation: the origin of diversity and its seemingly orderly arrangement in a natural system, and the superb adaptation of all organisms to each other and to their surroundings. For the essentialist, there was the additional problem of how to reconcile the discontinuity displayed by species and higher categories with the general continuity of all phenomena of life. Finally, there were a number of rather embarrassing special problems that seemed to be in conflict with the concept of the wisdom and benevolence of the creator, such as the problems of extinction and of the existence of vestigial organs. Creationism became a less and less satisfactory solution. The stage was thus set for a revolutionary new departure, and it was only a question of time before some naturalist would have the courage and originality to propose a solution clearly in conflict with accepted dogma. This person was the French biologist Lamarck.

8 ✐ Evolution before Darwin

TO A MODERN biologist the time lag between the first provoca- tive intimations of evolutionism by Leibniz in his *Protogaea* (1694) and the definite proposal by Lamarck (1800) seems extraordinar- ily long. Buffon had teetered on the brink of evolutionism all of his life, and numerous other thinkers had adopted a temporalized version of the chain of being, but none of them had made the decisive step of converting the unbroken chain of a created se- quence of ever greater perfection into a line of descent.

LAMARCK

Jean Baptiste Pierre Antoine de Monet, Chevalier de Lamarck (1744–1829), was born into a poor noble family from northern France.[1] He joined the army at age 17, fought with great bravery a number of battles in the Seven Years War, resigned at 19, and lived from then on in Paris on a very small pension and whatever else he could earn by hack writing for dictionaries and the like. Eventually, he became very much interested in natural history, particularly botany, and finally wrote a four-volume flora of France, which was justifiably praised for the excellence of its de- scriptions. Shortly afterwards he was chosen by Buffon to be tutor and travel companion to his son. This gave him an opportunity to visit Italy and other European countries, the only traveling La- marck ever did in his life. In 1788 Buffon secured for him a po- sition as assistant in the Botanical Department of the Natural His- tory Museum, a post he held for the next five years. Lamarck published prodigiously during the nearly thirty years he was in- terested in plants, and there is no doubt that he believed at that time in well-delimited species "created in the beginning" and not subsequently changing. Several statements of his clearly indicate that his thinking at that period was that of an essentialist.

In 1793, in connection with the reorganization of France's

scientific institutions, Lamarck was appointed professor of the "inferior animals," or invertebrates, as we now call them with the name given to them by Lamarck. This new appointment was the decisive event of his life. With enormous energy he familiarized himself with the diverse assemblage of the animals which Linnaeus had lumped together under the name "worms." Even though Lamarck was already 49 years old when he began these new studies, it is evident that they had a revolutionary impact on his concepts. Up to that time he had adhered to a typically eighteenth-century mixture of deism and a synthesis of Newtonian and Leibnizian ideas. From Newton he had adopted a belief in the lawfulness of the universe and the conviction that all phenomena, not only of inanimate nature but also of "organized bodies," could be explained in terms of movements and forces acting upon matter. From Leibniz he had acquired an optimistic conviction of the perfect harmony of the universe, of plenitude and continuity. This synthesis led to numerous contradictions, and it seems evident that the adoption of evolutionism was the consequence of the attempt to resolve at least some of these contradictions.

Lamarck had grandiose plans for a universal "terrestrial physics" (including biology) and, in the pursuit of these aims, he dabbled in just about all branches of science. He made himself ridiculous by opposing Lavoisier's brilliant new discoveries of the chemistry of oxygen and by his meteorological predictions. He also wrote a geology (Lamarck, 1802a) that was virtually ignored by his contemporaries and translated into English only a decade ago.

It was one of the duties of his new professorship to give an annual course on the invertebrates. For several years Lamarck devoted the first lecture to a *Discours d'ouverture*. The manuscripts of these lectures (or at least some of them) were preserved and have, in part, been published in recent years (Lamarck, 1907). The *Discours* for the year 1799 still represents Lamarck's thinking as he had acquired it from the botanist A.-L. de Jussieu and the Linnaean school: species were unchanging, and there was no hint at the possibility of evolution. Next year's *Discours,* presented on May 11, 1800, unveils Lamarck's new evolutionary theories, already containing the essential points of his *Philosophie zoologique* (1809). It is evident that between 1799 and 1800 Lamarck had had a "conversion," as it would be called in the religious literature. What could have induced a man who was already 55 years old to abandon his previous world view and replace it by one that is so revolutionary that it had not been held by anyone previously?

Past endeavors to explain Lamarck's evolutionism have almost without exception been unsatisfactory, owing to a failure to separate Lamarck's ideas on evolutionary changes as such and his endeavors to explain the physiological and genetic mechanisms responsible for it. In the subsequent account, an attempt will be made to separate carefully these two aspects of Lamarck's evolutionary theory.

A determined effort will also be made to interpret Lamarck in the context of his contemporary intellectual milieu. Few other authors have suffered as much in the past from whiggish historiography as Lamarck (see Chapter 1). To be sure, he is one of the most difficult figures in the history of biology to discuss. This is why there are probably more different interpretations, in fact even descriptions, of Lamarck's thought than exist for any other figure. Even excluding obsolete accounts, one needs only to compare the recent presentations of Mayr, Hodge, and Burkhardt to appreciate this point. Lamarck had profound intellectual commitments to Descartes, Newton, Leibniz, and Buffon. However, he was also clearly affected in his thinking by his zoological material, particularly by the variation and fossil history of mollusks. Hodge (1971a,b) correctly stresses that one cannot and should not interpret Lamarck in terms of Darwin's evolutionary theory. Lamarck had no theory of an origin of species nor did he consider common descent. Most remarkably for an early nineteenth-century naturalist, he totally disregarded geographic distribution, a body of information that was one of the most powerful sources of Darwin's theory of common descent.

Lamarck's New Paradigm

Lamarck states that his new theory is needed in order to explain two well-known phenomena in the world of organisms. The first is that animals show a graded series of "perfection." Under increasing perfection Lamarck understood the gradual increase in "animality" from the simplest animals to those with the most complex organization, culminating in man. He did not assess perfection in terms of adaptedness to the environment or by the role an organism plays in the economy of nature but simply in terms of complexity. The other phenomenon in need of explanation is the amazing diversity of organisms which suggests "that anything which it is possible to imagine has effectively taken place." Seemingly he is referring to the principle of plenitude.

A further ingredient added by Lamarck is the actual transformation of species in a phyletic line. "After a long succession of generations . . . individuals, originally belonging to one species, become at length transformed into a new species distinct from the first" (1809: 38–39). Everywhere in his discussions Lamarck reiterates the slowness and gradualness of evolutionary change. "With regard to living bodies, it is no longer possible to doubt that nature has done everything little by little and successively" (p. 11). In a discussion of originally aquatic animals he states, "Nature led them little by little to the habit of living in the air, first by the water's edge, etc." (p. 70).

"These changes only take place with an extreme slowness, which makes them always imperceptible" (p. 30). "It is difficult to deny that the shape or external characters of every living body whatsoever must vary imperceptibly, although that variation only becomes perceptible after a considerable time" (p. 45). "An enormous time and wide variation in successive conditions must doubtless have been required to enable nature to bring the organization of animals to that degree of complexity and development in which we see it at its perfection" (p. 50). This is no problem because for nature "time has no limits and can be drawn upon to any extent" (p. 114).

Numerous students of Lamarck's work have asked themselves what new observations or new insights had induced Lamarck to adopt this new viewpoint in 1800. What apparently happened (Burkhardt, 1977) was that, in the late 1790s, Lamarck took over the mollusk collection of the Paris Museum after the death of his friend Bruguière. When he started to study these collections which contained both fossil and recent mollusks, he found that many of the living species of mussels and other marine mollusks had analogues among fossil species. Indeed it was possible, in many cases, to arrange the fossils of the earlier and more recent Tertiary strata into a chronological series terminating in a recent species. In some cases where the material was sufficiently complete, it was possible to establish virtually unbroken phyletic series. In other cases, he found that the recent species extended far back into the Tertiary strata. The conclusion became inescapable that many phyletic series had undergone a slow and gradual change throughout time. Probably no other group of animals was as suitable for bringing about such a conclusion as the marine mollusks. Cuvier, who at the same time studied fossil mammals which on the average evolve far more rapidly than marine mollusks, found that none of the fossil elephants or other fossil types had a living analogue and thus came to the conclusion that the earlier species had become extinct and had been replaced by entirely new species. The recognition of phyletic series was of particular importance to Lamarck because it solved for him a problem which apparently had disturbed him for a long time, the problem of extinction.

347

Extinct Species

Ever since the study of fossils had become more intense, it had become apparent that many of the fossil species are quite unlike the living ones. The ammonites, so abundant in many Mesozoic deposits, are one conspicuous example. The situation became more acute when fossil mammals were discovered in the eighteenth century, such as mastodons in North America and mammoths in Siberia. Finally, Cuvier described entire faunas of fossil mammals from various horizons of the Paris basin. The more sober naturalists and students of fossils eventually accepted the fact that the earth had been inhabited in former eras by creatures that had since become extinct, and not all of them at the same time. Blumenbach, for instance, recognized an older period of extinction, mostly concerning marine organisms like bivalves, ammonites, and terebratulas, and a more recent extinction, concerning organisms with surviving relatives, like the cave bear and the mammoth. Herder already spoke of multiple earth revolutions, and other authors spoke of catastrophes, all of them resulting in extinctions. For other naturalists the concept of extinction was unacceptable for various ideological reasons. It was as inconceivable to the natural theologians as to the Newtonians, for whom everything in the universe is governed by laws. It also violated the principle of plenitude because the extinction of a species would leave a void in the fullness of nature. Finally, it violated concepts of a balance in nature which would not provide any causes for the occurrence of extinction (Lovejoy, 1936: esp. 243, 256).

The view that extinction was incompatible with the omnipotence and benevolence of God was very widespread throughout the eighteenth century. During a discussion of fossils John Ray stated in 1703: "It would follow, That many species of Shell-Fish are lost out of the World, which Philosophers hitherto have been unwilling to admit, esteeming the Destruction of any one Species a dismembring of the Universe, and rendring it imperfect; whereas they think the Divine Providence is especially concerned to secure and preserve the Works of the Creation" (*Physico-Theological Discourses*, 3rd. ed., 1713: 149).

Most of the philosophers of the Enlightenment and the first half of the nineteenth century were deists. Their God was not allowed to interfere with the universe, once he had created it. Any such interference would be a miracle, and which philosopher could afford to support miracles after what Hume and Voltaire had said about them? This created a formidable dilemma. Either one had to deny the occurrence of extinction, which is what Lamarck did (more or less), or else one had to postulate a law established at the original time of creation to account for the steady disappearance and appearance of new species through geological time. But how could such a law "of the introduction of new species" operate without being "special creation"? This was the (never fully articulated) objection Darwin raised against Lyell, who postulated such a law. But let us go back to the endeavors to "explain away" extinction.

In the course of the seventeenth and eighteenth centuries, four explanations were advanced for this disappearance of fossil species, none of them involving "natural extinction."

One was that extinct animals are those that were killed by Noah's Flood or some other great catastrophe. This explanation, which became rather popular in the first half of the nineteenth century, was quite incompatible with Lamarck's gradualism. Furthermore, since so many of the "lost species" were aquatic, a destruction by the flood seemed quite irrational.

A second explanation was that the supposedly extinct species might well be surviving in as yet unexplored portions of the globe: "There are many parts of the earth's surface to which we have never penetrated, many others that men capable of observing have merely passed through, and many others again, like the various parts of the sea bottom, in which we have few means of discovering the animals living there. The species that we do not know might well remain hidden in these various places" (Lamarck, 1809: 44).

Finally, some explained extinction by saying that it is the work of man. And this explanation was particularly chosen for the large mammals like mammoth and mastodon.

These three explanations still left much if not most of the

problem of extinction unsolved. The discovery of fossil species analogous to still living ones, therefore, afforded Lamarck the long-sought-for solution to a major puzzle. "May it not be possible . . . that the fossils in question belonged to species still existing, but which have changed since that time and have been converted into the similar species that we now actually find?" (1809: 45). In other words, extinction is only a pseudo-problem. The plenitude is nowhere disrupted, the strange species which we only find as fossils still exist but have changed to such an extent that they are no longer recognizable except where we have a continuity of fossil horizons and, as we would now say, an exceedingly slow evolutionary rate. Evolutionary change, then, was the solution to the problem of extinction. Furthermore, a study of evolution was another way to demonstrate the harmony of nature and the wisdom of the creator.

349

When drawing these conclusions, Lamarck at once noticed that this explanation was eminently logical for another reason. The earth has been forever changing during the immense period of time during which it had existed. Since a species must be in complete harmony with its environment and, since the environment constantly changes, a species must likewise change forever in order to remain in harmonious balance with its environment. If it did not, it would be faced with the danger of extinction. By introducing the time factor Lamarck had discovered the Achilles heel of natural theology. It would be possible for a creator to design a perfect organism in a static world of short duration. However, how could species have remained perfectly adapted to their environment if this environment was constantly changing, and sometimes quite drastically? How could design have foreseen all the changes of climate, of the physical structure of the surface of the earth, and of the changing composition of ecosystems (predators and competitors) if the earth was hundreds of millions of years old? Adaptations under these circumstances can be maintained only if the organisms constantly adjust themselves to the new circumstances, that is, if they evolve. Although the natural theologians, good naturalists that they were, had clearly recognized the importance of the environment and the adaptations of organisms to it, they had failed to take the time factor into consideration. Lamarck was the first to have clearly recognized the crucial importance of this factor.

Lamarck's new evolutionism had strong support from his earlier geological studies (Lamarck, 1802a). Like all Leibnizians, Lamarck was a uniformitarian as, indeed, were most naturalists in the eighteenth century. He postulated an immense age for the earth and, like Buffon, he envisioned continuous changes occurring during these immense time spans. Things changed constantly but they changed exceedingly slowly. This picture of a gradually changing world fitted extremely well with an evolutionary interpretation. However, it formed a complete contrast to the steady-state world of Hutton, which did not include any directional change and was thus not hospitable to evolutionary explanations.

Evolutionism, of course, was even less compatible with essentialist thinking, that is, with the belief in unchanging, discontinuous types. For an essentialist, changes in the earth's fauna could be explained only by catastrophic extinctions and new creations, a view represented in the writings of Cuvier and his followers. Lamarck was uncompromisingly opposed to catastrophism in any form, as is evident from his zoological writings as well as from his *Hydrogéologie* (1802a: 103).

Even though his new theory of transformation had solved several problems, it still faced some formidable puzzles. If Lamarck had been an uncritical adherent of Bonnet's chain of being with its gradual, unbroken transition from inanimate matter to the most perfect being, all Lamarck would have needed to do was to apply his species-transition principle to the *scala naturae*. But Lamarck was by no means a strict follower of Bonnet, as much as he believed in a gradation of perfection.[2] Even in his earlier writings Lamarck emphasized that there is no transition between inanimate nature and living beings. And even though Lamarck, the creator of the concept of biology, was a strong proponent of the essential unity of animals and plants, nevertheless he denied any gradation between the two kingdoms.

However, the difference between Lamarck and Bonnet was even more profound. The comparative anatomical researches at the Paris Museum, particularly in the 1790s, had revealed more and more discontinuities among the various morphological types, the vertebrates, mollusks, spiders, insects, worms, jellyfish, infusorians, and so on. Contrary to Bonnet, they do not form a graded series of species. "Such a series does not exist, rather I speak of an almost irregularly graduated series of the principal groups [*masses*] such as the great families; a series which assuredly exists among the animals as well as among the plants; but which, when the genera and particularly the species are considered, forms in many places lateral ramifications, the endpoints of which are truly isolated" (*Discours* XIII: 29). The picture of a linear chain is progressively replaced in Lamarck's writings by that of a branching tree. In 1809 he recognized two entirely separate lineages of animals, one leading from the infusorians to the polyps and radiarians, the other one, containing the majority of animals, arising from worms which had originated by spontaneous generation. By 1815 Lamarck recognized an even larger number of separate lineages.

The process of branching was seen by Lamarck as a process of adaptation and not, as in the case of Darwin and later evolutionists, primarily as a process capable of producing diversity of species. For the diversity of organic life had become a nagging scientific problem for those who no longer believed in a designed and created world. Spontaneous generation seemed to be the only conceivable alternative to special creation, in order to account for the origin of new phyletic lines (Farley, 1977). In order that "living bodies be truly the productions of nature, nature must have had and must still have the ability to produce some of them directly," said Lamarck (1802b: 103). Yet, knowing of the work of Redi and Spallanzani and in contrast to Maupertuis, La Mettrie, and Diderot, Lamarck denied that organic molecules could combine into complex animals like elephants, even under conditions of greater warmth in past periods of the earth. "It is exclusively among the infusorians that nature appears to carry out direct or spontaneous generations which are incessantly renewed whenever conditions are favorable; and we shall endeavor to show that it is through this means that she acquired power after an enormous lapse of time to produce indirectly all the other races of animals that we know." Once these lower organisms have originated, the known processes of evolution provide for their further development toward ever greater perfection. "Nature began and still begins by fashioning the simplest of organized bodies, and that it is these alone which she fashions immediately that is to say, only the rudiments of organization indicated in the term spontaneous generation" (1809: 40). However, Lamarck also accepted unquestioningly the spontaneous generation of intestinal worms and thought that they were the basis for the evolution of many higher organisms. The shift from one type of organism to a more complex one, he thought, is accomplished by the acquisition of a new faculty, this, in turn, being due to the acquisition of a new structure or organ (see below).

Was Lamarck the First Consistent Evolutionist?

Long lists of "early evolutionists" are recorded in some histories of biology. Indeed, H. F. Osborn, in his *From the Greeks to Darwin* (1894), fills an entire book with accounts of such forerunners of Darwin. As we have seen in Chapter 7, closer analysis fails to substantiate these claims. The forerunners either had theories of "origins" or of the unfolding of immanent potentialities of the type. A true theory of evolution must postulate a gradual transformation of one species into another and ad infinitum. No such ideas are found in the writings of de Maillet, Robinet, Diderot, and others who supposedly had influenced Lamarck. Several of Lamarck's forerunners, for instance Maupertuis, had postulated an instantaneous origin of new species. Linnaeus, in his later writings, was much impressed by the possibility of an unlimited production of new species by hybridization. Buffon had considered the possibility of the transformation of a species into a closely related one, but had emphatically rejected applying the same conclusion to a possible transformation of entire families. For all of these forerunners, nature was basically static. Lamarck replaced this static world picture by a dynamic one in which not only species but the whole chain of being and the entire balance of nature was constantly in flux.

Buffon had still stressed the immense gap between animals and man. Lamarck resolutely bridges this gap by considering man the end product of evolution. In fact, his description of the pathway by which our anthropoid ancestor became humanized is startlingly modern: "If some race of quadrumanous animals, especially one of the most perfect of them, were to lose by force of circumstances, or some other cause, the habit of climbing trees and grasping the branches with its feet in the same way as with its hands in order to hold on to them, and if the individuals of this race were forced for a series of generations to use their feet only for walking and to give up using their hands like feet there is no doubt . . . that these quadrumanous animals would at length be transformed into bimanous, and that their thumbs on their feet would cease to be separated from the other digits when they use their feet for walking" and that they would assume an upright posture in order "to command a large and distant view" (1809: 170). Lamarck here presented his view on the origin of man with far more courage than Darwin fifty years later in the *Origin*. Man "assuredly presents the type of the highest perfection that nature could attain to: hence the more an animal organization approaches that of man the more perfect it is" (p. 71). Since evolution is a continuing process, man will continue to evolve. "This predominant race, having acquired an absolute supremacy over all the rest, will ultimately establish a difference between itself and the most perfect animals, and indeed will leave them far behind" (p. 171). Even though man has now acquired certain characteristics not found in any animal, or at least not to a similar degree of perfection, man, nevertheless, shares most of his physiological characteristics with the animals. These characteristics, very often, are more easily studied in animals than in man, and in order to achieve a full understanding of man, it is therefore "necessary to try to acquire knowledge of the organization of the other animals" (p. 11). Aristotle had justified his study of the natural history of animals by the same argument.

353

Lamarck's Mechanisms of Evolutionary Change

Lamarck recognized two separate causes as responsible for evolutionary change. The first was an endowment which provides for the acquisition of ever greater complexity (perfection). "Nature, in successively producing all species of animals, beginning with the most imperfect or the simplest, and ending her work with the most perfect, has caused their organization gradually to become more complex." The causation of this trend toward ever greater complexity is derived "from powers conferred by the 'supreme author of all things' " (1809: 60, 130). "Could not His infinite power create an order of things which gave existence successively to all that we can see as well as to all that exists but that we do not see?" Or, as he stated it in 1815, nature "gives to animal life the power of progressively complicated organization." Clearly, the power of acquiring progressively more complex organization was considered by Lamarck an innate potential of animal life. It was a law of nature that did not require special explanation.

The second cause of evolutionary change was a capacity to react to special conditions in the environment. If the intrinsic drive toward perfection were the only cause of evolution, says Lamarck, one would find an undeviating single linear sequence toward perfection. However, instead of such a sequence, in nature we encounter all sorts of special adaptations in species and genera. This, says Lamarck, is due to the fact that animals must always be in perfect harmony with their environment, and it is the behavior of animals which reestablishes this harmony when disturbed. The need to respond to special circumstances in the environment will, consequently, release the following chain of events: (1) Every considerable and continuing change in the circumstances of any race of animals brings about a real change in their needs ("besoins"); (2) every change in the needs of animals necessitates an adjustment in their behavior (different actions) to satisfy the new needs and, consequently, different habits; (3) every new need, necessitating new actions to satisfy it, requires of the animal that it either use certain parts more frequently than it did before, thereby considerably developing and enlarging them, or use new parts which their needs have imperceptibly developed in them "by virtue of the operations of their own inner sense" ("par des efforts de sentiments intérieures").

354

Lamarck was neither a vitalist nor a teleologist. Even the trend toward "progressively complex or perfect organization" was not due to any mysterious orthogenetic principle but was the contingent by-product of the behavior and the activities required to meet new needs. Hence increasing perfection and the response to new requirements of the environment were only two sides of a single coin.

The crucial difference between Darwin's and Lamarck's mechanisms of evolution is that for Lamarck the environment and its changes had priority. They produced needs and activities in the organism and these, in turn, caused adaptational variation. For Darwin random variation was present first, and the ordering activity of the environment ("natural selection") followed afterwards. Hence, the variation was not caused by the environment either directly or indirectly.

In order to provide a purely mechanistic explanation of evolutionary change, Lamarck developed an elaborate physiological theory based on the ideas of Cabanis and other eighteenth-century physiologists, invoking the action of extrinsic excitations and the movement in the body of "subtle fluids" caused by the effort to satisfy the new needs. Ultimately these physiological explanations were Cartesian mechanisms, which were, of course, utterly unsuitable.

Relatively few of Lamarck's ideas were entirely new; what he did was order them into new causal sequences and apply them to evolution. But no one, so far, has made a consistent effort to trace them to their original sources. One of the key elements in Lamarck's theory, the claim that efforts to satisfy needs play an important role in modifying an individual, can be traced to Condillac and Diderot. Behavior caused by needs is a key factor in Condillac's explanation of animal behavior (1755); and Diderot, in *Le rêve de D'Alembert* (written in 1769) said quite simply that "the organs produce the needs, and conversely the needs produce the organs" (p. 180). This is all Lamarck needed to explain the ascent from one type of organism to a more perfect one. He considered this mechanism so powerful that he thought it could even produce new organs: "New needs which establish a necessity for some part really bring about the existence of that part as a result of efforts."

355

Even though the higher taxa may appear to be separated from each other by major gaps, this is all merely a matter of appearance, for "nature does not pass abruptly from one system of organization to another." When discussing the ten classes of invertebrates recognized by him (1809: 66), Lamarck insists dogmatically that "races may, nay must, exist near the boundaries, halfway between two classes." If we cannot find these postulated intermediates, it is due to their not yet having been discovered, either because they live in some remote part of the world, or owing to our incomplete knowledge "of past animals" (p. 23). With the reference to "past animals" and the statement that "existing animals . . . form a branching series" (p. 37), Lamarck seemed to have been rather close to the concept of common descent, but he never developed it. He was satisfied with having developed a mechanism that could explain the bridging of the gap between the higher taxa.

The idea that an organ is being strengthened by use and weakened by disuse was, of course, an ancient one, to which Lamarck gave what he considered a more rigorous physiological interpretation. Still, he considered this one of the cornerstones of his theory, and dignified it as his "First Law." "In every animal which has not yet passed beyond the limit of its development, the more frequent and sustained use of any organ gradually strengthens, develops, and enlarges that organ, and gives it a strength proportional to the length of time it has been used; while the constant disuse of such an organ imperceptibly weakens and deteriorates it, progressively diminishing its faculties until it finally disappears" (p. 113). This principle of use and disuse is, of course, still widespread in folklore and, as we shall later see, played a certain role even in Darwin's thinking.

The second auxiliary principle of evolutionary adaptation is the belief in an inheritance of acquired characters. This is formulated by Lamarck in his "Second Law": "Everything which nature has caused individuals to acquire or lose as a result of the influence of environmental conditions to which their race has been exposed over a long period of time—and consequently, as a result of the effects caused either by the extended use (or disuse) of a particular organ—[all this] is conveyed by generation to new individuals descending therefrom, provided that the changes acquired are common to both sexes, or to those which produce the young" (p. 113).

356

Lamarck nowhere states by what mechanism (pangenesis?) the inheritance of the newly acquired characters is effected. As was shown by Zirkle (1946), this concept was so universally accepted from the ancients to the nineteenth century that there was no need for Lamarck to enlarge upon it. He simply placed this principle in the service of evolution. Curiously, when Lamarckism had a revival toward the end of the nineteenth century, most of those who had never read Lamarck in the original assumed that Lamarckism simply meant a belief in the inheritance of acquired characters. Thus Lamarck received credit and blame for having originated a concept that was universally held at his time.

Before leaving the explication of Lamarck's paradigm, let me stress that it did not contain two beliefs which are frequently ascribed to him. The first is a direct induction of new characters by the environment. Lamarck himself rejected this interpretation by saying,

> I must now explain what I mean by this statement: *The environment affects the shape and organization of animals,* that is to say that when the environment becomes very different, it produces in the course of time corresponding modifications in the shape and organization of animals.
>
> It is true, if this statement were to be taken literally, I should be convicted of an error; for, whatever the environment may do, it does not work any direct modification whatever in the shape and organization of animals. (p. 107)

Even in the case of plants, which have no behavioral activities like those of animals "and consequently no habits *per se,* great alterations of environmental circumstances nonetheless lead to great differences in the development of their parts; so that these differences produce and develop some of them, while they reduce and cause the disappearance of others. But all this is brought about by the changes occurring in the nutrition of the plant, in its absorption and transpiration in the quantity of heat, light, air, and humidity, which it habitually receives." In other words, the changes in structure are produced by the internal activities of the plant in connection with its response to the environment, as in a plant that grows toward the light.

The second belief erroneously ascribed to Lamarck has to do with the effect of volition. Hasty readers of Lamarck's work have, almost consistently, ascribed to Lamarck a theory of volition. Thus, Darwin speaks of "Lamarck nonsense of . . . adaptations from the slow willing of animals" (letter of January 11, 1844, to J. D. Hooker). In part the misunderstanding was caused by the mistranslation of the word *besoin* into "want" instead of "need" and a neglect of Lamarck's carefully developed chain of causations from needs to efforts to physiological excitations to the stimulation of growth to the production of structures. Lamarck was not as naive to think that wishful thinking could produce new structures. For a full understanding of Lamarck's thinking it is important to know that Lamarck was not a vitalist but accepted only mechanistic explanations. He was not a dualist, and there is no reference to any duality of matter and spirit in his work. Finally, he was no teleologist, not recognizing any guidance of evolution toward a goal, predetermined by a supreme being.

A detailed analysis of Lamarck's explanatory model shows that it was remarkably complex. It made use of such universally accepted beliefs as the effect of use and disuse and the inheritance of acquired characters, it accepted spontaneous generation for the simplest organisms, as anyone was able to demonstrate any day for the production of infusorians from hay soaked in water (fully accepting the demonstration by Spallanzani and Redi that spontaneous generation was impossible for higher organisms), and used the physiological ideas of Cabanis and others on the interaction between the excitation of subtle fluids by efforts and the consequent effects on structures. Lamarck's paradigm was highly persuasive to the layperson, who held most of the beliefs of which it was composed. This is the reason why some of the Lamarckian ideas continued to be accepted so widely for almost a hundred years after the publication of the *Origin*.

The Difference between Lamarck's and Darwin's Theories

There has long been a futile controversy whether or not Lamarck was a "forerunner" of Darwin (Barthélémy-Madaule, 1979).[3] Darwin himself was quite explicit in denying any benefit from Lamarck's book, "which is veritable rubbish . . . I got not a fact or idea from it." In a more charitable moment, he stated, "But the conclusions I am led to are not widely different from his; though the means of change are wholly so" (Rousseau, 1969). It will help *358* the understanding of Darwin's theory by stating some of the components of an evolutionary theory.

The fact of evolution. The simple question here is whether the world is static or evolving. Even those who postulated an unfolding of immanent potentialities of essences believed ultimately in the unchanging nature of the essences. Lamarck's theory was in striking contrast to these static or steady-state theories. There is no doubt that he deserves credit for having been the first to adopt a consistent theory of genuine evolutionary change. Lamarck further postulated gradual evolution and based his theory on the assumption of progressive uniformitarianism. In all these respects, he clearly was Darwin's forerunner.

The mechanism of evolution. Here Lamarck and Darwin could not have differed more. The only component (not original with Lamarck) which these authors had in common was that both believed—although Darwin much less so—in the effect of use and disuse (soft inheritance).

A primary interest either in diversity or in adaptation. There is a fundamental and rarely sufficiently emphasized difference among evolutionists whether diversity (speciation) or adaptation (phyletic evolution) holds first place in their interest. Darwin came to the study of evolution through the problem of the multiplication of species (as encountered in the Galapagos!). The origin of diversity was, at least in the beginning, his primary interest. Evolution was common descent. This leads to an entirely different way of looking at evolution from that of the student of phyletic evolution (Mayr, 1977b).

Changes in time (the vertical dimension) are usually adaptive as seen by a Darwinian. Lamarck never explicitly articulated a concept of adaptation, but the entire causal chain of evolution postulated by him inevitably had to result in adaptation. Since the evolutionary force described by him was not teleological but materialistic, it produced adaptation by natural means. For the Darwinian, adaptation is the result of natural selection. For Lamarck adaptation was the inevitable end product of the physiological processes (combined with an inheritance of acquired characters) necessitated by the needs of organisms to cope with the changes in their environment. I can see no other way to designate his theory of evolution than as adaptive evolution. The acquisition of new organs and of new faculties was clearly an adaptive process. Accepting his premises, Lamarck's theory was as legitimate a theory of adaptation as that of Darwin. Unfortunately, these premises turned out to be invalid.

359

Lamarck in Retrospect

When, after 1859, Lamarck was rediscovered, following a long period of neglect, the term "Lamarckism" was usually given to a belief in soft inheritance. And the more decisively soft inheritance was refuted, the more "Lamarckism" became a dirty word. As a consequence, Lamarck's contribution as an outstanding invertebrate zoologist and pioneering systematist was entirely ignored. Equally ignored was his important stress on behavior, on the environment and on adaptation, aspects of biology almost totally neglected by the majority of the contemporary zoologists and botanists whose taxonomy was purely descriptive. No writer prior to Lamarck had appreciated as clearly the adaptive nature of much of the structure of animals, particularly in the characteristics of families and classes. More than anyone before him, Lamarck made time one of the dimensions of the world of life.

During the most whiggish period of biological history writing, Lamarck was mentioned only for his erroneous ideas, for his belief in soft inheritance, in innate perfectibility, and in speciation by spontaneous generation. It is time he receives credit for his major intellectual contributions: his genuine evolutionism which derived even the most complex organisms from infusorian or wormlike ancestors, his unflagging uniformitarianism, his stress of the great age of the earth, his emphasis on the gradualness of evolution, his recognition of the importance of behavior and of the environment, and his courage to include man in the evolutionary stream.

To determine what real impact Lamarck had on the subsequent development of evolutionary thought is exceedingly difficult (Kohlbrugge, 1914). He was almost totally ignored in France, he was admired by Grant in Edinburgh and made widely known in England through Lyell's criticism (which made Chambers an evolutionist!), but he was apparently more widely read in Germany than elsewhere. He was quoted and extensively used by Meckel, and, in spite of his simultaneous insistence on natural selection, by Haeckel. All this helped the acceptance of evolutionism. Yet, this popularity of Lamarckian ideas eventually became an impediment. It helped to delay for some 75 years after 1859 the general acceptance of the Darwinian explanatory model and of hard inheritance.

360

From Lamarck to Darwin

Lamarck's *Philosophie zoologique* (1809) signifies the first breakthrough of evolutionism. Yet, it required another fifty years before the theory of evolution was widely adopted. One must conclude that the creationist-essentialist world picture of the seventeenth and eighteenth centuries was far too powerful to yield to Lamarck's imaginative but poorly substantiated ideas. Nevertheless, the existence of a groundswell of evolutionary thought is unmistakable. The gradual improvement of the fossil record, the results of comparative anatomy, the rise of scientific biogeography, and many other developments in biological science contributed toward making evolutionary thinking ever more palatable. But this did not mean that it made Lamarck's eighteenth-century explanatory theories more acceptable.

Hence, one must make a sharp distinction between the acceptance of evolution and the adoption of a particular theory explaining its mechanism. This is particularly necessary as we encounter more and more explanations of evolution the further we progress into the nineteenth century. It is not always easy to understand the differences among these various theories, since some authors combined several of them or at least several of their components. It may help if I tabulate, at this time, the most important theories of evolution and specify how they differ from each other. Each of them was held by numerous supporters from the time of Darwin (or Lamarck) to the evolutionary synthesis.

One can perhaps recognize six major theories (some with several subdivisions):

(1) A built-in capacity for or drive toward increasing perfection (autogenetic theories). This was part of Lamarck's theory. It was widely supported, for example, by Chambers, Nägeli, Eimer

(orthogenesis), Osborn (aristogenesis), and Teilhard de Chardin (omega principle).

(2) The effect of use and disuse, combined with an inheritance of acquired characters.

(3) Direct induction by the environment (rejected by Lamarck, but adopted by E. Geoffrey Saint-Hilaire).

(4) Saltationism (mutationism). The sudden origin of new species or even more distinct types (Maupertuis, Kölliker, Galton, Bateson, de Vries, Willis, Goldschmidt, Schindewolf).

361

(5) Random (stochastic) differentiation, with neither the environment (directly or through selection) nor internal factors influencing the direction of variation and evolution (Gulick, Hagedoorn, "non-Darwinian evolution").

(6) Direction (order) imposed on random variation by natural selection (Darwinism in part, neo-Darwinism).

Theories (1), (2), and (3) had substantial support for well over a hundred years after Lamarck. Saltationism (4) is now refuted as the normal mode of speciation or of the origin of any other new types. It has, however, been substantiated for special cases (polyploidy and certain chromosomal rearrangements). The extent to which random differentiation (5) occurs is highly controversial at the present moment. Nevertheless, it is almost universally agreed that most evolutionary and variational phenomena can be explained by theory (6), in conjunction with (5).

The controversies between the supporters of these six theories have often been misinterpreted by nonbiologists as controversies over the validity of the theory of evolution as such. It is for this reason that I have focused so early on the existence of these different explanatory theories, even though in the immediate post-Lamarckian period the main argument was over evolution as such. Actually, at first, most of the new evidence in favor of evolution that began to accumulate through the first half of the nineteenth century was simply ignored. However, the reaction to these new facts was rather different in France, Germany, and England, the three major European countries in which biological research was cultivated.

The study of the developments in these countries is particularly important in order to refute the idea that evolutionism was a direct continuation of the liberated, materialistic, and often atheistic thinking of the Enlightenment. The facts do not support such an interpretation. The Enlightenment ended, so to speak, with the French Revolution (1789), and the next seventy years saw not only a great deal of reaction, particularly in England and France, but also new developments that were as important for the rise of evolutionary thought as the philosophizing of the Enlightenment.

FRANCE

362   In France, the scene during the quarter century after Lamarck was clearly dominated by Cuvier, even though he survived Lamarck by only three years. The only attempt to express less than orthodox ideas was made by Étienne Geoffroy Saint-Hilaire (1772–1844), the great comparative anatomist. There was a total absence of an evolutionary interpretation in all of Geoffroy's earlier anatomical work.[4] However, when studying certain Jurassic fossil reptiles from Caen in northern France in the late 1820s, Geoffroy was surprised to find that they were not members of such typically Mesozoic types as *Plesiosaurus* as he had expected but were closely related to the living gavials (crocodilians). This suggested to him the possibility of an actual transformation of the Jurassic crocodilians because "the environment is all powerful in modifying organized bodies." He further developed this idea in an essay published in 1833 in which he apparently tried to explain why different animals are different from each other in spite of the unity of plan. Here he attempted to give a physiological explanation by invoking an effect of the environment on respiration which, in turn, necessitates a drastic change in the environment of the "respiratory fluids" resulting in a profound impact on the structure of the organism. In contrast to Lamarck, Geoffroy does not invoke a change of habits as the intermediary that changes the physiology. For him the environment causes a direct induction of organic change, a possibility which had been definitely rejected by Lamarck. Even though the neo-Lamarckians at the end of the century held such direct induction in high esteem, it would be more appropriate to designate this hypothesis as "Geoffroyism," as was indeed done by some authors. The environmental influence, according to Geoffroy, was effected during the embryonic stage, and to substantiate this thesis, Geoffroy carried out extensive experiments with chick embryos.

The thesis that Geoffroy in his later years had become an evolutionist is still controversial. It has been ably argued by Bourdier (1969). Geoffroy did not believe in common descent but he believed that living species which had descended from antediluvian species by uninterrupted generation had, during this period, become considerably modified through external influences.

Geoffroy had a number of other ideas of interest to the evolutionist. He admitted that some of the environmentally induced modifications would be more useful than others. Those animals which acquire deleterious modifications "will cease to exist, and will be replaced by others, the forms of which had changed to correspond to the new circumstances." He expressed here a typically pre-Darwinian theory of elimination (see below).

363

There are a number of reasons why Geoffroy's evolutionary speculations had little lasting impact. Geoffroy, a deist, was religiously conservative, and his theory was not a theory of common descent but rather one of an activation of an existing potential in a given type. Some of his statements were rather contradictory, and the sudden transformation by saltation, as proposed by him, of lower egg-laying vertebrates into birds was rather a strain on the theory of the emergence of evolutionary potentials. His endeavor to make this credible by saying that such a drastic change could be caused by an equally drastic and sudden change of the environment was not at all convincing.

More damaging, probably, was the crushing defeat experienced by Geoffroy's major anatomical thesis, the extension of the unity of plan to the entire animal kingdom (Chapter 10).

Cuvier

No one in the pre-Darwinian period produced more new knowledge that ultimately supported the theory of evolution than Georges Cuvier (1769–1832).[5] It was he who placed the study of the invertebrates on a new basis by discovering, so to speak, their internal anatomy. It was he who founded paleontology and clearly demonstrated for the Tertiary strata of the Paris basin that each horizon had its particular mammalian fauna. More importantly, he showed that the lower a stratum was, the more different the fauna was from that of the present. It was he who proved extinction conclusively, since the extinct proboscidians (elephants) described by him could not possibly have remained unnoticed in some remote region of the world, as was postulated for marine organisms. He, more than anyone else, deserves to be considered the founder of comparative anatomy, little being added to the methods and principles worked out by him until after the publication of the *Origin*. Given this background and experience, one might have expected him to become the first proponent of a thoroughly sound evolutionary theory. In actual fact, Cuvier throughout his life was wholly opposed to the idea of evolution, and his arguments were so convincing to his contemporaries that even after his rather early death evolutionism was unable to assert itself in France for the next half century.

What facts or ideas were responsible for Cuvier's stubborn opposition? It is often stated that his firm adherence to Christianity precluded a belief in evolution, but a careful study of Cuvier's work does not support this interpretation (Coleman, 1964). He nowhere refers to the Bible in scientific argument, and his own interpretation of the past history is frequently in conflict with scripture. Thus he accepts several floods before the Mosaic one and states that there was no animal life in the early history of the earth. Cuvier never used the marvels of the world to demonstrate the existence and benevolence of the creator, as was done by the natural theologians; indeed, he quite deliberately never mixed science and religion. His theism never intrudes into his writing, except perhaps in the Academy debate of April 5, 1832.

A different ideological commitment seems to have been far more important. Cuvier had spent the most impressionable years of his youth at the Karlsschule in Stuttgart and had there become steeped in essentialism. This was reinforced by his subsequent studies in animal classification. Quite in contrast to Buffon, Lamarck, and other followers of Leibniz, Cuvier forever stressed discontinuity. His dismemberment of the *scala naturae* into four "embranchements" is characteristic of his attitude (see Chapter 4). He even insisted that it was impossible to establish any gradations within the embranchements. Each of them contains four classes "without forming a series or enjoying any incontestable rank" (1812). Even though some members of a group might display greater overall complexity, this was not necessarily true for every structure, and organisms that on the average were far simpler might be highly complex in certain structures. Cuvier, quite rightly, could see no evidence for the "steady increase of complexity or perfection" claimed by the adherents of the *scala naturae*. On the contrary, he saw everywhere discontinuities and irregular specialization.

His essentialism is reflected in his species concept (see Chapter 6). At first, his species definition strikes one as quite biological:

"A species comprehends all the individuals which descend from each other, or from a common parentage, and those which resemble them as much as they do each other." But then he constantly stresses that only superficial characters are variable. "There are [other] characters among the animals which resist all influences, either natural or human, and nothing indicates that time has more effect on them than climate or domestication." Triumphantly, he calls attention to the fact that the mummified animals from the Egyptian tombs which were many thousands of years old were quite indistinguishable from the living representatives of these species.[6] Even though Cuvier is aware of geographic variation, he stresses that it does not affect the basic characters of species: if we study a widespread species of wild herbivore and compare individuals from poor or rich habitats or from hot or cold climates, only such nonessential features as size and color may vary while the essential features of the important organs and body relations remain the same throughout.

365

Actually, like all other members of the Paris school, Cuvier had only a minimal interest in species. He, the paleontologist and comparative anatomist, was interested in the major types, but in his work he never really came into contact with the species problem. Even in his later work with fishes, he never looked at them from the populational point of view. He simply never studied the kind of evidence which later converted Darwin and Wallace to evolution.

Cuvier was the first geologist to stress the drastic nature of many of the breaks in the sequence of geological strata. He discovered that successive faunas might be first marine, then terrestrial, then again marine, and perhaps again terrestrial. There were obviously repeated invasions of the ocean, and these were not merely temporary floods. "We are therefore forcibly led to believe, not only that the sea has at one period or another covered all our plains, but that it must have remained there for a long time, and in a state of tranquillity . . . . These repeated eruptions and retreats of the sea have neither been slow nor gradual; most of the catastrophes [Cuvier himself in most cases used the milder term "revolutions," though most English translations use 'catastrophes'] which have occasioned them have been sudden; and this is easily proved especially with regard to the last of them." He then cites the case of the mammoths that have been found frozen in Siberian ice. "Preserved with their skin, hair, and flesh down to our own times. If they had not been frozen as soon as killed, putrefaction would have decomposed the carcasses." And yet they occurred in areas not previously arctic. However, it is not only the fauna which demonstrates the cataclysmic nature of these changes, but also geology: "The breaking to pieces and overturnings of the strata, which happened in former catastrophes, show plainly enough that they were sudden and violent like the last."

The limited fossil record available in Cuvier's time supported at first the conclusion that each of these catastrophes involved the total destruction of the fauna existing in the area where the catastrophe occurred. No species was known to continue through several geological horizons, at least none of the mammals Cuvier was most familiar with. Later it was found by Cuvier and Brongniart (1808) that the faunal change in successive horizons of the same formation was gradual, the main difference being that the most abundant species of a particular horizon were less common or rare in the two adjacent horizons. This discovery permitted tracing the horizons over considerable geographical distances. There are implications in Cuvier's writings that he considered the catastrophes more or less local events which permitted repopulation from unaffected areas. The recent discovery of the extraordinarily different Australian fauna had reinforced Cuvier's belief that entirely different faunas could simultaneously exist in different parts of the globe. Cuvier never speculated as to the cause of the catastrophes but implied that he considered them natural events such as earthquakes, volcanic eruptions, major floods, drastic climatic changes, and mountain building (which was beginning to be considered by geologists). It can be seen that Cuvier's catastrophism was quite "gentle" compared to that of such followers of his as Buckland, d'Orbigny, and Agassiz.

The universal manifestations of discontinuity seemed to Cuvier quite incompatible with an evolutionary interpretation. Lamarck and Geoffroy had invoked a steady occurrence of spontaneous generation to explain discontinuities. This, thought Cuvier, violated all the available evidence. Everything indicated that living things can only come from other living things. Harvey's "Omne vivum ex ovo" ("all life from eggs") was also his motto.

Most importantly, a concept of evolution was quite incompatible with Cuvier's concept of the harmonious construction of each organism. Each species had been created by divine will, and each was assigned from the beginning its special place in the economy of nature from which it could not depart. Fish, for example, were designed for an aqueous environment: "This is their place in the creation. They will remain there until the destruction of the present order of things" (*Histoire naturelle poissons*, I: 543). There was no scale of perfection for Cuvier, since each animal was perfectly adapted to its particular station in nature. He would have gladly endorsed Darwin's admonition: "Never say higher or lower!" These considerations inspired Cuvier to establish his famous correlation principle (see Chapter 10) which led him to such generalizations as that herbivores always had hoofs and no carnivore would have horns. Only certain combinations of form and function are possible, and only these were realized in nature. It was unthinkable   *367* for him that a new habit could induce structural changes. In particular, he rejected the idea that changed habits could affect the simultaneous alteration of many parts of the body and maintain the complex, harmonious interrelations of all organs. Furthermore, Cuvier held the idea that structure has primacy over function and habit, and that only a change in structure might necessitate a change in function.

Cuvier and Variation

Cuvier was too good a naturalist not to be aware of variation and this posed for him the problem of reconciling it with his essentialism. He does so by recognizing two levels of variability. One is manifested in the ephemeral reaction of an organism to such environmental factors as temperature and nutrient supply. Such variation does not affect essential characters, and Cuvier implied, if one wants to express it in modern terms, that this variation was nongenetic, that is, it did not affect the essence of the species. The most superficial characters he considered to be the most variable ones.

Of an entirely different nature would be variation of the essential organs, such as the nervous system, heart, lungs, and viscera. These organs, according to him, were completely stable in their configuration within classes and embranchements. They had to be stable because any variation in any of the major organs would produce unbalances with disastrous effects. Among the stable characters were also those that distinguish species, particularly fossil and living species: "And as the difference between these [fossil] species and the species which still exist is bounded by certain limits, I shall show that these limits are a great deal more extensive than those which now distinguish the varieties of the same species; and shall then point out how far these varieties may be owing to the influence of time, of climate, or of domestication" (*Essay*, 1811: 5–6).

Cuvier's claims of a complete constancy of organs and their proportions in the higher taxa of animals were completely unsupported by any investigations. If such studies had been undertaken by Cuvier, he would have discovered that, contrary to his claims, there are considerable differences in the relative size and configuration of the vital organs in related species, genera, and families. But even if he had found such differences, as he must have during his dissections, he probably would have merely gone back to his basic principle that each animal had been created to fill its assigned place in nature.

Much of Cuvier's argument is specifically directed against the evolutionary theories of Lamarck and Geoffroy, rather than against evolutionism in general. He particularly objects to the vague claim of evolutionary continuity so often made by Lamarck. To say that "this [species of] animal of the modern world descends in a direct line from this antediluvial animal, and to prove it by facts or by legitimate inductions, is what is necessary to do and, in the present state of knowledge, no one would even dare to attempt it" (Cuvier and Dumeril, 1829). On another occasion he stated, "If the species have changed by degrees, we should find some traces of these gradual modifications; between paleotherium and today's species we should find some intermediary forms: This has not yet happened." If Lamarck had been a more astute opponent, he could have probably pointed to a series of Tertiary mollusks that answered this requirement. As far as Cuvier's fossil mammals are concerned, the fossil record was, of course, far too incomplete for the demonstration of a series, and furthermore many of the fossils represent phyletic side branches that have since become extinct. This argument, of course, could not have been used by Lamarck because he did not acknowledge extinction.

In his controversy with Geoffroy and the *Naturphilosophen*, Cuvier was victorious because he realized that there are two kinds of similarities. On one hand, there is similarity due to the unity of the type (now referred to as homologies), on the other hand there is the similarity, well illustrated by the wings of bats, birds, pterodactyls, and flying fishes, which is due to similarity of function. "Let us then conclude that if there are any resemblances between the organs of the fish and those of other classes, it is only insofar as there are resemblances between their functions," said Cuvier. Curiously, when it came to animals that belong to the same anatomical type, let us say different species of fishes, Cuvier emphasized only their differences and ignored completely any resemblance that was not clearly due to similarity of function. He never asked why the various species of the same type were so similar in their basic structure. Thus, Cuvier ignored the powerful, comparative anatomical evidence for evolution.

Even more remarkable is his failure to draw the conclusions from the fossil record that now seem so obvious. This is the more surprising since Cuvier had an excellent understanding of the fossil record and asked very shrewd questions. He insisted that fossils could not be spontaneous products of the rocks but had to be the remains of formerly existing organisms. In contrast to Lamarck, he fully appreciated the importance of extinction: "Numberless living beings have been the victims of these catastrophes . . . Their races even have become extinct, and have left no memorial of them except some fragment which the naturalist can scarcely recognize." He realized how important the fossils were for an understanding of the earth's history. "How could one fail to see, that to the fossils alone is due the birth of the theory of the earth; that without them one would perhaps never have dreamt that there could have been successive periods in the formation of the globe and a series of different operations." He did not invoke any supernatural processes to account for the replacement of these faunas. "I do not pretend that a new creation was required for calling our present races of animals into existence. I only urge that they did not anciently occupy the same places, and that they must have come from some other part of the globe." The fossils provide the investigator with many problems:

369

> Are there animals and plants peculiar to certain beds, and found in no others? Which species appear first, and which follow? Do these two kinds of species sometimes accompany one another? Is there a constant relationship between the antiquity of the beds, and the resemblance or nonresemblance of the fossils with living creatures? Is there a similar climatic relationship between the fossils, and the living forms most closely resembling them? Have these animals and plants lived in the places where their remains are found, or have they been transported there from elsewhere? Do they still live somewhere today, or have they been wholly or partially destroyed?

He himself provided partial or complete answers to most of these questions. Yet, ultimately Cuvier denied that there is any evolutionary progression from a given fauna to that in the next higher stratum or, more generally, that there is a progression throughout the series of strata. Such a denial was possible as long as the stratigraphy of other regions or continents was unknown, and one could postulate that the new faunas were due to immigration from other areas. But further geological exploration showed that the fossil sequence was quite similar in all parts of the world. There were characteristic Paleozoic, Mesozoic, early and late Tertiary faunas (to use modern stratigraphic terminologies).

As we have seen, it was Cuvier himself who demonstrated that the fossils occurring in the highest strata belonged to species or genera that still have living representatives but that the fossils become increasingly dissimilar to modern forms, the deeper one penetrates into the geological sequence. In the Mesozoic strata, one finds a rich representation of peculiar reptiles without modern relatives (like dinosaurs, plesiosaurs, or pterodactyls), while mammals do not turn up until higher in the sequence. When they do, the first types are entirely different from living species. Yet, Cuvier refused so firmly to recognize any taxa of animals as higher or lower that the fossil sequence did not convey to him any evolutionary message.

Cuvier simply refused to face the issue. The progression of faunas through geological time had become so well established that a causal explanation had to be advanced. There seemed to be only two alternatives, either the older faunas evolved into the younger ones—an option Cuvier adamantly refused to accept—or else new faunas are created after each catastrophe. To admit this would have introduced theology into science, and that Cuvier also objected to. So Cuvier adopted the ostrich policy and ignored the bothersome problem.

As far as man is concerned, Cuvier accepted the Cartesian pronouncement that man was qualitatively different from all animals. In contrast to Aristotle and the early anatomists, he rejected the idea that zoology consists of a comparison of the ("degraded") animals with ("perfect") man. The study of man was something quite apart from the study of the four embranchements of the animals. Man was so unique that one would not expect to find him in the fossil record. Indeed, when Cuvier died (1832), no hominid fossil had yet been found, nor, in fact, not even any primate fossil, the first (*Pliopithecus*) not being discovered until 1837.

Cuvier's concept of the living world was, on the whole, internally consistent in spite of some contradictions and some important blind spots. It would have required a truly innovative mind to abandon the essentialistic paradigm and use the new facts to develop a replacement. Cuvier was not that person. As Coleman pointed out, Cuvier was by nature a conservative, an adherent of the status quo. Although extraordinarily well-informed, industrious, lucid in thought and exposition, he was not an intellectual revolutionary. After his death, facts piled up in rapid succession that made the nonevolutionary interpretation ever more implausible. Yet those who followed in Cuvier's footsteps, for instance Agassiz, Owen, Flourens, and d'Orbigny, were less cautious and more dogmatic than Cuvier. This induced them to support a veritable orgy of catastrophism. As far as Cuvier himself was concerned, he won every battle with his evolution-minded opponents, and he did not live long enough to realize that he had lost the war.

371

ENGLAND

The situation in England during the first half of the nineteenth century was in many ways fundamentally different from that in either France or Germany. Natural science, for instance, was totally dominated by geology; between 1790 and 1850 no other country in the world made as splendid a contribution to geology as Great Britain. At the same time, she was unique in the close alliance between science and Christian dogma. Most teaching of scientific subjects at English universities was done by ordained ministers, and well-known scientists continued the tradition established by Newton, Boyle, and Ray, to occupy themselves both with scientific and theological studies.

Piety led the physicist to a rather different emphasis on the manifestations of the hand of the creator from that of the biologist. The order and harmony of the universe made the physical scientist search for laws, for wise institutions in the running of the universe, installed by the creator. All was causal in nature, but the causes were secondary causes, regulated by the laws instituted by the primary cause, the creator. To serve his creator best, a physicist studied his laws and their working.[7]

The naturalist-biologist also studied the works of the creator, but his emphasis was on nothing so mechanical as the movement of falling bodies or of the planets while circling the sun. Rather, he concentrated on the wonderful adaptations of living creatures.

These could not be explained as easily in the form of general laws such as gravity, heat, light, or movements. Nearly all the marvelous adaptations of living creatures are so unique that it seemed vacuous to claim that they were due to "laws." But what could be the explanation of these wonderful adaptations? It rather seemed that these aspects of nature were so special and unique that they could be interpreted only as caused by the direct intervention of the creator. Consequently, the functioning of organisms, their instincts, and their manifold interactions provided him with abundant evidence for design, and seemed to constitute irrefutable proof for the existence of a creator. How else could all the marvelous adaptations of the living world have come into existence?

The materials of the two groups of investigators led them to rather different approaches. The God who made laws at the time of creation and then abdicated his authority, so to speak, in favor of secondary laws was far more remote than the God of the naturalist, who left the imprint of his design on every detail of living nature. Deism, a belief in a rather impersonal god of laws but not of revelations, was—one might say—almost a logical consequence of the developments in physics. The naturalists, on the other hand, adopted a kind of faith generally referred to as "natural theology."[8] It considered the seeming perfection in the adaptations of all structures and organic interactions as evidence for design. All of nature was the finished and unimprovable product of divine wisdom, omnipotence, and benevolence. What better way could there be to pay homage to one's creator than to study his works? For John Ray, the study of nature was the true "preparative to Divinity." Indeed, the study of the wonders of nature was the favorite preoccupation of countless country parsons throughout England.

British natural theology was distinguished from that on the continent in several ways. German physico-theology was man-centered. God had created the world for man's benefit, and it was the role of every creature to be useful to man. Man could not appear on the globe until the creation was ready for him. British natural theology stressed far more the harmony of all nature, and this led to the study of design in all mutual adaptations. The greater longevity of British natural theology can perhaps be explained by its more appealing conceptualization. While the wave of deism and enlightenment swept away physico-theology on the continent, it retained in England its full vigor in the eighteenth century (in spite of Hume's criticism) and rose to a new crescendo

372

in the first half of the nineteenth century with Paley's (1802) *Natural Theology: Or, Evidences of the Existence and Attributes of the Deity Collected from the Appearances of Nature,* and with the eight *Bridgewater Treatises* (1833–1836). The eight authors used various scientific subjects to demonstrate with commendable erudition and complete seriousness "the Power, Wisdom, and Goodness of God, as manifested in the Creation." Science and theology were so much a single subject to many scientists of the period, like the geologists Sedgwick, Buckland, and Murchison or the naturalist Agassiz, that even their scientific treatises were exercises in natural theology. This is true even for Lyell's *Principles of Geology.*

373

Particularly surprising for a late twentieth-century scientist is the willingness of the natural theologian to accept "supernatural evidence" on an equal footing with natural evidence. Not only was creation accepted as a fact but so was the subsequent intervention of God in his world as it pleased him.

However, the alliance between natural theology and science ultimately led to difficulties and contradictions. The canons of objective science came ever more often in conflict with attempts to invoke supernatural intervention. More specifically, the argument from design found it increasingly difficult to reconcile the occurrence of vestigial organs, of parasites and pestilence, and of devastating catastrophes like the Lisbon earthquake with design by a benevolent creator. As we shall later see, much of Darwin's argument in the *Origin* makes use of such contradictions. Various auxiliary hypotheses proposed to explain the fossil sequence and the pattern of world-wide geographic distribution could delay the downfall of natural theology temporarily but could not prevent its ultimate demise.

The critique came not only from science but also from philosophy. Hume in his *Dialogues Concerning Natural Religion* (1779) showed clearly that there was neither a scientific nor a philosophical basis for natural theology, and Kant in his *Critique of Judgment* (1790) rejected naive teleology. But this left an explanatory vacuum since science, prior to natural selection, had no satisfactory explanation for adaptation, considering that Lamarck's speculations were altogether unconvincing. In fact, many pious scientist-philosophers, like Lyell, Whewell, Herschel, and Sedgwick, seemed to have been positively afraid of a natural explanation, fearing that this would destroy the basis of morality. This is perhaps the major reason for the continuing survival of natural theology in Britain right up to the publication of the *Origin.* Natural theology played a peculiarly ambiguous role in the history of evolutionism. Darwin's most determined opponents were natural theologians, and yet the biological adaptations so lovingly described by them supplied some of the most convincing evidence for evolution as soon as one substituted natural selection for design.

Progressionism

374 The inclusion of geology into natural theology was a peculiarly British development (Gillispie, 1951). It attempted to reconcile the newer findings of geology and paleontology with the story of Genesis and the concept of design. The two pieces of evidence always cited as demonstrating the agreement of geological events with the Mosaic account were first the absence of man (the last act of creation) from the fossil record; and second the evidence of a Great Deluge, "a universal deluge," of the whole earth.

That there had been only a single flood began to be questioned already in the eighteenth century (by Blumenbach). It became increasingly implausible when the fossil record established one fossil fauna after another, almost always separated from the preceding one by a complete gap. In Cuvier's rather mild explanation (see above), the destruction of these faunas was referred to as "revolutions," but among his followers the concept of ever repeated catastrophes became dominant. Although Cuvier had sidestepped the issue of faunal replacement, some of his followers asserted emphatically that a brand new creation had taken place after each catastrophe, and that each succeeding creation reflected the changed conditions of the world. This concept has been designated *progressionism* (Rudwick, 1972; Bowler, 1976). In a way it is a creationist reshaping of the *scala naturae*.

The nature of the faunal progression with time was only gradually understood. Cuvier's findings dealt primarily with Tertiary changes of mammalian faunas. When the great fossil reptiles were discovered (at first mostly marine forms), it was realized that they lived at an earlier period (now called Mesozoic) than the mammals (the discovery of a Jurassic mammal near Oxford, England, therefore caused great consternation). The preceding late Paleozoic rocks contained the fossils of fishes, and still earlier deposits contained only invertebrates. The heated controversies of the period dealt with the causation (in creationist terms) of the progression with the question which types were "lower" or "higher," and whether man was or was not the endpoint of the progression. Each author had his own ideas. Louis Agassiz and some of his followers had the peculiar idea that each new creation (after the preceding catastrophe) reflected God's current concept of his creation, and that the succession of faunas represented the gradual maturation of the plan of creation in God's mind. It did not occur to him what a blasphemy this interpretation really was. It insinuated that God, time after time, had created an imperfect world, that he completely destroyed it in order to do a better job the next time, but failed again and again until his most recent creation.

375

Lyell and Uniformitarianism

For generations it has been an accepted dogma of British historians that, as first claimed by T. H. Huxley, "the doctrine of uniformitarianism when applied to biology leads of necessity to Evolution." Since Charles Lyell was the great champion of uniformitarianism,[9] it was concluded that Darwin's evolutionary thought was directly derived from Lyell. How dubious this claim is becomes evident when we realize how strongly the uniformitarian Lyell was opposed to evolution. It is only in recent years that the weakness of Huxley's argument has been pointed out by Hooykaas, Cannon, Rudwick, Mayr, Simpson, and others. Nevertheless, the geological arguments of the 1820s and 30s were of fundamental importance for the shaping of the mind of those biologists for whom the history of life on earth posed a problem. A discussion of uniformitarianism, even though primarily of concern to geology, is an indispensable prerequisite of a discussion of the birth of Darwin's evolutionary ideas.

The terms "uniformitarianism" and "catastrophism" were coined by the British philosopher William Whewell in 1832 in a review of Lyell's *Principles of Geology*. The terms referred to two opposing schools of geologists. Actually they were quite misleading because the principal issue was not the occurrence (or not) of catastrophes but rather the question whether the findings of geology support the steady-state world theory of Hutton and Lyell or the directionalist theory of most other geologists including progressionists and catastrophists. The major thesis of the directionalists was that life on earth had been changing through geological time. This was a rather new concept, being the result of Cuvier's fossil discoveries in the Paris basin and of other recent showings that successive geological horizons often have drastically different faunas; that more often than not they were separated from each other by sharp breaks, and that the lower (earlier) faunas consist largely or entirely of extinct types. Furthermore, it stated that these changes were progressive, as indicated by the sequence invertebrates-fishes-reptiles-mammals. The existence of a progressive sequence was also supported by the botanical stratigraphy of Adolphe Brongniart, who distinguished three periods: the first (Carboniferous) being characterized by primitive cryptogams, the second (Mesozoic) by gymnosperms (and a reduced number of cryptogams), and the third (Tertiary) by the beginning dominance of the angiosperms. The "highest" types, both of animals and plants, appeared last in the earth history. The existence of such a progression was denied by Lyell or, when he admitted it, explained as part of a cycle due to be reversed eventually (Ospovat, 1977).

The term "uniformitarianism" designates an even more complex set of theories than the term "catastrophism." In fact, the term conceals a bundle of at least six concepts or causations.

An attempt is made in table 1 to tabulate the most conspicuous differences among the opposing camps. As table 2 shows, Lyell held alternative *a* in all cases but one; but among the catastrophists, one finds different mixes of the various alternatives. Interestingly, it seems to me that Darwin was closer in his paradigm to Lamarck than to Lyell. I must warn the reader, however, that my classification is somewhat subjective and that other assignments are possible.

Although most of the six components of uniformitarianism, here distinguished, are of interest primarily only to geologists, a few words may be said about them as an explanation of the adopted categories in table 1.

(1) Naturalism. Without exception, all the participants in the controversy were devout Christians, and the only issue of disagreement was the extent to which they envisioned God to intervene in the workings of His world. In both camps there were some who thought that after creation only secondary causes were in operation. Obviously, all creations, whether a single original one or multiple creations after each catastrophe, were the direct work of the creator. For Lyell, all geological processes in the world were the results of secondary causes, not requiring the invoking of supernatural interventions. Lyell's critics reproached him for not

consistently applying this principle to the introduction of new species, a process which, Lyell's denials notwithstanding, had all the earmarks of ad hoc special creation.

(2) Actualism. This principle states that the same causes (physical laws) have operated throughout geological time, since the immanent characteristics of the world have always remained the same. The most important consequence of this postulate is that, as stated by Lyell in the subtitle of the *Principles,* it is legitimate to "attempt to explain the former changes of the earth's surface by reference to causes now in operation."

377

(3) Intensity of causal forces. Lyell and other extreme uniformitarians postulated that the intensity of geological forces was the same at all times, and that the time factor would account for apparent cases of increased intensity at certain periods. Some of his opponents thought that owing to the cooling of the earth there was a steadily decreasing intensity of geological phenomena, such as volcanism and orogeny. It is not quite clear whether or not certain authors supported a third possibility, that of an irregular increase or decrease in the intensity of geological phenomena.

(4) Configurational causes. This term, introduced by Simpson (1970), refers to the possibility that different constellations of the same factors can have drastically different results, a possibility entirely overlooked by the straight uniformitarians. The change of the earth's atmosphere from a reducing to an oxidizing one, also the irregular occurrence of ice ages as well as all the effects of plate tectonics on the size of the land masses and the extent of shallow shelf seas, and finally, the extent of volcanism, fall in this category. As a consequence, the now-prevailing physical conditions on earth do not necessarily reflect accurately the conditions that prevailed at earlier stages of the earth's history. A problem like the origin of life was insoluble as long as configurational causes were ignored. Lyell allowed for one configurational cause, the effect of the change in the position of land masses on climate (Ospovat, 1977).

(5) Gradualism. For most authors prior to the rise of catastrophism, historical changes on the face of the earth were believed to have been gradual. This was the opinion of Leibniz, Buffon (in part), Lamarck, and of most of Darwin's so-called forerunners. To uphold gradualism became more difficult after the discovery of the frequency of stratigraphic breaks. The greatest merit of Lyell's uniformitarianism was that it continued to emphasize the grad-

378

Table 1. Components of uniformitarianism.

PHENOMENON OR PROCESS	UNIFORMITARIAN VIEW	CATASTROPHIST OR DIRECTIONALIST VIEW
(1) Theological aspects of causes (*naturalism* or *supernaturalism*)	(a) Naturalistic (even if originally divine, now always due to secondary causes) (b) On the whole naturalistic, but allowing for occasional divine intervention	(c) Always allowing for direct divine intervention
(2) Causes through geological time (*actualism*)	(a) Same causes (physical laws) operative at all times	(b) Different causes operating in early history of earth
(3) Intensity of causal forces	(a) Always at same intensity as at present	(b) Irregular, varying in geological time (c) Steadily decreasing with geological time
(4) Configurational causes	(a) The same at all times	(b) Different in certain former geological periods
(5) Rate of change (*gradualism*)	(a) Many gradual but some rather drastic (saltational)	(b) Many truly cataclysmic changes
(6) Directional change of the world	(a) Rejected; world always in a steady-state condition, at most changing cyclically	(b) Yes; world changing through history in a more or less directional manner

ualness of geological changes, in spite of the new findings. Both Lyell and later Darwin were fully aware of the fact that earthquakes and volcanic eruptions could produce rather drastic effects but that they were nevertheless smaller by several orders of magnitude than the catastrophes postulated by some geologists. Yet, modern geological research has demonstrated that certain events in the past history of the earth virtually qualify as catastrophes (Baker, 1978; Alvarez et al., 1980).

Table 2. Supporters and opponents of evolutionism and their adoption of various components of uniformitarianism.

AUTHOR	COMPONENTS OF UNIFORMITARIANISM					
	1	2	3	4	5	6
Lyell	b	a	a	a	a	a
Darwin	a	a	a	?b	a	b
Lamarck	a	a	?b	b	a	b
Agassiz	c	a	?b	b	b	b

Solid lines: Darwin agreeing with Lamarck versus Lyell. *Broken lines:* Darwin agreeing with Lyell versus Lamarck.

(6) *Directionalism.* Lyell had adopted from Hutton the concept of a steady-state world, so popular among the pre-Socratic philosophers: "No vestige of a beginning, no prospect of an end," as Hutton had put it (1795). Lyell's opponents concluded that all the evidence indicated a directional, if not progressional, component in the history of the earth. This rather than any of the other five points was the basic difference of opinion between Lyell and the so-called catastrophists (Rudwick, 1971; but see Wilson, 1980).

What Does Darwin Owe to Uniformitarianism?

The various recent analyses indicate that, no matter how great Darwin's intellectual indebtedness to Lyell was, uniformitarianism (sensu Lyell) was actually more of a hindrance in the development of his evolutionism than a help.[10] Gradualism, naturalism, and actualism were the prevailing concepts from Buffon to Kant and Lamarck. The most distinctive part of Lyell's specific uniformitarianism was his steady-state (and cyclical) theory, and this was definitely quite irreconcilable with a theory of evolution.

Lyell was not a mere geologist; perusal of the *Principles of Geology* shows how well informed he was on biological subjects, including biogeography and ecology ("struggle for existence"). When speaking of biological matters he spoke with authority, yet, in retrospect, it is quite apparent that his creationism and essentialism led him into conflicts and inconsistencies.

Lyell had had his major training in the law and tended in his scientific controversies to give an extreme picture if not a caricature of opposing viewpoints. Thus he tended to attack individual errors in the accounts of the catastrophists and ignore their otherwise so substantial evidence for directional changes, such as the fossil content in the sequence of geological strata.

Apparently, he thought that his opponents postulated a fossil sequence strictly in terms of the *scala naturae* and triumphantly considered this to be refuted by the discovery of fossil mammals (at Stonesfield) in Jurassic strata ("in the age of reptiles") without realizing that these were triconodonts (ancestral mammals), thus fitting quite well into a directional series. He rightly refuted Lamarck's theory of an inherent trend toward perfection, but overlooked the fact that Lamarck had also postulated a second type of evolution, a continuous adjustment to the ever-changing environment ("circumstances"), which would, of course, inevitably result in continuous evolutionary change. For Lyell, an essentialist, such continuous evolution would make no sense at all.

Lyell's writings have long been completely misinterpreted owing to T. H. Huxley's erroneous claim that his uniformitarianism would inexorably lead to Darwinism and owing to Whewell's misleading labels "uniformitarianism" and "catastrophism." Lyell's steady-state world was not a completely static world, but one undergoing eternal cycles, correlated with the movements and climatic changes of the continents. Extinction was a necessary consequence of the changing world becoming unsuitable for certain species. And, of course, in a steady-state world the lost species had to be replaced through the "introduction" of new species. Since the loss of species by extinction and their replacement by new introductions occurred at a steady rate, Lyell insisted that he was following strict uniformitarian principles.

What was of foremost importance for the history of evolutionism was not Lyell's uniformitarianism but the fact that he shifted the emphasis from Lamarck's vague speculations on progression, growing perfection, and other aspects of "vertical evolution" to the concrete phenomena of species. The question, What are the causes for the extinction of species? led to all sorts of ecological questions. These, and the question, How are the replacement species introduced? were encountered by Darwin when he read the *Principles of Geology* during and after the voyage of the *Beagle.* As a result of Lyell's writings, these questions became the center of Darwin's research program.

This Lyell-Darwin relationship illustrates in an almost textbook-like fashion a frequently occurring relationship among scientists. It is the counterpart to "forerunner." It has often and

380

rightly been stated that Lamarck, even though a genuine evolutionist, was not really Darwin's forerunner. Darwin did not in any way build on the Lamarckian foundations; rather he built on the Lyellian foundations. But one can hardly call Lyell a forerunner of Darwin's, because he was adamantly opposed to evolution, he was an essentialist, he was a creationist, and his whole conceptual framework was incompatible with that of Darwin. And yet he was the first who clearly focused on the crucial role of species in evolution and stimulated Darwin to choose that way to solve the problem of evolution, even though this was done by showing that Lyell's proposed solutions were wrong. What does one call a person who shows the path even though he is not a forerunner in the conventional sense? In an analogous fashion my own work on geographic speciation and the biological species was stimulated by opposition to Goldschmidt's (1940) proposed solution of speciation through systemic mutations. There are literally scores of cases in the history of science where a pioneer in posing a problem arrived at the wrong solution but where opposition to this solution led to the right solution. *381*

I have analyzed at an earlier occasion (Mayr, 1972) the set of ideas and beliefs that had prevented an earlier acceptance of evolution. It consisted of natural theology and a very literal creationism together with essentialism. Paradoxically, within this framework the advance of scientific knowledge necessitated an ever-increasing recourse to the supernatural for explanation. For instance, the succession of faunas discovered by the stratigraphers necessitated abandoning the idea of a single creation. Agassiz was not afraid to postulate 50–80 total extinctions of life on earth, and an equal number of new creations. Even such a sober and cautious person as Charles Lyell frequently explained natural phenomena as due to creation. And this removed the facts of evolution from the realm of scientific analysis. Nothing, of course, is impossible in creation. "Creation," said Lyell, "seems to require omnipotence, therefore we can not estimate it."[11]

Chambers' Vestiges

After Lyell had crushed Lamarck in his *Principles,* evolutionism seemed to be quite absent from the thinking of British scientists. The rejection was universal, ranging from philosophers like Whewell and Herschel to geologists, anatomists, and botanists. Seemingly, there was happy contentment with the view of natural theology of a world created by a skillful designer. In this peaceful Victorian scene a bomb exploded in 1844 which thoroughly shook up the educated world of Britain—the publication of the *Vestiges of the Natural History of Creation*.[12] The contents of this volume were so heretical that the author had taken every possible precaution to remain anonymous. Everyone speculated who he might be and the guesses ranged from Lyell and Darwin to the Prince Consort! The reaction was something colossal. The Woodwardian Professor of Science at Cambridge and President of the Geological Society, Adam Sedgwick, was utterly outraged. He required no less than four hundred pages of print to state all of his objections, the flavor of which may be sampled by these sentences: "The world cannot bear to be turned upside down and we are ready to wage an internecine war with any violation of our modest principles and social manners . . . it is our maxim that things must keep their proper places if they are to work together for any good . . . if our glorious maidens and matrons may not soil their fingers with the dirty knife of the anatomist, neither may they poison the strings of joyous thought and modest feelings by listening to the seductions of this author, who comes before them with . . . a false philosophy." With such advertising, it is not surprising that the *Vestiges* had magnificent sales, eleven editions being necessary between 1844 and 1860, the sales in the first ten years (24,000 copies) greatly exceeding those of Lyell's *Principles of Geology* or Darwin's *Origin of Species* (9,500) in a corresponding ten-year period after publication.

The identity of the author was not revealed until after his death in 1871. He turned out to be Robert Chambers, the well-known editor of *Chambers' Encyclopedia*, an author of many popular books and essays. Even though Chambers was widely read and well-informed, he was very much of a layperson with all the deficiencies implied by this designation. Yet, it was he who saw the forest where all the great British scientists of his period (except for the nonpublishing Darwin) only saw the trees. Curiously, it was deism, rather than atheism, which led Chambers to postulate evolution. When there is a choice, he said, between special creation and the operation of general laws instituted by the creator, "I would say that the latter is greatly preferable as it implies a far grander view of the divine power and dignity than the other." Since there is nothing in inorganic nature "which may not be accounted for by the agency of the ordinary forces of nature," why not consider "the possibility of plants and animals having likewise been produced in a natural way." He rejects the suggestion that the origin and development of life is beyond our power of inquiry. "I am extremely loath to imagine that there is anything in nature which we should for any reason refrain from examining . . . and feel assured that our conception of the divine author of nature could never be truly injured by any additional insight we might gain into His works and ways."

His investigations finally led him to advance "the Principle of Progressive Development as the simplest explanation—as an explanation involving slow and gradual movement, such as we usually see in nature—as an explanation appealing to and allying itself with science instead of resting on a dogmatic assumption of ignorance." Chambers perceived two things very clearly from the available evidence: (1) that the fauna of the world evolved through geological time, and (2) that the changes were slow and gradual and in no way correlated with any catastrophic events in the environment.

Even though Chambers made some disparaging remarks about Lamarck, his thesis was in many ways the same as Lamarck's original theory, a gradual perfecting of evolutionary lines. Except for also postulating evolution, he was in no way a forerunner of Darwin.

Chambers marshaled his evidence as follows:

(1) The fossil record shows that the oldest strata have no organic remains; then follows an era of invertebrate animals; next a period during which fishes were the sole vertebrate forms in existence; next, a time when reptiles occur but not yet any birds or mammals, and so on.

(2) In all the major orders of animals, there has been a progression from simple to complex, "the highest and most typical forms always being attained last."

(3) The fundamental unity of organization is shown in every major group of animals, as revealed by the study of comparative anatomy.

(4) The facts of embryology, as worked out by von Baer, show that the embryos tend to go through stages resembling their more primitive relatives.

Although Chambers' discussions are full of errors and misconceptions, he displays an amount of common sense in his consideration of the evidence that is sadly lacking in the writings of the contemporary antievolutionists. When analyzing the arguments of the paleontologist Pictet, at that time still opposed to evolution, Chambers exclaims, "We can only wonder that a man learned in the subject can see such a difficulty in accepting faunal change by natural law."

What Chambers had really done was to apply the principles of uniformitarianism to organic nature. The hierarchy of animals as reflected in the natural system made no sense to him unless one adopted evolution. Here, as in the discussion of the fossil record, his arguments were remarkably similar to those of Darwin's *Origin.* Like Darwin, he constantly reiterated how many phenomena, for instance rudimentary organs, could be explained as the product of evolution but made no sense in terms of special creation. When seeing all this evidence, "the author embraced the doctrine of Progressive Development as a hypothetic history of organic creation."

All this sounds eminently reasonable, and yet Chambers was shredded to pieces by his critics, including some of the most distinguished British scientists of the day. T. H. Huxley, for instance, wrote a review so savage that, apparently, he himself later regretted it. The critics had no trouble pointing out that the evolutionary mechanisms which Chambers suggested were quite absurd. He relied on the universal and frequent occurrence of spontaneous generation. Recapitulation was one of the building stones of his theory, and the whole concept of progressive development was based on an analogy with "generation," that is, with the ontogeny of an individual. Like many dilettantes, Chambers was unbelievably gullible and backed up his belief in spontaneous generation by all sorts of folklore myths. Yet, every once in a while he makes some exceedingly shrewd guesses. He admits, for instance, that spontaneous generation may no longer occur. One of the reasons for this might be that it is "a phenomenon . . . as expressly and wholly a consequence of conditions which, being temporary, the results were temporary also." This, of course, is now the accepted explanation for the conditions at the time of the origin of life.

Even though Chambers was the only pre-Darwinian nineteenth-century British evolutionist, he is much too unimportant a figure to deserve further discussion. Still, he converted quite a few people to evolutionary thinking. The most important of these was A. R. Wallace, but Herbert Spencer, apparently, was also influenced. In Germany he made the philosopher Arthur Schopenhauer believe in evolution, and in the United States the poet and essayist Ralph Waldo Emerson. There is no doubt that it was through Chambers that a lot of people became accustomed to the thought of evolution. Even Darwin admitted, "In my opinion [the publication of the *Vestiges*] has done excellent service in calling in this country attention to the subject and in removing prejudices." It was valuable to Darwin for the additional reason that the critics of the *Vestiges* supplied Darwin with the standard catalogue of objections to evolution, objections which Darwin took great care to answer in the *Origin*.

The historian of science can derive two rather far-reaching generalizations from the *Vestiges:* First, Chambers, an ignorant layperson, saw a complex phenomenon rather clearly while all the contemporary, greatly better qualified specialists were distracted by seeming discrepancies (except for Darwin, who—however—for twenty years was withholding his findings). Second, Chambers saw and described the evolutionary process quite well, even though his explanations were not merely wrong but often positively childish. The frequently made claim that one cannot develop a scientific theory unless one has also worked out the explanation is clearly invalid. Darwin, of course, is another example. He postulated unlimited genetic variability and made it the basis of his theory of natural selection, even though all of his thoughts about the theory of inheritance were badly mistaken and insufficient.

What was perhaps most surprising about Chambers was his uniqueness on the British scene. He had almost no supporters, only Owen not being totally negative (Millhauser, 1959, 202). Indeed, all the more renowned British scientists of the period were at that time openly opposed to evolution, not only natural theologians like Buckland, Sedgwick, and Whewell but also Darwin's friends Lyell, Hooker, and Huxley. Even though more evidence in favor of evolution piled up all the time, as Lovejoy has argued so persuasively, the climate of opinion of England was so strongly opposed to evolution that no naturalist considered it seriously. It needed a substantial effort to change the climate of opinion, not the dabblings of a dilettante like Chambers, and this effort did not come until 1859.

Spencer

Herbert Spencer is often cited as having anticipated Darwin in propounding a theory of evolution, but there is little validity in this assertion.[13] Evolution, for Spencer, was a metaphysical principle. The vacuousness of Spencer's theory is evident from his definition: "Evolution is an integration of matter and concomitant dissipation of motion; during which the matter passes from an indefinite, incoherent homogeneity to a definite, coherent heterogeneity; and during which the retained motion undergoes a parallel transformation" (1870: 396). The stress on matter, movement, and forces in this and other discussions of evolution is a typical example of an inappropriate eighteenth-century-type physicalist interpretation of ultimate causations in biological systems, and has nothing to do with real biology. What little Spencer knew of biology in 1852 when he published his first essay on evolution was based on Chambers' *Vestiges* and on Lyell's refutation of Lamarck. Like Chambers, Spencer derived his concept of evolution from an analogy with ontogenetic development, the growth of the individual organism. It was transferred by him from these teleonomic phenomena to a teleological principle affecting the principles of progress adopted by Condorcet and other philosophers of the Enlightenment.

386

Spencer's ideas contributed nothing positive to Darwin's thinking; on the contrary, they became a source of considerable subsequent confusion. It was Spencer who suggested substituting for natural selection the term "survival of the fittest," which is so easily considered tautological; it was likewise he who became the chief proponent in England of the importance of the inheritance of acquired characters (in his famous controversy with Weismann).[14] Worst of all, it was he who became the principal spokesman for a social theory based on a brutal struggle for existence, misleadingly termed social Darwinism (Hofstadter, 1955).

It would be quite justifiable to ignore Spencer totally in a history of biological ideas because his positive contributions were nil. However, since Spencer's ideas were much closer than Darwin's to various popular misconceptions, they had a decisive impact on anthropology, psychology, and the social sciences. For most authors in these areas, for more than a century after Darwin, the word "evolution" meant a necessary progression toward a higher level and greater complexity, which is what it had meant to Spencer rather than to Darwin. This must be stated emphatically in order to dispel a long-standing myth. Unfortunately, there are still a few social scientists who ascribe this Spencerian type of thinking to Darwin.

GERMANY

The rise of evolutionism in Germany proceeded along fundamentally different lines from that in France or England, for several reasons. Natural theology had reached its height in Germany already in the eighteenth century under the influence of Christian Wolff and Hermann Samuel Reimarus, and had been far more deistic than the "interventionist" natural theology of the British. Instead of cultivating natural theology, Germany from Herder to the 1840s passed through an exuberant period of romanticism. This was an optimistic movement, seeing development and improvement everywhere, a striving toward higher levels of perfection, thus fostering ideas derived from the *scala naturae* and from the concept of progress so popular among the philosophers of the Enlightenment. This movement gave rise to a special branch of philosophy, *Naturphilosophie*. It is perhaps still not yet fully appreciated to what an extent various romantic movements, particularly *Naturphilosophie*, were a rebellion against the reductionism and mechanization of Newtonianism. No one has made this clearer than Goethe in his many writings, particularly in his *Farbenlehre*. To reduce every phenomenon and process in the organic world to movements and forces, or to heat and gravity was, quite rightly, unacceptable to most naturalists, who posited numerous alternatives. For instance, they could fall back on natural theology and explain everything in terms of creation and design. Those who did not want to invoke God to explain everything in nature elaborated a new view of nature, strongly influenced by Leibniz, which stressed quality, development, uniqueness, and usually a finalistic component. The excesses of Schelling and Oken could not have been received with such enthusiasm if it had not been for the prevailing disgust with the "heartless" mechanization of Newtonianism. *Naturphilosophie* was largely a backlash against a naive mechanistic interpretation of complex organic phenomena, quite inaccessible to such a simple-minded physicalist interpretation. Since all the best-known representatives of *Naturphilosophie*—Schelling, Oken, and Carus—ultimately were essentialists, they were quite unable to develop a theory of common descent. Yet, they all talked a great deal about development, meaning one or the other of two quite different processes: either an unfolding of a pre-existing potentiality (rather than any modification of the type itself) or else a saltational origin of new types either by spontaneous generation from inanimate material or from existing types. Much of this literature, particularly the work of Oken, is fantastic if not ludicrous. Most conclusions are based on analogies, often ridiculously far-fetched analogies.

It is exceedingly difficult to evaluate this literature and its lasting impact. Some historians have concluded that it retarded the coming of evolutionism to Germany, others that it prepared the ground for it and was responsible for the fact that Darwin and evolution were accepted more readily in Germany than in any other country. This much is certain, that there is a remarkable contrast between Germany and pre-Darwinian England. While in England not a single reputable scientist believed in evolution, in Germany such a belief was apparently widespread. The embryologist von Baer claimed in 1876 that in his 1828 work, "I have emphatically expressed myself as opposed to the then dominant theory of transmutation" (p. 241). In 1834 he repeated that he could "find no probability that all animals have developed from one another through transformation," although elsewhere in the same lecture he records with approval the idea, previously expounded by Buffon and Linnaeus, that the species of a genus "may have developed from a common original form."[15]

In J. F. Meckel's (1781–1833) great handbook of comparative anatomy (1821: pp. 329–350) a considerable number of pages is devoted to the subject of evolution, in particular to the origin of new species. He lists four possible mechanisms: (1) a frequent occurrence of spontaneous generation; (2) an inner drive toward change; (3) a direct effect of the environment; and (4) hybridization. What is most remarkable in his account is that Meckel takes it completely for granted that evolution is due to natural processes, so much so that God or creation is nowhere mentioned. Nature has taken over the role of God. How drastically different from the contemporary atmosphere in England!

Various historians (for example, Potonié, Schindewolf, and Temkin)[16] have rescued the names of numerous early German evolutionists from oblivion. Frankly, it is difficult to evaluate fairly the writings of Kielmeyer, Tiedemann, Reinecke, Voight, Tauscher, Ballenstedt, and other authors who published between 1793 (Kielmeyer) and 1852 (Unger). They are a peculiar mixture of sound ideas and absurdities. Often they seem to reflect the writings of Buffon, Herder, Lamarck, Geoffroy, and Cuvier, but the sources are never mentioned. It would require a very careful, comparative analysis to determine what is valuable and original in the writings of these authors. Since none of them seems to have had any noticeable influence in the ensuing decades, it seems questionable whether such an analysis would be worth the effort. At any rate, it is evident that these authors belong more to the genre of Chambers than to that of Darwin.

Considering the seeming universality of evolutionary thinking in Germany during the first half of the nineteenth century, it is quite puzzling that this background did not lead to the elaboration of a substantial theory of evolution by even a single German biologist. The mystery is heightened by the fact that at that time no other European country had a more competent group of zoologists and comparative anatomists than Germany, including Authenrieth, von Baer, Blumenbach, Burdach, Döllinger, Ehrenberg, Emmert, Heusinger, Kielmeyer, Leuckart, J. Müller, Pander, Rathke, Reichert, Rudolphi, Siebold, Tiedemann, and Wiedemann. The reasons for this failure are manifold. Most importantly, German philosophy at that period was strongly dominated by essentialism and this affected everyone's thinking. Typological thinking was reinforced by an admiration for Cuvier which is quite evident in the writings of some of the outstanding comparative anatomists of the period.

A second reason is that the evolutionism of the *Naturphilosophen* was so speculative and at the same time so sterile that it produced a violent reaction and induced the best zoologists to concentrate on straight descriptive work, as is very conspicuous in the writings of Leuckart, Ehrenberg, Müller, and Tiedemann. This reaction went so far that, when Weismann was a student in the 1850s, evolution was never mentioned at his university. The great excitement about evolution of the 1820s had been completely forgotten by then.

This rejection of speculation was reinforced by two additional considerations. The more the naturalists studied nature, the more they were impressed by the universality of beautiful adaptations. Since the mechanistic spirit of the period did not permit the acceptance of a teleological or supernatural explanation, one was forced, following the example of Kant's *Critique of Judgment,* to adopt an agnostic attitude. Finally, the 1830s, 40s, and 50s were a period of unprecedented developments in experimental biology, including physiology, cytology, and embryology. As a result, the leading German biologists of the period devoted all their efforts to the study of functional processes. Here they were able to apply successfully explanatory models made popular by the physical sci-

389

ences, models which would have been quite inappropriate when applied to evolution. A revival of evolutionism could not come from the laboratory; it had to be initiated by students of natural populations and species, as was the case in England. Alas, Germany's leading young naturalists, Kuhl and Boie, had succumbed to tropical diseases in the East Indies, or, like Illiger, had died of tuberculosis at a young age (Stresemann, 1975).

In spite of Weismann's statement, evolutionism was not completely dead in Germany in the 1850s. Bronn (Schumacher, 1975) wrote several essays on evolution, although ultimately rejecting it. Hermann Schaaffhausen (1816–1893), the co-discoverer of the Neanderthal skull (Temkin, 1959), stated clearly, "The immutability of species which most scientists regard as a natural law is not proved, for there are no definite and unchangeable characteristics of the species, and the borderline between species and subspecies is wavering and uncertain. The entire creation appears to be a continuous series of organisms affected by generation and development." He specifically rejects the argument that the living animals could not have descended from those of earlier periods because we now see no transformation of species. Since such a transformation requires "hundred thousands of years," Schaaffhausen points out, it would be unrealistic to expect that one could observe it directly.

Unger

Among Darwin's many forerunners, few merit mention more than the Viennese botanist Franz Unger (1800–1870). In his *Attempt of a History of the Plant World* (1852), he devotes a special chapter to evolution under the title "The Origin of Plants; Their Multiplication and the Origin of Different Types." He states (p. 340) that the simpler aquatic and marine plants preceded the more complex plants:

> It is in this marine vegetation consisting of thallophytes, particularly algae, that one must look for the original germ of all kinds of plants that have successively originated. There is no doubt that this empirically reconstructed pathway can be theoretically pursued still further backward until one finally gets back to an Urpflanze, in fact to an original cell which has given rise to the entire vegetable world. How this plant or rather cell ultimately originated is even more hidden from us than the fact of its existence. This much, however, is certain,

that . . . it must have designated the origin of all organic life and thus the representative of all higher development.

He continues by saying that on first sight one would expect a constancy of species, since parents always produce offspring of their own kind. However, this would require that all new species would have to originate by some process of spontaneous generation like the *Urpflanze*. Since all the evidence contradicts such a possibility, "no other alternative remains but to look for the source of the entire diversity in the plant world itself, not only of species but also of genera and higher categories." He adds very perceptively that there are too many regularities in the relationship of species to assume that the origin of new species could be due to purely external influences. "This indicates clearly that the cause for the diversity of the plant life cannot be an external one but must be internal . . . in one word, each newly originating species of plants . . . must originate from another one." As soon as one accepts this, the whole plant kingdom becomes a single organic unit. "The lower as well as the higher taxa appear then not as an accidental aggregate, as an arbitrary mental construct but united with each other in a genetic manner and thus form a true intrinsic unit" (p. 345). He then raises various other evolutionary questions such as whether a species undergoes a metamorphosis as a whole to become a new species, or whether only one or a few individuals change to become the ancestral stock of the new species. Indeed, the source of the variation that gives rise to the new species is evidently of great concern to him. Gregor Mendel was Unger's student, and has reported that it was Unger's pondering over the nature and source of variation leading to the origin of new species which had started him off on his genetic experiments (Olby, 1967).

THE PRE-DARWINIAN LULL

From the date of publication (1809) of Lamarck's *Philosophie zoologique* on, no one discussing species, faunas, distributions, fossils, extinction, or any other aspect of organic diversity could afford any longer to ignore the possibility of evolution. And it was not ignored, as the frequent references to Lamarck or to "development" clearly document. It was because he was aware of the "threat" of evolutionism that Lyell devoted so many chapters of the *Principles of Geology* to its refutation. In fact, the years from 1809 to 1859 are of fascinating interest to the historian of ideas.

Here was a legitimate theory, that of a dynamic, evolving world; here was an ever mounting pile of evidence in favor of this heterodox new theory; and here was an ever increasing number of authors who hesitantly referred to the possibility of evolutionary change. In view of these developments, Arthur Lovejoy has asked the intriguing question, "At what date can the evidence in favor of the theory of organic evolution . . . be said to have been fairly complete?" (1959a: 356). The answer, of course, depends on the strength of the resistance. One could go so far as to say that Cuvier's findings (1812) of the increasing taxonomic distinctness of the mammalian fossils in the Paris basin with increasing geological age should have been irrefutable proof of evolution for anyone but a special creationist. Lovejoy (1959a) and Mayr (1972b) have shown that by the 1830s and 40s there was abundant other evidence that should have led to the same conclusion. This includes the facts of geographic variation (for example, Gloger, 1833) which refute the constancy of species (later one of Darwin's major pieces of evidence). Every newly discovered biogeographic fact pointed in the same direction. The persistence of certain types of animals, like the brachiopod *Lingula* and certain mollusks, through many geological periods right back to the Silurian refuted the occurrence of universal catastrophes. The discovery that not all species hybrids are sterile helped to refute the claim of the complete isolation of species. The existence of rudimentary or abortive organs was in conflict with a creationist explanation of perfect design, as Chambers rightly pointed out. The "unity of type" (discovered by the comparative anatomists), the homology of the mammalian middle earbones (Reichert, 1837) and the discovery of other homologies among classes of vertebrates, the presence of gill arches in the embryos of terrestrial vertebrates (and other facts of comparative embryology), and much of the other evidence so convincingly used by Darwin in 1859 but discovered well before that date all supported the evolutionary theory. Referring to this evidence in at least twenty places in the *Origin,* Darwin uses exactly this argument. This evidence, he says, makes perfect sense only if we adopt evolution, but it would indicate the existence of an extraordinarily capricious creator if one adopts the creationist explanation.

Actually, as we have seen, a considerable number of authors arrived at this conclusion before Darwin. And yet, the leading authorities in zoology, botany, and geology continued to reject evolution. Since Lyell, Bentham, Hooker, Sedgwick, and Wollaston in

392

England and their peers in France and Germany were highly intelligent and well-informed scientists, one cannot attribute their resistance to stupidity or ignorance. The mounting evidence for evolution from the fields of biogeography, systematics, stratigraphy, and comparative anatomy did not "reduce [their own] hypothesis to a grotesque absurdity," as Lovejoy thought it should have done, but was somehow reconciled by them, either with a stable, recently created world, or with a steady-state cyclical world, or with a series of catastrophes. One can explain this attitude only by assuming—and all the evidence seems to support this assumption—that the opponents of evolution found it easier to reconcile the new facts with their established conceptual framework than to adopt the new concept of evolution. What was needed for the victory of the new ideas was a cataclysmic event that would sweep the boards clean. This event was the publication, on November 24, 1859, of the *Origin of Species* by Charles Darwin.

9 ✎ Charles Darwin

IN SPITE OF THE valiant efforts of various philosophers and such perceptive biologists as Lamarck, the concept of a created and essentially stable world continued to reign supreme until one man, Charles Darwin (1809–1882), destroyed it once and for all. Who was this extraordinary man, and how did he come to his ideas? Was it his training, his personality, his industry, or his genius that accounts for his success? This has been argued vigorously ever since there has been a historical literature on Darwin.[1]

Charles Darwin was born on February 12, 1809, at Shrewsbury, Shropshire, England, the second son and fifth of six children of Dr. Robert Darwin, an eminently successful physician, who, in turn, was the son of Erasmus Darwin, the author of *Zoonomia*. His mother, the daughter of Josiah Wedgwood, the celebrated potter, died when Charles was only 8 years old, and Darwin's elder sisters tried to fill her place. Our understanding of Darwin's youth and maturation is severely handicapped by the fact that almost all we know about it is taken from his *Autobiography* (1958), a set of reminiscences written for the benefit of his children and grandchildren when Darwin was 67 years old. Unfortunately, this document is not at all reliable, not only because his memory occasionally let him down but also because it was written with that exaggerated Victorian modesty that induced Darwin to belittle his own achievements and the value of his education. Biographers all too readily have tended to accept his words at face value, particularly where Darwin made disparaging remarks about his own abilities, and then wondered how such an uneducated dullard could have become the architect of perhaps the greatest intellectual revolution of all time.

One will never understand Darwin unless one appreciates the truth of his statement, "I was a born naturalist." Every aspect of nature intrigued Darwin. He loved to collect, to fish and hunt, and to read nature books like Gilbert White's *Natural History of Selborne*. As is the case for so many other young naturalists, school was nothing but a burden to him, and this was still largely true for his university years. Since natural history, or for that matter any science, was not a legitimate subject for study in the England of Darwin's youth (in fact, not until the 1850s), his father sent young Charles to the University of Edinburgh when he was only 16 years and 8 months old, to study medicine like his older brother Erasmus, who had gone there one year earlier. Medicine bored and appalled him, and the same was true for the lectures in some other subjects such as geology (by the famous Robert Jameson). Even though Darwin was utterly bored by most subjects (and this was equally true for his Cambridge years), he was conscientious enough to pass his various examinations with reasonably good marks.

395

The myth is forever repeated that Darwin became a naturalist through his experiences on the *Beagle*. The facts contradict these claims. The Darwin who joined the *Beagle* in 1831 was already an unusually experienced naturalist. I rather suspect he would have surpassed any contemporary new Ph.D. in biology in his knowledge of all kinds of organisms. He had an amazing knowledge not only of insects, which were his special group, but also of mammals, birds, reptiles, amphibians, marine invertebrates, fossil mammals, and plants. This expertise is evident not only from his pre-*Beagle* letters but also from his correspondence with J. S. Henslow during the first months on the *Beagle*. The facility with which he rattles off the names of the genera and families of organisms he collected is positively staggering. To be sure he made a few misidentifications, but this was quite excusable considering the limited knowledge of the period and the lack of an adequate library and reference collection on the *Beagle*.

Where did Darwin get that remarkable education that he must have had? The importance of keeping journals and extensive notes on his observations and collections he may have learned already at Shrewsbury grammar school, or later from Grant at Edinburgh or from Henslow and Sedgwick at Cambridge. His eager reading of the natural-history literature, as well as his contacts with geologists, botanists, entomologists, and other naturalists during his university years, were a far better preparation for his future career than would have been a thorough instruction in anatomy and other medicine-related subjects that was, for instance, the education of T. H. Huxley. While at Edinburgh Darwin actively participated in a local natural-history society (Plinian Society), where he himself presented some ideas and discoveries; he collected and studied marine life in tidepools under the guidance of the zoologist Robert Grant; he visited the local museum and met its curator; he took lessons in bird skinning; in short, he took his natural history very seriously. Only a few professions were suitable for a wealthy, middle-class boy, and the family was in a quandary when Darwin's total lack of interest in medicine became apparent.

This was the age of Paley and natural theology, the age when the professors of botany and geology at Oxford and Cambridge were theologians. Therefore the family, quite logically, decided that Charles should study for the ministry. He agreed, with the reservation that he would become a country parson, his ideal presumably being the Vicar of Selborne.

Darwin arrived at Cambridge in January 1828 and received his B.A. in April 1831. He endured a curriculum of classics, mathematics, and theology, which to him must have seemed intolerably dull, with sufficient fortitude to rank as the tenth among those who did not go in for honors. It left him enough time for his favorite pastimes of riding, hunting, natural-history collecting, and spending animated evenings with congenial friends so that he always looked back at his Cambridge life with much pleasure. "But no pursuit at Cambridge was followed with nearly so much eagerness or gave me so much pleasure as collecting beetles" (1958: 62). This hobby, begun at Shrewsbury, became an all-consuming passion. It cemented a friendship with W. Darwin Fox, a second cousin, at that time also at Christ College. Fox introduced him to entomology in the widest sense of the word and became one of Darwin's favorite correspondents in later years.

The most important factor of his life in Cambridge was his friendship with the Professor of Botany, the Reverend John Stevens Henslow. Henslow, in addition to being deeply religious and thoroughly orthodox, was an ardent naturalist. He not only had open house on Friday evenings for undergraduates interested in natural history but "during the latter half of my time," says Darwin, "I took long walks with him on most days; so that I was called by some of the dons, 'the man who walks with Henslow.'" From him Darwin absorbed a great deal of knowledge in botany, entomology, chemistry, mineralogy, and geology. In Henslow's house he met William Whewell, Leonard Jenyns, and others with whom he later corresponded.

Many people belong to one of two extreme classes of learners, the visual and the auditory. In his autobiography (pp. 63–64), Darwin recites several experiences which document that he had the superb visual memory of a good naturalist and taxonomist. Since Darwin clearly was of the visual type, he never got much out of lectures. "There are no advantages and many disadvantages in lectures compared with reading" (p. 47). It is therefore not without justification when Darwin later claimed that he was "self-taught," for he got his real education from observing and reading. To mention the books that impressed him as a young man is, therefore, as important or more so than mentioning the professors whose lectures he attended in Edinburgh and Cambridge. After reading White's *Natural History of Selborne*, Darwin "took much pleasure in watching the habits of birds, and even made notes on the subject. In my simplicity I remember wondering why every gentleman did not become an ornithologist" (p. 45). At Cambridge he was most impressed by the logic and clarity of the writings of Paley on Christian theology, but he also read his *Natural Theology*, which is an excellent introduction into natural history and the study of adaptation. Two books were particularly influential during the last year at Cambridge, Humboldt's *Personal Narrative* and Herschel's *Introduction to the Study of Natural Philosophy*. Darwin read them avidly and "no one or a dozen other books influenced me nearly so much as these two" (p. 68). From Herschel he learned a good deal about the methodology of science, and both books "stirred up in me a burning zeal to add even the most humble contribution to the noble structure of natural science" (p. 68). Reading Humboldt raised in him the ambition to become an explorer, preferably in South America, an ambition which, most unexpectedly, Darwin was soon able to satisfy.

    Since Darwin had not entered Cambridge until after Christmas, he had to make up two terms after his B.A., and Henslow persuaded him to devote these to a study of geology. He also arranged for Darwin to accompany Adam Sedgwick, Woodwardian Professor of Geology, on a geological field trip to Wales, on which Darwin learned a great deal about geological mapping. When he returned home, he found an invitation to join the next voyage of the *Beagle* as a naturalist. The objections of Charles' father were overcome by the counterarguments of Josiah Wedgwood, Darwin's uncle, who felt that "the pursuit of Natural History, though certainly not professional, is very suitable to a clergyman."

    All of Darwin's biographers agree that his participation in the voyage of the *Beagle* was the crucial event in his life. When the *Beagle* left Plymouth on December 27, 1831, Darwin was 22 years old, and on the return to England on October 2, 1836, five years later, he was a mature naturalist. When he left the *Beagle,* he was better trained and more experienced than almost any other of his contemporaries. The voyage gave Darwin a far more thorough and diversified experience than he could have acquired in any other way.[2] Yet it must be remembered that it required a person of Darwin's abilities and character to benefit as much from this opportunity as he did. It required a person with immense enthusiasm, a superb ability to make observations, great endurance, the stamina to work endless hours, the dedication to maintain an orderly and methodical set of notes, and perhaps most of all an unquenchable curiosity about the meaning of every natural phenomenon he encountered. All this was bought at a price. Life on board the *Beagle* was uncomfortable to the extreme, particularly since Darwin was exceptionally susceptible to seasickness. He was flat on his back the first three weeks of the voyage, as sick as a dog. There was a strong impulse when he sighted the first land to pack up his things and get back to terra firma, but Darwin overcame this temptation and stayed on board even though throughout the entire voyage (which was scheduled to last two years but lasted five) he was intolerably seasick every time the weather was bad.

Although Darwin had joined the *Beagle* as a naturalist, he was best prepared as a geologist, and it was geology on which he concentrated during much of the time. He took with him the first volume of Lyell's *Principles of Geology* which had just come out; the second volume containing Lyell's arguments against Lamarck and evolution reached him at Montevideo in October 1832. The two volumes gave Darwin a thorough grounding in uniformitarianism but also raised numerous doubts in Darwin's mind, as became apparent in later years. On the *Beagle* Darwin was daily challenged to make observations and to place them in the framework of a meaningful interpretation. He who described himself in his autobiography as an incorrigible loafer was just about the hardest working member of the crew. His cramped quarters forced him into extreme tidiness and Darwin himself ascribes to the *Beagle* discipline his methodical system of filing his notes. His intention of becoming a clergyman, he said, "died a natural death when, on leaving Cambridge, I joined the *Beagle* as naturalist" (*Auto.:* 57). Indeed, the letters which Darwin had sent home to Henslow and to his family as well as parts of his journals and the specimens he had shipped home had caused a sufficient stir so that young Darwin was already famous when he returned to England. There was

398

no longer any objection to his choosing quite formally the career of a naturalist.[3]

After leaving the *Beagle* in October 1836, Darwin first went to Cambridge to sort and distribute his collections, but on March 7, 1837, he moved to London. In January 1839, he married his cousin Emma Wedgwood, and in September 1842, the young couple moved to a country house in the small village of Down (Kent), 16 miles south of London, where Darwin lived until he died (on April 19, 1882). He visited London rather infrequently and traveled in England only to attend a few scientific meetings and to visit health resorts. After 1827 he never again crossed the channel to visit the continent.

The move to the country was necessitated by the state of Darwin's health, which had begun to deteriorate soon after he had settled in London. The symptoms were severe headaches, almost daily spells of nausea, intestinal upsets, sleeplessness, irregularities of heart rhythm, and periods of extreme fatigue. After Darwin had passed his thirtieth year, there were often long periods when he was unable to work more than two or three hours a day, and he was sometimes incapacitated for months on end. The exact etiology of this illness is still controversial (Colp, 1977), but all the symptoms indicate a malfunctioning of the autonomous nervous system. Some, if not all, of these symptoms are widespread among hardworking intellectuals. It is almost unbelievable that in spite of his constant illness Darwin was able to produce such a great volume of work. He made this possible by adopting an extraordinary working discipline, by taking refuge in a country retreat where he was protected against service on committees, offices in societies, and teaching obligations, and last but not least, by having a devoted wife to care for him.

Until a few years ago, all we knew about Darwin were his published works, a somewhat expurgated autobiography, and two rather carefully selected sets of his letters.[4] Since the Darwin jubilee in 1959, a veritable "Darwin industry" has developed. Every year, some two or three new volumes on Darwin and some aspects of his work are published, in addition to numerous journal articles. The mining of the rich treasure of unpublished Darwin notes, manuscripts, and letters (mostly at Cambridge University Library) continues, and the total Darwin literature can no longer be fully comprehended by a single person. Furthermore, the new material has not at all helped to dispel differences of interpretation; in fact, it has probably raised more new questions than it has answered old ones. Lack of space precludes any possibility of either a judicious analysis of these controversies or an attempt at a balanced resolution. Instead, my own treatment must by necessity be eclectic and subjective. I will try, however, to present in a logical sequence my own interpretation of the major questions of the Darwin literature. But before taking up the problem of Darwin's conceptual development, it is necessary to clarify the concept of evolution. We will never understand how Darwin became an evolutionist nor the nature of the opposition against him until we have disentangled the multiple threads woven into Darwin's theory of evolution.

400

DARWIN AND EVOLUTION

A retrospective survey of the various terms and definitions for evolution proposed since 1800 reveals quite clearly the ambiguities and uncertainties that have bedeviled evolutionists almost up to the present (Bowler, 1975). Would it be helpful to say, "Evolution is the history of the living world"? Not particularly, because discontinuous special creation would also be covered by this definition, and more importantly, because the definition fails to specify that organic evolution includes two essentially independent processes, which we might call transformation and diversification. The definition widely adopted in recent decades—"Evolution is the change of gene frequencies in populations"—refers only to the transformational component. It tells us nothing about the multiplication of species nor, more broadly, about the origin of organic diversity. A broader definition is needed which would include both transformation and diversification. Transformation deals with the "vertical" (usually adaptive) component of change in time. Diversification deals with processes that occur simultaneously, like the multiplication of species, and can also be called the "horizontal" component of change manifested by different populations and incipient species. Although Darwin was aware of this difference (Red Notebook, p. 130; Herbert, 1979) unfortunately he subsequently did not sufficiently stress the far-reaching independence of these two components of evolution, and this has been the cause of several of the post-Darwinian controversies. Two post-Darwinians, however, made a clear distinction between the two modes. Gulick (1888) used the term *monotypic evolution* for transformation and *polytypic evolution* for diversification. Romanes (1897:

21), who adopted Gulick's terminology, also referred to transformation as "transformation in time" and diversification as "transformation in space." Both Gulick and particularly Romanes appreciated that these were two very different components of evolution, an insight that was largely forgotten again after 1897, until Mayr (1942) and others revived it during the evolutionary synthesis.

Lamarck was almost exclusively interested in transformational (vertical) evolution. He stressed change in time and development from the lower to the more perfect groups. Darwin, by contrast, was far more interested in diversification (horizontal evolution), particularly during the early years of his career. The two founders of evolutionism thereby established two traditions that are still with us (Mayr, 1977b). Most evolutionists have concentrated on only one of the two components and have displayed rather little understanding of the other one. The leaders of the new systematics, for instance, were almost entirely concerned with the origin of diversity, while paleontologists, until quite recently, concerned themselves almost entirely with aspects of vertical evolution, that is, with phyletic evolution, with evolutionary advance, and with adaptational shifts and the acquisition of evolutionary novelty. Comparative anatomists and most experimental biologists had similarly restricted interests. They did not inquire into the nature of species as reproductively isolated populations nor into the mechanisms by which such reproductive isolation is acquired; in other words, they completely ignored populational evolutionism and the problem of the multiplication of species.

Darwin's Conceptual Development

The question when and how Darwin became an evolutionist has been much debated. Since the shift from a strict belief in creation to one in evolution requires a profound conceptual—indeed, ideological—reorientation, one must consider Darwin's attitude toward Christianity. No fundamentalist can develop a theory of evolution, and the changes in the nature of Darwin's faith are, therefore, highly relevant for our understanding of his conversion to evolutionism.

It is evident that Darwin grew up with orthodox beliefs; not until much later in life did he realize that his father had been an agnostic, or as Darwin called it, a skeptic. Darwin's favorite reading was Milton's *Paradise Lost,* which he took with him on all ex-

401

cursions during the voyage of the *Beagle.* Before going to Cambridge to study divinity, he read a number of theological treatises. "And as I did not then in the least doubt the strict and literal truth of every word in the Bible, I soon persuaded myself that our [Church of England's] creed must be fully accepted." Among his favorite reading at that time were also several volumes by the natural theologian Paley. "And taking [Paley's premises] on trust, I was charmed and convinced by the long line of argumentation." When on the *Beagle,* says Darwin, "I was quite orthodox, and I remember being heartily laughed at by several of the officers (though themselves orthodox) for quoting the Bible as an unanswerable authority on some point of morality" (*Auto.:* 85).

By implication, his orthodoxy included a belief in a created world tenanted by constant species. The scientists and philosophers with whom Darwin had the greatest amount of contact in Cambridge and London—Henslow, Sedgwick, Lyell, and Whewell—held essentially similar views. Prior to 1859, none of them reaffirmed more frequently or positively the constancy of species than did Lyell (even though he rejected the recency of the earth).

Darwin abandoned Christianity in the two years after his return to England. In part this was caused by a more critical attitude toward the Bible (particularly the Old Testament), in part by his discovery of the invalidity of the argument from design. For when Darwin had found a mechanism—natural selection—that could explain the gradual evolution of adaptation and diversity, he no longer needed to believe in a supernatural "watchmaker." With his wife and many of his best friends remaining devout theists, Darwin expressed himself rather carefully in the autobiography, but finally concluded: "The mystery of the beginning of all things is insoluble to us, and I for one must be content to remain an agnostic" (*Auto.:* 94).[5]

In his scientific writings, Darwin deals with the problem only a single time, in the concluding sentences of *The Variation of Animals and Plants under Domestication,* published in 1868. Here he states rather bluntly that we have a choice of either believing in natural selection or believing that "an omnipotent and omniscient creator ordains everything and foresees everything. Thus we are brought face to face with a difficulty as insoluble as is free will and predestination" (p. 432; see also Gruber, 1974). This much is certain, that by the time Darwin started to work out his collections, his Christian faith had been weakened sufficiently so that he could abandon a belief in the fixity of species.

And at this point the species problem became the focal point of Darwin's biological interests.

The Origin of New Species

Darwin called his great work *On the Origin of Species,* for he was fully conscious of the fact that the change from one species into another was the most fundamental problem of evolution. The fixed, essentialistic species was the fortress to be stormed and destroyed; once this had been accomplished, evolutionary thinking rushed through the breach like a flood through a break in a dike.

403

Curiously, the origin of species had not been a scientific problem before the eighteenth century. As long as no real distinction was made between species and varieties, and as long as it was widely believed that the seeds of one kind of plant could produce plants of other species—that is, as long as the whole concept of "kinds" of organisms was vague—speciation was not a serious problem. It became so only after the taxonomists, particularly Ray and Linnaeus, had insisted that the diversity of nature consists of well-defined, fixed species. Since species at that time were essentialistically defined, they could originate only by a sudden event, a saltation or "mutation" (as it was later called by de Vries). This, for instance, was the explanation put forward by Maupertuis: "Could we not explain in this way how from only two individuals the multiplication of the most various species could have resulted? Their first origin would have been due simply to some chance productions, in which the elementary particles would not have kept the order which they had in the paternal and maternal animals: each degree of error would have made a new species; and by repeated deviations the infinite diversity of animals which we know today would have been produced" (1756: 150–151).

Darwin was not the first to be concerned with the origin of diversity, but the pre-Darwinian solutions were nonevolutionary. According to the natural theologians and other theists, all species and higher taxa had been created by God, while Lamarck attributed it to a *deus ex machina,* spontaneous generation. Each evolutionary line, according to him, was the product of a separate spontaneous generation of simple forms which subsequently evolved into higher organisms. This postulate left just about everything unexplained.

What all the essentialists from Maupertuis to Bateson appreciated was the fact that if the species is typologically defined, then instantaneous speciation through a drastic mutation is one of the only two conceivable methods of speciation. That such instantaneous speciation can actually occur (through polyploidy) was not proven until the second decade of the twentieth century. The only other possible form of speciation within the essentialistic paradigm is speciation by hybridization, as proposed by Linnaeus (Larson, 1971: 102). After Linnaeus had found three or four natural hybrids and had named them new species, he became possessed by the idea that all species had originated by hybridization. In the course of the 1760s and 1770s his views became increasingly bizarre and in the end he thought that God had created only the orders of plants, and that all taxa of the categories below the order, down to the species, had resulted from "mixing," that is, hybridization.

404

This conclusion was vigorously opposed by Linnaeus' contemporaries. The plant hybridizer Kölreuter made numerous species hybrids in the 1760s but demonstrated that, contrary to Linnaeus' claims, these hybrids were not stable (Chap. 14). In later hybrid generations, he observed much segregation and a gradual but inevitable breakdown of the supposedly new species. This was a great relief to the essentialists, for it would have been quite unthinkable that one could produce a new *eidos* by mixing or fusing two previously existing ones.

A modern is apt to forget that prior to Darwin virtually everybody was an essentialist. Each species had its own species-specific essence and thus it was impossible that it could change or evolve. This, for example, was the cornerstone of Lyell's thought. All nature, according to him, consists of constant types, each created at a definite time. "There are fixed limits beyond which the descendants from common parents can never deviate from a certain type." And he stated emphatically: "It is idle . . . to dispute about the abstract possibility of the conversion of one species into another, when there are known causes, so much more active in their nature, which must always intervene and prevent the actual accomplishment of such conversions" (1835, II: 162). Yet, one searches Lyell's *Principles* in vain for a citation of such causes. It was simply impossible to adopt evolutionary thinking until the dogma of the constancy of species was destroyed. Lyell as well as his "catastrophist" opponents showed that it is quite conceivable to reconcile the fossil record with an essentially nonevolutionary concept of the history of the earth.

A realization of the dominance of essentialistic thinking helps to solve another puzzle. Why had all the attempts of the preceding 150 years, from Leibniz to Lamarck and Chambers, to develop a substantial theory of evolution been such failures? These failures are usually attributed to the lack of a reasonable explanatory mechanism. This is in part true, but that this is not the whole truth is indicated by the fact that the majority of the biologists who accepted the theory of evolution after 1859 simultaneously rejected Darwin's proposed explanatory mechanism, natural selection. What had made them evolutionists was not that they now had a mechanism but that Darwin had demonstrated the evolutionary potential of species and had thus made possible the theory of common descent, which explained so successfully almost everything about organic diversity that had been previously puzzling. The destruction of the concept of constant species and the posing and solving of the problem of the multiplication of species were the indispensable basis of a sound theory of evolution.

This new way of approaching the problem of evolution Darwin did not owe to Lamarck or any of the other of his so-called forerunners. They were all concerned with vertical evolution, with improving perfection, with evolution in the grand style. Rather, it was Lyell, the antievolutionist, who made the crucial contribution by making the reductionist move of dissecting the evolutionary movement into its elements, the species.[6] Lyell felt that one would never be able to come to firm conclusions concerning the history of organic life as long as one formulated the argument in terms of such generalities as progression and trends toward perfection, as Lamarck had done. Organic life, said Lyell, consists of species. If there is evolution, as claimed by Lamarck, species must be its agents. Thus, the problem of evolution cannot be solved by vague generalities but only through the study of concrete species, their origin, and their extinction. This led him to ask some very specific questions: Are species constant or mutable? If constant, can each species be traced to a single origin in time and space? Since species become extinct, what limits their life span? Can the extinction and the introduction of new species be currently observed and attributed to currently observable environmental factors?

Lyell thus admirably posed the right questions, questions which Darwin and Wallace pondered over in the ensuing decades. Lyell himself, being a dyed-in-the-wool essentialist, consistently came up with the wrong answers to his questions. For him it was types that originated and types that died out. Extinction and origination of species were two sides of the same coin. He never understood, at least not until Darwin and Wallace pointed it out to him, that the evolution of a new species population is a totally different process from the extinction of the last survivors of a dwindling species.

By the 1820s almost all geologists had come to agree that many species had become extinct over the course of time and that they had been replaced by new species. Several competing theories were proposed to account both for the extinction and for the introduction of new species. Some geologists believed that the extinctions had been catastrophic, in the most extreme case with God repeatedly destroying his entire previous creation, as Agassiz believed. Or did species die out individually either because their life span had run out or because conditions had become unsuitable for them? It was most important for the development of Darwin's theories that Lyell had opted for the last of these alternatives and had thus directed attention to ecology and geography and their contribution to the history of faunas and floras.

Lyell's *Principles of Geology* was Darwin's "bible" as far as the problem of evolution was concerned. There is abundant evidence that throughout most of the *Beagle* voyage Darwin accepted Lyell's conclusions without questioning. Lyell started from the same two observations as Lamarck: species live in a constantly (but slowly) changing world, and species are extremely well-adapted to their station in life. Since Lamarck believed that species could not become extinct, he concluded that they must undergo constant evolutionary change in order to remain adapted to the changes in their environment. Lyell, as an essentialist and theist, believed that species are constant and cannot change, therefore they cannot become adapted to the changes in their environment and must become extinct.

Lyell's explanation of extinction is reasonably plausible. He contributed one important thought, subsequently particularly developed by Darwin: it is not only the physical factors of the environment that can cause extinction but also competition from other better-adapted species. This explanation was of course in agreement with the concept of the struggle for existence, as it was widely held prior to Darwin's reading of Malthus.

Lyell was far less successful in his attempts to explain the replacement of the extinct species. In order to uphold his principle of uniformitarianism, he postulated that new species are introduced at an essentially constant rate, but he failed completely either to provide any evidence for such an introduction of species or to suggest any mechanism. Thus he laid himself open to the criticism of a German reviewer of the *Principles* (Bronn), who accused Lyell of having abandoned the principle of uniformity with respect to organic life. Lyell (1881) attempted to defend himself in a letter to his friend Herschel by saying that some unknown intermediate causes might be responsible for the introduction of new species. However, the description of the process by which new species are introduced is quite irreconcilable with any conceivable secondary causes: "Species may have been created in succession at such times and at such places as to enable them to multiply and endure for an appointed period and occupy an appointed space on the globe." The repeated choice of the word "appointed" indicates that for Lyell each creation was a carefully planned event (Mayr, 1972b). Such a frank appeal to the supernatural worried even Lyell a little, and he took considerable solace in Herschel's pronouncement: "We are led by all analogy to suppose that [the creator] operates through a series of intermediate causes and that in consequence the origination of new species, could it ever come under our cognizance, would be found to be a natural in contradistinction to a miraculous, process." As a mathematician and astronomer Herschel did not realize that except for evolution (and, as we now know, some chromosomal processes) there are no intermediate causes that could produce constant species at the right time and in the right place. Indeed, what Herschel and Lyell postulated was exactly the kind of miracle which they overtly rejected. Elsewhere, of course, Lyell admitted frankly that he adhered to "the perpetual intervention hypothesis" with respect to the concept of the creation (Lyell, 1970: 89). No wonder Darwin gave so much space in the *Origin* to the rejection of the special-creation hypothesis (Gillespie, 1979).

It is quite impossible to develop an evolutionary theory on the foundation of essentialism. Essences, being nonvariable in space and time, are nondimensional phenomena. Since they lack variation, they cannot evolve or bud off incipient species. Lyell thought he had solved the problem of the introduction of new species by pointing out that they will occupy vacant stations (niches). As an essentialist (and just like Linnaeus), he thought of speciation in terms of the introduction of a single pair that would be the progenitor of the new species. There are reasons to believe that Darwin prior to March 1837 held similar typological ideas. This is indicated by his description of the origin of the second *Rhea* species in South America. Progress in the speciation problem was not achieved until naturalists discovered that species taxa are dimensional phenomena. Species have an extension in space and time; they are structured and consist of populations which, at least in part (when they are isolated), are independent of each other. Thus, contrary to Lyell's insistence, species vary and each isolated species population is an incipient species and a potential source of the origin of diversity. According to Lyell's thesis, the vacant mockingbird niche on the Galapagos would be filled by the "introduction" (by whatever means) of the mockingbird species on the Galapagos. However, that each island had its own species was not explicable by Lyell's mechanism. Isolation and gradual evolution would explain it. This is the lesson Darwin learned from the Galapagos avifauna.

Darwin Becomes an Evolutionist

A great deal of research has been conducted in recent years to reconstruct, step by step, Darwin's "conversion." What Darwin himself says on the timing of his becoming an evolutionist is rather misleading. He starts the introduction of the *Origin of Species* with these sentences: "When on board H.M.S. *'Beagle'*, as naturalist, I was much struck with certain facts in the distribution of the inhabitants of South America, and in the geological relations of the present to the past inhabitants of that continent. These facts seemed to me to throw some light on the origin of species—that mystery of mysteries, as it has been called by one of our greatest philosophers." This implies, as does a similar statement in the autobiography, that he had become an evolutionist during the South American phase of the *Beagle* voyage. However, this is not substantiated by his journals. Indeed, when collecting on the Galapagos, he labeled the collections from the different islands simply "Galapagos," quite unaware of the phenomenon of geographic variation.[7] He should have seen the truth when the governor of the Galapagos told him that the tortoise of each island was recognizably different from those of the other islands, but this observation was not enough. Yet, what Darwin had seen in the Galapagos puzzled him sufficiently to pen these prophetic comments on the homeward voyage of the *Beagle* (June? 1836): "When I see these islands in sight of each other and possessed of but a scanty stock of animals, tenanted by these birds but slightly differing in structure and filling the same place in nature, I must suspect they are varieties . . . If there is the slightest foundation for these re-marks, the zoology of the Archipelagos will be well worth exam-ining: for such facts would undermine the stability of species" (Barlow, 1963).

It was not until March 1837, when the celebrated ornitholo-gist John Gould, who was working up Darwin's bird collections, told him of the specific distinctness of the mockingbirds (*Mimus*) collected by Darwin on three different islands in the Galapagos that Darwin finally recognized the process of geographic specia-tion. Apparently it was not until a good deal later that he learned that some of the finches also were restricted to certain islands. As a result, as Darwin stated in the *Origin*, "when comparing . . . the birds from the separate islands of the Galapagos archipelago, both with one another, and with those from the American mainland, I was much struck how entirely vague and arbitrary is the distinc-tion between species and varieties" (p. 48). It became clear to Dar-win that many populations (as we would now call them) were in-termediate between species and variety, and that particularly species on islands, when studied geographically, lacked the con-stancy and clear-cut delimitation insisted on by creationists and essentialists. Darwin's species concept was thus shaken to its foun-dations.

The spring of 1837 was one of the busiest in Darwin's life, and it was not until summer that he began to follow up on his conversion to evolutionism. In his journal he wrote: "In July [1837] opened first notebook on 'Transmutation of Species'—Had been greatly struck from about month of previous March on character of South American fossils—and species on Galapagos archipelago. These facts (especially latter) origin of all my views."

His encounter with Gould in March 1837 was the watershed in Darwin's thinking.[8] The destruction of the concept of constant species had a domino effect. Suddenly everything appeared in a new light. What had seemed so puzzling about his observations on the *Beagle* now seemed accessible to explanation: "During the voy-age of the *Beagle* I had been deeply impressed by discovering in the Pampean formation great fossil animals covered with armour like that on the existing armadillos; secondly, by the manner in which closely allied animals replace one another in proceeding southward over the continent; and thirdly, by the South American character of most of the production of the Galapagos archipelago, and more especially by the manner in which they differ slightly on each island of the group; none of these islands appearing to be very ancient in a geological sense. It was evident that such facts as these, as well as many others, could be explained on the supposition that species gradually become modified; and the subject has haunted me" (*Auto.:* 118–119).

The aspect of evolution that was clearly of the greatest interest to Darwin was the question of species and, more broadly, questions of the origin of diversity: the comparison of fossil with living faunas, of tropical and temperate zone faunas, of island and mainland faunas. Evidently, Darwin approached the problem of evolution in an entirely different manner from Lamarck, and problems of the evolution of diversity continued to dominate Darwin's thinking and interests.

It would be misleading to claim that from this point on Darwin had a clear picture of speciation. As shown by Kottler (1978) and Sulloway (1979), Darwin vacillated about speciation a great deal throughout his life. In particular there are indications that he might have thought that speciation on islands is different from speciation on mainlands. Like certain biologists even today, Darwin seemed to have considerable difficulty in visualizing barriers on the mainland which could isolate incipient species and believed that his principle of "character divergence" would overcome this difficulty.

Two extreme interpretations concerning the development of Darwin's theory of evolution can be found in the literature, both of them clearly erroneous. According to one, Darwin developed his theory in its entirety as soon as his conversion to evolutionism had happened. The other extreme is to say that Darwin constantly changed his mind and that later in life he completely abandoned his earlier views. The truth that seems to emerge from recent researches and from the study of Darwin's notebooks and manuscripts is that at first (in 1837 and 1838) in rapid succession Darwin adopted and rejected a series of theories, but that he more or less retained the overall theory he had developed by the 1840s through the rest of his life, even though he somewhat changed his mind concerning the relative importance of certain factors (such as geographic isolation and soft inheritance), without completely reversing himself. In fact, his statements on evolution in the sixth edition of the *Origin* (1872) and in the *Descent of Man* (1871) are remarkably similar to the statements in the essay of 1844 and in the first edition (1859) of the *Origin,* all contrary claims notwithstanding.

Geographic Speciation

Darwin and Wallace initiated an entirely different approach to the problem of the origin of species than did any of their "forerunners." Instead of comparing taxa in the time dimension, they compared contemporary taxa in the geographical dimension, that is, they compared populations and species which replace each other geographically. Actually, the concept of geographic speciation was by no means entirely novel in 1837 when it occurred to Darwin. Buffon was perhaps the first to call attention to the fact that when *411* one proceeds from one country to another far distant one, one finds that many species of the first country are represented in the distant country by similar ones. For instance, when one compares the mammals of Europe with those of North America, it is a real problem to decide whether the beaver, the bison, the red deer, the lynx, the snow-shoe hare, to list just a few examples, belong to the same species in the two countries or to different species. The same problem is posed by species of birds, insects, and many plants.

A few decades after Buffon the great zoologist Peter Simon Pallas (1741–1811) found similar pairs of vicarious forms when comparing the European and Siberian faunas. Closer study revealed that the more distant forms were often connected with each other by a graded chain of intermediates. The principle of geographic variation was discovered by these and similar studies, a principle which greatly helped to destroy the essentialist species concept. However, it was not until 1825 that Leopold von Buch drew what seemed to be the logical conclusion from these observations:

> The individuals of a genus strike out over the continents, move to far-distant places, form varieties (on account of the differences of the localities, of the food, and the soil), which owing to their segregation [geographical isolation] cannot interbreed with other varieties and thus be returned to the original main type. Finally, these varieties become constant and turn into separate species. Later they may again reach the range of other varieties which have changed in a like manner, and the two will now no longer cross and thus they behave as 'two very different species'. (pp. 132–133)

Von Buch most perceptively focused on the crucial aspects of geographic speciation: the spatial segregation of populations, their gradual change during isolation, and the concurrent acquisition of species-specific characteristics (most importantly isolating mechanisms) which would permit such a new species to return to the range of the parent species without mixing with it. In the beginning this was very much Darwin's theory of speciation, as is evident from his notebooks and his early essays.[9] Indeed, throughout his life Darwin thought that geographic isolation was an important component in much of speciation. This is substantiated by some statements in the *Origin:* "Isolation, by checking immigration and consequently competition [not to mention swamping!], will give time for any new variety to be slowly improved; and this may sometimes be of importance in the production of new species" (p. 105).

412

Speaking of species on oceanic islands, Darwin says, "A very large proportion are endemic, that is, have been produced there, and nowhere else. Hence, an oceanic island at first sight seems to have been highly favorable for the production of a new species" (p. 105). Clearly, the new species thus evolving on an island must have descended from immigrants: "It is an almost universal rule that the endemic productions of islands are related to those of the nearest continent, or of other near islands" (p. 399). And speaking of archipelagos, he says, "The really surprising fact in this case of the Galapagos archipelago and in a lesser degree in some analogous instances, is that the new species formed in the separate islands have not quickly spread to the other islands" (p. 401).

The origin of species—that is, the multiplication of species—is such a key problem in Darwin's theory of evolution that one would surely expect it to be the exclusive subject of one of the fourteen chapters of the *Origin.* But this is not the case. The discussion of speciation forms part of chapter IV ("Natural Selection"; pp. 80–130), a chapter dealing primarily with the causation of evolutionary change and with divergence. When reading this chapter, one is struck by the insufficiency of the analysis. Although Darwin does not say this in so many words, he virtually implies that geographical isolation and natural selection are alternate mechanisms for the production of species. Curiously, this seeming confusion has never been properly analyzed by a modern historian. Not surprisingly, it confused many readers of the *Origin,* including Moritz Wagner, and the confusion continues into the present. How else could Vorzimmer (1965: 148) have said, "Natural selection is the term Darwin gave to the process of speciation, as described by him." Darwin's ambiguity is the more surprising since speciation is the most characteristic phenomenon of "horizontal" evolution, while natural selection is the driving force of "vertical" evolution. His species book was apparently to have *Natural Selection* as the title, and it is under this title that the manuscript was eventually published (in 1975), while the 1859 abridgement was entitled (for short) *On the Origin of Species,* again implying the equivalence of the two terms. Speciation, for Darwin, was apparently always primarily an aspect of natural selection, as is also evident from some of his rejoinders to Wagner.

Before he became an evolutionist, while Darwin still adhered to Lyell's concept of an appointed life span for each species (with sudden origin and sudden death), he had been rather puzzled over the "introduction" of new species on continents. Thus, when he discovered a second species of *Rhea* (South American ostrich) on the flat, featureless plains of Patagonia, he thought its origin must have been due to a "change not progressive [that is, not gradual]: produced at one blow if one species altered" (Darwin, 1980: 63).

413

In the years immediately after he had become an evolutionist, Darwin explained speciation not only on islands but also on continents as being made possible by geographical barriers, like oceans, rivers, mountain ranges, and deserts (cf. *Essay,* 1844). Furthermore, he postulated that parts of continents (for instance, South Africa) might have experienced periods of rapid sinking, during which they had been converted temporarily into archipelagos, thus providing the needed isolation (*Origin:* 107–108), until they were subsequently again elevated. From his notebooks we now know how widely Darwin had accepted at that time the necessity of geographical isolation for speciation.

One is rather surprised, therefore, when one discovers to what an extent Darwin later reversed himself in *Natural Selection* (written 1856–1858) and in the *Origin* (1858–1859). He is now quite ready to accept sympatric speciation for many continental species, owing to some sort of ecological, habitat, seasonal, or behavioral specialization. He applied this mechanism particularly to species whose ranges were slightly overlapping or simply in contact with each other ("osculating"). Such distributions are today called *parapatric.* They are common, particularly in the tropics, and are now interpreted as zones of secondary contact of previously isolated species or incipient species. Darwin, on the other hand, took it for granted that these distribution patterns had developed in situ. "I do not doubt that many species have been formed at different points of an absolutely continuous area, of which the physical conditions graduate from one point to another in the most insensible manner" (*Nat. Sel.:* 266). As he explains elsewhere, he thought that one variety would develop at one end of the chain of populations, another one at the other end, and finally, an intermediate variety in the narrow zone where the two major varieties meet. Since the two major varieties would occupy a larger area than the intermediate variety, they would soon outcompete it, in a strictly typological manner, and cause its extinction. This would cause a clear discontinuity between the two major varieties and speciation would have been completed. As he said in the *Origin* (p. 111): "The lesser differences between varieties become augmented into the greater difference between species" (see also pp. 51–52, 114, 128).

414

Darwin's basic oversight was that he failed to partition isolation into extrinsic geographical-ecological barriers and intrinsic isolating mechanisms. This is well illustrated by a statement in *Variation* (1868, II: 185).

"On the principle which makes it necessary for man, whilst he is selecting and improving his domestic varieties, to keep them separate, it would clearly be advantageous to varieties in a state of nature, that is to incipient species, if they could be kept from blending, either through sexual aversion, or by becoming mutually sterile." He completely overlooked the fact that here he was dealing with two entirely different principles. Races of domestic animals are developed in strict spatial (minigeographical) isolation, while Darwin does not explain at all how the genetic differences could be built up in nature that would lead to sexual aversion or mutual sterility.

Darwin ignores this same difficulty when he lists cases where members of the same variety preferably pair with each other (*homogamy*) when two different "varieties" are brought together. He cites 13 cases (*Nat. Sel.:* 258) where he thinks that such preferential mating has been observed. Actually, when closely analyzed, none of the cases supports this contention. Omitting inappropriate cases (such as overlap outside the breeding season), each of the "varieties," now partly kept segregated by behavioral isolation, had clearly originated during a preceding period of spatial isolation during which the genetic isolation had been built up. This Darwin did not see, because at that time he did not appreciate the efficacy of ecological (vegetational) barriers, including those caused by the Pleistocene advances in the ice caps.

This much is certain, that a rather drastic change in Darwin's thinking had occurred between 1844 and 1856, when Darwin began writing his *Natural Selection.* When I attempted to trace the reasons for Darwin's later downgrading of the role of isolation (Mayr, 1959b), it was prior to the discovery of the *Notebooks on Transmutation,* and my analysis was onesided and incomplete. I attributed Darwin's uncertainties to four factors: (1) his ambiguous use of the term "variety," both for individual variants and for subspecies (populations). Of 24 usages of the term in the *Origin,* 8 refer to individual variants, 6 to geographical populations, and 10 to both (or are ambiguous); (2) his morphological species concept (in contrast to his earlier, biological concept); (3) his frequent confounding of the process of multiplication of species and that of phyletic evolution; and (4) his desire to find a single-factor explanation (curiously seeing natural selection somehow as an alternative, instead of as an accessory, to isolation).

Sulloway (1979) accepts the importance of these factors but points to four additional developments in the period from 1844 to 1859 that influenced Darwin's thinking: (1) his taxonomic work on barnacles, where he found a morphological species concept more practical than a biological one; (2) certain tactical considerations in making his conclusions more palatable to his peers, including the conceptualization of the (incipient) species as a competitor rather than a reproductive isolate; (3) the transfer of his ideas from birds and mammals to invertebrates (including uniparental ones) and to plants; and (4) his increasing attention to the *principle of divergence,* which he held responsible for the causation of diversity at higher taxonomic levels.

All four factors tended to strengthen Darwin's tendency to see in species something that is *different* (rather than reproductively isolated) and to see no need for isolation for the achievement of this difference. Hence, genuine geographical isolation would not be necessary. However, "some degree of separation must be . . . advantageous. This may arise from a selected individual with its descendants, as soon as formed even into an extremely slightly different variety, tending to haunt a somewhat different station, breeding at a somewhat different season, and from like varieties preferring to pair with each other" (*Nat. Sel.:* 257; *Origin:* 103). The typological frame of his mind is well documented in the statement: "If a variety were to flourish so as to exceed in numbers the parent species, it would then rank as the species and the species as the variety, or it might come to supplant and exterminate the parent species, or both might coexist, and both rank as independent species" (*Origin:* 52). Several of Darwin's statements implying sympatric speciation seem to be paraphrases of similar statements in the contemporary botanical literature (for example, Herbert, 1837). The influence of the botanists is not altogether surprising, since in the 1840s and 50s Darwin probably had more contact with botanists than with zoologists.

It is evident that Darwin was rather uncertain as to the actual role of isolation during the speciation process. In this he was not alone. Owen, in a review of the *Origin,* stated: "Isolation, says Mr. Darwin, is an important element in the process of natural selection but how can one select if a thing be isolated?" Darwin, of course, had not actually said such things, but it is true that he had treated geographic speciation in his chapter on natural selection. Hopkins, another critic, proposes a process of sympatric speciation by homogamy: "If it could be proved that there is a predominant tendency in the more perfect and robust of each species to pair with individuals like themselves for the transmission of their kind, then the necessary existence of natural selection as an operative cause must be admitted." What puzzled Darwin's critics consistently, and, for that matter, even Huxley and other of Darwin's friends, was how the interfertility of members of a species, including that of intraspecific varieties, could be converted into sterility. Darwin had brought this criticism unto himself by constantly stressing that varieties gradually turned into species but nowhere giving convincing illustrations of the gradual process of geographic speciation.

Even though Darwin never abandoned the concept of geographic speciation entirely, it is even less emphasized in the sixth edition of the *Origin* (1872) than in the first. A weakening of his reliance on geographic isolation is also indicated by his correspondence with Wagner, Weismann, and Semper. Darwin more and more treated speciation as a process of adaptation, an aspect of the principle of divergence, completely omitting any reference to the need for the acquisition of reproductive isolation. As Ghiselin (1969: 101) has rightly said, "There is no solid evidence that [when writing the *Origin*] he conceived of species as reproductively isolated populations." His own empirical observations had shown him time after time that islands were a favorite place for the origin of new species, but Darwin no longer considered how important spatial isolation is for the genetic building up of isolating mechanisms. It was this which eventually led to his drawn out controversy with Moritz Wagner (see Chapter 11).

Darwin's major ideas on speciation and evolution had crystallized in the course of a few years (1837–1839), even though he continued to modify them. By 1844 he was ready to compose a major essay of 230 handwritten pages which contains the gist of what eventually appeared as the *Origin*.[10] Darwin himself was so convinced of the importance of this manuscript that he gave instructions to his wife to have it published in case of his death. Yet, the only person to whom he dared to show this subversive document was the botanist Joseph D. Hooker. Fifteen more years passed by before Darwin finally published his theories, and the wait undoubtedly would have been even longer if it had not been for an event that will now be described. Thinking that the whole world was antievolutionary, Darwin felt no urgency to publish his views. But he misjudged the situation. The enormous success of Chambers' *Vestiges* should have warned him that there was far more interest in evolution than he thought, and that someone might independently arrive at similar ideas. And, indeed, there was such a person—Alfred Russel Wallace (1823–1913).

ALFRED RUSSEL WALLACE

The extraordinary coincidence of another naturalist coming up with an interpretation of evolution that was remarkably similar to that of Darwin has been a source of amazement ever since 1858. In most respects, both men were about as different as two people can be: Darwin, the wealthy gentleman, with many years of college education, a private scholar, able to devote all of his time to research; Wallace, a poor man's son with only a lower middle-class background (a very important factor in Victorian England), without any higher education, never particularly well-to-do, always having to work for a living, for the longest time in the exceedingly dangerous profession of a collector of birds and insects in fever-ridden tropical countries. But they agreed in some decisive points. Both of them were British, both had read Lyell and Malthus, both were naturalists, and both had made natural-history collections in tropical archipelagos. More about Wallace will be said later in connection with the description of his independent discovery of the principle of natural selection, but the part he played in forcing Darwin to speed up the publication of his species book must be reported here.[11]

Wallace left grammar school at the age of 13, and served as an assistant to his brother, a surveyor, for the next seven years. Roaming through moors and mountains in his survey work, Wallace became an enthusiastic naturalist. First he collected plants, but after becoming a friend of a dedicated entomologist, Henry Walter Bates, he added butterflies and beetles to his interests. Even more so than Darwin, Wallace received his most important stimulation from books.[12] Darwin's *Journal of Researches* and Humboldt's *Personal Narrative* inspired the two young naturalists to depart in April 1848 for the Amazon Valley, with the rather well-formed purpose "to gather facts, as Mr. Wallace expressed it in one of his letters, towards solving the problem of the origin of species, a subject on which we had conversed and corresponded much together" after reading the *Vestiges* in the fall of 1845 (Bates, 1863; vii). The mighty tributaries of the Amazon River dissect the entire basin into forest islands, so that many groups of species are distributed parapatrically as on an archipelago. Reminiscing about this more than fifty years later, Wallace wrote, "Ever since I had read the *Vestiges of Creation* before going to the Amazon, I continued at frequent intervals to ponder on the great secret of the actual steps by which each new species had been produced, with all its special adaptations to the conditions of its existence . . . I myself believed that [each species] was a direct modification of the preexisting species through the ordinary process of generation as had been argued in the *Vestiges of Creation*." Since Wallace was not an orthodox Christian, he had much less trouble accepting the evolution of species than Lyell or Agassiz.

How far the facts of the distribution of Amazonian species helped Wallace to crystallize his ideas we shall never know. Leaving Bates behind four years later, Wallace, on his return to England, was struck by catastrophe. The ship on which he traveled caught fire (August 6, 1852) and sank, with his entire magnificent collection and most of his journals, notes, and sketches. Yet, from memory Wallace pointed out (1853) that the distribution of each of numerous closely related species of monkeys, poorly flying birds, and butterflies was bordered by the Amazon and its tributaries. Undaunted by the crushing experience of the loss of nearly all the fruits of his four years in South America, Wallace at once made plans for a new expedition, carefully selecting the Malay archipelago as the most suitable place for a study of the origin of species (McKinney, 1972: 27). He left England in early March of 1854 and less than a year later (in February 1855) he wrote his

celebrated paper "On the Law Which Has Regulated the Introduction of New Species." To his friend Bates, with whom he had
evidently discussed evolution before and during their stay on the
Amazon, he wrote, "To persons who have not thought much on
the subject I feel my paper on the succession of species will not
appear so clear as it does to you. That paper is, of course, only
the announcement of the theory, not its development."

What Wallace had really tried to do was to solve the problem
of Lyell's "introduction of new species." We now know from his
unpublished notebooks (McKinney, 1972) that Wallace had rejected already by 1854 Lyell's assertion that species only vary within
certain limits and had come to the conclusion of a continuous very
slow change of the organic world over exceedingly long periods
of time. However, although the rejection of the constancy of species would permit him to adopt Lamarckian vertical evolution, it
did not solve the problem of the replacement of extinct species.
The introduction of new species continued to remain a puzzle,
and it is to this puzzle that Wallace addressed himself. As he clearly
stated in his 1855 paper, it was geography, that is, his distributional observations in Amazonia and the Malay Archipelago, which
gave him the answer: "The most closely allied species are found
in the same locality or in closely adjoining localities and . . .
therefore the natural sequence of the species by affinity is also
geographical." And this observation leads him to the law: *Every
species has come into existence coincident both in space and time with a
pre-existing closely allied species.* By stating either "in the same locality or in closely adjoining localities," Wallace obscured the strictly
geographical localization of incipient species, something Wagner
had seen much more clearly. Nevertheless, the process of splitting
a parental species into two or more daughter species, when read
backwards, leads automatically to the concept of common descent
and of phylogenetic trees. In short, Wallace had boldly sketched
a theory of evolution on an empirical basis, namely, the distribution pattern of closely related species.

Darwin and Wallace, thus, had introduced an entirely new
approach to evolutionism (although on a Lyellian foundation),
geographical evolutionism. Instead of attempting to solve the
problem of the origin of diversity via the origin of new major
types of organisms or through a comparison of taxa in the time
(vertical) dimension, they compared contemporary taxa in the
geographical dimensions, that is, they compared populations and
species which replace each other geographically.

How did Wallace's 1855 publication affect Darwin's thinking and activities?

DARWIN'S PROCRASTINATION

During the twenty years after 1837 Darwin never talked about evolution. What he was interested in was the species problem and, in the letters to his friends, he referred to his forthcoming work as "the species book." Can species change, and can one species be transmuted into another species? These were the concrete questions asked by Darwin, and in order to be able to answer these questions convincingly Darwin felt he had to collect an overwhelming amount of evidence. Had not Lamarck and Chambers also proposed the occurrence of evolution, without gaining any converts?

Considering that Darwin became an evolutionist in 1837, and conceived his theory of natural selection in September 1838, one would think that he would rush this, the most important theory in biology, to the printer as quickly as possible. Instead, he postponed publication for twenty years and was forced into action only by circumstances. Why this incredible procrastination? There are a number of reasons. At first Darwin was committed to giving priority to his geological researches, which were well advanced and belonged to the *Beagle* reports. But in 1846, when Darwin had satisfied his geological obligations, he started to work on barnacles (Cirripedia) and devoted the next eight years of his life to this subject, instead of getting on with his species book. This necessitates asking a number of questions. First, was Darwin ready in 1846 to start writing his species book? The answer clearly is that he was not, as he himself stated repeatedly in his letters and as is evident from the fact that he continued assiduously to collect facts. Even some of his basic ideas had not yet fully matured—for instance, his "principle of divergence," which apparently occurred to Darwin only in the 1850s.

The second question is, Why did not Darwin at least concentrate on getting the still-needed material for the species book instead of investing such an inordinate amount of time in his work on the barnacles? A study of the contemporary scene makes me suspect that Darwin was literally afraid to publish his views. The intellectual climate in England was not at all favorable for the reception of Darwin's theory. Chambers' *Vestiges*, published in 1844,

420

was savagely cut to pieces by all reviewers in spite of its deistic sentiments. The leading scientists of England, including Darwin's closest friends, Lyell, Hooker, and (at that time) even Huxley, were almost unanimously opposed to evolution. But it was not evolution as such that was so difficult to defend but rather its purely materialistic explanation by natural selection. It has been well described by Gruber (1974: 35–45) how clearly Darwin realized what a storm of protest this theory would provoke, and indeed, as we shall presently see, virtually no one in England accepted natural selection after the publication of the *Origin* except Wallace, Hooker, and a few other naturalists.

The third question is, Why did Darwin devote so much time to such a seemingly insignificant group as the Cirripedia? The answer to this is presumably threefold. First of all, it is quite obvious that Darwin did not have the slightest intention of investing eight years in this group when he first started working on a peculiar genus of barnacles he had collected in Chile. However, since he was not deeply committed to any other project, he found it convenient for a full understanding of this Chilean genus to study near and distant relatives and, finally, to prepare a monograph of the whole group. Also, Darwin felt that it would add weight to his opinions if he could establish his reputation as a systematist. The subsequent award of the Copley medal of the Royal Society for this work is evidence for the correctness of Darwin's reasoning. Finally, he found that working on the barnacles helped him understand variation, comparative morphology, the species concept, and the incompleteness of the geological record. There is little doubt that Darwin's barnacle studies greatly added to his sophistication and competence, and, as Ghiselin said, "The completed work was nothing less than a rigorous and sweeping critical test for a comprehensive theory of evolutionary biology" (1969: 129). Still, this does not explain why Darwin devoted the enormous period of eight years to this project. Here one can only suspect that Darwin felt that he had a tiger by the tail. He was unable to find a proper cutoff point; and being always seemingly close to completing the monographs, he would have had to throw away a great deal of investment if he had stopped earlier. Yet it is clear that Darwin did not start the barnacle work with the idea that this would be an excellent way to gain the experience which, in retrospect, he indeed acquired from his studies of this group.

Although Darwin published nothing about species and speciation during the 21 years between March 1837 (when he first understood speciation) and August 1858 (when the Linnean Society paper was published), we have learned from his notebooks and correspondence that the species problem was constantly on his mind. Darwin knew that the origin of species was the key to the problem of evolution, but he was still wavering concerning the meaning of species and the process of speciation.

By 1854 Darwin had essentially completed his barnacle work and was beginning to concentrate on sorting his species notes. One would think that the publication of Wallace's "Introduction of New Species" paper (1855) would have stirred him to action, but this did not happen. Darwin did not react to this pioneering paper until two years later, and then only because Wallace himself wrote to him, puzzled why there had been so little reaction. On May 1, 1857, Darwin answered, "I can plainly see that we have thought much alike and to a certain extent have come to similar conclusions . . . I agree to the truth of almost every word of your paper . . . I am now preparing my work for publication, but I find the subject so very large that . . . I do not suppose I shall go to press for two years" (*L.L.D.:* 95–96).

There was, however, one person who was rather thoroughly shaken up when reading Wallace's paper—Charles Lyell. As recently as 1851, in a major address, Lyell had vigorously rejected any concession to evolutionary thinking. But in the period from December 1853 to March 1854 he visited Madeira and the Canary Islands, primarily to study volcanism, but here he experienced in person what von Buch, Darwin, and other naturalists had previously described—the extreme localization of every island species of animals: "The Madeiras are like the Galapagos, every island and rock inhabited by a distinct species," he wrote in his journal (Wilson, 1970). While working on his observations and collections after his return to England, Lyell on November 26, 1855, read Wallace's paper, and it is evident that Wallace's theory greatly excited him. He started at once a series of notebooks on the species question, recording the results of his readings and his uncertainties. Eventually, he decided to visit Darwin at Down House and get the full story on Darwin's researches. Darwin, realizing how much his own ideas were in conflict with those of Lyell, had not discussed the problem of the origin of species with him as he had with Hooker. On April 16, 1856, Darwin gave Lyell a full report on his ideas. Although Lyell apparently was still not yet convinced, he nevertheless strongly urged Darwin to publish his ideas lest he be scooped by someone else. With the main reason for his hesitation now removed, Darwin began one month later, in May 1856, to write his big species book.

Two years later, in June 1858, when Darwin had completed the first draft of ten and a half chapters, the roof fell in on him. He received a letter from Wallace accompanied by a manuscript entitled "On the Tendency of Varieties to Depart Indefinitely from the Original Type." In his letter, Wallace said that if Darwin thought his paper sufficiently novel and interesting, he should send it to Lyell and, presumably, submit it for publication (the original Wallace letter is no longer in existence). Darwin forwarded Wallace's paper on June 18 to Lyell, with a letter saying, "Your words have come true with a vengeance—that I should be forestalled . . . I never saw a more striking coincidence; if Wallace had my manuscript sketch written out in 1842, he could not have made a better short abstract! . . . so all my originality, whatever it may amount to, will be smashed."

It is a well-known story how Lyell and Hooker presented Wallace's paper to the Linnean Society of London on July 1, 1858, together with extracts from Darwin's 1844 *Essay* and from a letter Darwin had written to Asa Gray on September 5, 1857. The issue of the *Proceedings* with these various papers was published August 20. It is interesting and significant that neither Darwin nor Wallace made any attempt in these papers to demonstrate evolution. They were primarily concerned with the mechanism of evolution. Darwin starts with a long discussion of varieties and their production, Wallace with a discussion of the balance of nature caused by the struggle for existence. In his case, it was quite logical because Wallace's 1858 paper was quite clearly a follow-up of his 1855 paper, in which he had firmly come out in favor of evolution.

The Publication of the *Origin*

The joint publication of Wallace's and Darwin's papers proposing the revolutionary theory of evolution by natural selection had astonishingly little effect. The President of the Linnean Society, in his annual report for 1858, stated, "The year . . . has not, indeed, been marked by any of those striking discoveries which at once revolutionize, so to speak, the department of science on which they bear." The ornithologist Alfred Newton claimed thirty years later to have been an exception, and to have found in the papers "a perfectly simple solution of all the difficulties that had been troubling me for months past" (Newton, 1888), and indeed he persuaded H. B. Tristram (1859) to interpret the substrate adaptations of larks as due to natural selection.

Since there seemed to be no end in sight as far as publication of the big book was concerned, Lyell and Hooker urged Darwin to write a short abstract for one of the journals. To make a long story short, the "abstract," prepared between July 1858 and March 1859, became the *Origin,* with its 490 text pages. Although Darwin still insisted that it was only an abstract, he finally gave in to the demands of the publisher, John Murray, to omit the word "abstract" from the title. The volume was published November 24, 1859, and the entire edition of 1,250 copies was at once subscribed to by the retail trade. There were no major revisions in the next three editions (1860–1866), quite a few changes in the fifth edition (1869), and still more, including a new chapter, in the last (sixth) edition (1872). By that time, Darwin was so busy with his other interests, particularly his botanical researches and his work on behavior, that he undertook no further revisions of the *Origin.* His later publications, particularly *The Expression of the Emotions in Man and Animals* (1872) and *The Effects of Cross and Self-Fertilization in the Vegetable Kingdom* (1876), were so pioneering and so outstanding that (it has been rightly said) they, together with his theory of coral reefs and the barnacle monographs, would have made Darwin famous even if he had not proposed evolution by natural selection. The claim by one of his denigrators that Darwin had escaped into these later researches after having been defeated by the opponents of his theory of evolution is quite absurd.

It has often been remarked how extraordinary it is that none of the great zoologists—whether physiologist, embryologist, or cytologist—made any contribution at all to evolutionary theory and that, at least during the nineteenth century, they all rather completely misunderstood the whole problem of evolution. How remarkable, it is also said, that two such "rank amateurs" as Darwin and Wallace found the solution!

There are several answers to explain this peculiar phenomenon but the simplest, no doubt, is that physiologists, embryologists, and indeed most experimental biologists deal with functional phenomena and come face to face with evolution only very indirectly. The naturalist, however, is constantly confronted by evolutionary problems. No wonder that this is what he is most interested in; no wonder that his constant attention to this problem places him in a much better position to ask the right questions and to find answers and solutions than the experimental biologist.

424

Finally, Darwin and Wallace were not amateurs, but, as naturalists, highly trained professionals.

This may explain why Bernard, Helmholtz, and Hertwig so utterly failed, as far as evolution is concerned. It fails to answer, however, why Owen, von Baer, Ehrenberg, Leuckart, or any of the other great systematists and comparative anatomists of the nineteenth century were so blind. There are probably multiple reasons for their failure. In the case of Owen and Agassiz it was unquestionably too strong a conceptual commitment to alternate interpretations; in the case of leading German zoologists like J. Müller, Leuckart, and so on it might have been a counterreaction to the unbridled speculation of the *Naturphilosophen*. What little speculating these zoologists did was related to the theory of morphology and to the information content of ontogenetic development. They were not interested in larger questions. More importantly, none of them was truly a student of natural populations.

425

10 ◈ Darwin's evidence for evolution and common descent

DARWIN WAS fully aware of the revolutionary nature of his work. He knew that it would encounter massive resistance and that, in order to prevail, he would have to overwhelm his opponents. This is why he had devoted twenty years to the accumulation of evidence and to the perfection of the logic of his proofs. The strategy he adopted, to discuss first the mechanism of evolution and only in the later chapters of the *Origin* the evidence supporting the thesis of evolutionary change, would probably not be adopted by many contemporary textbook writers, but it was consistent with the prevailing philosophy of science of his day (Hodge, 1977).

Not all those who studied the *Origin* in the past have realized that it does not deal with a monolithic theory of evolution but actually with a whole set of more or less independent theories, each of which will be analyzed in detail below (see Chapter 11). They include Darwin's theories of speciation, common descent, gradual evolution, and natural selection, in addition to the basic theory that the world of life is not static but evolving and so are the species of which it is composed. Darwin had to present evidence for each of these theories and argue against all potential alternatives. Most importantly, he had to try to refute the ideology of creationism, still dominant in mid-nineteenth-century Britain, even though often camouflaged under different names. This is why Darwin said of the *Origin* (p. 459), "This volume is one long argument" (see also Gillespie, 1979). It is impossible to give a complete abstract of everything presented by Darwin on the 490 pages of the *Origin,* but I shall try to describe what kind of evidence Darwin considered as supporting his theses, and how this fit with the biological knowledge of his day. I shall begin with the problem of an evolving world. As we have seen, Darwin was not the first to advance a theory of evolution, but he was the first not only to propose a feasible mechanism, namely, natural selection (see Chapter 11), but also to bring together such overwhelming evidence that within ten years after 1859 hardly a competent biologist was left who did not accept the fact of evolution.

THE EVIDENCE FOR THE EVOLUTION OF LIFE

The basic, direct evidence for evolutionary change is twofold: for horizontal evolution the nonconstancy of species as revealed by geographical researches, and for vertical evolution the fossil record, as revealed by geological researches. I have above already discussed Darwin's interpretation of the problem of the nonconstancy (multiplication) of species and shall now turn to the fossil record.

427

The Incompleteness of the Fossil Record

In the *Beagle* and post-*Beagle* years Darwin was primarily a geologist. He had read Lyell's *Principles of Geology* systematically and with enthusiasm, and was thus thoroughly familiar with the geological problems of the earth's history. Being at that period the most flourishing branch of natural history, geology had made tremendous strides during the first half of the nineteenth century. There was no longer any doubt that the earth was millions of years old, but was it really old enough to have permitted the development of the enormous diversity of the living world by gradual evolution, as required by Darwin's theory? Wouldn't it be necessary to postulate the occurrence of saltational evolution?

Fossils were used both to refute the theory of evolution, as by Cuvier, Agassiz, Bronn, and all British geologists, or to support it, as by Chambers and Wallace. It was therefore only natural that Darwin devoted two chapters in the *Origin* to the geological evidence in favor of evolution. From his earliest writings on, Darwin had adopted the strategy of anticipating and answering all possible objections to his theories before they were even raised. The objections raised by the geologists were so numerous and so formidable that Darwin devoted all of chapter IX to their refutation.

Let me begin with the problem of the age of the earth. Lyell, following Hutton, had postulated an earth of unlimited age. Darwin thought in terms of several thousand million years. In order to avoid circular reasoning, Darwin tried to prove his point with the help of purely geological data. He presents concrete figures on the enormous thickness of the geological strata, the slowness by which they are deposited, the slowness by which erosion takes place, all of it providing impressive geological evidence for the immense age of the world. Darwin was satisfied that it gave him enough time to allow for the production of any observed evolutionary phenomenon, even under the assumption of slow, gradual evolution. His actual figures were on the large side but of the right order of magnitude. For instance, he calculated that it might have taken 300 million years for the denudation of the Weald in Britain, while the best present estimate is 70–140 million years.

428

While Darwin was wrong by at most a factor of two to four, the contemporary physicists were wrong by several orders of magnitude. William Thomson (later Lord Kelvin), by calculating the rate of cooling of a body the size of the earth (while receiving radiant heat from the sun), concluded that the earth could not have been more than 100 million years old, and most likely only 24 million years old (Burchfield, 1975). This, of course, would not have been nearly enough time for the gradual evolution of the entire known animal and plant life. Kelvin's claims should have driven Darwin to abandon slow gradual evolution and to adopt instead evolution by large variations ("sports," that is, macromutations). Actually Darwin was so sure of his observations that, in response to Jenkin's critique, he ascribed in later years even less importance to sports than in 1859. Here was a clear confrontation of biological and physical evidence. For a physicist it was unthinkable that he could have overlooked some important factor, and so he simply concluded that the biological theory was wrong. Darwin, although greatly disturbed by the findings of the physicists, continued to be convinced of the validity of his biological findings and inferences, and finally concluded: "I feel a conviction that the world will be found rather older than Thomson makes it." The biologist, of course, was right. By allowing for radioactivity, then unknown, the physicist's estimate of the earth's age had to be enlarged by two orders of magnitude to about 4.5 billion years, more than enough for biological evolution. Darwin has sometimes been unjustly accused of having accepted, like Hutton and Lyell, an infinite age of the earth. This he did not do. He postulated several thousand million years, which turns out to be just about right.

There are a few physicists and mathematicians, however, who are still unhappy about the chronology adopted by the Darwinians. Some of the world's most distinguished physicists (including Niels Bohr and Wolfgang Pauli) expressed doubt to me that the accidental process of random variation and selection could produce within less than four billion years the great diversity of the world of life and the marvelous mutual adaptations of organisms to each other. When the arguments of a representative group of physicists and mathematicians were carefully scrutinized by a group of evolutionists, it became apparent that the physical scientists had a rather oversimplified understanding of the biological processes involved in evolution. Being typologists, they had failed to take the uniqueness-producing qualities of recombination sufficiently into consideration. Furthermore, they thought in terms of "tandem evolution," that is, the advance from one homozygous genotype to another, forgetting that genetic change in a species during evolution can proceed simultaneously at thousands, if not millions, of gene loci. In short, Darwin's prophetic estimates were once more confirmed, and the criticism by the physical scientists was shown to have been based on assumptions that are inappropriate for biological systems (Moorhead and Kaplan, 1967).

429

Perhaps the greatest advance made by geology in the fifty years prior to the *Origin* was in the recognition, delimitation, and naming of the geological ages, from the oldest—Sedgwick's Cambrian and Murchison's Silurian—to the Tertiary ones to whose chronology Lyell had made particularly important contributions.[1] These researches clearly demonstrated that each of the successive formations is characterized by a distinctive assemblage of fossil species and that the history of this succession had been essentially the same in all parts of the world. There was a rather acrimonious controversy whether the succession of faunas did or did not represent a progression, but in due time it became clear that fishes first occurred in the Silurian, reptiles in the Carboniferous, mammals in the Triassic, and placental mammals in the very latest Cretaceous. In broad outlines this had become evident by the 1850s, although much accuracy was added after 1859.

The replacement of floras and faunas, as well as the seeming progression, was explained by catastrophists like Agassiz in non-evolutionary terms. For an evolutionist like Wallace (1855) all this indicated a "gradual . . . change of organic life." He was further impressed by such facts as that "in each period, there are peculiar groups, found nowhere else, and extending through one or several formations . . . Species of one genus, or genera of one family occurring in the same geological time are more closely allied than those separated in time . . . [And all geographical and geological facts indicate that] no group or species has come into existence twice." There is nothing haphazard about the history of life on earth.

Yet only Chambers (largely based on misinformation), Darwin, and Wallace seemed to be able to see the fossil evidence as documenting evolution. Lovejoy (1959a) chided the geologists for being so blind, but one must realize that evolution, prior to 1859, meant evolution à la Lamarck and Chambers, that is, a *scala naturae*-like, steady, largely linear advance from "primitive" to more complex. Hence the fact that the earlier known fishes, the Placoderms, were highly complex, or that some primitive (nonplacental) mammals were found in the Jurassic, the Age of Reptiles, was considered refutation of evolution. That strata as well as organisms were sometimes misidentified added to the confusion. Except for indicating a general progression of floras (no angiosperms before the Cretaceous) and faunas, the geological record was almost more of an embarrassment to the evolutionist than a help. If the major groups of animals and plants had evolved slowly, it was believed that one should find connecting links between them. Actually, none were then known. Even *Archaeopteryx*, the almost perfect link between reptiles and birds, was not discovered until two years after the publication of the *Origin*. Darwin's opponents asked other embarrassing questions: Why are there such sharp breaks between the major geological periods? Do they not support catastrophism rather than evolutionism? Why are most of the major phyla fully formed already in the lowest fossil-bearing strata? Why are so many of the extinct types, like ichthyosaurs, pterodactyls, and dinosaurs, aberrant types that do not fit into any reconstructed evolutionary sequence?

Not surprisingly, chapter IX of the *Origin* is defensive from beginning to end. It opens at once with the most serious question of his opponents: Why are "specific forms . . . not being blended together by innumerable transitional links?" (p. 279). The reason, says Darwin, is that the geological record is far too imperfect for the preservation of such forms, and Darwin supplies one piece of evidence after the other to substantiate his claim. The geological researches of the last hundred years have thoroughly vindicated Darwin's assertion of the imperfection of the geological record. As it is preserved, it is a story of discontinuity. In Darwin's day it provided far better support for those who postulated the sudden origin of new types and species (saltationism) than for gradual evolution by natural selection. Indeed, the gaps in the record, in spite of the discovery of numerous "missing links" since Darwin,

430

are still so numerous and so large that the theory of an origin of new types by saltation (macromutation) was upheld by some paleontologists (for example, Schindewolf) and geneticists up to the 1940s, and by some paleontologists even today.

In spite of the absence of decisive evidence, Darwin had found the right answer by consistently treating the origin of new species as the key to the solution of evolutionary problems, a lesson the Galapagos Islands had taught him; he "reduced" all macroevolutionary problems to the species level, and to variation at the species level. As a consequence, the chapter on the geological record, somewhat unexpectedly, contains quite a few astute observations on speciation (pp. 297–298).

What is most impressive in Darwin's treatment of the fossil record is that he deals with it consistently as a biologist. Whenever possible he supplies ecological answers for puzzling phenomena of the geological record, in this respect following Lyell's lead. To the question why rich and diversified groups so often turn up so suddenly in the fossil record, he answers that this might be due not only to the imperfection of the fossil record but also to adaptive shifts: "It might require a long succession of ages to adapt an organism to some new and peculiar line of life, for instance to fly through the air; but that when this had been effected, and a few species had thus acquired a great advantage over other organisms, a comparatively short time would be necessary to produce many divergent forms, which would be able to spread rapidly and widely throughout the world" (p. 303). The fossil histories of birds, bats, or other organisms that invaded drastically different adaptive zones have fully substantiated Darwin's thesis.

Darwin was particularly anxious to find reasonable explanations for the sudden appearance of what seemed to be entirely new groups of organisms in the geological sequence, because this phenomenon had been cited by Agassiz, Sedgwick, and the Swiss paleontologist Pictet as an argument against the theory of gradual evolution. In addition to the shift of adaptive zone, Darwin lists several other reasons why the geological record is so imperfect (pp. 287–302); space forbids listing all of them. In tropical forests, for instance, the immediate decay of dead animals and plants prevents fossilization except under special circumstances such as burial by volcanic ash or mud. In continental areas with little erosion and sedimentation, there is often a total absence of sedimentary, fossil-containing deposits (for example, in large parts of Africa for the Tertiary or for certain stages of the Triassic or Permian in

431

many parts of the world). One other important cause of the loss of potential fossil deposits—unknown, of course, to Darwin—is the disappearance of continental shelves at the frontal edge of advancing "plates," as demonstrated by plate tectonics.

The best evidence for the fact that a group may be in existence without leaving any trace in the fossil record is provided by some living types that are almost or totally nonexistent as fossils; for instance, the agnath fishes (lampreys and hagfishes) are not known between the Paleozoic and the present. The coelacanth fishes, quite flourishing between the Devonian and the earlier Mesozoic, were thought to have become extinct in the Cretaceous (more than 70 million years ago) until a living species (*Latimeria*) was rediscovered in 1937 in the Indian Ocean.

Among all the sudden origins of faunas, none bothered Darwin more than the sudden appearance of most major phyla of animals in the lowest fossil-bearing rocks. Where could they have come from? In the eighty years after 1859, this difficulty became even more serious. Wherever new strata were explored, the earliest types invariably turned up in the Cambrian while nothing was found in the pre-Cambrian strata. Yet, the Cambrian is only about 600–650 million years old, while the earth as a whole is now believed to be 4.5 billion years old. Certainly a large part of the geological column is older, indeed much older than the Cambrian. The fact that there are rich faunas of trilobites, brachiopods, and other fossils in the oldest fossil-bearing strata but no trace of their common ancestors in still older strata forces Darwin to admit: "The case at present must remain inexplicable" (p. 308). Here, as always, Darwin was honest in admitting a difficulty, and a difficulty it still is, even today. To be sure, the fossil record has now been extended backward as far as about 3.5 billion years, thanks to the researches of Barghoorn, Schopf, Cloud, and other investigators, but virtually all these older fossils are microorganisms and, in strata older than one thousand million years, prokaryotes (Schopf, 1978). We have no choice but to conclude that the marvelous radiation of the invertebrates was indeed a comparatively "sudden" event in the late pre-Cambrian, between 700 and 800 million years ago. Presumably, a whole series of factors contributed to this outburst: There may have been a change in the chemistry of the oceans, diploidy and genetic recombination may have become more frequent, and there may have been changes in the ecosystem (such as the origin of predatory types). Perhaps we shall never know.

432

Inferences from the Fossil Record

Having attempted to answer in chapter IX all the embarrassing questions that his opponents might ask, Darwin was ready in chapter X to apply to the geological record the question he asked about all other aspects of diversity and adaptation: "Whether the several facts and rules relating to the geological succession of organic beings better accord with the common view of the immutability of species, or with that of their slow and gradual modification, through descent and natural selection" (p. 312).

Actually, Darwin upholds his own theory not only against the thesis of the immutability of species but also against orthogenetic theories such as Lamarck's and against catastrophism (or saltational theories). The chapter contains a particularly masterful application of the hypothetico-deductive method. Darwin not only presents the geological evidence but he also develops some rather general evolutionary principles. He stresses that "the variability of each species is quite independent of that of all others" (p. 314). As a result of this and of some other factors, each species has its own rate of evolution, and these rates can be either very slow or very fast. The same is true for higher taxa. "Genera and families follow the same general rules in their appearance and disappearance as do single species" (p. 316). This stress on the individuality of taxa and on the uniqueness of the evolutionary behavior of each taxon was a very unorthodox view in an age that was dominated by the thinking of physical scientists. They believed in general laws that could be expressed with mathematical precision, and they expected rates of evolution to be the same in all evolving organisms. This is emphatically denied by Darwin. "I believe in no fixed law of development, causing all inhabitants of a country to change either abruptly or simultaneously or to an equal degree" (p. 314).

Extinction. There were few aspects of the geological record that fitted Darwin's theory better than extinction. Lamarck, as we remember, considered extinction an impossibility. From Cuvier on, the incessant extinction of species and of whole higher taxa could no longer be denied, not even by those geologists who did not support extinction by catastrophes. Yet, if one denies evolution, extinction is an embarrassment. Why should the creator have produced so many vulnerable species? Why does he have to replace them? And by what process does he introduce the numerous new species to fill the vacant places in the economy of nature?

For Darwin, extinction was the necessary concomitant of evolution. With the world constantly changing, some species would find conditions no longer suitable, with the result "that species and groups of species gradually disappeared, one after another, first from one spot, then from another, and finally from the world" (p. 317). Biological factors, however, are even more important, says Darwin, than physical factors. "The improved and modified descendants of a species will generally cause the extermination of the parent species" (p. 321). Furthermore, a species may also be exterminated "by a species belonging to a distinct group." When an entire major group disappears, like the trilobites or ammonites, extinction is a slow and gradual process, sealed by the extinction of the last surviving species. "We need not marvel at extinction," says Darwin, "for it accords well with the theory of natural selection" (p. 322). But for Lyell it had accorded well with a theory that had a strong emotional appeal.

434

It is only in the last twelve pages of chapter X that Darwin presents the decisive evidence on evolution as revealed by the study of the fossil record. His conclusions can be summarized in a number of broad generalizations:

(1) All fossil forms can be fitted "into one grand natural system," even such extinct types as the ammonites (which are cephalopods) or the trilobites (which are arthropods).

(2) As a general rule, the more ancient a form is, the more it differs from living forms.

(3) Fossils from any two consecutive formations are far more closely related to each other than are the fossils of two remote formations.

(4) The extinct forms on any given continent are closely related to the living forms of that continent, as in Australia where the extinct Tertiary mammals, like the living ones, are mostly marsupials; and in South America, where the extinct Quaternary fauna prominently contains armadillos and sloths, like the modern fauna. To this phenomenon Darwin gave the term "the law of succession of types."

The evidence presented by Darwin in chapters IX and X is summarized by him in this statement: "Thus, on the theory of descent with modification, the main facts with respect to the mutual affinities of the extinct forms of life to each other and to living forms seem to me explained in a satisfactory manner and they are wholly inexplicable on any other view" (p. 333).

Since paleontology is the only biological science that can study macroevolutionary phenomena directly, the theory of evolution was a tremendous boon to paleontology. That evolution had occurred and that groups of related taxa are derived from a common ancestor was almost universally accepted by paleontologists soon after 1859. By contrast, Darwin's two other theories—gradual evolution and natural selection—were widely, indeed almost universally, rejected by paleontologists, as will be discussed in later chapters.

THE EVIDENCE FOR COMMON DESCENT

Once Darwin had abandoned the concept of the constancy of species, there was no longer any obstacle in the path of the theory of common descent. If an ancestral species of cats could give rise to several species, than it was conceivable, indeed logical, to derive all cats from a common ancestor. And since cats, weasels, dogs, and bears had a great deal in common, it was a legitimate hypothesis to derive all of them from a common ancestor that gave rise to all carnivorous mammals. Thus, common descent, when consistently applied, tied the whole organic world together. The enormous diversity of plants and animals which, up to that time, had seemed so chaotic and totally incomprehensible to the human mind suddenly began to make sense. This thought was at the same time so exciting and so satisfying that Darwin expressed it like a capstone, in the last sentence of the *Origin:* "There is a grandeur in this view of life, with its several powers, having been originally breathed into a few forms or into one; and that . . . from so simple a beginning endless forms most beautiful and most wonderful have been, and are being, evolved."

The failure of Lamarck, Chambers, and other earlier evolutionists to focus on the species had prevented them from discovering the concept of common descent which, except for natural selection, is perhaps the most heuristic concept developed by Darwin. A very large proportion of the phenomena of organic nature that before 1859 had seemed arbitrary and capricious acquired a logical pattern when explained as being due to common descent. Most of the arguments of chapters VI and X–XIII of the *Origin* are based on the demonstration that certain phenomena are more easily explained as due to common descent than due to special creation.[2]

Darwin was a great admirer of the philosophers John F. W.

Herschel and William Whewell, who based their philosophy and methodology of science on Newton. Whenever possible Darwin tried to apply their principles in his own writings. This included the admonition to look everywhere in natural phenomena for laws, and in particular to look for mechanisms or causes that were able to explain phenomena in widely different areas (Ruse, 1975b; Hodge, 1977). The theory of common descent must have delighted Darwin in this respect more than anything else he ever proposed, for the number of phenomena it was able to explain. This includes the Linnaean hierarchy, patterns of distribution, the facts of comparative anatomy, and indeed virtually all the facts that are now usually cited as substantiating evolution. Even the cell theory acquired a new significance because it explained why animals and plants, in other respects so different, are composed of the same basic components, cells, as a heritage of their common ancestors.

436

Common Descent and the Natural System

The adherents of the concept of the *scala naturae,* as we have seen, believed in a steady progression from the most simple to the most perfect organisms. Lamarck's theory of evolution was largely based on this concept. Yet, the more the knowledge of plants and animals advanced, the less did the similarities and differences of organisms conform to this pattern. Instead, organisms usually fell into well-defined and frequently rather isolated groups, like mammals, birds, and reptiles, that could not be arranged in a linear sequence from simple to perfect. On the other hand, nearly all taxa of organisms were clearly more similar to some than to other taxa. It was on the basis of this principle of degrees of similarity that naturalists from Aristotle on had grouped organisms, resulting, since the seventeenth and eighteenth centuries, in the Linnaean hierarchy (see Part II). As stated by Darwin, "From the first dawn of life, all organic beings are found to resemble each other in descending degrees, so that they can be classed in groups under groups. This classification is evidently not arbitrary like the grouping of stars in constellations" (*Origin:* 411). But what was the cause for the apparent pattern, what was the nature of the apparent constraints? To say, as did Louis Agassiz, that it reflected the plan of the creator explained nothing.

Everything, however, became clear as soon as one made the assumption that the members of a taxon are the descendants of a common ancestor. Darwin illustrates this in his famous diagram opposite to p. 116 of the *Origin*. This principle of common descent explains why "species descended from a single progenitor [are] grouped into genera; and the genera are included in, or subordinate to, subfamilies, families, and orders, all united into one class. Thus, the grand fact in natural history of the subordination of group under group . . . is in my judgment fully explained" (p. 413). And indeed it was.[3]

Two matters, however, need to be stressed at this time. The first is that Darwin, when proposing the theory of common descent, had found the solution of the great problem of the "natural system" that had exercised systematists for more than one hundred years. The inclusive hierarchy of groups under groups is a necessity if species are descended from common ancestors. Reciprocally, as Darwin continues to emphasize, the fact of the hierarchy of organisms is extremely powerful evidence in favor of his theory. There simply is no other possible explanation for the hierarchy unless one wants to postulate an extremely capricious creator. In the end Darwin reiterates that "descent is the hidden bond of connection which naturalists have sought under the term of the 'Natural System' " (p. 433). Indeed, every systematist since Darwin has accepted—or at least paid lip service to—the fact that any system of classification must be consistent with the theory of evolution, that is, that every recognized taxon must consist of descendants of a common ancestor.

The question is sometimes asked, Did Darwin become an evolutionist because he wanted to explain the Linnaean hierarchy? or more broadly, What is the causal connection between evolution and classification? Looking at the work of Lamarck or Cuvier will give us an answer to these questions. The excellent classifications of a Pallas, Latreille, Ehrenberg, or Leuckart did not lead to the establishment of evolutionary theories, nor did those of Cuvier or Agassiz. They all took the Linnaean hierarchy for granted but explained it in static terms, for it is quite possible to explain the best "natural" classification in terms of essentialism. Nor does the acceptance of evolution lead necessarily to a causal explanation of the Linnaean hierarchy. Most of the early evolutionists, like Lamarck, thought in terms of the *scala naturae* and tried, as much as possible, to list the higher taxa in an ascending lineage of growing perfection.[4] A tentative answer to the questions posed above would be that a knowledge of the Linnaean hierarchy alone would not automatically lead to the conception of the theory of evolution by common descent, but also that mere evolutionary thinking (like that of Lamarck and Meckel) without a full understanding of the Linnaean hierarchy would likewise fail. Darwin was in the possession of both ingredients.

Man and Common Descent

Combining all animals into a single hierarchy (phylogenetic tree) of common descent at once raised the problem of the position of man. Linnaeus (1758), without making any noise about it, had included man in his mammalian order Primates and had made it very clear in various of his writings how close he thought man was to the anthropoid apes. I do not have the space to present the evidence that has accumulated since that time, particularly through comparative anatomy, which shows the essential similarity of man and anthropoids. It is well known how proud Goethe was to have discovered the intermaxillary bone in man, the absence of which had been considered a diagnostic feature of *Homo*. However, all Darwin said in the *Origin* (p. 488) was, "Light will be thrown on the origin of man and his history." It was not until 1871 that Darwin was ready to state without reservations that man had arisen from ape-like ancestors. This had already been proclaimed in the 1860s by T. H. Huxley and Ernst Haeckel, and was soon accepted by most knowledgeable biologists and anthropologists.

The claim—or, one might say more correctly, the demonstration by science—that man was not a separate creation but part of the mainstream of life caused a tremendous shock. It was in conflict with the received teachings of the Christian church and even with the tenets of most schools of philosophy. It ended the reign of the anthropocentric world view and necessitated a reorientation of man's position toward nature. At least in principle, it provided a new basis for ethics, and in particular for conservation ethics (White, 1967). The shockwaves of the "dethroning" of man have not yet abated. Depriving man of his privileged position, necessitated by the theory of common descent, was the first Darwinian revolution. Like most revolutions, it went at first too far, as reflected in the claim made by some extremists that man is "nothing but" an animal. This is, of course, not true. To be sure, man is, zoologically speaking, an animal. Yet, he is a unique animal, differing from all others in so many fundamental ways that a separate science for man is well-justified. When recognizing this, one must not forget in how many, often unsuspected, ways man re-

438

veals his ancestry. At the same time, man's uniqueness justifies, up to a point, a man-directed value system and man-centered ethics. In this sense a severely modified anthropocentrism continues to be legitimate.

Common Descent and Patterns of Geographical Distribution

The first sentences in the *Origin* are, "When on board H.M.S. 'Beagle', as naturalist, I was much struck with certain facts in the distribution of the inhabitants of South America . . . These facts seemed to me to throw some light on the origin of species—that mystery of mysteries." In chapters XI and XII Darwin again refers to these distributional facts, and does so also in his autobiography. Two phenomena in particular had impressed Darwin, the fact that the fauna of the temperate parts of South America consisted of species closely related to those of the tropics of the same continent rather than to temperate zone species of other continents; and second, that the faunas of islands (Falklands, Chiloe, Galapagos) were closely related to those of the adjacent parts of continental South America rather than to those of other islands. The history of the "introduction" of these faunas thus seemed more important than the ecology of their area of occurrence. Distribution was clearly not random, but what factors exactly determined it?

439

This was by no means a new question, and it is necessary to give a short survey of the history of biogeography to understand why Darwin in the *Origin* asked the kinds of questions he did. With the help of the modern understanding of these problems, we can phrase more precisely the distributional problems with which the naturalists were particularly concerned in the eighteenth and nineteenth centuries: Is the seeming relatedness of the species in a local fauna (monkeys in the tropics, bears in the temperate zone) caused by the environment or by a common history? Are disjunct distributions due to multiple creations or due to a secondary separation of a previously continuous range or, alternatively, due to long-distance colonization?

The ancients knew already that there were regional differences in the distribution of animals and plants, and ascribed the occurrence of certain species to climatic factors, while they attributed discontinuities, such as between the Indian and African elephants, to former connections (Hippocrates, Aristotle, Theophrastus, and others). When the idea began to spread that the earth is a globe rather than a flat disk, it raised new problems, for instance the question of the possible presence of humans (antipodes) on the other side of the globe. Free speculation about such questions was no longer possible after the Church usurped total dominance over the western mind, and zoogeographic problems were posed in biblical terms. This made the problem of different faunas and floras much more formidable. Since according to the Bible all life had descended from the inhabitants of the Garden of Eden or, more precisely, from the survivors of Noah's Flood, their descendants must have spread out from the place where the Ark had set down, supposedly on Mount Ararat. This interpretation precluded a purely static conception of patterns of distribution, since it was based on the occurrence of dispersion and migration.

A dispersal from Mount Ararat appeared credible when only the faunas of Europe and the adjacent part of Africa and Asia were known. The discovery of the entirely new continent of America and the realization by the end of the sixteenth century that it had a rich fauna that was drastically different from anything known in the Old World caused great consternation. The further discovery of the faunas of central and southern Africa and the East Indies, and finally the even more unique Australian fauna, raised ever more formidable questions for the pious biogeographer. A dispersal of an immutable animal life from a single center of creation over the entire world became more and more a logical impossibility.[5]

The botanist J. G. Gmelin (1747) was apparently the first to suggest that a creation of species had taken place all over the world. The biblical story of the Garden of Eden and of Noah's Ark was quietly superseded by various theories of *centers of creation*. Some authors still postulated an origin from a single pair; others had each species originate in the number of individuals characteristic for that species and all over its present range.

No one in the eighteenth century had as great an influence on the development of biogeography as Buffon who, therefore, has been called the father of zoogeography. In his violent antagonism to Linnaeus he refused to classify animals on the basis of shared characters and chose instead the "practical" system of arranging them according to their country of origin. In other words, he grouped them into *faunas*. The faunal lists which he thus obtained enabled him to draw all sorts of conclusions—for instance, that the fauna of North America was derived from that of Europe.

Buffon (1779) postulated both historical and ecological causal factors (Roger, 1962). When the earth began to cool off, life was first created in the far north, because the more tropical regions were still too hot for animal life. As the earth gradually cooled off, the northern fauna moved with the decreasing temperature toward the tropics, and a new northern fauna originated, presumably in Siberia. The fauna that had been able to occupy South America was protected by mountains on the isthmus of Panama from being invaded by the new northern fauna, and this is why "of all the animals of the southern parts of our continent not a single one is found in the southern parts of [America]" (p. 176). In the old world "not a single large and prominent species is known in the tropical parts [*terres du Midi*] that had not previously existed in the north" (p. 177). Since Buffon believed that faunas are the product of the country, he is rather puzzled by the amount of difference in the tropical faunas of the two continents because "the species produced by the creative [*propre*] forces of the southern regions of our continent ought to resemble [*auroient dû ressembler*] the animals of the southern regions of the other continent," but, as already stated, not a single species of the two tropical regions is the same.

What Buffon proposed was that when "born," a fauna is the product of the district where it originated, but that it can and will disperse as the climatic conditions change. When born, species are created according to definite laws, each species adapted to its climatic zone, and this is the reason why we can observe tropical faunas, desert floras, arctic faunas, and so forth. As nature has made the climate for the species, said Buffon, so it has made the species for the climate: "The earth makes the plants; the earth and the plants make the animals" (Buffon, 1756, VI).

The fossil and subfossil proboscideans and other distributional data had a dominant impact on the historical component of Buffon's theories. Where his "product of the country" beliefs came from is less clear, but I suspect that his Newtonian philosophy was responsible. Origins had to be due to certain forces.

The explanatory conflict revealed in Buffon's writings continued until 1859. Even though every traveler described the drastic differences between faunas and floras, this was subconsciously unacceptable for those who felt that distributions should display design just like everything else in God's created world. Thus the tropical faunas of different continents or of different islands "ought to resemble" each other, as Buffon had said, but they did not. In pre-evolutionary days there was no explanation for this failure of expectation.

Buffon was not alone in his emphasis on historical factors. Linnaeus (in 1744) derived all plants from a mountainous tropical island from which they had spread all over the world (Hofsten, 1916). Remarkably advanced ideas were published by the zoologist E. A. W. Zimmermann (1778–1783). The distribution of mammals, he demonstrated, is not sufficiently explained by climate but is clearly influenced by the history of the earth. Indeed, the distribution of animals provides evidence of changes on the surface of the earth. When two countries that are now separated by an ocean have different mammal faunas even though they have the same climate, then they must always have been separated. However, when such countries have similar or the same species, then it is legitimate, he said, to infer a former connection. He lists islands, like Great Britain, Sicily, Ceylon, and the Greater Sunda Islands, which formerly must have had continental connections, and postulates a former connection of North America with northern Asia. With some justification, Zimmermann is considered by some authors the founder of historical biogeography. C. F. Willdenow (1798) was the first botanist to explain discontinuous ranges of species as the result of a secondary interruption of previously continuous ranges.

Alexander von Humboldt, in his younger years, had wanted to write "a history and geography of plants or historical information on the gradual dispersal of plants over the whole globe (1805)." But when he finally published his *Ideas on a Geography of Plants* it was almost entirely devoted to floristics and plant ecology. His interest was evidently in the present distribution of plants and their dependence on the physical factors of the environment. By then, he had come to consider questions of origins as insoluble.

The rapid advances of biogeographical knowledge toward the end of the eighteenth and the beginning of the nineteenth centuries raised new difficulties. More and more cases were discovered where related species like the beaver in Eurasia and North America had adjacent but separated ranges, or where even the same species occurred at widely separated localities, such as plants of the Alps being also in the Pyrenees, in the mountains of Scandinavia, or even in the Arctic lowlands. The explanation of such disjunct distributions became one of the major issues of biogeography in the first half of the nineteenth century (von Hofsten, 1916).

When the two Forsters discovered European plants in Tierra del Fuego on Cook's second voyage, they concluded at once that similar climates had led to the production of similar species (1778). (By contrast, the very same distribution was one of Darwin's favorite illustrations of the extraordinary dispersal power of plants.)

The emphasis on historical factors that can be found in the writings of Buffon, Zimmermann, Willdenow, and other eighteenth-century writers can no longer be perceived in the writings of biogeographers of the early nineteenth century. As the faunas and floras became better known, and, in particular, after the strangeness of the Australian biota was discovered, the main emphasis was placed on the uniqueness of the biota of various places (Engler, 1899; 1914). Each flora and fauna had been introduced at a definite center or focus of creation. Alphonse de Candolle (1855; 1862) recognized twenty botanical regions (not including the separate floras of islands), each presumably a separate center of creation.

443

Those who, like Louis Agassiz (1857: 39), believed in an entirely static world could conceive of no limitations to the creative power of God and proposed, therefore, that species had been separately created in every disjunct portion of the range of a species; he thus carried the theory of multiple centers of creation to its logical extreme. When Agassiz, in the 1850s, wrote about biogeography, his uncompromisingly fundamentalist interpretation seemed like a throwback to a long past period.

The emphasis on regional difference and centers of creation also dominates the writings of Lyell, to whom Darwin owes much of his thinking about biogeography (Hodge, 1981). Not surprisingly, Darwin still adhered to a creationist interpretation of distributions when he was on the *Beagle*. When studying the impoverished animal life of some terraces, he states: "It seems a not very improbable conjecture that the want of animals may be owing to none having been created since this country was raised from the sea" (Darwin, 1933: 236). Local creation under the influence of the local environment (particularly climate) was Darwin's interpretation at that time.

Happenings in the twenty-three years between Darwin's return from the *Beagle* and the publication of the *Origin* had profoundly affected biogeographical theory. The so-called catastrophists, no matter how wrong they might have been in most of their other claims, made the important point that the face of the earth had undergone rather drastic changes which, if one assumes that the biota are in harmony with their environment, could not have helped but greatly affect distributions. This was strikingly and unexpectedly confirmed by Agassiz's Ice Age theory. With much of northern Europe covered by ice, and the climate of the remainder profoundly influenced by this icecap, drastic shifts in the zones of vegetation and their inhabitants were inevitable. Two authors who used this new insight to convert static biogeography into a dynamic, developmental science were Edward Forbes and Alphonse de Candolle. In a major monograph Forbes (1846) attempted to explain the distribution of the flora and fauna of the British Isles as a product of the recent geological history. He postulated that each species had a single center of origin and that all discontinuous ranges were the result of secondary disruptions of former continuity. He explained the composition of the British biota as being due to post-Pleistocene colonization by southern and eastern elements. He emphasized that, in addition to purely physical barriers like oceans and mountain ranges, there are also climatic and vegetational barriers as, for instance, those which separate the Alpine flora of the European mountains and the closely allied Arctic flora. Darwin had reached similar conclusions in manuscript but did not publish them until thirteen years later.

Forbes differed from Darwin in two important respects. Impressed by geological change and underestimating dispersal abilities of animals and plants, he was a great land-bridge builder and in particular a proponent of a now sunken former mid-Atlantic continent, Atlantis. More importantly, Forbes retained a belief in the immutability of species, and when he found related species in different areas, he ascribed this to separate creations rather than to evolutionary differentiation during isolation. This is a typical instance of the phenomenon so well described by Thomas Kuhn of the reluctance of authors to abandon a long-familiar paradigm.

No other pre-Darwinian author paid as much attention to the problem of "disjunct species" (his terminology) as the botanist Alphonse de Candolle (1806–1893). He defined disjunct species as plants which live in separate areas that are sufficiently isolated so that a current dispersal from one to the other area seems impossible. In an early paper (1835) he had still accepted the multiple creation of disjunct species, but in his great *Géographie botanique raisonnée* (1855) he had decidedly shifted to a historical explanation of divided ranges, emphasizing that the current geographical and climatic conditions play only a secondary role.[6] Rather, different dispersal opportunities in former periods must have been responsible. Although de Candolle's plant geography is a splendid analysis of the origin of distributional discontinuities and the first fully consistent attempt by a plant geographer to explain present distributions as a product of history, he was unable to provide a comprehensive explanation of the history of faunas and floras, not yet having accepted evolution. After the publication of the *Origin* he suggests that "the theory of a succession of forms by deviations of anterior forms" might be regarded as "the most natural hypothesis" to explain disjunctions (1862).

445

Darwin's Explanation of Geographical Distribution

It was Darwin who took the decisive step to free biogeography from the restrictions imposed on it by creationist presumptions.[7] Prior to 1859 there had been essentially two theories concerning the origin of biota (ignoring at this moment subsequent migrations). The theists proposed that each species had been separately introduced by creation and that there were in principle as many centers of creation as there are species or disjunct species areas. This explanation implied a creator of extraordinary capriciousness, an implication acceptable only to an extreme fundamentalist. The deists and the natural theologians who believed in a designed world believed that the creation and introduction of new species had to obey certain laws and was caused by appropriate forces. Consequently they expected to find similar ("related") species in all hot tropical regions, in all arid and desert regions, on all mountains, and on all islands. But this, of course, is not at all what the biogeographers found, as Darwin pointed out again and again. And this failure of the two existing theories induced him to introduce a third causal theory, distribution as a result of common descent.

By postulating common descent of related species and of members of the same higher taxon, Darwin was able to draw far-reaching conclusions on the former distribution and movement of these taxa. He presents his evidence in chapters XI and XII of the *Origin,* chapters which are a delight to read owing to their methodological rigor and the logic of the argument. Darwin no longer had to ask, Is this species where it is because the Creator placed it there? Unfettered by such religious constraints he was able to ask questions such as: Why does the fauna or flora of a given district have its particular composition? Why are the biota of certain districts similar, and those of others dissimilar? What determines the faunal composition of islands? Or, what are the causes of disjunct patterns of distribution?

By asking these questions, Darwin became the founder of causal biogeography. Indeed, his whole interest centered in questions of causation, and we find very little descriptive biogeography in these chapters. In the actualist tradition, Darwin insisted on interpreting distributions in terms of the present configuration of the continents and opposed any reckless construction of land bridges, in contrast to Forbes and to most biogeographers of the ensuing eighty years. In this respect, as in so many others, Darwin was far closer to modern thinking than his contemporaries and early followers.

His argument is essentially two-pronged. On the one hand he attempts to refute previously held invalid beliefs and on the other he attempts to introduce new causal theories. He begins by supporting "the view that each species was first introduced within a single region . . . He who rejects it . . . calls in the agency of a miracle" (*Origin:* 352), precisely what Asa Gray had said of Agassiz's theory of multiple creations. The fact that the British Isles and the European continent have so many species in common while Europe has not a single species of mammal in common with either South America or Australia, argues Darwin, fits the laws of dynamic biogeography but is inexplicable under the theory of special creation.

According to the theory of "laws of creation" one should expect biota to be the immediate product of local climate. Darwin thoroughly refutes this theory: Whether we compare climatically similar tracts of Europe and North America or "in the southern hemisphere, if we compare large tracts in Australia, South Africa, and western South America between latitudes 25° and 35°, we shall find parts extremely similar in all their conditions, yet it would not be possible to point out three faunas and floras more utterly dissimilar" (p. 347). The same can be shown for forested regions, for islands, and for the oceans. Thus, there is no indication whatsoever for the introduction of constant species according to appropriate laws.

According to Darwin's causal theory of biogeography, patterns of distribution, particularly discontinuities, can be explained rather simply by making one of two possible assumptions: either (1) the taxon in question has the dispersal ability to cross barriers,

446

as a mountain species that is able to cross lowlands to colonize another range of mountains, or (2) the discontinuous ranges are the remnants (relics) of previously continuous ranges. The postulate of descent from common ancestors together with the two stated assumptions permits the explanation of any pattern of distribution without any recourse to supernatural agencies. Thus, it becomes the main task of the biogeographer to study the nature of barriers and of the dispersal abilities of animals and plants. "Barriers of any kind, or obstacles to free migration, are related in a close and important manner to the differences between the productions of various regions" (p. 347). Darwin did not envision barriers as purely physical obstacles because there is a close inverse relation between the efficiency of barriers and the dispersal facilities of species and, furthermore, because he considered the ranges of competing species also as constituting powerful dispersal barriers.

447

Darwin knew that the correct evaluation of dispersal is the key problem in the explanation of patterns of distribution (pp. 356–365). He was the first to approach these problems by ingenious experiments which showed that the dispersal power of organisms, particularly of plant seeds, is much greater than previously believed, and that there is no particular need to invoke land bridges to explain much transoceanic dispersal. The one factor which he grossly underrated was the power of wind and air currents to transport not only seeds but also small animals.

Darwin, like Forbes (though he arrived at this conclusion independently), places great stress on the effect of the glacial period on present distribution (pp. 365–382). He deals with it on a worldwide basis, attempting to explain the presence of northern elements in the southern hemisphere and on tropical mountains. Disjunct distributions are of crucial importance in his chain of argument when he reasons by analogy from the distribution of disjunct populations of the same species to the distribution of allied species of the same genus, and so on up the hierarchy of categories.

Most of chapter XII is devoted to a discussion of the inhabitants of oceanic islands (pp. 388–406). Darwin points out that creationists are utterly unable to explain why there are so few species on oceanic islands, or why certain groups of animals, such as terrestrial mammals, urodele amphibians, and true freshwater fishes, are consistently absent from them. The strange imbalance of the biota of oceanic islands and the striking difference between the faunas of continental and oceanic islands are inexplicable "on the view of independent acts of creation" and "seem to me to accord better with the view of occasional means of transport" (p. 396). This also explains why invariably the inhabitants of oceanic islands are most closely related to those of the nearest continent, which induces Darwin to ask the creationists, "Why should the species which are supposed to have been created in the Galapagos archipelago, and nowhere else, bear so plain a stamp of affinity to those created in America?" (p. 398).

448   Darwin, who forever looked at natural-history phenomena from a biological point of view, was fully aware that successful dispersal involves two capacities: the ability to get to a new location, and the ability to colonize it successfully. "We should never forget that to range widely implies not only the power of crossing barriers but the more important power of being victorious in distant lands in the struggle for life with foreign associates" (p. 405). He finally summarizes his findings with typical Victorian indirectness as follows: "I think the difficulties in believing that all the individuals of the same species, wherever located, have descended from the same parents, are not insuperable" (p. 407).

In biogeography, as in so much else of his work, Darwin was far ahead of his contemporaries, and biogeographical science did not really catch up with him until the 1940s, even though a few advanced authors in the intervening years were strictly Darwinian biogeographers.

Biogeography after 1859

Scientific biogeography, as it exists today, had its beginnings in chapters XI and XII of the *Origin of Species*. Considerations of space forbid giving an extensive treatment of the rich history of the ensuing 120 years. An attempt shall, however, be made to mention some of the major trends.[8]

Regional Biogeography. An interest in comparing the faunas and floras of different regions goes back to the seventeenth century. With Buffon and Linnaeus it was a major preoccupation and so it was with various biogeographers like de Candolle, Swainson, and Schmarda in the first half of the nineteenth century. The publication of P. L. Sclater's classification of the world into (six) zoogeographic regions on the basis of the distribution of birds (1858) was, however, the beginning of a new period.

Darwin was never particularly interested in regional biogeography. Apparently he considered this method of dealing with distributional phenomena as too static and descriptively taxonomic. Nevertheless, in the first sixty years after the *Origin*, the regional treatment preoccupied most biogeographers. The bible of this school was A. R. Wallace's authoritative two-volume work, *The Geographical Distribution of Animals* (1876). Although everyone agreed that the major zoogeographic regions coincided more or less with the major continental land masses, different schemes were proposed for combining these into "regions" according to the group of organisms on which the geographical classification was based. Students of mammals were impressed by the similarity of the mammal fauna of Eurasia and that of North America, combining them into a Holarctic Region. Students of birds, in contrast, discovered much relationship between the birds of North and South America and some of them proposed to separate a Neogaea from the Old World (Paleogaea) (see Mayr, 1946a). For the botanist, still other delimitations seemed more natural. For instance, the plants of the entire region from the Malay peninsula to New Guinea and the Pacific islands belong to a single flora while, for animals, there is a conspicuous break between a western Indo-Malayan element and an eastern Australo-Papuan element, separated from each other by a north-south line between New Guinea and the Greater Sunda Islands. The exact location of this line remained controversial for three-quarters of a century until it was recognized that "Wallace's Line" between Borneo and Celebes reflects the edge of the Asiatic continental shelf, while "Weber's Line" between Celebes and the Moluccas is the line of faunal balance (Mayr, 1944b).

 Unsatisfied with this coarse-grained analysis, regional biogeographers, beginning with de Candolle, devoted much effort to attempts to produce a fine-scaled classification of subregions and *biotic districts*, endeavors that have continued to the present day. On the whole, such studies remained on a descriptive level and contributed little to generalization.

 A rebellion against this static approach was started by E. R. Dunn (1922), who proposed instead a causal analysis of faunas. G. G. Simpson (1940; 1943; 1947) became the leader of this new movement, particularly as far as mammals were concerned, and Mayr for birds.[9] Simpson showed that there are different kinds of bridges connecting land masses (for example, "corridors," "filter bridges") and stressed in particular the statistical element in considering the probability of dispersal across water. This, actually, was a return to Darwin's classical framework of causal bio-

geography which had been neglected by Wallace and his followers. Dispersal is the key problem in this approach.

History of Continents and Means of Dispersal. The two great bones
of contention, with respect to Darwinian biogeography, are the
former history of continents and their connections and, second,
the means of active and passive dispersal in various animal groups.

As far as continental connections are concerned, three major
schools can be recognized. One of them continued Forbes' pro-
clivity of postulating land bridges and formerly existing islands
and sunken continents. Distributional discontinuities were explained by the existence of former land bridges between Europe
and North America, between Africa and South America, between
South America and Australia, between Madagascar and India, between Hawaii and Samoa, and so forth. There is no ocean that
was not crisscrossed by land bridges during the heyday of this
school. The authors of these land bridges had one thing in common: they had a very low opinion of the dispersal abilities of animals and plants.[10]

However, not all land bridges were without geological support. All biogeographers agreed that islands situated on continental shelves, the so-called continental islands such as Great Britain,
Ceylon, and the Sunda Islands, once had had continental connections, as had already been stated by Zimmermann and by Forbes.
A land bridge across the Bering Strait between North America
and northeast Asia was also universally accepted. The land-bridge
builders, however, went much further and, completely disregarding all geological considerations, often proposed land bridges to
explain the occurrence of species on islands that are nothing but
the peaks of volcanic cones rising from ocean deeps.

Reckless land-bridge building was vigorously opposed by all
those biogeographers who continued Darwin's Lyellian tradition,
which postulated an essential permanence of continental masses
and ocean basins, only admitting occasional rises and falls of sea
level as during the Pleistocene glaciations. A. R. Wallace sided with
Darwin in the opposition to land bridges (Fichman, 1977). The
reaction against land-bridge building was particularly strongly expressed by Matthew (1915) and Simpson (1940) but also by Mayr
(1941; 1944a), Darlington (1957), and several plant geographers
(Carlquist, 1974).[11] The biogeographers of this school have two
things in common. They are averse to accepting any changes in
the outlines of the continents not validated by geology, and they
had an even greater faith than Darwin in the ability of most kinds
of animals and plants to cross seemingly formidable water gaps.

A third school arose after the publication in 1915 of Wegener's continental-drift theory. This theory, although supported by a number of biogeographers, was at first not very successful, for two reasons. First, the geophysicists unanimously opposed it because they were unable to discern any forces which could account for such large-scale movements of parts of the earth's crust as postulated by Wegener. Second, those biogeographers who adopted continental drift misused it badly, invoking it primarily to explain late Tertiary and Pleistocene phenomena. The resistance of biogeography to the continental-drift theory, as originally proposed, was not reactionary but soundly based on the then-existing information.

Continental drift took a new lease on life in the 1960s through the development of the theory of plate tectonics.[12] This theory has its greatest success in explaining distribution patterns that originated in the Jurassic and Cretaceous, for instance the distribution of the major groups of freshwater fishes, but it still leaves many open questions. According to plate tectonics, for instance, Australia and Antarctica were attached to South America until early Tertiary. Later, Australia separated from Antarctica and drifted northward, coming only rather recently in contact with the outliers of the Asian continent. Why, then, does the bird life of Australia, with the possible exception of a few small groups, consist almost entirely of Asiatic elements? The history of the Pacific also is still controversial. Madagascar, India, and southeast Asia pose additional puzzles.

The two major mistakes made by some recent biogeographers is a failure to recognize that different higher taxa have established their present pattern of distribution at different geological ages (when the position and distances of the various plates were different from today and from that existing during the major dispersal period of other higher taxa), and second, that the distribution pattern of a group is profoundly affected by its dispersal ability. Groups with a relatively low active dispersal ability, like most terrestrial mammals, true freshwater fishes, or earthworms, have very different patterns from those of easy dispersers like freshwater plankton, ballooning spiders, birds, and some groups of insects. A specialist who sweepingly generalizes on the basis of his familiarity with a single group of organisms is apt to arrive at unbalanced conclusions.

In a way, the continental-drift theory is a synthesis of the theory of the permanence of oceans and continents and the theory of land bridges. Although the major land masses (plates) are still

451

considered permanent, their positions and connections are changing in the course of time, even though these changes occur so slowly that the reconstructed outlines of the continents in the middle of the Tertiary are not strikingly different from the present ones. As far as the reconstructed history of the distribution of mammals and birds is concerned, accepting plate tectonics necessitated less of a revision of the conclusions of the permanence-of-oceans school than one might have expected. It affects primarily the interchange of the older Holarctic element between Eurasia and North America (across the North Atlantic rather than Bering Strait) and the origin of the older Australian fauna (South America via Antarctica). Plate tectonics required more revision for the interpretation of the distribution of groups that had their major dispersal prior to the mid-Cretaceous.

452

Discontinuities. The explanation of the origin of discontinuities has continued to be one of the most controversial subjects in biogeography. One can distinguish two kinds of discontinuities, primary and secondary. A primary discontinuity originates when colonists reach an isolated area and succeed in establishing a permanent population there. For instance, when Scandinavian insects and plants dispersed to Iceland in the post-Pleistocene period, such colonization, it is now quite certain, took place across a large watergap. This is a typical case of a primary discontinuity.

Secondary discontinuities originate owing to the fractionation of an originally continuous range through a geological, climatic, or biotic event. The blue magpie (*Cyanopica cyanea*) occurs in eastern Asia (Transbaicalia to China and Japan) and has a completely isolated colony in Spain and Portugal. Clearly, this pattern of distribution could not possibly have become established by long-distance dispersal but resulted from the break-up of an originally more or less continuous Palearctic range owing to the Pleistocene deterioration of the region between the two isolates. Unfortunately the situation is not always so clear, leading to arguments as to whether or not long-distance dispersal could account for the discontinuity or, on the contrary, whether there is evidence for a former physical continuity.

When the exuberance of land-bridge building had abated, indeed when it was considered rather disreputable to postulate any land bridge that was not well-documented geologically (particularly in the 1940s and 1950s), the extraordinary ability of many groups of organisms to colonize exceedingly isolated places was discovered. The entire fauna and flora of the Hawaiian Islands,

to mention only one case, is the product of transoceanic colonization, even though this was apparently facilitated by the availability of some, now drowned, stepping stones in the eastern
Pacific. However, a reaction against too great a reliance on long-
distance dispersal developed in the wake of plate tectonics. Perhaps there was a continental connection, it was said, wherever
there is now a wide ocean. Indeed, it is now known that Africa
and South America were still connected in the earlier Cretaceous,
and that Europe and North America had a transatlantic connection as late as the Eocene.

453

A somewhat eccentric biogeographic theory was proposed in
the late 1950s, "vicariance biogeography," which, so far as I can
understand it, stresses former continuities and downgrades the
importance of long-distance dispersal.[13] Quite logically, it found
its chief support among ichthyologists, because primary freshwater fishes have a particularly low dispersal ability. Actually, it does
not seem that vicariance biogeography has introduced any new
principles, since the occurrence of secondary discontinuities was
already well-known to Forbes, Darwin, Wallace, and other pioneers of biogeography (von Hofsten, 1916). Darwin in particular
was fully aware of the two causes for disjunction.

Faunal Elements. Barriers come and go. The rise of the isthmus of Panama connecting North and South America about five
million years ago, the establishment of the Bering Strait bridge,
and the Pleistocene lowering of sea level and temperature with
the advancing ice fronts are a few examples of the elimination or
production of barriers. As a result, partly isolated faunal areas
alternate between periods of high isolation, providing an opportunity for the production of endemics, and periods of faunal interchange. Biotas, for this reason, are not homogeneous but consist of various biotic elements, differing in the time of immigration.
The oldest traceable element, if not known from elsewhere at an
earlier date, is usually called the autochthonous element of the
area, actually simply meaning that its earlier history is unknown.
On the basis of various patterns of autochthonous radiation and
invasion by extraneous faunal elements, Mayr (1965b) distinguished six types of faunas. This classification stresses that, in addition to an original old element (usually unanalyzable), there are
faunal elements that can be classified according to the time of
arrival. This methodology allows for a dynamic interpretation that
is more realistic than typologically assigning faunas to the permanent pieces of the earth's crust recognized by plate tectonics.

Ecological Biogeography. The factors of the environment which influence distribution were of great interest to Darwin. In a way, one might say, this interest was a return to the traditions of Buffon, Linnaeus, and Humboldt, except that now the study of these factors was firmly based on evolutionary principles. Such factors were the major theme of Wallace's *Island Life* (1880). Recent ecobiogeography is again paying particular attention to a component of the environment that had first been emphasized by Lyell and was considered by Darwin to be of greater influence on the distribution of species than any other: competition. He felt that the presence or absence of a competing species determines the success of colonizations and is, more than anything else, responsible for extinction. This emphasis on competition, never entirely dormant after Darwin and quite prominent in the writings of Wallace, Simpson, and Mayr, has experienced a renaissance in the work of David Lack and the school of Hutchinson-MacArthur on species diversity. A mathematical model was proposed in MacArthur and Wilson's *Theory of Island Biogeography* (1967) in which the vague ideas of Darwin and his followers are formalized and quantitized. This publication has proved immensely stimulating and has resulted in numerous, precise, biogeographical analyses by such authors as Diamond, Cody, and Terborgh.[14] The emphasis in this research is on the colonizing powers of individual species, on the interaction of species in determining species diversity at concrete localities, and on the causes of extinction of individual species. Only the very first beginnings have been made in the comparison of the effect of these factors on groups of animals and plants that differ in their dispersal facilities, reproductive strategies, life expectancies, physiological tolerances, genetic systems, and other attributes that can affect colonizing power and competitive ability. There are still rather drastic differences of interpretation in the evaluation of these factors, and since this is the hallmark of all active fields of research, one is justified in believing that this will long remain an active branch of biogeography.

Unfortunately, those who have published in this area have often confused two subject matters. The word *biogeography* means the science dealing with the distribution of organisms, while *ecological biogeography* means the effect of ecological (environmental) factors on distribution. However, the geographic variation of the adaptations of organisms to their environment is known as *geographical ecology*. The first major book in this field was Semper's *Natürliche Existenzbedingungen der Thiere* (1880). A more recent one was Hesse's *Tiergeographie auf ökologischer Grundlage* (1924), in spite of its misleading title. The question that is most important in this field is what adaptations enable an animal or a plant to exist in certain climatic zones, and particularly in such special or stressful environments as the Arctic, deserts, brackish water, the deep sea, caves, or hot springs. This geographical ecology merges directly into ecological physiology as represented, for instance, by the work of Schmidt-Nielsen (1979).

Morphology as Evidence for Evolution and Common Descent

455

Among the lines of evidence for evolution, morphology was ranked very high by Darwin. He said of it, "This is the most interesting department of natural history, and may be said to be its very soul" (*Origin:* 434). Why did Darwin think morphology was so important? We cannot answer this question without a short review of the history of this field.

Morphology is the science of animal and plant form. Just where it belongs within the theoretical framework of biology has always been controversial and, to some extent, continues to be so. Rather remarkable were the frequent attempts, beginning late in the eighteenth century, to establish a "pure morphology," more or less independent of biology, a science that would appeal equally to the biologist, the mathematician, and the artist. It is quite impossible to understand the complex history of morphology unless one realizes that the term is used to designate several independent and rather different developments.

Two of these deal with proximate causes: (1) the morphology of growth, including all processes of growth and development that can be formulated mathematically, particularly allometric growth; and (2) functional morphology, the description of structures in terms of the functions they serve.

Three others concern ultimate causations: (3) idealistic morphology, that is, the explanation of form as a product of an underlying essence or archetype; (4) phylogenetic morphology, the derivation of form from that of a common ancestor (or quite often the tracing back of form to that of the reconstructed common ancestor); and (5) evolutionary morphology, which views form either as response to environmental needs (Lamarck-type explanations) or as adaptation produced by selection pressures.

In view of these many different ways of looking at form (and there are others not here mentioned), it is obvious that a unified treatment of form is quite impossible. In particular, those aspects of morphology that deal with proximate causations belong either to physiology or to embryology and will not here be treated.

The center of interest in morphology from the Greeks to the eighteenth century had been in human anatomy.[15] Yet the anatomy of a Galen or Vesalius was simply an auxiliary discipline of physiology, based on the observation that a painstaking study of structure (preferably combined with experiment) can reveal a great deal about bodily functions. Not surprisingly, anatomy was considered a branch of physiological medicine from the Greeks to the Renaissance.

A new trend began to develop in the sixteenth century when animals were dissected not merely in order to contribute to the understanding of the function of the parts of the human body but also as part of the great revival of interest in nature. Belon's (1555) famous illustration comparing the skeleton of a bird with that of man was an early indication of this new interest. As more and more animals were dissected and compared with each other—and this included not only vertebrates but also insects (Malpighi, Swammerdam) and marine invertebrates, zoologists began to remember the pioneer in this field, Aristotle. Indeed, in his great biological works Aristotle had laid a substantial foundation for a science of morphology.

Three of Aristotle's ideas in particular had a lasting effect. The first is the clear recognition that there are groups of animals that are joined together by a "unity of plan." All warm-blooded terrestrial quadrupeds, for instance, not only are characterized by hair and other external features but also resemble each other in heart, lungs, liver, kidneys, and virtually all other internal organs. Aristotle established a similar unity of plan for other groups of vertebrates and for several taxa of invertebrates such as crustaceans and some of the mollusks. He took it for granted that animals sharing the same plan have equivalent parts, parts that we would now call homologous. However, owing to his primary interest in function, he made no distinction between similarities due to what we would now consider common descent and those due to function. And this confusion persisted for another two thousand years.

Aristotle was also keenly aware of certain correlations. He observed, for instance, that no animal has both tusks and horns. If one part of an animal was enlarged, as compared to other similar ones, this would be compensated by the reduction in another part.

For, as Aristotle said, "Nature invariably gives to one part what she subtracts from another." This thought was taken up again by Goethe and later elaborated in Geoffroy's "loi de balancement" (see Chapter 7).

A third Aristotelian concept important in the history of morphology is, of course, that of the *scala naturae*. Those who revived an interest in comparative anatomy in the seventeenth and eighteenth centuries were very much impressed by the unity of plan and attempted to establish similarities as, for instance, in the extremities of various kinds of mammals, even though some dig underground like moles, others swim like whales or fly like bats. These efforts suffered from the same weakness as those of Aristotle that no further analysis was made of what "similarity" meant. As a result, some of the comparisons were rather ridiculous, such as when the botanist Cesalpino compared the roots of plants with the stomach of mammals, the stem with the heart, and so forth, because the equivalent organs had similar functions.

The discovery of ever new types of animals and plants in exotic countries and of new internal structures revealed by the comparative studies of anatomists steadily added to the seemingly unlimited diversity of the living world. Yet, there were glimpses of underlying patterns, documented particularly by a seeming unity of plan in certain groups of organisms. This was used by morphologists to bring order into the living universe just like the laws of Galileo, Kepler, and Newton had brought order into the physical universe. Any structure or phenomenon that resembled, even in the slightest, something similar in a different organism was used at once to draw far-reaching analogies. Linnaeus was a past master of analogy, rather charmingly displayed in his description of the flower (Ritterbush, 1964: 110).

This trend reached its culmination in the idealistic morphology of the German *Naturphilosophen*. It is no coincidence that this movement was ushered in by a poet, Johann Wolfgang Goethe (1749–1832), for in a way it was a fusion of Plato's essentialism with aesthetic principles. A search for an underlying *eidos* induced Goethe to propose that all organs of the plant are nothing but modified leaves. Goethe took his studies very seriously and was the person who in 1807 introduced the term "morphology" for this field. He was as much interested in animals as in plants and did, himself, quite a few dissections to learn about the structure of vertebrates. These, together with his theoretical ideas, led him to assert "that all the more perfect organic natures, as which we

457

consider fishes, amphibia, birds, mammals, and at their highest point Man himself, are all formed according to an *Urbild* [archetype], which varies only more or less in its basically constant parts, and which still daily develops and becomes modified by reproduction" (Goethe, 1796). As Lovejoy and others have shown, these ideas had nothing to do with evolution, but some of Goethe's ideas were vague anticipations of principles later formulated by Geoffroy.[16]

458 Lorenz Oken (1779–1851) was the most imaginative but also the most fantastic representative of idealistic morphology. The more bizarre of his comparisons are now charitably forgotten, but one of his ideas, although largely erroneous, preoccupied morphology for the next fifty years. Like Goethe in his leaf theory, Oken compared not only the "same" structure in different organisms but also different structures in the same organism, particularly those serially arranged in different segments like, for instance, the vertebrae. This led him to the famous theory that the skull was composed of fused vertebrae. Although Oken later turned out to be wrong in this particular case, the approach as such was actually very productive in arthropod morphology, helping to homologize mouth parts and other cephalic appendages with extremities.

Morphology in this pre-evolutionary period was searching desperately for an explanatory theory. Under the influence of the then dominant philosophy of essentialism, it finally combined the observation of certain types of structure (unity of plan) with the concept of Plato's *eidos,* postulating that organisms represent a limited number of archetypes. Morphologists were looking for the real essence, the ideal type, or, as the Germans called it, the *Urform,* under the great observed variability. The period when idealistic morphology, as it was called, flourished was quite short in zoology, Richard Owen being the last serious representative (1847; 1849), even though a few attempts to revive it were made in this century.[17] In botany, in spite of an early and vigorous opposition by Schleiden, Hofmeister, and Goebel, a school of idealistic morphology has survived to the present day. Alexander Braun (1805–1877) was its early leader and Agnes Arber and W. Troll were recent representatives. Indeed, there is a strong element of this philosophy in the writings of many of the plant morphologists of the last generation (for instance, Zimmermann and Lam).

When idealistic morphology originated early in the nineteenth century, as pointed out by Bowler (1977b) and Ospovat (1978), it was quite a radical departure from orthodox natural theology, according to which each structure of an organism was designed purely for the sake of utility for a particular species, to provide greatest adaptation. But then why should the anterior extremity of a mole (digging tool), a bat (wing), a horse (running leg), and a whale (paddle) have essentially the same structure, while the wings of insects, birds, and bats, all serving the same function, have very different structures? This made no sense at all under the theistic concept that every creature, in all of its details, was specifically designed to fill a particular niche in nature or was the result of pure adaptation to its environment. The more the comparative anatomists and paleontologists learned, the less the theistic, ad hoc explanation of designed adaptation fitted the facts. A deistic mode of ascribing structure to natural laws that would produce types and be accountable for the unity of the type attempted to escape the contradiction. As it turned out this idealistic morphology concept of structural variation ultimately provided a perfect stepping stone to the theory of common descent (see also McPherson, 1972; Winsor, 1976b).

459

The satisfaction which idealistic morphology gave by supplying an ordering principle was more than offset by two great weaknesses. Not being based on evolution, it made little effort to discriminate between structural similarities due to common descent (homologies) and those due to similarity of function (analogies) and thus often produced highly heterogeneous assemblages. More importantly, being devoid of explanatory capacity, it was not at all able to account either for the origin of archetypes or for their mutual relationships. What satisfaction idealistic morphology gave was primarily esthetic, which is why it had such appeal during the romantic period in the first half of the nineteenth century.

Cuvier

The most important event in the history of morphology was perhaps the founding of the Paris Natural History Museum by Buffon. It was the world center of morphological research during the ensuing one hundred years. Daubenton, who did the anatomical work for Buffon's *Histoire naturelle*, stressed the unity of the plan but otherwise limited himself largely to description. An entirely new spirit is found in Vicq-d'Azyr's work (1748–1794) (Russell, 1916). He was the first anatomist to adopt a consistent comparative approach. Unlike Daubenton's work, which was confined to the study of external morphology and the principal visceral organs (lungs, stomach, and so on), Vicq-d'Azyr's interest encompassed all anatomical systems, not just a few selected ones. But perhaps his greatest achievement was to have created a close bond between anatomy and physiology. Cuvier's functional approach was clearly the result of Vicq-d'Azyr's influence.

While virtually all outstanding anatomists prior to him had been physicians by training, Georges Cuvier (1769–1832) was first and foremost a zoologist.[18] His stress on physiology was not motivated by an interest in human physiology but by his conviction that structure could be understood only through the study of its relation to function. Description for Cuvier was necessary to provide the raw material for broad generalizations. The two morphological generalizations for which Cuvier is most famous are the *principle of the correlation of parts* and the *principle of the subordination of characters.*

According to the principle of the correlation of parts, each organ of the body is functionally related to every other organ, and the harmony and well-being of the organism results from their cooperation. "It is in this mutual dependence of the functions and the aid which they reciprocally lend one another that are founded the laws which determine the relations of the organs and which possess a necessity equal to that of metaphysical or mathematical laws, since it is evident that the seemly harmony between organs which interact is a necessary condition of existence of the creature to which they belong and that, if one of these functions were modified in a manner incompatible with the modifications of the others, the creature could no longer continue to exist" (Coleman, 1964: 68). This principle enabled Cuvier to explain the gaps found between various animal groups, particularly between his four great embranchements. Intermediate organisms would have to possess combinations of organs that would be disharmonious, and they would not be viable.

In a practical application of this principle, Cuvier stated that on the basis of only a small part of a fossil (he thought mostly in terms of mammals), one could reconstruct the entire organism. As he said: "At the sight of a single bone, of a single piece of bone, I recognize and reconstruct the portion of the whole from which it would have been taken. The whole being to which this fragment belongs appears in my mind's eye" (Bourdier, 1969: 44). Although this is surely a fertile heuristic working rule, it has also serious limitations. Thus, it tricked Cuvier into identifying the skull of a Chalicothere as that of a horse and its foot (claws) as that of a sloth, being unaware of the existence of the fossil family of Chalicotheres which has this unusual combination of characteristics.

Cuvier had such an exalted concept of the perfection of the correlation of parts that this was one of the major reasons why he could not conceive of any evolutionary change. Actually, he never undertook any studies of variation of the correlation of parts either within species or within higher taxa which would have shown him at once that the correlation is not nearly as perfect as he claimed.   *461*

Cuvier's second great principle, although in some ways just an application of his first one, is the subordination of characters. This is basically a taxonomical principle, which permitted him to establish hard and fast rules by which to recognize and rank the higher taxa of animals (see Chapter 4). The two principles together permitted Cuvier to demonstrate the nonexistence of a gradual chain of being and to replace it instead by his four great embranchements (phyla) which had no special connections with each other.

Buffon's unity of plan became in Cuvier's hands the type concept. This continued to dominate the teaching of zoology for a hundred years after Darwin, as is evident from any elementary textbook published during that period. There were two reasons why Cuvier's influence on morphology was so impressive and lasted so long. The first is that his sober empirical approach, free of all metaphysical speculations, appealed to an age that was rebounding from the excesses of *Naturphilosophie*. The second reason was Cuvier's biological approach. His was an adaptational morphology, which stressed the functional meaning of all structures in relation to the mode of living of each organism. It was, one can say, an almost ecological approach. At the same time it was felicitously combined with the recognition that all adaptive variation was constrained by the unity of the type.

There were, however, some important questions that were rather conspicuously sidestepped by Cuvier. The first was, How far does the unity of the type extend? Is there not as much difference within some of his embranchements, for instance among the Radiata, as between them? Far more bothersome was another question, What is the meaning of these four types, and what is their origin? Why are there exactly four types and not, rather, ten or only a single one? The question of the origin and the meaning of the great morphological types remained a deep concern of the comparative anatomists for decades to come. It was, of course,

Darwin who succeeded in answering the questions that Cuvier had bequeathed to his followers.

Geoffroy Saint-Hilaire

The interests of the other great French morphologist of the period, Étienne Geoffroy Saint-Hilaire (1772–1844), even though he was a colleague and friend of Cuvier's for nearly forty years, developed in a very different direction.[19] In contrast to Cuvier, he was almost exclusively a morphologist and, as is indicated in his great theoretical work, *Philosophie anatomique* (1818), his ideal was the establishment of a pure morphology. He carried comparison and the establishment of homologies to a much greater refinement than did his predecessors. Man was no longer the great type with which everything else was compared. Indeed, Geoffroy extended the systematic comparison right through the entire class of vertebrates.

Two principles which he established are guidelines for decisions on homology to the present day. One is the *principle of connections,* which states that, when one is in doubt as to the homology of structures in widely different organisms, say a fish and a mammal, "the sole general principle one can apply is given by the position, the relations, and the dependencies of the parts, that is to say, by what I name and include under the term of connections." This principle, says Geoffroy, is a sure guide when a structure is greatly modified by a functional transformation, for "an organ can be deteriorated, atrophied, annihilated, but not transposed." For instance, the humerus will always lie between the shoulder articulation, and the bones of the lower arm (radius and ulna). An auxiliary principle is that of "composition," which states that all homologous structures are composed of the same kinds of elements and this facilitates identifying specific components in a series of elements, let us say individual hand bones. The entire modern method of establishing homologies throughout the vertebrate series or throughout the arthropods is ultimately based on Geoffroy's method. Geoffroy's reputation would be shining even more brightly if he had not also promoted several other rather fantastic ideas.

There is little doubt that he was strongly influenced by the writings of Oken and other German *Naturphilosophen* and idealistic morphologists. This induced him, in opposition to Cuvier, to extend the unity of plan to all animals, both vertebrates and inver-

462

tebrates. He thus adopted Goethe's great ideal of a single proto-type for the whole animal kingdom. Geoffroy and some of his younger friends claimed that one could "homologize" (as we would now call it) the anatomy of the squid, a mollusk, with that of a vertebrate by turning the squid upside down and partly inside out. As Geoffroy put it, "Every animal is either outside or inside his vertebral column." In a public debate in Paris before the Academy of Sciences on February 15, 1830, this theory was decisively refuted by Cuvier. Geoffroy had made no distinction between similarities due to relationship and those due to function (convergence). Cuvier summarized his demonstration by stating that "the cephalopods do not form a connection with anything. They are not the result of evolution from other kinds of animals and they have not led to the development of any kind of animals superior to them." This totally demolished Geoffroy's claim that he could reduce the four branches of the animal kingdom recognized by Cuvier to a single one.[20]

463

In contrast to Cuvier, who believed that function determines structure, Geoffroy held that structure determines function. If changes of structure occur, says Geoffroy, they will cause changes in function. "Animals have no habits but those that result from the structure of their organs; if the latter varies, they vary in the same manner all their springs of action, all their faculties and all their actions" (Russell, 1916: 77). The bat is forced to live in the air as a result of the modification of its hand. The thoroughly unbiological assumption that structure precedes function was curiously revived by the mutationists after 1900: Cuénot, de Vries, and Bateson claimed, between 1900 and 1910, that organisms are exposed to the mercy of their mutations but that some mutations "preadapt" them to new behaviors and adaptive shifts.

Geoffroy's writings are full of original ideas. He was the author of the "Loi de balancement," according to which the amount of material available during development is limited so that, if one structure is enlarged, another one has to be reduced in order to maintain an exact equilibrium. "The atrophy of one organ turns to the profit of another; and the reason why this cannot be otherwise is simple, it is because there is not an unlimited supply of the substance required for each special purpose." Roux's "struggle of the parts" was a later revival of this thought (also stated by Goethe in 1807) and supported in our time by Huxley and Rensch, except now in terms of selection pressures.

Richard Owen

Owen (1804–1892) was the last great idealistic morphologist of the pre-Darwinian period.[21] His major work, *On the Archetype and Homologies of the Vertebrate Skeleton* (1848), was an attempt to produce an internally consistent theory of morphology. It was an eclectic system incorporating Cuvier's teleology, Geoffroy's principle of connections, Oken's idea of the serial repetition of parts, and some aspects of Lamarck's dual nature of evolution (translated into static terms). The concept of an archetype was carried by him to an extreme, the vertebrate archetype being strictly segmental, including even the entire skull. His endeavor to determine the homology of every bone in the vertebrate skull caused him to provide an elaborate nomenclature of these bones, much of which is still in use long after Owen's theories have been forgotten.

Another terminological proposal of Owen's likewise had a lasting impact. One of the greatest weaknesses in the work of the idealistic morphologists was that their conclusions were largely based on similarities revealed by comparison. However, they failed to make a terminological distinction between those analogies that were due to similarity of function and those others that seemed to be of a different and more fundamental kind, already known to Cuvier (Chapter 7). Owen separated the two as follows: "Analogue. A part or organ in one animal which has the same function as another part or organ in a different animal"; and "Homologue. The same organ in different animals under every variety of form and function." The difficulty, of course, was to determine which was the "same" organ, and this is where Geoffroy's principle of connections was particularly helpful.

Homology and Common Descent

The idealistic morphologists were completely at a loss to explain the unity of plan and, more particularly, why structures rigidly retained their pattern of connections no matter how the structures were modified by functional needs. As Darwin rightly said, "Nothing can be more hopeless than to attempt to explain this similarity of pattern in members of the same class, by utility or by the doctrine of final causes" (*Origin:* 435). The real explanation, says Darwin, is as simple as the egg of Columbus. All mammals, birds, or insects share the same morphological type, resulting in an extraordinary anatomical similarity, because they all have descended from a common ancestor from whom they have inherited this structural pattern. Natural selection will be constantly at work to modify the components of this pattern so as to make them most efficient for the functions they have to serve but this does not require destruction of the basic pattern.

Darwin thus replaced the archetype of idealistic morphology by the common ancestor. As a consequence homology was redefined by the Darwinians: "Attributes of two organisms are homologous when they are derived from an equivalent characteristic of the common ancestor." Darwin himself never clearly gave this definition but it is implied in his discussions. Owen, lacking an explanation for the existence of homologies, was forced to define homology in terms of Geoffroy's principle of connections. To retain this awkward definition in evolutionary biology would have been absurd and this is why modern students (Simpson, Bock, Mayr) have redefined homology in terms of derivation from the common ancestor. To prove that this definition is met in a particular case, all sorts of evidence must be utilized including that of connections.[22] One important aspect of the evolutionary redefinition of homologous is that it is applicable not only to structural elements but to any other properties, including behavioral ones, which might have been derived by inheritance from a common ancestor.

There is one curious aspect about Darwin's treatment of morphology in the *Origin*, in the light of the prevailing thought of 1859. He repeatedly emphasizes the point that natural selection provides the answer to all the listed morphological questions. Actually, it is the theory of common descent with modification that provides the answers, as agreed to by all evolutionary morphologists of the ensuing period, while the observed phenomena did not at all shed a decisive light on the nature of the forces responsible for the modification. This is why post-Darwinian morphologists so often explained morphological changes by use and disuse or direct influences of the environment combined with an inheritance of acquired characters instead of by natural selection.

In view of Darwin's enormous stress on the importance of morphology, one is surprised how little (pp. 434–439) he says on the subject in the *Origin*. In part this is due to the fact that he had by implication stated his evolutionary-morphological principles already in his barnacle monograph (Ghiselin, 1969: 103–130); in part the explanation is also that this is a subject which Darwin had not yet reached in his big manuscript (*Natural Selection*), when he abandoned it in 1858 to write the *Origin*. Hence, during the rush of preparing the manuscript of the *Origin*, all he could do

465

was to sketch out the barest outline of the problems of morphology. It was left to his followers, particularly Gegenbaur, Haeckel, and Huxley, to fill the gap.

Morphology after 1859

Morphology somewhat fell into disrepute after Cuvier and Geoffroy, particularly in Germany and France. Either it was considered simply a handmaiden of (medical) physiology, or it was denounced as being purely descriptive (not making use of experiments), or on the contrary it was considered as too speculative, particularly as practiced by the *Naturphilosophen*.[23] The field was in the search of a new identity at the time when the *Origin* was published. Darwin's theory of common descent gave new meaning to morphological research, particularly in zoology, as indicated by the fact that in the decades (one is tempted to say, in the century) after 1859 the emphasis of evolutionary biology was almost exclusively on phylogeny. It is most instructive to compare the first edition of Gegenbaur's great textbook of comparative zoology (published in 1859 just before the *Origin*) with the second edition, published eleven years later. There is remarkably little difference except that terms like "morphological type" or "archetype" were replaced by "common ancestor" (Coleman, 1976).

What Geoffroy and Owen had started—the search for the homology of even the most insignificant element of anatomy—was carried out with ever-increasing enthusiasm and extended to all phyla of the animal kingdom. The leading zoologists, from Haeckel and Huxley on, thought about nothing as much as about phylogeny and about the reconstruction of common ancestors. In fact, from 1859 to about 1910 most of zoology was actually comparative anatomy and phylogeny. This intensive activity produced a marvelous descriptive knowledge of the animal kingdom and led to the discovery of many previously unknown types of animals, including new classes and even phyla. As descriptive as much of this work was, the triumphs of this methodology should not be underrated. What could be more fascinating than the derivation of the mammalian middle-ear bones from reptilian jaw elements, or the limbs of the tetrapods from paired rhipidistian (fish) fins, or the muscles that move the eyeballs from segmental muscles, to mention some interesting homologies of the vertebrates. Perhaps even more intriguing and in part still controversial are homologies among the invertebrates, particularly the segmental appendages (extremities, mouth parts, and so on) of the arthropods.

As far as the theory of phylogenetic morphology was concerned, much of this comparative research still reflected pre-evolutionary thinking. The arguments to a very considerable extent were still the arguments and questions of Geoffroy and Cuvier, translated into evolutionary terms. For instance, there was still the problem of the subordination of characters, or as it was now called, the weighting of characters. When it came to establishing the phylogeny of the invertebrates, it was still argued "which character had primacy," the presence and form of the coelom, as was long insisted upon by the British zoologists, or the ontogeny of the mouth (Protostomia, Deuterostomia), as believed by the Vienna school.

467

All sorts of methodological weaknesses became apparent during these controversies, inducing many zoologists to transfer their attention from problems of ultimate causation to those of proximate causation. One school, arising from embryology, attempted to produce a physiological, if not completely mechanical, explanation of animal form (His). Roux's *Entwicklungsmechanik* was the logical culmination of this trend. Another school stressed the functional aspects of structure, an approach that was particularly fruitful for structures that have to do with locomotion (Böker, 1935; Gray, 1953; Alexander, 1968). The most distinguished representative of purely functional morphology was d'Arcy Thompson (*On Growth and Form*, 1917). It is perhaps not a coincidence that this work contains a lengthy introductory statement in which Darwinism (natural selection) is rejected. What united His, Roux, and d'Arcy Thompson was that they only saw the proximate causation of form, not only ignoring but virtually denying the evolutionary causation. As Raup (1972:35) has rightly criticized it, "In terms of modern evolutionary biology, Thompson was contending that the genetic make-up of an evolving organism is . . . [so] plastic, that it can be altered completely as part of species level adaptation to immediate functional problems." That natural selection is responsible for incorporating in the genetic program the growth constants responsible for geometrically interesting forms of snails, ammonites, foraminifera, is, of course, fully realized in the more recent literature.

The fact that the explanation of adaptation was one of the main interests of Darwinian biology was almost totally ignored by the post-Darwinian morphologists. Phylogeny, homology, and the reconstruction of the common ancestor—conceptually very little different from Owen's archetype—described their sphere of interests for the hundred years after 1859. In fact, authors like Naef, Kälin, Lubosch, and Zangerl virtually returned to the principles of idealistic morphology. Almost the only exception was Hans Böker (1935; 1937), who in a superb functional-evolutionary morphology asked all the right questions, as seen in hindsight, concerning the adaptive value of structures and their changes, but based his interpretations unfortunately on the wrong evolutionary philosophy (neo-Lamarckism). As a result his visionary study failed to have any effect.

468   It was not until the 1950s that a new movement got started which sometimes refers to itself as *evolutionary morphology.* Instead of adopting the backward look to the common ancestor so characteristic of classical comparative anatomy, the representatives of the new school start with the ancestor and ask what evolutionary processes were responsible for the divergence of the descendants. Why and how did the ancestral type give rise to new morphological types? To what extent was a change in niche occupation or, indeed, the invasion of an entirely new adaptive zone responsible for the anatomical reconstruction? What was the nature of the selection pressures? Was behavior the pacemaker of the ecological shift? What was the nature of the population in which the decisive shift occurred? These are the kind of questions asked by this school. This approach takes all that for granted which the previous generation still had to establish: phyletic sequences, homologies, and the probable structure of the common ancestor. Evolution for them is not merely genealogy but the totality of the processes that are involved in evolutionary change. The new approach is clearly a borderline field, since it has built bridges both to ecology and to behavioral biology. The new questions which this approach has opened up promise to keep morphology busy and exciting for many years to come.[24]

The solution to perhaps the greatest problem of morphology requires a bridge to genetics, a bridge which at this time cannot yet be built. I am referring to the origin and the meaning of the great anatomical types, already known to Buffon under the name "unity of plan." Within the mammalian *Bauplan,* for instance, such strikingly different functional types evolved as whales, bats, moles, gibbons, and horses, without any essential change of the mammalian plan. Why is the chordate type so conservative that the chorda still is formed in the embryology of the tetrapods and gill arches still in that of mammals and birds? Why are the relations of structures so persistent that they can form the basis of Geoffroy's principle of connections? Clearly this is a problem for developmental physiology and genetics, indicated by such terms as the cohesion of the genotype or the homeostasis of the developmental system, terms which at this time merely conceal our profound ignorance.

A new frontier was opened up when morphological studies were expanded to include microstructures. The study of cells revealed that they were built exactly the same way in animals and plants (except for the presence of chloroplasts in the cells of green plants), providing the first convincing evidence for the monophyly of animal and plant kingdoms. At the same time the study of the cells of lower organisms revealed the drastic break between higher organisms (eukaryotes), which have well-developed nuclei and mitosis, and lower organisms (prokaryotes such as bluegreen algae and bacteria), which lack nuclei and well-organized chromosomes.

When the analysis was carried still one step further, to the morphology of the macromolecules, a still further frontier was opened which permitted an endless array of new kinds of investigations. It is now possible, for some of the better analyzed macromolecules (like cytochrome C), to construct phylogenetic trees from the lowest eukaryotes to the highest animals and plants, sometimes even including the prokaryotes. Not surprisingly, these studies consistently confirm the results of macromorphological studies, but molecular phylogeny is sometimes able to shed light on previously obscure lines of relationship.

Embryology as Evidence for Evolution and Common Descent

The last area which supplied Darwin with evidence for evolution was embryology. Darwin (*Origin:* 442) lists five sets of facts in embryology that are exceedingly puzzling unless one adopts the theory of descent with modification. He placed great value on "the leading facts in embryology, which are second in importance to none in natural history" (p. 450), as well as on his own interpretation of these facts. "Hardly any point gave me so much satisfaction when I was at work on the *Origin,* as the explanation of the wide difference in many classes between the embryo and the adult animal, and of the close resemblance of the embryos within the same class. No notice of this point was taken, as far as I remember, in the early reviews of the *Origin*" (*Auto.:* 125). In letters to Gray and Hooker, he likewise complains that neither his reviewers nor his friends had paid attention to his embryological arguments,

469

even though they are "by far the strongest single class of facts in favor of" evolution.

Embryology gave Darwin one of his strongest anticreationist arguments. If species had been created, their ontogeny should lead them by the most direct route from the egg to the adult stage. But this is not at all what one finds, since quite extraordinary detours are usually encountered during development. "There is no obvious reason why, for instance, the wing of a bat, or the fin of a porpoise, should not have been sketched out with all the parts in proper proportions, as soon as any structure became visible in the embryo" (*Origin:* 442). Why should the embryos of land-living vertebrates go through a gill-arch stage? Why should young baleen whales develop teeth, and the higher vertebrates have a notochord? These are only a few of the countless embryonic structures that can be understood only as part of the phyletic heritage.

How did Darwin explain these developmental detours? His interpretation was based on his ideas on the origin of variation. He believed that "the adult differs from its embryo, owing to variations supervening at a not early age, and being inherited at a corresponding age. This process, whilst it leaves the embryo almost unaltered, continually adds, in the course of successive generations, more and more differences to the adult" (p. 338). In other words, Darwin bases his conclusions on the assumption that the most recent evolutionary acquisitions are due to variations that had occurred very late in ontogeny. Consequently, embryos which have not yet reached the ontogenetic stage in which these variations show ought to be more similar to each other than the adult individuals of different groups of animals that had become distinct owing to diverse new acquisitions. "Thus, community in embryonic structure reveals community of descent" (p. 449). The younger the embryos, the more similar they ought to be to each other, and by investigating and comparing embryos, one should be able to find clues to common descent. This is, says Darwin, how it was discovered that the Cirripedia belong to the crustacean class. A study of embryology often provides helpful clues on phylogeny. For instance, "The two main divisions of cirripedes, the pedunculated and sessile, which differ widely in external appearance, have larvae in all their several stages barely distinguishable" (p. 440).

In order to strengthen his argument that similarities in ontogeny are indications of common descent, Darwin—true to his method—refutes a conceivable alternate explanation. Someone might claim that the special characteristics and similarities of embryos are ad hoc adaptations for the larval existence. This is indeed possible, says Darwin, if the larvae "are active and have been adapted for special lines of life" (p. 439). But, he continues, "we cannot, for instance, suppose that in the embryos of the vertebrata the peculiar loop-like course of the arteries near the bronchial slits are related to similar conditions,—in the young mammal which is nourished in the womb of the mother, in the egg of the bird which is hatched in the nest, and in the spawn of a frog under water" (p. 440).

If Darwin's complaint that his embryological evidence in favor of evolution was overlooked is justified, it is in part because everyone's attention was distracted by a controversy of long standing. It is necessary, therefore, to consider the history of embryological thought.[25] As far back as the Greeks, it had been recognized that there was some sort of parallel between the seriation of stages in the growing embryo and the seriation of organisms from the lowest to the highest, later known as the *scala naturae*. Aristotle, for instance, classified organisms into those with a nutritive soul (plants), a nutritive and sensitive soul (animals), and those also with a rational soul (man). During the development of the embryo, these three kinds of souls, he postulated, come successively into operation. This vague idea became far more concrete toward the end of the eighteenth century, particularly with Bonnet, who carried the belief in a great chain of being to its greatest heights.

The study of this parallelism led to certain conclusions concerning the relation between ontogeny and the animal series which was formulated by Meckel (1821, I: 345) as follows: "The development of the individual organism obeys the same laws as the development of the whole animal series; that is to say, the higher animal, in its gradual evolution, essentially passes through the same *permanent* organic stages which lie below it." These developments were due to a "tendency, inherent in organic matter, which leads it insensibly to rise to higher states of organization, passing through a series of intermediate states."

It would lead to a complete misinterpretation of these ideas if one did not fully realize that there was no implication of evolution in this idea of a parallelism between the stages of ontogeny and the stages of perfection in the (static!) ladder of being. The word "evolution" still had the old meaning of the unfolding of an existing potential of the type. The French anatomist Étienne Serrès, a pupil of Geoffroy, had similar ideas. He considered "the

471

whole animal kingdom . . . ideally as a single animal which . . .
here and there arrests its own development and thus determines
at each point of interruption, by the very state it has reached, the
distinctive characters of the phyla, the classes, families, genera,
and species" (1860: 833).

All the proponents of the parallelism between ontogeny and
the *scala naturae* were essentialists. For them the *scala naturae* consisted of a seriation of types, and they believed they could discover
the same sequence of types in ontogeny. The end of ontogeny was
"the permanent stage," the temporary halting point of Serrès. The
theory of a parallelism between the stages of ontogeny and the
stages of the *scala naturae* was later called the Meckel-Serrès law.
In post-Darwinian days, when the conceptualization on which it
was based (*scala naturae,* essentialism, *Naturphilosophie*) had been
superseded, the Meckel-Serrès law was often grossly misrepresented. The most misleading of the misinterpretations was the replacement of the words "permanent stage" by the word "adult."
Most of those who had adopted the Meckel-Serrès law were competent embryologists and knew perfectly well that no stage in the
development of mammalian or chick embryo was "identical with"
(terminology actually used by one of their opponents!) an adult
reptile or fish. But since mammals and birds have no gills and
breathe through lungs, the gill arches represented the fish stage
on the *scala naturae.* To the best of my knowledge none of the
members of the Meckel-Serrès school had ever claimed that the
ontogenetic stages represented the adult stages of the lower types.
Nor did they hold the beliefs in causation and chronology that
became associated with the term "recapitulation" in post-Darwinian days.

It must further be remembered that the 1820s and 30s saw
the climax in the great controversy between the adherents of a
single *scala naturae* (or a single type for the whole animal kingdom) and the Cuvierians, with their thesis of four entirely independent embranchements. K. E. von Baer (1792–1876), who more
or less independently had arrived at views similar to those of Cuvier, not only held that each of the animal phyla had its own ontogeny but rejected the whole idea of a parallelism between ontogeny and level of organization.

He devoted a major part (the fifth scholium) of his famous
animal embryology (1828) to this refutation. In this scholium he
presents a parody of Lamarck's ideas and rejects evolution in any
form whatsoever; he rejects any idea of an animal series, stating

472

that all animals are grouped around a certain number of archetypes which coincide with Cuvier's four embranchements and very specifically rejects "the prevalent notion that the embryo of higher animals passes through the permanent forms of the lower animals." He repeats his conclusion for the vertebrates: "The embryos of the Vertebrata pass, in the course of their development, through the permanent forms of no known animals whatsoever."

To take the place of these refuted ideas, he proposes his own laws of individual development:

(1) That the more general characters of the large group of animals to which the embryo belongs appear earlier in the development than the more special characters.

(2) From the most general forms, the less general are developed and so on, until finally the most special appear.

(3) Every embryo of a given animal form, instead of passing through the state of the other definite forms, rather becomes separated from them all.

(4) Fundamentally, therefore, the embryo of a higher form never resembles the adult of any other animal form but only its embryo.

What actually happens in ontogeny, says von Baer (I: 153), can be summarized as follows: "There is gradually taking place a transition from something homogeneous and general to something heterogeneous and special." It is this statement which inspired Spencer's theory of evolution, but it is, of course, rather misleading as far as ontogeny is concerned. Why should the fish-like gill arches be something "homogeneous and general" in mammalian ontogeny? The same question applies to the teeth of the baleen whale embryos and other instances of recapitulation. Von Baer presumably considered these characteristics to be integral parts of the archetype and hence "general."

When Darwin began to read up on embryology after 1838, he had the choice between the theory of parallelism of the *Naturphilosophen* and von Baer's theory of straight-line differentiation. In his "Sketch" of 1842, he seems to be close to von Baer's position, stating that at an early stage of ontogeny "there is no difference between fish, bird, etc. etc., and mammal . . . it is not true that one passes through the form of a lower group." And he reaffirms in 1844 "that the young mammal is at no time a fish . . . or that the embryonic jellyfish is at no time a polype."

In the 1840s and 50s Louis Agassiz expanded the Meckel-

Serrès laws into a threefold parallelism by a progressionist interpretation of the fossil record: The stages of the embryo repeat
not only the scale of perfection as observed among existing types
but also the fossil succession: "It may therefore be considered as
a general fact . . . that the phases of development of all living
animals correspond to the order of succession of their extinct representatives in past geological times. As far as this goes, the oldest
representatives of every class may then be considered as embryonic types of their respective orders or families among the living"
474   (1857; 1962: 114). This thought greatly intrigued Darwin, as is
evident from his comments in the *Origin* (p. 338); "Agassiz insists
that ancient animals resemble to a certain extent the embryos of
recent animals of the same classes; or that the geological succession of extinct forms is in some degree parallel to the embryological development of recent forms. I must follow Pictet and
Huxley in thinking that the truth of this doctrine is very far from
proved. Yet, I fully expect to see it hereafter confirmed . . . For
this doctrine of Agassiz accords well with the theory of natural
selection." Perhaps also on the basis of his work on the Cirripedia
Darwin now seems to have moved considerably closer to the
Meckel-Serrès doctrine. But, as usual, Darwin was rather cautious
in his generalizations.

The same cannot be said of his exuberant follower Ernst
Haeckel, who transformed the Meckel-Serrès statement of parallelism into an evolutionary law. In 1866 he published his *biogenetic
law* (theory of recapitulation), according to which "ontogeny is a
concise and compressed recapitulation of phylogeny, conditioned
by laws of heredity and adaptation." Fritz Müller had independently come to a similar conclusion (1864): Ontogeny repeats
phylogeny because phylogeny is the cause of the ontogenetic
stages! Consequently, an analysis of ontogeny will tell us all about
phylogeny, that is, about common ancestry. If true, it would be a
most wonderful heuristic principle.

With Darwin's silent blessings (1872: 498) and Haeckel's enthusiasm, the theory of recapitulation was immensely popular and
successful in the three or four decades after 1870. It led to a
splendid flowering of comparative embryology and was responsible for many spectacular discoveries, for instance, by Kowalewsky,
that the tunicates are chordates[26] and that the relationship of the
major phyla of the animal kingdom is rather different from previous conceptions (the Protostomia-Deuterostomia phylogeny).
Embryology also became an indispensable tool in establishing oth-

erwise uncertain homology. By the end of the century, various excesses as well as a growing interest in proximate causation led to disenchantment and to the eventual rejection of recapitulation, particularly in its extreme form.

The question has been asked recently, How could recapitulation have received such unqualified acceptance in the Haeckelian period in spite of von Baer's cogent arguments against the Meckel-Serrès law? Had von Baer's writings been overlooked? Certainly not, because he was widely quoted (Ospovat, 1976). Furthermore, his arguments did have considerable weight, because most authors (including Darwin) rejected the claim that ontogeny is the recapitulation of the *adult stages* of the ancestors. The majority of phylogenists adopted a mild version of recapitulation which merely states that the embryo during ontogeny passes through a series of stages which correspond to those of the ancestors, as is indeed often correct. Most of von Baer's arguments refuting the thesis that embryos pass through the adult stages of the ancestors were not applicable to the modified version. Indeed, the difference between the opposing theories was much less than usually claimed.

475

Von Baer's laws were not widely adopted because they were largely descriptive and sterile from the explanatory point of view, while the thesis of recapitulation was wonderfully heuristic; because von Baer's attempt to refute the parallelism of ontogeny and the animal series was part of a more extensive argument against evolution, hence after 1859, interpreted as part of von Baer's anti-evolutionism; because von Baer believed in a teleological, necessary progression from lower to higher and from homogeneous to heterogeneous; and because the claim that ontogeny always goes from simple to more complex could easily be refuted for the most conspicuous cases of recapitulation. Also, von Baer's interpretation was pervaded by the spirit of *Naturphilosophie*, which by 1866 had become quite unfashionable, even though still upheld by Serrès and a few idealistic morphologists.

When Haeckel's biogenetic law lost its appeal, attempts were made to return to von Baer's laws (for example, de Beer, 1940; 1951), but it was evident that this was not the right solution either. It was unavoidable that one had to reject both recapitulation and von Baer's laws.

How does the modern biologist explain the presence of gill arches in the ontogeny of mammals? To be frank, until the physiology and biochemistry of developmental systems is better understood, only a tentative answer is possible. One can suggest that the genetic program for development consists of a set of such complex interactions that it can be modified only very slowly. This is demonstrated particularly convincingly for the so-called vestigial organs, for instance the remnants of the posterior extremity of whales whose ancestors entered water around 55 million years ago. Darwin's thesis that evolutionary new acquisitions are superimposed on the existing genetic structure, even though frequently attacked, has a correct nucleus. Once the genetic basis of a structure is thoroughly incorporated into the genotype and forms part of its total cohesion, it can be removed only at the risk of destroying the entire developmental system. It is less expensive to keep the complex regulatory system of mammalian embryogenesis intact, even though (as a by-product) it produces unneeded gill arches, than to break it up and produce unbalanced genotypes. Our understanding of developmental regulation is far too incomplete to rule out the possibility that late evolutionary acquisitions are indeed "added" to the genotype more loosely than characteristics inherited from remote ancestors. We do not have a recapitulation of ancestral types, but we do occasionally have in ontogeny the recapitulation of individual ancestral characters and developmental pathways. How they are to be identified and how to explain their developmental physiology are matters of current discussion.

Chapter XIII completes Darwin's presentation of the evidence for evolution by common descent. Two aspects of this marshaling of fact and argument are particularly noteworthy. One is the ever-repeated emphasis that all the known facts of natural history are perfectly consistent with evolution by common descent, but that many of them cannot be reconciled at all with creation. The other aspect is that Darwin's theory settled at once numerous arguments in all branches of biology which had seemed hopelessly puzzling for many generations. It is this capacity of the theory of evolution which has induced biologists to refer to it as the greatest unifying theory in biology. The areas which had already supplied Darwin with the most telling evidence for evolution—paleontology, classification, biogeography, morphology, and embryology—have continued to supply the most convincing proofs of evolution up to modern times.[27] Almost the only recent, but in fact highly important, addition is molecular biology.

11 ⌁ The causation of evolution: natural selection

BY THE SUMMER of 1837 Darwin was a convinced evolutionist. It had become clear to him that species are modifiable and that they multiply by natural processes. But how these changes occur and what factors are responsible for the transformation of species was at first very puzzling to him. Luckily for the historians, Darwin put down all of his speculations and brain waves in little notebooks, and the rediscovery of these notebooks has permitted a reconstruction of the rather circuitous pathway of Darwin's hypothesizing. Like Lyell, Darwin had speculated on the introduction of new species on the *Beagle,* when he was still a creationist, and by necessity had adopted a saltationist model (for example, for the origin of the second species of South American *Rhea,* or "ostrich"). In these early speculations Darwin was dealing with pairs of sympatric species occurring in the plains of Patagonia. Here Darwin neither could see isolation nor, in the case of successive species, could he readily apply Lyell's explanation of the filling of a vacant niche by a new species. He could find no evidence for a change of climate, hence no need for extinction of the earlier species. Yet extinction had occurred in the case of the giant llama, its place now being filled by the guanaco. Kohn (1981) and others have described this stage in Darwin's thinking well.

In July of 1837 Darwin started the first of four notebooks labeled B, C, D, E by him and referred to as *Notebooks on Transmutation* (de Beer, 1960). The thoughts recorded in these notebooks reflect in the most wonderful manner the tortuous path by which some fifteen months later Darwin arrived at his theory of evolution by natural selection. Since this is a highly complex theory, as we will see, it could not be conceived at a single moment, even though Darwin remembers a definite date when he experienced a major illumination. In his autobiography (1958: 120) he telescopes the slow and involved development of the theory into a single moment described in a memorable passage:

> In October [actually September 28] 1838, that is, fifteen months after I had begun my systematic inquiry, I happened

to read for amusement Malthus on Population, and being well prepared to appreciate the struggle for existence which everywhere goes on from long-continued observation of the habits of animals and plants, it at once struck me that under these circumstances favorable variations would tend to be preserved, and unfavorable ones to be destroyed. The result of this would be the formation of new species. Here, then, I had at last got a theory by which to work.

478 Just exactly what happened on September 28, 1838? From his notebooks it is clear that it was one particular sentence in Malthus which set the intellectual avalanche in Darwin's mind in motion: "It may safely be pronounced, therefore, that population, when unchecked, goes on doubling itself every 25 years, or increases in a geometrical ratio."

The causal chain of the theory of natural selection is very logical, as will be shown below. Darwin, however, did not reach it in any simple manner but rather by developing and subsequently rejecting a series of alternate theories. Yet, he was able to retain valid components of the rejected theories and use them when eventually constructing the theory of natural selection. Even that theory was not conceived and completed on a single day. Schweber (1977) attributes much of the change in Darwin's thinking to his reading of Brewster and Quetelet in the two to three months before the Malthus episode. Kohn (1981) also seems to think that most of the theory was assembled by the end of September 1838 (but under different influences than suggested by Schweber). Hodge (1981) makes it plausible that the most decisive change in Darwin's thinking occurred in November 1838. Ospovat (1979), by contrast, thinks that Darwin's concept of natural selection and the nature of adaptation was still rather immature in 1838 and required many more years of maturation to the form recorded in the *Origin* (1859), that is, the form in which the world encountered it. On one point all these authors are in agreement, namely, that the theory evolved slowly and piecemeal. Indeed, even in his later writings Darwin is often inconsistent when referring to selection and he makes statements occasionally that are incompatible with other statements made almost simultaneously.

In the three years after his return from the *Beagle* voyage Darwin read perhaps as much nonbiological literature as he did books and papers on animals and plants (Herbert, 1974; 1977; Manier, 1978). It is evident that Darwin did not live in an intellectual vacuum but was at all times in active contact with the ideas that formed the zeitgeist of his period. Not surprisingly, this has led to the question to what extent Darwin's new ideas had originated, so to speak inevitably, as the product of his scientific findings, and to what extent he had simply adopted or modified ideas existing among his contemporaries. Biologists, on the whole, tend to minimize external influences, while nonbiologists, historians of ideas, and social historians tend to go to the other extreme.

The name "Malthus" has induced a school of social historians to propose the thesis that it was Malthus's social theory which gave Darwin his theory of evolution by natural selection (see below), an interpretation that is vigorously opposed by the biological historians. However, they, in turn, have major disagreements of interpretation among themselves, as I have pointed out. The reason for this is the extraordinary complexity of Darwin's explanatory paradigm. In the physical sciences the crucial component of a new theory is usually provided by a single factor, whether it be gravity, relativity, the discovery of the electron, or what else it might be. Biological theories, particularly those in the field of evolutionary biology, are, by contrast, highly complex. Darwin's theory of evolution by natural selection, for instance, has eight major components, several of which can again be subdivided, as we will see. More importantly, it is the interpretation of the interaction of its components that is usually decisive in a biological theory. In order to determine exactly what Darwin owes to Malthus, it is necessary to dissect carefully Darwin's explanatory model. The nature of his model can be reconstructed from the first five chapters of the *Origin* entitled "Variation under Domestication," "Variation under Nature," "Struggle for Existence," "Natural Selection," and "Laws of Variation."

THE LOGIC OF THE THEORY OF NATURAL SELECTION

Darwin's theory consisted of three inferences based on five facts derived in part from population ecology and in part from phenomena of inheritance.

Fact 1: All species have such great potential fertility that their population size would increase exponentially (Malthus called it geometrically) if all individuals that are born would again reproduce successfully.

Fact 2: Except for minor annual fluctuations and occasional major fluctuations, populations normally display stability.

Fact 3: Natural resources are limited. In a stable environment they remain relatively constant.

Inference 1: Since more individuals are produced than can be supported by the available resources but population size remains stable, it means that there must be a fierce struggle for existence among the individuals of a population, resulting in the survival of only a part, often a very small part, of the progeny of each generation.

These facts derived from population ecology lead to important conclusions when combined with certain genetic facts.

Fact 4: No two individuals are exactly the same; rather, every population displays enormous variability.

Fact 5: Much of this variation is heritable.

Inference 2: Survival in the struggle for existence is not random but depends in part on the hereditary constitution of the surviving individuals. This unequal survival constitutes a process of natural selection.

Inference 3: Over the generations this process of natural selection will lead to a continuing gradual change of populations, that is, to evolution and to the production of new species.

The question that the historian of science must ask is which of these facts were new with Darwin, and if none were new, why others before him had not made the same inferences? He must also ask in what sequence Darwin acquired his various insights and why Malthus's reference to the exponential increase of populations proved to be so crucial for the final putting together of Darwin's logical framework.

Before we analyze Darwin's theory in detail, a few statements must be made about Darwin's frame of mind in the critical period from 1837 to 1838. His general reading had convinced him of the importance of the gradual nature of all changes. He emphatically rejected sudden origins. *Natura non facit saltus* (nature makes no leaps) was as much his motto as it had been Lamarck's. It was also fully consistent with Lyell's anticatastrophism (see Chapter 7).

The second point to be kept in mind is Darwin's original preoccupation with diversity. Darwin always had a theory about everything, and long before he conceived the theory of natural selection he had a theory on the formation of species on islands. His theory of speciation was that, if a group of animals is isolated

480

from the main body of the species population, they gradually become different under the impact of the new conditions until they become a different species. With reference to his early theory, Darwin proclaimed "my theory, very distinct from Lamarck's" (B: 214), referring to his understanding of Lamarck's "evolution by willing." Actually, his theory seems to have been very much the same as later neo-Lamarckian theories of change caused by the local environment (Ruse, 1975a: 341). It was a strictly typological theory, the isolated species population responding to the new conditions equally and as a whole. Rather amusingly, later in life, long after he had given up this theory, Darwin accused Wagner (quite wrongly) of similar beliefs and emphasized "that neither isolation nor time by themselves do anything for the modification of species" (*L.L.D.*, II: 335–336). Many statements could be cited from Darwin's notes (Ruse, 1975) illustrating Darwin's early theory, but I shall give only two. "According to this view, animals on separate islands ought to become different if kept long enough apart, with slightly different circumstances" (B: 7). "As I have before said, *isolate* species, especially with some change, probably vary quicker" (B: 17).

481

The years 1837 and 1838 were unquestionably the intellectually most exciting period in Darwin's life. He read an enormous amount, not only geology and biology but a great deal of philosophy and "metaphysics."[1] It was in these years that Darwin moved sharply toward agnosticism, that his population thinking began to develop, and that he relied far less on soft inheritance (Mayr, 1977a). Some of this is directly reflected in his notebooks, some of it can only be inferred. It was a period of drastic reorientation for Darwin, and it is not surprising that by the end of 1838 many facts and concepts with which Darwin had long been familiar took on an entirely new meaning.

THE MAJOR COMPONENTS OF THE THEORY OF NATURAL SELECTION

There is probably no more original, more complex, and bolder concept in the history of ideas than Darwin's mechanistic explanation of adaptation. A number of scholars have attempted to reconstruct the steps by which Darwin arrived at his final model.[2] They attempted to place a whole series of facts and ideas into a new setting. Instead of following this more or less chronological method of analysis (for which I refer to the listed literature), I will take up the major concepts of which Darwin's theory is composed, and attempt to analyze their history before Darwin, as well as in Darwin's thinking.

Fertility

The exuberant fertility of living organisms was a favorite theme of authors writing about nature. To mention only authors with whom Darwin was thoroughly familiar, one finds fertility referred to in the writings of Buffon, Erasmus Darwin, Paley, Humboldt, and Lyell. Darwin was particularly impressed by the incredible reproductive rate of protozoans, about which he learned from the writings of C. G. Ehrenberg (Gruber, 1974: 162). Two factors are perhaps mainly responsible for Darwin's failure to incorporate this information in his evolutionary theorizing at an earlier date. One is that Darwin apparently did not see that organisms with relatively few offspring—like birds and mammals—had, potentially, the same exponential rate of increase as microorganisms. The other one, as we shall see below, is that within the essentialist framework high fertility is indeed irrelevant. If all individuals are identical, it does not matter what percentage of them is killed off before reproduction. Only after some others of his ideas had sufficiently matured did fertility become an important component of his theory.

Human fertility had been a concern of social thinkers for many generations, and Malthus did not claim any originality on this question. Indeed, he directly refers to Benjamin Franklin as the author of the calculations which gave him the idea of the geometric increase. Buffon and Linnaeus (Limoges, 1970: 80) long before had presented some calculations showing how quickly the world would be filled by a single species if it reproduced itself without checks. And Paley (1802: 540), one of Darwin's favorite authors, had already stated that "generation proceeds by a geometrical progression . . . [while] the increase of provision . . . can only assume the form of an arithmetic series." Had Darwin forgotten that he had once read this in Paley (who, in turn, presumably had gotten it from the first edition of Malthus)?

The Struggle for Existence and the Balance of Nature

In the two generations before Darwin, a rather drastic change was beginning to take place in man's interpretation of the harmony of nature. The natural theologians had taken up a theme already popular among certain Greek philosophers that the interaction between animals and plants and their environment displayed a beautiful harmony. Everything was ordered in such a way that it was in balance with everything else. If one species became a little too common, then something would happen to bring it back to its earlier level. The idea of a well-ordered interdependence of the various forms of life was evidence of the wisdom and the goodness of the creator (Derham, 1713). To be sure, predators destroy prey. But predators, once created, must live. Prey providently has been so designed as to provide reproductive surplus and thus sustenance for predators. The seeming struggle for existence is only a surface phenomenon; it nowhere disturbs the basic harmony of it all. The harmony of nature is so great that species can neither change nor become extinct, or else the harmony would be disturbed; nor do they need to improve, because there is no higher level of perfection.

483

The concept and even the term "struggle for existence" is quite old, being frequently referred to in the seventeenth and eighteenth centuries, as Zirkle (1941) has shown. However, this struggle was considered by Linnaeus (Hofsten, 1958), Kant, Herder, Cuvier, and many others as, on the whole, a relatively benign affair serving to make the necessary corrections in the balance of nature. As the knowledge of nature increased, an opposing interpretation, in which the fierceness of the struggle for existence was beginning to be recognized, acquired increasing cogency and popularity. It is indicated in some of the writings of Buffon and in a few statements of Linnaeus, it is expressed in some of the writings of the German historian Herder, and it was strongly emphasized by de Candolle, from whom Lyell took it over when giving a lengthy exposé of the severity of the struggle for existence. It was in the writings of Lyell that Darwin first encountered the concept of struggle for existence, not in Malthus.

The concept of the unchanging harmony of the designed world had, of course, become quite untenable when the fossil record became known, showing how many species had disappeared; and also when the studies of the geologists revealed how greatly the world had changed through the ages. Lamarck attempted to rescue the concept of the benign balance of nature by denying extinction and by explaining the disappearance of types by evolution. Accepting such an interpretation spelled the end of the belief in a static world.

Adaptation, insofar as such a concept existed, could no longer be considered a static condition, a product of a creative past, and became instead a continuing dynamic process. Organisms are doomed to extinction unless they change continuously in order to keep step with the constantly changing physical and biotic environment. Such changes are ubiquitous, since climates change, competitors invade the area, predators become extinct, food sources fluctuate; indeed, hardly any component of the environment remains constant. When this was finally realized, adaptation became a scientific problem. After 1837 Darwin's interests increasingly shifted from problems of diversity to problems of adaptation.

484　　What Darwin attempted to do was to analyze in more detail the factors that lead to the struggle for existence. The struggle is of course the result of his facts 1, 2, and 3 above, that is, the checks which the limited resources place on the potential growth of populations.[3] That there are various factors which stabilize human populations had been emphasized by authors since the seventeenth century, and perhaps earlier. In 1677 Matthew Hale listed five major checks on the increase of human populations: epidemics, famines, wars, floods, and conflagrations. Linnaeus (Gruber, 1974: 163) gave this graphic account: "I know not by what intervention of nature or by what law man's numbers are kept within fitting bounds. It is, however, true that most contagious diseases usually rage to a greater degree in thickly populated regions, and I am inclined to think that war occurs where there is the greatest superfluity of people. At least it would seem that, where the population increases too much, concord and the necessities of life decrease, and envy and malignancy towards neighbours abound. Thus it is a war of all against all."

The struggle for existence, Linnaeus's vivid description notwithstanding, rarely takes the form of actual combat. Ordinarily it is simply competition for resources in limited supply. In the days of essentialism, competition was usually described—particularly when applied to animals and plants—as competition among species. The crucial event in Darwin's mind, when reading Malthus's statement on fertility, was that he finally fully realized how important is the competition among *individuals* of the same species, and *how entirely different the consequences of this competition are from typological competition among species.*[4]

Quantification was greatly stressed by the philosophers of science (such as Herschel and Whewell) and statisticians (Quetelet) in Darwin's day. It has therefore been suggested by several authors (for example, Schweber, 1977) that Malthus's statement made such a strong impression on Darwin because it was expressed in quantitative terms ("geometrical ratio"). It is indeed possible that this added to the attractiveness of Malthus's statement for Darwin, although the "law of natural selection" is anything but either a quantitative or predictive law. This explains Herschel's later reference to natural selection as the "law of the higgledy-piggledy," a definition which illustrates well what that philosopher thought of qualitative nondeterministic generalizations.

Several recent authors have shown the gradual change in the decades before 1838 in two concepts, the nature of the struggle of existence (from benign to fierce) and the actors in competition (from species to individuals), but we do not yet have a thorough stepwise analysis. A certain amount of awareness of intraspecific competition existed already before Darwin without, however, seriously affecting the concept of the balance of nature. But this is what the reading of Malthus did to Darwin: "Even the energetic language of Decandolle does not convey the warring of the species as inference from Malthus—increase of brutes must be prevented solely by positive checks, except that famine may stop desire" (D: 134). Darwin quite rightly remarks that up to then people had always thought that animals had as many offspring as they "needed." That reproductive rate is largely independent of the vacancies in the economy of nature was a thought that was quite incompatible with the natural theologian's concept of the balance of nature. The teleological idea that members of a species had as many offspring as they needed was given up very slowly, and had to be combatted by David Lack even in recent years.

Artificial Selection

In his autobiography, as well as in his correspondence, Darwin stated a number of times that he had long been convinced of the importance of artificial selection, but that it was not until after he read Malthus that he realized how to apply this conviction to evolution. For instance, "I came to the conclusion that selection was the principle of change from the study of domestic production; and then, reading Malthus, I saw at once how to apply this principle" (letter to Wallace, 1858). This sequence has been questioned by Limoges and other recent writers, since Darwin apparently never used the word "selection" in his notebooks prior to his reading of Malthus (instead he spoke of "picking") and because the writers could not see how the study of domestication could have influenced Darwin's thinking. However, Wood (1973) and

485

Ruse (1975a) have shown that Darwin read widely in the literature of the animal breeders in the year before his theory formed, and that his underlining of crucial sentences in pamphlets by Sebright and Wilkinson (read in the spring of 1838) shows how clearly Darwin understood the principles of artificial selection, and how important he considered them.

In this connection, it is well to remember that Darwin's college friends at Cambridge, reputed to be mostly interested in riding and hunting, were the sons of country squires and had undoubtedly a considerable interest in agriculture and animal breeding. Otherwise one might well question how Darwin at this early stage would have come to discover the importance of animal breeding for his scientific interests!

Darwin owed to the breeders good ideas and bad ones. The firm belief that the mere placing of animals or plants under the conditions of domestication would increase their variability was one of the bad ideas. Fortunately, Darwin acquired from the breeders also some exceedingly valuable concepts. The most important among these was, of course, the emphasis on the individuality of every member of a herd. It was this insight rather than the practice of artificial selection which gave Darwin the key component of his theory of natural selection.

Many years after the Malthus episode, Darwin stated repeatedly that he had arrived at the concept of natural selection by analogy with artificial selection. Neither the notebook entry of September 28, 1838, nor other portions of his notebooks support this recollection. Even though his reading of the animal-breeding literature undoubtedly gave Darwin several important insights, there is much to suggest that he developed the analogy only several months later, when it occurred to him that artificial selection was a splendid experimental confirmation of natural selection. Hodge (1981) thinks that this occurred in November 1838 during a visit to Shropshire.

Darwin's new model of natural selection was purely deductive, and in order to conform to the prescriptions of the leading philosophers of his time (Comte, Herschel, and Whewell), Darwin felt that he had to provide proof for the validity of his theory—preferably, as is customary in the physical sciences, experimental proof. But how could one experiment with evolution, since evolutionary change is so slow? It is at this point that Darwin remembered the activities of the animal breeders. Artificial selection, Darwin concluded, was the greatly accelerated analogue to natural selection. It supplied the experimental proof he needed so badly. But later in life artificial selection became so important in his thinking that Darwin even thought it had given him the original inspiration for natural selection, which does not seem to have been the case.

Population Thinking and the Role of the Individual

The realization of the uniqueness of every individual was perhaps the most revolutionary change in Darwin's thinking in 1838. This uniqueness was, of course, and always had been part of his daily experience. Who is not aware that no two human beings are identical, nor any two dogs or horses? The individuality of every member of his herd is taken for granted by every animal breeder. This is what gives him the opportunity to change the properties of his herd through the deliberate selection of certain individuals to serve as sires and dams of the next generation. And yet, just because it is so commonplace, this individuality had been largely ignored by philosophers. Once Darwin had been made aware of the importance of the uniqueness of individuals, everything he did in the next twenty years reeinforced this new insight. His taxonomic studies of the barnacles were particularly convincing. Darwin found the individual variability so great that again and again he was in doubt whether two specimens were variants of a single species or two different species. This variability was not limited to the external morphology but affected also all the internal organs. Whether the contemporary emphasis on the political individual ("the rights of the individual") or that of certain schools of philosophers (Schweber, 1977) also contributed to Darwin's thinking is uncertain, and for me, rather questionable.

It was the "discovery" of the importance of the individual which led Darwin from typological to population thinking. It was this that made him realize that the struggle for existence due to competition, so vividly described by Malthus, was a phenomenon involving individuals and not species. By introducing population thinking, Darwin produced one of the most fundamental revolutions in biological thinking. As stated in Chapter 2, it is a peculiarly biological concept, alien to the thinking of the physical scientist. Adoption of population thinking is intimately tied up with a rejection of essentialist thinking. Variation is irrelevant and therefore uninteresting to the essentialist. Varying characters are "mere accidents," in the language of essentialism, because they do

487

not reflect the essence. It is most interesting to read in the essays of Darwin's critics (Hull, 1973) how puzzled they were by Darwin's stress on variation as the all-important feature of life. As far as the philosophical literature is concerned, this lesson still has been learned only by a few. Toulmin's recent volume (1974) is a conspicuous exception. Those who still question the power of natural selection invariably still use essentialist arguments. Darwin's own shift to population thinking was gradual and slow. His language is still quite typological in many of his post-1838 discussions.

488

THE ORIGIN OF THE CONCEPT OF NATURAL SELECTION

Ask any biologist what concept is most characteristically connected with Darwin's name and he will answer: natural selection. This was the great new principle Darwin introduced into biology, in fact into all of man's thinking. Yet, it has been claimed again and again that the concept was not at all new with Darwin but that it had already been often proposed, from the days of the Greeks on (see Zirkle, 1941, for example). In order to substantiate or refute the legitimacy of this claim, it is important to make a clear distinction between two processes that are consistently confused in the literature. I shall call the first one the process of *elimination.* It is the concept of the existence of a conservative force in nature which eliminates all deviations from the "normal," all those individuals that do not possess the perfection of the average type. Such elimination is quite compatible with essentialism. On the other hand, it is obvious that for the essentialist there can be no selection because the essence is unchanging and all variants are merely "accidents," such as the occurrence of monstrosities and other "degradations" of the type. Biological change was usually referred to by the term "degradation" in the seventeenth and eighteenth centuries. If a major degradation proved to be viable, it constituted a new "type." Indeed, the whole *scala naturae* was originally presented as a descending scale of diminishing perfection (degradations). Most degradations, however, are not viable; they are unable to survive or reproduce and are eliminated, thereby restoring the purity of the type. An elimination of clearly inferior or totally unfit individuals indeed does take place all the time and is part of natural selection. It is recognized in modern evolutionary biology as "stabilizing selection" (Schmalhausen, 1949; Waddington, 1957; Dobzhansky, 1970).

In his historical study Zirkle (1941) listed numerous cases of "natural selection before the origin of species," beginning with Empedocles. Virtually all the older cases cited by him merely describe elimination. This includes, for instance, Lucretius, Diderot, Rousseau, Maupertuis, and Hume. In the case of Prichard, Spencer, and Naudin the improvement is ascribed to "Lamarckian" forces, such as use and disuse, the exercise of faculties, or the influence of the environment, while elimination constantly removes the inferior types.

Eiseley (1959) vigorously promoted the thesis that Edward Blyth had established the theory of evolution by natural selection in 1835 and that Darwin surely had read his paper and quite likely had derived a major inspiration from it without ever mentioning this in his writings. The subsequent discovery of Darwin's notebooks has permitted the refutation of Eiseley's claims. More importantly, Blyth's theory was clearly one of elimination rather than of selection. His principal concern is the maintenance of the perfection of the type. Blyth's thinking is decidedly that of a natural theologian, for whom all occurrences of variation "are among those striking incidences of design which clearly and forcibly attest the existence of an omniscient great first cause." Everything attests design and the perfect balance of nature (Schwartz, 1974). Darwin quite likely had read Blyth's paper but paid no further attention to it since it was antievolutionary in spirit and not different from the writings of other natural theologians in its general thesis. In later years, Blyth became one of Darwin's valued correspondents.

There are two major reasons why a concept of natural selection was so alien to the western mind prior to the nineteenth century. One is the all-pervading prominence of essentialism which made any idea of a gradual improvement impossible. All it could permit was a sudden origin of new types and the elimination of those that are inferior. In discussions of adaptation in the writings of the naturalists, it is only species that are compared with each other, never individuals. The second reason was the equally pervading acceptance of a global teleology according to which the creator's design automatically effected perfection. It would have been held impious if not heretical to search for any mechanism that would improve on this perfection. In natural theology there

489

simply was no opportunity for an improvement by natural selection.

The process of natural selection, as conceived by Darwin, is fundamentally different from the process of elimination of the essentialists. The concept of a static type is replaced by that of a highly variable population. New variations are produced continuously, some of them superior and some of them inferior to the existing average. Since one sees this type of variation in every human population, it is difficult to understand why population thinking was so rare prior to Darwin, and why it took so long to be generally accepted after Darwin. Population thinking was virtually nonexistent prior to 1800. Even such a vigorous antiessentialist as Lamarck thought only in terms of (identical) individuals, not in terms of variable populations consisting of uniquely different individuals. Natural selection would have made as little sense to Lamarck as to the strictest essentialist.

Up to the present day many authors fail to understand the populational nature of natural selection. It is a statistical concept. Having a superior genotype does not guarantee survival and abundant reproduction; it only provides a higher probability. There are, however, so many accidents, catastrophes, and other stochastic perturbations that reproductive success is not automatic. Natural selection is not deterministic, and therefore not absolutely predictive. This was perceptively pointed out by Scriven (1959), but still causes problems to philosophers raised in an essentialistic tradition. On the other hand, the evolutionary theory permits numerous probabilistic predictions (Williams, 1973a).

Following his usual strategy, Darwin advances a number of arguments showing that the interpretation of the essentialists and natural theologians is not valid. There is room for improvement in all species. He documents this (*Origin:* 82) by the success of so many introduced species over the native fauna and flora. If the natives had been perfect, they would not have succumbed so easily. Therefore "extremely slight modifications in the structure or habits of one inhabitant would often give it an advantage over others."

Natural selection would of course be helpless if there were not abundant intraspecific variation: "Unless profitable variations do occur, natural selection can do nothing" (p. 82). Darwin places great stress on the occurrence of useful variations. Since among domestic animals variations occur that are useful to man, he asks, "Can it, then, be thought improbable . . . that other variations useful in some way to each being in the great and complex battle of life, should sometimes occur in the course of thousands of generations? If such do occur, can we doubt (remembering that many more individuals are born than can possibly survive) that individuals having any advantage, however slight, over others, would have the best chance of surviving and procreating their kind?" (pp. 80–81). This leads him to the following definition. "This preservation of favourable variations and the rejection of injurious variations, I call Natural Selection" (p. 81). Variation and its inheritance belong to the subject matter of genetics and Darwin's assumptions and theories will be analyzed in detail in Chapter 16.

Interestingly, among the factors that control natural selection, Darwin, following Lyell, always thought that biotic factors—the interaction among competing species and their relative frequency—was more important than the physical environment. Accordingly, "any change in the numerical proportions of some of the inhabitants, independently of the change of climate itself, would most seriously affect many of the other species " (p. 81). Darwin also fully realized a fact forgotten by many later authors: that not only the adult phenotype is a target of selection. "Natural selection will be enabled to act on and modify organic beings at any age, by the accumulation of profitable variations at that age, and by their inheritance at a corresponding age" (p. 86). "In social animals it will adapt the structure of each individual for the benefit of the community" (p. 87).

Darwin's Debt to Malthus

Social historians have from time to time advanced the thesis that the theory of evolution by natural selection was inspired by the social and economic situation of England in the first half of the nineteenth century. This thesis is based on the logic that natural selection is the result of the struggle for existence, about which Darwin, it was claimed, had learned from Malthus. More broadly, the claim is made that Darwin's theory was a product of the industrial revolution with its fierce competition, misery, poverty, and struggle for survival; or else a product of the replacement of feudalism (monarchy) by democracy. Is there any validity to these claims? Young and others have made valiant efforts to substantiate the thesis that Darwinism is the product of Malthusianism.[5] Some of these writers do not even bother to partition Darwinism into its several components, even though they all agree that the concept of natural selection "arose from an interest in racial, national and class forms of war and conflict" and that "Darwin's principles were the application of social-science concepts to biology" (Harris, 1968: 129). Unfortunately all those who have defended this thesis have confined themselves to such broad, general statements. By contrast, all the serious Darwin students who have thoroughly analyzed the sources of Darwin's theory (most recently Herbert, Limoges, Gruber, Kohn, Mayr) agree that Malthus's influence on Darwin was very limited ("one sentence") and highly specific. What Darwin and Wallace had taken from Malthus was the "populational arithmetic," but not his political economy. The Marxist claims "that Darwin and Wallace were extending the laissez-faire capitalist ethos from society to all nature to make a Weltanschauung out of the new captains' of industry's utopia of progress through unfettered struggle" is not supported by any evidence whatsoever (Hodge, 1974). To be sure, Darwin did not live in an ivory tower; he must have seen what went on in England all around him; he read all the relevant literature (Schweber, 1977; Manier, 1978), and this might have facilitated his acceptance of certain ideas. Yet, if the theory of natural selection were the logical and necessary consequence of the zeitgeist of the industrial revolution, it should have been widely and enthusiastically adopted by Darwin's contemporaries. Actually, just the opposite is true: Darwin's theory was almost universally rejected, indicating that it did not reflect the zeitgeist.

As my analysis of the eight components of Darwin's theory has revealed, none of them was original with Malthus and all of them had been encountered by Darwin in his earlier reading, most of them repeatedly. The struggle for existence was forever talked about from the Greeks to Hobbes, Herder, de Candolle, and Lyell, although no one stressed its intensity more than Malthus. The various checks on overpopulation were widely discussed in the literature. Individuality (population thinking) was a concept wholly alien to Malthus, and without it, of course, natural selection is unthinkable. Why then did the reading of Malthus's comment on the potential geometric increase of populations have such an impact on Darwin? The answer is that Darwin read him at a moment when some of his other thinking had matured to a point where high fertility had acquired a new meaning.

There is now good evidence from Darwin's notebooks for a considerable shift in his thinking in the half year prior to September 1838. Under the impact of studying the writings of the animal breeders, Darwin was beginning to be converted from essentialistic to population thinking. In his earlier notes, Darwin applied variation, competition, and extinction quite typologically to species or to incipient species (for example, to the varieties of mockingbirds). The discussions of the breeders made him appreciate for the first time the enormous importance of individual variation. In his third notebook, just a few pages (D: 132) before his famous Malthus statement (D: 135), he emphasizes that individual variation makes "every individual a spontaneous generation." This is when Darwin suddenly saw that there is competition not only among species but indeed among individuals, and it was this individual variation which made natural selection possible.

493

The ironic aspect of his "debt to Malthus" is that Darwin uses the new insight to come to conclusions diametrically opposed to those of Malthus. Malthus' principal argument had as its object to refute the claims of Condorcet and Godwin of an unlimited perfectibility of man. By adding the ingredient of population thinking, Darwin arrived precisely at the opposite conclusion to Malthus. It is even more ironic that Malthus was fully aware of the successes of breeders due to artificial selection: "I am told that it is a maxim among the improvers of cattle that you may breed to any degree of nicety you please . . . and that some of the offspring will possess the desirable qualities of the parents in a greater degree" 1798: 163). However, Malthus uses this very statement only to refute it, at least as far as unlimited perfectibility is concerned. It was as unthinkable for him as for Lyell to accept any transgressions of the limits of the type. For both of them all individuals are essentialistically alike. It is thus evident, to repeat once more, that the role of Malthus was very much that of a crystal tossed into a saturated fluid. If Darwin at this moment had read Franklin's pamphlet or some of the natural-history literature stressing super-fecundity and its consequences, it is quite likely that it would have electrified him just as much as did the sentence in Malthus. It was a clear case of the "prepared mind" seeing something that he had not seen when he was not yet prepared.

Some sociologists have also construed a debt of Darwin to Spencer. There is no basis for such a claim. Darwin's theories of evolution were essentially completed by the time Spencer (1852) first thought of evolution. Furthermore, Spencer's ideas, with their reliance on finalistic principles and Lamarckian inheritance, were totally irreconcilable with Darwinian evolution. As Freeman has rightly concluded (1974: 213), "The theories of Darwin and Spencer were unrelated in their origins, markedly disparate in their logical structures, and differed decisively in the degree to which they depended on the supposed mechanism of Lamarckian inheritance and recognized 'progress' as 'inevitable.' " The misconception that Spencer's evolutionism was the same as Darwin's has been a great handicap to anthropology and to sociology.

A. R. Wallace and Natural Selection

494   The enormous resistance which Darwin's theory encountered in the ensuing eighty years proves conclusively how difficult it is to put its eight components together properly. It is not like many of the discoveries in the physical sciences, where at a given period the same discovery is often made simultaneously by several people because they were looking for the missing piece in a jigsaw puzzle (Merton, 1965). That a second person, without knowing of Darwin's work, would come up with the same theory of evolution by natural selection would seem utterly improbable. This theory was so novel, in such a contrast to anything that anybody had ever thought before, that it took nearly another hundred years before it was generally accepted. That another person, among the relatively few people thinking about evolution, would come up with essentially the same theory, at the same time, was totally unexpected, and yet it happened.[6]

The story about Darwin receiving Wallace's essay in June 1858 (see Chapter 9) raises numerous questions. Was Darwin justified in writing to Lyell, "I never saw a more striking coincidence; if Wallace had my ms. sketch written out in 1842, he could not have made a better short extract! Even his terms now stand as heads of my chapters"? Was Wallace's theory really so nearly identical with Darwin's? How did Wallace assemble the pieces of his theory? Did he arrive at it by the same steps as Darwin, or by a process of convergence?

We must remember that Wallace had been convinced of evolution since about 1845, and that in 1855 he had published his evidence for speciation. From that time on he searched for the factors that are responsible for evolutionary change. It is relevant to stress here once more the important influence of Lyell's *Principles of Geology.* Wallace had read Lyell's superbly argued case against the modification of species as carefully as had Darwin. A good deal in the similarity of the arguments of Darwin and Wallace is clearly due to the fact that both men attempted to refute the very specific points raised by Lyell. By the concreteness of his antievolutionary objections, Lyell had prepared the ground for specific counterarguments (McKinney, 1972: 54–57).

Even though constantly thinking about these problems, Wallace's ideas apparently made little progress from 1855 until a memorable day in February 1858. "At that time I was suffering from a rather severe attack of intermittent fever [malaria] at Ternate in the Moluccas, and one day while laying on my bed during the cold fit, wrapped in blankets, though the thermometer was at 88°F., the problem [of how species transformation occurs] again presented itself to me, and something led me to think of the 'positive checks' described by Malthus in his *Essay on Population,* a work I had read several years before and which had made a deep and permanent impression on my mind" (Wallace, 1891: 20).

As in the case of Darwin, the illumination was sudden, and caused by thinking about Malthus's *Essay on Population.* A close reading of Wallace's 1858 essay "On the Tendency of Varieties to Depart Indefinitely from the Original Type" shows, however, that the parallelism is not complete.

Wallace states his thesis with extraordinary clarity: "There is a general principle in nature which will cause many varieties to survive the parent species, and to give rise to successive variations, departing further and further from the original type" (1858: 54). The language in which this observation is presented is rather typological; Wallace's conclusion, however, clearly contradicts Lyell's claim that "varieties have strict limits, and can never vary more than a small amount away from the original type."

The most important aspect of Wallace's analysis is that he carefully stayed away from the quagmire of the morphological controversy on species and varieties but based his conclusion on a rather strictly ecological argument. He concluded that population size of a species is not at all determined by fertility but by natural checks on potential population increase. An enormous number of animals must die each year to keep the number constant, and "those that die must be the weakest—the very young, the aged, and the diseased—while those that prolong their existence can only be the most perfect in health and vigour—those who are best able to obtain food regularly and avoid their numerous enemies. It is, as we commenced by remarking, 'a struggle for existence', in which the weakest and the least perfectly organised must always succumb" (pp. 56–57).

In this early part of Wallace's discussion his emphasis is on regulation of population size, on stabilizing selection (elimination), and on competition between species. He is now "in a condition to proceed to the consideration of varieties, to which the preceding remarks have a direct and very important application." In the ensuing discussion Wallace applies the term "variety" to variant individuals, that is, individuals within a population that do not share the same properties. If a species produces a superior variety, "that variety must inevitably in time acquire a superiority in numbers" (p. 58.).

496   Curiously, Wallace's account is marred by the same weaknesses as Darwin's. There is still a good deal of typological thinking, particularly in reference to the nature of varieties, and like Darwin he still accepts the effectiveness of use and disuse, a process rather universally believed in at that period. Like Darwin, Wallace rejects "the hypotheses of Lamarck" in a formulation that indicates that it was taken straight from Lyell. Instead, Wallace explains the short, retractile talons of the cat tribe and the elongated neck of the giraffe in strict selectionist terms. Indeed, Wallace emphasizes strongly that the acquisition of new adaptations is completely consistent with the interpretation that they are the result of selection. He concludes his essay with these words:

> We believe we have now shown that there is a tendency in nature to the continued progression of certain classes of *varieties* further and further from the original type—a progression to which there appears no reason to assign any definite limits . . . This progression, by minute steps, in various directions, but always checked and balanced by necessary conditions, subject to which alone, existence can be preserved, may, it is believed, be followed out, so as to agree with all the phenomena presented by organised beings, their extinction and succession in past ages, and all the extraordinary modifications of form, instinct, and habits which they exhibit. (p. 62)

Let us now try to compare Wallace's chain of argument in more detail with that of Darwin.[7] Both of them started out with the species problem or, as Wallace himself said, in a retrospective account in 1908, with the thought of "the possible causes of the change of species." Nevertheless, Wallace's own analysis was in some ways more a study of population ecology than a study in speciation (which he presumably thought he had dealt with adequately in 1855). Wallace, quite in contrast to Darwin, relates the problem of evolution very directly to man. What had long puzzled him, who for eight years had lived among natives, was what the checks were that "kept all savage populations nearly stationary." "These checks [enumerated by Malthus]—disease, famine and accidents, war, etc.—are what keep down the population, and it suddenly occurred to me that in the case of wild animals these checks would act with much more severity, and as the lower animals all tended to increase more rapidly than man, while their population remained on the average constant, there suddenly flashed upon me the idea of the survival of the fittest" (Wallace, 1903: 78).

As in the case of Darwin, the crucial component of the theory was the recognition of individuality. Precisely fifty years later (in 1908) Wallace reports this as follows: "Then there flashed upon me, as it had done twenty years before upon Darwin, the *certainty*, that those which, year by year, survived this terrible destruction must be, on the whole, those which have some little superiority enabling them to escape each special form of death to which the great majority succumbed—that, in the well known formula, the fittest would survive. Then I at once saw, the ever present variability of all living things would furnish the material."

As stated, there are subtle differences between the interpretations of Wallace and of Darwin. Wallace apparently had been far more impressed by the general thesis of Malthus and particularly by the enormous annual losses which keep populations at a steady level, the "positive checks." Population thinking, apparently, had different sources in the two authors of natural selection. It was animal breeding and taxonomic work in Darwin; it was the study of human populations and taxonomic work in Wallace. Wallace had a low opinion of the value of studying domestic varieties and concluded that we cannot draw "inferences as to varieties in a state of nature" by observing domestic animals. For this and other reasons he did not use the term "selection" in his essay and apparently was always somewhat unhappy about this term.

In spite of these finer differences in approach, Wallace entirely agreed with Darwin in the final conclusion: unlimited variability in populations exposed to drastic regular decimation must lead to evolutionary change. In due time, however, the thinking of Wallace diverged increasingly from that of Darwin, as far as natural selection was concerned. For instance, by 1867 Wallace renounced any reliance on use and disuse, and in the 1880s he was one of the first to endorse enthusiastically Weismann's rejection of any inheritance of acquired characters. Wallace did not believe in a separate category of sexual selection and least of all in "female choice" (Chapter 12). He also believed that reproductive isolating mechanisms were strictly the result of selection. Yet, he lost his nerve when it came to applying this consistent selectionism to man, since he considered it impossible that natural selection could have endowed primitive man with such a large brain as well as with his capacity for morality. Some superior power must have been responsible for this (Wallace, 1870).

498 Wallace did not return from the East Indies until 1862, four years after his paper had been read at the Linnean Society. He was never jealous of Darwin, indeed he always was a great admirer of his, even though they later disagreed on the answers to certain problems. Wallace eventually acquired fame on his own account, particularly through his splendid *Malay Archipelago* and his *Geographical Distribution of Animals* (1876), the classic of zoogeography for the next eighty years.

Forerunners of Natural Selection

Two methods of attack are used most frequently against a new theory; the first one is to claim that the new theory is wrong, the second one is to claim that it is not new. True to the second part of this tradition, after the publication of the *Origin* one claim after the other was made of priority in publication of the concept of natural selection.[8] Because an essentialist simply cannot conceive of evolutionary change through natural selection, all claims prior to 1800 are inadmissible for this reason alone. There are, however, a number of proposals of genuine natural selection prior to Wallace and Darwin in 1858.

William Charles Wells (1757–1817), an English doctor who had lived in South Carolina for a while, advanced in 1818 the theory of natural selection in a sort of postscript to an essay dealing with a human color variant (Wells, 1818). Wells stated, as had quite a few others before him, that negros are far more resistant to tropical diseases than whites. Conversely, negros are far more susceptible to the diseases of the temperate zone. "Regarding then as certain that the negro race is better fitted to resist the attacks of the diseases of hot climates than the white, it is reasonable to infer that those who only approach the black race will be likewise better fitted to do so than others who are entirely white." This, he says, is indeed true for mulattoes. He then refers to the practice of animal breeders, "When they find individuals possessing in a greater degree than common, the qualities they desire, couple a male and female of these two together, then take the best of their offspring as a new stock and in this way proceed till they approach as near the point in view as the nature of things will permit. But what there is done by art seems to be done with equal efficiency, though more slowly, by nature in the formation of varieties of mankind, fitted for the country which they inhabit." He declares that this is the way in which human races develop in the different climatic zones of the world.

499

Although Wells clearly proposes a theory of evolution by natural selection, it is only evolution of adaptation to local climates within a species and at that for man only; the principle is never applied to genuine evolution, to the multiplication of species, to a development of higher taxa or to common descent.

The person who has the soundest claim for priority in establishing a theory of evolution by natural selection is Patrick Matthew (1790–1874). He was a wealthy landowner in Scotland, well educated, very well read, and well traveled (Wells, 1974). His views on evolution and natural selection were published in a number of notes in an appendix to his work *On Naval Timber and Arboriculture* (1831). These notes have virtually no relation to the subject matter of the book, and it is therefore not surprising that neither Darwin nor any other biologist had ever encountered them until Matthew brought forward his claims in an article in 1860 in the *Gardeners' Chronicle*. Matthew's background was very much the same as that of many of Darwin's friends among the landed gentry, having to do with animal and plant breeding. He enunciates clearly that success in this endeavor depends on the selection (a word he uses repeatedly) of the best suited individuals. Indeed, the main thesis of his book is that this principle should also be applied to tree culture. The choice of words indicates that he had read Erasmus Darwin, Lamarck, Malthus, and Lawrence. He clearly adopts a theory of evolution and, quite remarkably, evolution by common descent. "Are they [species] the diverging ramifications of the living principle under modification of circumstances?" He considers gradual evolution as much more probable than "total destruction and new creation" (catastrophism). He rejects Linnaeus's origin of species by hybridization and believes that "the progeny of the same parents, under great difference of circumstance, might, in several generations, even become distinct species, incapable of co-reproduction" (p. 384).

The remarkable similarity of Matthew's thinking to that of Darwin is clearly indicated in his statement:

> The self regulating adaptive disposition of organized life may, in part, be traced to the extreme fecundity of Nature, who, as before stated, has, in all the varieties of her offspring, a prolific power much beyond (in many cases a thousandfold) what is necessary to fill up the vacancies caused by senile decay. As the field of existence is limited and pre-occupied, it is only the hardier, more robust, better suited to circumstance, individuals, who are able to struggle forward to maturity from the strict ordeal by which Nature tests their adaptation to her standard of perfection and fitness to continue their kind by reproduction . . . The breed gradually acquiring the very best possible adaptation of these to its condition which it is susceptible of, and when alteration of circumstances occurs, thus changing in character to suit these as far as its nature is susceptible of change. (p. 385)

Patrick Matthew undoubtedly had the right idea, just like Darwin did on September 28, 1838, but he did not devote the next twenty years to converting it into a cogent theory of evolution. As a result it had no impact whatsoever.

Prichard, Lawrence, and Naudin have also been mentioned as having anticipated Darwin, but their statements are weak and inconclusive when compared to those of Matthew. They either refer to the improvement of the human species or to choosing plant varieties, but the appreciation of the possibilities of selection are not utilized for the development of a theory of evolution.

A careful indication of sources of ideas was not the tradition of the times. Lamarck, for instance, hardly ever cited the authors whom he had used. Not surprisingly, it has therefore been claimed repeatedly that Darwin had been aware of these forerunners and had used their findings without giving due credit, but not a shred of evidence supports these claims. There is every reason to believe that Darwin did not know the relevant writings of Wells or Matthew, and that the statements of Lawrence, Prichard, and Naudin, if known to Darwin, were too vague and too unrelated to a theory of evolution by common descent to have attracted his attention. It is true that authors whose writings Darwin had used are rarely cited in the *Origin* by name, but this is due to the fact that he considered this work an abstract and that he would provide detailed references in his more complete work. Now that *Natural Selection* has been published (1975), it is much easier to determine what earlier publications Darwin had used, and what he had not.

This emerges even more clearly from his notebooks and other manuscript material, and they document convincingly that Darwin had been unaware of the writings of either Wells or Matthew.

THE IMPACT OF THE DARWINIAN REVOLUTION

The Darwinian revolution has been called, for good reasons, the greatest of all scientific revolutions. It represented not merely the replacement of one scientific theory ("immutable species") by a new one, but it demanded a complete rethinking of man's concept of the world and of himself; more specifically, it demanded the rejection of some of the most widely held and most cherished beliefs of western man (Mayr, 1972b: 988). In contrast to the revolutions in the physical sciences (Copernicus, Newton, Einstein, Heisenberg), the Darwinian revolution raised profound questions concerning man's ethics and deepest beliefs. Darwin's new paradigm, in its totality, represented a most revolutionary new *Weltanschauung* (Dewey, 1909).

501

The sweeping nature of the changes proposed by Darwin is best documented by listing some of the more philosophical implications of Darwin's theories:

(1) The replacement of a static by an evolving world (not original with Darwin).

(2) The demonstration of the implausibility of creationism (Gillespie, 1979).

(3) The refutation of cosmic teleology.

(4) The abolition of any justification for an absolute anthropocentrism by applying the principle of common descent to man.

(5) The explanation of "design" in the world by the purely materialistic process of natural selection, a process consisting of an interaction between nondirected variation and opportunistic reproductive success which was entirely outside the dogma of Christianity.

(6) The replacement of essentialism by population thinking.

To this list must be added various philosophical-methodological innovations, such as the consistent application of the hypothetico-deductive method (Ghiselin, 1969; Ruse, 1979a), a new evaluation of prediction (Scriven, 1959), and the bringing of the study of ultimate (evolutionary) causations into science (Mayr, 1972b).

How ready was the world to accept these revolutionary new concepts or, to put it differently, how much time did it take before Darwin's thought was adopted? The impact of the *Origin* was unprecedented. Except for Freud, perhaps no other scientist has been translated so extensively, reviewed so frequently and in such detail, and had so many books written about him. All the numerous quarterlies and review journals of the period had extensive reviews, and so did most of the religious or theological periodicals. So rich is this literature that a secondary literature has developed which deals with this review literature (for example, Ellegard, 1958; Hull, 1973). Another category of literature deals with Darwin's impact and gradual acceptance in various parts of the world. No other phase in the history of biology has been described by historians with more loving detail than the battles resulting from Darwin's theory (Kellogg, 1907; Vorzimmer, 1970; Glick, 1974; Conry, 1974; Moore, 1979).

502

The nature of the opposition to Darwin can be understood much better if one realizes the general attitude toward evolution in the middle of the nineteenth century. Prior to Darwin, the consideration of evolution was thought to be part of the realm of philosophy. Indeed, virtually all of those who had speculated on evolution were theologians or other nonbiologists who basically had no competence whatsoever to deal with this complex biological subject matter. Even Lamarck, Darwin's most distinguished forerunner, failed to marshal systematically facts to support his evolutionary speculations or to provide a detailed analysis of the possible mechanisms of evolution. In line with the concepts of the period, he entitled his work *Philosophie zoologique* (1809), and indeed it was a philosophy rather than a zoology. Darwin was the first author to deal with the subject of evolution strictly scientifically. He supported his thesis by a massive body of facts, and this rich evidence changed the situation quite fundamentally. As long as the discussions about evolution were conducted on a philosophical basis, the argument could be stated in metaphysical terms. The publication of the *Origin* made this approach impossible, once and for all time. Darwin showed, implicitly and explicitly, that there are three and only three possible explanations for the diversity of the living world and for the ingenuity of its adaptations. This challenge forced every thinking reader of his detailed and perceptive analysis into the uncomfortable position of having to make a choice in favor of either one or the other of these three possible explanations.

The first one is that of a continual creation, involving the perpetual intervention of the creator in replacing species and faunas that had become extinct, and in creating ever new adjustments and adaptations. Lyell and Sedgwick were among the many scientists who up to a point adopted this explanation. It included the belief that every feature of every species was specially created to adapt the species for the environment into which it was placed. It is probable that in 1859 this theistic explanation of the world was still the majority opinion, at least in Great Britain. However, such a "perpetual intervention hypothesis," as Lyell called it (Wilson, 1970: 89) was too extreme even for many devout scientists; even Lyell and Agassiz had misgivings.

503

This led them to a second, deistic theory of evolution: a belief in the existence of teleological, evolutionary laws, ordained at the time of creation, that would lead to ever greater perfection and adaptation and would guarantee an orderly replacement of faunas in the geological sequence. It would account for any other kind of order and regularity found in nature (Bowler, 1977b; Ospovat, 1978). Lamarck, who originally seems to have adhered to such a theory, eventually realized the absence of a consistent trend toward ever greater perfection. The difficulties increased as biological knowledge grew. Much of Darwin's argumentation in the *Origin* was directed toward uncovering irregularities in the patterns of distribution of faunas and floras and of morphological trends which defied any interpretation in terms of progressive laws.

No other phenomenon was more awkward for the deists than the production of new species to replace those that had been lost by extinction. That God was responsible for their appearance was taken for granted. To ascribe their creation to a miracle was, of course, unacceptable to scientists like Herschel and Whewell. Thus, whistling in the dark, they ascribed their origin to "intermediate causes" or "causal laws" governing the introduction of new species instituted by the creator (see Chapter 9). How could such laws possibly operate? There were actually only three possibilities, (1) special creation, which would be a miracle, (2) spontaneous generation, an origin that was not respectable scientifically, at least as far as higher organisms are concerned, nor would it explain the perfect design of each species, or (3) derivation from other species, and that would be evolution. Lyell, unlike Gray, was not ready to accept evolution by natural selection as an "intermediate cause." Herschel and Lyell did not know enough natural history to appreciate that there are no conceivable mechanisms by which such

secondary laws could be implemented without coming into conflict with the laws of physics and chemistry.[9] It was precisely this
insight which induced Darwin to postulate the third of the three
possible explanations, a strictly nonteleological mode of evolution
in which random variation is converted into directional trends and
into adaptation by natural selection, without any recourse to
supernatural forces, not even at the beginning.

It is not admissible to judge the debate following the publication of the *Origin* within the framework of modern thinking.
One must remember how powerful in the 1850s and 1860s the
hold was which creationism still held, particularly in England. Virtually all of Darwin's peers were creationists, most of them even
rather orthodox theists, who found nothing unscientific in invoking supernatural forces in their arguments. Hopkins, one of Darwin's reviewers, called Darwin unscientific for postulating that the
trilobites, a group of extinct fossil invertebrates which suddenly
appeared in the fossil record, were derived from ancestors still
unknown as fossils. Yet the same Hopkins did not hesitate to believe that the trilobites had been created at the time when they
first showed up in the fossil beds (Hull, 1973).

It is evident that a creationist interpretation was, to those who
believed in a personal god, as legitimate (in fact even more so) as
a so-called scientific explanation. The battle about evolution (and
particularly natural selection) was not a purely scientific controversy; rather it was a struggle between two ideologies, natural theology and objective science. I will not here report on the struggle
between religion (church) and science,[10] since this volume deals
with biological thought. However, since creationism, at least in
England, was a dominant "scientific" school in the 1850s, Darwin
had to adopt the bold strategy of showing for one natural phenomenon after another that it could be explained quite reasonably as the product of evolution but that it did not fit at all what
one would expect from the action of a wise, benevolent, and all-
powerful creator: "Why, it may be asked, has the supposed creative force produced bats and no other mammals on remote islands?" (*Origin:* 394). Here and in about thirty other places in the
Origin Darwin argues that a given phenomenon is consistent with
evolution or with common descent but makes no sense when ascribed to "a special act of creation" (p. 55). Again and again, Darwin repeats, "On the view that each species has been independently created, I can see no explanation."

Darwin's Five Theories

The rich literature on the impact of the *Origin* is unfortunately badly flawed since it takes insufficiently into consideration that Darwin had actually proposed five largely independent theories. As a result, when a historian or philosopher speaks of Darwinism one rarely knows whether he means evolution as such, man's descent from the apes, natural selection, or what. The word "Darwinism" has continued to change its meaning over the years. In the period immediately after 1859 it referred most often to the totality of Darwin's thinking, while it strictly means natural selection for the evolutionary biologist of today. Darwin himself contributed to this ambiguity by referring in the *Origin* to the theory of evolution ten times as "my theory" but designating natural selection only three times as "my theory." Indeed, there is considerable evidence to indicate that Darwin considered all components of his evolutionary theory as a single indivisible whole. One can infer this from the fact that he mingles in many chapters of the *Origin* seemingly unrelated subjects. For instance, in the first chapter the causes of variability, the problem of species vs. varieties, and artificial selection. In the second chapter he deals with variation in nature and with the species problem. The next two chapters deal with mechanisms of evolution (struggle for existence, natural selection), speciation, divergence of characters, extinction, and the theory of common descent.

In spite of the beauty and brilliance of individual discussions, much of the organization of the *Origin* strikes the modern reader as rather chaotic. It is this, probably, which has induced many readers of the *Origin* to complain that it was a "difficult" book. Although I find his arguments not completely convincing, Hodge (1977) makes a case that Darwin had a clear tripartite organization in mind, and that he more or less followed it.

Many subsequent writers adopted Darwin's belief that common descent, gradualness, and natural selection are an indivisible single paradigm, and this belief induced these writers to treat these subjects jointly when discussing the post-1859 fate of "Darwinism." Actually, a much clearer picture emerges if five strands of Darwin's thought are treated separately. That they do not constitute an indivisible whole is demonstrated by the fact that so many evolutionists accepted some of Darwin's theories but rejected others (table 3).

505

Table 3. The composition of the evolutionary theories of various evolutionists. All these authors accepted a fifth component, that of evolution as opposed to a constant, unchanging world.

	COMMON DESCENT	GRADUAL-NESS	POPULATIONAL SPECIATION	NATURAL SELECTION
Lamarck	No	Yes	No	No
Darwin	Yes	Yes	Yes	Yes
Haeckel	Yes	Yes	?	In part
Neo-Lamarckians	Yes	Yes	Yes	No
T. H. Huxley	Yes	No	No	(No)
de Vries	Yes	No	No	No
T. H. Morgan	Yes	(No)	No	Unimportant

The full title of Darwin's work, *On the Origin of Species by Means of Natural Selection, or the Preservation of Favored Races in the Struggle for Life,* has added to the erroneous impression that only a single theory is involved. The fact that Darwin treated speciation (in chapter IV) under natural selection has reenforced this interpretation, but it is quite wrong. Let me illustrate with an example that speciation and natural selection are two independent processes. A population could become established on an island and could (theoretically) eventually become so different from the parental population merely by random genetic processes (genetic drift) that it is no longer able to reproduce with the parental population, that is, it would have speciated (without any contribution at all of natural selection). The same virtual independence of two evolutionary phenomena is true for patterns of geographic distribution. They are largely due to accidents of dispersal superimposed on geographical and geological processes that also go on independent of natural selection. To imply, as Darwin did, that natural selection explains patterns of distribution is quite misleading.

Let me now attempt to particularize the various theories of which Darwin's evolutionary paradigm is composed.

Evolution as Such

The theory that the world is not constant, but the product of a continuing process of evolution was, of course, not new with Darwin. Yet, in 1859, in spite of the writings of Lamarck, Meckel, and Chambers, the majority opinion still was that the world is stable. A number of rather peculiar compromises, such as progressionism, were defended between 1800 and 1859, in order to avoid having to accept evolution. However, the massive evidence which Darwin presented was so convincing that within a few years every biologist became an evolutionist, even Owen, Mivart, and Butler in England, who opposed Darwin's other theories. Agassiz, a holdout to the end, died in 1873. France was actually the only country where evolutionism as such had to battle for acceptance (Conry, 1974; Boesiger, 1980). For many biologists of today, evolution is no longer a theory but simply a fact, documented by the changes in the gene pools of species from generation to generation and by the changes in the fossil biota in accurately dated geological strata. Current resistance is limited entirely to opponents with religious commitments.

Evolution by Common Descent

Darwin, curiously, was the first author to postulate that all organisms have descended from common ancestors by a continuous process of branching. When accepting the splitting of a parental species into several daughter species, he was led to the concept of common descent almost by necessity. By tracing the ancestry up into the higher taxa Darwin was led to consider all living beings "as the lineal descendants of some few beings which lived long before the first bed of the Silurian system" (*Origin:* 488) and that life had "been originally breathed into a few forms or into one" (p. 490).

A continuing multiplication of species thus could account for the total diversity of organic life. By reducing the problem of origins to a single one, that of the first origin of life, common descent made spontaneous generation, a process quite incompatible with Darwin's continuistic ideas, superfluous. Even though the solution of this ultimate problem was clearly outside the capacities of contemporary science Darwin could not resist speculating about it (*L.L.D.,* III: 18).

The theory of common descent greatly facilitated the acceptance of evolution, as Darwin himself said in the *Origin,* because of its power to explain so much in comparative anatomy, biogeography, systematics, and other areas of biology that had previously been puzzling. Even Lyell and the botanist George Bentham, who had originally been opposed to it, had by 1868 adopted the theory of common descent.

It has become customary in the literature of the history of biology to refer to the "Darwinian revolution."[11] This term, which I myself have used, is ambiguous, however, since the total corpus

507

of Darwin's thought ushered in several intellectual revolutions. Particularly well defined are two. The first one is that by including man in the phyletic tree of common descent, Darwin took away from man his privileged position in nature assigned to him in the Bible and in the writings of virtually all philosophers. This, so to speak, dethroned man. It was a truly revolutionary concept, quite different from considering man the pinnacle of the chain of being. The second revolution was natural selection (see below).

508   ### Gradualness of Evolution

Darwin's insistence that evolution is entirely gradual encountered almost as much resistance as his theory of natural selection. This had both empirical and ideological reasons. A gradual change from one type to another is simply unthinkable for an essentialist (typologist). Lyell and others insisted that the potential for variation of a species has fixed limits which no amount of selection could transcend. Every species was separated from every other species by an unbridgeable gap, and if one wanted to postulate evolution, one would have to postulate the sudden origin of new types by a saltation. This is why Lyell postulated the steady occurrence of the "introduction of new species," a discontinuous process. Darwin's theory that populations are the *locus operandi* of speciation and that this permits every grade of intermediacy between geographical varieties and species undermined the essentialistic argument.

On the other hand, certain empirical findings seemed to support the essentialist position. Comparative anatomists, with few exceptions, stressed the fundamental differences between the structural plans of the higher taxa, differences which, they said, could not be explained by gradual evolution. Paleontologists, likewise, insisted on the sudden origin of new types in the fossil record and on the total absence of any intermediate types. Wherever one looked in nature, its most impressive aspect was discontinuity.

Experimental biologists, all of them solidly essentialists, found it particularly difficult to comprehend gradual evolution. Unused to thinking in terms of variable populations, they could not conceive of the origin of anything new except through the saltational production of a deviating individual, a hypothetical process, later called macrogenesis. Nägeli, His, Kölliker, W. H. Harvey, Mivart, Galton, and other distinguished authors publishing in the 1860s, 70s, and 80s supported macrogenesis. Still, until the 1890s this was a minority opinion. Obviously, macrogenesis is incompatible with gradual natural selection, and Darwin never had any use for it.

Darwin, more clearly than any of his opponents, saw that the observed discontinuities are, so to speak, artifacts of history. He explained the gaps between genera and still higher taxa by the dual processes of character divergence and extinction, an explanation that is now universally adopted. Competition and the invasion of new niches and adaptive zones leads to steady divergence, but the extinction of intermediate types and connecting links is responsible, more than anything else, for the observed discontinuities between the higher taxa. Thus, such breaks are secondary artifacts rather than a reflection on the original process of the formation of the taxa.

The source of Darwin's strong belief in gradualism is not entirely clear. In part it was evidently the result of observation, such as the gradual differences among Galapagos mockingbirds and finches or the historically documented continuity among the most aberrant races of dogs, pigeons, and other domestic animals. But as Gruber (1974) points out, there may have been a metaphysical component in Darwin's belief. As a result of studying the writings of the theologian Sumner (1824: 20), Darwin had come to the conclusion that all *natural* things evolve gradually from their precursors, while discontinuities, such as sudden saltations, are indicative of a *supernatural* origin, that is, indicative of intervention by the creator. All of his life Darwin took great pains to reconstruct a gradual evolution of phenomena that at first sight seemed clearly the result of sudden origins.

Natural Selection

In spite of occasional flashes of insight among his forerunners and the simultaneous proposal by Wallace, there is no question that it was Darwin who established the theory of evolution by natural selection, who supported it by numerous well-chosen examples and carefully reasoned arguments, and who, by tying it in with an equally well-substantiated theory of evolution, brought it to the attention of the western world. By explaining "design" in nature as the result of a purely nonteleological, materialistic process, the theory of natural selection eliminated the need for any global teleology. Darwin's theory provided a causal explanation of the seemingly perfect order in living nature, that is, of the adaptation of organisms to each other and to their environment. Clearly the theory of natural selection was the most revolutionary concept

advanced by Darwin. By providing a purely materialistic explanation for all phenomena of living nature, it was said it "dethroned God." Rightly the theory of natural selection can be designated a second Darwinian revolution.

THE RESISTANCE TO NATURAL SELECTION

510   If a modern biologist speaks of Darwinism, natural selection is the
component of Darwin's paradigm he has in mind. Darwin realized
from the beginning that it was the most revolutionary of his ideas;
not surprisingly, it was the one his opponents reacted to most violently, beginning with Herschel, who called it the "law of the
higgledy-piggledy," and Sedgwick, who considered it a "moral
outrage." Natural selection was the component of Darwinism that
offended Darwin's opponents most deeply ("it dethrones God"),
and which, not unnaturally, is even today resisted most strenuously. Darwin's friend Asa Gray, a devout Christian, was one of
the very few Darwinians who succeeded in reconciling natural selection with a belief in a personal god. And it was not only theologians, philosophers, and lay people who opposed this thesis but,
up to the evolutionary synthesis of the 1930s and 40s, even the
majority of biologists (Mayr and Provine, 1980).

Even Darwin's friends and sympathizers were, at best, luke-
warm about natural selection.[12] Natural selection was not stressed
in a single one of the favorable reviews of the *Origin* that were
published after 1859 (Hull, 1973). The endeavor to explain the
world, including all living organisms, in a purely materialistic
manner was distasteful to most of Darwin's supporters. Lyell never
accepted natural selection, and when he finally accepted evolution, he most often referred to it as "Lamarck's theory," to the
great annoyance of Darwin.

T. H. Huxley, Darwin's bulldog, was a staunch defender of
natural selection throughout Darwin's life, and yet Poulton (1908)
provides good evidence that he "was at no time a convinced believer in the theory he protected." Huxley was a morphologist,
physiologist, and embryologist, and the evolution of the organic
world was to him equivalent to the evolution (as he called it) of
the chick embryo in the egg (*L.L.D.*, II: 202). Natural selection
did not fit too well with this concept, and in a historical paper
about Darwin ("The Coming of Age of the Origin of Species,"

1893: 227–243), Huxley makes no reference at all to natural selection. When using the word "Darwinism," Huxley applied it as
often as not simply to the theory of evolution by common descent.
There are indications that he was by no means sure that the theory of natural selection would prove ultimately valid. This is implied by his words, "Whatever may be the ultimate fate of the
particular theory put forth by Darwin . . ." Huxley thought that
major saltations could achieve what gradual evolution by selection
could not (see Chapter 11).

The only solid support Darwin received for natural selection
was from the naturalists. In the first place there was of course his
co-discoverer, Wallace, who was an even more unreserved champion of selectionism than Darwin himself. He drew the line only
when it came to man and his mind. Wallace's South American
companion Henry W. Bates made important contributions to selectionism, and so did Fritz Müller in Brazil (another naturalist;
see below). The botanists, on the whole, were opposed to selection, but Darwin's friend J. D. Hooker always expressed himself
on selection in the sense of Darwin and later so did Thiselton-
Dyer. Abroad, nobody was as convinced a selectionist as August
Weismann, at least after 1880. In fact, as we shall see, he was
perhaps the first evolutionist to ascribe evolutionary change exclusively to natural selection. It is clear from Weismann's biography
and from his research on butterflies that he had been an ardent
naturalist all his life.

It is always said that the 1858 publication of the Darwin–Wallace theory of natural selection had at first been totally ignored.
This is not correct. The ornithologist Alfred Newton describes
how he and his friends had been arguing for years about the origin of species and what an exciting surprise the Darwin–Wallace
issue of the Linnean journal had been: "I sat up late that night to
read it . . . I went to bed satisfied that a solution had been found"
(1888: 241). Newton, in turn, called the paper to the attention of
Canon Tristram, who in a careful study of desert larks interpreted their cryptic coloration as the result of natural selection,
one month before the publication of the *Origin* (Tristram, 1859:
429). Indeed, he describes in full detail under what conditions
paler individuals and individuals with longer bills would be favored by selection. Owen mentioned the Darwin–Wallace paper
favorably, in 1858 in a presidential address but turned against
selection after the publication of the *Origin*.

Selectionism probably had its best support in the 1880s after

511

Weismann had refuted the inheritance of acquired characters and had converted Lankester, Thiselton-Dyer, and others (see Chapter 12). By the 1890s it lost again much support, and it was not until the occurrence of the evolutionary synthesis of the 1930s and 40s that the theory was finally adopted by virtually all biologists. Since the criticism of natural selection was almost universal, it is impossible to try to review it. An exception will be made for one critique because it is always said to have been particularly effective.

512   No other attack on Darwin's theory has drawn more attention than that of the physical scientist and engineer Fleeming Jenkin (1867). This is in part due to Darwin's own statement, "Fleeming Jenkin has given me much trouble, but has been of more real use to me than any other essay or review" (letter to Hooker, 1869, *M.L.D.*, II: 379). When read by a modern reader, Jenkin's review seems anything but impressive. It is based on all the usual prejudices and misunderstandings of the physical scientists. Though Jenkin admits that "all must agree that the process termed natural selection is in universal operation," what he understands by natural selection is actually the essentialist's process of elimination. If Jenkin had understood that reproductive success is the basic principle of natural selection, he would not have written: "The tendency to produce offspring more like their superior parents than their inferior grandfathers can surely be of no advantage to any individual in the struggle for life. On the contrary, most individuals would be benefitted by producing imperfect offspring, competing with them at a disadvantage."

Jenkin agrees with Darwin and most contemporaries that "two distinct kinds of possible variation must be separately considered: *First* that kind of common variation . . . [which is referred to as an individual variation] . . . and *secondly*, that kind of variation which occurs only rarely, and may be called . . . briefly a 'sport', as when a child is born with six fingers on each hand."

As far as individual variation is concerned, Jenkin, like Lyell, Owen, and all essentialists, asserts that natural selection would soon exhaust the available reservoir of such variation. Individual variation, he insists, can never pass beyond the confines of a definite "sphere" of variation. It can never step outside "the type." Selection can make a dog run faster or improve his olfactory faculty but it can never make it into something that is not a dog. He repeats again and again "that no species can vary beyond defined limits." This widespread assumption is not only an automatic consequence of essentialistic thinking but it represents also the experience of animal and plant breeders who had found that the available variation in a closed flock or herd is soon exhausted by intense artificial selection.

This ignores, of course, that the situation in nature is radically different because the reservoir of variation is continuously replenished by gene flow and mutation. In small, closed populations, continued natural selection can be effective only if new genetic variation is produced abundantly. Like the early Mendelians, Jenkin postulated an enormous "mutation pressure," with natural selection actually contributing nothing to the evolutionary change. Owing to his complete inability to understand natural selection, he asserts again and again that its efficiency is limited to the cases "in which the same variation occurs in enormous numbers of individuals . . . [it] does not apply to the appearance of new organs or habits."

513

Here Jenkin approaches the core of his criticism. Even if one were to admit the gradual improvement of species by the selection of individual variation, this does not help us at all, says Jenkin, "as the origin of species requires not the gradual improvement of animals retaining the same habits and structure, but such modification of those habits and structure as will actually lead to the appearance of new organs." It is obvious that Jenkin, like Mivart, was particularly impressed by the difficulty of explaining the origin of new organs. As an essentialist he could not imagine that this could occur in any way other than by saltation, and this leads him to turn his attention to the second type of variation.

Darwin had occasionally referred in the *Origin* to "sports," or as he also called them, "single variations," because, as he said, they offer "such simple illustrations" (*L.L.D.,* II: 289). One might suggest that new structures which carry species beyond their normal sphere of variation arise by sports. This, says Jenkin, is highly improbable for a number of reasons, but particularly because when a sport breeds, his "progeny will on the whole be intermediate between the average individual and the sport." In other words, Jenkin postulated the universal occurrence of what in the later genetic literature was called "blending inheritance." This claim was particularly astonishing considering that Jenkin had chosen families with six-fingered individuals as typical illustrations of sports. It was known since Maupertuis and Réaumur that polydactyly (six-fingeredness) is inherited without any intermediacy. Darwin could have easily refuted Jenkin by pointing out that six-fingered individuals do not have children with five-and-a-half fingers and grandchildren with five-and-a-quarter fingers, nor are the descendants of albinos half-pigmented. The animal breeders had reported literally countless cases where such sports, by back crossing, had become standard breeds, such as the Ancon sheep referred to by Darwin (*Origin:* 30). If Jenkin's claim of intermediacy had any validity, all such sports would have been quickly extinguished by back crossing.

514   The fact that Darwin did not use this argument confirms that Darwin himself was rather confused on the problem of variation (see also Chapter 16). As it was, he meekly accepted Jenkin's argument of blending, and it induced him to stress even more than before the unimportance of sports for evolution. What Darwin also failed to see was that the same argument of blending would hold for individual variation if it reflected genuine genetic change. Vorzimmer (1963; 1970) has rightly pointed out that Jenkin's review had only minimal effect on Darwin, earlier claims by historians notwithstanding. It is quite erroneous, in my opinion, to cite Jenkin's review as a brilliant and devastating criticism of Darwin. In fact, it contains more erroneous assumptions and misleading conclusions than the parts of the *Origin* which it attacks. Particularly weak in his argument are his inappropriate analogies of biological processes with physical phenomena, as for instance the comparison of evolutionary change with the flight of a cannonball. To a modern reader, it is astonishing that such physical scientists as Haughton, Hopkins, and Jenkin thought that by applying the thinking of the physical sciences, they could cope with such extraordinarily complex phenomena, unparalleled in the inanimate world, as the evolution of biological systems (Hull, 1973).

Reasons for the Strength of Resistance to Selection

Considering how rapidly the theory of evolution was accepted by biologists, it is puzzling how reluctant they were to adopt natural selection. It was not until the "evolutionary synthesis" (see Chapter 12) of the 1930s that natural selection was accepted by the majority of the biologists as the exclusive direction-giving mechanism in evolution. And even then natural selection remained so alien a concept to philosophers and nonevolutionists that, to the present day, evolutionists have to make great efforts to demonstrate the efficacy of selection to nonevolutionists.

The opposition, of course, was not total. Almost all opponents allowed for some selection but claimed that major evolutionary phenomena and processes could not be explained by it. Darwin himself, as we know, allowed for some nonselectional processes, such as the effect of use and disuse; still, for him selection was by far the most important mechanism of evolutionary change. Most of his opponents considered its importance minor, if not negligible.

What were the factors that contributed to the extraordinary strength of the antiselectionist resistance? It seems that it cannot be ascribed to any one factor but that it was due to the great range of opposing arguments. No one has yet tabulated and analyzed all the objections that were raised, but the more important ones can be found in the writings of Kellogg (1907), Delage and Goldschmidt (1912), Plate (1924), Hertwig (1927), Tschulok (1929), and various French authors such as Caullery, Cuénot, Vandel, and Grassé. Arguments by philosophers were stated by Cassirer (1950) and Popper (1972). What follows is a partial listing of the major factors contributing to the resistance to natural selection.

Threat to the Argument from Design

Explaining the perfection of adaptation by materialistic forces (selection) removed God, so to speak, from his creation. It eliminated the principal arguments of natural theology, and it has been rightly said that natural theology as a viable concept died on November 24, 1859. This is what so outraged not only theologians but all those naturalists for whom natural theology was their basic *Weltanschauung*. For them the theory of natural selection was thoroughly immoral. This is what Sedgwick meant by his anguished outcry, "The pretended physical philosophy of modern day strips Man of all his moral attributes." By repudiating final causes, Sedgwick continued, Darwin's theory "indicates a demoralizing understanding on the part of its advocates. What is it that gives us the sense of right and wrong? of Law? of duty? of cause and effect?" (Hull, 1973). God had given purpose to the world and the moral world order was part of his purpose. If you replace this purpose by the automatic process of natural selection, you not only remove the creator from our concept of the world but you also destroy the foundation of morality.

Sedgwick's outcry thus reveals that much more was involved than the refutation of Paley's concept of designed adaptation. This is made even clearer by K. E. von Baer's (1876) opposition to Darwin. Von Baer was a committed teleologist. The organic world was not only *zweckmässig* (a term very popular with Kant meaning well-adapted) but also *zielstrebig* (goal-directed). Owing to the existence of goal-directedness, so he claims, adaptation precedes the formation of new structures, while according to Darwin adaptation is the result of the formation of structures through natural selection (1876: 332). For a teleologist a trend toward greater perfection, to ever-increasing harmony, is inherent in nature. As Agassiz also stressed, indications of an underlying plan could be seen everywhere. Such a plan could be effected only through the existence of laws, and several such "laws" were proposed in the pre-Darwinian period, like MacLeay's quinarianism as the basis of classification, or Edward Forbes' law of polarity to explain distribution, or Agassiz's three-fold parallelism between ontogeny, fossil progression, and morphological progression (Bowler, 1977b).

The acceptance of evolution made the problem of a well-ordered world particularly acute. If the world had been created in a single instant (or in six days) and had remained constant ever since, its harmony could be explained as the product of a well-conceived plan. The maintenance of order, however, became a serious problem in an evolving and continuously changing world. For the early evolutionists (the *Naturphilosophen,* Lamarck, and Chambers) it was axiomatic that evolution was an "upward" movement. From raw matter and the simplest organisms (Infusoria) there was a steady progression, culminating in the evolution of man. The acceptance of cosmic teleology was thus a necessary corollary to the acceptance of evolution. Being able to explain a temporalized *scala naturae* was thus bought at the price of now having to explain final causes. In fact, so impressive was the image of progressive evolution that even after it had ceased to be a problem for those who accepted the theory of natural selection, it largely retained its credibility not only among a surprisingly large segment of the biological community but particularly among lay people and theologians. The fight against cosmic teleology ("necessity") was the principal object of Monod's *Chance and necessity,* and the same objective has been explicit or implicit in the writings of all evolutionists who have dealt with so-called progressive evolution (for example, Simpson). Yet, to convince someone who is not familiar with the evolutionary mechanisms that the world is not predetermined and—so to speak—programmed seems hopelessly difficult. "How can man, the porpoise, birds of paradise, or the honey bee have evolved through chance?" is the standard question one receives surprisingly often even today. "Does not a world without purpose also leave man without purpose?" it is asked. The acceptance of natural selection thus seems to pose a serious metaphysical dilemma.[13]

The situation in the 1860s and 1870s was aggravated by battles between liberals and conservatives among the theologians (with the liberals trying to accommodate Darwin's thought) and between church and state. For some evolutionists, none more so than Haeckel in Germany, the main significance of evolution and the refutation of all finalism was that this served as a pacemaker of materialism. As Weismann stated it (1909: 4–5), "The principle of selection solved the riddle, how it is possible to produce adaptedness [*Zweckmässigkeit*] without the intervention of a goal-determining force."

Hence natural selection not only eliminated the need for a designer but it also spelled the end of cosmic teleology (finalism). Eventually it became clear that the term "teleological" had been applied to a very heterogeneous medley of phenomena, some of which, but not cosmic teleology, are valid scientific processes (see Chapter 2).[14] Finalism died a slow death, even in evolutionary biology, and was revived by some post-Darwinian evolutionists under the concept of *orthogenesis,* or related concepts (see below).

The publication of the *Origin* produced a decisive change in the relation between science (biology) and religion, particularly in England. Up to 1859 creationism, natural theology, creationist idealistic morphology, and other explanatory theories in which God played an important role were considered legitimate scientific theories. In the controversies, scientist opposed scientist. After 1859 religious arguments quickly disappeared from the statements of scientists, and, as Gillispie (1951) has pointed out perceptively, the controversy became one between organized religion (churches) and scientists.

The Power of Essentialism

Natural selection is meaningless to an essentialist, for it can never touch the underlying essence; it can only eliminate deviations from the type. For him natural selection is by and large only a negative process, able to eliminate the unfit but unable to play a constructive role. Lyell specifically referred to the "purely eliminative power of natural selection" and postulated that it would require some truly creative natural forces to produce the highest plants, animals, and man.

It has been claimed that natural selection, although rejected by vitalists (which indeed it was), was, by contrast, accepted by most mechanists. The facts do not substantiate this claim. Virtually all experimental biologists were mechanists and yet until rather recently, that is, until the evolutionary synthesis, they rejected natural selection almost unanimously. Only those who had adopted population thinking accepted it. The embryologists, in particular, who were always working with a given individual organism and until recent times never studied populations, found it most difficult to understand natural selection. This is very evident in the writings of T. H. Morgan and of E. B. Wilson, who, according to Muller (1943: 35), as late as the 1930s "was not yet quite ready to admit that higgledy-piggledy can provide an adequate explanation of organic adaptations."

518

It is one of the paradoxes of the field that several well-known experimental biologists who were thoroughly familiar with selectionism nevertheless used essentialist arguments in their evolutionary analyses. This is true, for instance, for two such distinguished biologists as Waddington and Monod. It was characteristic of the arguments of the physicists and mathematicians attending the Wistar conference (Moorhead and Kaplan, 1967).

The Ambiguity of the Term "Selection"

Darwin himself was never entirely happy with the term "selection," many of his supporters disliked it, and his opponents criticized and ridiculed it. What he later called natural selection, Darwin called "wedging" on September 28, 1838: "One may say there is a force like a hundred thousand wedges trying [to] force every kind of adapted structure into the gaps in the oeconomy of nature" (D: 135). He adopted the term "selection" in early 1840, when the analogy with the artificial selection of the breeders occurred to him (Ospovat, 1979).

Limoges (1970; 144–146) points out quite correctly that there was considerable doubt in the post-Darwinian literature about the nature of natural selection. Is it an agent, a process, or the result of a process? The greatest weakness of the term is that it implies someone who selects. Darwin's critics were quite outraged by his uninhibited personification of nature. Wherever the natural theologian might have invoked God, Darwin invokes Nature: "Nature cares nothing for appearances, except in so far as they may be useful to any being. She can act on every internal organ, on every shade of constitutional difference, on the whole machinery of life" (*Origin:* 83). "Natural selection is daily and hourly scrutin-

izing, throughout the world, every variation, even the slightest"
(p. 84). Had not Darwin abolished the God of the Bible only to
replace him by a new god, Nature?

Unhappiness among his friends about the term "natural selection" induced Darwin to adopt Spencer's metaphor "survival of
the fittest" in the later editions of the *Origin.* This was rather unfortunate, because now the objection was raised that the entire
theory of natural selection rested on a tautology: "Who survives?
The fittest. Who are the fittest? Those that survive." Darwin, of
course, never said anything of the sort. All he said was that among
the countless variations that occur in every species, some that are
"useful in some way to each being in the great and complex battle
of life should sometimes occur in the course of thousands of generations" (*Origin:* 80), and "that individuals having any advantage,
however slight, over others, would have the best chance of surviving and of procreating their kind" (p. 81). There is nothing circular in this statement. Williams (1973b) and Mills and Beatty
(1979) have analyzed the logical basis of Darwin's argument and
have likewise concluded that no tautology is involved (but see
Caplan, 1978).

519

Repeated attempts were made in the ensuing years to find a
better term than either natural selection or survival of the fittest,
but none was successful. Darwin himself thought of "natural preservation," but even that term fails to bring out the creative component of natural selection, which is due to the alternation between genetic recombination and reproductive success, an aspect
of natural selection emphasized by Julian Huxley, Dobzhansky,
and other recent evolutionists. The modern generation of biologists has been so thoroughly habituated to the term "natural selection" that it no longer experiences the misgivings of the Darwinian period.

Evolution Due to Accident

Darwin's theory rejected uncompromisingly the existence of
any finalistic factor in the causation of evolutionary change, and
this strengthened the resistance of many to natural selection. Most
of his contemporaries could see only a single alternative to teleological determination, this being accident. Indeed, until modern
times many scientists and philosophers have rejected selection,
saying that it was unthinkable that "the marvelous harmony of
organisms" could all be due to accident. Those who raised this
objection overlooked the fact that natural selection is a two-step

process. At the first step, the production of genetic variability, accident, indeed, reigns supreme. However, the ordering of genetic variability by selection at the second step is anything but a chance process. Nor is selection, as has sometimes been claimed, something that is intermediate between chance and necessity, but something entirely new that escapes the dilemma of a choice between these two principles. No one has stated this better than Sewall Wright (1967: 117): "The Darwinian process of continued interplay of a random and a selective process is not intermediate between pure chance and pure determination, but in its consequences qualitatively utterly different from either."

It is remarkable how generally it is overlooked that with natural selection Darwin had introduced an entirely new and revolutionary principle which is not at all vulnerable to the objection that his theory relies entirely on accident. Darwin himself apparently forgot this occasionally, since he confessed at one time that he was greatly bothered by "the extreme difficulty or rather impossibility of conceiving this immense and wonderful universe . . . as the result of blind chance or necessity" (1958: 92), as if these were the only two available options.

Natural selection has been particularly puzzling to the physical scientists, because it is so different from physical theories and laws. It is neither strictly deterministic nor predictive but probabilistic with a strong stochastic element. Whether one likes or dislikes such an undisciplined process is irrelevant. The fact is that it occurs in nature and that it is of overwhelming importance for the fate of genotypes.

Methodological Objections to Natural Selection

Scientific objections to the theory of natural selection were not the only ones to be raised. It must be remembered that in the *Origin* the principles and the methodology of evolutionary biology were presented to the world for the first time. Nearly all of Darwin's opponents were mathematicians, engineers, physicists, philosophers, theologians, or other kinds of scholars whose knowledge of biology was deplorably limited. Nevertheless, they felt that evolution was a subject of sufficient importance to justify anybody's participation in its discussion. Being unable to come up with scientific arguments, they resorted instead to the claim that Darwin had violated the canons of sound scientific methodology (Hull, 1973). They said that his work was speculative, hypothetical, inferential, and premature. They also criticized his conclusions on the basis that they had not been reached by induction, which they claimed was "the only sound scientific method." Furthermore, again and again, the theory of evolution was declared to be unacceptable because it was not based on experiment (as late as 1922 by Bateson). Comparative-observational evidence was not scientific, it was said; it had to be experimental.

These criticisms are all based on the assumption, now fully understood to be wrong, that phenomena and processes containing time-generated information must be studied by the same methods as purely functional processes. More broadly, it assumes that the methods that had been found useful in the physical sciences (with its very limited universe of phenomena) are fully sufficient for all sciences. The critics who accused Darwin of not following proper scientific methods and of not supplying ironclad proof did not realize that science in the mid-19th century was undergoing a methodological revolution. Darwin's consistent application of the hypothetico-deductive method (Ghiselin, 1969) greatly helped to establish the respectability of this method and led to a revision of the criteria needed to establish the validity of a theory (see Chapter 2). It was Darwin more than anyone else who showed how greatly theory formation in biology differs in many respects from that of classical physics (Hull, 1973; Hodge, 1977; 1981).

Historical narratives can only rarely (if at all) be tested by experiment. However, one can "speculate" about them, as Darwin would have said, that is, one can formulate hypotheses based on observations, which can then be tested by further observations. And this is what Darwin did incessantly. Darwin's speculating was a well-disciplined procedure which he used, as does every modern scientist, to give direction to the testing of further observations and, where feasible, to the planning of experiments.

By far the most important departure in Darwin's methodology was that he demonstrated the legitimacy of why-questions. Evolutionary causations can be analyzed *only* by asking why-questions. "Why is a leaf insect green?" is not a search for final causes but for past (or current) selection pressures. "Why are the animals of the Galapagos more closely related to the animals of South America than to those of other Pacific islands?" again is a perfectly legitimate scientific question. The hypothetical answer that the fauna must have reached the islands by transoceanic colonization permitted all sorts of predictions, for instance that this fauna came most likely from the nearest source area (South America) or

521

that nonflying animals (unless they had special means of dispersal) would have much greater difficulties in reaching islands than flying animals; indeed, terrestrial mammals are rare or absent from true islands, but bats reach most of them.

By his new methodology Darwin transferred the whole realm of ultimate causations from theology to science. He was fully aware of what he was doing. For one set of phenomena after another, he asked, "Is this better explained by (special) creation or as the result of evolution by common descent?" (Gillespie, 1979).

Absence of Proof

Even some of Darwin's most ardent supporters admitted that the theory of natural selection was based almost entirely on deductive reasoning. His opponents called this method pure speculation and demanded inductive or experimental proof. Almost the only thing Darwin could offer was analogy with artificial selection. But, as T. H. Huxley admitted, no animal breeder had ever succeeded in producing a new, reproductively isolated species by selection. And the most aberrant races of dogs and pigeons were called "pathological" by Kölliker, who quite rightly insisted that they would never be able to maintain themselves in nature.

The discovery of mimicry by H. W. Bates (1862) came as a godsend, and Darwin at once wrote a joyous and highly laudatory review of it. What Bates had observed was that each species or geographic race of the unpalatable (if not poisonous) heliconid butterflies was associated where it occurred with one or several species of edible butterflies which mimicked them in coloration (Batesian mimicry). But more than that! When a heliconid species varied geographically (and most of them do, often very strikingly), its mimicking satellites underwent exactly the same changes as their unpalatable models. Bates (1862: 512) quite rightly concluded that this type of variation could only be due to "natural selection, the selecting agents being insectivorous animals, which gradually destroy those sports or varieties that are not sufficiently like [their models] to deceive them." The geographical variation of the butterflies, some of it being very gradual, showed furthermore that the mimicry was not acquired by major saltations but gradually by natural selection. Later genetic analysis confirmed this conclusion.[15]

Bates' work, a most brilliant piece of natural-history research, was soon confirmed by other investigators. Wallace discovered a similar situation in Indomalayan papilionid butterflies and new instances of various kinds of mimicry are discovered each year. The most important extension of the mimicry principle was due to Fritz Müller (1879), who showed that mutual mimicry may also occur among unpalatable, poisonous, or venomous animals such as wasps and snakes (Müllerian mimicry). Since their potential predators apparently have to learn, at least in part, which color pattern to avoid, it pays for a guild of owners of warning coloration to adopt a single pattern in a given region. There is a selective advantage for any member of the guild to have this warning color pattern. Not surprisingly, in line with the demands of natural selection, all species belonging to a single Müllerian complex vary geographically in a parallel manner (Turner, 1977).

Much research in evolutionary biology, particularly after 1930, has been devoted to the endeavor to establish the selective value of various attributes of plants and animals (see Chapter 12).

Impossibility of Falsification

According to Popper, only such theories are scientific that can be "falsified." Several philosophers opposed to natural selection have stated that it is impossible to falsify any claims made on behalf of natural selection. Here one must make a distinction between the theory of natural selection as such and the application of natural selection to specific cases. As soon as one deals with specific cases, one can make predictions that can be falsified in principle, by testing them against various assumptions. It is, of course, also true that the exclusive reliance on falsifiability is questioned by several contemporary philosophers. Finally, since few if any neo-Darwinians claim that every component of the phenotype and every evolutionary change is the result of ad hoc selection, the nonfalsifiability argument has not much force.

Ideological Resistance

Inevitably, the concept of natural selection was eventually applied to man. This resulted in various excesses (such as racism) but also in the counter claim that an assumption of genetic differences of selective significance in man was in conflict with the principle of equality. Extreme egalitarianism led to the development of strongly environmentalist schools, particularly in American anthropology and in behaviorist psychology. As noble as these movements were in their basic ideology, and perhaps as necessary as they were to combat racism and social prejudice, the major claims of these schools were not substantiated by any concrete evidence but were based on an unbiological concept of equality. The situation was made worse when Lysenkoism raised its ugly head in the USSR and when certain Marxist groups in the western countries decided to attack genetics and to promote environmentalism. Some of the attacks against sociobiology in recent years originated from the same ideology. Linking Darwin's name with Herbert Spencer's social Darwinism was also detrimental for the acceptability of natural selection (Freeman, 1974; Nichols, 1974; Hertwig, 1921; Greene, 1977; Bannister, 1979).

524

Empirical Objections

The students of diversity raised some observational objections to natural selection. On the basis of the survival of superior individuals and the gradual change of populations, one would expect complete continuity in nature, they claimed. What one actually found was nothing but discontinuities: All species are separated from each other by bridgeless gaps; intermediates between species are not observed. How could the sterility barrier between species have possibly evolved by gradual selection? The problem was even more serious at the level of the higher categories. Higher taxa, like birds and mammals, or beetles and butterflies, are far too distinct from each other, the skeptics said, to permit the explanation of their origin through gradual evolution by natural selection. Furthermore, how can selection explain the origin of new structures like wings, when the incipient new organs can have no selective value until they are large enough to be fully functional? Finally, what is the role of the very small differences among the individuals of a population, seen in all gradual evolution (including geographic variation), when, it was said, the differences are far too small to be of selective significance? The defenders of gradual evolution had to be able to refute these objections and had to provide evidence in favor of a rather formidable list of prerequisites of their theory:

(1) Availability of an inexhaustible supply of individual variation

(2) Heritability of individual variation

(3) A selective advantage of even the slightest variation to be of evolutionary significance

(4) No limits in the response to selection

(5) An explanation by gradual variation of major evolutionary novelties and the origin of higher taxa

Neither Darwin nor his supporters were at first able to supply this evidence. As a result the traditional objections were raised again and again, up to recent times, most forcefully by Schindewolf (1936), Goldschmidt (1940), and some French zoologists (Boesiger, 1980). It was not until the period of the new systematics that Rensch, Mayr, and others demonstrated the populational origin of the discontinuities (Mayr, 1942; 1963) and that the geneticists supplied the evidence on the variation needed to permit natural selection to be effective.

525

ALTERNATE EVOLUTIONARY THEORIES

The acceptance of the theory of evolution created a dilemma for all those who rejected Darwin's explanatory principle of natural selection. What else could be the factor (or factors) controlling evolution if it was not natural selection? Many alternate explanations were proposed in the eighty years after 1859 and were in fact far more popular during that period than natural selection. In order not to misrepresent the climate of opinion, let me emphasize that natural selection was not totally condemned. Many biologists admitted, "Of course, natural selection occurs, but it cannot be the exclusive causal factor in evolution because too many evolutionary phenomena cannot be explained by it." One must therefore remember that the mere acceptance of a certain amount of selection did not make an author a Darwinian if he simultaneously accepted the existence of other controlling factors in evolution. The three theses of Darwin and the neo-Darwinians that were found to be particularly unacceptable by the anti-Darwinians were gradualism, the rejection of soft inheritance, and the rejection of finalism (teleology). One can classify, therefore, the various anti-Darwinian theories according to which of these three components of Darwin's theory they opposed specifically. They will therefore be discussed under these three headings: (1) saltationist theories, (2) neo-Lamarckian theories, and (3) orthogenetic theories (Kellogg, 1907; Mayr and Provine, 1980).

Saltationist Theories
Early theories that opposed Darwin's gradualism (His, Kölliker, and others) have already been discussed (see above). They had relatively few followers in the 1860s to 1880s. After 1894 saltationist theories gained rapidly in popularity and were dominant

early in the century under the name "mutationism." The role of these theories in the twentieth-century controversies will be discussed in Chapter 12.

Neo-Lamarckian Theories

The most determined and most successful opposition to Darwinism was posed by theories usually combined under the name "neo-Lamarckism."[16] The paradoxical aspect of this designation is that the most fundamental component of Lamarck's theory—that there is a finalistic element in evolution leading phyletic lines of organisms to ever greater perfection—was not the principal thesis of neo-Lamarckism. Admittedly, however, neo-Lamarckism shared with Lamarck two major concepts: that evolution is "vertical" evolution, consisting in an improvement of adaptation (neglecting or disregarding altogether the origin of diversity), and second, that acquired characteristics of an individual can be inherited (soft inheritance). Neo-Lamarckism, thus, can be considered as much a theory of inheritance as a theory of evolution, and a discussion of soft inheritance is therefore included in Chapter 16.

The idea that the environment exerts a decisive influence on the characteristics of an organism goes back to ancient folklore. The idea was exceedingly popular among philosophers, particularly in the period preceding and during the Enlightenment (Locke, Condillac). Among English writers David Hartley (1749) is a good example of an extreme environmentalist. That "changes in the conditions of life" greatly contribute to the "variation of the type" was accepted by Buffon, Linnaeus, Blumenbach, and Lamarck, all of whom also accepted to a greater or lesser degree that such acquired characters could be inherited. Blumenbach, for instance, believed that the dark-skinned human races had been derived from light-skinned races by the action of the strong sunlight in the tropics on man's liver. Darwin was no exception (see Chapter 16). He always believed in some effect of use and disuse and its inheritance, and he adopted the theory of pangenesis to account for it. However, he attributed to it only a minor role in evolution, as compared to selection.

"Neo-Lamarckism" covers an exceedingly heterogeneous group of ideas. No two neo-Lamarckians had quite the same views, but it would lead too far to describe their various theories in detail. One of them, designated as Geoffroyism, ascribes evolutionary change to the direct influence of the environment. Even though

Lamarck had expressly rejected the occurrence of such direct induction, those who, late in the nineteenth century, believed in such a process were included among the neo-Lamarckians. Many naturalists believed in it as a process co-existing with natural selection. They argued, for instance, that gradual geographic variation could not be explained except through environmental induction. Geoffroyism had many followers, particularly in the early decades of the twentieth century, as an "opposition party" to mutationism with its reliance on discontinuous saltations as the only source of evolutionary change. Environmental induction seemed to be the only way to account for gradual variation as it was everywhere observed by naturalists.

527

Concepts related to use and disuse, combined with an inheritance of acquired characters, were prominent among the neo-Lamarckians' theories. This is true for Cope's "law of growth and effort." An organ that had become more useful in a new environmental situation would have its growth enhanced in each generation and thus become ever better adapted to its environment. This is, of course, very similar to some of Lamarck's ideas. The suggested mechanism for such a process was "that the germ cells carry a record of the past efforts of the growth force in a manner analogous to memory" (Bowler, 1977a: 260). Here Cope had a mechanism that would produce adaptation naturally without recourse to design or supernatural forces. Most American evolutionists prior to 1900 were neo-Lamarckians.

Many neo-Lamarckian theories invoke mental forces. This started with Lamarck's own "efforts" to satisfy "needs" (erroneously interpreted as "willing" to produce new structures); "consciousness" is mentioned by Cope and other neo-Lamarckians, and reaches its climax in Pauly's psycho-Lamarckism, which had a considerable influence on Boveri and Spemann (Hamburger, 1980). What characterized all neo-Lamarckian theories was the postulate that something experienced by one generation could be transmitted to the next generation and become part of its heritage. Consequently, all neo-Lamarckians supported the inheritance of acquired characters. As long as the nature of the genetic material was unknown, neo-Lamarckism explained adaptation far better than the haphazardous process of random variation and selection. As soon as minimutations and recombination were recognized as the genetic material of evolution and the assumption of soft inheritance had been refuted, the conversion of the younger neo-Lamarckians to Darwinism took place quite rapidly.

Orthogenetic Theories

The third set of anti-Darwinian theories, likewise originating in remote history, is based on the concept that somehow evolution is due to a finalistic component.[17] Although the *scala naturae* was static and although the author of Genesis was not at all thinking of evolution when he had God create man on the sixth day of creation, a necessary sequence from lower to higher is nevertheless implicit in both cases. Indeed, the assumption of some sort of cosmic teleology was very widespread among philosophers and in many religions.

Erasmus Darwin considered "the faculty of continuing to improve" as one of the basic properties of life itself: "Would it be too bold to imagine, that in the great length of time, since the earth began to exist, perhaps millions of ages before the commencement of the history of mankind, would it be too bold to imagine, that all warm-blooded animals have arisen from one living filament, which THE FIRST GREAT CAUSE endued with animality, with the power of acquiring new parts, attended with new propensities, directed by irritations, sensations, volitions, and associations; and thus possessing the faculty of continuing to improve by its own inherent activity, and of delivering down those improvements by generation to its posterity, world without end!" (1796, I: 509).

For Lamarck, evolution was clearly a movement to ever greater perfection, and the progressionists among the geologists also saw an upward trend in the creation of each new fauna and flora, a trend that would keep organic life perfectly adapted to the changing conditions of the earthly environment (Agassiz, 1857; Bowler, 1974b). It is immaterial whether the presumed mechanism was a set of "laws" that would automatically guarantee perfect adaptation or the constant immediate attention of the creator, the end-product was the same: an inexorable movement toward perfection.

Teleological thinking was extremely widespread in the first half of the nineteenth century. For Agassiz and other progressionists the sequence of fossil faunas simply reflected the maturation of the plan of creation in the mind of the creator. Theistic as well as deistic philosophers needed to uphold the universal operation of final causes in nature, since it was one of the most important, if not the only piece of evidence for the existence of a creator. Theists like Sedgwick and K. E. von Baer saw purpose everywhere in nature. In a review of Darwin's *Origin* von Baer wrote, "My goal is to defend teleology" because "natural forces must be coordinated or directed. Forces which are not directed— so-called blind forces—can never produce order . . . If the higher forms of animal life stand in a causal relationship to the lower, developing out of them, then how can we deny that nature has purposes or goals?" Louis Agassiz, likewise, sarcastically rejected the efficacy of blind forces. Even Darwin had originally accepted finalism, hence his remarkable statement (*Notebook* B, p. 169): "If all men were dead, then monkey make men.—Man makes angels." As Herbert (1977: 199, 200) points out, the study of geographic variation soon made Darwin give up any orthogenetic notions. Comparing vicarious species, he found no evidence for necessary, built-in progressive trends. And after he had adopted natural selection, he had no longer any need for a finalistic principle.

Among the numerous defenders of a finalistic principle in evolution, Nägeli (1865; 1884) and Eimer (1888) developed the most elaborate theories. These were based either on the assumption that a perfecting principle is immanent in all organic life or else that the (genetic) constitution places a constraint on all organisms in such a manner that evolution can proceed only in a more or less rectilinear direction. Eimer, adopting a term first proposed by Haacke, called the perfecting principle *orthogenesis;* other biologists and philosophers coined different names for essentially the same postulated force in evolution: Berg, *nomogenesis;* Henry Fairfield Osborn, *aristogenesis;* and Teilhard de Chardin, the *omega principle.* A belief in some sort of intrinsic, direction-giving force was particularly widespread among paleontologists, who everywhere saw evidence for evolutionary trends, extending over millions, if not tens or hundreds of millions, of years. The thesis, widespread among anthropologists, that human evolution goes by necessity through a definite series of stages likewise belongs here (White, 1959).

When it came to explaining the causation of the orthogenetic principle, there was much vagueness and uncertainty among its defenders. Some of them saw in evolution simply the unfolding of the potential of a basically unchanging essence, "evolution" in the most literal sense. It was, one might say, an application of the principle of preformation (in embryology) to evolution. This was essentially Louis Agassiz's thinking and it was endorsed as recently as 1914 by the geneticist Bateson. Others referred to mysterious laws which cause orthogenetic evolution: "Evolution of organisms is the result of certain processes inherent in them, which are based upon law. Purposive structure and action are thus a fundamental property of the living being" (Berg, 1926: 8), a statement which, of course, explains nothing. Eimer attempted to get away from an immanent teleological principle by suggesting that the environment directed variation, but the adequate response of the organism still rested on an immanent teleological capacity.

The Darwinians rejected any internal direction-giving mechanism or purposive principle for a number of reasons. First, because the defenders of orthogenesis were quite unable to provide any reasonable mechanism consistent with a chemico-physical explanation. Second, because a detailed study of such trends invariably revealed numerous irregularities and sometimes even complete reversals (Simpson, 1953). Finally, because when evolutionary lines split, the daughter lines may have very different trends, an occasional one even reversing the previous trend. Again this is incompatible with an integral mechanism. The observation that larval and adult stages of metamorphosing insects and marine organisms often display entirely different trends was rightly cited by Weismann and Fritz Müller as a further argument against orthogenesis.

In due time all theories defending orthogenesis were refuted, but this does not justify ignoring this literature. The major representatives of orthogenesis, whether paleontologists or other kinds of naturalists, were keen observers and brought together fascinating evidence for evolutionary trends and for genetic constraints during evolution. They were right in insisting that much of evolution is, at least superficially, "rectilinear." In horses, the reduction of the toe bones and the changes in the teeth are well-known examples. In fact, the study of almost any extended fossil series reveals instances of evolutionary trends. Such trends are of importance to the evolutionist because they reveal the existence of continuities that are worth exploring, and have therefore been given much attention in the current evolutionary literature.

Trends may have a dual causation. On the one hand they may be caused by consistent changes of the environment, such as the increasing aridity of the subtropical and temperate zone climate during the Tertiary. This set up a continuing selection pressure which resulted in the toe and tooth evolution of the horses. A response to such a continuing selection pressure is what Plate had in mind when he introduced the term "orthoselection" (1903). On the other hand, trends may be necessitated by the internal cohesion of the genotype which places severe constraints on the morphological changes that are possible.[18] Hence, evolutionary trends are readily explained within the explanatory framework of the Darwinian theory and do not require any separate "laws" or principles.

EVOLUTIONARY PROGRESSION, REGULARITIES, AND LAWS

The Darwinians had considerable difficulty in making it clear to their opponents that to deny the existence of an internal perfecting principle did not mean a denial of observed evolutionary progression. Denial of a progression from the infusorians to the angiosperms and vertebrates might well imply a rejection of evolution altogether. Darwin, fully aware of the unpredictable and opportunistic aspects of evolution, merely denied the existence of a lawlike progression from "less perfect to more perfect." It was in this spirit that he once reminded himself "never to say higher or lower."

Of course, Darwin did not follow his own advice and referred in the *Origin* frequently to evolutionary progress (pp. 149, 336–338, 388, 406, 441, and 489). This was necessary not only to refute Lyell's concept of a steady-state world but also to counter a newly developed school which denied any difference in perfection between the simplest and the most complex organisms. Ehrenberg, for instance, claimed that there is no structural advance from the lowest organisms, the infusorians, to the highest, the vertebrates. All of them have the necessary structures to perform all animal functions. All are "perfect." This curious claim completely ignored the fact that there is a tremendous advance from the diffuse nerve fiber of a coelenterate to the magnificently evolved central nervous system of a cetacean or primate. Ehrenberg's claim, of course, had strong antievolutionary implications. Lyell likewise tended to deny any aspect of progression in the sequence of faunas from the lowest fossiliferous beds to the present, with the single exception of the recency of man. Obviously these claims constituted an implicit denial of improvement through natural selection. Darwin realized that "naturalists have not as yet defined to each other's satisfaction what is meant by high and low forms," and yet, he continues, "The most recent forms must, on my theory, be higher than the more ancient; for each new species is formed by having had some advantage in the struggle for life over other and preceding forms" (*Origin:* 337).

Indeed, the series of morphological and physiological innovations that have occurred in the course of evolution can hardly be described as anything but progress. I think of such phenomena as photosynthesis, eukaryoty (organization of the nucleus), multicellularity (metazoans, metaphytes), diploidy, homoiothermy, predation, and parental care, to mention only a few of such evolutionary innovations made in the three billion years since the origin of the first prokaryotes. On almost any measure one can think of, a squid, a social bee, or a primate, is more progressive than a prokaryote. Yet, the word "progressive" implies a linearity which one does not find. Nor does one find only a single sequence since there is progressive evolution in plants, in arthropods, in fishes, in mammals, indeed in almost any group of organisms, each lineage displaying a very different expression of progress.

A careful analysis of everything Darwin wrote on evolutionary progress shows that he did not contradict himself. What he objected to was finalism, that is, a belief in an intrinsic drive to perfection, controlled by "natural" laws. Where Darwin encounters improvements in the course of evolution, he finds that they can be easily explained as the a posteriori results of variation and natural selection. Evolutionary progress, where it occurs, is not a teleological process, a conclusion in which all evolutionary biologists agree with Darwin.

The main objection raised by the anti-Darwinians has always been that the living world is full of progressive trends but that it is inconceivable they should be caused by random variation and natural selection. The Darwinian answers, Why not? After all, any improvement, any new structural, physiological, or behavioral invention made in any gene pool can lead to evolutionary success and thus to progress as it is traditionally defined. This Darwin had already seen quite clearly.

A far more intransigent problem has always been how to define progress. Here no two authors agree. Complexity is certainly not necessarily a measure of progress, for in many evolutionary lines the oldest members are the most complex, and progress has consisted in simplification. Almost no one has been able to sidestep entirely Lamarck's measure of progress, a comparison with man. When Julian Huxley (1942) makes "control of the environment" a measure of progress, there is no doubt that this sets man on a pinnacle well above any other organism, even though termites, bees, and some other organisms have been reasonably successful in controlling their environment. Independence of the environment is perhaps a better measure, another good one being the capacity of the nervous system to store and utilize information. Open behavior programs must surely be considered as more progressive than rigidly closed ones.

In spite of all these manifestations of evolutionary advances, Darwinians, on the whole, have been rather reluctant to talk of evolutionary progress. They seem to be afraid that this might be interpreted as an endorsement of the existence of teleological factors. Also, there seems to be an intellectual contradiction between the spirit of progress and the materialistic means (struggle for existence) by which it is achieved. Finally, the staggering frequency of extinction adds to the questionable value of any temporary progress seemingly achieved by any evolutionary line. When all these difficulties are kept in mind, it becomes apparent why the definition of evolutionary progress is so difficult, if not impossible.[19]

533

Not all, and perhaps only the smallest amount, of evolution consists of progress. Much of the genetic change caused by natural selection simply serves to maintain a status quo. In order to keep step with the evolutionary (genetic) changes of one's competitors, enemies, sources of food, and even of the physical environment, a population must change from generation to generation. This is what Van Valen (1973) has referred to as the "Red Queen Principle" ("One must run in order to be able to stay at one's place"). More importantly, "Any gain in fitness by one unit of evolution is balanced by losses in fitness by others." This is true at various levels. Macromolecules, for instance, regularly replace amino acid residues in order to maintain an optimal interaction with their molecular milieu. When an organism (population or species) falls behind in its race for maintaining the optimal balance, it faces extinction.

In many cases success simply consists in becoming different or more different, thus reducing competition. Darwin (*Origin:* 111) saw this clearly when he proposed the principle of character divergence. It promotes continuous change, but not necessarily progress. Indeed, it has induced endless phyletic lineages to enter evolutionary deadend streets.

A denial of evolutionary progress does not necessarily mean that the process of evolution is chaotic. That it is not has been stressed by a number of authors who have recognized evolutionary laws (for example, Rensch, 1960). One particular regularity, the one which relates ontogeny to phylogeny, has been considered by numerous authors from Haeckel (1866) and Severtsov (1931) to living authors. This subject is a terminological and conceptual jungle into which Gould (1977) has brought some light and order. Two trends are encountered most frequently: (1) an addition of new characters late in ontogeny, and (2) a shift in the maturation of the gonads, with the result that the organism either reproduces in an immature or larval stage (neoteny) or that it postpones adulthood (retardation). Obviously these various "life history strategies" are selected for the greater reproductive success which these shifts in maturation guarantee. Although these processes are particularly important in plants (Stebbins, 1974; 1979) and invertebrates, man has often been described as a fetalized ape (Bolk, 1915). There are, however, several ways by which the changes of human life history (as compared to that of apes) can be described and no consensus has yet been achieved (Starck, 1962).

Any shift in the adaptive zone which an organism occupies will initiate new evolutionary trends. For instance, there are well-known trends in the restructuring of cave animals and of parasites. In plants there are trends leading from trees to perennial herbs and to annuals. There are trends involving modes of reproduction and the nature of the karyotype. All this demonstrates that the mere fact that variability as such is undirected does not preclude the possibility that natural selection converts such variability into more or less regular trends. New trends may originate when organisms invade new adaptive zones or when the environment undergoes a change (including the occurrence of new predators or new competitors). Each new level of complexity in organic systems favors the beginning of new trends (Huxley, 1942; Stebbins, 1969; 1974).[20]

12 ✍ Diversity and synthesis of evolutionary thought

THE DIVERSITY of opinion among evolutionists in the eighty years following the publication of the *Origin of Species* is quite extraordinary. Each branch of biology had its own tradition, and so did each country. Germany embraced evolutionism quickly and rather completely (*L.L.D.*, III: 88). Ernst Haeckel, Germany's most enthusiastic evolutionist, both helped and hindered the spread of Darwinian thinking. He did a very effective job of popularizing Darwinism, but used it at the same time as a weapon against all forms of supernaturalism, particularly Christianity, thereby provoking counterattacks in which evolutionism was equated with materialism and immorality. This could not prevent the spread of evolutionary thinking as such, but it was an important factor in the almost universal rejection in Germany of the theory of natural selection.[1]

Evolution by descent with modification was rather generally adopted also in England within the decade after the publication of the *Origin,* at least by the biological community. Natural selection, by contrast, was largely found unpalatable. It was accepted only by a few naturalists—Wallace, Bates, Hooker, and some of their friends, and later by Poulton, Meldola, and other entomologists—but not by a single experimental biologist.[2] Of decisive importance for subsequent developments in Britain was the fact that Ray Lankester was converted to selectionism by reading some of Weismann's essays. He enthusiastically supported the invitation issued to Weismann to give lectures in England. It was Ray Lankester who founded a school of selectionism at Oxford, represented through several generations by E. S. Goodrich, Julian Huxley, G. de Beer, and E. B. Ford. There was nothing like it at Cambridge or at University College, London, until R. A. Fisher and J. B. S. Haldane began to publish.

In the United States, in spite of Asa Gray's enthusiastic support, evolutionism had a much harder time. There being only a small professional class of biologists and paleontologists, much of the controversy was conducted by writers, theologians, and philosophers. Still, with the death of Agassiz in 1873 and the acceptance of evolution by Dana in 1874, all resistance to evolution ended among the professionals.[3] The theory of natural selection, by contrast, encountered continuing resistance. The situation was complicated by the temporary popularity of Spencer's so-called social Darwinism.[4] Partly as a reaction to it and partly as a result of traditional American egalitarianism, an extreme environmentalism developed in psychology and anthropology which minimized, if not denied altogether, any genetic contribution to the differences among human individuals. Up to a point this tradition is alive even at the present day. It is evident that, except for details of chronology, the history of the acceptance of Darwin was essentially the same in Germany, Britain, and the United States. Evolution was accepted quickly but natural selection at first by only a minority.

In France the resistance to Darwin was much greater than in any other major western country. Not a single leading French biologist came out in favor of selection after 1859 and even evolutionism as such did not begin to spread until the 1870s.[5] The first chair of evolutionary biology at the Sorbonne was created for Giard in 1888. When evolution was finally adopted in France in the 1880s and 90s, it was in the form of neo-Lamarckism, which at the same period enjoyed considerable popularity also in the United States and in Germany. Although natural selection was temporarily supported by an occasional author and was adopted by Teissier and l'Héritier in the 1930s, its more general acceptance did not take place in France until after 1945 (Boesiger, 1980).

Paradoxically (considering the subsequent success of Lysenko), in the period up to the end of the 1920s there was probably no other country in which Darwinism, including the theory of natural selection, was as widely accepted as in Russia. Originally this was primarily for political reasons but in part it was also due to the flourishing condition of population systematics in Russia (Adams, 1968). The impact of this situation on the further development of population genetics will be discussed below.

NEO-DARWINISM

As is described in Chapter 16, Darwin, although largely a champion of hard inheritance, still allowed some scope for the effects of use and disuse and other aspects of soft inheritance. As the knowledge of cytology and particularly that of chromosomes grew, several authors began to question any inheritance of acquired characters. These doubts were expressed rather casually and did not receive much notice. The rejection of soft inheritance made no real headway until Weismann, in 1883 and 1884, published his germ-track theory and proposed a complete and permanent separation of soma and germ plasm.[6] The total rejection of any inheritance of acquired characters meant a rejection of all so-called Lamarckian, Geoffroyian, or neo-Lamarckian theories of evolution. In fact, it left only two conceivable mechanisms of evolution: saltation (evolution owing to sudden, major departures from the existing norm) and selection among minor variants. Weismann adopted an uncompromising selectionism, a theory of evolution designated by Romanes (1896) as neo-Darwinism. It may be defined as the Darwinian theory of evolution without recourse to any kind of soft inheritance. Indeed, Weismann accepted most other components of Darwin's theory except pangenesis, now no longer needed.

537

The elimination of soft inheritance, which up to that time had been considered a major source of individual variability, obliges the evolutionist, said Weismann, "to look for a new source of the phenomenon, upon which the processes of selection entirely depend." His knowledge of cytology permitted him to name the particular phenomenon most likely to supply the needed genetic variability. It was the process now called "crossing over." If such a rebuilding of chromosomes during gamete formation (meiosis) did not exist, genetic variation (except for occasional new mutations) would be limited to a reassortment of the parental chromosomes. By contrast, chromosomal recombination has the consequence that "no individual of the second generation can be identical with any other . . . [in every generation] combinations will appear which have never existed before and which can never exist again." No one before Weismann had understood the extraordinary power of sexual recombination to generate genetic variability.

The importance of recombination in evolution was at first rather neglected in the genetic literature which, written in terms of bean-bag genetics (see Chapter 13), presented evolution by the formula "mutation and selection." Actually, the genotypes which are the target of selection are the immediate product of recombination rather than of mutation. It was not until the work of C. D. Darlington (1932; 1939)[7] and Stebbins (1950: chap. 5) that the evolutionary significance of systems of recombination ("genetic systems") was fully appreciated.

538   Much has been written about Weismann's genetical and cytological theories, but the development of Weismann's ideas on evolution has been rather neglected by historians. Until such an analysis has been made, only a few tentative statements can be offered. In 1872 Weismann entered into the controversy between Moritz Wagner and Darwin on the role of geographic isolation and displayed a rather remarkable lack of understanding of the problem. Certain comments made later in the 1870s indicate that Weismann at that time still believed in soft inheritance. It was not until 1883 that he categorically rejected soft inheritance, and in the next few years he emphasized the role of recombination (*amphimixis*). It was in these years that he developed the theory, until quite recently almost universally adopted, that the selective advantage of sex is its capacity to multiply genetic variability at a high rate, and thus provide more abundant material for selection. Weismann was the first to ask questions about the regulation of the life span (age at time of death) by natural selection (see also Korschelt, 1922). More broadly, he introduced an entirely new way of looking for the meaning, that is, the selective value, of every aspect, morphological or otherwise, of organisms. Everything in the living world was for him the product of the "Allmacht der Naturzüchtung" (power of natural selection).

After Weismann had passed the age of sixty, however, he began to become somewhat uncertain about the ability of selection to control evolutionary trends unaided, and he proposed the principle of "germinal selection," admitting the improbability "that the adaptations *necessary* for the existence of organisms could originate by *accidental* variations." He postulated therefore the occurrence of "directed variation . . . which is caused and guided by the conditions of life of the organisms" (1896: iv). Weismann categorically rejected any inner (orthogenetic) drives, and postulated instead that the selection of certain characters, let us say longer tail feathers in a bird, favors simultaneously those genotypes that have a tendency to vary the length of the tail feathers. What Weismann did was to make a distinction between a given genetic variant and the capacity of the organism to produce variants of the given character, and to point out that both can be selected for. His thinking was ambivalent, however, and he conceded that the extraordinary similarity of model and mimic in certain butterflies "can not be due to 'accidental', but must be due to *directed* variation which is caused by the utility itself" (1896: 45). Weismann now admits that "the Lamarckians were right, when they insisted that what up to now one has exclusively admitted as selection, i.e. the selection of individuals, was not sufficient to explain all phenomena" (1896: 59). Random variation, ordered by selection, was thus no longer considered by Weismann as sufficient.

539

The evolutionary phenomena that troubled Weismann, such as similar trends of variation in many species of the same genus, or the gradual reduction of useless or rudimentary organs (such as the loss of eyes in cave animals), no longer pose serious problems to the evolutionary geneticist of today. The harmonious integration of the genotype places definite constraints on possible genetic variation and this, as well as selection for or against certain regulatory "genes," can account for all observed "orthogenetic trends." These constraints and regulations are the modern equivalent of Weismann's germinal selection.

Weismann's impact on evolutionary biology was far-reaching. He forced every biologist to take a position on the problem of the inheritance of acquired characters. By insisting that there is only one direction-giving force in evolution, selection (even if slightly watered down by his belated theory of germinal selection), he forced his opponents to produce evidence supporting their opposing theories. For the next fifty years most evolutionary controversies dealt with the problems which Weismann had posed with such uncompromising clarity. Furthermore, by his imaginative genetic theories he prepared the ground for the rediscovery of Mendel, an event which ultimately led to the solution of the evolutionary problems that had stumped Weismann.

THE GROWING SPLIT AMONG THE EVOLUTIONISTS

The evolutionists presented a rather solid front as long as they still had to convince the world of the fact of evolution. This was largely true until about 1882, the year of Darwin's death. In the next twenty years, however, more and more events took place which sowed seeds of dissension among them. The first of these was Weismann's uncompromising rejection of any inheritance of acquired characters. The reaction which this provoked was a hardening of the claims of the neo-Lamarckians.

Far more important, although this was probably not fully recognized at the time, was the growing disciplinary radiation of biology. The rise of evolutionism after 1859 coincided with an increasing break-up of zoology and botany into special fields, such as embryology, cytology, genetics, behavioral biology, ecology, and others. Many of these new disciplines of biology were primarily experimental in their approach, and this resulted in the development of an ever widening gap between the experimental biologists on one hand and those others (mostly zoologists, botanists, and paleontologists) who had been raised as naturalists and worked with whole organisms. The experimentalists and the naturalists not only differed in their methods but tended also to ask different questions. Both groups were interested in evolution, but had very different approaches and stressed different aspects of evolution. The experimental evolutionists, most of them originally embryologists, entered the newly developing field of genetics. Their interest was the study of proximate causations, with particular emphasis on the behavior of genetic factors and their origin. Bateson, de Vries, Johannsen, and Morgan were typical representatives of this camp. Several of them had a strong interest or background in the physical sciences and in mathematics. The naturalists, by contrast, were interested in ultimate causations; they tended to study evolutionary phenomena in nature and were particularly concerned with problems of diversity. Paleontologists, taxonomists, naturalists, and geneticists spoke different languages and found it increasingly difficult to communicate with one another.

The naturalists, from the beginning, were particularly fascinated by diversity, its origin and meaning. The species problem was at the center of interest for the taxonomists, whereas evolutionary trends and the origin of the higher taxa intrigued the paleontologists and comparative anatomists. By contrast, diversity was almost totally excluded from evolutionary discussions of the pre-synthesis geneticists. They were only concerned with transformational evolution. Their focus was entirely on genes and characters, and on their changes (transformation) in time. They wrote as if they were unaware that there are taxa, and that they (different populations, species, and so on) are the real actors on the evolutionary stage. Even a phenomenon like adaptive radiation, as Eldredge says correctly (1979: 7), "is viewed as a problem of divergent anatomical specializations among a series of related organisms, rather than as a spectrum of discrete species occupying a diverse array of ecological niches." The emphasis was on transformation, not on diversity. It was this total neglect of diversity, or at best its *deus ex machina* explanation by de Vriesian mutations or Goldschmidt's hopeful monsters, which left the naturalists completely dissatisfied.

541

The disagreement affected almost any aspect of the interpretation of evolution. The three questions that were disputed particularly heatedly were: (1) whether all inheritance is hard (as Weismann believed) or whether some is soft; (2) whether mutation, selection, induction by the environment, or intrinsic tendencies are the principal direction-giving factors in evolution; and (3) whether evolution is gradual or saltational. Kellogg (1907) has well described how many different combinations of opposing interpretations were held by different evolutionists. The disagreement was polarized by the rediscovery of Mendel's rules in 1900, which induced the early Mendelians to use the particulateness (discontinuity) of the genetic factors as evidence for the importance of saltational processes in evolution, particularly in the origin of species. From that date on, one could speak of two camps of evolutionists, the Mendelians and the naturalists (Mayr and Provine, 1980).[8]

The interpretation of evolution by either camp was an unfortunate mixture of valid ideas and of misconceptions. The naturalists had erroneous ideas on the nature of inheritance and of variation, while the experimental geneticists, dominated by typological thinking, ignored the existence of populations, and concerned themselves with the frequency of genes in closed gene pools. They ignored the problems of the multiplication of species, of the origin of higher taxa, and of the origin of evolutionary novelties. Both camps were quite unable to understand—and therefore to refute effectively—the arguments of their opponents.

The two camps also represented different research traditions. The naturalists very much continued in the original Darwinian tradition of studying natural populations and paying particular attention to the problem of the origin of diversity. Most importantly, they continued Darwin's tradition of asking questions about ultimate causation. Any question concerning the "why?" of an adaptation or of any other biological phenomenon was answered prior to Darwin with, "It is due to design," or "It is the result of natural law laid down by the Creator." Both answers excluded the given phenomenon, for all practical purposes, from scientific analysis. Darwin's theory of natural selection provided the first rational approach to the study of ultimate causes, and such causes were the foremost interest of the naturalists.

The experimental geneticists, by contrast, had derived much of their methodology and thinking from the physical sciences. They were convinced that their methods were more objective, more scientific, and hence superior to the "speculative" approach of the evolutionary naturalists. T. H. Morgan (1932), for instance, was convinced that only the experimental method would permit "an objective discussion of the theory of evolution, in striking contrast to the older speculative method of treating evolution as a problem of history."

The inability to understand the arguments of the opponents was aggravated by the fact that experimentalists and naturalists dealt, on the whole, with different levels in the hierarchy of natural phenomena. The geneticists dealt with genes while the naturalists dealt with populations, species, and higher taxa. It is only in rather recent years that the difficulty of transferring the findings and conclusions of one hierarchical level to another one, particularly to a higher one, has been fully recognized (Pattee, 1973). Furthermore, the geneticists, on the whole, worked with the one-dimensional system of a single gene pool, while there were multidimensional components of geographical space and time in the considerations of the naturalists. The issue, however, which separated the two camps most decisively was whether evolution was gradual or saltational.

Gradual Evolution or Saltations?

Any group of individuals of a species, from the lowest sexually reproducing organisms to man, shows individual variation. This means that the individuals differ from each other in size, proportions, intensity of coloration, and many other characteristics that can either be measured or graded. This variation is also referred to as continuous variation, because one extreme end of the curve of variation usually grades imperceptibly into the other extreme, let us say the smallest and the largest individual, if a large enough population sample is available.

A seemingly entirely different type of variation is represented by the occurrence of an occasional individual that falls way outside the norm of variation of a species population. Is such an individual perhaps a new species? According to the creationist dogma, which was so powerful from the sixteenth to the eighteenth centuries, all new species have been created "in the beginning," that is, at the time of the original creation recorded in Genesis. The spontaneous occurrence of an occasional individual that fell way outside the normal range of variation of the known species was referred to as discontinuous variation, thus a source of considerable puzzlement. Was this perhaps evidence for continuing creation, as postulated by Saint Augustine, or did this indicate a far greater plasticity of the species essence than had hitherto been accepted?

Darwin had a rather low opinion of the evolutionary importance of discontinuous variation. The occurrence of variant individuals who in some character rather strikingly differ from their parents and siblings, and—indeed—from all other members of their population, is mentioned in the *Origin* only casually. When Fleeming Jenkin attacked him in 1867, Darwin reduced the number of references to such variants even more. As he wrote to Wallace (*L.L.D.* III: 108), "I always thought individual differences more important; but I was blind and thought that single [discontinuous] variations might be preserved much oftener than I now see is possible or probable . . . I believe I was mainly deceived by single variations offering such simple illustrations, as when man selects." For one like Darwin, who was always searching for causal explanations, such unique saltations were singularly unsatisfactory. They were seemingly accidents of nature and most authors who wrote about them never even attempted an explanation.

When one reads Darwin's discussions of variation, one senses that he felt that it would be easier to explain ordinary continuous variability. His theory of natural selection was based on the assumption of an unlimited supply of individual variation and this in turn was based on his observation that every individual is uniquely but very slightly different from every other one. He refers to these individual variants again and again: "We have many slight differences which may be called individual differences, such as are known frequently to appear in the offspring from the same parents . . . these individual differences are highly important to us, as they afford material for natural selection to accumulate . . . I believe mere individual differences are amply sufficient."

Darwin's thesis that the gradual accumulation of very slight variants by natural selection was the mechanism of evolution was not popular among his contemporaries. He was criticized not only for failing to explain the causation of this continuous variation but also for ignoring or at least underestimating the widely accepted importance of discontinuous variation. T. H. Huxley, who retained considerable allegiance to essentialism all his life, disagreed with Darwin's downgrading of saltations. In his famous *Times* review (April 1860) he remarked: "Mr. Darwin's position might, we think, have been even stronger than it is, if he had not embarrassed himself with the aphorism 'Natura non facit saltum', which turns up so often in his pages. We believe . . . that nature does make jumps now and then, and a recognition of the fact is of no small importance."

Huxley was not alone in this opinion. Among those who accepted evolution after 1859 were not a few who were far more impressed by the occurrence of sudden mutations than was Darwin. Botanists and horticulturalists, in particular, cited numerous cases, more or less in the same category as Linnaeus's Peloria (see Chapter 6), where a strongly deviant type suddenly originated. Nevertheless, Darwin and his friends (such as Asa Gray) continued to deny that such aberrant types were of evolutionary importance. By the end of the 1880s this apparently had become the prevailing opinion. Darwin's tendency to equate discontinuous variation with the production of monstrosities and his argument that complex new adaptations could not possibly be acquired by a single sudden jump seemed to have carried the day. Weismann (1892) was as convinced a gradualist as Darwin: "An abrupt transformation of a species is inconceivable, because it would render the species incapable of existence" (2nd ed., p. 271). But more and more other evolutionists concluded that gradual variation was insufficient to account for the ubiquitous discontinuities observed between species and between higher taxa.

One who was particularly unhappy over the emphasis on the role of gradual change in evolution was the British zoologist William Bateson (1861–1926), who later played such a decisive role in the rise of genetics. His first important work was on the embryology of the hemichordate *Balanoglossus,* work done in the laboratory of the American zoologist William K. Brooks. There Bateson became interested in the problem of evolution and particularly in the role of variation, without which natural selection is meaningless: "Variation, whatever may be its cause . . . is the essential phenomenon of Evolution. Variation, in fact is Evolution. The readiest way then, of solving the problem of Evolution is to study the facts of Variation" (1894: 6). As far as Darwin's reliance on continuous variation as the basis of evolution was concerned, Bateson, like T. H. Huxley before him, objected to the "gratuitous difficulties which have been introduced by this assumption" (p. 15). "Species are discontinuous: May not the Variation by which Species are produced be discontinuous too?" (p. 18). He repeats this suggestion in his conclusion: "Discontinuity . . . has its origin not in the environment, nor in any phenomena of Adaptation, but in the intrinsic nature of organisms themselves, manifested in the original discontinuity of Variation" (p. 567). At that time, curiously, Bateson's interest in variation was entirely evolutionary rather than genetic. He assembled an enormous amount of material (598 pages) on variation in natural populations and its possible role in speciation in his *Materials for the Study of Variation.*

To be sure, many variants were simply monstrosities. Bateson, however, concentrated on those deviations from the norm that were of the magnitude of species differences. From this evidence Bateson concluded "that the Discontinuity of which Species is an expression has its origin . . . in the intrinsic nature of organisms themselves, manifested in the original Discontinuity of Variation" (p. 567). "It suggests that the Discontinuity of Species results from the Discontinuity of Variation" (p. 568). Bateson did not think in terms of populations but in terms of discrete types, and he did not change this interpretation to the end of his career (see his Toronto address of 1922). Discontinuous variation, thus, was for him the key to evolution, and this is why he started his program of work on inheritance (see Chapter 16).[9]

The events of the ensuing years indicate that Bateson's argument decisively influenced the thinking of many of his contemporaries. At the turn of the century two works appeared that promoted even more vigorously the thesis that new species originate by sudden saltation. The Russian botanist S. Korschinsky (1899; 1901), further developing a thesis proposed by Kölliker in 1864, asserted that all organisms have the capacity to produce occasionally an offspring which differed discontinuously from other members of the species ("heterogenesis"). Going beyond Darwin (1868),

545

who had reported many such cases among cultivated plants, Korschinsky emphasized that the deviations from the type were not always drastic but might represent any grade of difference from the normal condition. The production of such deviant individuals was not caused by the environment but was due to an intrinsic potentiality.

De Vries's Mutationism

Saltationism received its greatest boost from de Vries's mutation theory (1901; 1903). Like Bateson, de Vries started from the assumption that there are two kinds of variation. Among these "the ordinary or so-called individual variability can not . . . lead to a transgression of the species border even under conditions of the most stringent and continued selection" (1901: 4). Hence speciation must be due to the spontaneous origin of new species by the sudden production of a discontinuous variant. "The new species thus originates suddenly, it is produced by the existing one without any visible preparation and without transition" (p. 3).

Unfortunately, de Vries's argument was entirely circular: he called any discontinuous variant a species, hence species originate by any single step that causes a discontinuity. The origin of species, he says, is the origin of species characters (p. 131). De Vries had no concept of populations or of species as reproductive communities. He was a strict typologist. His theory of evolution thus was based on the assumptions (1) that continuous, individual variation is irrelevant, as far as evolution is concerned, (2) that natural selection is inconsequential, and (3) that all evolutionary change is due to sudden, large mutations and, furthermore, that species have mutable and immutable periods. De Vries describes how from 1886 on he had studied variable species in the surroundings of Amsterdam in order to find one that was truly mutable. "I have taken into culture in the course of years far more than 100 such species, but only a single one lived up to my hopes" (p. 151). All other species, he said, were in an immutable period. The only mutable species was *Oenothera lamarckiana*.

One can only shake one's head when one reads de Vries's *Die Mutationstheorie*. This brilliant physiologist and geneticist, whose 1889 book on intracellular pangenesis was, prior to 1900, the most sensible and prophetic discussion of the problems of inheritance, violates all the canons of science in his *Mutation Theory*. Not only are most of his conclusions circular, but he builds his entire theory on a single exceptional species, postulating without the shadow of a proof that the "far more than 100 other species" which did not behave like *Oenothera* happened to be "in an immutable period." He finally concludes (p. 150) that species do not *originate* by the struggle for existence and natural selection, but are *exterminated* by these factors.[10]

In spite of its evident shortcomings and the vigorous opposition by leading naturalists (for example, Poulton, 1908), de Vries's work dominated the thinking of biology from 1900 to 1910. As Dunn (1965a: 59) rightly has said, "In some ways the publication of the first volume of de Vries's great work in 1901 made a greater impression on biology than the rediscovery of Mendel's principles." The leading textbook of genetics during the de Vries era (Lock, 1906: 144) summarizes the thinking of the Mendelians in the statement: "Species arise by mutation, a sudden step in which either a single character or a whole set of characters together become changed." T. H. Morgan, at first (1903), was most enthusiastic about de Vries's theory. The Mendelians thought that such evolution by mutation refuted gradual evolution by selection. Consequently Bateson claimed that "the transformation of masses of populations by imperceptible steps guided by selection is, as most of us now see, so inapplicable to the fact that we can only marvel both at the want of penetration displayed by the advocates of such a proposition, and at the forensic skill by which it was made to appear acceptable even for a time" (1913: 248). Johannsen was even more opposed to any role of selection in evolution.

To show how totally he rejected Darwin's theory of natural selection, Bateson remarked condescendingly: "We go to Darwin for his incomparable collection of facts [but reject his theoretical explanations] . . . for us he speaks no more with philosophical authority. We read his scheme of Evolution as we would those of Lucretius or of Lamarck" (1914: 8). In this rejection of Darwin, Bateson went far beyond de Vries, who insisted that his theory was a modification of Darwin's, not a replacement for it.

R. A. Fisher (1959: 16) appraised the situation quite rightly when he said, "The early Mendelians could scarcely have misapprehended more thoroughly the bearings of Mendel's discovery . . . on the process of evolution. They thought of Mendelism as having dealt a death blow to selection theory, a particulate theory of inheritance implied [to them] a corresponding discontinuity in evolution." As a result, the opinion was widespread that Darwinism was dead. This led Nordenskiöld in his otherwise so authoritative *History of Biology* (1920–1924) to the statement: "The [Darwinian] theory has long ago been rejected in its most vital points . . . the objections made against the theory on its first appearance very largely agree with those which far later brought about its fall." Morgan, among others, thought that mutation pressure alone could achieve everything that Darwin had ascribed to natural selection.

What upset the naturalists most was the frequent assumption made by Mendelians (for example, by de Vries) that so-called individual or fluctuating variation lacked a genetic basis. This was of decisive importance for the evaluation of geographic races, some of which were considered by the Darwinians to be incipient species. De Vries had to reject the concept of geographic speciation because it was in direct conflict with his mutation theory (speciation by genetically different individuals). He stated this quite clearly with respect to geographic races in man:

> The variability exhibited by man is of the fluctuating kind: whereas species arise by mutation. The two phenomena are fundamentally different. The assumption that human variability bears any relation to the variation which has or is supposed to have caused the origin of species is to my mind absolutely unjustified . . . Favorable and unfavorable conditions of life, migration to a different climate and so forth affect the fluctuating characters of man to no small extent. But only for a time; as soon as the disturbing factor is removed, the effect which it produced disappears. The morphological characters of the race on the other hand are not in the least affected by such influences. New varieties do not arise by this means. Since the beginning of the diluvial period man has not given rise to any new races or types. He is, in fact, immutable, albeit highly variable. [Another of de Vries's claims without any basis of fact!] (1901, I: 155–156)

The interpretation of evolution by the early Mendelians can be summarized in the following conclusions:

(1) Every change in evolution is due to the occurrence of a new mutation, that is, of a new genetic discontinuity. Hence the moving force in evolution is mutation pressure.

(2) Selection is an inconsequential force in evolution, at best playing a role in eliminating deleterious mutations.

(3) Since mutation can explain all evolutionary phenomena, individual variation and recombination, neither of which produces anything new, can be disregarded. Most continuous individual variation is nongenetic.

The naturalists were dismayed. All that they had discovered and described since Wallace's pioneering paper of 1855 was ignored by the experimentalists. As Rothschild and Jordan said (1903: 492), "Whoever studies the distinction of geographic varieties closely and extensively, will smile at the conception of an origin of species *per saltum.*" The naturalists saw gradualness everywhere and they all believed, at least to some extent, in natural selection. Poulton (1908: xviii), not without justification, ridiculed the mutationists by stating, "Mutation without selection may be left to those who desire to revive Special Creation under another name."

549

But the naturalists also had misconceptions. For instance, they were so sold on gradualness that they even belittled Mendelian inheritance. To be sure, discontinuous characters might obey the Mendelian rules, they admitted, but, after all, such characters were of little evolutionary consequence. Gradual, quantitative characters, the only ones of importance in evolution, did not follow the Mendelian rules laid down by de Vries and Bateson, the naturalists claimed, hence one would have to search for another solution. And they found this solution either in some orthogenetic capacity of evolutionary lines or in neo-Lamarckism. Natural selection, even though acknowledged by most of them as an evolutionary force, was not the major factor in evolution. Rather the naturalists continued to believe in the effects of use and disuse, a direct induction by the environment or other manifestations of soft inheritance. Up to the 1920s and 1930s, virtually all the major books on evolution—those of Berg, Bertalanffy, Beurlen, Böker, Goldschmidt, Robson, Robson and Richards, Schindewolf, Willis, and those of all the French evolutionists, including Cuénot, Caullery, Vandel, Guyénot, and Rostand—were more or less strongly anti-Darwinian. Among nonbiologists Darwinism was even less popular. The philosophers, in particular, were almost unanimously opposed to it, and this opposition lasted until relatively recent years (Cassirer, 1950; Grene, 1959; Popper, 1972). Most historians likewise rejected selectionism (Radl, Nordenskiöld, Barzun, Himmelfarb).

In various confrontations between the two camps there was no evidence of a willingness to compromise; all the argument was directed toward trying to prove that the other camp was wrong. At a meeting between geneticists and paleontologists at Tübingen in 1929, the paleontologists adopted the worst possible strategy (Weidenreich, 1929). Instead of concentrating on the evolutionary phenomena which the geneticists, particularly the Mendelians, had not been able to explain, they concentrated on trying to prove the existence of an inheritance of acquired characters, a subject which they were in no manner whatsoever qualified to discuss. And yet, there were numerous evolutionary problems that were not at all explained by the "changes in the frequency of genes" concept of evolution of the geneticists, such as highly uneven rates of evolution, the basic constancy of major structural types, the absolute discontinuities between them, and the problem of the multiplication of species.

550

When the controversy between the two camps had started (in the 1890s and the early 1900s), both camps held ideas that were incompatible with the ideas of the other camp. More importantly both camps supported certain explanations that could be refuted by the other camp. But it was impossible to recognize this until both camps had clarified and, in part, considerably revised their own ideas. In order to be able to understand the eventual resolution of the conflict, it is necessary to describe the advances made in both camps, in evolutionary genetics (after about 1906) and in evolutionary systematics (from the post-Darwinian period to the 1930s). These advances eventually made a reconciliation of the two opposing camps possible and led to a synthesis of the valid components of the two research traditions.

ADVANCES IN EVOLUTIONARY GENETICS

What de Vries, Bateson, and Johannsen had been doing was only one of several possible kinds of evolutionary genetics and did not outlive the first decade of the century. The teachings of Bateson's opponents, the biometricians (Provine, 1971; see Chapter 16) were even more short-lived. The simplistic ideas of these pioneers were subjected to a rather radical revision by a new generation of geneticists. Schools that originated in experimental zoology, like that of T. H. Morgan at Columbia, stayed closest to the original evolutionary ideas of Mendelism, stressing mutation and the discontinuous independence of individual genes (Allen, 1968). But other geneticists who had entered genetics from natural history or from animal or plant breeding, like Nilsson-Ehle in Sweden, East, Jones, Jennings, Castle, and Payne in the United States, and Baur in Germany, made findings which showed that there is no conflict between the genetic evidence and either natural selection, gradualness of evolution, or population thinking.

A detailed history of these findings is presented in Chapter 17. Those of the greatest importance for the interpretation of evolution may be summarized in these statements:

(1) There is only one kind of variation, large mutations and very slight individual variants being extremes of a single gradient.

(2) Not all mutations are deleterious; some are neutral and some are distinctly beneficial.

(3) The genetic material itself is invariant (constant), that is, there is no soft inheritance.

551

(4) Recombination is the most important source of genetic variation in populations.

(5) Continuous phenotypic variation can be explained as the result of multiple factors (polygenes) together with epistatic interactions and is not in conflict with particulate inheritance.

(6) A single gene may affect several characters of the phenotype (pleiotropy).

(7) Experimental as well as observational data demonstrate the effectiveness of selection.

These findings completely refuted the antiselectionist, saltational evolutionary theories of de Vries and Bateson. Curiously, this by no means spelled the end of saltationism, which continued for several decades to have substantial support, as for instance by the geneticist Goldschmidt (1940), the paleontologist Schindewolf (1950) (and other, particularly German, paleontologists), the botanist Willis (1922; 1940), and some of the philosophers. Eventually it was universally accepted that an origin of species and higher taxa through individuals does not occur, except in the form of polyploidy (principally in plants). The phenomena which the adherents of macrogenesis had used as support could now be readily explained in terms of gradual evolution. Particularly important for the reconciliation was the recognition of the importance of two previously neglected evolutionary processes: drastically different rates of evolution in different organisms and populations, and evolutionary changes in small, isolated populations. It was not until the 1940s and 50s that well-argued defenses of macrogenesis disappeared from the evolutionary literature in the wake of the evolutionary synthesis.

Evolution by mutation pressure, a concept popular from Bateson and de Vries to Morgan, also lost ground after 1910 but did not disappear entirely, being revived recently by the proponents of "non-Darwinian evolution." A belief in mutation pressure was losing ground not only because the sentiment in favor of selection continued to grow stronger, particularly in the 1920s, but also because of the discovery of reverse mutation. Steady evolutionary change through mutation would be possible only if there was a cascading of mutations all in the same direction. However, if the probability of mutating from *a* to *a'* is no greater than that of the reverse mutation from *a'* to *a,* then no evolutionary trend can develop. After Morgan had discovered *eosin* in 1913, a reverse mutation from *white-eye,* more and more reverse mutations were discovered, and in many cases the frequency from wild type to mutant was no greater than the reverse from mutant to wild type, as shown by Muller and Timofeeff-Ressovsky (Muller, 1939). The assumption that directional evolution (evolutionary trends) could be caused by mutation pressure was made exceedingly improbable by these findings.

Perhaps the most important contribution made by the new genetics was the decisive refutation of soft inheritance. This had been delayed again and again by claims of an experimental substantiation of an inheritance of acquired characters. Some of these claims were based on experimental error, others quite clearly were fraudulent (Burkhardt, 1980). It is of extraordinary interest how often experimenters when they are convinced of the to-be-expected outcome of their tests, "produce" data which they were unable to get in their actual experiments. This psychological phenomenon has also been observed in other areas of experimental biology (such as cancer research and immunobiology).

Although the final disproof of soft inheritance had to wait for the demonstration, by molecular genetics, that information acquired by the proteins cannot be transmitted back to the nucleic acids (and this was not proven until the 1950s), nevertheless the geneticists showed (and this was accepted by such naturalists as Sumner, Rensch, and Mayr) that all the phenomena of gradual evolution and adaptational variation that had previously been cited as evidence for an inheritance of acquired characters could be interpreted in terms of constant genes. Contrariwise, all endeavors to demonstrate soft inheritance were failures (see Chapter 17).

From the beginning some geneticists were more interested in the mechanics of inheritance, others in the evolutionary aspects. Those who wanted to understand the genetic basis of evolution increasingly appreciated that evolution was a population phenomenon and that it had to be studied as such. A field began to develop that later was designated as population genetics. Workers interested in statistics, like Yule, Pearl, Norton, Jennings, Robbins, and Weinberg, made the first important contributions to this field. We still lack a good history of this period, but it seems that these authors had already arrived at many of the later conclusions of population genetics. Most of their findings were published in technical journals and did not become as widely known as they deserved. Regrettably, the naturalists were largely unaware of this work.[11]

Eventually it became customary to designate as *population genetics* that brand of genetics that investigates the changes of gene frequencies in populations. The term "population genetics" is actually ambiguous, because two largely independent research programs were involved. One is represented by *mathematical population genetics* connected with the names R. A. Fisher, J. B. S. Haldane, and Sewall Wright. Its "populations" were statistical populations, and research in this field could be done with pencil and paper, later with a calculator, and now with the computer. The other population genetics dealt with actual populations of living organisms studied in the field and in the laboratory. The history of *that* branch of population genetics has not yet been written. It is represented by the work of Schmidt (*Zoarces*), Goldschmidt (*Lymantria*), Sumner (*Peromyscus*), Langlet (*Pinus*), Baur (*Antirrhinum*), Chetverikov, Timofeeff-Ressovsky, Dobzhansky (*Drosophila*), Cain, Sheppard, Lamotte (*Cepaea*), and Ford and Sheppard (*Panaxia, Maniola*), to mention the names of some of the many students of the distribution of genes in natural populations and their changes in time. To distinguish it from mathematical genetics, Ford (1964) has appropriately designated this field as *ecological genetics.*

Mathematical population genetics had its beginnings in the controversy between the Mendelians (Bateson, in particular) and the biometricians (Weldon, Pearson). The biometricians, although quite correctly stressing the importance of continuous variation as the material of natural selection, had assumed inheritance to be blending. The early Mendelians, cognizant of the particulate nature of inheritance, had stressed discontinuous variation. The main development of evolutionary genetics was to show that there is no conflict between particulate (nonblending) inheritance, continuous variation, and natural selection.[12]

The basis of all mathematical population genetics is the so-called Hardy-Weinberg equilibrium principle, established in 1908.

553

It states that two alleles (*a* and *a'*) will remain at the same frequency in a population from generation to generation unless their frequency is affected by immigration, mutation, selection, nonrandom mating, or errors of sampling. (For a history of the discovery of this principle, see Provine, 1971: 131–136.) Much of mathematical population genetics of the ensuing thirty years dealt with the question of how the genetic composition of populations of various sizes is affected by different rates of mutation, different selection pressures, and errors of sampling.

554   The first question to be studied was how effective selection is when the selective advantage of a new allele introduced into a population is only slight. The British mathematician H. T. J. Norton worked this out for different selection intensities of genes occurring at different frequencies (1915). To the surprise of almost everybody he was able to show that even rather small selective advantages or disadvantages (less than 10 percent) led to drastic genetic changes in relatively few generations. This finding greatly impressed J. B. S. Haldane (who published a series of researches on the mathematics of selection in the 1920s) and the Russian naturalist-geneticist Chetverikov. The conclusion that alleles only slightly differing in selective value could replace each other rather rapidly in evolution later induced several neo-Lamarckians (Rensch and Mayr, for example) to abandon their belief in soft inheritance. For it was now evident that phenomena like climatic races and other environment-correlated adaptations could be interpreted in terms of selection acting on multiple alleles and genes.

Beginning with 1918 R. A. Fisher (1890–1962) published a series of papers on the mathematics of gene distributions in populations. These researches dealt with the partitioning of genetic variance into an additive portion (caused by alleles or independent genes with similar effects) and a nonadditive portion (epistasis, dominance, and so forth), with the conditions under which balanced polymorphism would be maintained, with the role of dominance, and with the rate at which a favorable gene would spread in populations of different sizes. Some of his findings, as on balanced polymorphism, are now so well established that we can hardly comprehend that someone had to be the first to work this out. Others of his researches are so fertile in conclusions that it is only in the last decades that they have been fully applied.

Fisher's most important conclusion was that much of continuous variation, at least in man, is due to multiple Mendelian factors rather than to environmental influences. His stress on genes with small phenotypic effects was a major contribution to the coming reconciliation between geneticists and naturalists. Like most mathematical geneticists, Fisher tended to minimize the effects of an interaction among gene loci.

Fisher always thought in terms of large populations, and although he was fully aware of the existence of errors of sampling, he thought that, owing to the selective differential of competing genes and to recurrent mutation, such errors of sampling would be in the long run of little evolutionary consequence, as is indeed true for large populations. Another geneticist, Sewall Wright (b.1889), disagreed with Fisher on this point, and this revived an old argument which, as a matter of fact, is not entirely settled to this day. The first person to advance the thesis that much evolutionary change is simply a result of chance variation was J. T. Gulick (1872). He was led to this thesis when observing the incredible diversity of local populations of Hawaiian landsnails (*Achatinella*) and their seemingly haphazard variation in the absence of definable differences in environmental factors. From that date on, the thesis that much variation is selectively neutral was proposed again and again. Fisher (1922: 328) called such random variation the *Hagedoorn effect* after two Dutch investigators who had brought together a great deal of evidence in its support. Their thesis (like that of Gulick) was based on the assumption that much of such variation is effectively neutral as far as selection is concerned. Fisher, by contrast, thought that most allelic polymorphism in populations was due to a superiority of heterozygotes.

Sewall Wright, a student of William E. Castle, had worked since 1914 on color inheritance and on the effects of inbreeding in guinea pigs. This work had persuaded him that "effective breeding populations" (later called demes) even in wild animals often were of sufficiently small size to make errors of sampling a nonnegligible factor. Although gene flow from adjacent populations would usually prevent the random fixation of genes, there would nevertheless be sufficient "genetic drift" to favor gene combinations that would be unlikely to occur in large populations. In his first major account Wright (1931a) expressed himself in a way that sounded as if he were proposing genetic drift as an alternative mechanism to natural selection, and this caused considerable confusion. Through Dobzhansky's book (1937) Wright's thesis became widely known among evolutionists, and a tendency developed in the 1940s and 50s to ascribe to genetic drift almost any puzzling evolutionary phenomenon. Drift played an important role in the writings of Dobzhansky and also in Simpson's (1944) concept of the "inadaptive phase" of quantum evolution. Eventually a reaction to the liberal invoking of genetic drift set in, as described by Mayr (1963: 204–214).

As a student of Castle, Wright had had considerable contact with naturalists and was particularly interested in the 1920s in the researches of F. B. Sumner (Provine, 1979). As a result, Wright tended to think in terms of natural populations and was aware of the changing fitness values of genes. "Genes favorable in one combination are . . . extremely likely to be unfavorable in another" (1931: 153). Unfortunately, he made little use of this insight in his equations and graphs, where he deals almost exclusively with single genes and with constant fitness values. Wright, like Chetverikov, was greatly impressed by pleiotropic effects: "Since genes *as a rule* have multiple effects . . . it is probable that in time a gene may come to produce its major effects on wholly different characters than at first" (1931a: 105). Owing to his background, Wright, among the mathematical geneticists, was in his thinking closest to the thought of the naturalists. That he regarded species as aggregates of populations was a prerequisite for his later collaboration with Theodosius Dobzhansky (1900–1975).

Chetverikov

A rather different school is represented by the population genetics that originated in Russia, primarily through the work of Sergei S. Chetverikov (1880–1959) and his students. Russia differed in its traditions quite significantly, not only from the United States but also from Western Europe. Natural selection had been much more widely accepted (prior to the 1920s) than elsewhere, and natural history seems to have had a much higher prestige and influence at the universities. Even today most zoology students, in Moscow for instance, spend their summers at biological field stations or elsewhere doing fieldwork. Also, in the USSR there were a number of genetics institutes (two in Leningrad, one in Moscow), and it would seem that in the 1920s the number of geneticists in the USSR was as large as in all of the rest of continental Europe together. Chetverikov was the head of the Department of Genetics from 1924 to 1929 in Koltsov's Institute of Experimental Biology in Moscow. He was a passionate butterfly specialist, still describing a new species from the Ural mountains when he was 76 years old. He was equally interested in evolution, publishing in

1906 a seminal article on population fluctuations, the evolutionary importance of which (particularly the bottlenecks) no one had previously fully appreciated. From the early 1920s on, Chetverikov taught genetics and became the leader of a large informal group of genetics enthusiasts. In 1929 he had to leave Moscow for political reasons and was no longer able to continue his genetic researches (Adams, 1968; 1970; 1980a).

Owing to his own background as a naturalist, Chetverikov answered the questions and objections of the anti-Mendelians far more effectively than did either Morgan or the mathematicians. In one of the most important publications in the whole history of evolutionary biology (1926) he sets himself "the goal of clarifying certain questions on evolution in connection with our current genetic concepts" (p. 169). First he showed that there is a complete, almost "imperceptible transition from mutations that have completely normal viability" to mutations of ever lower viability and even lethality. The claim that all mutations are deleterious is not true. Indeed, as was later shown by Dobzhansky and others, mutations may occur that are of higher fitness than wild type. Chetverikov saw clearly, as had Fisher and others before him, that a new mutant always first appears as a heterozygote and that, if it is recessive, it may long remain concealed in the population (unless lost by errors of sampling), because only homozygotes will be exposed to selection. He therefore came to the conclusion "that a species, like a sponge, soaks up heterozygous mutations, while remaining from first to last externally (phenotypically) homozygous" (p. 178). Thus a great deal of concealed genetic variability ought to exist in every species. To test his hunch, he trapped 239 wild *Drosophila melanogaster* females near Moscow, and brother-sister mated their offspring. In this small sample he found no less than 32 loci which segregated for visible recessives, thus confirming his supposition. No one before him had suspected the amount of concealed variation in a wild population. His students, particularly N. V. Timofeeff-Ressovsky, B. L. Astaurov, N. P. Dubinin, and D. D. Romashov, rigorously began to analyze genetic variation in wild populations, integrating it with experimental research in the laboratory. Dobzhansky, although not himself a member of the group (he worked with Philipchenko in Leningrad), keenly followed these researches, which were one of the influences on his later *Drosophila* researches.

For Chetverikov, changes in populations were not the result of mutation pressure but of selection. Basing his argument on

Norton's table (1915), he concluded that "even the slightest improvement of the organism [a slightly superior gene] has a definite chance of spreading throughout the whole mass of individuals comprising the freely crossing population (species)" (1961: 183). It does not matter whether the new gene is dominant or recessive, nor whether its selective advantage is 50 percent or only 1 percent; "the complete replacement of a gene by a better-adapted one always proceeds . . . to an end." In contrast to Fisher and Haldane, who devoted most of their efforts to proving the effectiveness of selection, Chetverikov, in line with the Russian tradition, took selection far more for granted. This enabled him to turn to other problems.

Most of Chetverikov's conclusions were eventually also reached independently by Fisher, Haldane, and Wright and entered the evolutionary literature of the west primarily from them. Where Chetverikov was way ahead of the western group was in his much clearer recognition of the evolutionary importance of gene interaction. He emphatically rejected "the former notion of the mosaic structure of the organism consisting of various independent genes" and concluded that "each inherited trait . . . is determined by not just some one gene but by their whole aggregate, their complex." No gene has a constant fitness value, because "the very same gene will manifest itself differently, depending on the complex of the other genes in which it finds itself" (p. 190). The phenotypic expression of each gene is determined by its "genotypic milieu."

Chetverikov had based these conclusions on the discovery, particularly by the Morgan group, of pleiotropic gene action, that is, the effect of a gene on several components of the phenotype (see Chapter 17). His student Timofeeff-Ressovsky had discovered important manifestations of pleiotropy (1925). By contrast, the mathematical population geneticists, especially Fisher and Haldane, concentrated, for reasons of simplicity of a first approach, on the behavior of individual genes. In their equations as well as in their graphs, they illustrated the increase or decrease in the frequency of individual genes under the effects of selection, mutation, and errors of sampling. Genetics textbooks in the 1940s and 50s suggested laboratory exercises in which genes were represented by beans of several colors, placed in a bag, mixed and reassembled for each generation, according to certain experimental specifications. Since any interaction among genes was excluded in these exercises, Mayr (1959d) dubbed the kind of genetics which ignored gene interaction as "bean-bag genetics." Unfortunately, too much of mathematical population genetics at that time was bean-bag genetics. Even an author like Sewall Wright, who was fully cognizant of the importance of gene interaction, dealt in his calculations and illustrations almost exclusively with the behavior of individual genes. As a result, it was not until the 1950s and later that Chetverikov's concept of the genotypic milieu was fully incorporated into the thinking of evolutionary biologists.

In spite of the fact that Russian language publications were hardly ever read outside the USSR, the work of Chetverikov's school was not entirely unknown in England and the United States. Not only Chetverikov's 1927 article but at least three papers by Timofeeff-Ressovsky were published in English or German language publications, and a complete translation of Chetverikov's 1926 paper was available in Haldane's laboratory. After leaving the USSR, both Timofeeff-Ressovsky and Dobzhansky further helped to spread the ideas of the Chetverikov school. There is no doubt that it made a substantial contribution to the evolutionary synthesis.

Chetverikov and the mathematical population geneticists completed the destruction of the genetic theory of evolution of the Mendelians. They confirmed the importance of selection and the nonexistence of mutation pressure; they established the genetic basis of gradual Darwinian evolution and confirmed the nonexistence of soft inheritance. Finally, they showed that there is no conflict between the discontinuity of genes and the continuity of individual variation. An important foundation was thus laid for a bridge to the camp of the naturalists, who had rejected all along the de Vriesian macromutations and mutation pressure and had likewise emphasized the importance of gradual evolutionary changes and of natural selection.

ADVANCES IN EVOLUTIONARY SYSTEMATICS

The rapid advances made in evolutionary genetics were paralleled by similar advances in systematics or, more broadly, by advances in the understanding of organic diversity by naturalists. As a matter of fact, the type of population genetics conceived by Chetverikov involved little more than the transfer of concepts and methodologies to genetics that had existed in systematics for more than a hundred years. I am referring to the study of different geo-

559

graphic races of a species casually discussed by Buffon (for North American animals) and Pallas (for Siberian animals) and fully developed by Gloger.[13] The more perceptive taxonomists since that time have given much thought to innate differences among different populations, particularly geographical races of species.

Such population differences are referred to by Linnaeus (1739), Buffon (1756), Blumenbach (1775), Pallas (1811), von Buch (1825), and Gloger (1827, 1833). It was generally known to foresters from Sweden, Germany, and France from the middle of the eighteenth century on (Langlet, 1971). The fact that pines and rhododendrons from different altitudes in the Himalayas greatly differ in frost hardiness was discovered by Hooker (1853) and quoted by Darwin (1859: 140). It was soon recognized that this variation was closely correlated with the nature of the environment, and the term *climatic race* was introduced in the middle of the nineteenth century. In botany this was extended to a study of edaphic factors, the combination of edaphic and climatic factors being responsible for the development of *ecotypes* (Turesson, 1922). More geographical in treatment was the work of Baur on *Antirrhinum* populations in Spain (Schiemann, 1935; Stubbe, 1966). In zoology, these interests led to the studies of Schmidt (1917) on fishes, of Goldschmidt on *Lymantria,* and of Sumner on *Peromyscus.* None of this material was as suitable, however, for a detailed genetic analysis as *Drosophila.* It is important to recognize that Chetverikov's work consisted largely in the application of a classical *Fragestellung* to a new and particularly suitable material.

The development of population systematics that could easily be translated into population genetics was a major contribution of the naturalists. They were continuing a tradition, to which Darwin himself had belonged, which was concerned with the study of natural populations, with variation within populations, and with the changes on geographical gradients from population to population. They adopted the population as the unit of evolution, rather than the phyletic line favored by the comparative anatomists and paleontologists. The naturalists were the only biologists who studied isolation and the role of geographical as well as of individual variation. Except for the animal breeders they were the first ones to understand individuality and to base their methodology on this knowledge, resulting in the admonition to collect "series" or to make "mass collections." This, in turn, led to the application of statistics, and Galtonian statistics at that, which stresses variance rather than mean values. Unfortunately, no adequate history of evolutionary natural history is so far available, although some of the developments are described in Stresemann's *Ornithology* (1975) and in historical comments made by Mayr (1963).

The most important contributions made by the naturalists were conceptual ones. A true understanding of natural selection, speciation, and adaptation was not possible until population thinking had displaced typological thinking. The population thinking of the naturalists had a particularly important impact on Chetverikov and his school. But it was not the naturalists alone who helped to spread this concept. A second source of population thinking were, as in Darwin's day, the animal and plant breeders. Those geneticists, like Castle, East, Emerson, and Wright, who had the closest contact with breeders also avoided most successfully the pitfalls of typological thinking. Among the naturalists it led to the new concept of races as variable populations, each with a different geographical history. It led to the development of the biological species concept, and ultimately it culminated in the so-called new systematics, more correctly called population systematics (see Chapter 6).

561

It was the naturalists who solved the great species problem, a problem which the geneticists either side-stepped altogether or answered unsuccessfully in a typological manner. The naturalists showed that species are not essentialistic entities, to be characterized morphologically, but that they are aggregates of natural populations that are reproductively isolated from each other and fill species-specific niches in nature. A full understanding of the nature of species could not be achieved until a number of further insights had been acquired, such as the distinction between taxon and category, and the realization that the word "species" is a relational term like the word "brother" and that, philosophically speaking, a species taxon is an individual, the members of a species being "parts" of this individual. The truth of this assertion becomes evident when one considers that the genes of all the members of a species are components of the same gene pool (Ghiselin, 1974b; Hull, 1975; see also Chapter 6).

Speciation

The new understanding of the nature of populations and of species enabled the naturalists to solve the age-old problem of speciation—a problem that had been insoluble for those who looked for the solution at the level of genes or genotypes. At that level the only solution is instantaneous speciation by a drastic mutation or other unknown processes. As de Vries (1906) had stated, "The theory of mutation assumes that new species and varieties are produced from existing forms by certain leaps." Or as Goldschmidt (1940: 183) had stated, "The decisive step in evolution, the first step toward macroevolution, the step from one species to another, requires another evolutionary method [that is, the origin of hopeful monsters] than that of sheer accumulation of micromutations." The naturalists realized that the essential element of the speciation process is not the physiological mechanism involved (genes or chromosomes) but the incipient species, that is, a population. Geographic speciation, consequently, was defined by Mayr in terms of populations: "A new species develops if a population which has become geographically isolated from its parental species acquires during this period of isolation characters which promote or guarantee reproductive isolation when the external barriers break down" (Mayr, 1942: 155).

The most important conceptual advance was a clear formulation of the problem. In order to explain speciation it is not sufficient to explain the origin of variation or of evolutionary changes within populations. What must be explained is the origin of reproductive isolation between populations. Speciation, thus, is not so much the origin of new types as the origin of effective devices against the in-flow of alien genes into gene pools.

This insight had a history of more than a hundred years. The first person to state that speciation is in most cases "geographical" was von Buch (1825). The concept was strong in Darwin's 1837–38 notebooks and in his 1842 and 1844 essays (Kottler, 1978; Sulloway, 1979), as well as in Wallace's 1855 paper. But it rather weakened in later years (see Chapter 9). Beginning with the 1850s Darwin thought that speciation, particularly on continents, could also occur without strict geographical isolation, and this involved him in a heated controversy with Moritz Wagner.

The Role of Isolation

Moritz Wagner (1813–1887), a celebrated explorer, collector, and geographer, had devoted three years (1836–1838) to the exploration of Algeria. Here he found that each species of flightless beetles (*Pimelia* and *Melasoma*) invariably was restricted to a stretch of the north coast between two rivers descending from the Atlas mountains. As soon as one crossed a river, a different but closely related species appeared (Wagner, 1841: 199–200). Wagner was able to confirm such an isolating capacity of rivers during his later travels in western Asia and extended it by comparing faunas at either side of mountain ranges (for example, the Caucasus) or, in the case of montane species, for major peaks separated by valleys such as the great volcanoes of the Andes. This led him to the conclusion:

> The formation of a genuine variety which Mr. Darwin considers to be an incipient species, will succeed in nature only where a few individuals transgress the limiting borders of their range and segregate themselves spatially for a long period from the other members of their species . . . the formation of a new race will never succeed in my opinion without a long continued separation of the colonists from the other members of their species . . . Unlimited crossing, the uninhibited cross fertilization of all individuals of a species will always result in uniformity and will turn back to the original condition any variety the characters of which have not been fixed through a series of generations.

563

All this sounds like a rather reasonable description of the process of geographic speciation. Unfortunately, Wagner combined this with some peculiar ideas on variation and selection. He felt that the isolation of the founder population resulted in increased variability, and he also thought that it was only in such an isolated population that natural selection truly had an opportunity to operate (Sulloway, 1979).

This was too much for Darwin, who not only insisted quite rightly that natural selection and evolutionary change could take place without isolation but also implied rather clearly that isolation was not a necessary condition for species formation. Darwin concludes his rejection of Wagner's thesis with the emphatic statement: "My strongest objection to your theory [of geographic speciation] is that it does not explain the manifold adaptations in structure in every organic being" (*L.L.D.*, III: 158), as if speciation and adaptation were mutually exclusive phenomena. Perhaps Darwin was forced into this extreme stand by Wagner's claim, "Organisms which never leave their ancient area of distribution will never change" (1889: 82), a statement obviously not strictly true but perhaps closer to the truth than was thought in the first 75 years after it was made.

In due time Weismann was drawn into the controversy. He published (1872) a rejoinder to Wagner which is perhaps the weakest of Weismann's otherwise so outstanding publications. Wagner's original question, "Can species multiply without geographic isolation?" was changed into the question, "Is isolation itself the factor which is responsible for the changes in isolated populations?" and "Is isolation necessary for varieties to become constant?" As in Darwin's writings, the question of the acquisition of·reproductive isolation is nowhere mentioned, and the entire emphasis is on the degree of morphological difference. How little Weismann and his contemporaries understood what really the essentials of the problem of the multiplication of species are may be illustrated by this quotation. "In this it is quite unimportant how they [endemic species occurring in isolated areas] originated, whether by amixia in a period of variation or by natural selection, which tried to adjust the immigrants to the new environmental conditions of the isolated area. The change can even have been caused by influences which had nothing to do with the isolation, as for example the direct influence of the physical environment or the process of sexual selection" (1872: 107).

Wagner remained very much alone in his insistence on the importance of geographic isolation. A. R. Wallace completely sided with Darwin and concluded "that geographical or local isolation is by no means essential to the differentiation of species, because the same result is brought about by the incipient species acquiring different habits or frequenting a different station; and also by the fact that different varieties of the same species are known to prefer to pair with their like, and thus to bring about a physiological isolation of the most effective kind." Needless to say, Wallace brought forth no proof whatsoever for any of these assertions.

The ironic aspect of the controversy between Darwin and Wagner is that both of them consistently argued past each other. Wagner insisted that reproductive isolation could ordinarily not be acquired without geographic isolation. Darwin, at that time very much fascinated by the principle of divergence, answered "that neither isolation nor time by themselves do anything for the modification of species" (*L.L.D.*, II: 335–336), as though Wagner had denied the occurrence of phyletic evolution. In his entire correspondence with Wagner, Semper, and Weismann it is quite evident that Darwin failed to understand how difficult a problem the acquisition of reproductive isolation is.

One of the major difficulties was that most of those who joined the controversy in the ensuing years—Romanes, Gulick, and even Wallace (Lesch, 1975)—made no clear separation of geographical and reproductive isolation, nor of individual and geographical variation, and often dealt with speciation as if it was the same as natural selection.[14] The confusion is particularly painful in the writings of Romanes, who invented the misleading term "physiological selection" for reproductive isolation. There still is no completely critical analysis of this literature, but one can, on the whole, recognize two camps, those who followed Darwin in not clearly distinguishing the two kinds of isolation (among them, Weismann, Semper, Romanes, Gulick, and Wallace) and those who, following Wagner, considered geographical isolation as a factor *sui generis* and indispensable for speciation (for example, Seebohm, K. Jordan, D. S. Jordan, Grinnell, a good many entomologists such as H. W. Bates and perhaps Meldola and Poulton, as well as Kerner and Wettstein among the botanists).

565

After 1900 the theory of speciation through geographic isolation suffered an almost complete eclipse, because in the theory of mutationism (as developed by Bateson and de Vries) isolation was no longer considered necessary. Owing to the efforts of D. S. Jordan, K. Jordan, Stresemann, Rensch, Mertens, and other taxonomists, the importance of geographical isolation during speciation was not forgotten altogether. Yet as late as 1937 Dobzhansky included both the intrinsic genetic factors and the extrinsic geographical barriers in his list of isolating mechanisms. It was one of the major theses of Mayr's *Systematics and the Origin of Species* (1942) that there is a fundamental difference between the two kinds of isolating factors and, as Wagner and K. Jordan had previously insisted, that geographical isolation is a prerequisite for the building up of intrinsic isolating mechanisms. A further conceptual clarification was achieved by a populational definition of isolating mechanisms (Mayr, 1970: 56). However, even today some authors confound the mechanisms of speciation—genes, chromosomes, and so forth—with the location of the populations involved in speciation (that is, whether the populations are sympatric or allopatric), not realizing that the two aspects are independent of each other and both are by necessity involved simultaneously. Since 1942 the importance of geographic speciation, as worked out by the naturalists, has not been denied. The main question that has remained controversial is the relative importance of alternate processes, such as instantaneous speciation (by polyploidy and other chromosomal reorganization) and sympatric speciation.

One further contribution to evolutionary thought made by the naturalists was their recognition of the adaptive nature of geographic variation within species. This strongly reenforced the belief in gradual evolution. It had been observed by perceptive naturalists long before 1859 not only that different populations in many species differ from each other (geographic variation) but also that much of this variation is gradual and correlated with factors of the environment—that it is adaptive (Gloger, 1833; Bergmann, 1847). The intensive study of such climatic variation by Allen (1870s), Sumner (1920s), and Rensch (1920s and 30s) provided powerful support in favor of Darwin's thesis of the gradualness of evolutionary change as well as of the importance of the environment (Mayr, 1963: 309–333). Similar but less systematic studies were made in plants, particularly through the transplantation of northern individuals of tree species to southern latitudes, experiments which confirmed climate-correlated geographic variation (Langlet, 1971; Stebbins, 1979). However, in the period during which the early Mendelians insisted that genetic variation is drastic and discontinuous, such adaptive geographic variation was considered by most naturalists (prior to the early 1930s) as important evidence in favor of soft inheritance (Rensch, 1929).

566

THE EVOLUTIONARY SYNTHESIS

Through the first third of the twentieth century the gap between the experimental geneticists and the naturalists seemed so deep and wide that it looked as if nothing would be able to bridge it. The distinguished German biologist Buddenbrock said in 1930: "The controversy . . . is as undecided today as it was 70 years ago . . . neither party had been able to refute the arguments of their opponents and one must assume that this situation is not going to change very soon" (p. 86). The members of the two camps continued to talk different languages, to ask different questions, to adhere to different conceptions, as is abundantly evident from the contemporary literature (Mayr and Provine, 1980).

How could this stalemate be broken? How could both camps be persuaded to admit that some of their assumptions were erroneous or else—particularly in the case of the experimentalists—that their explanatory framework omitted important components? Two conditions had to be met before the two camps could unite: (1) a younger group of geneticists had to arise who took an interest in diversity and in the populational aspects of evolution, and

(2) the naturalists had to learn that the genetic interpretation of this second generation of geneticists was no longer opposed to gradualism and natural selection.

When this state had been reached, a meeting of the minds came quite suddenly and completely in a period of about a dozen years, from 1936 to 1947. It was in these years that biologists of the most diverse subdivisions of evolutionary biology and from various countries accepted two major conclusions: (1) that evolution is gradual, being explicatory in terms of small genetic changes and recombination and in terms of the ordering of this genetic variation by natural selection; and (2) that by introducing the population concept, by considering species as reproductively isolated aggregates of populations, and by analyzing the effect of ecological factors (niche occupation, competition, adaptive radiation) on diversity and on the origin of higher taxa, one can explain all evolutionary phenomena in a manner that is consistent both with the known genetic mechanisms and with the observational evidence of the naturalists. Julian Huxley (1942) designated the achievement of consensus on these points as *the evolutionary synthesis*. It required that the naturalists abandon their belief in soft inheritance and that the experimentalists give up typological thinking and be willing to incorporate the origin of diversity in their research program. It led to a decline of the concept of "mutation pressure," and its replacement by a heightened confidence in the powers of natural selection, combined with a new realization of the immensity of genetic variation in natural populations.

This tells us what happened during the synthesis, but it does not tell us how it came about. There is now rather general agreement that the reconciliation was the work of a handful of evolutionists who were able to build bridges among different fields and remove misunderstandings.

The Architects of the Evolutionary Synthesis

What qualifications did an evolutionist have to have in order to be able to function as a bridge builder? First of all, he had to be more than a narrow specialist. He had to be willing to become acquainted with areas of biology outside his own field of specialization, and to learn the new findings in these other fields. He had to be flexible, able to discard earlier ideas, and able to accept new ones. For instance, Sumner, Rensch, and Mayr, who had originally believed in soft inheritance, adopted a strict neo-Darwinian interpretation after becoming acquainted with the new genetic findings. What is still lacking is a critical analysis of the writings of the architects of the synthesis. What, if any, were their new ideas? Was it a rich assembly of facts that had the decisive impact? Was it the focusing of attention on concrete evolutionary phenomena (speciation, adaptive radiation, evolutionary trends, and so on) that was particularly effective? Which new genetic insights were most helpful in eliminating misunderstandings? What was the particular role played by each of the bridge builders? None of these questions (and there are many others) has yet been fully answered. Evidently, only a beginning has been made in the study of the evolutionary synthesis (Mayr and Provine, 1980).

568

If we define as the architects of the synthesis those authors who in major publications actually constructed bridges among various fields, six names in particular come to our mind: Dobzhansky (1937), Huxley (1942), Mayr (1942), Simpson (1944; 1953), Rensch (1947), and Stebbins (1950). It must be stressed that there were numerous other evolutionists who had helped to "clear the terrain" so that the bridges could be built and who had supplied important building materials. This includes, first of all, Chetverikov and Timofeeff-Ressovsky in the USSR; Fisher, Haldane, Darlington, and Ford in England; Sumner, Dice, Sturtevant, and Wright in the United States; Baur, Ludwig, Stresemann, and Zimmermann in Germany; Teissier and l'Héritier in France; and Buzzati-Traverso in Italy. Two multiauthor volumes have also contributed to the synthesis: Heberer's edited volume, *Die Evolution der Organismen* (1943), and Julian Huxley's *The New Systematics* (1940).

When one looks at the ten to twelve people most active in the synthesis, one finds that each of them occupied his own special niche. Mentioning the names Dobzhansky, Simpson, Mayr, Rensch, Huxley, and Stebbins makes this very evident. Yet, they all had one thing in common: they had recognized the communication gap between the various evolutionary schools, and had attempted to bridge it by reconciling the gene-frequency approach of T. H. Morgan, R. A. Fisher, and others with the population thinking of the naturalists.

As astonishing as its sudden arrival was the rapidity with which the synthesis spread through evolutionary biology. At an international symposium at Princeton, New Jersey, held January 2–4, 1947, in which representatives of the most diverse fields and schools (except hardline Lamarckians) participated, there was universal and unanimous agreement with the conclusions of the synthesis. All participants endorsed the gradualness of evolution, the preeminent importance of natural selection, and the populational aspect of the origin of diversity (Jepsen, Mayr, and Simpson, 1949). Not all other biologists were completely converted. This is evident from the great efforts made by Fisher, Haldane, and Muller as late as the late 1940s and 50s to present again and again evidence in favor of the universality of natural selection, and from some reasonably agnostic statements on evolution made by a few leading biologists such as Max Hartmann.

There is complete agreement among the participants of the evolutionary synthesis as well as among historians that it was one particular publication that heralded the beginning of the synthesis, and in fact was more responsible for it than any other, Dobzhansky's *Genetics and the Origin of Species* (1937). As L. C. Dunn rightly said in the preface, the book symbolized "something which can only be called the Back-to-Nature Movement." The very first chapter was devoted to organic diversity, and other chapters covered variation in natural populations, selection, isolating mechanisms, and species as natural units. Dobzhansky had successfully integrated the naturalist's profound understanding of evolutionary problems with the knowledge which he had acquired in the preceding dozen years as an experimental geneticist. Truly he was the first to build a solid bridge from the camp of the experimentalists to that of the naturalists.

The evolutionary synthesis settled numerous old arguments once and for all, and thus opened the way for a discussion of entirely new problems. It was clearly the most decisive event in the history of evolutionary biology since the publication of the *Origin of Species* in 1859. Yet, historians and philosophers of science have been puzzled over just exactly how the synthesis fits into the theory of scientific advance. It was definitely not a revolution, since it was clearly only the final maturation of Darwin's theory of evolution. But does it even deserve the epithet "synthesis"? This I assert emphatically.

I described above the radically different thinking and preoccupations of the two camps of evolutionary biologists, that of the experimental geneticists and that of the population naturalists. They truly represented two very different "research traditions," as Laudan (1977) has called it. Laudan observes that "there are times when two or more research traditions, far from mutually undermining one another, can be amalgamated, producing a synthesis which is progressive with respect to both the former research traditions" (p. 103). What happened in evolutionary biology from 1936 to 1947 was precisely such a synthesis between two research traditions that had previously been unable to communicate with each other. There was no victory of one paradigm over another, as described in Kuhn's theory of scientific revolutions, but rather "an exchange" of the most viable components of the two previously competing research traditions. For this reason it would be incorrect to state that the synthesis was merely the acceptance by the naturalists of the newer findings of genetics. This would ignore the numerous concepts contributed by the naturalists: population thinking, the multidimensionality of the polytypic species, the biological species concept (with the species defined as a reproductively and ecologically autonomous entity), the role of behavior and of change of function in the origin of evolutionary novelties, and the entire emphasis on the evolution of diversity. All these concepts are indispensable for a full understanding of evolution, and yet they had been virtually absent from the conceptual framework of the experimental geneticists.

In the short run, it was perhaps the refutation of a number of misconceptions that had the greatest impact on evolutionary biology. This includes soft inheritance, saltationism, evolutionary essentialism, and autogenetic theories. The synthesis emphatically confirmed the overwhelming importance of natural selection, of gradualism, of the dual nature of evolution (adaptation and diversification), of the populational structure of species, of the evolutionary role of species, and of hard inheritance. Even though this amounted to a drastic narrowing down of the options available to an evolutionist, it left many problems unsolved. These problems fall into two categories, indicated by these two questions: (1) What is the meaning of a given phenomenon (selection, gradual evolution, biological species, and so on)? and (2) How does a given evolutionary principle or phenomenon actually operate in an individual case, and what new problems does this raise (as applied to selection, isolation, the production of variation, stochastic processes, and so on)?

13 ✐ Post-synthesis developments

A NUMBER OF reasonably well-delimited periods can be recognized in the history of evolutionary biology. The main concern of
evolutionists in the period from 1859 to about 1895 was the proof
of evolution and the establishment of the various lines of common
descent. Phylogenetic research was the major preoccupation of
evolutionists. From about 1895 to the beginning of the evolutionary synthesis (1936), controversies within the field dominated research and writing. The great questions of the period were: Is
evolution gradual or saltational? Is inheritance soft or hard? Is
genetic change due to mutation pressure or selection pressure?
The period from 1936 to the 1960s was dominated by the evolutionary synthesis and the working out of the finer details of the
new insights. The populational approach dominated all investigations and there was a new interest in diversity, particularly at
the level of populations and species; the adaptational aspects of
variation were analyzed as due to selection forces, but all genetic
interpretations were dominated by the gene-frequency concept.

Later developments in evolutionary biology are diffuse. They
include a strong interest in the stochastic components of variation
and a recognition of the diversity of the genetic material (in the
form of various types of DNA). Broad contacts were established
with ecology and behavioral biology, and the study of the evolution and evolutionary role of macromolecules became an ever more
important branch of evolutionary biology. As a result of all these
developments, the study of evolution has become a highly diversified science. More than that! The expansion of evolutionary
thinking to all branches of biology led to a tearing down of the
walls between evolutionary biology and other fields of biology, to
the extent that it is now impossible to say whether such fields as
evolutionary ecology, evolutionary ethology, and molecular evolution should be included with evolutionary biology or with the
adjacent fields with which they have been fused. What is perhaps
most significant is that it has finally become respectable for a bi-

ologist to ask why-questions without being suspected of being a teleologist.

The unified interpretation of the evolutionary process had a highly beneficial impact on the standing of evolutionary biology in the whole field of biology. By eliminating all interpretations that signaled an implicit conflict with physico-chemical explanations (namely, those theories that were vitalistic or teleological), evolutionary biology became far more respectable than it had been during the preceding period, when it was maligned by the experimentalists as "speculative." The new insight which emerged from the elucidation of the structure of DNA in 1953—that living matter consists of two fundamentally different components, a historical one (the genetic program) and a functional one (the translated proteins)—necessitated at once an extension of the causal analysis of all biological phenomena to the historical component. This led to the recognition that it is not only legitimate but indeed essential to any reasonably complete biological analysis to include a study of the evolutionary history of all components of living organisms. This extension of evolutionary thinking has affected every branch of biology.

572

Evolutionary biology certainly represents a splendid example of shifting interests and research programs in a field of science. My simplified description conceals, however, that hardly any line of approach is ever completely terminated, even when more promising new approaches open up, nor does it mention that the roots of any new approach usually reach back several decades prior to the date when it became productive. Every new technique and every switch of an investigator from one field of competence to another is likely to initiate new approaches. Obviously it is impossible to represent adequately the full complexity of the advances in evolutionary biology or for that matter in any field of science.

In 1946 a special society was established in the United States to cultivate the study of evolution: and in 1947 Ernst Mayr founded *Evolution,* a journal devoted to research in evolutionary biology. *The American Naturalist,* which in the 1930s had been largely converted into a journal of experimental biology, returned after the synthesis to its specialization in evolutionary biology. Other periodicals devoted to evolution were founded in the United States and in other countries. The number of new textbooks on evolution has grown steadily, and so have courses in evolutionary biology given at colleges and universities. The literature has grown to such an extent that frequent review papers are now necessary.

This prodigious activity poses a serious problem for the historian. It has now become quite impossible to analyze the recent advances anywhere near adequately. The best I can do is to sketch the major outlines of some recent researches and to mention at least some of the unanswered questions that appear particularly intriguing to the current generation of evolutionists. For a listing of the relevant literature I must refer to the contemporary journals and some recent textbooks.[1] Let me begin by mentioning evolutionary problems that have occupied, in recent years, the attention of population genetics and of molecular biology.

573

POPULATION GENETICS

Population genetics since the 1930s had as its major objective to test the conclusions of mathematical population genetics in the field and in experimental populations in the laboratory. This work was dominated by the definition of evolution as "the change of gene frequencies in populations." Outstanding in this research tradition is the important series by Dobzhansky and associates entitled *The Genetics of Natural Populations* (1938–1976), dealing largely with *Drosophila pseudoobscura* and its sibling species (Lewontin et al., 1981). What Dobzhansky attempted to determine were numerical values for selection pressure, gene flow, effective population size, frequency of lethals and other concealed recessives, and other factors of potential evolutionary significance.[2] Of particular advantage in this research was the fact that this species, like most other species of *Drosophila*, is rich in paracentric chromosomal inversions (recognizable in the banding pattern of the giant salivary chromosomes), each with a somewhat different geographic range. Dobzhansky found that the relative frequency of a given inversion may vary not only geographically but also seasonally and in some cases through a series of years. A number of regularities indicated that the frequency was controlled by selection, and this was confirmed experimentally. Mayr (1945) attempted to interpret the gene arrangements as ecotypes adapted in such a way that the bearers of different inversions could utilize different local niches. This has been subsequently confirmed by Coluzzi et al. (1977) for gene arrangements in mosquitoes (*Anopheles*). What is most remarkable is that bearers of different gene arrangements not only have a different fitness in different niches but also the behavioral capacity to search out the right niche.

A major technological advance in the study of *Drosophila* populations was made by Teissier and l'Héritier, who invented "population cages" in which populations of *Drosophila* of various sizes and of various genetic heterogeneities could be continued for many generations without the input of new alien genes; keeping these cages under different conditions of temperature and food supply, one could test different genes and gene combinations for relative fitness and calculate selection pressures. Dobzhansky and others soon took up this technique, which is now utilized in numerous genetics laboratories, with all sorts of modifications. It provided an excellent method for the experimental study of natural selection in populations.

574

MOLECULAR BIOLOGY

Many of the findings of biochemistry, ever since there has been such a branch of biology, were important for evolutionary biology, even though this was not at first recognized. One might mention here the discovery of nuclein by Miescher in 1869, the work of Nuttall in immunology, that of Garrod on inborn errors in metabolism, of Landsteiner on blood groups, and the later work of Beadle and Tatum. Yet, the phoenix-like rise of molecular biology did not really start until the discovery of the structure of DNA in 1953. This, at first, had little impact on previously established evolutionary concepts. Of the greatest immediate importance was the finding that the translation from nucleic acids to peptides and proteins is a one-way street (the "central dogma"). This discovery supplied the last and most conclusive proof for the impossibility of an inheritance of acquired characters.

The extraordinary precision and reliability of the replication of the germ plasm during each nuclear division was, until recently, not a conceptual problem. Essentialists took it for granted and believers in soft inheritance considered it irrelevant. Biophysicists, however, were rather puzzled by the almost error-free performance of the complicated replication process. An occasional error, of course, was recorded, and this is what the geneticist calls a mutation. For the evolutionist, the margin of error was not particularly disturbing since he knows what immense numbers of gametes and zygotes are lost anyhow prior to or during development. What was unexpected was the discovery of repair mechanisms which permit the subsequent "repair" of errors in replication. The existence of these mechanisms poses problems concerning the definition of "rate of mutation," but they do help to explain the infrequency of observed errors of replication.

The finding that the genetic code is, on the whole, identical in all organisms, including the prokaryotes, was an important addition to the evidence that all life on earth, as it exists now, can be traced back to a single origin. This and some other discoveries of molecular biology have helped to simplify and unify biology, but there have been other findings that may require certain modifications in genetic theory or at least in our understanding of genetic processes.

Most of the early analysis in molecular biology was done on viruses and bacteria and, following Occam's razor, the assumption was made that the findings in prokaryotes could be applied to the eukaryotes without modification. Recent studies indicate, however, that this assumption is not necessarily valid. In particular, it is now evident that the eukaryote chromosome has a very complex structure, radically different from the simple tandem DNA double helix of the prokaryotes. Instead, the DNA is intimately associated with various proteins, particularly histones, with which it forms molecular aggregates (nucleosomes) of various sizes, which seem to differ functionally. At the present time these researches are primarily of interest for physiological genetics, but there can be little doubt that eventually the knowledge of the organization of DNA in the eukaryote chromosomes may give answers to various not yet solved evolutionary problems, such as the control of evolutionary trends, the stability of the phenotype in many evolutionary lines, rapid shifts to new evolutionary grades during genetic revolutions, and others. It is quite possible that we stand on the threshold of major discoveries.

When Nirenberg and Matthaei succeeded in 1961 in breaking the genetic code, it was widely believed that the last major problem of molecular biology had been solved. Actually the frequency of altogether unexpected discoveries has since accelerated. The major bearing of these discoveries, up to now, has been on aspects of gene physiology, but there is little doubt that all these aspects of genes are also of evolutionary importance, as will surely become apparent when the molecular processes are fully understood.

The structures that control genetic processes are of submicroscopic dimensions, and molecular biologists have been extraordinarily ingenious in developing new techniques that permit inferences on molecular structures and processes and on their variation. In fact at this point more is being learned about molecular evolution by the application of new techniques than by the development of new concepts. One of the most important of these techniques, first employed by Clem Markert, is that of starch gel electrophoresis.[3] Soluble proteins migrate in a gel in an electric field different distances, depending on their size and their electrical qualities, and can thus be separated from each other. Each protein can be made visible in the gel by a different staining technique. By this method the genotype of an individual can be determined directly without any breeding analysis. In fact, 20, 30, or even more than 70 gene loci can be analyzed simultaneously for the existence of alleles. The method makes possible what no method had permitted before, the determination of the level of heterozygosity of individuals and of populations. It also permits the comparison of different geographic populations of a species, and of related species, to determine what fraction of the spectrum of alleles is the same or different. The greatest weakness of the method is that it reveals the variation only of "structural" (enzyme) genes. A second weakness is that it cannot separate alleles that have the same electric load and it underestimates therefore the number of alleles. By using additional methods (heat degradation, change in pH), further alleles are often discovered. Since only a few enzymes have been analyzed thoroughly, it is still controversial what fraction of genetic variability is overlooked during the conventional electrophoresis method.

The elegance of the technique and the ease by which it can be applied even by a nonbiochemist have led to a veritable explosion of studies of enzyme variation ever since Hubby and Lewontin (on *Drosophila*) and Harris (on man) first used this technique in 1966 to estimate heterozygosity in individuals and populations. The number of discoveries made possible by this technique is large: new sibling species, a quantification of degrees of difference between closely related and more distant species, a correlation (or not) of enzyme change with speciation events, a correlation (or not) of geographic variation of enzymes with climatic and other environmental factors, and many others.

One of the conclusions drawn from these studies, more or less confirmed by the behavior of other macromolecules, is that there are regularities in the rate at which molecules change over geological time, that is, the rate at which amino acid residues are being replaced in evolution. It has therefore been postulated by

certain authors (first by Pauling and Zuckerkandl and later particularly by Sarich and Wilson) that one can utilize this regularity to
construct a "molecular clock" and infer the date of the branching
point between two evolutionary lines from the degree of difference of homologous molecules (Wilson et al., 1977).

At the present time there are considerable discrepancies between the chronologies of the branching points calculated by the
molecular clock and those calculated by paleontologists on the basis of the (admittedly scanty) fossil record. And there is other evidence that the concept of the molecular clock must be applied
with caution. For instance, the same molecule may change during
the same geological time interval more rapidly in some phyletic
lines than in others. Also, it seems that the rate of change may
occasionally be drastically slowed down in certain phyletic lines.
The molecular distance between man and chimpanzee, for instance, is smaller than that between certain species of *Drosophila.*

The other difficulty is that the concept of the molecular clock
implies a built-in regularity, one might almost say an autonomy,
of the changes. The clock has sometimes been described in such
terms as "the occurrence of one mutation every two million years."
Such a formulation is of course utterly misleading; mutations at
the same gene loci occur quite frequently, but are consistently
eliminated by errors of sampling or by natural selection until the
molecular milieu has changed sufficiently to favor a change in the
three-dimensional structure of the molecule. In other words, molecular clocks are governed by natural selection, not by rates of
mutation. This has been shown for many macromolecules, but for
none more convincingly than for hemoglobin. Here the replacement of even a single amino acid among more than three hundred
can be highly deleterious. Thus, sickle-cell anemia is caused by the
replacement in the beta chain of hemoglobin of a single glutamic
acid residue by valine. In man more than two hundred different
hemoglobin mutations are now known (discovered as "private"
blood types), and even though in many cases they are not the
cause of a pronounced blood disease, not a single one of these
mutations has succeeded in becoming either fixed or polymorphic
in the hominid ancestry. That these mutations are being discriminated against by natural selection is indicated by the fact that our
distant relative, the chimpanzee, has a hemoglobin almost identical with ours in spite of the documented high mutation rate of
hemoglobin.

The explanation of the molecular clock phenomenon is pre-

577

sumably that each macromolecule regularly interacts in the cell with some 10 to 25 other macromolecules. However, when some of these other molecules evolve, by answering specific selection forces, these changes sooner or later produce a selection pressure on the original molecule to replace an amino acid residue in order to achieve the best possible fit with its genetic milieu and restore a steady state.

Kinds of DNA

578

Because all genes consist of DNA it was assumed after 1953 that all genes were basically identical in their function and evolutionary characteristics. The researches of the last two decades have revealed that this is not the case. There are a number of categories of genes such as enzyme genes, genes for structural (non-soluble) proteins, regulatory genes, and perhaps many more kinds of which so far we have not even an inkling. A higher organism may have enough DNA in its nucleus for about five million genes, and yet genetic research finds evidence for only about 10,000, or at most 50,000 traditional (enzyme) genes. They (together with other kinds?) are among the so-called unique sequences, but there are also several classes of "repetitive DNA" and much seemingly "silent" DNA, the function of which is quite puzzling. Much of the DNA that is not accounted for by the enzyme DNA evidently has regulatory functions. Only the very first beginnings have been made in the study of the differences in the evolutionary behavior of the various kinds of genes (Davidson and Britten, 1973; 1979).

New discoveries in molecular genetics have been following each other since the late 1960s, and particularly since 1975, at such breakneck speed that it is quite impossible for a nonspecialist to keep up with them. Furthermore, some of these discoveries were so unexpected that their interpretation is still altogether controversial. These discoveries concern the structure of the eukaryote genome. For instance, it was found that some genes—transposable genes—may change their position on the chromosome. Even more surprising was the discovery that many genes include sequences ("introns") that are not transcribed into the messenger RNA (mRNA) but are excised during the transcription process and the remaining parts of the genes ("exons") are subsequently "spliced" together into the functional mRNA. Two questions arise: How could such a peculiar system have ever evolved? Are the introns merely inert ballast or do they have a still unknown function? The teleological answer that the seemingly functionless DNA is stored up "in order to have it available in future times of need" is altogether unsatisfactory. One interpretation that is rather popular at the moment (Orgel and Crick, 1980) is that this extra DNA is, so to speak, parasitic, the organism being helpless to prevent its replication and accumulation. Although valid arguments in favor of this hypothesis exist, it is intuitively distasteful to a Darwinian. Surely natural selection, a Darwinian would say, should be able to come up with a defense mechanism against such an expensive type of parasitism. Considering how very little is known about the functioning of gene regulation, it is premature to write off the introns as genetically inert. For all we know, it might be very important to keep certain segments (exons) of the gene separated from each other prior to translation. Indeed there is now evidence that the introns help to regulate gene splicing.

579

Equally puzzling is the fact that closely related species or genera may drastically differ from each other in their repetitive DNA and in other components of the genome without much visible morphological change and sometimes even without a loss of the ability to hybridize. How this may affect evolutionary potential is still entirely unknown. Ever since the pioneering work of Mirsky and Ris (1951), it has been known that different groups of organisms have differing amounts of DNA in their cells (nuclei). The smallest amounts are found in prokaryotes and fungi, the largest amounts in urodeles, lungfishes, and some groups of plants. Some regularities are known (nearly all of them with exceptions), such as that annual plants usually have less DNA than related perennials or trees. Species with slower growth rates (longer developmental times) tend to have more DNA than their relatives. The enormous differences in amount of DNA in different taxa seems to support the idea that much of the excess DNA cannot be of very high selective significance. Further evolutionary speculations, however, are premature as long as our knowledge of gene regulation in eukaryotes is as rudimentary as it is today.

Evolutionists since Lamarck have been familiar with the principle of "mosaic evolution," which says that different components of the phenotype may evolve at highly unequal rates. It is now being discovered that such inequality of evolutionary rate is also true for molecular evolution. Wilson and colleagues (1974), for instance, believe that enzyme genes in mammals and anurans (such as frogs) evolve at about the same rates, while the regulatory genes which control morphological evolution change at a much higher rate in mammals than in frogs. In South American mimicking butterflies, the genes controlling color patterns show very strong geographic and virtually no individual variation, while the enzyme genes of these species show very high individual and virtually no geographic variation (Turner, Johnson and Eames, 1979). Also a strong difference in variability between enzyme genes and other protein genes was found by recent investigators. Finally, the genes that control speciation seem to vary quite independently of the enzyme genes. Here is a new frontier of evolutionary biochemistry which I rather suspect will produce major surprises in the near future. This much is evident already: that different groups of genes seem to answer to different selection pressures and follow their own evolutionary pathways. The results of the study of one group of genes, let us say enzyme genes, can not be generalized to apply to all categories of genes. This seems equally true for response to selection pressure, to variability (level of heterozygosity), and to molecular clocks. Chromosomal changes also have very different evolutionary rates in different organisms. The karyotype seems to be exceedingly stable in some groups and to change very rapidly in others, for instance in certain groups of mammals.

Each set of genes may play a different role in evolution. Enzyme gene differences apparently accumulate at a fairly regular rate and are thus ideal yardsticks for molecular clocks. Speciation events seem to be largely independent of enzyme genes. The reason why there are different kinds of genes is that they have different functions; however, our understanding of these functions is as yet very elementary.

Chetverikov's concept of the genetic milieu is beginning to acquire a new meaning. The study of the action of genes, it is now being realized, must be supplemented by the study of the interaction of genes. A pioneering discussion of the functioning of genotypes was Lerner's *Genetic Homeostasis* (1954), in which massive evidence for the importance of gene interaction was presented. Dobzhansky's research on "synthetic lethals" reenforced this thinking. He showed that certain genes or chromosomes could convey superior fitness in some combinations, and be lethal in combination with other chromosomes. This spelled the end of the faith in constant fitness values of genes, even though these findings, in the absence of an analysis of the causes of such relativity, are only a starting point in a new area of research (see Mayr, 1963, chapter 10; see also Mayr, 1974; Carson, 1977).

The study of molecular evolution has revealed the surprising fact that most macromolecules of higher organisms can be traced right back to the prokaryotes. And yet a prokaryote may have only a fraction ($^1/_{10,000}$) of the amount of nucleic acid of a higher organism. Where do all the other genes come from?

The first geneticists who speculated about this were apparently members of Morgan's group (Metz, 1916; Bridges, 1918). The sophisticated researches of Sturtevant, Bridges, and Muller revealed that new genes arise when new pieces of chromosome are inserted into an existing chromosome. This is either effected by unequal crossing over or by a major chromosomal mutation, particularly a translocation. The analysis of salivary chromosomes in *Drosophila* provided a welcome opportunity to confirm the occurrence of duplications inferred on the basis of purely genetic evidence. In other cases whole chromosomes may be added to the genome (owing to nondisjunction) or the chromosome set as a whole may be duplicated (by the process of polyploidy). The work of the early pioneers on gene duplication has been greatly expanded in recent years (for example, Ohno, 1970). The evolutionary advantage of small-scale duplications is that they interfere much less with the normal functioning of the genome than do sometimes major translocations or the addition of whole chromosomes (as in Down's syndrome) or chromosome sets. Small duplications are, thus, more easily incorporated into the gene pool. The duplicated genes can assume new functions and by divergent mutation become increasingly more different from their sister gene. It has been questioned whether such duplication can lead to the production of entirely new proteins, but the evolutionary history of far too few macromolecules is known to permit such sweeping conclusions. It is, however, quite possible, if not probable, that the most important classes of macromolecules were invented in the very early history of life.

581

Origin of Life

When Darwin in 1859 proposed the theory of common descent, he realized that at the beginning there had to be a "first life," and he expressed it in the somewhat biblical sentence of life "having been originally breathed into a few forms or into one" (*Origin:* 490). This was a most daring formulation, since at that time the differences among the numerous kinds of organisms appeared far too great for a single origin. Even after the students of phylogeny had succeeded in tracing animals and plants back to simple algae and flagellate ancestors, a common origin of the prokaryotes (bacteria and relatives) and of the eukaryotes (higher organisms) seemed still altogether unlikely. And yet this is now well established through the researches of molecular biology. Not only the general chemical similarity of all forms of life, but specifically the fact that the genetic code is identical everywhere (including the prokaryotes), leaves no longer any doubt that life, as it is now found on earth, originated only a single time. And there are now sound theories concerning the origin of the eukaryotes (Margulis, 1981). All organisms now living on earth unquestionably descended from a single ancestral stock. If there had been several independent origins of life, all the others succumbed to the competition of the stock which now rules the world.

An origin of life from inanimate matter would be spontaneous generation. It so happens that, precisely at the time when the theory of common descent was proposed by Darwin, the concept of spontaneous generation was under particularly heavy attack, owing to the experimental refutation of such a possibility by Pasteur and others (Farley, 1974). This posed a real dilemma for the evolutionists, and Darwin stated with resignation (1863): "It is mere rubbish, thinking at present of the origin of life; one might as well think of the origin of matter." But then, of course, being the inveterate speculator that he was, he mused in 1871: "It is often said that all the conditions for the first production of a living organism are now present, which could ever have been present. But if (and oh! what a big if!) we could conceive in some warm little pond, with all sorts of ammonia and phosphoric salts, light, heat, electricity, etc. present, that a proteine compound was formed, ready to undergo still more complex changes, at the present day such matter would be instantly devoured or absorbed, which would not have been the case before living creatures were formed" (*L.L.D.*, III: 18).[4]

The reason why the problem of the origin of life was so intractable for several generations after 1859 was that the whole question had to reformulated. One thought typologically of a living species suddenly arising from inanimate matter, and one thought of the earth as if its atmospheric and other environmental conditions had remained constant throughout all geological ages. These assumptions had to be thoroughly revised. The botanist Schleiden (1863) was apparently the first to suggest that an origin of life, "a first cell," might have been possible under the entirely different atmospheric conditions of the young earth. This has now been thoroughly substantiated. The young earth is now believed to have had a reducing atmosphere, consisting mostly of water vapor, methane, and ammonia. Free oxygen, which would oxidize and thus destroy any possible precursors of life, was virtually absent at the time when life originated on earth (about 3.5–3.8 billion years ago). The oxygen which, from about 1.9 billion years ago on, began to accumulate on earth was produced by the photosynthetic organisms which had evolved by then.

The second revision concerns life. Here the essentialistic concept of its sudden origin had to be replaced by the evolutionary concept of gradualism. We now realize that the origin of life was as gradual as the origin of man. Just as *Homo sapiens* is connected with the lower primates by a series of intermediate hominids, so did life have a series of precursors. Such intermediate molecular stages between inanimate matter and well-organized living beings are not now found in nature. They would not be able to survive in an oxidizing atmosphere and exposure to the enormous variety of microorganisms that subsist on organic molecules. In a reducing atmosphere, ultraviolet radiation and lightning can indeed produce organic compounds such as purines, pyrimidines, and amino acids that serve as the building stones of life. This was experimentally demonstrated by Miller (1953), following a suggestion by Urey. Haldane (1929) and Oparin (1924) had already previously suggested scenarios to explain how the gap between inanimate matter and life could be bridged. Fox (1977) has made very imaginative contributions toward the solution of this problem. Curiously, the findings of molecular biology have complicated the task of explanation rather than simplified it. Polypeptide chains (proteins), even in the simplest organisms, are assembled from amino acids under the guidance of a nucleic acid genetic program. Indeed, there is now such a complete "symbiosis" between nucleic acids and proteins that it is difficult to imagine either being able to function without the other. How then could the first proteins have been assembled and replicated without nucleic acids, and how could nucleic acids have originated and been maintained in the primeval "organic soup" if they had no other meaning than to control the assembly of proteins? (See Chapter 10, for a further discussion of this problem.)

The problem of the origin of life, that is the reconstruction of the steps from simple molecules to the first functioning organism, is one that poses a keen challenge to the students of molecular evolution.[5] A full realization of the near impossibility of an

583

origin of life brings home the point how improbable this event was. This is why so many biologists believe that the origin of life was a unique event. The chances that this improbable phenomenon could have occurred several times is exceedingly small, no matter how many millions of planets in the universe.

The preceding short review of recent advances in molecular biology indicates the close connection between research in molecular biology and in evolutionary biology. The vital interest of the molecular biologist in evolution is documented by the founding of a journal for molecular evolution and by a series of recent symposia and review volumes (for example, Ayala, 1976). As the evolutionist would say, the study of the evolution of molecules has become an important branch of evolutionary biology.

584

The assertion is sometimes made that, in addition to the Darwinian theory of evolution, there is now a "molecular theory" of evolution. The validity of this claim is doubtful. Two of the more important evolutionary phenomena that occur at the molecular level—hard inheritance (as espoused from Weismann, 1883, to the Morgan school) and mutation (de Vries, 1901; Morgan, 1910a)— were accepted, at least in principle, decades before the rise of molecular genetics. At this time it is still uncertain whether some of the recent discoveries of molecular genetics (repetitive DNA, gene splicing, wandering genes) do or do not require any revision of the synthetic theory of evolution. Presumably, the new discoveries enlarge only the amplitude of genetic variation that is available for the exercise of natural selection, as well as forming constraints on the action of natural selection.

I have used molecular biology as an illustration of the increasingly close relationship between evolutionary biology and other branches of biology. An equally active interaction has developed between evolutionary biology and many other biological disciplines. At the present time evolutionary questions seem to dominate the field of ecology. They are also of great importance in behavioral biology. This is well illustrated in recent textbooks of ecology and of animal behavior.

Even though the evolutionary synthesis did not solve all the problems of evolutionary biology, at least it created a united front. A glance at the current evolutionary literature shows how much disagreement in the interpretation of certain specific problems of evolution still exists. Yet, the opposing viewpoints do not question any of the basic theses of the synthetic theory; they merely have different answers for some of the pathways of evolution. I will attempt to demonstrate the nature of these disagreements by discussing some of the open questions in three major areas of evolutionary biology: the theory of natural selection, the problem of speciation, and the processes of evolution above the species level (macroevolution).

NATURAL SELECTION

The strong resistance to natural selection which had characterized the post-Darwinian period and Mendelism was largely broken by the synthesis. The resistance had been so powerful because it was the one thing which all anti-Darwinians had in common, for the neo-Lamarckians opposed selection as fervently as did the mutationists. The most widely known selection experiments of the first third of the century were those of Johannsen. Having received much of his training in chemical laboratories, he approached his task in the most unbiological manner. In order to have suitable experimental material, he first attempted to establish homogeneous classes, "pure lines." Not surprisingly, such samples of genetically identical individuals, the result of many generations of inbreeding, did not respond to selection. From this Johannsen concluded (1915: 609, 613) that selection cannot produce a deviation from the mean in self-fertilizing species, "and even the most careful experiments with cross-fertilizing plants and animals confirm most convincingly our interpretation of an inability of selection to achieve more than a mere isolation or separation of previously existing constitutionally different organisms: selection of different individuals creates nothing new; a shift of the biological type in the direction of selection has never been substantiated!" He finally concludes that it is "completely evident that genetics has deprived the Darwinian theory of selection entirely of its foundation, and . . . the problem of evolution is still an entirely open question" (p. 659). This conclusion was widely accepted among the experimentalists, and even T. H. Morgan (1932) stated, "The implication in the theory of natural selection that by selecting the more extreme individuals of a population, the next generation will be moved further in the next direction, is now known to be wrong." As late as 1936, two distinguished British zoologists, G. C. Robson and O. W. Richards, concluded, "We do not believe that natural selection can be disregarded as a possible factor in evolution. Nevertheless, there is so little positive evidence in its favour . . . that we have no right to assign to it the main causative role in evolution." No wonder that in this intellectual climate of the 1920s and 30s the Darwinians had to devote so much effort to the refutation of antiselectionist arguments.

The skepticism of the anti-Darwinians was not entirely unjustified. Direct proof for the occurrence of natural selection in nature and even in the laboratory was very scanty almost up to the middle of the twentieth century. The demonstration by Bumpus (1896) of differential mortality in sparrows, as a result of an ice storm, was for several decades the only evidence, and thus forever cited by selectionists. To make matters worse, the Darwinians themselves were rather divided about selection in the pre-synthesis period. As we saw, most of them, following Darwin's lead, accepted some soft inheritance, such as use and disuse. Wallace clearly was the most consistent of the early selectionists and the first to endorse Weismann's thesis that there is no soft inheritance and hence an "Allmacht der Naturzüchtung." In fact Wallace ascribed even the origin of isolating mechanisms strictly to selection, in conflict with Darwin, who could not visualize such a sympatric process. Modern students of speciation tend to agree with Darwin. But Weismann and Wallace were very much in the minority in their unconditional support of natural selection. Most other evolutionists held various reservations. (For objections that were raised against the efficacy of natural selection, see Chapter 11; for more detailed presentations see Kellogg, 1907; Mayr and Provine, 1980; and numerous volumes in the anti-Darwinian literature.)

Many factors played a role in the changing climate of opinion concerning natural selection. The following were perhaps the most important ones:

(1) The actual demonstration of the effectiveness of selection in selection experiments in the laboratory as well as in the work of numerous animal and plant breeders. Experiments carried out in nature, such as those of Kettlewell on industrial melanism (Ford, 1964), were particularly convincing. The introduction of the method of population cages by Teissier and l'Héritier in the 1930s (see above), a technique soon adopted by Dobzhansky and other *Drosophila* workers, gave rise to an active program of experiments on natural selection under different conditions of temperature, humidity, food supply, crowding, and competition of different genetic stocks.

(2) The refutation of the existence of soft inheritance by ge-

neticists, which left virtually no alternative than to explain gradual evolution by natural selection.

(3) The refutation of the claim that most attributes of organisms are without selective value. Even Haldane (1932: 113) had admitted, "There is no doubt that innumerable characters [of animal and plant species] show no sign of possessing selective value, and, moreover, these are exactly the characters that enable a taxonomist to distinguish one species from another." Eventually it was shown by various investigators, for instance Rensch and particularly E. B. Ford's Oxford group, that many of the characters that had previously been called "neutral" have a selective value when carefully investigated.

587

(4) The calculations of Norton, Haldane, Fisher, and others showing that even very slight selective advantages are important when continued over many generations.

(5) The spread of population thinking, in particular the demonstration by the new systematists that discontinuities among species and higher taxa could be explained as originating gradually through geographic speciation and extinction, hence not requiring saltations.

In his *Genetics and the Origin of Species* (1937) Dobzhansky devoted an entire chapter of 43 pages to the subject of natural selection. What made his presentation particularly effective was that he treated selection not merely as a theory but as a process that could be substantiated experimentally. Furthermore, he showed that there is no conflict between gradual adaptive geographic variation (as reflected in Rensch's climatic rules, for instance) and selection. This removed any need to take refuge in Lamarckian explanations, as the naturalists had been forced to do previously, owing to the arguments of the mutationists. Mayr (1963: 182–203) analyzed in detail many of the problems which selectionism had raised in the preceding decades. Among these problems five will be singled out for further discussion.

Kinds of Natural Selection

There are several ways in which kinds of natural selection can be classified. One of these is based on the portion of the curve of variation to which selection pressure is applied. *Stabilizing selection* refers to selection directed against both tails of the curve of variation; this corresponds to the "elimination" of the essentialists, that is, all deviations from the "normal" are discriminated against. *Di-*

rective selection is said to occur when one tail of the curve is favored and the other is discriminated against by natural selection, resulting in a steady advance of the mean value of the curve. *Diversifying (disruptive) selection* favors both tails of the curve over the mean, resulting in a bimodal curve, as is found in species with mimetic or other forms of polymorphism.

The Probabilistic Nature of Natural Selection

Essentialists have had great difficulty in understanding that selection is a statistical rather than an all-or-none phenomenon. The philosopher Charles Sanders Peirce saw this perhaps more clearly than his contemporaries and remarked that even though natural selection may fail in an individual case, "variation and natural selection . . . in the long run will . . . adapt animals to their circumstances." Mayr (1963: 184) likewise emphasized the probabilistic nature of selection. Even though philosophers may still refer to "survival of the fittest," biologists no longer use such deterministic language.

The Target of Selection

By adopting the formulation "evolution is caused by mutation and selection," some geneticists contributed to a widely held misconception. This formulation has been interpreted to mean that the mutated gene is the actual target of selection. By contrast, naturalists from Darwin on and the more perceptive geneticists have always emphasized that not genes but whole organisms—potentially reproducing individuals—are the unit of selection. This means that the effects of recombination and of gene regulation, as well as the capacity of developing phenotypes to respond to the environment, are as important for selection as is mutation, indeed quantitatively actually more important by several orders of magnitude. A difficulty arose, however, when Fisher (1930) and other mathematical geneticists chose the gene to be the unit of selection and attributed to each gene a definite fitness value. Fitness was redefined as the contribution a given gene makes to the gene pool of the next generation (see also Haldane, 1957). This, in turn, led to a very questionable definition of evolution ("change of gene frequencies in populations") and to the legitimate criticism that frequency changes in single genes left many, in fact most, evolutionary phenomena unexplained. Some, if not most, of the current criticism of the theory of selection consists of attacks on the un-Darwinian assumption of genes as units of selection.

588

This must be emphasized, because it demonstrates how misleading and confusing the concept of "internal selection" is, which has been promoted by several recent authors. It is quite impossible to partition selection into two parts, one caused by the external environment and another one caused by the internal factors of physiology and development. Such a partition is impossible because the result of selection is determined by the interaction between the external environment and the physiological processes of the organism as a whole. There is no internal selection. All developmental and regulatory processes contribute to the fitness of an individual either favorably or unfavorably, but this is evaluated when the individual is exposed to the external environment (including competition from individuals of the same or other species). Darwin already was fully aware of the importance of these internal factors, as is evident for instance from his discussion of correlation (*Origin:* 143–150). When a modern author still ascribes to the Darwinians the obsolete formula "mutation and selection," it is not surprising that he considers it as insufficient to explain appropriate evolutionary response. No one who still uses this formulation can possibly understand the actual causation of evolutionary change. Leading evolutionists have rejected the mutation as the target of evolution for more than forty years.

The Product of Selection as a Compromise

Since the phenotype as a whole is the target of selection, it is impossible to improve simultaneously all components of the phenotype to the same degree. Selection cannot produce perfection, for in the competition for reproductive success among members of a population, it is sufficient to be superior and not at all necessary to be perfect. Furthermore, every genotype is a compromise between various selection pressures, some of which may be opposed to each other, as for instance, sexual selection and crypsis, or predator protection (Endler, 1978). Owing to the cohesion of the genotype, it is sometimes not possible to improve one component of the phenotype without damaging some other component. After each shift into a new adaptive zone, certain adaptations to the previous zone become liabilities. The aquatic mammals had to reduce and eliminate so far as possible all specific adaptations for a terrestrial mode of living. The bipedal hominoids are still burdened by their quadrupedal past.

What evolutionists had long called the compromises of evolution is called by modern ecologists the optimization process of

589

evolution. There is a cost to every evolutionary advance (such as running faster, having more offspring, utilizing a new source of food) and selection determines whether or not the added advanage is worth the cost. The result is that the phenotype is often a patchwork of features that were specifically selected for a particular function (or as the answer to a particular selection pressure) and others that are the by-product of the genotype as a whole and are simply tolerated by selection. From Darwin's time on, naturalists have asked themselves into which of the two categories they should class differences among species. For instance, is the difference in the striping of Burchell's and Grevy's zebra a result of different selection pressures in the different parts of Africa where these species originated, or, as is more likely, was there simply a selection for striping to which the genotypes of the two species responded differently? As long as certain geneticists believed in an independent fitness of every gene, each having one optimal fitness value, one could believe that every aspect of the phenotype was the appropriate response to ad hoc selection. But the fact that the individual as a whole is the target of selection and, furthermore, that many (if not all) genes are interacting with each other, sets severe limits for the response of the phenotype to selection. This is why man still has an appendix, a vulnerable sacroiliac joint, and poorly built sinuses. Gregory (1913; 1936) called the totality of the ad hoc adaptations the *habitus* and the tolerated remnants of the past the *heritage*.

The conclusion that not every detail of the phenotype is shaped by ad hoc selection is reenforced by a phenomenon which Bock (1959) has designated as *multiple pathways*. For instance pelagic marine invertebrates have a great diversity of mechanisms by which to stay afloat in the water: gas bubbles, oil droplets, or an enlargement of the body surface. In each case natural selection, which is always opportunistic, made use of that part of the available variation that led most easily to the needed adaptation.

The misguided atomistic-reductionist strategy of dissecting an organism into as many parts as possible and demonstrating the selective value of each of these pieces has brought the whole concept of adaptation somewhat into dispute. To that extent, some of the objections to natural selection brought forth by opponents of selectionism (for example, Grassé, 1977a) were perfectly valid. Selection is probabilistic, errors of sampling in small populations inevitably have stochastic effects, and the integration of the organism as a whole always places severe constraints on the response of individual features. To be sure, organisms on the whole are well adapted to their environment, because those that were not had too low a reproductive success to survive. But this does not mean that every aspect of the phenotype of an organism is optimal in its construction and functional efficiency.

Selection as a Creative Force

Selection, for an essentialist, is a purely negative factor, a force which eliminates deleterious deviations from the norm. Darwin's opponents, therefore, insisted in the spirit of essentialism that selection could not create anything new. By saying this, they revealed that they had neither understood the two-step process of selection nor its populational nature. The first step is the production of an unlimited amount of new variation, that is, of new genotypes and phenotypes, particularly through genetic recombination rather than by mutation. The second step is the test to which the products of the first step are subjected by natural selection. Only those individuals that can pass this scrutiny became contributors to the gene pool of the next generation. Chetverikov, Dobzhansky, and others have rightly stated that this back and forth between genetic recombination and the selection of a highly limited number of progenitors of the next generation is indeed a creative process. It provides in each generation a new starting point and a new opportunity to take advantage both of new environmental and of new genetic constellations.

591

Unresolved Issues in Natural Selection

The interpretation of the five selectionist problems here discussed is relatively uncontroversial. By contrast some other problems still give rise to unresolved disagreements among evolutionary biologists. Some of these problems shall now be discussed.

Variability and Natural Selection

For the last fifty years two schools have disagreed with each other on the level of genetic variability in natural populations. For H. J. Muller and most classical geneticists, each allele had a different selective value, one of them, normally the "wild type," being the "best" and thus the prevailing gene in the population. He considered it the function of natural selection to eliminate other, inferior, alleles, the supply of which is continuously replenished by mutation. It is concluded from this reasoning that most individuals in a population should be homozygous at most loci, since the burden of deleterious recessives (the "genetic load") would otherwise become too great. Muller, Crow, and their associates were the most vigorous proponents of this traditional view.

The other school, of which Dobzhansky was the leader (also Mather, Lerner, Mayr, B. Wallace, and their students) considers the genotype a harmoniously balanced system of many genes with the heterozygotes often superior to the homozygotes of any of the alleles. Furthermore, this school denies absolute fitness values of genes so that several alleles could be the "best," depending in each case on their genetic milieu and on the prevailing external selection pressures. The thinking of the balance school had its beginning in Chetverikov's concept of the genetic milieu, a concept which was expanded in the theory of the genotype as a balanced system (Dobzhansky, 1951; Mather, 1943).

To determine the frequency of concealed recessives in a population with the help of the classical techniques of genetic analysis was not feasible, because only one locus at a time could be made homozygous. It was therefore impossible to settle the argument between the "classical" and the "balance" school. Finally, in 1966, the application of the enzyme electrophoresis method enabled Hubby and Lewontin to establish for *Drosophila* and simultaneously Harris for man an amazingly high level of allelic polymorphism. They found, and this was abundantly confirmed by later investigators, that even a single individual may be heterozygous for 10 percent or more of its loci, and a species for 30 to 50 percent. It thus seemed as if the issue had been clearly decided in favor of Dobzhansky's balance theory. It also seemed a vindication of Darwin's faith in the existence of a virtually inexhaustible supply of genetic variation.

However, as is the case with most new lines of research, the study of enzyme variability raised more new questions than it answered. Why do certain species have a much higher level of variability than others? What is the relation between the level of variability and the ecology of a species? What portion of the variability is maintained in the population by selection and what other part by chance (the mutation of virtually neutral alleles)? What relation is there between the variability of enzyme genes and the variability of the other DNA of the genotype? In the endeavor to answer these questions the study of the variability of enzymes by the electrophoresis method is now one of the most active fields of evolutionary genetics (Lewontin, 1974; Ayala, 1976; Ayala et al., 1974b).

The most controversial problem concerning this high genetic variability is its source. One would expect that errors of sampling and selection pressure against inferior homozygotes would drastically reduce the level of allelic variability. How can four, six, or even more than ten alleles at a single locus be simultaneously maintained in a population?

Random-Walk Evolution

When the enormous genetic variability of natural populations was discovered in the 1960s, the thesis of a selective neutrality of much of this variation was once more advanced. The proponents of this theory—King and Jukes (1969) and Crow and Kimura (1970)—refer to genetic change owing to stochastic processes (essentially neutral mutations) as "non-Darwinian evolution," a term which is quite misleading since Lamarckism, orthogenesis, and mutationism are also forms of non-Darwinian evolution. Others have called it, perhaps more appropriately, "random-walk evolution." Ever since then, an active controversy has been going on concerning the proportion of observed genetic variability in natural populations which is due to selection to that which is due to chance. Curiously, ideological commitments seem to play a role in this controversy, since Marxists, on the whole, attribute a greater role to random-walk evolution than non-Marxists. My own feeling is that selection is far more important than admitted by the promoters of non-Darwinian evolution, but that indeed there is a random component in much of the variation at some gene loci.

What has become rather probable is that a selective superiority of heterozygotes alone would not be able to maintain such high levels of genetic diversity. But there are other factors that favor genetic diversity (Mayr, 1963: 234–258). In the case of polymorph snails and insects, a rare phenotype is somewhat protected against predators because the predator's "search image" has become conditioned to the more common phenotype (apostatic selection) (Clarke, 1962). It has also been shown (first by Petit and Ehrman, 1969) that females of many species have a mating preference for males with rare genotypes; this also helps to prevent the loss of rare genotypes from populations. Other cases of variable selective values have been discovered and it seems probable that frequency-dependent selection is a rather important mechanism by which the genetic variability of populations is maintained.

Evidence is now also accumulating that different genotypes are not only superior in different subdivisions of the species' niche but may also have a preference for such subniches and the ability to find them. This agrees with the finding that genetic diversity is usually greater in diversified than in simpler habitats (Nevo, 1978; Powell and Taylor, 1979). A further mechanism by which genetic variability is maintained is defense against parasites and pathogens, as Haldane pointed out long ago (1949). High genetic variability in immunity-giving genes (antibody producers, and so forth) protects populations against devastating losses because the pathogens will be unable to cope with rare immune genes. Finally, if epistatic interactions are important, as we believe they are, genes at low frequency may be maintained because they are of high selective value in certain combinations. Considering how many of such selection-controlled mechanisms have now been discovered, all of which permit the diploid gene pool to store genetic variability, one is forced to conclude that much of the observed genetic variability of populations may well be the result of natural selection.[6]

594

The Cost of Selection

Haldane (1957) and Kimura (1960) made some calculations which showed how "expensive" it is to replace one allele in a large population by a selectively superior one. They concluded from this that evolution would have to proceed very slowly, that is, on relatively few loci simultaneously, or else the total mortality would be forbiddingly high. This conclusion was in apparent conflict with well-established rapid rates of evolutionary change, as for instance in freshwater fishes, as well as with the high level of heterozygosity in most natural populations. Obviously Haldane had made some unrealistic assumptions. Mayr (1963: 262), and later several other authors (Lewontin, 1974), called attention to the kind of simplifying assumptions made by Haldane. For instance, in a species in which only a small fraction of all the offspring reproduce owing to density-dependent competition, there is so much mortality in every generation anyhow that weighing down this "expendable surplus" with deleterious homozygotes is no great burden. More important is the fact that Haldane's calculations pertain to large populations, while rapid evolutionary changes happen most frequently in small populations (see below). Haldane may indeed be right for large, populous species. This is indicated by the evolutionary inertia of such species as revealed by the fossil record, but his calculations are not valid for small, particularly for founder populations, the very populations in which most of the crucial evolutionary events seem to take place.

Natural Selection as a Unitary Phenomenon

As long as the theory of natural selection was severely questioned, not much thought was given to possible subdivisions of natural selection. Now, with the validity of natural selection firmly established, new questions have come to the fore, for instance whether or not there is a process that might be called group selection, and whether or not it is legitimate to differentiate sexual selection from natural selection, as Darwin had done. Both questions have led to extensive controversies and it is necessary to say a few words to explain the nature of the argument.

595

Group Selection

The thesis that the individual is the principal unit of selection has been challenged by some evolutionists who postulate a process of *group selection* (Wynne-Edwards, 1962). Those who support this kind of selection claim that there are phenomena that could not possibly be the result of individual selection. They refer in particular to characteristics of entire populations, such as aberrant sex ratios, rates of mutation, distance of dispersal and various other mechanisms favoring either in-breeding or out-breeding in natural populations, and degrees of sexual dimorphism. Such differences among populations, say the proponents of group selection, can be established only when an entire population (deme) is favored over other demes because it differs in its genetic constitution for the stated factor. Whether, and to what extent, such group selection actually occurs is still actively discussed in the current literature, but the general consensus is that most of such cases can be interpreted in terms of individual selection, except perhaps in social animals (Lack, 1968; Williams, 1966).

The controversy concerning group selection has led to the realization that there are indeed uncertainties concerning various aspects of selection. Evolutionists have become aware that a number of rather different phenomena had been lumped together in the past and that they will not fully understand the workings of selection until they have partitioned the field into its components.

Sexual Selection

As early as the late eighteenth century some animal breeders had suggested that females show a preference for more vigorous males and that this explains sexual dimorphism. The process by which an individual gains reproductive advantage by being more attractive to individuals of the other sex was designated by Darwin as *sexual selection*. Darwin clearly distinguished it from natural selection (*sensu stricto*), which conveys superiority in general fitness (environmental tolerance, resource utilization, predator thwarting, disease resistance, and so forth). Darwin's interest in sexual selection was already evident in early handwritten notes (around 1840), but he allotted less than three pages to the subject in the *Origin* (1859: 87–90). Yet, in the *Descent of Man* (1871) the discussion of sexual selection occupies more pages than the evolution of man. However, nothing demonstrates Darwin's intense interest in the subject better than his extended correspondence with Wallace on the causation of sexual dimorphism (Kottler, 1980). The Darwin-Wallace correspondence was the inauguration of a controversy concerning the significance of sexual selection that is not yet at an end. (For a review of the early phase of the controversy, see Kellogg, 1907: 106–128.) Darwin's effort to keep sexual selection distinct from natural selection encountered strenuous objections. In 1876 even Wallace abandoned sexual selection and so did most experimental biologists in the ensuing years since, like T. H. Morgan, they were interested only in proximate causations (for example, what hormones or genes are responsible for sexual dimorphism). The recognition of sexual selection was entirely rejected by the mathematical population geneticists, who considered evolution as a change in gene frequencies and defined fitness simply as the contribution of a gene to the gene pool of the next generation. Since this definition, indeed, applies equally to natural and sexual selection, any distinction between the two kinds of selection is obliterated.

In recent years the individual has been reinstated as the principal target of selection and it has become respectable to revive Darwin's concept of sexual selection (Campbell, 1972). Admittedly, Darwin had included under sexual selection aspects of sexual dimorphism that would be better listed under natural selection, such as some aspects of male aggressiveness. What is left, however, are all those aspects of male adornment (and song) which Darwin had explained as being due to "female choice." Even though the principle of female choice was defended by most naturalists for the last hundred years, it was rejected by the majority of biologists and by virtually all nonbiologists for the reason that it credited females with a discriminating ability "which they could not possibly have." Recent studies of ethologists and other field naturalists, however, have proved conclusively that females not only in vertebrates but also in insects and other invertebrates are usually very "coy," by no means accepting for copulation the first male they encounter. Indeed, the selection of the male which finally is admitted for copulation is often a very protracted process. Female choice in these cases is an established fact, even where the criteria are not yet known on the basis of which the females make their choice.

This strongly contrasts with males, who usually are ready to mate with any female and quite often do not even discriminate among females of their own and of other species. The reasons for this drastic difference between males and females were pointed out by Bateman (1948) and further elaborated by Trivers (1972), on the basis of the principle of investment. A male has sufficient sperm to inseminate numerous females, and his investment in a single copulation is therefore very small. A female, by contrast, produces relatively few eggs, at least in species with female choice, and may furthermore invest much time and resources in brooding the eggs or developing the embryos and in taking care of the brood after hatching. She may lose her entire reproductive potential by making a mistake in the selection of her mate (for instance, by producing inferior or sterile hybrids). The principle of female choice explains also a number of other phenomena that were previously puzzling, for instance, why polymorphism in species of butterflies with Batesian mimicry is usually limited to females. Females would discriminate against males that deviate too far from the species-specific image of the mating partner (releasing mechanism).[7]

There is now a justified tendency to interpret sexual selection rather broadly as any morphological or behavioral characteristic that gives a reproductive advantage.[8] Mayr (1963: 199–201) called attention to the potentially "selfish" aspects of some kinds of natural selection, specifically those which enhance the reproductive success of individuals without adding to the general adaptedness of the species. Hamilton (1964), Trivers (1972), and Dawkins (1976) have shown how widespread this kind of sexual selection is and how profoundly it affects animal behavior and evolutionary trends. Wilson (1975) has reviewed much of the relevant literature. Reproductive selfishness would seem a milder equivalent of the struggle for existence than "nature red in tooth and claw" made proverbial by the social Darwinists.

597

During the 1880s and 90s when social Darwinism was confused with real Darwinism, cooperation and altruism were often cited as evidence for the evolution of human ethical tendencies that could not possibly have been the product of natural selection. This claim overlooked the fact that cooperating, particularly in social organisms, may be of selective advantage. Darwin had already recognized this when he said, "I use the term Struggle for Existence in a large and metaphorical sense, including dependence of one being on another" (1859: 62).

598   The problem of altruism and its evolution which Haldane raised in 1932 is now again the focus of attention. Altruism is usually defined as an activity that benefits another individual (the "recipient") to the seeming disadvantage of the altruist. Haldane pointed out that an altruistic trait would be favored by natural selection if the beneficiary was sufficiently closely related, so that his survival benefitted the genes which he shared with the altruist. For instance, if there is 1 chance in 10 that an altruistic act would cost the life of the altruist, but the beneficiaries were the children, siblings, or grandchildren of the altruist, with all of whom he shares more than 10 percent of his genes, selection would favor the development of altruism. This particular form of selection has also been designated as *kin selection,* and the fitness which refers to all the carriers of the same (or similar) genotype is known as *inclusive fitness.* Haldane's rather simple theory has since been elaborated by Hamilton, Trivers, Maynard Smith, G. C. Williams, Alexander, West-Eberhard, and many others and has become part of sociobiology.[9]

Sociobiology, broadly speaking, deals with the social behavior of organisms in the light of evolution. There is little argument that much if not most of social behavior in animals has a strong genetic component. The part of sociobiology that is being attacked is that which deals with man: Can man's social behavior be compared with that of animals? To what extent is the social behavior of man part of his primate heritage? These are among the questions that have been asked. Much of the argument seems to be semantic. For instance, E. O. Wilson and other sociobiologists have been accused of preaching the genetic determinism of behavior. This does not represent their views accurately. All they have said, and one can argue about the validity of this claim, is that much of man's social behavior has a genetic component. But that is not the same as genetic determinism. It must be remembered that a behavior may be controlled by "closed" or "open" programs and that even open programs have a considerable genetic component. The profound differences in social behavior among human groups, some of them closely related, show how much of this behavior is cultural rather than genetic.

The Evolutionary Significance of Sex

Several recent authors have been puzzled about the possibility that there may be a conflict between Weismann's theory of sex and the principle of reproductive success. A species with uniparental reproduction can generate twice as many reproducers as a sexually reproducing species that "wastes" half of its zygotes on males. Accordingly, one would expect natural selection to favor uniparental reproduction (for example, parthenogenesis) over sexual reproduction (Williams, 1975; Maynard Smith, 1978). Uniparental reproduction indeed is widespread both among plants and animals, and yet it is of much lower frequency than sexual reproduction. No explanation has yet been advanced that would satisfactorily explain this puzzle. Undoubtedly, in the long run, sexual selection is superior because it provides an escape in case of any major change in the environment. However, in the short run in relatively stable environments one would expect the doubled fertility of uniparentals to carry the day. Perhaps one should again invoke the principle of the "expendable surplus": even in sexually reproducing organisms there is already a sufficiently large expendable surplus; a doubling of it would not be of any particular selective advantage. Moreover, there is little doubt that abandoning sexuality cuts down drastically on future evolutionary options. Evolutionary lines that switch to uniparental reproduction most likely become extinct sooner or later, and with this also any mechanisms that would permit such a switch. What are left are strictly sexual lineages unable to switch to uniparentalism but able to fill the niches vacated by extinct uniparental lineages. Sexual reproduction is, of course, obligatory wherever there is the possibility for a second parent to participate in parental care. There are many other correlations between sexuality, behavior, and niche utilization (Ghiselin, 1974a). It has long been known that there is a regular alternation between sexual and parthenogenetic generations among a large assemblage of organisms (certain parasites, freshwater plankton, aphids) and that the shift from one to the other state is closely correlated with changes in the environment.

Natural selection is indeed often puzzling and the modern evolutionist is as perplexed about the selective aspects of some

599

natural phenomena as were Darwin and Wallace. Considering how useful an organ the human brain is, the question is sometimes asked, Why did not selection produce as large a brain in all organisms? Yes, why? Or to turn this argument upside down, what selection pressure gave Neanderthal man a brain as large as that of Darwin, Einstein, or Freud? It was this inability to account for the large brain of our primitive ancestors which made Wallace doubt that selection could account for the origin of man *as man.* What Wallace overlooked is that the crucial moment in all selection is an emergency or catastrophe. An organ or function is usually not altered by selection during normal times; rather, it is selected at a time when it represents the tail end of the curve of variation and permits its carrier to survive in an emergency when the other thousands or millions of individuals of the species succumb. "Catastrophic selection," as Lewis (1962) has emphasized quite rightly, is a very important evolutionary process.

600

MODES OF SPECIATION

Darwin, as the first avowed representative of population thinking, stressed the gradualness of the process of geographic speciation (see Chapter 11). The Mendelians emphatically denied Darwin's conclusion, assuming instead, as it was stated by de Vries, that "new species and varieties are produced from existing forms by certain leaps." The mode of speciation was a major bone of contention between the naturalists and the Mendelians (Mayr and Provine, 1980). Comparative anatomists, students of phylogeny, and even experimental geneticists thought of evolution strictly in "vertical" terms, and the phyletic line was for them the unit of evolution. It was a major contribution of the new systematics to have adopted the population as the unit of evolution and to base the explanation of speciation on this concept. New species, the new systematists asserted, originate when populations become isolated, and this thesis was defended by Mayr (1942) with the support of rich detail. Nothing was said, at first, about the size of such isolated populations, except that Wright (1932) called attention to the fact that genetic drift might occur in small and very small populations as a result of errors of sampling.

The theory of geographic speciation was based primarily on birds, butterflies, and other wide-ranging insects, some groups of snails, and other groups of animals with well-defined patterns of geographic variation. Geographic speciation is so well established in these groups, and the sequence of steps through which isolated populations become differentiated is so abundantly documented, that no doubt could remain after 1942 that geographic speciation is an important and presumably the prevailing mode of speciation in animals.

Since the number of insuperable geographical barriers (mountains, water, and so forth) on continents is limited, some other kind of barriers must be responsible for the evidently active speciation on continents. Various authors (see Mayr, 1942) suggested that it might be vegetational barriers or other kinds of uninhabitable terrain. Keast (1961) showed the truth of this assumption very convincingly for Australian birds. Haffer (1974) found that the alternation of humid and dry Pleistocene periods in the Amazon basin had been responsible for the extremely active speciation of birds there, and Williams and Vanzolini showed the same for reptiles and Turner for butterflies. The effectiveness of any vegetational belt as a barrier depends on the dispersal facilities of a given species. For flightless grasshoppers or subterranean mammals, even a remarkably narrow zone of ecologically unsuitable terrain can be an effective dispersal barrier. A failure to recognize vegetational barriers has led some authors (White, 1978) to postulate nongeographical speciation mechanisms.

In textbook illustrations until rather recently, geographic speciation was usually shown in a diagram in which a widespread species was cut in half by a geographical barrier. The two halves, while isolated from each other, would in time become so different that they would react to each other as different species when contact was reestablished at a later period. More detailed studies of the distribution patterns of speciating groups, and particularly of species that seemed to have originated recently, suggest a different solution, however. When Mayr in the 1940s and 50s worked out the geographic variation of South Sea island birds, he was impressed how frequently the most peripheral population was the most divergent one, often reaching such a level of distinctness that it had been ranked as a distinct species or even genus. Mayr recorded in 1942 several cases of highly distinct allopatric "genera" which distributionally were nothing but far-distant subspecies. At that time his emphasis was on taxonomic questions (how to classify such populations), but he continued to speculate on the

601

causal background of this phenomenon. Being aware of the frequency of founder populations beyond the periphery of the solid species range, it finally occurred to him that such founder populations would be the ideal place for a drastic genetic reorganization of the gene pool in the absence of any noticeable gene flow and under conditions of a more or less strikingly different physical and biotic environment (Mayr, 1954).

There were two reasons why Mayr postulated the importance of founder populations. One was the observation that aberrant populations of species almost invariably are peripherally isolated and that, more often than not, the most aberrant population is the most distant one, like *D. tristrami* (San Cristobal) in the genus *Dicaeum*, *D. galeata* (Marquesas) in *Ducula*, and literally scores of similar examples listed by Mayr (1942; 1954). By contrast, the amount of geographic variation in contiguous species ranges is usually minor. The other reason, pointed out by Haldane (1937; 1957), is that large, widespread populations—in fact all more populous species—are evolutionarily inert, because new alleles, even favorable ones, require very long periods of time to spread through the entire species range. Genetic homeostasis (Lerner, 1954) strongly resists any changes in a large, undivided gene pool. The facts of geographic isolation do not seem to support Sewall Wright's model, according to which the most rapid evolution takes place in large species consisting of only partially isolated demes. In fact, populous, widespread species often continue virtually unchanged through the fossil record from the time of their first appearance until the time of their extinction. By contrast, the evolutionary fate of peripherally isolated populations is often very different. They are usually founded by a small number of individuals, indeed often by a single fertilized female, and contain only a small fraction of the total genetic variability of the parent species. This, Mayr postulated, would lead to greatly increased homozygosity and to a change in the fitness value of many genes in a drastically changed genetic milieu. Many epistatic interactions would be quite different from what they had been in the parental population. Mayr postulated therefore that such founder populations are particularly well situated to undergo a drastic genetic restructuring, sometimes amounting to a veritable "genetic revolution" (Mayr, 1954). That such founder populations might undergo drastic genetic changes can hardly be questioned. The brilliant researches of Hampton Carson (1975) on speciation in the genus *Drosophila* on the Hawaiian Islands have convincingly substantiated Mayr's theory. There is little doubt observationally that rapid speciation is most easily accomplished in very small populations.

That chromosomes may play an important role in speciation had been recognized since the first quarter of the century. Almost a third of the first edition of Dobzhansky's *Genetics and the Origin of Species* (1937) is devoted to a discussion of chromosomal phenomena, and chromosomes played an even more important role in the botanical literature. To be sure, de Vries's *Oenothera* "mutations" were eventually shown to be mostly chromosomal rearrangements and not a normal mechanism for speciation. However, soon afterwards polyploidy was discovered, a process in which, by a doubling of the chromosome set, new species can originate in a single step (Stebbins, 1950; Grant, 1971). The discussion of the role of chromosomes in speciation, however, has suffered from two misconceptions.

603

The first misconception is the assumption by some specialists of one group of organisms that their findings extend to all organisms; that this is not legitimate has been pointed out by a number of investigators. The claim, for instance, that *all* speciation is due to chromosomal reorganization was refuted by Carson (1975), who showed that active speciation in Hawaiian *Drosophila* can take place without any visible change in the chromosomes. Since these Hawaiian species can be analyzed in great detail with the help of their giant salivary chromosomes, whatever structural chromosomal changes may occur during speciation must be very minute. In other groups of organisms, closely related species often differ strikingly in their karyotype by inversions (paracentric or pericentric), translocations, Robertsonian fusions or fissions, or other changes of chromosome structure. Different groups of organisms specialize in different mechanisms of chromosomal change (Mayr, 1970: 310–319; White, 1974).

The second misconception was the assumption that chromosomal speciation is an alternative to geographic speciation. Actually, chromosomal and geographic speciation represent two entirely different dimensions. The kind of chromosomal differences one finds as differentiating closely related species (in contrast to those characterizing chromosomal polymorphism) almost invariably reduce the fitness of the heterozygotes, owing to all sorts of disturbances during meiosis. Such chromosomal rearrangements would have little chance of establishing themselves in a large population where they would have to pass through many generations of heterozygosity. It is only in a small founder population with a high level of inbreeding that they would have a chance to pass quickly through heterozygosity to the fitter homozygous condition of the new chromosomal type. What is true for chromosomal rearrangements is also true for new epistatic balances of genes, for the acquisition of new isolating mechanisms, and for new endeavors in niche utilization. They all are more easily acquired during the passing through the bottleneck of a founder population than by a slow process of selection in a large, populous species. There is no contradiction in saying that a certain new species originated by geographic *and* chromosomal speciation. The term "peripherally isolated" becomes somewhat ambiguous in species with low population density and very much reduced dispersal facility. In such cases a species may consist of numerous more or less isolated colonies and a new isolated colony may be founded in a previously vacant portion in the midst of the species range. Yet even such a founder population would go through the same steps of inbreeding and homozygosity as if it were isolated beyond the periphery of the species range.

Some evidence is now beginning to be found that ease of speciation is largely (and negatively) correlated with population size and that rapid speciation is not necessarily limited to founder populations. A drastic reduction of population size as has, for instance, occurred in many Pleistocene refuges also speeds up speciation, as was demonstrated by Haffer (1974) and others for the forest refuges in the Amazon basin. However, such species are apparently never even nearly so aberrant as are some of those that originated in peripheral isolates.

A second major controversy in the field of speciation concerns the old argument between Darwin and Wagner of the 1860s and 70s as to whether geographic isolation is necessary at all (Mayr, 1963; Sulloway, 1979). Again and again mechanisms were suggested that would permit the division of a single deme into two reproductively isolated ones (sympatric speciation) without any extrinsic barriers to gene flow. Three mechanisms of sympatric speciation have been suggested most frequently, (1) diversifying (disruptive) selection, which would pull apart a bimodal gene distribution, (2) allochronic speciation, owing to a drifting apart of the breeding seasons, and (3) colonization of a new host, in the

case of host-specific species. Sympatric speciation by host specialization is a concept that was highly popular from Darwin to the
evolutionary synthesis and again enjoys considerable popularity
(Bush, 1974). However, as I pointed out in 1942, even though
conditioning to new hosts is potentially an important method of
sympatric speciation in monophagous species, and in particular in
plant feeders, the occurrence of such speciation is subject to many
limitations, and the frequency of cases where it is more probable
than is geographic speciation is still an open question (White, 1978).
I rather suspect that here also the problem is distorted when a
strict separation is made between geographic speciation and speciation by colonization of a new host. It is evident that the shift to
a new host is much easier in a small founder population—in nature or in the laboratory—than within the continuous range of a
large populous species.[10]

605

The greatest unsolved problem in speciation research remains that of the genetic basis of speciation. To describe the process of speciation, one still relies in the main on inferences from
patterns of distribution. It will not be possible to resolve the controversies on the frequency and validity of the various possible
modes of speciation until we have acquired a better understanding of the underlying genetic processes. As recently as 1974 (p.
159) Lewontin said quite rightly, "We know virtually nothing about
the genetic changes that occur in species formation." This is, unfortunately, largely still true today. The older literature (Jameson,
1977) has been made virtually obsolete by the discovery of the
heterogeneity of the DNA.

At first, it was thought that a comparison of the frequency of
enzyme alleles in populations before and after speciation would
provide a decisive answer. Such investigations attempted, rather
in the spirit of bean-bag genetics, "to construct a quantitative theory of speciation in terms of genotypic frequencies" (Lewontin,
1974: 159). However, all the evidence gathered through this line
of research indicated that shifting frequencies of enzyme alleles
are not a primary agent in speciation. For instance, the degree of
difference in isozymes among closely related species varies greatly
in different genera. The passing of the species threshold does not
seem to coincide with a drastic shift in gene frequencies. This was
interpreted by some authors as constituting a falsification of Mayr's
theory of genetic revolutions in founder populations. This it would
be indeed if the enzyme genes were the primary genetic mecha-

nism for reproductive isolation.

It is becoming increasingly probable that there are special genetic mechanisms or regulatory systems that control the degree of reproductive isolation. Such mechanisms may be limited to a relatively small number of genes or to a restricted portion of the karyotype (Carson, 1976). They may be found in the various new kinds of DNA (such as middle repetitive DNA) that have been identified in recent years. The rapid and quite unexpected discoveries in molecular genetics make it seem probable that a major revision of our interpretation of the genetics of speciation may be in the offing.

If only a limited portion of the DNA controls the reproductive isolation between species, it would be possible that rather few mutational steps or limited karyotypic restructuring could initiate the speciation process. This would be far easier in a founder population consisting of a few individuals than in a widespread populous species. On the other hand, the multiplicity of isolating mechanisms separating most species indicates that full species rank is acquired in most cases only through a protracted process. Since speciation is gradual, that is, extending over a number of generations even in founder populations, one would not expect that it would happen through a single mutation. And all the indications are that it does not. But just exactly what happens during speciation is still a riddle. Carson (1976: 220) suggests that it is "a shifting internal balance of gene interactions in which a strong role is played by regulatory genes."

What Mayr's theory had left altogether unexplained was the irregularity of genetic revolutions. They occur in some but not in all peripherally isolated founder populations. Why? Much progress has been made in the understanding of the genome since 1954. It is now evident that some parts of the DNA, the isozyme genes, are less affected by the genetic revolution than others (presumably some of the regulatory systems). Templeton (1980) has speculated on some of these factors and, in particular, why genetic revolutions occur only under certain conditions.[11] It is still too early, considering our limited knowledge of the role of various classes of repetitive DNA and other recently discovered aspects of the genotype, to provide a definitive explanation. However, all the recent researches have provided further evidence in support of Mayr's theory that decisive evolutionary events occur most often, by way of genetic revolutions, in peripherally isolated founder populations.

MACROEVOLUTION

A third major area of activity after the evolutionary synthesis, in addition to natural selection and speciation, was macroevolution. Macroevolution has been defined in various ways: evolution above the species level, evolution of the higher taxa, or evolution as studied by paleontologists and comparative anatomists. By about 1910, paleontology, particularly invertebrate paleontology, owing to its success in determining stratigraphy, had become very much preoccupied with geological questions, a preoccupation which resulted in a loss of interest in evolutionary history. The study of macroevolution prior to the evolutionary synthesis was conducted by the paleontologists without any effective connection with genetics. Only very few paleontologists were strict Darwinians, accepting natural selection as the dominant agent in evolution. Most paleontologists believed either in saltationism or in some form of finalistic autogenesis. Macroevolutionary processes and causations were generally considered to be of a special kind, quite different from the populational phenomena studied by geneticists and students of speciation.

607

All this changed dramatically with the evolutionary synthesis. Its major effect was to discredit some of the beliefs most widely held previously among students of macroevolution. Important assumptions that were now rejected include the following:

(1) that major saltations are indispensable in explaining the origin of new species and higher taxa;

(2) that evolutionary trends and the continuous improvement of adaptations require the existence of autogenetic processes; and

(3) that inheritance is soft.

It was a major achievement of Rensch and Simpson to be able to show that an explanation of the phenomena of macroevolution does not require the acceptance of any of these three theories, and that in fact the phenomena of evolution above the species level are consistent with the new findings of genetics and microsystematics. Obviously, this conclusion had to be based on inference, consisting of morphological, taxonomic, and distributional evidence, since higher taxa were at that time—and, except for molecular evidence, are still today—inaccessible to genetic analysis.

In defense of paleontology it must be said that although saltationists and defenders of autogenetic processes were very much in the majority, there were also quite a few gradualists and some defenders of natural selection. As early as 1894, W. B. Scott had vigorously defended the gradualness of evolutionary change against Bateson. Even though in all species there is more or less pronounced variation around the "normal," says Scott, new departures in phylogeny do not come from the extreme variants but rather from a gradual shifting of the normal (p. 359). Osborn and other advocates of orthogenesis also supported gradual evolution against saltationism.

608

Natural selection likewise had its defenders. Even though most paleontologists agreed that natural selection was insufficient to explain the phenomena of macroevolution, there were some rather vigorous supporters of natural selection, such as Dollo, Kovalevsky, Abel, Goodrich, and Matthew. However, it is not clear from their writings whether they considered natural selection alone as sufficient to explain all evolutionary phenomena. The writings of these and other contemporary macroevolutionists have not yet been sufficiently analyzed to determine this.

Simpson in the introduction to *Tempo and Mode in Evolution* (1944) stated that his work was an attempt to achieve a synthesis between paleontology and genetics. Building a bridge between the two fields was made doubly difficult by the almost exclusive attention of geneticists to changes in gene frequencies, based on the assumption that nonadditive gene effects were of negligible importance. This restriction was adequate in the interpretation of only some macroevolutionary problems (such as evolutionary trends) but not of others (for example, the origin of diversity).

The synthesis between genetics and paleontology occurred in two steps, so to speak, represented by these questions: (1) Are there macroevolutionary phenomena that are clearly in conflict with a genetic interpretation of the Darwinian theory? (2) Can all the laws and principles of macroevolution be developed simply by studying gene frequencies in populations? Eventually it became evident that both questions had to be answered no.

The first task of the Darwinian macroevolutionists was to refute the claim of the anti-Darwinians that there are macroevolutionary phenomena which are in conflict with the formula "genetic variation and natural selection." This refutation was successfully carried out by Rensch and Simpson. Both of them, and also Julian Huxley, showed that there is no need to invoke a mysterious autogenetic factor to explain evolutionary trends but that increases in the size of the entire body, changes in the proportions of individual structures (such as teeth), the reduction of certain parts (for example, toes in horses, eyes in cave animals), and other long-continued evolutionary regularities can readily be explained by natural selection. It has since been pointed out that genetic as well as functional constraints reinforce the effectiveness of natural selection in controlling trends (Reif, 1975).

Various authors as far back as Geoffroy Saint-Hilaire have proposed evolutionary "laws." In every case it was shown that the law in question can be expressed in terms of natural selection. This includes, for instance, Dollo's so-called "law of irreversibility," which states that structures that had been lost in evolution can never be reacquired exactly in the same way. This finding is an obvious consequence of the fact that the genotype is constantly changing during evolution, and that if the need again arises for a structure that had been previously lost, the structure will be generated by a very different genotype than that which had produced the original organ, and thus the new structure will not be identical with the previously lost one (Gregory, 1936).

609

Most evolutionary phenomena relate to complex structures, organ systems, whole individuals, and populations. No approach was less able to lead to a full explanation than the reductionist one of expressing everything in terms of gene frequencies. Such reductionism, however, is not at all demanded by neo-Darwinism. Much of the objection of the anti-Darwinians became pointless as soon as the exclusive reliance on the reductionist approach was abandoned.

Simpson was particularly interested in rates of evolution. He showed that some evolutionary lines change rapidly, others exceedingly slowly, while the majority of lines have an intermediate rate. Furthermore, he showed that in the course of evolution a phyletic line may either accelerate or slow down its rate. The most rapid rate of evolutionary change was designated by Simpson as *quantum evolution,* which he defined as "the relatively rapid shift of a biotic population in disequilibrium to an equilibrium distinctly unlike an ancestral condition" (1944: 206). Simpson thought that this explained the well-known observation that "major transitions do take place at relatively great rates over short periods of time and in special circumstances" (p. 207). From the context of his discussion in 1944 and in his later writings (1949: 235; 1953: 350; 1964b: 211) it is evident that what Simpson had in mind primarily was a great acceleration of evolutionary change within a phyletic line. Simpson's thinking was clearly influenced by Sewall Wright's (1931) model of an inadaptive phase of genetic drift followed by natural selection. Extreme changes in evolutionary rates are, of course, well documented in the fossil history. Bats apparently originated from insectivores within a few million years but experienced no further major structural modification in the 50 million years since then. The shift from thecodont reptiles to *Archaeopteryx* likewise required relatively few million years, but the class of birds as a whole has not been materially modified since the appearance of the first modern birds more than 70 million years ago. Such drastic changes in rates of evolution do not in the least imply a conflict between the origin of the bat or bird morphotype and the Darwinian theory.

610

The problems connected with rates and trends of evolution could be interpreted in terms of the geneticists' formula that evolution is a change in gene frequency. However, this is a meaningless formulation as far as most other problems of macroevolution are concerned, and is one of the reasons why genetics made such a relatively small contribution to the solution of macroevolutionary problems. This inappropriate formulation is also responsible for the considerable time lag between the synthesis and an adequate treatment of some of these problems.

Evolutionary Novelties

One of the most frequently raised objections to Darwin's gradualism was that it was unable to explain the origin of "evolutionary novelties," that is, of entirely new organs, new structures, new physiological capacities, and new behavior patterns. For instance, how can a rudimentary wing be enlarged by natural selection before it enables its possessor to fly? it was asked. In fact, how can any incipient organ be favored by natural selection until it is fully functional? Darwin (1859; 1862) provided the answer to this question by pointing out that a change in function of a structure is the key element in the solution of this problem. His solution was rather generally ignored until it was further developed by Dohrn (1875), by Severtsov (1931), and by Mayr (1960).

During such a shift of function a structure always passes through a stage when it can simultaneously perform two functions, like the antenna of a *Daphnia* which is a sense organ as well as a swimming paddle. This duality of function is made possible because the genotype is a highly complex system which always produces certain aspects of the phenotype that had not been directly selected for but are simply "by-products" of the selected genotype. Such by-products are then available as the machinery for new functions. This is what permits an anterior extremity (with a patagium) of a tetrapod to function as a wing, or a lung in a fish as a swim-bladder. There are numerous "neutral aspects" in the phenotype of any organism that are "permitted" by natural selection (not selected against) but that had not been specifically selected for. Such components of the phenotype are available to take on new functions. Shifts in function are also known for macromolecules and for behavior patterns, such as when feather preening becomes a courtship display in certain ducks.

611

As Severtsov showed, an intensification of function is often all that is needed to permit a structure to adopt a seemingly new function. In this manner, for instance, the anterior extremity of a walking mammal is converted into the digging shovel of a mole, the wing of a bat, or the flippers of a whale. All that is needed as the starting point for the development of eyes is the existence of light-sensitive cells. Natural selection will then favor the acquisition of any needed auxiliary mechanism. This is why photo-receptors or eyes have evolved independently more than forty times in the animal kingdom (Plawen and Mayr, 1977). In most cases, no major mutation is necessary in order to initiate the acquisition of the new evolutionary novelty; sometimes, however, a phenotypically drastic mutation seems to be the first step, as in the case of mimetic polymorphisms, but once this step is made, minor modifying mutations will accomplish the finer calibration (Turner, 1977). The crucial factor, however, in the acquisition of most evolutionary novelties is a shift in behavior.

Behavior and Evolution

Behavior was for Lamarck an important evolutionary mechanism. The physiological processes initiated by behavioral activity ("use versus disuse"), combined with an inheritance of acquired characters, were for him the causes of evolutionary changes. After the invalidity of this proposed evolutionary mechanism had been demonstrated by genetics, the mutationists went to another extreme. According to them, major mutations generate new structures, and these "go in search of an appropriate function." The modern evolutionist rejects both interpretations. For him, changes in behavior are indeed considered important pacemakers in evolutionary change. However, the chain of causations is quite different from that envisioned by Lamarck or by the mutationists. The modern interpretation is that changes in behavior generate new selection forces which modify the structures involved.

Mayr (1974a) showed that different kinds of behavior play different roles in evolution. Behavior that serves as communication, for instance courtship behavior, must be stereotyped in order not to be misunderstood. The genetic program controlling such behavior must be "closed," that is, it must be reasonably resistant to any changes during the individual life cycle. Other behaviors, for instance those that control the choice of food or habitat, must have a certain amount of flexibility in order to permit the incorporation of new experiences; such behaviors must be controlled by an "open" program. New selection pressures, induced by changes in behavior, may lead to morphological changes facilitating the occupation of new ecological niches or adaptive zones. For instance, Bock (1959) showed that the primitive woodpeckers, which had switched to the behavior of climbing on tree trunks and branches, still had essentially the ancestral foot structure. However, the new habit created selection forces in several lines of woodpeckers which led to various highly efficient specializations of foot and tail structure adapted to more efficient climbing. Many if not most acquisitions of new structures in the course of evolution can be ascribed to selection forces exerted by newly acquired behaviors (Mayr, 1960). Behavior, thus, plays an important role as the pacemaker of evolutionary change. Most adaptive radiations were apparently caused by behavioral shifts.

Phylogenetic Research

Classical phylogenetic research was almost entirely looking toward the evolutionary past. It asked: What was the structure of the common ancestor, and how can he be reconstructed through a study of the homologous features of his descendants? To document the validity of Darwin's theory of common descent was the primary objective of this discipline. Its major interest was to determine for isolated types and phyletic lines where they should be placed on the phylogenetic tree. Common descent was the main emphasis of comparative anatomical research from T. H. Huxley and Gegenbaur to Remane and Romer.

Dissatisfied with the diminishing returns which this approach produced, a group of younger evolutionary morphologists began to ask why-questions. They developed a new methodology by turning the evolutionary tree upside-down, so to speak, that is, by making the common ancestor the starting point of their inquiry.

They asked: Why did the lines diverge that originated from a common ancestor? What factors permitted certain descendants to enter new niches and adaptive zones? Was a change in behavior a crucial component in the adaptive shift? The emphasis in this new approach was clearly on the nature of the selection forces. Severtsov, Böker, Dwight Davis, Bock, von Wahlert, and Gans were among the pioneers in this new evolutionary morphology. Their approach built a bridge between morphology and ecology, leading to the establishment of a new borderline field which is still in its youth and on the threshold of further interesting developments. *613*

A few of the more interesting results of these researches may be mentioned. One was the refutation of the concept of the "harmonious development of the type," a major dogma of idealistic morphology. When *Australopithecus* was discovered, for instance, the anatomist Weidenreich remarked to me that it could not be ancestral to man. It could not be a link between anthropoid and man, owing to its "disharmonious type" (advanced pelvis and extremities, primitive brain and face).

Actually, the concept of the harmonious development of the type had been refuted many times before. When studying the structure of *Archaeopteryx,* the link between reptiles and birds, de Beer (1954) called attention to the fact that this connecting link was already much like the later birds in certain features (such as feathers, and wings) while still a reptile in others (its teeth and tail). He designated this type of unequal evolutionary rates as *mosaic evolution.* Even then this was not a new discovery. The same principle had been discussed in considerable detail by Abel (1924: 21), who, in turn, had learned it from Dollo (1888), who was profoundly influenced by Lamarck (1809:58): "In fact, the organs that have little importance or are not essential to life are not always at the same stage of perfection or degradation; so that if we follow all the species of a class we shall see that some one organ of any species reaches its highest perfection, while some other organ, which in that same species is quite undeveloped or imperfect, reaches in some other species a high state of perfection." Our reasoning today is very different from that of Lamarck, but his observation of highly unequal rates of evolution of different structures and organ systems was entirely sound.

Key Character
What is most interesting about the unequal evolution of the type is that one particular feature, a *key character,* is so often involved in the new departure. In the case of the evolution of birds

from reptiles, it was the development of the feather, which almost certainly preceded flight. In the case of the evolution of land-living reptiles from aquatic amphibians, it was internal fertilization. The search for the key character is a major objective in the study of the evolution of higher taxa. In the evolution of man, for instance, a series of key characters were involved in the transition from the arboreal anthropoid stage to that of *Homo sapiens.* Upright posture, a manipulating hand, toolmaking, the hunting of big ungulates, and a language-based system of communication are suggested successive key features.

614

The anatomists of the school of idealistic morphology always stressed the conservative nature of the type. There is indeed something extraordinarily conservative about the ensemble of features which make up the vertebrate type or the mammalian type or the avian type. It is now evident that much evolution is virtually restricted to the key character and a few other characters correlated with it. A bat, in its entire structure, is still very much of an insectivore except for the flight adaptations (including those involving the sense organs). Even a whale is still very much of a mammal except for its adaptations for life in the seas. In turn there is hardly any mammalian character that cannot be directly followed back to the reptiles. The "unity of the type" clearly has a genetic basis with the interaction of genes and regulating genes serving as a conservative, if not almost inert, element.

Grades

One of the most characteristic features of macroevolution is the relative rapidity with which shifts into new adaptive zones occur, as that from insectivores to bats or from reptiles to birds. When a phyletic line enters a new adaptive zone, as when birds entered the zone of flight, it undergoes at first a very rapid morphological reorganization until it has reached a new level of adaptation. Once it has achieved this new *grade,* it can radiate into all sorts of minor niches without any major modifications of its basic structure. For instance, all birds are anatomically remarkably similar to each other, being merely variations on a theme. The importance of the phenomenon of grades has been known for a long time (see Bather, 1927) and was again stressed by Huxley (1958).

The clear recognition that there are highly unequal rates of evolution, as particularly emphasized by Simpson (1953), alternating with periods of remarkable stability as indicated by the term "grade," is important both for the theory of classification (see Chapter 5) and for an interpretation of the relations between evolution and ecology.

The evolutionary morphology of animals is still in the early stages of its development. Its greatest achievement is, perhaps, that of considerable conceptual clarification. This includes the clear distinction between the functioning of a structure and the biological role of a structure in relation to the organism's environment. The concept of preadaptation has been redefined to express the potential of a feature to adopt new functions and new biological roles. Bock (1959) developed the concept of multiple pathways and Mayr (1960) clarified the concept of multiple functions. The major emphasis of the new thinking is on the biological meaning of the structural, physiological, and behavioral features of organisms and on the pathways by which selection forces can gradually modify such features.[12] Darwin would have been most gratified over the final conclusion of all these researches: even the most drastic structural reconstructions proceed gradually, particularly when populations (including founder populations) enter new habitats and carve themselves new niches.

In spite of the most determined efforts of botanists, the reconstruction of the phylogeny of plants has lagged behind that of animals, primarily for two reasons. (1) The fossil record of most plant groups is infinitely poorer than that of animals, particularly since remnants of the diagnostically important reproductive system of plants are much scarcer than those of the vegetative system. (2) The differences in the internal anatomy (vascular structures) of the orders of angiosperms are far smaller than the differences in the internal anatomy of the 24 phyla of animals. However, the study of fossil pollen and of various chemical constituents and macromolecules of plants is beginning to open up entirely new dimensions of understanding. Owing to the difficulties encountered by the plant morphologists, it has been possible only within the last decade or two to undertake the kind of causal investigations of plants that have been conducted by the evolutionary animal morphologists. The pioneering work in this new causal morphology is Stebbins' (1974a) study of the evolution of flowering plants. He searches for the adaptive significance of every structure by asking, "What kinds of ecological conditions and environmental changes would have been most likely to have given rise to the observed morphological differences?" This stress on the adaptive significance of characters is radically different from

615

the approach of the traditional taxonomist, who was only interested in clues to common descent. The same adaptive feature can, of course, be acquired repeatedly in unrelated lines through convergence, a fact which is troublesome for the classifier but a valuable source of information for the student of evolutionary causes. Another landmark study in evolutionary plant morphology is the work of Carlquist on convergent adaptations (such as woodiness) of island plants (1965) and on ecological strategies of xylem evolution (1975).

616   An even more recent frontier is the study of the evolution of microorganisms. This is being advanced on two fronts. One is the study of microfossils, initiated by Barghoorn, Cloud, and Schopf, while the other is the comparative study of macromolecules and metabolic pathways of fungi, protists, and prokaryotes. Unfortunately, lack of space precludes even mentioning the host of exciting problems opened up by these researches.

The Origin of Macroevolutionary Diversity

There is one aspect of macroevolution that was neglected to an extraordinary degree in the hundred years after Darwin: the origin of higher taxa, or, to say it in different words, the origin of macroevolutionary diversity. Even during and after the synthesis this problem was neglected by the paleontologists, who might discuss adaptive radiation but did not at all come to grips with the problem of how the taxa originated that radiated into different niches and adaptive zones. This neglect had a number of reasons (not yet analyzed by anyone), two of which I would like to single out.

The first one, of course, was the essentialistic thinking universal among morphologists but most apparent in the schools of idealistic morphology. These anatomists were greatly impressed by the conservative nature of the ensemble of features that make up a morphological type or archetype, whether it is the mammalian, the vertebrate, or the arthropod type. Once such a type had evolved, as Schindewolf (1969) and other paleontologists were quite right to stress, it was virtually immune to major restructuring. Furthermore, intermediate stages between one type and another, either still living or in the fossil record, were very rare or absent. The gene-frequency approach of population genetics was quite unable to supply any solution to this problem of origination.

The second reason for the stalemate in the study of the origin of new types was the concentration of paleontologists on straight-line phyletic evolution, that is, on the "vertical" component of evolution. All the great pre-synthesis leaders of paleontology—Cope, Marsh, Dollo, Abel, Osborn and Matthew—concerned themselves primarily with evolutionary laws, evolutionary trends, and the evolution of adaptation. All this would lead to better adaptation but not to greater diversity. How new diversity originated was either explained in terms of essentialistic saltations or it was not mentioned at all. The latter was true even for Simpson (1944; 1953), whose evolutionary (that is, vertical) species definition made it difficult for him to analyze the problem of the branching of phyletic lines.

617

Curiously, the answer had been available since the synthesis (Mayr, 1942; 1954) but was ignored by the paleontologists until used by Eldredge and Gould (1972) in their model of so-called *punctuated equilibria.* They pointed out that when one looks at the geological record, one finds that most fossils belong to wide-spread, populous species that show little change in the time dimension until they become extinct. A certain proportion of lineages undergoes a process of vertical phyletic evolution (Gingerich, 1976) in which the species of one time level evolve into descendant subspecies or species at the next time level. Far more frequently the extant species are supplemented by—or the extinct species are replaced by—new species that suddenly turn up in the fossil record. In the classical literature this sudden introduction of new species was usually ascribed to instantaneous saltations. Eldredge and Gould, however, accepted Mayr's interpretation that such new species had originated somewhere in an isolate (peripheral or not) and were able to spread far and wide if they were successful. This interpretation of the "introduction of new species" (as Lyell had called it 150 years earlier) agrees well with the fossil record (Boucot, 1978; Stanley, 1979). That such an origin of new types is not pure speculation is documented by the origin of new minor types in peripheral isolates in the living fauna.

In one respect Gould and Eldredge differ fundamentally from Mayr. They maintain that punctuated equilibria are produced by discontinuities of such size that they correspond to Goldschmidt's hopeful monsters: "Macroevolution proceeds by the rare success of these hopeful monsters, not by continuous small changes within populations" (Gould, 1977: 30). What Goldschmidt had postulated, and this seems to be endorsed by Gould, is the production of new species or higher taxa by a single step through a single individual. Mayr, by contrast, considers evolution in founder populations a populational process, which is gradual evolution on the human time scale (Bock, 1979). It appears to be saltational only when measured on the geological time scale. Undoubtedly regulatory genes are participating in these changes or are largely responsible for them, but this does not require saltations.

618

What is crucial is the fact that prior epistatic and regulatory systems are broken up during a genetic revolution in the founder population, making room for new ones. This greatly facilitates and speeds up the acquisition of new adaptations. These are, of course, not acquired by single steps, and selection for their improvement continues. It may even be accelerated by the establishment of descendant founder populations. It is unknown and presumably variable whether such an evolutionary shift requires a few, scores, hundreds, or thousands of generations, but it is certainly by several orders of magnitude faster than the traditional phyletic evolution described in the paleontological literature as requiring millions of years. Even so, evolution through changes in founder populations is not a process of saltation but one of gradual evolution. The most important departure in the new way of thinking is to treat it as a populational phenomenon.

In a few contemporary situations there are fortunate constellations of geography and ecological opportunities that permit us to demonstrate the gradual, step-by-step origin of such macroevolutionary origins. The Hawaiian archipelago, in which the various islands of the chain were colonized from the west (Kauai) to the east (Hawaii), provides a graphic illustration of such almost gradual evolutionary steps. This has been demonstrated by Bock (1970) for species and genera of honeycreepers (Drepanididae) and by Carson and Kaneshiro (1976) for *Drosophila*.

The continuing introduction of new species by the process of geographic speciation (Stanley, 1979) is made possible because simultaneously there is a steady loss of species by extinction. Extinction, thus, is the counterpart to speciation, as was realized already by Lyell, and a problem of equal importance, particularly for the ecologist.

Extinction

When one observes with what extraordinary faithfulness a mimicking species may copy even rather incidental characteristics of its model, one gains the conviction that nothing is impossible to selection. This, however, is contradicted by the frequency of extinction in nature. When such highly successful orders and phyla of animals as the trilobites, the ammonites, or the dinosaurs became extinct, why was natural selection unable to reconstruct even a single species in these large taxa in such a way as to permit its survival? As a matter of fact, the ammonites *had* gone through at least four previous periods of mass extinction, during which a single lineage survived in each case and gave rise to a new adaptive radiation. However, at the last of these "crashes" not a single species had the proper constellation of genes to be able to cope successfully with the environmental challenge it encountered, whatever it was.

619

Extinction, as is becoming clearer all the time, is a highly complex problem. The dinosaurs became extinct only when the last of scores or hundreds of species had become extinct. The question thus is, Why did this whole higher taxon become obsolete? A look at the history of the phyla and orders of plants and animals shows that they differ greatly in their proneness for extinction. Indeed, one can establish definite regularities in the pattern of extinction, as shown by Van Valen (1973). It is my own conviction that extinction is somehow correlated with the cohesion of the genotype. Surely, the rate of mutation ought to be approximately the same in different species of organisms. However, some of them have a genotype that is so well integrated, and thus has become so inflexible, that it can no longer produce the departures from the traditional norm that might permit a major switch in resource utilization or an answer to the challenge of a competitor or pathogen. These, of course, are only words until we have learned more about the structure of the eukaryote genotype and its regulatory system.

The diversity of a fauna or flora depends on the equilibrium between speciation and extinction events. Our vastly increased knowledge of fossil biota has made it possible in recent years to trace species diversity through geological time. The analyses show that there are periods of an exponential increase of diversity, such as in the earliest Cambrian and in the Ordovician; periods of steady state when the diversity remains about the same for millions, if not hundreds of millions, of years; and periods of massive extinction (Sepkoski, 1979). What is perhaps most interesting is the extraordinary stability of certain ecological associations. Instead of a gradual enrichment of such faunas, the species diversity remained the same for entire geological periods and turnover was largely due to a 1:1 replacement of extinct by newly colonizing species.

The Ordovician "species explosions" may have been due to a replacement of generalists by specialists; more recent changes, particularly in the oceans, may have been due to plate movements, the extent of shallow-shelf seas, and climatic events (including ice ages). The pioneering recent researches are clearly only a beginning.

620

There have been a number of periods of mass extinction, as at the end of the Permian and again at the end of the Cretaceous. Indeed, the end of the Paleozoic and of the Mesozoic are defined by these mass extinctions. There have been numerous suggestions concerning an extraterrestrial causation of the extinction, such as the passing of the earth through a cloud of cosmic dust. Others explained it by drastic changes of climate, in turn caused by plate tectonics. The discovery that at the border between Cretaceous and Tertiary there is a greatly enriched deposit of iridium has induced Alvarez and colleagues (1980) to hypothesize that the earth was struck by an asteroid, with the dust cloud blocking out sunlight for several years. As appealing as this theory is on first sight, it raises numerous unanswered questions, such as how does one account for the survival of mammals, birds, angiosperms, nondinosaurian reptiles, and so forth? Clearly the study of extinction remains a wide open frontier.

THE EVOLUTION OF MAN

No other thought was as distasteful to the Victorian imagination than that man could have descended from the apes. Even if evolution could be demonstrated for all other organisms, surely man with all of his unique human characteristics must have been specially created. Even A. R. Wallace refused to credit natural selection with the evolution of man, much to Darwin's dismay. Actually, as the anatomists knew very well, man is remarkably similar to the anthropoid apes in his morphology. This is why Linnaeus had unhesitatingly included him among the Primates. Within a few years after the publication of the *Origin,* Haeckel in Germany (1866; 1868) and T. H. Huxley in England (1863) published volumes in which man was postulated to have descended from the apes. Even Lyell (1863) eventually admitted at least the antiquity of man, and Darwin in 1871 published a major work, *The Descent of Man,* in which the problems of human evolution were discussed in considerable detail.

In the meantime (actually already before the publication of the *Origin*) the first fossil hominids were found, in particular Neanderthal man (1856). Haeckel, with his usual romantic imagination, went even so far as to reconstruct the "missing link" between man and apes, naming him *Pithecanthropus*. The search for this missing link was unexpectedly soon crowned with success, when a Dutch Army doctor and amateur anthropologist, E. Dubois, found the skull of *Pithecanthropus* (now included in *Homo*) *erectus* in Java in 1891. The number of new finds of fossil man has increased steadily since that time, none of them more important than the Taung child (*Australopithecus africanus*) described by Dart from South Africa in 1924. Numerous subsequent finds of australopithecines by Broom, the Leakeys, and others have permitted a reconstruction of this remarkable creature. In its pelvis and posterior extremity it hardly differs from modern man; in its dentition and face it is somewhat intermediate between apes and man; and in its brain (about 450 cc as compared to 1500 cc in modern man), it is still essentially on the ape level.

621

Additional finds made in southeast Asia, Ethiopia, Kenya, and Tanzania now permit reconstructing an almost unbroken chain from the oldest *Australopithecus* (*afarensis*) through *A. africanus*, *Homo habilis*, *H. erectus*, to *Homo sapiens*. Chronological as well as morphological considerations suggest that *A. africanus* was a polytypic species, isolated populations of which gave rise both to the robust *Australopithecus robustus* (a side line) and to *Homo habilis*. It is most unlikely that we will ever recover enough fossils to determine where the isolates were located in which these species evolved nor what caused their divergence from *A. africanus*. *Australopithecus robustus*, which coexisted with *Homo habilis*, became extinct more than one million years ago. Although *Australopithecus* can now be traced back to about four million years ago, it is still controversial how many million years earlier this hominid line had branched off the line that leads to the African apes, the chimpanzees and gorillas. The ultimate decision depends a good deal on where one places the fossil *Ramapithecus* and whether one considers it to be ancestral only to the hominids or also to the African apes, or a side branch. It seems increasingly probable that a shift from an ape-like ancestor (*Ramapithecus?*) to the hominid condition occurred very rapidly and perhaps as recently as only 5 to 7 million years ago. Only further fossil discoveries can give us certainty.

What is amazing is the extraordinary similarity between man and the great African apes in molecular characteristics and chromosome structure. Here is an evident case of mosaic evolution, where some segments of the genotype (the basic macromolecules) have remained conservative while other segments, those controlling general anatomy and in particular the central nervous system, have evolved at an exceedingly rapid rate. However, the crucial fact that the hominid line branched off from the line leading to the African apes is now no longer in doubt.

622

What is far more important than the uncertainties of chronology is our growing understanding of the steps that led from the anthropoid to the human condition. The assumption of upright posture when our ancestors descended from the trees was apparently the first and perhaps the most decisive step. It freed the anterior extremity for the function of manipulation, which permitted the carrying of objects and far more extensive tool use and eventually tool manufacturing than found in any ape. The hunting of big game and the development of a true language were apparently other major steps in the evolution of man. To characterize man by such criteria as consciousness, or by the possession of mind and of intelligence, is not very helpful, because there is good evidence that man differs from the apes and many other animals (even the dog!) in these characteristics only quantitatively. It is language more than anything else that permits the transmission of information from generation to generation and thus the development of nonmaterial culture. Speech, thus, is the most characteristic human feature. It is often said that culture is man's most unique characteristic. Actually, this is very much a matter of definition. If one defines culture as that which is transmitted (by example and learning) from older to younger individuals, then culture is very widespread among animals (Bonner, 1980). Thus, even in the evolution of culture there is not a sharp break between animal and man. Though culture is more important in man, perhaps by several orders of magnitude, the capacity for culture is not unique with him but a product of gradual evolution.

One of the most surprising discoveries of anthropological research has been the rapidity with which *Homo* evolved. Even allowing for the concomitant increase in body size, the growth of the hominid brain from 450 to 1600 cc was remarkably fast. Perhaps equally remarkable is that once the *Homo sapiens* stage had been reached (more than 100,000 years ago), no further noticeable increase in brain size occurred. Why primitive man should have been selected for a brain of such perfection that 100,000 years later it permitted the achievements of a Descartes, Darwin, or Kant, or the invention of the computer and the visits to the moon, or the literary accomplishments of a Shakespeare or Goethe, is hard to understand. But then, of course, man will always be a puzzle to man.

Eugenics

The recognition that natural selection, and natural selection alone, had raised man from the level of an ape to that of a human being suggested to Galton soon after Darwin's death that one might apply this principle of selection in order to achieve a biological improvement of man. This utopian scheme, to which he gave the name *eugenics,* found at first many adherents. In fact, a large number of geneticists and other biologists agreed in their writings that it was a noble idea to improve mankind by facilitating the reproduction of the "best" members of the species and by preventing the reproduction of individuals who had genetic diseases or were otherwise inferior. Actually, two kinds of eugenics must be distinguished. Negative eugenics endeavors to reduce the number of deleterious genes in a population by preventing the reproduction of carriers of dominant genes and by reducing the reproductive rate of heterozygous carriers of recessives (where such heterozygotes can be diagnosed). Positive eugenics strives to enhance the reproductive capacity of "superior" individuals (Haller, 1963; Osborn, 1968). When one reads the writings of these early believers in eugenics, one is impressed by their idealism and humanity. They saw in eugenics a means to go beyond the improvements made possible by education and a rise in the standard of living. No political bias was at first attached to eugenics, and it was supported by the entire range of opinion from the far left to the far right. But this did not last long. Eugenics soon became a tool of racists and of reactionaries. Instead of being applied strictly to population thinking, it was interpreted typologically; soon, without the show of any evidence, whole races of mankind were designated as superior or inferior. In the long run it led to the horrors of Hitler's holocaust.

As a consequence it has become almost impossible, since 1933, to discuss eugenics objectively. This does not invalidate the fact, however, that it was through natural selection that man reached humanity, and it is equally true that we know of no method other than selection to improve the human genotype. Nevertheless, to apply artificial selection to man is impossible, at least for the time

623

being, for a number of reasons. The first one is that it is quite unknown to what extent nonphysical human characteristics have a genetic basis. Second, human society thrives on the diversity of talents and capabilities of its members; even if we had the ability to manage the selection, we would not have any idea for what particular mixture of talents we should strive. Finally, the concept that people are genetically different, even were it scientifically even better established than it is today, is not acceptable to the majority of western people. There is a complete ideological clash between the concepts of egalitarianism and eugenics. We must remember that the principles of the United States Constitution are based on the writings of the leaders of the Enlightenment, whose ideals were magnificent but whose knowledge of biology was deficient, to put it mildly. As Bateson said many years ago, "Not even the patristic writings contain fantasies much further from physiological truth than those which the rationalists of the 'encyclopedia' adopted as the basis of their social schemes" (1914: 7). At the present time, eugenics is a dead issue and will remain so until populational thinking is more widely adopted and until we know far more about the genetic component in human characteristics.[13]

624

If we would ask what the most characteristic aspect of current evolutionary research is, we would have to use the term *interaction*. In the reductionist phase the attention was on the action and the fitness of single genes; now more and more attention is paid to the interaction of genes, to regulatory mechanisms, and to the genotype as an active system. Studies of the fitness of the isolated individual are supplemented by studies of kinship selection, inclusive fitness, reciprocal altruism, parent-offspring relations, and so forth. The study of the evolution of plants and animals is enriched by the study of their coevolution (Ehrlich and Raven, 1965). The evolution of herbivores cannot be understood except as a response to the evolution of their food plants. This was long understood, as shown by the frequent references to contributions made by the shift from browsing to grazing in the evolution of horses and other temperate-zone mammals during the Tertiary. Most insect evolution from the Cretaceous on is intimately connected with the evolution of the angiosperms. Studies of the evolution of social systems and of ecosystems focus strongly on the effects of interactions. All this is, of course, an obvious consequence of natural selection. Natural selection is exerted by the environment, and the environment of an individual consists not only of inanimate nature but also of other individuals of the same species and of individuals of other species (both plants and animals). In the last analysis, thus, most studies of interactions during evolution are nothing but an expansion in the application of natural-selection research. This is splendidly demonstrated in modern textbooks of evolutionary biology (Futuyma, 1979), behavior (Alcock, 1980), and ecology (Rickleffs, 1978).

UNSOLVED PROBLEMS IN EVOLUTIONARY BIOLOGY

The evolutionary biologist is often asked what the unsolved problems of his field are. It turns out that few of them deal with basic principles, since an alternative to Darwinism has become ever more improbable the more we have learned about life. As for problems, one could perhaps mention the question, What proportion of the observed variability of life is the product of selection and what other part is due to stochastic processes? More specific problems are the origin of life (how nucleic acids and polypeptides became associated), the origin of viruses, the details of the conversion of prokaryotes into eukaryotes, the functioning of the eukaryote chromosome, the classification of the various kinds of DNA (structural, regulatory, repetitive, and so on) and their respective roles in evolution and speciation, the relationship and phylogeny of the major types of plants and invertebrates, the respective roles of intra- and interspecific competition in evolution, the evolution of different kinds of behavior and their role as pacemakers in evolution, and the reasons for the extraordinarily high frequency of extinction (why is natural selection so helpless to prevent it?). Any specialist can add to this list. A particularly rich field for investigation is the pluralism (multiple pathways) found in evolution. For almost any challenge posed by the environment, different evolutionary lines have found different answers. What constraints do the different answers (such as the external skeleton of arthropods vs. the internal skeleton of vertebrates) impose on the future evolution of these lineages? The whole field of evolutionary constraints is still virtually untouched. The fusion of evolutionary biology with ecology, behavioral biology, and molecular biology has raised an almost endless number of new questions. Yet, to repeat, there is little likelihood that any new discoveries will force a major modification of the basic theoretical framework that was arrived at during the evolutionary synthesis.

EVOLUTION IN MODERN THOUGHT

The frequency and often violence of controversies in the camp of the evolutionists have confused some nonbiologists. As a result they have become skeptical of the whole concept of evolution or at least of the Darwinian principle of natural selection. The question is therefore legitimate as to what role evolution and Darwinism play in modern thought. It is perhaps fair to state at the outset that no well-informed biologist doubts evolution any longer. In fact, many biologists consider evolution not a theory but a simple fact documented by the change of gene pools from generation to generation and by the changes in the sequence of fossils in successive accurately dated geological strata. It is probably equally fair to say that the vast majority of well-informed lay people accept evolution as readily as the fact that the earth circles the sun and not the reverse. Whatever opposition to evolution survives today is restricted to persons with religious commitments. Certain fundamentalist sects still insist on the unquestioned acceptance of the literal story of Genesis in spite of the falsification of this story by the overwhelming evidence of science. A rational debate between scientists and fundamentalists is impossible because one camp rejects supernatural revelation, the other camp scientific fact.

More interesting is the occasional resurgence of antiselectionism. The most prominent authors of the antiselectionist literature are usually journalists, jurists, writers, and philosophers, and their arguments are based on such an ignorance of the facts of genetics, systematics, biogeography, ecology, and other branches of biology that a rational debate is impossible. What is disturbing, however, is that a few serious and well-qualified scientists have accepted the arguments of the lay antiselectionists and have also proclaimed that the formula "variation and selection" cannot fully explain evolution. Such scientifically qualified antiselectionists are a very small minority. Their arguments are usually based on a failure to recognize the probabilistic nature of selection, on a failure to realize that the individual as a whole is the target of selection, and on the failure to appreciate the numerous constraints encountered by selection. No special effort is made by evolutionists to refute these authors because the counterarguments have all been stated in the literature many times and in considerable detail.

These minor controversies have been unable to delay, much less stop altogether, the impact of evolutionary thinking on all

spheres of human thought. Evolutionary thinking is no longer restricted to biology, and there is no field of human endeavor with
a historical component that has not adopted evolutionary thinking
and evolutionary methodology. We now use the word "evolution"
very freely, beginning with the evolution of the universe up to the
evolution of human society, the evolution of languages, the evolution of art forms, and the evolution of ethical principles.

The indiscriminate application of the term "evolution," however, has led to some unfortunate formulations, if not absurdities.
Nonbiologists who favor the evolutionary conceptualization are
often unaware of the Darwinian or neo-Darwinian theory and may,
for instance, promote orthogenetic schemes, such as the theory
that human culture automatically passes through a series of stages
from that of the hunter-gatherer to that of the urban megalopolis.
Teleological principles have been very popular among those who
have used evolutionary language outside of biology, but when these
teleological schemes were refuted, it was thought that this refuted
the whole concept of evolution. A study of such literature demonstrates rather painfully that no one should make sweeping claims
concerning evolution in fields outside the biological world without
first becoming acquainted with the well-seasoned concepts of organic evolution and, furthermore, without a most rigorous analysis of the concepts he plans to apply. Evolutionary thinking is indispensable in any subject in which a change in the time dimension
occurs. However, there are many "kinds" of evolution, depending
on the nature of the causes that are responsible for the change,
on the nature of the constraints, and on the nature of the success
of the changes. The appropriate analysis of the different kinds of
so-called evolution in different areas has not yet been undertaken.
Nevertheless, there is no doubt that applying evolutionary principles has greatly enriched many areas of human thought.

627

III

Variation and
Its Inheritance

E VEN PRIMITIVE people are
well aware of two aspects of living nature: an immense variability
within each species and a tendency for characteristics of parents
to be transmitted to their offspring. Philosophers and scientists have
attempted explanations for inheritance from the days of the pre-
Socratics to the end of the nineteenth century, but it was only in
the year 1900, when Mendel's work was rediscovered, that a maturation of concepts permitted the establishment of genetics as an
autonomous science of inheritance. Another fifty years passed by,
however, before biologists fully understood what aspect of inheritance is most significant, that is, the existence of a genetic program. It constitutes the most fundamental difference between living organisms and the world of inanimate objects, and there is no
biological phenomenon in which the genetic program is not involved. Geneticists, not without justification, have therefore claimed
that genetics is the most basic of all biological disciplines.[1]

The particular importance of genetics is that it deals with a
level in the hierarchy of biological phenomena that bridges the
gap between those parts of biology that deal with whole organisms, as do systems and most of evolutionary biology, and those
that deal with purely molecular phenomena. It therefore has contributed to the unification of biology by showing that the genetic
processes in animals and higher plants are exactly the same. More
importantly, genetics helped to solve the problems of the mechanisms of evolution and development. A comprehension of the
fundamental principles of inheritance is a prerequisite for a full

understanding of virtually all phenomena in all other branches of biology, whether physiological, developmental, or evolutionary biology. Much of the acceleration in the progress of biology in the twentieth century is due to a better understanding of the mechanisms of inheritance. Likewise, many of the controversies in biology during the first half of the twentieth century were due to the difficulty of integrating the findings and concepts of genetics into the older, previously established branches of biology. At the same time, an important contribution to the maturation of genetics was made by the introduction of concepts from neighboring fields, concepts that had been previously absent from genetics. Such concepts came from systematics (population thinking), information theory (program), and biochemistry.

One might even raise the question whether the intensive interaction between genetics and the other branches of biology has not led to the disappearance of genetics as a separate science. Population genetics has become a branch of evolutionary biology, the study of gene action has become part of molecular biology, and the developmental aspects of genetics have become the domain of developmental biology. Some particularly partisan champions of genetics have gone to the other extreme and said that fundamentally all parts of biology are branches of genetics. Considering the fact that the genetic program in some way or another is involved in all biological activities (even where open programs control a certain action), this claim is not altogether as absurd as it may seem. These conflicting points of view serve to emphasize the central and integrating role of genetics in biological thought.

The universality of genetic phenomena is, by necessity, the cause of a considerable heterogeneity in the science of genetics. The study of the origin of new genetic programs (mutation, recombination), of their transfer to the next generation (transmission genetics), of the behavior of genetic factors in gene pools (population genetics), and of the translation of genetic programs into phenotypes (physiological or developmental genetics) requires separate disciplines, some dealing with proximate, others with ultimate, causations. The most frequent interactions of any of these disciplines are often not with each other but with branches of biology outside of genetics, such as systematics, embryology, physiology, or biochemistry.

My treatment here is focused on transmission genetics and its components, that is, on the units of inheritance (genes), their changes (mutations), their arrangement (in chromosomes), their

reassortment (recombination, errors of sampling), and their transfer to the next generation. The history of those aspects of genetics
that are most important for the causal explanation of evolution,
such as population genetics, have already been treated in Chapters 11 and 12. Finally, physiological genetics is inseparable from
developmental biology and will be treated in a separate volume.
Each of these branches of genetics has its own conceptual framework and its own history.

Although some historians, such as Barthelmess (1952) and
Stubbe (1965), unhesitatingly begin their history of genetics with
the ideas on inheritance held by the ancients, other historians feel
that "to trace the origins of genetics to a time before the beginnings of modern science seems . . . to lose sight of its essence"
(Dunn, 1965: xiv). In this argument I side with Barthelmess and
Stubbe. From primitive man on, people had ideas on inheritance,
on the causes of resemblances, and on the origins of new kinds of
organisms and new characteristics. Most of these ideas were erroneous, but even if one were to assert that not a single one of
the concepts and beliefs of Hippocrates or Aristotle had survived
to modern times, the historian of ideas nevertheless must study
them carefully. Before one can understand the ground on which
new ideas grow, one must understand how the old ones were
modified or why they were eliminated. Furthermore, ideas on inheritance very often were part of more universal ideologies, like
animism, atomism, essentialism, creationism, physiological mechanism, or holism, and certain genetic theories cannot be understood if one is not aware of this philosophical background. Mendelian genetics did not confront a vacuum but rather already
available theories of inheritance, such as pangenesis, blending inheritance, or the various theories of multiple determinants. One
will never fully appreciate the intellectual force of the Mendelian
revolution unless one knows what the existing theories were which
it displaced.

14 ⌀ Early theories and breeding experiments

EVERY INDIVIDUAL in a sexually reproducing species (except identical twins) is unique. This amount of uniqueness is far greater than that found in the world of inanimate objects. Although at the macro-level one finds also unique "individuals" (such as planets or·volcanoes) and unique systems (galaxies and weather systems), the most abundant of all individuals, the components of matter (molecules, atoms, elementary particles), are never unique. Most of the laws of the physical sciences are based on this lack of uniqueness (see Chapter 2).

A correlate of individuality in organisms is variation; any living group consisting of unique individuals by necessity displays variation. The origin and nature of variation in living organisms were not understood until the twentieth century, and the lack of an established theory of variability was a great impediment for nineteenth-century biology. It was the weakest link in the chain of argument in Darwin's theory of natural selection, a fact of which Darwin himself was keenly aware. Indeed, it worried him all his life.

That variation, or parts of it, is somehow connected with inheritance must have been dimly appreciated even by primitive man. That an offspring may resemble its parents or grandparents in certain traits was, of course, ancient knowledge. All animal and plant breeding is based on an awareness that some qualities are inherited. Any endeavor toward improvement of a breed, whether by selection or by cross-breeding, was implicitly based on the postulate of inheritance. Even the role of the sexes in fertilization was well understood in certain cultures. The Assyrians, at least as far back as 2000 B.C., fertilized the flowers of female date palm trees with pollen shed by the flowers of male trees.

Nevertheless, the nature of inheritance and its mechanism remained rather a mystery. The early observations of the primitive naturalists and agriculturalists, as well as the speculations of physicians and philosophers, raised numerous questions, most of which were debated until the beginning of the twentieth century. Perhaps there is no other area of biology in which the refutation of erroneous ideas and dogmas was more important for an advance of understanding than in genetics. Some of the erroneous ideas or dogmas were:

(1) that a parental pneuma, rather than a transfer of gross physical substance, is the agent of inheritance;

(2) that only one of the parents transmits the genetic elements (opposed by Buffon and Kölreuter);

634

(3) that the contribution made by the father is quantitatively and qualitatively different from that of the mother (from Aristotle to Linnaeus);

(4) that the environment and the activities of the body (use and disuse) have a strong determining influence on the genetic material (inheritance of acquired characters);

(5) that there are two sharply distinguished kinds of inheritable properties, those that vary discontinuously (by saltation) and those that vary continuously by infinitesimally small gradations;

(6) that the characters (properties) themselves are directly inherited, rather than the potential for their formation (in the form of the genetic program);

(7) that the genetic contributions of both parents fuse in the offspring (blending inheritance).

This is only a small sample of widely held misconceptions concerning inheritance. Inheritance, being such a conspicuous phenomenon, became the subject of a diversified folklore-"science," the remnants of which can be found among laypersons even today. Animal breeders, for instance, insisted that if a female of a pure race had once been inseminated by a male of a different race or by a mongrel, her "blood" would be permanently impure so that she could no longer be used for breeding purposes. This belief often was also applied to mankind, particularly in the racist literature. It was also widely believed that a single offspring could be fathered by several males so that a child of a female who had received several males in the period during which she had conceived would combine the characteristics of these several fathers. There was also the belief in a great plasticity of the genetic material; for example, it was believed that any accidents of the mother, like being frightened by a snake, might affect the fetus.

One of the most characteristic aspects of traditional ideas on inheritance, when seen in retrospect, was the frequent incompatibility of simultaneously accepted views. A belief in an invisible constant essence was combined with a belief in strong environmental influences of all sorts or with a recognition of a differential contribution of the two parents. Strictly quantitative concepts (such as "strength of the father's influence") were held alongside purely qualitative ones (the inheritance of individual features, as in Plato's eugenics). The inheritance of somatic damages (mutilations) was almost universally believed in, even though one could readily see that a warrior who had lost an arm did not produce armless children, not to mention the genetic ineffectiveness of thousands of years of circumcision among the Jews.

635

THEORIES OF INHERITANCE AMONG THE ANCIENTS

Despite thoughtful and critical analysis by a number of Greek philosophers, still no unified theory of either variation or inheritance was developed by the ancients, and the ideas of these philosophers differed widely from one another. However, a principle of inheritance was quite generally accepted, continuing the tradition of the *Iliad* and other epics, where an inheritance of the heroic qualities of the father by the son was taken for granted. Yet, the Greek philosophers had only the vaguest ideas on how the characteristics of the parents were transmitted to their offspring. Two authors who had the greatest influence in subsequent centuries on the thinking concerning generation and inheritance were Hippocrates and Aristotle.[1]

The famous physician Hippocrates (ca. 460–377 B.C.) taught that "seed material" came from all parts of the body to be carried by the humors to the reproductive organs (*De generatione*, sections 1 and 3). Fertilization consists in the mixing of the seed material of father and mother. That all parts of the body participate in the production of seed material is documented by the fact that blue-eyed individuals have blue-eyed children and baldheaded men have children that become baldheaded. If parts of the body are unhealthy, the corresponding part in the offspring may also be unhealthy.

The idea of such a panspermy or pangenesis had apparently been first expressed by Anaxagoras (ca. 500–428 B.C.) and had representatives at least to the end of the nineteenth century, including Charles Darwin (see Chapter 16). If one believes in the effect of use and disuse or any other form of an inheritance of acquired characters, as did just about everybody from Hippocrates to the nineteenth century, one is virtually forced to accept such a theory. Characteristic for the theory of pangenesis is also the alternation between the formation of the body (phenotype, soma) and through it the formation of seed stuff (sperm, genotype) which then directly through growth is converted again into the body of the next generation, a concept essentially maintained *636*   until first challenged in the 1870s and 1880s (Galton, Weismann).

Aristotle

None of the ancients had a deeper interest in questions of generation than Aristotle, who devoted one of his major works (*De generatione*) to this problem. He discussed variation and inheritance also in other writings, as in *De partibus*. Aristotle was altogether opposed to the atomistic interpretation of inheritance by Hippocrates and his forerunners. How could it explain the inheritance of characteristics that could not produce seeds, such as dead tissues, like nails or hair, or behavioral characteristics, such as voice or locomotion? Also characters may be transmitted by the father at an age when the character is not yet shown, such as baldness or premature graying of the hair. Aristotle likewise rejects the possibility that the sexual product of the male is a tiny preformed animal, as was later believed by some seventeenth- and eighteenth-century authors.

Aristotle's theory of inheritance was a rather holistic one. He held, as had some of his forerunners, that the contribution of male and female were somewhat different. The semen of the male contributes the form-giving principle (*eidos*), while the menstrual blood (*catamenia*) of the female is the unformed substance that is shaped by the *eidos* of the semen. He compares the effect of the semen to that of the carpenter's tools on wood. The "female always provides the material, the male provides that which fashions the material into shape; this in our view, is the specific characteristic of each of the sexes: that is what it means to be male or to be female."

This statement would suggest a marked difference in the role of semen and catamenia, yet in other places Aristotle postulates a striving, almost a struggle, between male and female seed stuffs. When the male material predominates, a boy will be born. If it is only a partial victory, it may be a boy with the characteristics of the mother; or, if the parental strength is inferior to that of the grandparents, it will be a child with characters of the grandparents, and so forth.

What is very important in Aristotle's thinking is the role of the *eidos* of each individual. To be sure, each child has the characteristics of the species to which it belongs, but it also has its own specific individuality. A child of Socrates, says Aristotle, is apt to have the characteristics of Socrates.

It has been stated, not without justification, that Aristotle's separation of a form-giving principle (*eidos*) from the material that is being formed is not too different from the modern concept of the genetic program which controls the shaping of the phenotype (Delbrück, 1971). This ignores the fact that Aristotle's *eidos* was a nonmaterial principle; furthermore, it was rather consistently confused by subsequent authors with Plato's quite different concept of the *eidos,* with the result that Aristotle's concept was virtually ignored until after 1880. (Buffon's concept of a *moule intérieur* is superficially similar to Aristotle's *eidos,* but there seems to be no historical connection [Roger, 1963]. Buffon's mold was a strictly material entity.) The similarity of Aristotle's ideas to the modern ones was not recognized until 1970.

As in other areas of biology, the major contribution made by the Greeks was that they introduced an entirely new attitude toward inheritance. They considered it no longer as something mysterious, given by the gods, but as something to be studied and to be thought about. In other words, they claimed inheritance for science. Indeed, they were the first to pose many of the questions that were the subject of the great genetic controversies of the nineteenth and early twentieth centuries. And one of the schools of philosophy, the Epicureans, introduced a new concept, that of the existence of very small invisible particles, which later became a dominant concept of genetics.

For some two thousand years after the days of Aristotle and the Greek atomists, remarkably little was added to the subject of generation and inheritance that was new. This is equally true for the Alexandrian and the Roman period, and medieval discussion continued largely in terms of such classical models as remained available. Many of the questions that the Greeks had asked but were unable to answer decisively were the major questions occupying the new science of the Renaissance. Some of these questions, not all of them clearly articulated by the Greeks, were:

637

(1) What is the nature of fertilization? What is transmitted during copulation that is responsible for conception?

(2) Can living beings originate spontaneously, or is a sexual union always necessary for the production of new individuals?

(3) What are the respective contributions to the characteristics of a child made by the father and mother? Does the mother make a (we would now say "genetic") contribution in addition to serving as the nurse of the developing embryo?

638

(4) Where is the male semen formed—in a special organ, or throughout the body?

(5) How is the sex of the offspring determined?

(6) To what extent are the heritable characters affected by use or disuse, the environment, or other factors?

These and many other questions had to be answered—in fact, they had to be first formulated properly—before a science of genetics was possible.

NEW BEGINNINGS

When an interest in nature reawakened in the later Middle Ages, it encountered an entirely different spiritual and intellectual climate from that of the Greeks. The will of God and his power of creation were seen in every object and every process. The emphasis was on "origins," the generation of new individuals, rather than on the principle of continuity implicit in inheritance. This spirit, particularly well developed in the sixteenth century, is superbly described by Jacob (1970: 19–28). Spontaneous generation, the infusing of life into non-living organic matter, was considered as natural as regular reproduction. The production of monsters created hardly more wonder than that of normal beings. The conversion of the seeds or seedlings of one plant into those of another plant (heterogony) was considered an everyday occurrence. The origin of new beings was always a *generatio ab initio.* Since the emphasis was on the development that follows the original generation, this period of human thought is particularly important for the history of the field that, after about 1828, was designated as embryology.

One must remember that from the fifteenth to the eighteenth century biology as such did not yet exist. Rather there were two spheres of interest with only tenuous cross-connections, natural history and medicine (including physiology). Generation was primarily studied by professors of anatomy and by medical physiologists who investigated proximate causations and rarely asked questions concerning inheritance. Their interest was in developmental biology. By contrast, the students of natural history had as their major interest the diversity of nature, the result of ultimate causes.

Since all members of the species share in the same essence, inheritance was an obvious necessity and was not seen as a scientific problem. When considered at all, it was done within the framework of the species question. Variation, however, was very much in everyone's mind, particularly the naturalist's. The herbalists, botanists, hunters, and animal keepers all delighted in aberrant individuals. At first this related only to strikingly different "mutations" (see below), but eventually, as more and more specimens accumulated in the herbaria and museums, ordinary individual variation also became apparent and was studied. In due time this became an important source of evidence against the validity of essentialism.

From the Middle Ages to the nineteenth century the thinking of western man was completely dominated by essentialism (see Chapter 2). Since according to this philosophy all members of a species share the same essence (unaffected by external influences or occasional accidents), the study of nature is simply the study of species. So dominant was essentialistic thinking throughout the sixteenth, seventeenth, and most of the eighteenth centuries that no systematic investigations seem to have been made about the variation of individual characteristics. When naturalists encountered deviations from the typical expression of the species, they might recognize intraspecific "varieties" (quite typologically conceived), but these merited no special attention. With the emphasis so strongly on the species, it is not surprising that it was the species problem which gave rise to some of the earliest thoughts about inheritance—those of Linnaeus, Kölreuter, Unger, and Mendel.

Any study of the mechanisms of inheritance must be based on the crossing of individuals that differ in definite and seemingly constant characteristics. Variation, thus, is the primary problem to be explained by a theory of inheritance. Yet, an essentialist does not know how to handle variation. The conceptual dilemma for him is that "essentially" all individuals of a species are identical. As a result, different kinds of variation were hopelessly confused with one another to the end of the nineteenth and even into the twentieth century. This confusion was not resolved until population thinking had replaced essentialism in systematics and evolutionary biology. The nature of the difficulties is most easily illuminated by a historical survey. It will show how the heterogeneity of variation was gradually perceived and the differences among the components understood.

Linnaeus

As far as the essentialist is concerned, a species, by definition, has no essential variation. All variation is "accidental," not affecting the species essence (for detail see Chapter 6). A variant is not a different species; it is a "variety." Although variants and varieties had long been known to naturalists and horticulturalists, Linnaeus is generally credited with having formalized the concept of the variety. He had considerable contempt for varieties and made fun of the flower lovers who named them so enthusiastically. On the whole he considered varieties to be unimportant, reversible modifications caused by climatic or soil conditions. He also knew of monstrosities, which he likewise considered irrelevant. It never occurred to him to ask about the biological importance of much of variation. "Varieties are plants changed by some accidental cause" (*Phil. Bot.*, 1751: para. 158).

In his *Philosophia Botanica* (para. 158) Linnaeus characterized the variety as follows: "There are as many varieties as there are different plants produced from the seed of the same species. A variety is a plant changed by an accidental cause: climate, soil, temperature, winds, etc. A variety consequently reverts to its original condition when the soil is changed." Here a variety is defined as what we would now call a nongenetic modification of the phenotype. In his discussion of varieties in the animal kingdom (para. 259) Linnaeus indicates that he includes under the term "variety" not only nongenetic climatic variants but also races of domestic animals and intrapopulation genetic variants. When we carefully go through his writings we discover that under the name "variety" Linnaeus includes at least four entirely different sets of phenomena: (1) nongenetic modifications, owing to differences in nutrition, climate, cultivation, or other environmental influences on the phenotype; (2) races of domestic animals or cultivated plants; (3) genetic intrapopulation variants; and (4) geographic races, like the races of man.

As time went on and the heterogeneity of the phenomena that had been subsumed under the term "variety" was discovered, new terms were proposed for different kinds of varieties. Yet the elaborate terminology which resulted from these efforts (see Plate, 1914: 124–143) did not eliminate the problem, because it did not eliminate the underlying conceptual confusion. Most authors failed to distinguish clearly between (1) genetic and nongenetic variation; (2) continuous and discontinuous variation (see Chapter 16); and (3) individual and geographical variation. As a consequence, when different authors talked about "varieties," they often had entirely different phenomena in mind. The situation was aggravated by the fact that, beginning with Linnaeus, two different traditions developed which divided botanists and zoologists. When *641* zoologists spoke of varieties, they usually meant geographic races; when botanists did so, they usually meant cultivated varieties or intrapopulation variants. And yet this difference in tradition was the first indication of a sorting out of the different kinds of varieties.

MENDEL'S FORERUNNERS

It was in the Linnaean era that the first hesitating steps were taken that eventually led to the founding of genetics. Methodologically, there are two ways to learn about inheritance. One is the study of pedigrees. It is rather easy to follow conspicuous characteristics in the human species through several generations, and by this method Maupertuis was able in 1745 to record the occurrence of polydactyly (the presence of a sixth finger and toe), now known to be due to a dominant gene, through four generations. By a peculiar coincidence, Réaumur, at about the same time (1751), likewise demonstrated the dominant inheritance of polydactyly in man (Glass, 1959). This was soon followed by similar studies on hemophilia and color blindness. Although these pedigrees were well-known to biologists in the nineteenth century, they were not used as the basis of theories of transmission genetics.

The other method of investigating inheritance is by breeding. This method was employed by two schools, the species hybridizers and the animal and plant breeders, schools which had very different interests and objectives.[2]

The Species Hybridizers

Linnaeus is often described as a pedantic schoolmaster interested in nothing but artificial classifying. To be sure, he was quite fanatical in his endeavor to classify anything under the sun that showed variation. On the other hand, he often surprises the readers of his essays by his highly unorthodox thoughts on all sorts of natural-history subjects. As with any author in whose mind a rich fermentation of ideas is going on, he often promoted simultaneously, or at least consecutively, ideas that seemed strangely in conflict with each other. This is well illustrated by Linnaeus's change of mind concerning the nature of species. The constancy of species was the cornerstone of Linnaeus's early work and his statement (1735) "Tot sunt species . . ." is perhaps his best-known dogma (see Chapter 6). Yet later in life he played with the idea

642

(one can hardly avoid expressing it in these words) that natural species freely hybridize with each other. In one of his theses (Haartman, 1764; *Amoen. Acad.*, 3: 28–62) no less than a hundred putative species hybrids are listed, 59 of which are described in detail. In a prize essay (1760) dealing with the nature of sex in plants, written for the Academy of Sciences at St. Petersburg, Linnaeus described two hybrids claimed to have been produced artificially by hand cross-pollination. One was a hybrid goatsbeard (*Tragopogon pratensis* × *T. porrifolius*), the other a hybrid speedwell (*Veronica maritima* × *Verbena officinalis*).

It is irrelevant whether or not the plants that Linnaeus had produced were really offspring of the mentioned parental species (which is somewhat doubtful); what is important is that Linnaeus asserted here that a new constant species—that is, an entirely new essence—had been produced by the hybridization of two species. This claim was completely in conflict with all of the previous ideas of Linnaeus and other essentialists. The hybrid unless it had both essences, would have to have an intermediate essence, and if again hybridized with one of the parents or with another species, it would virtually produce a continuity of essences, a conclusion completely contradicted by the well-defined discontinuities among species found in nature. Nevertheless, Linnaeus himself was so convinced of the production of new essences that he gave new species names to both of his hybrids and entered them in his authoritative *Species Plantarum* (1753).

Linnaeus sent some of the seeds of his hybrid goatsbeard to St. Petersburg, where they were raised by the German botanist Kölreuter, who himself was occupied with crossing species of plants. The goatsbeards raised by Kölreuter in 1761, presumably the F_2 generation, showed considerable variability, utterly refuting Linnaeus's claim to have produced a constant new species.

Kölreuter

Joseph Gottlieb Kölreuter (1733–1806), like nearly all other biologists of the eighteenth century, obtained his education in a faculty of medicine (at the University of Tübingen). He received his degree in 1755 after seven years of study and went for the next six years to St. Petersburg, where he was appointed in the Academy of Sciences as a natural historian. Here he worked on, among other things, fertilization (pollination) of flowering plants and the production of hybrids. Since Kölreuter later was often viewed as a forerunner of Mendel, it is important to emphasize that he did not undertake his work on plant breeding with any purely genetic *Fragestellung*. What he was interested in were such problems as the biology of flowers and the nature of species.

643

His first successful cross was that between the two tobacco species *Nicotiana rustica* and *N. paniculata*. The hybrids grew extremely well, and "the keenest eye can discern no imperfection, from the embryo to the more or less complete formation of its flowers." Indeed, it appeared as if, like Linnaeus, he had succeeded in producing a new species. However, all efforts to pollinate the hybrid flowers with each other were in vain. Not a single seed was produced by the hybrid, while a normal flower would produce 50,000 seeds. This struck Kölreuter as "one of the most wonderful of all events that have ever occurred upon the wide field of nature." This finding was a great relief to him since it restored his faith in an essentialistic species concept. Again and again in the ensuing years Kölreuter crossed species belonging to a large number of different plant genera. Indeed, he undertook more than 500 different hybridizations involving 138 species. The findings were invariably similar. There was a striking reduction in the fertility, if not complete sterility, of the hybrids. In fact, when Kölreuter discovered normal fertility in some of his "species" crosses, he discarded these, saying they were obviously not good species. He was correct. He left an exact description of all of his crosses, and in retrospect we are able to agree with him; the discarded crosses were indeed crosses among intraspecific variants.

When he examined the pollen of the hybrid plants under the microscope, he found that in nearly all instances the individual grains were shrunken—indeed were mere empty husks. No wonder pollination was unsuccessful. He found well-formed pollen grains only in a few cases and was then able to produce some F_2 plants. Fertility was greater in back-crosses, that is, when he pollinated the hybrid plants with the pollen of either of the two parental species. Continuing such back-crossing for a number of generations, he finally obtained plants that were indistinguishable from the species to which the hybrids had been back-crossed. He described this result in a somewhat quaint language as being able to restore the original species.

In his other crossing experiments, for instance with various species of carnations (*Dianthus*), there was sometimes a much less drastic reduction of fertility and F_2 and F_3 generations were much more easily produced, but in principle the results were always the same. Each species was clearly protected to a lesser or greater degree by a sterility barrier. This, of course, had already been shown by Buffon in his studies on the mule and other animal hybrids, but it had not yet been generalized.

Another important finding made by Kölreuter concerned the appearance of the first- and second generation hybrids, and of back-crosses. He found that the F_1 hybrids were all more or less alike and in most of their characters intermediate between the two parental species. As it was often phrased, the characters of the parent species had become "blended" in the F_1 offspring. F_2 hybrids, on the other hand, showed a great deal of variability, and some resembled their grandparents more than their own F_1-generation parents. These findings were again and again confirmed in the ensuing hundred-year period between Kölreuter and Mendel, at least as far as species crosses are concerned.

Kölreuter belonged to the school which thought that a scientific explanation in biology had to be physical or chemical in order to be convincing. This is why he explained the difference between the F_1 and F_2 generations with the help of a chemical model. Just as an acid and a base form a neutral salt, said Kölreuter, thus in an F_1 hybrid the female "seed material" unites with the male "seed material" to form a "compound material." In the F_2 hybrids they do not combine in equal proportions, and a variety of offspring is produced, some of which resemble more one, some more the other, grandparent. He was unable to explain why this was so, but it is clear that he did not consider the combining of the parental "seed material" as a process of blending. Indeed, to my knowledge, no experienced plant breeder, except Nägeli, has ever asserted blending inheritance as an exclusive mechanism.

Kölreuter was aware that the F_2 hybrids in some crosses fall into three types, two of them like the two grandparental species and the third like the F_1 hybrid. However, with his focus on the species problem rather than on individual characters, Kölreuter found only few cases of such clear-cut segregation. His basic objective was to prove that the hybridization of two species does not produce a third species, and with few exceptions this conclusion is as valid today as it was two hundred years ago. The only exceptions are allotetraploids discovered 150 years after Kölreuter.

Reading Kölreuter's painstakingly detailed reports on his many crosses fills us with admiration not only for his industry but for the perceptiveness of his inquiry. His demonstration that flowers are sterile if pollen is prevented from reaching the pistil of the female flower proved conclusively that the male seed material was necessary for fertilization. By comparing numerous characteristics of hybrids with those of the two parental species, and by the production of reciprocal hybrids, he was the first to prove the equal contribution made by the two parents as documented by the intermediacy of the F_1 hybrids. He thus decisively established the significance of sex as well as of fertilization, two points that were still quite controversial in his day. Furthermore, he conclusively refuted preformation, whether of the ovist or spermist variety.

To a modern person it is self-evident that both father and mother contribute to the genetic endowment of a child. Curiously, this was not nearly so evident to former generations. The uncertainty goes back to the Greeks, where "chauvinistic males" ascribed the main character-forming qualities to the father and where, in the writings of Aristotle and others, the father supplied the form and the mother only the substrate to be shaped by the form. In the seventeenth and eighteenth centuries the problems became hopelessly entangled with the problem of development. Was there preformation (or even preexistence) of the germ or "epigenesis" of an unformed egg? The preformationists by necessity had to choose whether the preexisting embryo was located in the egg ("ovists") or in the sperm ("spermists"). Almost all the leading biologists of the seventeenth and eighteenth centuries (Malpighi, Spallanzani, Haller, Bonnet) were ovists and thus attributed most of the genetic potential to the female. Van Leeuwenhoek and Boerhaave were among the spermists, the former as co-discoverer of the spermatozoon understandably so.

How these extraordinarily well-informed and intelligent authors could have such one-sided theories is difficult to explain. All these authors must have been aware of the fact that in the human species every child shows a mixture of the traits of both parents. They knew that the mulattoes, the result of a cross between a white and a black person were intermediate. They knew that hy-

645

brids between species, like the mule between horse and ass, were intermediate between the two species. All these well-known facts and others clearly refuted not only a naive belief in preexistence (*emboitement*) but also any concept of a purely one-sided contribution by either male or female. And yet such observations shook neither the ovists nor the spermists, as if these authors kept these observations in two unconnected compartments of their brain.

Some of their contemporaries were more perceptive. Buffon saw clearly that both father and mother made a genetic contribution, but it was P. M. de Maupertuis, more than anyone else, who developed a theory of inheritance that can be considered as foreshadowing later developments (Glass, 1959; Stubbe, 1965). Maupertuis espoused a theory of pangenesis, based on the thought of Anaxagoras and Hippocrates, postulating particles ("elements") from both father and mother as responsible for the characters of the offspring. Most components of his theory can be found in the later theories of Naudin, Darwin, and Galton.

As fundamental as Kölreuter's findings were for the understanding of sexuality and reproduction in plants, it would be a mistake to consider him a forerunner of Mendel. Kölreuter always regarded the essence of the species in a monolithic manner. The very fact of the intermediacy of the F_1 hybrid which he found in most cases seemed to confirm to him his holistic interpretation. At no time did he partition the phenotype into individual characters and trace the fate of a given character in different combinations through several generations. This is precisely what was necessary to establish laws of genetics, as Mendel and de Vries were the first to recognize.

Kölreuter is to be admired not only for the importance of his findings on flower biology and the nature of hybrids but also because his experimental approach indicated an excellence of planning and execution unknown among his contemporaries. Alas, like many pioneers, he was too far ahead of the interests of his time and directed some of his most elegant experiments on the demonstration of what to us seems obvious, the sexuality of plants.

Kölreuter's results with species hybrids were in conflict with existing dogma to such an extent, and his findings so novel and revolutionary, that they were not accepted by his contemporaries. Learned volumes were published as late as 1812 and 1820 in which the existence of sexuality in plants was denied and the credibility of Kölreuter's experiments was questioned. In view of this situation, academies in Prussia and Holland offered prizes in the 1820s

646

and 1830s in order to settle the problem of hybridization in plants and its utilization for the production of useful varieties and species. This stimulated the work of Wiegmann, Gärtner, Godron, Naudin, Wichura, and other hybridizers, which has been well described by Roberts (1929), Stubbe (1965: 97–110), and Olby (1966: 37–54, 62–66). All of these investigations were in Kölreuter's tradition. They dealt with the sexuality in plants and with the nature of species.

Only some of the crosses were between Mendelian varieties within species but, as in the case of Kölreuter, the results, even when published, were not followed up. All these authors confirmed, again and again, Kölreuter's results, such as the intermediacy and relative uniformity of the F_1 generation, the increased variability of the F_2 generation (with clear indications of reversion to the parental species), the identity of reciprocal crosses, a contribution by both father and mother, (usually more or less equal) to the characters of the hybrid and the occasional occurrence of somatic hybrid vigor even in sterile hybrids. Clear-cut Mendelian segregation rarely occurred, even in the F_2, which is not surprising, because species differences often, if not usually, are highly polygenic. Furthermore, Kölreuter's *Nicotiana* species, as well as many of the species with which the other hybridizers worked, were polyploids and the number of chromosomes was often larger in one of the parents than in the other, and then the parent with the larger chromosome set predominated in the appearance of the hybrid.

647

Again and again it must be emphasized that these students were not engaged in an investigation of the laws regulating the inheritance of individual characters. They were interested in the essence of the species as a whole, and to some extent they understood this better than those who practiced the bean-bag genetics of the early Mendelian period. The great split in evolutionary biology that characterized the period between 1900 and the evolutionary synthesis of the 1930s can, to some extent, be traced back to some of the cross currents among the plant hybridizers early in the nineteenth century.

Gärtner

Carl Friedrich von Gärtner (1772–1850) was by far the most erudite and industrious of the pre-Mendelian species hybridizers. In his main work (1849) he summarizes the results of nearly 10,000 separate crossing experiments among 700 species yielding 250 different hybrids. Darwin said of this work that "it contains more valuable matter than that of all other writers put together, and would do great service if better known."

The piling up of the huge amount of information which Gärtner had assembled should have resulted in numerous generalizations, if the claims of the inductivists were valid. This, however, did not happen. Neither Darwin, who studied this work so carefully, nor any other contemporary saw any general laws emerge from Gärtner's facts. Actually, Gärtner asked himself the same kinds of questions that Kölreuter had asked nearly a hundred years earlier and was, on the whole, quite satisfied simply to describe the results of his crosses. Perhaps one can pay Gärtner a left-handed compliment by saying that he showed so conclusively what answers could and what others could not be gotten from these questions that he vacated the field for an entirely new approach. We know that Mendel, who owned a copy of Gärtner's book, studied it most carefully, and it is more than likely that this work helped him to phrase the new questions, which resulted in his spectacular breakthrough. Among the thousands of crosses which Gärtner carried out were a few dealing with intraspecific varieties of peas and maize. In these, as we shall presently see, Gärtner was indeed a forerunner of Mendel.

Gärtner was not the only German plant hybridizer of this period, but the others (such as Wiegmann or Wichura) likewise worked within the traditional framework and thus failed to add anything of significance to our understanding of inheritance.

Naudin

The French hybridizer Charles Naudin (1815–1899)[3] differed from Gärtner in having a very definite theory, but in his basic thinking he was not too far removed from him. He thought that it was an altogether unnatural process to bring the essences of two species together in the production of hybrids. This expressed itself in their sterility and in the reversion of the later generation hybrids to one or the other of the parental species. There was no blending of the parental essences. Furthermore, Naudin treated the species essence as a whole rather than as a mosaic of independent characters, as Mendel would do in his work. Some of Naudin's species were apparently merely Mendelian varieties (for instance those of *Datura*) and here Naudin apparently obtained clear-cut Mendelian ratios, which, however, were completely consistent with his interpretation of a perfect segregation of the parental essences. Even though the results of some of his crosses, such as the uniformity of the first and the variability of the second hybrid generation, were thus quite "Mendelian," Naudin was not a forerunner of Mendel either in theory or in methodology, as evidenced by his failure to look for repeatable ratios.[4] The same is true for his compatriot D. A. Godron (1807–1880, who is entirely concerned with the same questions (fertility of the hybrids, their return to the parental type, and so forth), as was Kölreuter nearly a hundred years earlier. As his other publications show, his major interest was in the nature of the species.

649

The Plant Breeders

An entirely different tradition developed side by side with the activities of the species hybridizers, that of the practical plant breeders. Their purely utilitarian aim was to improve the productivity of cultivated plants, to increase their resistance to disease and to frost, and to produce new varieties. Although they also made use of species crosses, their major interest was the crossing of varieties, many of them differing, as we would now say, in merely one or a few Mendelian characters. These plant breeders have a far better claim to be considered as direct forerunners of Mendel than the plant hybridizers.

The first of them was Thomas Andrew Knight (1759–1853), who worked particularly with varieties of fruit trees. He is of special interest to us for having recognized the usefulness of the edible pea (*Pisum sativum*) as genetic material because "the numerous varieties of strictly permanent habits of the pea, its annual life, and the distinct character in form, size and color of many of its varieties, induced me, many years ago, to select it for the purpose of ascertaining, by a long course of experiments, the effects of introducing the pollen of one variety into the prepared blossom of another" (1823). This special suitability of the edible pea was apparently well-known among the plant breeders (including Gärtner) and was undoubtedly the reason why Mendel eventually devoted most of his efforts to this species. Knight was a careful experimenter who always emasculated the flowers before applying pollen from different plants, and who used unpollinated or openly pollinated flowers as controls. He described both dominance and segregation (in back crosses), but did not count the different kinds of seeds he had obtained and thus did not calculate ratios.

Two contemporaries of Knight, Alexander Seton (1824) and John Goss (1820), confirmed dominance and segregation and established the true breeding character of what we would now call the recessives. Some of the experiments of these three breeders were contradictory because they did not appreciate that in the F_1 of the pea the appearance of the seed coat (transparent or opaque) is determined by the mother, while the color of the pea itself (the cotyledons) is determined by the genetic constitution of both parents. Gärtner at a later date, in crossing experiments with maize, encountered similar difficulties with the seed coat (pericarp) which contributed to his failure to consistently obtain strict Mendelian ratios. The confusion was resolved only many years later. The endosperm is formed through the fusion of two maternal and one pollen nuclei and may, therefore, display paternal characters, a phenomenon (later investigated by de Vries and Correns) called *xenia* by plant geneticists (Dunn, 1966).

650

The crucial difference between the species hybridizers and the numerous plant breeders (see Roberts, 1929) was that the latter often studied individual characters and followed their fate through a series of generations. A particularly successful application of this new methodology was made by the French agriculturalist Augustin Sageret (1763–1851). When crossing two varieties of the melon *Cucumis melo*, he arranged the characters in a set of five pairs:

Variety 1	*Variety 2*
Flesh yellow	Flesh white
Seeds yellow	Seeds white
Rind reticulated	Rind smooth
Ribs pronounced	Ribs barely indicated
Taste Sweet	Taste sweet/sour

The hybrids he obtained were not intermediate between the two parents; instead there was a close resemblance in each character either with one or the other parent. He came to the conclusion "that the resemblance of a hybrid to his two parents consists, in general, not in an intimate fusion of the various characters that are peculiar to either of them but rather in a equal or unequal distribution of the unchanged characters; I say equal or unequal because this distribution is far from being the same in all the hybrid individuals of the same provenance and there exists among them a very great diversity" (1826: 302).

In the description of his crosses he clearly designates the characters of one or the other parent as "dominant." No one before him had used this terminology quite as unequivocally. Sageret not only confirmed the phenomenon of dominance and discovered the independent segregation of different characters, but he was also fully aware of the importance of recombination. "One can not admire too much the simplicity of means with which nature has provided itself with the ability to vary its productions infinitely and to avoid monotony. Two of these means, uniting and segregating characters, in diverse ways combined, can lead to an infinite number of varieties." Sageret also recognized that ancestral characters occasionally turn up in these crosses, "the potential for which existed but the development of which had not before been favored." As we shall see, Darwin later on was greatly interested in such reversions. Sageret unfortunately never followed up his imaginative and innovative researches.

651

The question has often been asked in recent years why these plant breeders had stopped seemingly so near to the achievement of a genetic theory. Numerous answers have been given, most of them clearly inappropriate. An insufficient understanding of cytology is certainly not responsible because Mendel's explanation was not based on cytological theory, nor is it necessary.

The failure of these breeders to develop a genetic theory cannot be attributed to faulty technique because several of them were scrupulously careful in preventing unwanted pollinations and in providing control experiments. One has the feeling that they were quite satisfied merely to get clear-cut results. They simply failed to ask questions about underlying mechanisms; if they had asked them, as Mendel later did, they would indeed have had to add to their technique the careful counting of offspring and the calculation of ratios. In other words, ultimately their failure, if this is what we want to call it, consisted in not asking the decisive questions. They failed to do so because they did not think in terms of variable populations. A populational interpretation was a prerequisite for the new approach toward inheritance.

And yet by the 1850s a broad basis had been laid, both by the hybridizers and by the plant breeders. They had clearly established most of the facts needed for a genetic theory, such as the equal contribution of both parents, dominance, the relative uniformity of the first filial (F_1) generation, segregation (the increased variability of F_2), and the usual identity of reciprocal crosses.

The stage was set for the appearance, sooner or later, of an exceptional gifted individual, who would ask previously unasked questions and solve them with new methodologies. This individual was Gregor Mendel (see Chapter 16).

15 ✐ Germ cells, vehicles of heredity

652 BY SAYING that a child has inherited this or that characteristic from one of his parents, one postulates a process that provides for continuity between one generation and the next. Indeed, continuity is the essence of the entire concept of heredity. The Greeks already vaguely understood that sexual union was the key to the solution of the problem of heredity, but how the "genetic material" (as it was later called) was transmitted from one generation to the next was entirely a matter of speculation (see Chapter 14). Some of the proposed theories were highly improbable because the inheritance of physical and behavioral characteristics was far too precise and detailed to be explained in terms of "heat" or of "pneuma" or of other generalized physical forces, as proposed by most early philosophers. The school of Hippocrates seems to have come much closer to the truth when it explained inheritance as due to the transmission of seed stuff. A qualitative theory of inheritance was proposed by Lucretius, according to whom the qualities of hair, voice, face, and other parts of the body are determined by the mixture of atoms contained in the seed inherited from the ancestors. All observations on inheritance suggested that something qualitative-corpuscular was transmitted but, whatever it was, it was far too small to be seen by the naked eye. An entirely new branch of biology, cytology, first had to develop before the challenge of the nature of the genetic material could be met. The development of this new discipline was not possible until the microscope had been invented and applied to the study of cells.[1]

That eggs are necessary for the development of a new individual had long been evident, and that the male semen is also important was likewise already a widespread belief among the Greeks, and more or less conceded even by most so-called ovists in the seventeenth and eighteenth centuries. However, definite proof was not available until the 1760s. The similarity, if not identity, of the hybrids produced in reciprocal crosses (as in the work of Kölreuter) led to the inevitable conclusion that the genetic contributions of father and mother are equivalent. This insight raised new questions: How could eggs and spermatoza (or pollen grains) be equivalent in spite of their conspicuous differences in size and shape? Where in the body of male and female is the seed stuff produced that acts as the conveyor of the parental characteristics? And how is the seed stuff structured, to be able to transmit the complex characteristics of an individual to his offspring? These questions could not be answered until the cell theory had been established.

The discovery that all living organisms (strictly speaking, only the eukaryota) consist of cells and cell products was made possible by one of the greatest technological advances in the history of biology, the invention of the microscope. The earliest simple microscopes were apparently invented around 1590 by some Dutch spectacle makers, but it was not until 1665 that Hooke, in his *Micrographia,* described and illustrated some pores and box-like structures in a thin slice of a piece of cork. More and better pictures were published by Nehemiah Grew in the years 1672 to 1682 and by Malpighi in 1675 and 1679. What these authors saw were walls, as is evident from the word "cell," and nothing was said about the possible biological meaning of their discovery. Soon afterwards, students of animal tissues, particularly of embryos, like Swammerdam (1737), C. F. Wolff (1764), Meckel (1821), Oken (1805; 1839), and others described globules or bubbles. At this time it is no longer possible to determine which of the observed globules were real cells and which others artifacts (Baker, 1948; Pickstone, 1973). It took a century and a half after Hooke's first description before any real progress was made in the study of cells, a progress made possible by technological advances in the construction of improved microscope lenses.

In the meantime a number of authors—in part, perhaps, stimulated by atomistic speculations in the physical sciences—had begun to ask what were the ultimate components of the human (an animal) body. According to Hippocratic dogma, the body consists of liquids and solids, and Boerhaave and other anatomist-physiologists of the eighteenth century thought that these solids consisted of very small fibers.[2] Haller became a chief proponent of the fiber theory, which was also adopted by Erasmus Darwin. Although the fiber theory was wrong, it had the merit of focusing attention on the problem of the ultimate building materials of the body.

Since fibers, globules, or cells were promoted by different au-

653

thors, and since the findings of botanists and zoologists often seemed to be in conflict, there was an evident need for a unification of this area of biology. Bichat had recognized 21 different categories of animal tissue: did they have the same building stones, and if so, what were they? The search for some common element seemed particularly important in those days of idealistic morphology.

In the 1820s and 30s microscopes began to be manufactured in England, France, Germany, and Austria and soon became a regular part of the equipment of the better laboratories. The new instruments incorporated a number of recent improvements and stimulated microscopic research as nothing had done before. Not only did these investigations show how much of the claims of the eighteenth century had been based on artifacts but, more importantly, they showed that there was more to cells than a wall. Up to that time the word "cell" (as it was used by Haller and Lamarck) was little more than a name. It was largely seen as a structural element with the emphasis on the cell wall, and nothing was said of function. Only gradually did the improved instrumentation permit investigators to pay attention to the cell contents. Then it was found that living cells are not empty but filled with a sticky fluid which was called *sarcode* by the French zoologist Dujardin (1835) and *protoplasm* by Purkinje (1839) and von Mohl (1845). Protoplasm was far more than a technical term for the cell contents (excluding the nucleus). By going to the popular literature of the period, one can see that it was widely conceived as the "substance of life" in a rather vitalistic manner (see *Oxford English Dictionary,* under protoplasm"). It was considered the ultimate building material of everything living, and for nearly a hundred years it was interpreted as the real agent of all physiological processes.

Later, when biochemistry began to "dissect" the cell contents, it became apparent that there was no unitary substance deserving the name protoplasm, but it was not until the introduction of the electron microscope, after 1940, that it was realized what a complex aggregation of structures the cell contents are, with functions undreamed of by the early students of protoplasm. The name "protoplasm" has now virtually disappeared from the biological literature, and the aggregate of cellular structures and cell fluid outside the nucleus is now referred to by Kölliker's term *cytoplasm.* The importance of the cell wall was more and more deemphasized, Leydig (1857) and M. J. S. Schultze (1861) finally pointing out that animal cells had no cell wall, many of them being quite naked, covered only by a membrane.

The other conspicuous element of the cell contents is the nucleus. Although nuclei had been observed in plant cells and even in some animal cells at least as far back as the beginning of the eighteenth century, the British botanist Robert Brown (1773–1858) deserves credit for having been the first (1833) to consider the nucleus a regular component of the living cell. What its function was remained, however, quite uncertain, and its role in the cell was at first totally misinterpreted. Until the 1870s, cell and protoplasm were considered as almost synonymous and the nucleus as an unimportant cell component that was either present or absent. Indeed, it was believed to be absent from most cells during part of the nuclear cycle. This was an understandable conclusion since the spherical nucleus, surrounded by a membrane, disappears during cell division.

655

THE SCHWANN-SCHLEIDEN CELL THEORY

By the end of the 1830s the uncertainties about the cell crystallized into two major questions: What is the role of the cell in the organism? and, How do new cells originate? These questions received a preliminary answer in the Schwann-Schleiden cell theory.

The most influential cytologist of the period was the botanist M. J. Schleiden (1804–1881). With enormous enthusiasm he not only recruited the zoologist Theodor Schwann (1810–1882) for cytological research but also developed some of the leading young botanists of the century, like Hofmeister and Nägeli. It was he who induced young Carl Zeiss to start his subsequently so important optical firm, and gave him enough orders to insure success. (Zeiss, in due time, more than fully repaid his debt to biology by inventing better optical instruments, particularly microscopes.)

Schleiden belonged to the generation of young German biologists who reacted violently to *Naturphilosophie* and who attempted instead to explain everything in a reductionist physico-chemical way (Buchdahl, 1973). It was unthinkable for him to answer the question, "How do new cells originate?" by saying "From pre-existing cells." This would have been too much like preformation, a theory thoroughly in disrepute at that time. Consequently, Schleiden applied the principle of epigenesis to cell formation and es-

tablished, in 1838, a theory of "free cell formation." He suggested
that the first step in cell formation was the formation of a nucleus
by crystallization from granular material within the cell contents.[3]
This nucleus would grow and eventually form a new cell around
itself with the outer nuclear membrane becoming the cell wall
(Schleiden gave a detailed account, 1842: 191). New nuclei might
form within existing cells, or even crystallize within a formless organic fluid. A major controversy of the ensuing two decades was
whether or not such free cell formation took place, the eventual
answer being negative. Although he was shown to have been
wrong, Schleiden decisively advanced cytology by focusing attention on a problem and by proposing a succinct, testable theory.
More important, in the long run, was his insistence that the plant
consists entirely of cells and that all the highly diverse structural
elements of plants are cells, or their products.

656

In a classical publication, *Mikroskopische Untersuchungen über
die Übereinstimmung in der Struktur und dem Wachstum der Tiere und
Pflanzen* (1839), Schwann showed that Schleiden's conclusion also
applies to animals. (Independently this was also claimed by Oken
in 1839.) By examining embryonic animal tissues and by following
their subsequent development, he succeeded in demonstrating a
cellular origin even for tissues like bone, which—when full grown—
show no trace of such an origin. The fact that both animals and
plants consist of the same fundamental element, cells, was one
additional piece of evidence for the unity of life and was celebrated as one of the great biological theories, the cell theory. It
helped to give substance to the word "biology" (coined by Lamarck and Treviranus), until then a largely unfulfilled program.

As important as this new insight was, one is somewhat puzzled over the immense excitement caused by the cell theory. No
one yet really understood the cell and the functions either of the
nucleus or of cytoplasm. At that time, the name "cell theory" was
primarily applied to Schleiden's theory of free cell formation (Virchow, 1858), and perhaps the idea of purely physico-chemical formation of nuclei and cells (by crystallization) had considerable appeal in the prevailing climate of extreme physicalism and
reductionism.

A very different reason is suggested by the fact that Brücke
and others referred to cells as "elementary organisms." His thinking evidently was influenced by idealistic morphology. Just as
Goethe "reduced" all parts of the plant to the leaf, so did Brücke

reduce all parts of any organism to the cell. Indeed, Wigand (1846) called the cell the *eigentliche Urpflanze*.

Some of these statements (also by other authors saying similar things) have an almost vitalistic flavor. This led to a reaction among the physicalists, as evidenced by Sachs's (1887) declaration that the cell was a phenomenon of only secondary significance, since the formative forces reside throughout the organic substance. Others devalued the cell because for them the protoplasm was the basic substance of life. Clearly the cell did not fit into an explanatory model based on the universality of "forces." Claims like those of Sachs were vigorously opposed by E. B. Wilson in the introductory statement of *The Cell* (1896).

657

Regardless of what it meant to various authors, the cell theory contributed to the firmer establishment of the unity of the living world. Furthermore, it led to the concept of organisms as republics of elementary living units. "The characteristics and unity of life cannot be limited to any one particular spot in a highly developed organism (for example, to the brain of man)" (Virchow, 1971 [1858]: 40); rather, life is found in an equal manner in each cell. Somehow this was, at that time, considered a strong argument against vitalism. Whether or not Virchow was influenced by somewhat similar ideas of Oken remains to be studied.

For Schwann and Schleiden the cell still was primarily a structural element, but already in the 1840s other authors stressed the physiological, particularly developmental, nutritional, role of cells. As the knowledge of cells and their constituents (particularly the nucleus) grew, the meaning of the concept "cell theory" gradually shifted. Schleiden's theory had the immediate effect of stimulating exceedingly active research on dividing cells of animals and plants. In 1852 R. Remak (1815–1865) showed that the frog egg is a cell and that new cells in the developing frog embryo are formed by the division of previously existing cells. He emphatically rejected free cell formation. In this he was joined by Rudolf Virchow (1855), who showed for many normal and pathological animal and human tissues that every cell originated by division from a preexisting cell. He established "as a general principle, that no development of any kind begins *de novo*, and consequently [one must] reject the theory of [spontaneous] generation just as much in the history of the development of individual parts as we do in that of entire organisms" (Virchow, 1858: 54).

Kölliker, as well as several botanists, arrived at the same conclusion at about the same time, even though Schleiden's authority tended to delay its acceptance in botany. In 1868 (II: 370) Darwin was still uncertain whether or not free cell formation occurs. In due time Virchow's famous aphorism *omnis cellula e cellula* (1855)—"every cell from a pre-existing cell"—was accepted by everyone, even though the details of the process of division, particularly of the nucleus, were not understood at that time (see below, under "Mitosis").

658

With this new interpretation of the cell, the stage was set for a renewed consideration of the process of fertilization. If all parts of the body consist of cells, is this also true for the gonads (ovaries and testes)? And what about the male and female "seed stuff"? Does it consist of cells? How different are the sex cells of males and females? Such well-formulated questions were, of course, not asked at the beginning, but they were logical consequences of the cell theory, and it became apparent, in due time, that no viable theory of inheritance could develop until the role of cells in fertilization had been elucidated. It was in these decades that the concept of germ cells emerged.

THE MEANING OF SEX AND FERTILIZATION

That there are two different sexes in animals was known from the earliest times, since the analogy with man was inescapable.[4] Sexuality in plants, at least its almost universal occurrence, was a much later discovery. To be sure, sexuality is rather obvious in some dioecious plant species (that is, species in which one individual has only male, and another one only female, flowers) and this knowledge was used by the ancient Assyrians when they fertilized female date palms with pollen shed by the flowers of male trees (see Chapter 13).

After the Middle Ages, N. Grew (1672) speculated on the role of pollen as the agent of fertilization. However, it was not until Rudolf Jakob Camerarius (1665-1721) published his *De Sexu Plantarum Epistola* (1694) that the sexual nature of plant reproduction was firmly established. He clearly designated the anthers as the male sex organs, and emphasized that pollen is needed for fertilization, as he had discovered through experimentation. Camerarius was fully aware of the fact that sexual reproduction in plants was the exact equivalent of sexual reproduction in animals. He asked some exceedingly penetrating questions about the exact role played by the pollen grains during fertilization: "It would be most desirable . . . if we could learn from those who have access to microscopes, what is the content of pollen grains, how far they penetrate in the female apparatus, whether they reach the place intact where the seed [sperm] is received, and what issues from them when they burst" (1694: 30). This challenge was subsequently taken up by Kölreuter and other hybridizers but was not fully met until the work of Amici, Hofmeister, and Pringsheim (from 1830 to 1856; see Hughes, 1959: 59–60, and below). 659

Camerarius also recognized the role of wind in pollination and that seed set might occur under certain conditions even though pollination was prevented. His *Epistola* had a great impact on his contemporaries and was apparently responsible for the increasing number of attempts at experimental plant hybridization in the eighteenth century, culminating in the work of Linnaeus and Kölreuter (see Chapter 14 and Zirkle, 1935). Still, sexuality in plants continued to be widely denied well into the nineteenth century.

Even Kölreuter and Linnaeus did not sufficiently stress the universality of sexual reproduction in plants and the obligatory need for cross pollination in the majority of plant species. Nor was it generally realized that plants with "flowers" (as understood by the layperson) are invariably animal-pollinated. In 1795 Christian Konrad Sprengel (1750-1816) published a classical treatise on the pollination of flowers by insects which emphasized all these points, but his work was so far outside the standard thinking and interests of the period that it was almost completely ignored. The most remarkable aspect of this volume is that Sprengel carefully described the numerous mutual adaptations of flowers and insects that facilitate cross fertilization or make self-fertilization impossible. It was the first "flower biology," a fact duly appreciated by Darwin (*Origin:* 98; 1862). An obvious inference from Sprengel's work, although not made until more than a century later, is that individuals of sexually reproducing species are not types or pure lines but members of populations.

The Nature of Fertilization

After the cell theory had been established, it would have seemed obvious that one would have to ask whether this theory could be extended to eggs and spermatozoa. This was done very quickly for the spermatozoa, which von Baer still had considered to be parasitic worms in the semen. Kölliker showed already in 1841 that they are cells, as can be shown by the study of spermatogenesis. In the case of the egg, things went more slowly. Neither von Baer, who discovered the mammalian egg in 1827, nor Purkinje, who in 1830 discovered the large nucleus of the ovarian egg (which he called the germinal vesicle), were aware of the cellular nature of these structures. It was Remak who in 1852 showed that the frog egg is a single cell, and Gegenbaur who in 1861 extended this conclusion to the eggs of all vertebrates by demonstrating that the yolk granules were not cells.

660

With modern hindsight one would think that the nature of fertilization in animals would be speedily inferred, once it was realized that the egg is a cell and so is the spermatozoon: Of course, one would say, fertilization is the fusion of these two germ cells, giving rise to a new individual. Actually, it took many decades before this insight was achieved. The same conclusion should have been reached for pollination in plants, on the basis of the observations of Kölreuter, Amici, Mendel, and others. One suggestive observation after another was made between 1824 and 1873, only to be either ignored or to be interpreted in a manner which to us seems clearly contradicted by observation. As late as 1840, so progressive a botanist as Schleiden questioned sexuality in plants. But then, from 1873 to 1884, everything suddenly fell into place. Why had the phenomenon of fertilization caused such difficulties of interpretation for such a long time?

There are many reasons, but perhaps the most important one is that this phenomenon belongs both to functional and to evolutionary biology. Embryologists were impressed by the fact that an unfertilized egg may long remain dormant, and does not start to develop until it is fertilized. They assigned therefore to the spermatozoon a purely mechanical role, as is indeed true in the exceptional phenomenon of pseudogamy, corresponding to the dropping of a coin into a juke box. In contrast, those who were interested in heredity saw in fertilization a process resulting in a mixing of the paternal and maternal hereditary endowment. Not surprisingly, those holding such diverging interpretations of the meaning of fertilization subscribed to entirely different explanatory models. That fertilization has a dual significance, and that the two opposing interpretations were both correct, was not fully realized until the last two decades of the nineteenth century.

Since the controversy about fertilization is one of the most interesting ones in the history of biology, leading to a direct confrontation of the students of proximate and of ultimate causes, it would seem worthwhile to provide a short survey of the opposing arguments.

Much as they differed in detail, theories of fertilization from the Greeks to the early nineteenth century postulated that the mother supplied a single, more or less uniform, unit of matter, later designated as an "ovum," while the male supplied some potency, involving pneuma, heat, or some physical or vital forces that induced the development of the egg. As late as 1764 Wolff still assumed that pollen and animal semen merely served as refined nutrition necessary for the stimulation of growth and development of the embryo. Even von Baer's (1828) account has a remarkably Aristotelian flavor. In Aristotelian terms the female contributed the material, the male the efficient, formal, and final causes.

661

This interpretation appeared to be confirmed when it was discovered by Bonnet in 1740 that eggs of plant lice (aphids) can develop even without the presence of males (parthenogenesis). Clearly, the potential of the eggs to develop could be induced by some generative force exercised by the female herself. This was a shocking discovery to Bonnet's contemporaries, but the researches of the nineteenth century showed that such "virgin birth" was widespread in the animal kingdom, either as a seasonal phenomenon (as in aphids and rotifers) or as a permanent one (Churchill, 1979). In the 1840s a special kind of parthenogenesis (arrhenotoky) was discovered in the hymenoptera where unfertilized eggs give rise to haploid male offspring. The discoverer of this extraordinary process was Johann Dzierzon, a contemporary of Mendel and, like him, a Catholic priest of Silesian birth and a beekeeper. He substantiated his hypothesis (1845) that drones come from unfertilized eggs of the honeybee by ingenious crossing experiments between German and Ligurian bees. A process of uniparental reproduction (apomixis), analogous to parthenogenesis, is even more widespread in the plant kingdom, where it was the source of a great deal of confusion during the early period of genetics (see Mendel's *Hieracium* crosses, Chapter 16). The special role of parthenogenesis as an evolutionary strategy has been much discussed in recent years.[5]

One can observe again and again in the history of biology that problems have their ups and downs, and so it was with fertilization. After the splendid beginnings made by Camerarius, Köl-reuter, and Sprengel came a lull in the first quarter of the nineteenth century. When fertilization again began to attract attention in the 1830s and 40s, it was during a period of extreme physicalism. According to von Liebig, all chemical activity depended upon molecular agitation induced by the close contact of two substances and their constituent particles (Coleman, 1965). T. L. W. Bischoff (1847) articulated this concept more explicitly by applying it to fertilization: "The semen acts on contact through a catalytic force, that is, it constitutes a particular form of matter characterized by intrinsic movement which is conveyed to the egg . . . in which it causes the same or a similar organisation of the atoms."

There was no thought of a penetration of the spermatozoon into the egg nor of an effect of the gross movements of the spermatozoa. Everything was due to "molecular excitation." This interpretation fitted so beautifully into the mechanistic-reductionist dogma then prevailing in the schools of Schwann, du Bois-Reymond, and Ludwig that it was almost universally adopted. One of the major proponents of this theory was the great morphologist Wilhelm His, who wanted to reduce all biological phenomena to chemistry, mathematics, and above all mechanics. "The fertilized egg contains the excitation for growth. In this excitation is the entire contents of genetic transmission from the paternal as well as from the maternal side. It is not the form which is transmitted, nor a specific form giving stuff, but merely the excitation to induce the form giving growth, not the characters themselves but the beginning of a uniform process of development" (1874: 152). It was one of the tragedies of biology and biochemistry that His greatly influenced the thinking of his nephew, F. Miescher. This was in part responsible for the fact that Miescher completely missed the significance of his own discovery of nucleic acid (see Chapter 19). As late as 1899 Jacques Loeb could write: "The ions and not the nucleins in the spermatozoon are essential to the process of fertilization."

Owing to the powerful influence of the physical interpretation of fertilization, observation after observation was ignored even though they clearly contradicted the reductionist interpretation or at least demonstrated that it provided only a partial solution. Let us review the history of the discoveries which finally provided the key to the riddle of fertilization.

The Process of Fertilization

The first question to be answered was whether it was the seminal fluid as a whole, or the spermatozoa in the fluid, that effected fertilization. As early as the 1780s L. Spallanzani performed experiments which should have given him the right answer, but did not. Male frogs dressed in little panties, permeable for some of the seminal fluid but not for the spermatozoa, were unable to fertilize the eggs of the females with which they were mated. In 1824, two Swiss physiologists, J. L. Prévost and J. B. Dumas, published the results of a series of imaginative and decisive experiments on frogs which demonstrated conclusively that the spermatozoa were the fecundating element and the seminal fluid only the vehicle. The presence of spermatozoa inside of fertilized eggs was shown by M. Barry in 1843 for rabbits and by G. Newport in 1851 for the frog, though neither author observed the actual entry of a spermatozoon into an egg or was able to determine their subsequent fate. Their observations were therefore unable to dislodge the physical theories of fertilization. In 1854 Thuret showed for *Fucus* (a common kelp) that ciliated spermatozoids surrounded the egg and entered it. Having observed this, he was even able to perform artificial fertilizations.

The first conclusive description of the process of fertilization was provided in 1856 by N. Pringsheim for the freshwater alga *Oedogonium*. He actually observed the entry of the male gamete into the female oogonium and drew from this observation the correct conclusions that the first cell (the zygote) of the new organism is formed by the fusion of the male gamete and the female egg cell and that fertilization is achieved by a single spermatozoid. With the sexuality of cryptogams at that time still controversial, this observation likewise was ignored in spite of the decisiveness of Pringsheim's demonstration. Matters were not helped when in the 1850s and 1860s several authors emphasized that the vigor of a plant was enhanced if the germ was the product of fertilization by several pollen grains.

It was Mendel (letter of Nägeli, July 3, 1870) who set out to refute "the opinion of . . . Darwin that a single pollen grain does not suffice for fertilization of the ovule." (Darwin's opinion was largely based on a misinterpretation of the work of Naudin, who actually had adopted the "one pollen grain" hypothesis.) Mendel, experimenting with *Mirabilis jalappa*, obtained 18 well-developed seeds from fertilization with single pollen grains. "The majority of plants [raised from these seeds] are just as vigorous as those derived from free self-fertilization." This settled the problem unequivocally; alas, owing to Nägeli's lack of interest, this correspondence was not published until 1905 (Correns, 1905).

Other workers at the same period elucidated the sequence of steps during the process of fertilization in plants. In 1823 J. B. Amici saw how an isolated pollen grain put out a pollen tube, and by 1846 he was able to show that an egg cell in the ovule was stimulated, after the arrival of the pollen tube, to develop into an embryo. Neither he nor Hofmeister, who in 1849 confirmed the sequence of events, had any idea just what role the pollen tube played!

Botanists had been the pioneers in cell research in the first half of the nineteenth century. But after about 1850, when adequate methods of tissue fixation became available, animal cytologists took over the leadership.[6] The absence of a cell wall in animal cells made it much easier to concentrate on the nucleus and its changes during cell division, and furthermore, fertilization was not complicated by the phenomena of pollen tube, embryosac, and so on.

After Kölliker and Gegenbaur had proven the cellular nature of both egg and spermatozoon, and with the cell being redefined as a nucleus located in protoplasm, the stage was set for the question: What happens to the male and the female nucleus, and to the male and the female protoplasm, when the spermatozoon enters the egg?

For the 25-year period after 1850 the views on fertilization can be considered to reflect two alternate theories, the contact theory and the fusion theory. The physicalists who saw in fertilization a transmission of excitation thought that the mere contact of spermatozoon and egg cell was the essence of fertilization, and if one is only interested in the initiation of the cleavage of the fertilized eggs (that is, in proximate causation), this is a conceivable explanation. Nevertheless, it required considerable credulity and a complete lack of interest in ultimate causation to accept the thesis that a mere exchange of excitations could explain the combination of paternal and maternal characteristics in the newly formed individuals. The opposition to this view was strengthened by the results of the increasingly precise microscopic analysis of the process of fertilization. This eventually led to the abandonment of the contact theory of the physicalists.

The Role of the Nucleus

Even though it was widely appreciated by that time that the spermatozoon consisted largely of nuclear material, the conclusion that the nucleus is really the crucial element in fertilization was not generally accepted. The reason for the hesitation was the common belief that the spermatozoon dissolved once it had entered the egg. To be sure, some authors had seen two nuclei in freshly fertilized eggs, one author even observing that they fused, but they had failed to draw the conclusion that one of these nuclei was nothing but the spermatozoon reconverted into a nucleus.

665

Two technical developments greatly helped in bringing about the final solution. One was the realization that neither the mammalian nor the avian egg were favorable for fertilization studies. Zoologists, as a result, tested the eggs of numerous other kinds of organisms and eventually found that, depending on the particular problem to be solved—whether fertilization, mitosis, or chromosomal continuity—various other species were far more suitable. Even more important was the rapid development of microscopic technique.

Microscopes and their lenses were constantly improved, ultimately culminating in the introduction of the oil immersion lens in 1870. Wilhelm His invented the microtome (about 1866), and in the ensuing years new types of microtomes allowed the preparation of ever thinner sections. New methods of fixing the various kinds of biological materials were also discovered, and, finally, the invention of aniline dyes made an endless array of new stains available, many of them with a highly specific affinity for certain cell components or molecules. These technical advances increased the amount of visible microscopic detail by at least an order of magnitude.

Bütschli (1873; 1875) and Auerbach (1874) in nematodes and Schneider (1873) in a platyhelminth were perhaps the first persons to observe and understand that the nucleus of the zygote was formed by the fusion of the egg nucleus and a male nucleus derived from a spermatozoon, but their somewhat casual observations were not given due attention. It was Oskar Hertwig who once and for all established the nature of fertilization. Armed with excellent equipment, he studied in the spring of 1875 the process of fertilization in the Mediterranean sea urchin *Toxopneustes* (=*Paracentrotus*) *lividus*. The eggs in this species are small, with very little yolk, and thus transparent even at high magnifications. Both egg and sperm were simple to preserve, fix, and stain. Hertwig clearly demonstrated that the second nucleus observed in the egg right after fertilization was derived from a spermatozoon. He also showed that only a single spermatozoon is involved in fertilization. Finally, he showed that the male and the female nucleus fuse into a single nucleus and give rise, by division, to all the nuclei of the developing embryo. The nucleus of the fertilized egg cell (zygote) never disappears, and there is a complete continuity between it and all the nuclei in the newly developing organism, as had already been shown by Schneider, Bütschli, and others. As Flemming phrased it in a succinct aphorism: *Omnis nucleus e nucleo.*

666

Cytological research reached a level of activity during the 1870s and early 1880s previously unknown in any branch of science: "At that time it was not uncommon for the leading cytologists, most of whom worked in German laboratories, to publish up to seven papers in a year" (Hughes, 1959: 61). Hertwig's account (1876) still contained some errors and was therefore not at once accepted by the other leading students of fertilization (see the specialist literature for the claims of van Beneden and Strasburger). However, these errors were soon corrected, and Hertwig's valid observations confirmed by the superb investigations of Hermann Fol (1845–1892). He correctly described the two maturation divisions of the egg nucleus (see below) and was able, through enormous perseverance, to observe the actual penetration of a spermatozoon into the egg. He fully confirmed that the male nucleus fuses with the female nucleus and gives rise, as Hertwig had claimed, to the nuclei of all the cells of the new organism. Fol experimentally induced simultaneous fertilization by several spermatozoa and showed that this process always results in aberrant cleavage and nonviable larvae. Fertilization is always effected by a single spermatozoon (Fol, 1879), thus confirming Mendel's observations in plants. Nearly all the students of fertilization, in animals and in plants, henceforth agreed that the fusion of the nuclei was the decisive element.

These findings decisively refuted the claims of the physicalists that a transmission of excitation is the essence of fertilization. To be sure, natural and chemically induced parthenogenesis proved that the cleavage process could be induced in eggs without fertilization. But genuine fertilization always consists of the mixing of the substance contained in the nuclei of male and female gametes. The acceptance of this conclusion was only one manifestation of the spreading rebellion against the dogmas of physicalism in the second half of the nineteenth century. The excessive and rather paralyzing preoccupation with forces, movements, and quantities was replaced by a growing recognition of the importance of form and quality. A similar emancipation took place at about the same period in chemistry (Fruton, 1972). Yet, in the 1870s the fascination with "forces" was still so great that many cytologists devoted far more attention to the "locomotory apparatus" of the cell, aster, and spindle fibers than to nucleus and chromosomes. Others recognized clearly that a mixing of substances is the true nature of fertilization and that this insight raises an entirely new set of questions, as we will see in the next section. Most of all, it encouraged, indeed it required, a study of the microstructure of cell and nucleus.

667

THE MATERIAL BASIS OF VARIATION AND INHERITANCE

When the importance of variation began to be realized toward the end of the eighteenth and the beginning of the nineteenth centuries, questions concerning its causation began to be asked. Variation may affect any aspect of an organism, any of its so-called characters, whether morphological or nonmorphological. There had to be some underlying element, some physiological or chemical element that caused this variation. At first, it was not even evident what questions one should ask and it is only through hindsight that it is possible to phrase them in a precise form.

The questions to be answered were: Is the totality of characters (of a species) controlled by a single, uniform, species-specific substance or is each character determined by a separate particle that can vary independently? Is the genetic material "soft," so that it can change gradually during the lifetime of the individual or through generations, or is it "hard," that is, completely constant and only changeable through a sudden and radical alteration, a "mutation", as it was called subsequently? How are the hereditary particles formed in the body? Do the particles contributed by father and mother during reproduction retain their integrity after fertilization, or do they fuse completely?

These were the foremost questions concerning generation and inheritance that were asked in the second half of the nineteenth century. Some of the most brilliant minds in the whole history of biology puzzled over them and were able to greatly narrow down the number of possible answers. They came up with many ingenious hypotheses, some correct and many of them not, but again and again they found themselves up against seemingly irreconcilable contradictions. How could they have possibly guessed the ultimate answer to their questions, which was found almost a century later by molecular biology? The unprecedented answer was that the genetic material is merely a "blueprint," a program of information, not at all forming part of the developing organism, and chemically quite different from it. But it was a long way until this insight was finally achieved. Let us now go back to 1850.

668

Protoplasm had only recently been named and proclaimed to be the fundamental substance of the living organism when it was postulated (Brücke, 1861) that protoplasm could not possibly perform its functions unless it was composed of "ultimate units," some sort of structural elements. Indeed, all those who thought more deeply about inheritance realized that the cell as a whole could not serve as the basic element of inheritance. After all, each gamete is only a single cell, and how could such a single unit control, as a unit, the ten thousand or more differences in the characters of an individual?

Forerunners of the Gene Concept

From the 1860s to 1900 the nature of the structural elements in cytoplasm and nucleus was the subject of endless speculation,[7] most of it with little or no experimental or observational basis. This orgy of uninhibited speculation in the period between Spencer (1864), and Weismann (1892) was quite in contrast to the attitude of the preceding thirty years (1835–1864), which had been a relatively sober period, evidently in reaction to the preceding excessively speculative period of *Naturphilosophie* (from about 1800 to 1835). During the sober period many authors wrote purely descriptive accounts of the most exciting subjects and resolutely refused to make generalizations, even when they seemed to be staring them in the face.

In another sober period (after 1895) T. H. Morgan ridiculed Weismann as "the philosopher from Freiburg" and, with reductionism and positivism on the rampage, nothing but scorn was expressed for "speculation." Some of this criticism was deserved (see below). Yet, we are now inclined to think more kindly of the authors of these speculations, for they made one invaluable contribution: Even though their answers might have been wrong, they were the people who began to ask the right questions. How can one find answers if one doesn't know what questions to ask? Wrong theories very often bring new life to a stagnant field and new observations which they engender very often lead almost automatically to their own eventual refutation.

Virtually all these authors postulated that the body, including its cells, consisted of minute particles or corpuscles. These corpuscles had to play a dual role, in ontogenetic development and in inheritance. But this was the extent of the agreement. On every other point these authors disagreed with each other. There was vast disagreement as to the nature of these particles, their role in development, and their transmission from generation to generation. Each author would coin a new term for these particles and propose a new theory of development and inheritance.[8]

669

A capacity for self-replication had to be one of the properties of these particles. And this, at once, provided for a drastic difference from inanimate nature, where self-replication does not occur. The growth of a crystal, for instance, proceeds in an entirely different manner from the growth of a cell.

Finally, for evolution to occur, these particles had to have the capacity either to change continuously ("soft" inheritance) or to be nearly constant ("hard" inheritance). Complete constancy would preclude evolution, so the particles must be able to "mutate" occasionally, that is, to change from one constant state to another. Thus, a theory of transmission genetics was complete only if it provided simultaneously an explanation for the physical nature of these particles, for their location and arrangement in the cells, for their replication, and for their mutation. In a more or less complete manner, the theories of inheritance proposed in the ninety years between 1860 and 1950 attempted to find solutions to these problems.

The first general theory of inheritance and development was the purely deductive one proposed by the philosopher Herbert Spencer (1820–1903). It was strongly influenced by the phenomenon of regeneration, for instance the ability of certain animals to regenerate a lost tail. Spencer (1864) postulated the existence of *physiological units*, intermediate in size between cells and simple organic molecules. These units were thought to be self-replicating, species-specific, identical units (within a given individual). Spencer makes seemingly contradictory statements concerning the amount of difference among the units of different individuals of the same species. Differences among siblings are attributed to differences in the number of units contained in the respective gametes received from father and mother. The form of an organism is caused by the capacity of these units to place themselves adjacent to each other in a predetermined manner, just as molecules do when forming a crystal. Furthermore, the physiological units have the capacity to respond to the environment and thus bring about an inheritance of acquired characters.

670 The next major theory of inheritance was Charles Darwin's theory of pangenesis, published in 1868 in *The Variation of Animals and Plants under Domestication.* As de Vries (1889) correctly pointed out, Darwin's theory actually consists of two components, one being the hypothesis that the hereditary qualities of an organism are represented in the germ cells by a large number of individually different invisibly small particles, the so-called *gemmules.* These multiply by division and are transmitted from the mother cell to the daughter cells during cell division.

The most important part of this hypothesis is that it postulates the existence of an enormous number of *different* kinds of gemmules—a population of gemmules, so to speak—in contrast to the essentialistically conceived physiological units of Spencer, all of which were identical in a given individual. The second component of Darwin's theory, pangenesis, will be discussed below.

In the ensuing fifteen years various other authors postulated similar hereditary particles, like the plastidules of Ellsberg (1874) and Haeckel (1876), either all identical (like Spencer's) or individually different (like Darwin's), without adding any essentially new ideas.

By far the most ambitious and speculative theory of inheritance of that period, however, was proposed by the botanist Nägeli (1884). He asserted, more clearly than most others before him, that the protoplasm of an organism consists of two components, ordinary or nutritive protoplasm and "idioplasm," a name for that part of the protoplasm which is responsible for the genetic constitution of the organism. This separation was inferred from the observation that father and mother usually contributed about equally to the genetic constitution of the offspring even though the egg may have a mass that is more than a thousand times larger than that of the spermatozoon. Consequently, only a small fraction of the egg, approximately equal to the mass of the spermatozoon, can consist of idioplasm. One would think that this conclusion would have induced Nägeli to postulate that the idioplasm is restricted to the nucleus. Curiously, this is not the case; rather, his idioplasm consists of long strands which go from cell to cell (independent of the nucleus). Each strand consists of numerous groups of molecules (micelles), the cross-section through the strand being everywhere identical. Each strand has specific properties, and bundles of such strands control the properties of cells, tissue systems, and organs. Growth consists of the elongation of these strands without any change in their consistency.

Nägeli explains the activity of the idioplasm as due to differential states of excitation of different groups of molecules within the strands. This is why he called his speculations the "mechanical-physiological theory of evolution." He concludes several hundred pages of such claims with the modest statement, "The idioplasm theory . . . permits the only possible interpretation how inheritance and phylogenetic change can take place naturally, that is mechanically" (1884: 81). Barthelmess (1952) states that he has presented Nägeli's speculations in such detail because they are perhaps the most extreme example of the speculations of the period: "Today we are appalled at such a house of cards of fantasies and are amazed at the self-assurance with which the author claims that his is the only possible solution of the great riddle of organic evolution." Nevertheless, precisely because Nägeli had speculated about every conceivable aspect of the processes of inheritance and development, he exerted an enormous influence. Indeed, during the next twenty years one does not find a paper in this field that does not quote him extensively, and usually with considerable reverence. After all, the great Nägeli was one of the dominant figures of his day. Nevertheless, almost every detail of his theory was radically wrong and almost none of it based on any known fact. One point, which must be kept in mind when evaluating Nägeli's theory of inheritance, is that he was mostly interested in species hybrids where Mendelian segregation of characters is rare or absent. This is one reason why he could not understand Mendel's findings in peas (see Chapter 16).

The one idea of Nägeli's that had a truly constructive impact was his insistence on a strict separation of idioplasm from the remaining protoplasm. At about the same time at which Nägeli published, three other authors independently came to the same conclusion and inferred, additionally, that the genetic material was contained in the nucleus (see below). It is really quite incomprehensible why Nägeli failed to recognize the nucleus as the seat of his idioplasm. After all, the role of the nucleus in fertilization was

671

widely appreciated by 1884, and the relative equality of material and paternal idioplasm, which was the original impetus for his speculations, should have also suggested to Nägeli the role of the nucleus. Haeckel, on the basis of far less evidence, had concluded as early as 1866 (I: 287–288) "that the nucleus has to take care of the inheritance of the heritable characters, while the surrounding cytoplasm is concerned with accommodation or adaptation to the environment."

By about 1884 it had become reasonably well-established and accepted that fertilization in both animals and plants consists of the fusion of a maternal and paternal germ cell (gamete), that both gametes make an equal contribution to the formation of the new zygote, and that the crucial process is the fusion of the respective nuclei. The attention next shifted to the nuclei. Are they just an amorphous mass of germ substance, as was the unspoken assumption of the epigenesists—perhaps only the fuse which ignites the developmental process in the egg cell—or is the nucleus, in spite of its minute size, highly structured with its invisible microstructure holding the key to the extraordinarily precise and specific development that follows fertilization? If one envisions the nucleus as nothing but the initiator of cell development and cell division, one will assume that the nucleus is dissolved after it has done its job, to be newly formed before each new cell division or at least before gamete formation.

Since nearly all of the cytologists in the second half of the nineteeth century had been trained as physiologists or embryologists, the emphasis was on developmental problems, and there seemed no need to postulate a continuity of the nucleus. The genetic question of the transmission of the characters of the parental generation to the daughter generation was rarely asked.

The last remnants of a belief in "free cell formation" or in the *de novo* formation of nuclei were finally eliminated in the years 1875 to 1880, when five investigators—Balbiani, van Beneden, Flemming, Schleicher, and Strasburger—were able to follow continuously all the events during cell division. They proved three important facts: (1) that the division of the nucleus starts before cell division, (2) that there is a regular sequence in the changes of the nuclear material (see below), and (3) that the basic phenomena of nuclear and cell division are the same in the plant and animal kingdoms.

It became increasingly evident that the role of the nucleus is not purely physiological (that is, serving as an initiator of cell development in a purely physical sense). Instead it is a highly structured organ presumably patterned in a highly specific way. The question as to the nature of this pattern has continued to occupy the students of cells from this point on, and the final answers are not yet in.

Progress in this research was characterized by ever finer analysis. The steps involved were the shift from the individual as a whole to the cell, from the cell as a whole to the nucleus, and now from the nucleus as a whole to its major structural elements, the chromosomes.

673

CHROMOSOMES AND THEIR ROLE

Depending on the interests of the investigator, nuclear division was interpreted in two entirely different ways in the 25 years prior to the birth of genetics (1900).[9] For those primarily interested in development, the big question was; How can the undifferentiated egg cell, by simple division, give rise to the differentiated cells of nerve tissue, glandular tissue, epidermis, and the hundreds of other kinds of tissues recognized by histologists and physiologists? This consideration, for instance, dominated Weismann's theories. This group of investigators was primarily interested in proximate causes.

Those relatively few authors who were interested in transmission genetics were laying the groundwork that eventually led to the question, What are the mechanisms that effect a division of the nuclear material in such a way that exactly equal halves are conveyed to the daughter cells of a dividing cell? The questions the two groups of investigators asked were thus quite different. The embryologists asked, How can we interpret cell division as a mechanism that explains the differentiation of the phenotype? The transmission geneticists were, as we would now call it, interested in the accurate perpetuation of the genotype, that is, in the problem of inheritance. The interpretation of the transmission geneticists left the problem of differentiation entirely unsolved, while the answers proposed by the developmental geneticists raised some formidable, and, as later became apparent, insoluble difficulties for the interpretation of genetic transmission.

Evidently, not even the first step toward a resolution of the conflict could be made until there was a better understanding of what went on inside the nucleus during cell division, a process which I shall not attempt to describe in full detail even though the mechanics of cell division (mitosis) are among the most marvelous processes known in the living world.

Mitosis

674

A. Trembley (1710–1784) in the 1740s was the first person to describe cell division of a sort, the fission of a protozoan (Baker, 1952).[10] Cell division was also found, in the second half of the eighteenth century, in diatoms and other algae and was studied rather intensively by Ehrenberg in the 1830s in various protozoans. From the 1840s on, the process by which a somatic cell divides (named *mitosis* by Flemming in 1882) was described and illustrated more and more frequently (Wilson, 1896; Hughes, 1959).

When a cell divides, its nucleus also divides, and this was eventually recognized as the most important aspect of cell division. At first it was believed that the nucleus was simply filled with granular material which, during cell division, was equally distributed to the two daughter nuclei (direct cell division). However, as the quality of the optical equipment and of the microscopic techniques (such as staining) improved and hence the precision with which every phase of cell (and nuclear) division could be studied, the simple picture had to be revised. The nucleus at certain stages of mitosis seemed to be filled with threads, ribbons, or bands which, since they were heavily stained, were said by Flemming (1879) to consist of *chromatin*. Since the word *chromosome* was not proposed until 1888 (Waldeyer), each author used different terminologies and descriptions. What presumably were chromosomes was seen by Remak (1841), Nägeli (1842), Derbès (1847), Reichert (1847), Hofmeister (1848, 1849), and Krohn (1852), all of whom saw mitotic figures and sometimes provided good illustrations of metaphase plates. One must remember that these papers were published in the period of free cell formation, in which many authors thought that the nucleus dissolves and two new nuclei form anew from the cell sap.

The first person who observed the complexity of the reorganization of the nucleus was the French zoologist E. G. Balbiani (1825–1899). As early as 1861, in beautiful preparations, he pictured in exquisite detail all the stages of mitosis in a protozoan. Unhappily, however, Balbiani completely misidentified what he saw. Not realizing that each protozoan is but a single cell, he identified the nucleus as a testis and the chromosomes as spermatozoa.

As a result, this pioneering work was without any subsequent influence. Direct nuclear division was accepted by the majority of authors until the mid-1870s.

The technical advances in microscopy permitted the demonstration that the nucleus (and its chromatic contents) do not disappear between cell divisions but are maintained during the resting stage, even though in a different form. Furthermore, it permitted an accurate description of three major (and several minor) stages of mitosis, subsequently called prophase, metaphase, and anaphase (see Fig. 1).

675

The resting nucleus (between cell divisions) does not stain well, but there were indications that much of the nuclear material is organized in one or several thin threads, or a network of threads. As cell division approaches, the membrane around the nucleus disappears, the chromatic threads tend to condense and become more easily stained by the appropriate dyes. Eventually this material contracts into a few heavily staining bands, designated *chromosomes*. Every species normally has in each cell a constant number of such chromosomes—in man there are forty-six—and these arrange themselves during nuclear division in an "equatorial plate." It is at this stage (metaphase) that each chromosome seems to split into two. At first it was reported that they split transversely, and this error created considerable confusion. Ultimately it was clearly observed (Flemming, 1879) that they split longitudinally; the split evidently occurs before metaphase, that is, when the staining material (chromatin) is still in the uncondensed stage and almost impossible to observe. At the next stage the two halves of the split chromosomes separate from each other and move to the opposite poles of the dividing nucleus. Around each bundle of chromosomes (at the two poles) a new nuclear membrane is formed, and the chromosomes revert back to their threadlike and largely invisible resting condition.

It took a good many years and the labors of numerous investigators until these various stages of mitosis were recognized and correctly interpreted. The first reasonably accurate descriptions of mitosis were given in 1873 by three investigators: Schneider, Bütschli, and Fol. The importance of this process was recognized at once, and became the subject of furious activity, with the zoologist van Beneden and the botanist Strasburger making particularly important contributions. Eight years later a reviewer listed 194 papers (by 86 authors) on cell division and related topics, published from 1874 through 1878. No one, however, contributed as much to the correct description and interpretation of mi-

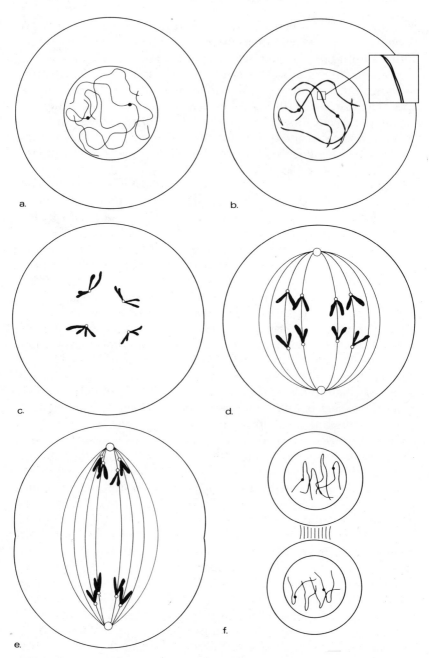

Figure 1. Stages of mitosis. (a) Early prophase. (b) Late prophase; each chromosome splits, apparently in early prophase, into two chromatids. (c) Polar view of metaphase. (d) Early, (e) late anaphase. (f) Telophase. The small circles in (a), (b), and (f) represent the nucleoli.

tosis as Flemming, who in 1882 provided a superb summary of the state of the field. Observations of mitosis in plants revealed that it proceeds exactly as in animal cells. This was one more piece of evidence for the unity of cellular processes in animals and plants. (It was more than half a century later before the pronounced difference between the cell division of eukaryotes and prokaryotes was discovered.)

Every new observation confirmed the existence of this exceedingly complex process of cell division. Why is such complexity necessary? Why do not cell and nucleus simply divide in half, as Remak had thought? This is the question which W. Roux asked himself in 1883. In an age in which only proximate causes were studied, culminating in Nägeli's mechanico-physiological speculations, Roux wanted to know about the ultimate causation. He ventured to ask a why-question: Why, he asked, do we have such a complex process if a simple division would do the job? He answers it by saying that a simple direct division of the nucleus would be quite sufficient if the nuclear material was homogeneous. However, if the nuclear material is heterogeneous, if it consists of innumerable particles each of which has a different genetic capactiy, then there is only one possible method by which this material can be divided in such a way that each particle is represented in both daughter cells. This method is to arrange all particles serially, like beads on a string, and then to split this string longitudinally, "so that each chromatin globule is divided into two halves, so that from each string of globules two adjacently placed strings result"

677

In the preceding years it had been repeatedly suggested (Balbiani, 1881; Strasburger, 1882) that in the resting nucleus all the chromatin is arranged as a single long thread. This is the observation on which Roux's hypothesis rested: "The mitotic figures . . . are mechanisms which allow the nucleus to be divided not only quantitatively but also according to the mass and nature of its individual qualities. The essential process in nuclear division is the halving of each of the maternal corpuscles; all other processes serve the object to transfer one of the daughter corpuscles to the center of one of the daughter cells, and the other to the center of the other daughter cell." This process guarantees that both daughter cells are identical, not only quantitatively but also qualitatively. They are identical in every respect.

This is Roux's thesis throughout his entire article of 19 pages, but then he gets cold feet. In a single sentence he capitulates to the possibility of unequal division: "Since the second cleavage division [of the frog egg] determines the front end and rear end of the embryo and, since one must assume that the different development of the anterior and posterior part is connected with an inequality of the material, it becomes probable that the nuclear material is divided into qualitative parts during the second division" (1883: 15). This contradicts his principal thesis since there is no difference whatsoever between the mitosis of the first and the second division.

678   A mechanism for equal division, Roux's major thesis, is of course precisely the modern interpretation of mitosis, an interpretation that was curiously ignored by all those, like Weismann, who in the ensuing years attempted to explain differentiation by an unequal distribution of nuclear material into daughter cells. Yet, as Wilson said (1896: 306), "Not a single visible phenomenon of cell-division gives even a remote suggestion of qualitative division. All the facts, on the contrary, indicate that the division of the chromatin is carried out with the most exact equality."

Roux's theory was frankly a speculation, but it was a speculation quite different from the speculations of Nägeli or Jacques Loeb. Roux asked a why-question in the tradition of Harvey, whose curiosity over the meaning of valves in the veins was a major contribution to his discovery of the circulation of blood. Indeed, Roux asked implicitly, What is the selective value of this complicated process? Nägeli and Loeb did not ask why-questions; rather, they attempted to interpret biological phenomena in a reductionist manner, in terms of physics and chemistry, and that kind of speculation was at that time far more respectable than the "teleological" or "Aristotelian" (as it was referred to) approach of Roux. Nevertheless, Roux's hypothesis, like Hertwig's theory of fertilization, was further evidence of the gradual emancipation of biology from purely physicalist interpretations.

The story of the nucleus was completed by 1880 because every cytological investigation had confirmed Flemming's aphorism. The center of attention, from this point on, shifted to the chromosomes. What do the chromosomes do during cell division?

From the Nucleus to the Chromosomes

In 1883 the Belgian cytologist van Beneden published a brilliant analysis of fertilization in the horse threadworm *Ascaris bivalens,* an unusually suitable organism because it has only four chromosomes. He was able to show that the gametes have only two chromosomes and that the fertilizing male nucleus did not fuse with the female nucleus in such a way as to produce a blending of their nuclear substance, but that the two chromosomes of the male nucleus merely join the two chromosomes of the egg nucleus, forming the new nucleus of the zygote with four chromosomes (which he calls chromatic globules). At the first cleavage division of the fertilized egg (zygote), each of the four chromosomes divides longitudinally as in every other mitotic division, and each daughter cell receives the same two paternal and two maternal chromosomes as had participated in fertilization (see Chapter 17).

Although van Beneden observed that exactly one half of the nuclear material of the new individual is derived from the father, the other half from the mother, he did not establish any connection between his observation and inheritance. Not being a theorist, he did not draw the obvious conclusions from his brilliant cytological demonstration. This was done almost simultaneously and independently of each other by four German biologists, although Galton (1876) had preceded all of them.[11]

Weismann (1883), in a penetrating analysis of the problem of inheritance, concluded that the nuclear material was the hereditary substance, which he designated by the name "germ plasm" (one year earlier than Nägeli's "idioplasm"): "Heredity is brought about by the transference from one generation to another, of a substance [germ plasm] with a definite chemical, and above all, molecular constitution."[12] In 1884 the leading animal cytologist, Oskar Hertwig, and the leading plant cytologist, E. Strasburger, reviewing the extensive literature of the past ten years, were able to demonstrate conclusively that no other interpretation was possible but that the nucleus is the carrier of inheritance. All of them gave Haeckel credit for having been the first to have postulated this role of the nucleus. Hertwig and Kölliker (1885) went one step further by asserting that the truly effective material in the nucleus was nothing but the peculiar chemical isolated by Miescher from cell nuclei, and named by him "nuclein." This, Kölliker declared emphatically, must be the material basis of inheritance.

Eventually it was even possible to demonstrate experimentally that the nucleus is the seat of the genetic material. Boveri (1889), in a series of ingenious investigations, fractured eggs of a sea urchin by vigorous shaking, and found that large nonnucleated egg fragments could be fertilized with sperm of a different, morphologically quite distinct species (and genus) of sea urchin. Even though this haploid hybrid egg fragment had only the paternal chromosome set in the maternal cytoplasm, it developed normally, but the resulting larva had chiefly paternal characteristics. Control larvae produced through simultaneous fertilization by the spermatozoa of two different species had intermediate morphological characters. Here was clear-cut experimental proof that the nucleus determines the characteristics of an organism.[13]

680

The study of the behavior of chromosomes during fertilization clarified the problem of the "fusion" of the gametes. Yes, there is a fusion of the gametes, indeed even a fusion of the nuclei of the gametes, but there is no fusion of the paternal and maternal chromosomes. This new insight opened up an entirely new field of research, the question of the joint or independent action of the paternal and maternal genetic material and the question of the feasibility of blending inheritance (see chapter 17).

It is necessary to look back to Schleiden's theory (1838) that each nucleus originates *de novo* in each new cell by a process of crystallization in order to appreciate the enormous advance made in these 45 years. Not only was the continuity of the nuclei now firmly established, but also established was the fact that the crucial aspect of the production of a new individual is the combining of the genetic elements in the nuclei of the male and female gametes. From 1884 on, the thesis that the nuclei play the predominant role in inheritance was no longer challenged, even though some scientists postulated for another fifty years or so that important components of the genetic material might also reside in the cytoplasm. What remained unclear was the connection between nucleus and cytoplasm. Was the nucleus the dominant element and the cytoplasm a two-way street enabling the cytoplasm to return material to the nucleus and thereby affect its heritable qualities?

The next fifteen years were years of speculation by Nägeli, de Vries, Weismann, and others. The cytological understanding was consolidated but no major new insights were added. Indeed, the knowledge of chromosomes and cell nucleus had reached a remarkable degree of maturity by the 1890s. Wilson's magistral book, *The Cell in Development and Inheritance* (1896; 1900), is convincing evidence for this. The time had come when the venerable problem of the inheritance of acquired characters had to be reviewed in the light of the new cytological knowledge.

16 ✍ The nature of inheritance

INTUITIVELY all students of nature felt the existence of some conflict or contradiction between the facts of inheritance ("She is exactly like her mother!") and those of variation. Inheritance implies continuity and constancy; variation implies change and divergence. When a breeder crossed animals or plants, he often encountered unexpected variants among the offspring. Even when one compared siblings with each other, one often found amazing variation. Eventually it became an important question, Where does the new variation come from? But it was not until Darwin had established his theory of natural selection that the source of variation became a key problem in biology. Natural selection can be effective only if there is an abundant supply of variation, a supply that must be forever renewable. How can this be reconciled with a belief in the constancy of inheritance?

The answer traditionally had been that inheritance is not necessarily constant, that not all of it is "hard." To be sure, in some characteristics an individual might closely resemble his father, or mother, or even a grandparent or earlier ancestor. All animal breeding is based on the fact that such hard inheritance exists. But, if inheritance were completely hard, there could be no variation. Hence, it was reasoned that there may be two, not mutually exclusive, sources of variation. Either some inheritance is soft, that is, susceptible to various influences, or else the genetic material is hard but has the capacity to generate new variation occasionally. Throughout the nineteenth and first third of the twentieth century, the question of soft inheritance and of the source of genetic variation remained controversial.

DARWIN AND VARIATION

One of the two cornerstones of Darwin's theory of natural selection was the postulate of the availability of an unlimited supply of variation. Each individual is unique and different from every other one: "These individual differences are highly important for us, as they afford materials for natural selection to accumulate" (*Origin:* 45). But where does this variation come from? What is its source? This question puzzled Darwin all his life. What a large role variation played in Darwin's thinking is documented by the fact that he devoted a 900–page work to the variation of animals and plants under domestication (1868). He had planned to write a corresponding work on variation in nature but never did so, being overwhelmed by the abundance of material. His enormous information on variation was condensed into the first two chapters (59 pages) of the *Origin*. Recent writers on Darwin (such as Ghiselin, 1969; Vorzimmer, 1970; authors of several journal articles) have fully appreciated the importance of variation. Inheritance, as such, and its laws were of much less immediate interest to Darwin than variation and its causation.

682

Even at the present time variation and its causation are not fully understood. In the middle of the nineteenth century, the subject was enveloped in great confusion. How difficult this subject is becomes apparent when one realizes how bewildered even Darwin was, who had been preoccupied with variation all his life and who had thought deeply about it. In retrospect it is obvious that much of it could not be clarified until after the rise of genetics (for example, the distinction between genotype and phenotype). However, some of the confusion was also due to the failure to apply population thinking consistently.

The most fascinating aspect of Darwin's confusions and misconceptions concerning variation is that they did not prevent him from promoting a perfectly valid, indeed a brilliant, theory of evolution. Only two aspects of variation were important to Darwin: (1) that it was at all times abundantly available, and (2) that it was reasonably hard. Instead of wasting his time and energy on problems that were insoluble at his time, Darwin treated variation in much of his work as a "black box." It was forever present and could be used in the theory of natural selection. But the investigation of the contents of the box, that is, the causes of the variation, occupied Darwin only occasionally and with little success (as in his pangenesis theory; see below). Fortunately for the solution of the major problems with which Darwin was concerned (for example, the success of individuals in the struggle for existence), a study of the contents of the box was unnecessary. It could be postponed until more auspicious times. One of the secrets of success in science is to select "soluble" problems (Medawar, 1967).

There were two aspects of variation that caused Darwin particular difficulties.

(1) The distinction between intrapopulational and interpopulational variation. Darwin, unfortunately, never made a clear distinction between individual and geographical varieties, and this, particularly after the 1840s, bedeviled his discussions of speciation (Mayr, 1959a; Kottler, 1978; Sulloway, 1979; see also Part II). It also affected his discussion of variability under domestication: "Individuals . . . of our older cultivated plants and animals . . . generally differ much more from each other than do the individuals of any one species or variety in a state of nature" (*Origin:* 7). Actually, individual (that is, intrapopulational) variation in strains, breeds, and races of domestic animals and plants is often extraordinarily low, and to have uniform breeds is one of the ideals of the breeder. What Darwin had primarily in mind was the range of variation within domesticated species as a whole, that is, interpopulational variation. However, where stabilizing selection is not exercised by the breeder, individual variation may increase.

(2) The belief that there are two entirely different kinds of intrapopulational variation.[1] Breeders and naturalists—in fact, anybody having to do with variation—believed until well into the first quarter of the twentieth century that there are two kinds of variation, discontinuous and continuous (also called individual) variation. Discontinuous variation is represented by all striking deviations from the "type," that is, by any variant not connected with the "normal type" by a graded series of intermediates; albinos would be one example of discontinuous variation. For an essentialist, anything new can originate only by a drastic deviation from the type—a mutation or saltation—and discontinuous variation therefore played an important role in the evolutionary theories of essentialists (see Part II).

Even though Darwin acknowledged the existence of discontinuous variation as a separate category, he considered it evolutionarily unimportant. In contrast to all earlier writers, he emphasized the universal prevalence and biological significance of individual or continuous variation. From where did Darwin derive this crucially important insight? He owed it primarily to the study of the works of the animal breeders who, from Bakewell and Sebright on, had stressed the uniqueness of individuals, a fact which made selection and an improvement of breeds possible. This lesson was reenforced by Darwin's taxonomic studies, where he

683

found, like other taxonomists before him, that no two individuals were entirely identical when examined carefully. It is this individual variation, as Darwin never failed to emphasize, which provides the raw material for selection and thus for evolutionary change. The nature of this continuous variation about which Darwin himself was very uncertain remained controversial until the genetic researches of Nilsson-Ehle, East, Baur, Castle, Fisher, and other geneticists settled the issue after 1910 (see Chapter 17).

684 Those who disagreed with Darwin raised two objections in particular. First, they claimed, up to Johannsen and later, that such continuous variation merely reflects the plasticity of the phenotype but is not heritable. Their second claim, going back to Lyell and earlier, was that such variation is severely limited, never being able to transgress the limits of the "type." Both of these claims were subsequently refuted, and the paramount importance of individual variation is today unchallenged. Furthermore, as we shall presently see, genetics eventually showed that there is no real difference between the genetic basis of continuous and discontinuous variation.

Merely to assert that there is abundant variation was not sufficient for Darwin. For him, a true child of his times, variability had to be ascribed to an identifiable cause. He did not believe in "spontaneous" variation. "I do not believe that variability is an inherent and necessary contingency, under all circumstances, with all organic beings, as some authors have thought" (*Origin:* 43). The most important causes of variation for Darwin were various influences on the reproductive system of the parents, particularly those caused by shocks or severe changes of the environment. These influences, he believed, do not produce specific variants or directional variations; rather, they simply increase the variability of the offspring, providing increased scope for the exercise of natural selection.

Occasionally, Darwin admits he had somewhat carelessly spoken of variation as having "been due to chance. This, of course, is a wholly incorrect expression, but it serves to acknowledge plainly our ignorance of the cause of each particular variation" (*Origin:* 131). Darwin's friend Hooker saw much more clearly that there was no need for a causal connection between special conditions and a particular type of variation occurring under these conditions. "I incline to attribute the smallest [individual] variation to the inherent tendency to vary; a principle wholly independent of physical conditions" (Hooker to Darwin, March 17, 1862; 1918, I: 37). For Darwin an "inherent tendency to vary" probably smacked of the same spirit as Lamarck's "inherent tendency toward perfection." To accept genetic variability simply as another manifestation of the imperfection of the organic world was not sufficiently "causal" for Darwin. Both Darwin and Hooker were rather vague on the actual subject of their disagreement. Were they referring to variation as a process or to variation as the product of this process? Also, what was meant by chance? In an age when only those processes were scientifically respectable that obeyed a "law," it was not admissible to acknowledge stochastic perturbations.

685

In spite of all his disclaimers, Darwin's opponents pounced upon his statements that variation was due to chance. In various forms ("Are mutations random?") this argument has continued virtually into our day. What the opponents of Darwinism failed to grasp is that Darwin and his followers never questioned the strict physico-chemical causation of all variations; they simply denied that it had a teleological component. Genetic variation is not a specific response to an adaptive need.

To sort out the gradual change in Darwin's thinking about the causes of variation is particularly difficult because it is correlated with a simultaneous change in his thinking about the causation of adaptation (natural selection) and about the nature of inheritance (whether soft or hard). Anyone who does not believe in natural selection is forced to rely on soft inheritance and, furthermore, is forced to postulate the existence of adaptive responses to environmental needs. Once Darwin had adopted natural selection as the mechanism of evolutionary change, all he needed was a process (or processes) that generates variability. However, if inheritance is normally hard (and, as we shall see, Darwin had reached this conclusion)—that is, if the characters of the parent are normally transmitted unchanged to the offspring—then it will require some special stimulus to make them vary. And since the offspring is a product of the reproductive system, that stimulus somehow must affect the reproductive system. This chain of arguments was completely logical.

And it seemed to Darwin that he had the evidence to support his argument. Seeing that there are in nature both highly variable as well as very uniform species, he concluded that there must be factors that could affect the variability of species. At this point he remembered domesticated species, like the races of dogs or the varieties of cabbage, each presumed to be derived from a single ancestral species. And he observed, that "the most favourable conditions for variation seem to be when organic beings are bred for many generations under domestication" (1844: 91). And what particular aspect of domestication is it that is responsible for this increase in variability? What is it that causes the normally so stable genetic constitution to vary? It is "simply due to our domestic productions having been raised under conditions of life not so uniform as, and somewhat different from those [of the] parent species" (*Origin:* 7). Darwin does not postulate, as might be read into this statement, a direct induction of new characters as a result of life in a different environment but merely that some factor, perhaps "an excess of food," increases genetic variability as such. And Darwin believes, furthermore, that this increased variability is due to the fact that male and female reproductive systems appear "to be far more susceptible than any other part of the organisation to the action of any change in the conditions of life" (*Origin:* 8; see also similar statements elsewhere in Darwin's writings).

686

The crucial difference between this interpretation and that of the adherents of soft inheritance is that Darwin's variation was not given any special direction by the environment or by any needs of the organism. Whatever directional trend is observed in evolution has a different cause: "It is the steady accumulation through natural selection . . . that gives rise to all the more important modifications of structure" (*Origin:* 170). Implicit in many statements scattered through Darwin's writings is the conviction that the genetic material is usually not affected by the environment. In this belief Darwin was virtually alone prior to the 1870s.

As is evident from his notebooks, Darwin had struggled with the problem of inheritance from the time he first started thinking about evolution, but he said relatively little about this subject in the *Origin.* What he did believe was that most individual variation is heritable. "Perhaps the correct way of viewing the whole subject, would be, to look at the inheritance of every character whatever as the rule, and non-inheritance as the anomaly" (*Origin:* 13). Obviously, natural selection cannot favor any nonheritable character, hence "any variation which is not inherited is unimportant for us" (*Origin:* 12). It was not until the *Variation of Animals and Plants under Domestication* (1868) that Darwin published his ideas on inheritance, in the form of his hypothesis of pangenesis. This I will take up presently, but in order to facilitate the analysis I want to mention that there are two components of Darwin's theory of inheritance on which the historians of science have not yet come to a complete agreement. The first is whether Darwin believed in blending or in particulate inheritance. Since the nature of this argument cannot be fully illuminated, except in terms of Mendelian inheritance, we will postpone its discussion until after the treatment of the rediscovery of Mendel (see Chapter 17). The second disagreement concerns the question to what extent Darwin believed in any form of soft inheritance and, in particular, in an inheritance of acquired characters.

687

SOFT OR HARD INHERITANCE

The belief that either the environment or "use versus disuse" (or both) would affect the heritable qualities of characters was universally held almost to the end of the nineteenth century (Zirkle, 1946) and by numerous biologists well into the twentieth century (Mayr and Provine, 1980). This belief is usually referred to by the words "inheritance of acquired characters," but this terminology is imprecise because the belief usually also included the postulate of a modifiability of the genetic material by general climatic and other environmental conditions (Geoffroyism) or by nutrition directly, without peripheral (phenotypic) characters necessarily serving as intermediaries. A belief in an effect on the offspring of various experiences or encounters of the pregnant mother is recorded in the Bible (Moses 1: 30) and was accepted in the teratological literature as a major cause of the birth of monsters. In these cases a nongenetic modification of the phenotype *is* often the correct explanation.

The basic concept underlying this belief is that the genetic material itself is pliable, or "soft." For this theory it does not matter whether the genetic material changes slowly or fast, nor whether it changes directly or via "acquired characters"; what matters is that the genetic material is believed not to be constant, not unchangeable, not "hard." Curiously, soft inheritance was so universally accepted, it was considered so axiomatic, that it was not until after 1850 that even the first attempts were made to justify it and to work out its mechanisms. Darwin, Spencer, and Haeckel were among the first who tried to do so (Churchill, 1976). Except for a few ignored pioneers, the first suggestions of the possibility of exclusively hard inheritance were made even later (see below).

Among the neo-Lamarckians at the end of the nineteenth century it was assumed that Lamarck was the father of the concept of "inheritance of acquired characters." Actually it was the standard concept in the eighteenth century, held by all outstanding biologists of the period, including Buffon and Linnaeus. Blumenbach, for instance, believed that the dark-skinned human races were derived from the light-skinned races by the action of the strong sunlight of the tropics on the liver. This resulted in a blackening of the bile, which in turn caused a pigment deposit in the skin. Those with more experience with human races than Blumenbach had little trouble in refuting him by pointing out, as did Herder, that white people living in the tropics as well as their children did not become black skinned, while the descendants of the African slaves remained dark skinned after many generations of life in the temperate zone; whenever a noticeable change of skin color was observed, it was due to the interbreeding of races. No one, however, demolished the claim of an influence of climate on the racial characteristics of man more thoroughly than Prichard (1813). He concluded "that the colour acquired by the parent on exposure to heat, is not imparted to his offspring, and has consequently no share in producing natural varieties." The same impotence of the climate could be demonstrated for animals. Species which were bred in menageries or zoological parks for many generations did not change their appearance in the slightest. In spite of such early indications of the invalidity of this concept, soft inheritance was tenaciously held onto by most authors. The only concession its adherents made to the opposing evidence was to assume that there is both hard and soft inheritance.

The acceptance of an unchanging essence, the basic credo of the essentialists, would seem to necessitate a belief in hard inheritance. It puzzles me therefore how universally the essentialists at that period were able to reconcile soft inheritance with the concept of an unchanging essence. They escaped the dilemma by defining all characters subject to soft inheritance as "accidental," their variation not affecting the essence. Louis Agassiz compromised by claiming that it was part of the intrinsic potential of the essence to be able to change, to respond to the environment, and even to be "prophetic." More consistent essentialists looked for instances (like migrating human races) where a change of climate was without permanent effect. C. F. Wolff, for instance, noticed with satisfaction that even though certain plants had changed dramatically when transplanted from western Russia (St. Petersburg) to Siberia, their descendants nevertheless returned to their original form when taken back to St. Petersburg. This proved to him that the influence of the external factors was unable to penetrate to the essential organization of the organism (Raikow, 1947; 1952). Here was a possible basis for a theory of hard inheritance, but it was not followed up.

The question of the existence of soft inheritance became of crucial importance as soon as the evolutionary theory was proclaimed. Was evolution due to an inheritance of acquired characters, as believed by Lamarck? The transplantation of plants into other environments, particularly from the lowlands to the mountains and back, became a favorite method after 1859 to test the influence of the environment, by such experimenters as Bonnier and Kerner. Yet, the method was not very satisfactory, because individuals of most lowland species were unable to tolerate the alpine climate, and when species were employed that had alpine ecotypes, it required extreme precautions to prevent a mixing up of the transplants and the local ecotypes; hence, Kerner's conclusion, "In no instance was any permanent or hereditary modification in form or color observed," had little influence. The matter was finally settled by the work of Clausen, Hiesey, and Keck from the 1930s on, but by that time a refutation of an inheritance of acquired characters was no longer necessary.

Darwin and Soft Inheritance

Throughout his life Darwin believed in both soft and hard inheritance, changing his opinion only on the relative importance of the two. In his early notebooks soft inheritance clearly dominated. He even recorded the possibility, though he was not fully convinced, of a paternal influence on later pregnancies in interracial crosses (B: 32, 181; C: 152) as well as the possibility that "wishing of the parents" might affect the offspring (B: 219). Most of his statements are sufficiently vague so that they can be equally interpreted as accounts of nongenetic changes or as effects on later generations (B: 3, 4; C: 68, 69, 70, 195, 220). Darwin, even at that time, definitely denied that drastic bodily changes such as mutilations could have genetic consequences (C: 65–66, 83; D: 18, 112).

In the *Origin*, more than twenty years later, Darwin no longer mentions any dubious folklore of the breeders and, having adopted natural selection as the causative agent in evolutionary change, relies in the main on hard inheritance. A careful reading of the work reveals, however, that Darwin still cites occasionally evidence seemingly in favor of soft inheritance. He accepted three potential sources of such variation. The first one, an effect of changes in the environment that induce increased variability by way of the reproductive system, can be reconciled with hard inheritance. The other two demand a belief in soft inheritance: the direct effect of the environment, and the effect of use and disuse.

690

Direct Effect of the Environment

The environment was one of the factors considered by Darwin as a possible cause of variation. Repeatedly in the *Origin* he states that "climate, food, etc. probably produce some slight and direct effect" (p. 85; similar statements are made on pp. 15, 29, 43, and 132). Darwin frequently commented on the great number as well as great variability of races of domestic animals and cultivated plants. He ascribed this increased variability to changed and particularly favorable conditions of life. Actually, in cultivated plants hybridization is the main source of increased variability (as Darwin was aware, to some extent), while in some breeds of domestic animals, on the contrary, it was the destruction of well-balanced epistatic systems by intensive inbreeding that was responsible (Lerner, 1954). Equally often, Darwin stresses that such direct effects "are of quite subordinate importance to the effects of natural selection" (p. 209). This unimportance of the "conditions of life" in producing new variation is also recorded on pages 10 and 134. Darwin actually expressed himself much more clearly in a letter to Hooker (*L.L.D.*, II: 274): "My conclusion is that external conditions do extremely little, except in causing mere variability. This mere variability (causing the child not closely to resemble its parents) I look at as very different from the formation of a marked variety or new species . . . The formation of a strong variety or species I look at as almost wholly due to the selection of what may be incorrectly called chance variations or variability." As Darwin made no clear distinction between genotype and phenotype, it is impossible, in virtually all the cases listed by him, to say whether he considered the variation induced by the environment to have been genetic or not.

There was perhaps no other period during which Darwin rated the direct influence of the environment as low as that during the writing of the *Origin*. But 1862, after having completed the first volume of *Variation,* he wrote to Hooker, "My present work is leading me to believe rather more in the direct action of physical condition"; and by 1878 he admitted, "I probably under-rated [the] power [of external conditions] in the earlier editions of the *Origin* (in Vorzimmer, 1970: 264). To Galton he wrote in 1875, "Every year I come to attribute more and more to such agency [modification 'by use and disuse during the life of the individual']."

691

The Effect of Use and Disuse

Among all the phenomena that Darwin considered as evidence in favor of soft inheritance, none was as important to him as the effect of use and disuse. It was the study of domestic animals which led him to adopt this belief. "There can be little doubt that use in our domestic animals strengthens and enlarges certain parts and disuse diminishes them; and that such modifications are inherited" (*Origin:* 134). Darwin was so strongly convinced of the importance of this factor that he devoted to its discussion a whole section (pp. 134–139) of chapter V of the *Origin.* As examples he discusses the reduction of wings in flightless birds, the loss of anterior tarsi in dung beetles, winglessness in beetles of Madeira (in part), the reduction of eyes in moles and other burrowing mammals, and the loss of eyes and pigment in cave animals. Regarding rudimentary organs in general, Darwin states, "I believe that disuse has been the main agency" (p. 454) in their production. The importance he attributes to this factor is indicated by the frequency with which he invokes it in the *Origin* as an agent of evolution (for instance on pp. 11, 43, 134, 135, 136, 137, 168, 447, 454, 472, 473, 479, and 480). Use and disuse, of course, is of importance only if one believes in an inheritance of acquired characters. This Darwin affirms repeatedly. He describes how the constant milking of cows leads to an inherited increase in the size of the udder. Darwin is quite positive: "Modifications [caused by use and disuse] are inherited" (p. 134).

The modern evolutionist has no difficulty in explaining all alleged effects of disuse as due to a relaxation of stabilizing selection, often reinforced by counterselection forces. Although Darwin realized that selection made a contribution to the production of rudimentary organs (p. 143), he was not prepared to go to the extreme of relying entirely on selection to explain rudimentation.

His thinking was still so much conditioned by pre-Darwinian

concepts that he sometimes interpreted observations in terms of use and disuse, which to us seem "obviously" due to natural selection. By careful measurements, Darwin found, for instance, "in the domestic duck that the bones of the wing weigh less and the bones of the leg more, in proportion to the whole skeleton, than do the same bones in the wild duck" (p. 11). Darwin, curiously, does not ascribe this to selection during domestication but assumes the modification to be, in part, nongenetic, corresponding to differences in plants growing in different soils, and in part due to "the domestic duck flying much less, and walking more than its wild parent" (p. 11). His experience with plants and plant breeders had persuaded Darwin to accept a far greater phenotypic plasticity in animals than is actually found.

692

There is one other line of evidence revealing Darwin's belief in soft inheritance. He thought that the genetic basis of a character would be strengthened if the organism was exposed to the same circumstances for a long time or if the structure was used steadily: "Variety when long in blood gets stronger and stronger" (C: 136). When considering whether "some actions become hereditary and instinctive and not others," he concludes, "Therefore it can only be those actions which many successive generations are impelled to do in the same way" (C: 171), and, "The longer a thing is in the blood the more persistent any amount of change and shorter time less so" (D: 17; also D: 13). It took another hundred years before it was recognized that this is the result of stabilizing selection.

The conclusion he drew from this was that the older a domestic breed or a geographical variety, the stronger was its influence in crosses. He refers to this as "Yarrell's Law," after William Yarrell, one of his animal breeder friends, from whom Darwin apparently had this generalization (C: 1, 121; D 7–8, 91). He admits however that this law does not always work (E: 35).

On the other hand, a character would be weakened if exposed to adverse conditions. He believed that "if we . . . were to cultivate, during many generations, the several races of the cabbage in very poor soil . . . that they would to a large extent or even wholly, revert to the wild aboriginal stock" (*Origin:* 15). Ideas like this were widely held at that period among animal and plant breeders.[2]

Several recent historians have accepted Darlington's claim (1959) that Darwin had recognized nothing but hard inheritance in the first edition (1859) of the *Origin* but "fell back on the assumption that acquired characters were inherited" after reading in 1867 Jenkin's critical review. This claim is erroneous in every detail, as was shown by Vorzimmer (1963; 1970) and others, and by the preceding analysis of Darwin's views on soft inheritance. Admittedly, Darwin in his later years conceded a little more influence to soft inheritance than he had in 1859, but it never became a major component of his interpretation. Whenever he compared the contribution to evolutionary change made by the inheritance of acquired characters to that made by natural selection, he always made it quite clear that he continued to consider selection as the decisively important factor.

693

Darwin's Theory of Pangenesis

Chapter XXVII of Darwin's *Variation of Animals and Plants under Domestication* (1868) is devoted to his "provisional hypothesis of pangenesis," as it is called in the chapter heading.[3] Darwin justifies his hypothesis because "it may be serviceable by bringing together a multitude of facts which are at present left disconnected by any efficient cause" (1868: 357). And in a chapter subheading Darwin lists "the facts to be connected under a single point of view, namely, the various kinds of reproduction—the direct action of the male element on the female—development—the functional independence of the elements or units of the body—variability—inheritance—reversion."

No simple theory could provide all the answers to this ambitious program, and Darwin's theory of inheritance, to which he himself somewhat misleadingly affixed the term "the hypothesis of pangenesis," is indeed a whole package of theories. The first of these is that the transmission of heritable qualities, as well as the guidance of development, is caused by individually different, very small, and hence invisible, particles, the so-called gemmules (see above). Each kind of cell in the body is represented by its own kind of gemmule; the mosaic of characteristics in hybrids is due to the mixing of parental gemmules; and the facts of reversion to ancestral characteristics, a phenomenon which greatly fascinated Darwin, is due to the activation of previously dormant gemmules.

As de Vries (1889) was the first to point out succinctly, this genetic theory of Darwin's, proposing that the various characters of an organism have separate, independent corpuscular bases, was the first well-rounded and internally consistent theory of inheritance. It permitted the explanation of a large number of observations, and it is a historical fact that all subsequent theories of inheritance, particularly those of Galton (1876), Weismann (1883–1892) and de Vries (1889), were greatly influenced by Darwin's theory. It permitted an explanation, not very different from the subsequent Mendelian one, of "prepotency" (dominance) and "reversion" (recessiveness), regeneration, and other genetic and developmental phenomena.

694   As stated so far, this theory was unable to account for an inheritance of acquired characters. How could the effect of use and disuse on peripheral organs (hands, skin, eyes, brain) be communicated to the reproductive organs? To account for this, Darwin proposed a "transportation hypothesis" (as de Vries later called it). In any stage of the life cycle cells may throw off gemmules, "which circulate freely throughout the system, and when supplied with proper nutriment multiply by self division, subsequently becoming developed into cells like those from which they were derived" (Darwin, 1868: 374). This circulation of the gemmules is the second part of Darwin's theory; it permits the gemmules to accumulate in the sexual organs or, in the case of plants, in buds. Finally, "in variations caused by the direct action of changed conditions . . . the tissues of the body, according to the doctrine of pangenesis, are directly affected by the new conditions, and consequently throw off modified gemmules, which are transmitted with their newly acquired peculiarities to the offspring" (pp. 394–395).

This is the pangenesis theory in the more restricted sense, and it is the theory Darwin's critics usually had in mind when referring to Darwin's pangenesis theory. The idea of a transport of germinal material from the body to the reproductive organs was by no means original with Darwin, and Zirkle (1946) was able to list ninety forerunners, from Hippocrates on (see also Lesky, 1950: 1294–1343). Darwin himself (1868: 375) refers to the somewhat similar theories of Buffon, Bonnet, Owen, and Spencer, while pointing out where his theory differs from theirs.

Darwin was quite reticent about his transportation theory, referring to it as "a mad dream" and "still born," yet thought that "it contains a great truth." It was, of course, soon refuted (see below). The ironical part is that the theory became quite unnecessary when fifteen years later Weismann rejected soft inheritance, basing his rejection on quite a formidable array of facts and theory. If there is no inheritance of acquired characters, then there is no need to postulate a migration of genetic material from the soma to the germ cells.

The Decline of Soft Inheritance

Darwin was one of the first authors to stress the prevalence of hard inheritance, but even he, as we saw, was unable to abandon soft inheritance altogether. Who then was the first author to do so unequivocally? Every preformationist ought to reject soft inheritance implicitly, but I am not aware that any author ever articulated this principle. It is sometimes said that Prichard in the first edition (1813) of his *Researches in the Physical History of Man* was the first to reject soft inheritance. Prichard, indeed, denied that climate is responsible for racial differences in man, but he still allowed for soft inheritance with respect to culture and other factors, and he accepted soft inheritance to a greater extent in later editions of his work. Lawrence (1819), even though he said, "The offspring inherit only their [the parents'] connate peculiarities and not any of the acquired qualities," nevertheless allowed for the origin of birth defects owing to influences on the mother, and gives other indications of an occasional belief in soft inheritance (Wells, 1971). So did every other author up to the 1870s. Perhaps the first person to deny the occurrence of soft inheritance categorically was His: "Until it has been refuted, I stand by the statement that characters can not be inherited that were acquired during the lifetime of an individual" (1874: 158). Weismann (1883), Kölliker (1885), Ziegler (1886), and others followed him in due time (Churchill, 1976).

The controversies between the pioneers of hard inheritance and their opponents (for example, Virchow) show how axiomatic the belief in an inheritance of acquired characters still was in the 1880s and to what extent it was supported by contemporary ideas on the nature of life.

Darwin's cousin Francis Galton (1822–1911)[4] rejected soft inheritance largely but perhaps not altogether. In the 1870s he developed some extraordinarily prophetic ideas on inheritance, which were· apparently completely· ignored by the contemporary biologists, in part because Galton published in nonbiological journals and in part because some of his most original thoughts were not published at all. This is, for instance, true for an explanation of the characters of hybrids, which Galton communicated to Darwin by letter on December 19, 1875. Here he proposes a typical Mendelian theory of particulate inheritance with the hereditary units not fusing but able to segregate (Olby, 1966: 72). Yet, he was not particularly interested in conspicuous, discontinuous characters (like red versus white in flowers). He was far more concerned with generalized characters, such as size or (in man) intelligence. In 1876 Galton published a detailed, well-rounded theory of inheritance in which he anticipated many of the ideas, including the reduction division, subsequently developed by Weismann and others.

Galton accepted Darwin's theory "of a multitude of organic units, each of which possesses its own proper attributes." Since he rejected Darwin's thesis of pangenesis (or at least the part called by de Vries the "transportation theory"), he concentrated on the fact that the entire potential of an organism is encased in the fertilized ovum. For this sum total of the genetic particles, he coins the term *stirp*, apparently the same as Weismann's germ plasm (1883) and Nägeli's idioplasm (1884). Like Darwin, Galton was much impressed by reversions to ancestral conditions and by the sudden appearance in individuals of characters not observed in the parents. He therefore concluded, as had Naudin (1865) before him, that "comparatively few of . . . the host germs [in the stirp] achieve development," the others remaining dormant, sometimes for generations (1876). He discussed the meaning of sex, and concluded that it functions to maintain genetic variability, that is, to prevent the loss of genes (as we would say it now). He stated that such a loss is very unlikely to occur when the fertilized egg is composed of the contribution of two parents. He appreciated the necessity of a reduction division of the nucleus; he developed (long before Weismann) a theory of germinal selection (1876: 334, 338). Like all his contempories except Mendel, he believed that each genetic determinant is represented in the stirp by numerous identical replicas; he discussed random fixation, and had many other interesting ideas. Unfortunately, in the Spencerian tradition, he thought of inheritance very much in terms of "movements and forces," and consequently his explanation of ontogeny as a result of inheritance is very unsatisfactory. (After 1885 Galton developed an entirely different theory of inheritance; see Chapter 18.)

The part of the stirp that is not used up during individual development is transmitted from generation to generation. Just how evolutionary change occurs is not made clear, and even though throughout the years Galton rejected soft inheritance, it is implied in the cryptic statement: "It may well happen that some species of germs may have failed in achieving development during very many generations, by the end of which time they may have become considerably modified" (1876: 338). He adopted this explanation because he accepted "the evidence that structural changes might react on the sexual elements" (p. 348), but he rejected Darwin's transportation hypothesis. To disprove it experimentally, Galton made blood transfusions among rabbits with different coat colors. He then inbred the transfused rabbits, but among their offspring there was never any deviation from the parental color, as there should have been if alien gemmules had been circulating in the blood, as Darwin had hypothesized. These experiments did not induce Darwin to give up his pangenesis hypothesis. He said rather angrily that the experiments merely show that the gemmules are transported by other means than by blood circulation. This possibility was decisively refuted by Castle and Phillips (1909), who transplanted the ovaries of an immature black guinea pig to an albino female whose own ovaries had been completely removed. She was then mated to an albino male and in three successive litters bore only black offspring.

697

Galton, a dilettante and maverick, pioneered in many areas. He was a strong proponent of population thinking, appreciating the uniqueness of the individual more clearly than any of his contemporaries. This led him to his discovery of the uniqueness (and hence absolute diagnostic value) of fingerprints and to his development of populational statistics (Hilts, 1973). Two of the major concepts of statistics were created by Galton: regression and correlation. He is perhaps most widely known as the founder of eugenics.

The 1870s were a period of transition. The attacks on soft inheritance were disorganized, relating only to certain aspects of that doctrine. In Darwin's theory of pangenesis the cell was still accepted as the unit of structure in the organism. Even Galton, who had the most progressive theory of inheritance, failed to connect it with the new findings of cytology. As a consequence, he was unable to lay a theoretical foundation for his speculations. He, like Darwin, failed to understand that entirely new problems arose as soon as it was recognized that the nucleus rather than the cell as a whole is the carrier of the genetic material. One now had to ask, what is the relation between the nucleus and the cytoplasm of the cell? Is there any input from the cytoplasm into the nucleus, and particularly into the nucleus of the germ cell?

It must be remembered that ideas about inheritance and its physical basis had been exceedingly vague until the 1870s. This all changed when the nucleus was recognized as the vehicle of inheritance and when the complex structure of the chromatin inside the nucleus was discovered. The elaborate architecture of the germ plasm did not look like a structure that would respond appropriately to general environmental influences like climate and nutrition. A finely organized chromatin structure seemed more compatible with hard than with soft inheritance. How reliable was the evidence that had so far been accepted as proving soft inheritance? Did the new evidence help to refute it? Unfortunately, both Galton and Darwin were unaware of the spectacular advances in cytology made at this period in Germany.

698

AUGUST WEISMANN

The first person who not only asked these questions quite unequivocally but also gave decisive answers to them was August Weismann (1834–1914), one of the great biologists of all time.[5] He was unique among all those who worked on cytology, development, and inheritance in the last century by being an uncompromising selectionist. His theory of evolution, which excluded all remnants of a belief in an inheritance of acquired characters or other kinds of soft inheritance, was designated as *neo-Darwinism* (Romanes, 1896).

From the point of view of scientific methodology, again, he was notable for his period in his careful, rational analysis of every problem he encountered. When he wanted to interpret a given phenomenon or process, he attempted to reason out all the possible alternate solutions. Almost invariably this included the solution that is now considered to be the right one. Owing to the insufficient and sometimes even faulty information available at his time, Weismann himself sometimes chose an alternative that is now rejected. This does not in the slightest diminish the magnitude of his intellectual achievement. He never made a hasty decision but always first surveyed the entire field of possible solutions. His was the first truly comprehensive theory of genetics and his theorizing prepared the way for the research of the entire next generation. As Correns said, the rediscovery of the Mendelian rules in 1900 was no great intellectual achievement after Weismann had paved the way.

As a youth, Weismann (born January 17, 1834, in Frankfurt) had been an enthusiastic collector of butterflies, beetles, and plants. He first studied medicine and even practiced it some years but then shifted to zoology (histology). Almost at once he was hit by a serious eye disease that made microscopic work impossible and forced him into partial retirement, which turned out to be a blessing. He switched from empirical to theoretical studies and devoted his time to thinking deeply about biological problems and their solution. Evolution through natural selection, the material basis of inheritance, and the mechanisms of development were the three interrelated areas of his concern. He saw more clearly than any of his contemporaries that the great controversy about the validity of Darwinism could never be settled without a comprehensive theory of inheritance.

699

His first major paper on inheritance was published in 1876, a whole series of important essays appeared in the 1880s, and finally, in 1892 he published his monumental *Keimplasma* (628 pages). Like all imaginative pioneers, Weismann was quite openminded and never hesitated to revise his theories when he thought this was required by new evidence. Unfortunately, his revisions, particularly those published after 1890, were not always improvements, when seen in the light of modern knowledge.

In a theory of heredity, proposed in 1876, Weismann explained inheritance as due to molecular movements, citing with approval von Helmholtz's statement (1871: 208) that "all laws must be reduced in the last analysis to laws of motion." When he rejected Darwin's pangenesis theory, it was because it was based on "stuff" rather than on motion, and not because it sponsored soft inheritance. Weismann at that time still believed in the "influence of external conditions on the inheritable evolutionary material" (1868: 12). Yet, his trust in soft inheritance was apparently weakening, for he tested it by numerous experiments in the years 1875–1880.

The genetic theory which Weismann proposed in 1883 and 1885 was not only vastly different from his first attempts but also truly comprehensive. It was dominated by two new insights. The first was that all the genetic material is contained in the nucleus. As stated quite explicitly by Weismann, his theory was "founded upon the idea that heredity is brought about by the transmission from one generation to another of a substance with a definite chemical and, above all, molecular constitution" (1889; Eng. trans.: 167). The second insight was a rejection of an inheritance of acquired characters in any form.

There are three ways to refute an inheritance of acquired characters. The first is to show that the mechanisms by which it is supposed to operate are impossible. This was primarily Weismann's approach. There is nothing in the structure and division of cells that would make an inheritance of acquired characters possible. In fact, in certain organisms (Weismann specifically cites hydroids) the future germ cells are segregated at very early larval stages after only a few cell divisions and are "put on ice," so to speak, until the reproductive process is initiated. There is no possible way by which the influences on the remainder of the organisms could be transmitted to the nuclei of the segregated germ cells.

700   This observation led Weismann in 1885 to his theory of the "continuity of the germ plasm,"[6] which states that the "germ track" is separate from the body (soma) track from the very beginning, and thus nothing that happens to the soma can be communicated to the germ cells and their nuclei. We now know that Weismann's basic idea—a complete separation of the germ plasm from its expression in the phenotype of the body—was absolutely correct. His intuition to postulate such a separation was faultless. However, among two possible ways for effecting this he selected the separation of the germ cells from the body cells, while we now know that the crucial separation is that between the DNA program of the nucleus and the proteins in the cytoplasm of each cell.

A second way to refute an inheritance of acquired characters is by experiment. If there is an inheritance of acquired characters, then something must be conveyed from the affected part of the body to the germ cells. The old theory of use and disuse, in which even Darwin believed mildly, could be tested by a total disuse of a structure (Payne's experiments); alternatively, if any body part sends gemmules to the germ cells, then amputation of this body part through many successive generations would result in a gradual size reduction of this organ. Finally, if changes of the phenotype in plants due to cultural conditions were heritable, selective breeding from the largest and smallest individuals of pure lines should produce progressive results (Johannsen, 1903). Beginning with Hoffmann and Weismann, such experiments were conducted up to the 1930s and 40s and the results were uniformly negative (see also Galton, Romanes, and Castle and Phillips). In other words, the theory failed every test of its validity.

The third way of refuting the theory of the inheritance of acquired characters is to show that the phenomena that are claimed to *require* the postulate of an inheritance of acquired characters can be explained equally well or better on the basis of the Darwinian theory. Much of the evolutionary literature of the 1920s, 30s, and 40s was devoted to this third approach (see Part II).

Weismann believed in an inheritance of acquired characters throughout the 1870s. Exactly what brought about his eventual conversion is not clear. Nor is it clear whether Weismann had first become convinced of the invalidity of the theory of an inheritance of acquired characters and then adopted the germ-track theory or vice versa. The fact is that he already cites in his 1883 paper so many lines of argument against soft inheritance that one can well imagine that this general conviction preceded the proposal of a specific mechanism. This interpretation is strengthened by the fact that Weismann was a strict selectionist already in the 1870s and presumably had simply no need for an additional mechanism.

701

Weismann's revolutionary rejection of soft inheritance encountered great hostility. It was attacked not only by the neo-Lamarckians, who reached the height of their influence in the 1880s and 1890s, but even by orthodox Darwinians, who continued to accept Darwin's occasional reliance on the effects of use and disuse (for example, Romanes, 1896; Plate, 1903). However, it was adopted by Lankester, Poulton, and Thiselton Dyer in England and had probably, up to the 1930s, more adherents in England than in Weismann's home country. Near universal acceptance did not occur until the 1930s and 40s, as a result of the evolutionary synthesis (Mayr and Provine, 1980).

Weismann's Theory of Inheritance

Having eliminated the complicating factor of soft inheritance, Weismann was ready to propose his own theory. When evaluating it, one must remember that like all the other German students of cytology and generation, Weismann was far more concerned with explaining the genetic control of development than the mechanism of transmission from generation to generation. He concluded that "the orderly changes . . . during embryogenesis must be the result of corresponding systematic changes in the idioplasm" (1892: 61). Some years later (1899: 21) he recalled that at the time he proposed his theory, "there were two alternatives to explain ontogenetic differentiation: (1) the hypothesis of a systematic and progressive dissection of the totality of the genetic potential contained in the germ plasm into ever smaller groups [to be segregated into different cells], or (2) the hypothesis that the determinants of all characters remain together in all the cells of the developing organisms but that each of them is tuned to respond to a specific stimulus which only activates this trait: A pure 'dissection' and a pure 'activation' theory. I decided in favor of the former, because on the basis of the facts available at that time it seemed to be the more probable one." As we all now know, it was the wrong choice.

702

Before presenting in detail Weismann's theory of inheritance, I call attention once more to the clarity with which he realized the difference between genotype and phenotype. Indeed, in some of his statements he comes very close to proposing that development is controlled by a genetic program. He rejects the idea postulated in Bonnet's evolution theory that the genetic determinants are the preformed rudiments of the to-be-developed parts themselves, and instead considers them "active living units which intervene [eingreifen] in the process of development in a specific manner, that is, in such a way that the character is produced which they have to determine" (1899: 23).

Since Weismann approaches the problem of inheritance from the point of view of the developmental physiologist, he attempts to explain the nature of the genetic material on the basis of its effect on ontogeny: "The chromatin is able to give a specific character to the cell in whose nucleus it resides. Considering that the thousands of cells of which an organism is composed possess very different characters, it is evident that the chromatin which controls them cannot be the same in every cell but must differ according to the nature of the cell" (1892: 43).

Weismann postulates an elaborate hierarchy of hereditary units that control ontogeny. The smallest is the *biophore,* each consisting of an aggregate of diverse molecules having the capacity for growth and replication. Each biophore controls a specific feature (property) of a cell. All living substance is composed of biophores (1892: 56–57). The number of possible kinds of biophores is unlimited, that is, it is as great as the number of possible combinations of molecules. Nucleus as well as cytoplasm are composed of biophores, even though the properties of the cytoplasm of a cell are determined by the nucleus.

Muscle cells, blood cells, and other components of the body are controlled by specific compounds of biophores which Weismann calls *determinants* and which represent the next higher rank of units in the hierarchy of particles. The determinants are genotypic units while the biophores carry out the physiological implementation. A single cell may contain numerous replicas of the same determinant (1892: 81). This is particularly true for the nucleus of the gamete. The crucial difference between Weismann's theory and the Mendelian theory of inheritance is that Weismann postulates that a single cell, including the gametes, may contain numerous replicas of the same determinant (1892: 81), while in the Mendelian theory there are only two (one from each parent). This single difference in the two postulates requires two entirely different theories of inheritance.

The determinants, in turn, are joined together in a phylogenetically acquired architecture into still higher units, the *ids* which, Weismann sometimes implies, are the same as the chromosomes. The germ plasm consists of several if not many ids, which, like the biophores, can grow and replicate. The rate of replication of each kind of unit is independent of that of the others.

The crucial components of Weismann's theory seem to be these:

(1) There is a special particle (biophore) for each trait.

(2) These particles can grow and multiply independent of cell division.

(3) Both nucleus and cytoplasm consist of these biophores.

(4) A given biophore may be represented by many replicas in a single nucleus, including that of the germ cell.

(5) During cell division the daughter cells may receive different kinds and numbers of biophores (unequal cell division).

As we now know, postulates (2) to (5) are wrong and are responsible for the fact that Weismann was not able to arrive at a correct theory of inheritance. By adopting an entirely different strategy, Morgan and his school were able to succeed where Weismann had failed. Instead of trying to explain the gene ontogenetically, they concentrated on the gene from a phylogenetic point of view; that is, instead of studying the unit of developmental genetics, they studied the unit of transmission genetics.

Weismann's ingenious theory was at once vigorously attacked, particularly by botanists who favored the activation theory of ontogeny (see above). The fact that in many kinds of plants a bud may be produced almost anywhere which can develop into flowers, as well as the fact that one can often reconstitute a new plant (with flower-producing germ cells) from a single leaf or other vegetative structure, completely refutes a strict separation of germ track and soma track. These and other experiments likewise prove that unequal nuclear division, that is, an unequal partitioning of

703

the genetic particles of the mother cell in the two daughter cells, cannot take place. Furthermore, as Roux (1883) had demonstrated so convincingly, the entire elaborate process of mitosis makes no sense unless one postulates an equal division of the germ plasm during cell division. Kölliker (1885), Oskar Hertwig (1894), and Driesch (1894) summarized particularly effectively the evidence against Weismann's "dissection" theory.

An Alternate Theory of Inheritance

704

The various criticisms led to a different interpretation of the genetic processes during ontogeny which incorporated two major new concepts relating to the connection between nucleus and cytoplasm and to the problem of differentiation.

Strasburger (1884), aware of the chemical difference of nucleus (nuclein) and cytoplasm, proposed that the nucleus remains at all times intact but produces molecular excitations "which are conveyed to the surrounding cytoplasm and there determine the metabolic processes of the cell and give it a species specific character." Wilhelm His and others adopted similar physical interpretations. Haberlandt (1887) suggested instead that the nucleus sends not vibrations but specific molecules to the cytoplasm and thus regulates its activities. De Vries (1910: 203) identified Haberlandt's molecules with enzymes. Unfortunately, Haberlandt never followed up this remarkable theory, which so nearly anticipates messenger RNA.

De Vries himself instead suggested that the genetic units, the pangens, migrate from the nucleus to the cytoplasm and thus determine the character of the respective cells. This suggestion was adopted by Weismann (Churchill, 1967). He was fully aware that not all genetic units can be functional at all times and in all cells. Nevertheless, he rejected an activation theory of the gene for two reasons. First, he thought that the activity of a cell was controlled by a determinant (aggregate of biophores) and he was unable to visualize what would happen to a cell if the determinant controlling it was inactivated. Furthermore, he simply could not think of a mechanism that would control the activation or deactivation of the hundreds of thousands of different determinants of an organism: "If one would assume that all determinants of the germ plasm are transmitted during ontogeny to all cells, then one would have to explain the entire differentiation of the body by an orderly inactivation of all determinants of a cell with the exception

of the single one which is specific for this particular kind of cell"
(1892: 86). It did not occur to him that each biophore (we would
now say "gene") could be activated and deactivated independently
or that the activity of the cell was due to an interaction between
the diffuse cell products in the cytoplasm and the products of the
activity of the nucleus. Weismann did not deny activation and inactivation but restricted it to the determinants rather than to the
biophores (1892: 100–101). His opponents accused him of believing in extreme preformationism. There is much justification in
this accusation. Complex characters are caused by pre-packaged
assemblages of biophores: determinants. The "eyes" on the feathers of a peacock could not possibly be produced by large numbers
of independent genes. They require a carefully packaged set of
determinants, said Weismann. His emphasis was entirely on structural elements. No allowance was made for rates of growth, developmental fields, temporary periods of activity or inactivity of
biophores, and so forth. This atomistic interpretation of the determination of traits in the activation theory contributed to its rejection.

705

The controversies stimulated by Weismann's elaborate theories came to deal more and more with problems of development
and, in a way, moved more and more away from a genuine theory
of inheritance. This is quite evident, for instance, in the work of
Oskar Hertwig (1898). Hugo de Vries was about the only author
who continued to focus on transmission genetics (see below).

The Meaning of Sex

The distribution of genetic factors during cell division was
not the only aspect of inheritance on which Weismann theorized.
Having thought deeply about these matters, he contributed several major new theories, one of which had to do with the controversial role of sexual reproduction. Why should a mother "waste"
half of her reproductive capacity by producing male offspring,
when females in parthenogenetic species can reproduce without
fertilization and thus double their reproductive potential? Weismann pointed out that there is no sound evidence in favor of any
of the previously proposed physiological theories of sexuality—for
instance, that sexual reproduction is a rejuvenation process. Rather,
said Weismann, sexual reproduction is the only way by which the
unlimited individual variation can be produced that is so characteristic of biological populations. During fertilization "two groups

of hereditary tendencies are, as it were, combined. I regard this combination as the cause of hereditary individual characters, and I believe that the reproduction of such characters is the true significance of sexual reproduction. The object of this process is to create those individual differences which form the material out of which natural selection produces new species" (1886: 279).

This was by no means a novel idea, because as far back as the 1780s Herder (1784–1791: 138) had stated most perceptively that "the most successful method by which nature combines in her species both diversity and constancy of form, is the creation and conjugation of two sexes. How wonderfully combined are the traits of both parents in the face and body build of the children; as if their souls had been poured into them in different proportions and as if the thousand-fold forces of their organization had been distributed among them, and how often do we find the traits of former generations in the children." But, of course, there is no biological significance to such variation unless one adopts also natural selection. In Darwin's thinking, curiously, sexual reproduction as a source of individual variation played only a very minor role. There is no doubt that Weismann was the major champion of the importance of sex as source of variation (see Chapter 11), even though Galton (1876: 333) had also recognized it.[7]

When we look at the total lifework of Weismann, we are awed by the diversity of the problems he analyzed and by the sound intuition with which, again and again, he suggested the correct interpretation. His only major mistake was to reject the activation theory, which forced him to adopt unequal cell division (which he called the "dissection theory") and a hierarchy of particles. In numerous essays, Weismann raised a great diversity of biological problems, some of which, like "What is the biological meaning of death?" had hardly ever been raised before. Inheritance and evolution were his main interests. E. B. Wilson, long ago, stated that the modern theory of genetics rests on the Weismannian foundation. In an age in which soft inheritance was at the height of its popularity, he was the champion of exclusively hard inheritance. In an age relying on physical forces, it was he who stressed particles and what might be called neo-preformationism. His theory of inheritance was based on the assumption of particulate inheritance; indeed, the theory of blending inheritance was specifically refuted by him (1892: 388, 544). It was he who emphasized that the units of inheritance are carried by the chromosomes and who predicted the occurrence of a reduction division (see Galton, 1876:

334, and Chapter 17). Weismann played an equally important role as an evolutionist by his uncompromising emphasis on natural selection (neo-Darwinism). Although the early Mendelians (including T. H. Morgan prior to 1910) rejected Weismann, his ideas ultimately prevailed, particularly where the application of genetics to evolution was concerned.

HUGO DE VRIES

The Dutch plant physiologist Hugo de Vries (1848–1935)[8] differed quite fundamentally from Weismann and the German cytologists in two ways. He had been educated in organic and physical chemistry, which permitted him to see the functional problems of inheritance in a different and much more meaningful light than contemporary zoologists and botanists. Furthermore, his main genetic interests were transmission genetics and the origin of organic diversity.

When studying de Vries's influence on our understanding of variation and inheritance, one must distinguish the impact of three publications: His *Intracellräre Pangenesis* (1889; quotations are from English translation of 1910), the report of his rediscovery of Mendel's rules (1900), and his *Mutation Theory* (1901–1903). The theory of intracellular pangenesis, which was published prior to 1892 and influenced Weismann's theory of inheritance, incorporated the same advances in the understanding of cells as Weismann's work had, yet differed by placing a primary emphasis on questions of transmission genetics. This brilliant and persuasive work, curiously, did not have the influence it deserved. Only after Weismann's theory had been refuted was it remembered how much closer de Vries had been to the later findings. Also, the mental preparation which *Intracellular Pangenesis* had given de Vries predestined him to become one of the rediscoverers of Mendel.

De Vries's primary interest in heredity was evolutionary and started, as with Unger and Mendel (see below), with the species problem. De Vries rejected the concept of the species "as a unit and the totality of its specific attributes as an indivisible concept" (1889: 11). "But if the species characters are regarded in the light of the theory of descent it soon becomes evident that they are composed of single factors more or less independent of each other." A study of organism leads inevitably "to the conviction of the composite nature of specific characters."

There were two important influences on de Vries's thinking: the year he had spent in the reductionist-mechanistic laboratory of Julius Sachs in Würzburg, and his close relations with the physical chemist Jacobus Hendricus van't Hoff in Holland. It is not surprising therefore that he wanted to carry the analysis down to the basic units of the living world. "The character of each individual species is composed of numerous hereditary qualities" based on factors that "are the units which the science of heredity has to investigate. Just as physics and chemistry go back to molecules and atoms, the biological sciences had to penetrate to these units in order to explain, by means of their combinations, the phenomena of the living world" (1889: 13).

708

Genetic Units

The various authors from Spencer to Weismann postulated three theories about the nature of genetic units. In a rather simplified manner, these theories may be stated as follows:

(1) Each unit has all species characters; it is, so to speak, an entire species homunculus (Spencer, Weismann's ids, Nägeli's idioplasm).

(2) Each unit has the features of a single cell (Darwin's gemmule, Weismann's determinant).

(3) Each unit represents a single species character or trait (de Vries's pangen, Weismann's biophore).

De Vries's 1889 theory differed from that of Weismann (1892) by giving each pangen independent existence and the capacity to be activated and to vary independently of the others (Weismann's biophores were tied together into determinants). De Vries (1889: 67–68) refutes with well-reasoned arguments Weismann's objections against the recognition of individual units for each hereditary trait. One can summarize de Vries's genetic theory in these statements:

(1) Inheritance is due to material bearers of hereditary qualities, to be called pangens.

(2) Every hereditary character has its special kind of pangen.

(3) The more highly differentiated an organism is, the more kinds of pangens it has.

(4) Each pangen can vary independently of any others.

(5) All nuclei contain the same pangens but only a very lim-

ited number of pangens are released into the cytoplasm of a given cell, all others remaining inactive in the nucleus of that cell.

(6) A given nucleus may contain many identical replicas of a given pangen.

(7) In order to become active, a pangen must move from the nucleus to the cytoplasm.

(8) There is no movement of pangens from the cytoplasm to the nucleus.

(9) There is no movement of pangens from one cell to another.

(10) Pangens always divide during cell division but may also divide between cell division so that a given pangen may be represented in the cytoplasm (as well as in the nucleus) by many identical replicas.

(11) The entire protoplasm of an organism consists of pangens.

(12) Occasionally a pangen changes and this "forms a starting point for the origin of varieties and species" (1889: 71). (This is the source of his later mutation theory; see Chapter 12.)

De Vries had every justification to claim that his theory was an excellent foundation for an experimental analysis of inheritance, and soon after the publication of his brilliant book (1889), he himself initiated such an experimental program. It was based on the thesis of the independent variation of each genetic unit; consequently "each one can of itself become the object of experimental treatment in our culture experiments" (1889: 69).

There is little doubt that de Vries's genetic theory is closer to current concepts than any other that preceded it. However, two of his major assumptions were drastically wrong: that the pangens themselves move from the nucleus to the cytoplasm; and that a given pangen could exist in the nucleus in multiple replicas. He thought that this explained dominance and quantitative characters. "If certain pangens are fewer in number than others then the character represented by them is only slightly developed; if there are very few, the character becomes latent" (1889: 72). De Vries shared this erroneous postulate with Weismann and all other authors in the 1880s and 90s who theorized about heredity. It is very obvious that it would be meaningless to calculate Mendelian ratios if one makes such an assumption. The crucial next step in the history of genetics was the overthrow of the "multiple replica theory" of genetic factors. The total refutation of blending was another one.

The period from the 1860s to the 1890s was a period of quite uninhibited speculation. We must conclude this whether we look at the writings of Spencer, Haeckel, and Darwin, or those of Galton, Nägeli, de Vries, and Weismann. This period also continued to be handicapped by erroneous concepts and the failure of an adequate discrimination between components in complex problems. This includes the failure to cleanly separate character transmission between generations from gene physiology (differentiation); it includes the failure to distinguish (except for de Vries) unit characters from the species essence; and the failure to discriminate between genotype and phenotype. And yet this period was an indispensable stage in the development of genetics. It was in this period that the right questions were first asked, that an interest in the corpuscular and chemical nature of the transmitted genetic material developed, and that the cytological foundation was laid without which no causal theory of inheritance could have been elaborated. At the end of the period almost all conceivable alternatives had been proposed, and the stage was set for the new insight or discovery that would permit unequivocal choices between competing theories. This decisive event was the rediscovery in 1900 of the work of Mendel. It gave rise with one stroke to an entirely new branch of biological science.

GREGOR MENDEL

It is one of the great ironies in the history of science that the answer to the problem of heredity had already been found while so many distinguished investigators searched for it so assiduously during the 1870s, 80s, and 90s. It had been published in the *Proceedings of the Natural History Society of Brünn* (Brno).[9] Father Gregor Mendel had given two lectures to that society on February 8 and March 8, 1865, in which he described the results of plant-breeding experiments he had conducted since 1856. The report, published in 1866, is one of the great classics of the scientific literature, a model scientific report, clearly setting out the objectives, concisely presenting the relevant data, and cautiously formulating truly novel conclusions. Who was this hidden genius, and why was his work neglected until 1900, when it was suddenly rediscovered?

Johann Mendel (1822–1884; the name Gregor was given to

him when he became a priest) was born in Austrian Silesia, the son of poor peasants. He was not at all the "obscure monk" he is sometimes described; even though he carried out his genetic experiments in Brünn in virtual intellectual isolation, Mendel had received an excellent education in the high schools of Troppau and Olmütz and, ultimately, for two years (1851–1853), at the University of Vienna, in order to be qualified to teach physics and other sciences at the high school level. He was, thus, actually a well-trained young scientist who had been educated in Vienna under some of the outstanding physicists and biologists of his time. Of special importance is the fact that Franz Unger, his professor of botany, had adopted a theory of evolution in 1852 which included the opinion that variants arise in natural populations which in turn give rise to varieties and subspecies until finally the most distinct of them reach species level (see Chapter 8). He thus implied that the study of varieties was the key to the solution of the problem of the origin of species. This idea apparently greatly stimulated his student Mendel. It is highly significant that, as in the case of Darwin, it was the species question which inspired Mendel in his work on inheritance, quite in contrast to the German embryologists and cytologists whose basic interest was the physiology of development. In his famous 1866 paper Mendel states that his time-consuming experiments were necessary in order to "reach the solution of a question, the importance of which cannot be over-estimated in connection with the history of the evolution of organic form." Evidently he wanted to test Unger's theory and this meant a study of varieties.

711

As a consequence of his evolutionary approach, Mendel adopted, as Thoday (1966) has correctly pointed out, the method of population analysis rather than the study of the single individual that is traditional in functional analysis. He analyzed large populations of offspring and was fully aware that it was "necessary to observe without exception all members of the series of offspring in each generation" (1866: 4). He literally analyzed tens of thousands if not hundreds of thousands of seeds and plants, his experiments requiring the work of eight planting seasons. Everything we know about Mendel indicates that he was an extremely meticulous person. He kept careful records of weather, sun spots, and other variable phenomena, and was fascinated with numerical relationships. This ideally predestined him for a populational approach to inheritance.

Of decisive importance for Mendel's success was the fact that he had been trained in physics as thoroughly as in biology (or more so). His favorite high school teacher had been a physicist, and physics seems to have been the major subject in his own teaching. In Vienna he took courses with the famous Doppler and with other physicists, and he even served for a time as a demonstrator at the Physics Institute of the University of Vienna. It is this experience which must have taught him to keep careful records of his experiments, to arrive at numerical generalizations, and to attempt a rudimentary statistical analysis. This approach, of course, was particularly suitable, indeed necessary, for a population analysis. Thus, although his concepts (population, evolution) came from biology, most of his methodology came from physics.

712

Owing to his excellent grasp of the botanical literature, and particularly his thorough reading of Gärtner (see above), Mendel was keenly aware of the extreme importance of selecting the right kind of plant for his experiments:

> Selection of the plant group for experiments of this kind must be made with the greatest possible care if one does not want to jeopardize all possibility of success from the very outset.
> The experimental plants must necessarily
> (1) Possess constant differing traits.
> (2) Their hybrids must be protected from the influence of all foreign pollen during the flowering period or easily lend themselves to such protection.
> (3) There should be no marked reduction in the fertility of the hybrids and their offspring in successive generations. (Mendel, 1866)

The latter point was of crucial importance, considering one great weakness in Mendel's conceptual framework: He had little idea what a species was. He designated the "forms" which he crossed sometimes as species or as subspecies or as varieties, because "in any event, the rank assigned to them in a classification system is completely immaterial to the experiments in question, just as it is impossible to draw a sharp line between species and varieties, it has been equally impossible so far to establish a fundamental difference between the hybrids of species and those of varieties" (p. 5).

Actually, there is indeed a drastic difference, as Kölreuter had intuitively appreciated better than Mendel. Differences among intrapopulation variants are usually single-gene differences and display uncomplicated Mendelian segregation, while the differences between species are often highly polygenic and fail to segregate cleanly.

As long as Mendel faithfully adhered to the third of his stated principles, he was safe. When at a later period he turned to other material, because devastations by the pea weevil (*Bruchus pisi*) made further work on peas impossible, Mendel ran into disturbing complications which seemed to undermine the generality of his previous findings. However, in 1856 he had fortunately decided to select the pea *Pisum sativum* and related forms as his experimental material owing to a large number of advantages of this species which had been appreciated by plant hybridizers from Andrew Knight on.

713

Owing to his uncertainty as to what a species is, Mendel used the term "hybrid" indiscriminately, both for actual species hybrids and for heterozygotes of a single gene. This has confused certain historians. Even though Mendel occasionally calls himself a hybridizer and in his paper often refers to Kölreuter, Gärtner, and other plant hybridizers, he himself does not at all belong to that tradition. As a student of Unger and of the problem of evolution, Mendel was concerned with single-character differences and not, like the hybridizers, with the species essence. To understand this fully is very important for the interpretation of Mendel's work. It is totally misleading to say that Mendel's conceptual framework was that of the hybridizers. It is precisely the breaking away from the tradition of the hybridizers that characterizes Mendel's thinking and constitutes one of his greatest contributions.

Another remarkable aspect of Mendel's work was that it quite evidently employed the hypothetico-deductive method. The entire planning of his experiments, the explanation of his method, as well as the choice of his material permit no other interpretation than that already early in his work Mendel had a well-formed theory in his mind and that his experiments actually consisted in the testing of his theory. His approach thus differed widely from that of both the earlier hybridizers like Gärtner, who through an inductive approach piled up mountains of results without getting to any conclusion whatsoever, and workers like Nägeli, who speculated wildly without ever attempting to test the validity of their speculations. The hypothetico-deductive approach, of course, was not new with Mendel; it had been adopted by perceptive investigators from the eighteenth century on, among both physicists and biologists, Darwin and Schleiden being typical examples.

Stripped to its essentials, Mendel's theory was that for each heritable trait, a plant is able to produce two kinds of egg cell and two kinds of pollen grain, each representing either the paternal or the maternal character (if they are different). Or, to express the same hypothesis in different words, each character was represented in the fertilized egg by two hereditary elements (and no more than two) one derived from the mother (from the female gamete) and one derived from the father (from the male gamete). (It is admittedly controversial to which extent Mendel and the early Mendelians thought in such terms.)

714   Exactly when this theory was formed in Mendel's mind we shall never know, because his voluminous notes and manuscripts were all burned either late in Mendel's life or after his death. All we can do is to conjecture. Most likely this theory came to Mendel about 1859 after some preliminary breeding, but was firmly in his mind during the later years of intensive breeding work.

Mendel's Findings

Mendel procured 34 more or less distinct varieties of peas from several seed dealers and subjected them to two years of testing. Of these varieties, 22 remained constant when self-fertilized, and these he planted annually throughout the entire experimental period. In these 22 varieties, seven pairs of contrasting traits were chosen for experimental testing. Two plants differing in a given pair of traits were hybridized, and the behavior of the trait was followed in the ensuing generations. The 22 varieties differed from each other by far more than the seven selected traits, but Mendel found the other traits unsuitable because either they produced continuous or quantitative variation not suitable for the study of the clear-cut segregation that he was interested in, or else they did not segregate independently.

The traits chosen were the following:

(1) whether the ripe seeds were smoothly round or angular and deeply wrinkled;

(2) whether the ripe seed (cotyledon) is yellow or green;

(3) whether the seed coat is white or gray;

(4) whether the ripe pod is smooth and nowhere constricted or deeply constricted between the seeds and more or less wrinkled;

(5) whether the unripe pod is green or vivid yellow;

(6) whether the flowers are proficient along the main stem or are terminal;

(7) whether the stem is long (6–7 ft.) or short (¾–1½ ft.).

What Mendel found is now familiar to every beginning biology student. He had chosen seven character pairs, one of which was always clearly dominant. In all of his experiments, therefore, the first hybrid population (F_1) was uniform and agreed with the character of one of the parents. Dominant were, for instance, round seeds, yellow seed coloration, gray coloration of the seed coat, green coloration of the unripe pod, long stems, and so on. Mendel introduced, probably independently of others such as Martini and Sageret who had used a similar terminology, the term *dominant* (*dominierend*) for this predomination of one character in the first hybrid generation, and *recessive* (*recessiv*) for the alternate characters.

When the F_1 hybrids were self-fertilized, producing an F_2 generation, the recessive character reappeared. In the case of seed shape, among 7,324 seeds collected from 253 self-fertilized hybrid plants, 5,474 were round and 1,850 wrinkled, giving a ratio of 2.96:1. In the case of seed color, 8,023 seeds collected from 258 hybrid plants gave 6,022 yellow and 2,001 green seeds, representing a ratio of 3.01:1. Mendel summarizes the results of the crossing of first generation hybrids as follows: "In this generation along with the dominant traits the recessive ones reappear in their full expression, and they do so in the decisively evident average proportion of 3:1, so that among each four plants of this generation three show the dominant and one the recessive character" (1866: 10).

Mendel did not stop at this point, but produced an F_3 generation by self-fertilizing a large number of plants of the F_2 generation. In his experiment with round versus wrinkled seeds, which had given him 75 percent round and 25 percent wrinkled in the F_2 generation, he found that all the plants raised from wrinkled seeds bred true for this character. Plants raised from round seeds showed segregation in the F_3. Among 565 plants raised from round seed, 193 yielded only round seeds, thus being constant for this trait; however, 373 plants produced both round and angular seeds in the proportion of 3:1. In other words, among the round seeds one-third bred true for this character, and two-thirds gave round and wrinkled seeds. Mendel carried most of his experiments through four to six generations, and the results were always the same. He had clearly discovered a law-like regularity.

What interpretation did Mendel give to his findings? The distinction between genotype and phenotype was not made until nearly fifty years later and the concepts of pangens and genes, and of chromosomes and other cell and nuclear elements, had not yet been developed. It would have been miraculous in the absence of such factual and conceptual assistance if Mendel in 1865 had created all of Mendelian genetics out of nothing. And he did not. Nevertheless, it was so natural to interpret his presentation in Darwinian and Weismannian terms that de Vries, Correns, and Bateson did just that automatically when they read Mendel's paper. None of them even attempted to question Mendel's priority. This "honor" was reserved for historians. Olby (1979) has recently suggested that "Mendel was no Mendelian." The validity of this claim depends entirely on one's definition of "Mendelian." If one needs to have adopted all the genetic findings made from 1900 to 1915 then, indeed, Mendel was no Mendelian. He did not name genes and assign them to definite loci. Throughout most of his paper he referred to the inherited characters in a language remarkably similar to that by which Bateson referred to "unit characters," as would anybody who does not make the genotype-phenotype distinction.

716

Considering that Mendel did not know any of the findings of cytology (most of which were made in the 1870s and 80s), how did he visualize the transport of the characters in the "Keim und Pollenzellen" (female and male gametes)? He postulated that the characters are represented by "gleichartige [identical] oder differierende [differing] Elemente." He does not specify what these "Elemente" are—who could have done so in 1865?—but considers this concept sufficiently important that he refers to these "Elemente" no less than ten times on pages 41 and 42 of the *Versuche.* Evidently they correspond reasonably well to what we would now call genes. Where Mendel differed from the later genetic interpretation is that he ascribed a different fate to the *gleichartigen* and *differierenden* elements. He thought that if they were identical, the homologous elements of male and female gamete would fuse completely after fertilization. This is why in the F_2 he wrote A and $a,$ instead of AA and aa. If the elements were different, he assumed that the association in the hybrid plant would be only temporary, to be dissolved again during the formation of the gametes of the hybrid plant (1866: 42).

Mendel summarizes his "hypothesis" (his word) of the behavior and the attributes of the elements by saying, "The distinguishing traits of two plants can, after all, be caused only by differences in the composition and grouping of the elements existing in dynamic interaction in their primordial cells" (p. 42; trans. Stern and Sherwood, 1966).

Where Olby and colleagues are correct is in refuting the assumption, universally made among geneticists and previously by Mayr, that Mendel had a clear picture of pairs of alleles which neatly separated during gamete formation. His description of the "Vereinigung gleichartiger Elemente" ("union of identical elements") through fusion falsifies this claim. The absence of a concept of gene loci with sets of alleles is confirmed by Mendel's description of polygenic color inheritance in a *Phaseolus* cross where the same recessive Merkmal a is postulated for the two simultaneously present Merkmale A_1 and A_2. In modern terminology the recessives at these two independent loci would have to be designated differently, as a_1 and a_2.

Then why did Correns, de Vries, and Bateson assign to Mendel the priority of having discovered Mendelism? The main reason, as Correns pointed out quite succinctly, is that after the cytological researches of the three preceding decades and after the genetic theorizing of de Vries (1889) and Weismann (1892), the 3:1 ratios could be explained in no other way than by assuming that during gamete formation there is a 1:1 segregation of the "Anlagen" for equivalent characters. Indeed, this was what Mendel had already almost (but not quite) postulated. He did postulate it for the "differierende Merkmale" (1866: 42), while for the "gleichartigen Merkmale" he merely postulated that each must be represented in the gametes. Mendel himself never says explicitly that they must be represented in the gamete by only a single element, but 3:1 ratios would not occur with such law-like universality if this were not the case. With the vastly expanded knowledge of cytology and inheritance available in 1900, Mendel's rediscoverers immediately took this for granted. The 3:1 ratios left them no other alternative.

Olby and the others who have recently questioned the nature of Mendel's contribution are thus right in insisting that Mendel did not, by a single stroke, create the whole modern theory of genetics. He did not have a theory of the gene, but neither did his rediscoverers, as Olby points out quite correctly (1979: 58). However, Mendel's various discoveries (segregation, constant ra-

717

tios, independent assortment of characters), combined with the new insights acquired between 1865 and 1900, led, one is tempted to say automatically, to the theory quite legitimately called Mendelian. Among Mendel's more important conclusions concerning a single set of characters are these:

(1) Dominant and recessive genes do not affect each other while associated in the heterozygote. Even if one were to cross round-seeded peas with wrinkled peas for a hundred generations, the round peas would remain as round as they were at the beginning and the same would be true for the wrinkled ones.

718

(2) Gametes always contain only the *Anlage* of one of the two alternate characters. This is as true for the gametes produced by heterozygotes as for those produced by homozygotes. Evidently, the determinants of the parental traits are separated prior to gamete formation. This explains the phenomena of segregation and of recombination, so well known to breeders.

(3) A plant produces thousands of egg cells and millions of pollen grains (or spermatozoa, in the case of animals) and the meeting of gametes with different genes is a matter of chance. When small samples are used, one has to expect deviations from the 3:1 ratios, but the range of these deviations is statistically predictable.

Important for the planning of his crosses was Mendel's conviction, experimentally tested by him, that "propagation in phanerogams is initiated by the union of one germinal and one pollen cell into a single cell" (1866: 41). The insight that only a single pollen grain is involved in fertilization was based on the work of Amici and other botanists which Mendel had evidently learned from Unger, whose excellent textbook of the anatomy and physiology of plants he owned and who had written elsewhere about this subject. It was a great handicap for Darwin to have accepted from the breeders the belief that egg cells are simultaneously fertilized by several male gametes.

Mendel now applied his new insight to crosses involving two pairs of characters. He found, for instance, that when a plant with round yellow seeds is crossed with a plant with wrinkled green seeds, one can obtain four different combinations in the F_2. For instance, in a given cross he obtained 350 round yellow, 108 round green, 101 wrinkled yellow, and 32 wrinkled green seeds, coming rather close to the expected 9:3:3:1 ratio. The conclusion was evident: each character is inherited independently of the other, and the dominant-to-recessive ratio is unaffected by the other character (1866: 42). Finally, Mendel made a cross involving three sets of characters, showing that all three were inherited independently.

By clearly focusing on individual characters and their behavior in subsequent generations, Mendel was able to arrive at certain generalizations. He formulated the "law of combination of different characters," now referred to as the *independent assortment of characters.* Correns (1900: 98) phrased it as follows: "In the gametes of an individual hybrid the Anlagen for each individual parental character are found in all possible combinations but never in a single gamete the Anlagen for a pair of characters. Each combination occurs with approximately the same frequency." It is self-evident, but needs special emphasis, that the laws of inheritance can be worked out only if the two parents differ from each other in their genetic constitution. This permits the demonstration of two important factors of inheritance. First, the equal contribution of both parents, and second, the maintenance of the integrity of differing elements (their "nonblending" in subsequent generations). Mendel stressed this in his correspondence with Nägeli: "I am inclined to regard the separation of parental traits in the progeny of hybrids in *Pisum* as complete, and thus permanent . . . I have never observed gradual transitions between the parental traits or a progressive approach toward one of them" (Correns, 1905).

719

In his smaller samples, Mendel had some rather pronounced deviations from the expected 3:1 or 2:1 ratios.[10] He was fully aware of the statistical nature of such sampling errors, and to compensate for them, in an age long before the existence of statistical tests of significance, he simply grew large populations of his crosses. Fisher (1936) raised the question of whether Mendel's results were not "too good," since the deviations from expectancy, as calculated by chi-square tests, are smaller than expected, he said. However, the internal evidence as well as everything we know about Mendel's painstaking and conscientious procedure make it quite evident that no deliberate falsification is involved. It is possible that Mendel threw away a few particularly deviant crosses, thinking that they had been falsified by foreign pollen; it is also possible that he continued repeating a certain cross until the numbers approached the expected ratio, not realizing that this introduced bias into his method, but it is most likely that the bias is introduced by the fact that pollen, during maturation, is produced in the form of tetrads and that this, particularly in cases of self-fertilization and limited amounts of pollen, may lead to results that are "too good" (Thoday, 1966). Furthermore, if germination in Mendel's plants was only eight or nine out of ten, as is usual in such experiments, it invalidates Fisher's chi-square calculations and brings Mendel's results right in line with those of other pea hybridizers (Weiling, 1966; Orel, 1971). Thus there was really nothing drastically wrong with Mendel's figures; indeed, Mendel was an almost pedantically precise recorder of data, as also shown by his work in meteorology.

720

Mendel's Most Significant Contribution

The almost explosive development of genetics after the rediscovery of Mendel's work suggests that there is something crucial in Mendel's findings that permitted the field to make a new start after more than thirty years of floundering with erroneous or at least premature speculations. What was this crucial component?

Dominance, reversion, the identity of reciprocal crosses, the uniformity of the first hybrid generation, and the variability of the second generation had been described by numerous authors before (Zirkle, 1951). Nor was his postulate new (in 1900) that there are certain elements (particles) that control characters. This was essentially Darwin's theory of gemmules, and more particularly de Vries's theory of pangens. Nor was the refutation of blending inheritance Mendel's decisive contribution. First of all, he himself believed in a fusion of the "gleichartigen Elemente," but, more importantly, de Vries and Weismann believed, at least in part, in particulate inheritance. Yet, Mendel greatly contributed toward the eventual eradication of the last remnants of a belief in blending. He emphasized that, if the factors derived from father and mother differed, they would never fuse but would invariably again separate during the formation of the germ cell. It was only a small step from this to the postulate that the "gleichartigen Elemente" would likewise remain discrete after fertilization. This independence, and separate existence, so to speak, of the genetic factors in the germ plasm simultaneously gave a great boost to a belief in hard inheritance.[11] I stressed above that although Mendel's method had been strongly influenced by physics, his conceptual framework was supplied by biology. Unlike the physicalists (His, Loeb, Bateson, Johannsen), inheritance for Mendel was not due to forces or excitations but due to concrete materials supplied by maternal egg cells and paternal pollen cells. The basis of inheritance was the *quality* of the transmitted parental matter. From Haeckel (1866) and Darwin (1868) on this was the standard assumption of all those who approached the problem of inheritance as naturalists or whole-organisms biologists.

What, then, was the outstanding contribution made by Mendel? When we carefully compare his theory of inheritance with those of Darwin, Galton, Weismann, and de Vries (1889), we discover two crucial differences. First, all these earlier authors postulated the existence of numerous identical determinants for a given unit character in each cell (each nucleus) and speculated, likewise, that many replicas of a single determinant might be transmitted simultaneously to the germ cells. If this were the case, no consistent ratios would be found in crosses. This assumption made the development of a clear-cut genetic theory almost impossible. The universality of the 3:1 ratio refuted the multiple-particle postulate. It is consistent only with a single-particle postulate. This was Mendel's greatest contribution. Mendel's other significant contribution was the discovery that these particles exist in sets—genes and their alleles, we would now say. Through this assumption it was possible to explain segregation and recombination. His inference that each character is represented in a fertilized egg cell by two, but only two, factors, one derived from the father and the other from the mother, and that these could be different, was the new idea which revolutionized genetics. What Mendel provided was an exceedingly simple theory which any amateur could easily test on a given set of alternate characters. Indeed, this is so simple that experiments of this sort are now done by teenagers in the science classes of some high schools. At the same time, Mendel's simple generalizations laid the foundation for the development of genetics after 1900.

The rediscoverers of Mendel rather concealed the true nature of Mendel's discovery by speaking of Mendel's three laws: (1) the law of segregation, (2) the law of dominance, and (3) the law of free assortment.

Phenotypic segregation in the F_2, of course, had been found by many pre-Mendelians from Kölreuter, Knight, and Sageret on. However, it was never as central in anybody else's work as in Mendel's nor previously applied to the genetic material itself ("die Elemente"). Emphasis on segregation was an effective way to counteract any leanings toward a belief in blending inheritance, but segregation alone is not the essence of Mendelism. If one has

721

multiple determinants for a simple character, as was believed by everybody but Mendel before 1900, one can have segregation without 3:1 ratios. What was crucial in Mendel's theory was his insistence that when the parents differ in a character, the elements or *Anlagen* for these characters remain discrete in the hybrids and separate again in the formation of the germ cells of these hybrids. This clearly is one of Mendel's decisive contributions, the other one being the inference, made necessary by the 3:1 ratios, that each character is represented in the germ cells by one, and only one, element.

722

Dominance, as we shall presently see, is not a "law." That there was dominance in all seven of Mendel's character pairs was evidently due to his deliberate choice of such characters.

Finally, free assortment is also not a valid "law," because it was discovered soon after 1900 that characters could be "linked," by having their determinants on the same chromosome (see Chapter 17). Mendel's "laws" may have been a helpful didactic device in the early days of Mendelism, but by now they have rather lost their usefulness and have been replaced by others.

Why Mendel's Work Was Ignored

The clarity of Mendel's writings, the simplicity of his theory, and the desperate need for such a theory at the time when it was published (1866) make it a disturbing puzzle why his work was so utterly ignored. The glib answer that the world was not yet ready for it is no answer. If Mendel was ready for it, why not some others? The question is sufficiently important for some of the basic tenets in the history of ideas to be studied more carefully. What then are the possible causes?

The first, of course, is that Mendel published very little. Of the immense amount of data which he must have accumulated between 1856 when he started his work and 1871 when he discontinued his crosses, he published only his lecture to the Brünn Natural History Society and one other short paper, on hawkweed crosses (1870). To put it mildly, Mendel was not a prolific author. From his correspondence with Nägeli (Stern and Sherwood, 1966) we know that he found the *Pisum* results completely confirmed in crossing experiments with *Matthiola annua, M. glabra, Zea,* and *Mirabilis,* work done in 1869. Alas, this was long before the days of the admonition, "Publish or perish," and Mendel never informed the world of this confirmation of his earlier discovery that had been announced in a single publication.

The *Verhandlungen* of the Brünn Society were sent to the libraries of 115 or more institutions, including the Royal Society and the Linnean Society in Great Britain. Mendel had forty reprints made of his paper, and we know that he sent them, presumably among others, to two famous botanists: A. Kerner von Marilaun at Innsbruck (well-known for his transplant experiments) and Nägeli, one of the leading botanists of his time and known to Mendel as a student of hybrids. This resulted in an active correspondence with Nägeli of which, unfortunately, only Mendel's letters have survived. It is quite evident that Nägeli either did not understand Mendel's thesis or, more likely, opposed it. Instead of encouraging Mendel, he apparently did just the opposite, nor did he invite Mendel to publish his results in one of the prestigious botanical journals where they would have come to the attention of others. Instead, he encouraged Mendel to test his theory of inheritance in the hawkweeds (*Hieracium*), a genus in which, as we now know, parthenogenesis (apomixis) is common, leading to results that are incompatible with Mendel's theory. In short, as one historian has put it, "Mendel's connection with Nägeli was totally disastrous." When Nägeli in 1884 published his great book on evolution and inheritance, he entirely failed to mention Mendel even a single time in a long chapter dealing with hybridization experiments. This is almost unbelievable since everything else in this chapter is far less significance than Mendel's work. Was Nägeli contemptuous of the Catholic priest in far away Moravia? Or was it simply scientific intolerance? Presumably it was the latter. It is rarely pointed out the Nägeli was one of the few biologists who subscribed to a theory of pure blending inheritance (Mayr, 1973: 140). During fertilization the maternal and the paternal idioplasms blend, according to him, owing to the *fusing* of the homologous strings of micelles into a single strand. To accept Mendel's theory would have meant, for Nägeli, a complete refutation of his own. Without studying Mendel's work as carefully as he should have done, Nägeli simply concluded that Mendel must be wrong (Weinstein, 1962).

Mendel's modesty did not help his case. After having been snubbed by Nägeli, he apparently made no effort to contact other botanists or hybridizers or to lecture at national or international meetings. He referred to his seven years of work involving more than 30,000 plants as "one isolated experiment"!

Mendel was fully aware of the fact that the situation in the pea is an unusually simple one. This no doubt is the reason why he chose this species as his principal material. Almost all the complications of chromosomal inheritance which have since been discovered were already present in one or the other of the species of experimental plants with which Mendel had worked. With the means available to him, he would surely have been stumped by the complications introduced by linkage, crossing over, and polyploidy. Indeed, the apomixis of *Hieracium* later thwarted him completely. Mendel therefore was under the impression that his findings were perhaps not true for all species of plants, and said himself, "A final decision can be reached only when the results of detailed experiments from the most diverse plant families are available" (1866: 2). Perhaps Mendel's attitude, in this instance, was adversely affected by his training in physics. The physicists (at least in Mendel's times) always searched for general laws. Hence, the "laws" which Mendel had found for peas would be valid only if they also applied to *Hieracium* and to all other plants. Did Mendel think his *Pisum* laws were not valid, because he found some other kinds of plant to which these laws did not seem to apply?

724

As I pointed out above, there was another weakness in Mendel's approach. When he decided that "the validity of the laws proposed for *Pisum* needs confirmation" (1866: 43), he turned to the hybridizing of species. Even though Mendel realized that this was not quite the same as hybridizing varieties (p. 39), nevertheless the work on species hybrids made him unsure and unwilling to promote his *Pisum* results as vigorously as they deserved. He was particularly puzzled by alleged constant species hybrids. In this Mendel was not alone. The nature of species was what hybridizers were most interested in and, prior to 1900, Mendel's crosses of species of beans (*Phaseolus*) and hawkweeds (*Hieracium*) were mentioned by the hybridizers (from Nägeli to Hoffmann and Focke) rather than his Mendelian ratios of varieties of peas.

For a long time after 1900 it was widely believed that continuous variation obeyed entirely different laws of inheritance than those of Mendel, and this might add another reason for the neglect of Mendel's work. After all, gradual continuous variation was widely considered after 1859 as the only variation of interest to the evolutionist.

Historians have determined that Mendel's work had been cited about a dozen times prior to 1900. The most important of these is Focke's great review book *Die Pflanzen-Mischlinge* (1881). Anyone subsequently working on hybridization consulted Focke, and nearly all those who referred to Mendel after this date stated that they had discovered the reference in Focke. Yet, Focke himself never realized the importance of Mendel's work and referred to Mendel's work in a way that would not encourage anyone to consult the original paper.

In 1864 Mendel was forced to discontinue his *Pisum* work owing to the heavy infestation of pea weevils and to exciting results in other plant genera. He abandoned all of his crossing work in 1871 after he had been elected abbot of his monastery and had become too preoccupied with administrative burdens. After Mendel had died of nephritis in 1884 at the relatively young age of 62, it took another sixteen years before the world appreciated the greatness of his discovery.

725

Finally, it should be mentioned that the rediscoverers of Mendel (particularly Correns), with their advanced understanding of cytology, read more into Mendel's account than is actually there. Heimans and Olby deserve credit for having pointed out deficiencies in Mendel's interpretation. This by no means diminishes Mendel's greatness. But by showing that his theory was not as complete and therefore not as fully explanatory as had been claimed by geneticists for three quarters of a century, Heimans and Olby make it easier to understand why the work was ignored for 34 years.

For reasons that are not at all clear, Mendel's age was not particularly interested in "pure" transmission genetics. Inheritance was generally considered only in connection with other biological phenomena, such as the species problem (and that of species hybrids), environmental induction (and the inheritance of acquired characters), differentiation during development, the consolidation of species characters in isolation and their breakdown ("blending") following the removal of the isolating barrier, and so forth. There has been much speculation as to what effect Mendel's paper would have had on Darwin had he read it. I agree with those who think it would have had little influence if any. It took many years (after 1900) before the "true Darwinians," as they liked to call themselves, understood that gradual evolution and continuous variation could be explained in Mendelian terms. Darwin, presumably, would have had the same difficulty. He knew Sageret's work, but apparently it did not help him to understand variation. And when it comes to the problems in which Darwin, as an evolutionist, was most interested, such as "the mysterious laws of correlation," the acquisition of reproductive isolation, and the establishment of the "cohesion of the genotype," even we are still very much in the dark, eighty years after the rediscovery of Mendel.

Without any knowledge of chromosomal cytology, without the theoretical analysis by Weismann, and without the benefit of the many other seminal discoveries made between 1865 and 1900, Mendel had discovered a new way of looking at the phenomena of inheritance, he emphasized the behavior of unit characters and used this insight to arrive at far-reaching generalizations. His achievement was one of the most brilliant in the history of science. Mendel was a dedicated scientist, reflected in the enthusiasm with which he reports his findings to Nägeli (April 18, 1867): "Every day from spring to fall, one's interest is refreshed daily, and the care which must be given to one's wards is thus amply repaid. In addition, if I should, by my experiments, succeed in hastening the solution of these problems, I should be doubly happy" (Stern and Sherwood, 1966).

His short treatise, "Experiments on Plant Hybrids," as Curt Stern has characterized it so well, "is one of the triumphs of the human mind. It does not simply announce the discovery of important facts by new methods of observation and experiment. Rather, in an act of highest creativity, it presents these facts in a conceptual scheme which gives them general meaning . . . [Mendel's classic] remains alive as a supreme example of scientific experimentation and profound penetration of data" (Stern and Sherwood, 1966: v).

17 ✐ The flowering of Mendelian genetics

DARWIN'S *VARIATION* (1868), de Vries's *Intracellular Pangenesis* (1889), and Weismann's *Germ Plasm* (1892) created an accelerating interest in the problem of inheritance.[1] Hugo de Vries and Carl Correns began in 1892 with systematic crossing experiments, and both published in 1899 important results of their experiments on xenia (endosperm formation by the pollen nucleus; see Dunn, 1966). Then, in the spring of 1900 one of the most extraordinary events in the history of biology took place with seemingly explosive suddenness, though it was actually only the climax of a long development. Three botanists—de Vries, Correns, and Tschermak—within a period of a few months published statements that they had independently discovered certain laws of inheritance, only to find, when checking the literature, that Mendel had anticipated them by thirty-five years. Ever since that memorable spring, suspicions have been expressed as to whether the statements made by the three rediscoverers were to be taken as literally true. The problem seems to be important enough to be looked into a little more closely.[2]

THE REDISCOVERERS OF MENDEL

De Vries in his 1889 *Intracellular Pangenesis* had clearly formulated the view that inheritance was to be dissected into unit characters, each inherited independently.[3] He had also outlined a program of experimentation. Being preoccupied at that time with physiological experiments, he did not seriously begin crossing experiments until 1892, at first concentrating on *Silene, Papaver,* and *Oenothera.* In 1894, among 536 F_2 plants of *Silene,* he found 392 hairy and 144 smooth ones (2.72:1). In an F_2 cross of poppies in 1895 he got 158 black to 43 white petal spots (3.67:1), and found in 1896 that the white spotted ones bred true. Other experiments in these years reinforced these findings. By the autumn of 1899 he had obtained clear segregation in over 30 different species and varieties. He was finally convinced that the segregation of alternate characters obeyed a general law and that he was justified in publishing his results. In March 1900, he submitted three papers within a few weeks of each other, describing his findings, two to the Academie des Sciences in Paris (to be presented at the session of March 26, 1900) and one to the German Botanical Society, received March 14 (see Krizenecky, 1965). The Paris papers were actually published a few days earlier (prior to April 21) than the German paper (April 25). In a footnote to the German paper, de Vries wrote: "I first learned of its [Mendel's paper] existence after I had completed the majority of my experiments and had deduced from them the statements communicated in the text." Olby (1966: 129), on the basis of a good deal of indirect evidence, concluded that de Vries might have read Mendel's paper as early as 1896 or 1897; Zirkle (1968) thought it was not until 1899 and Kottler (1979) found further evidence for the later date.

728

In his notes for lectures in these years de Vries still used his own terminology—active (A), latent (L)—instead of Mendel's dominant and recessive, and in a demonstration plate for students he used variable percentages for segregation (77.5%:22.5%, 75.5%:24.5%), as if he were not yet aware of the true causation of segregation. It is also worth mentioning that of his very numerous *Oenothera* crosses, he referred in his 1900 paper only to the *lamarckiana* × *brevistylis* cross, the latter being the only genuine gene mutation he had found in his *Oenothera* material.[4] As he clearly stated in his correspondence with Bateson, de Vries made a distinction between progressive and derivative characters, only the latter obeying the Mendelian rules.

De Vries says that he had found the reference to Mendel in the bibliography of an article published in 1892 which he apparently consulted a few years later and which induced him to read Mendel's original publication. There is no doubt that at that time he had already found segregation ratios which we would now interpret as 3:1 ratios, as well as the true-breeding of the recessives, but this does not necessarily mean that these findings had induced him to abandon his earlier erroneous notions. Like all other investigators of the 1880s, de Vries had originally believed that characters might be controlled by multiple particles (see Chapter 15). Ratios like 394 to 144, or 158 to 43, or 77.5%:22.5% do not mean anything if one believes in replicate factor determination. When using ratios, de Vries referred to 2:1 or 4:1 ratios (Kottler,

1979). Did the reading of Mendel's paper cause him to abandon his original theory and adopt Mendel's theory of a single element from each parent determining an individual character? We shall never know. As it is, we must accept de Vries's statement that he "had deduced" from his own experiments the law of segregation, just like Mendel had derived this law from similar results. By concentrating on the experimental analysis of unit characters, de Vries had certainly come very close to the solution. It was only a small step to abandon the last erroneous component (frequent pangene replication) of his earlier theory. Yet, Bateson failed to find the Mendelian explanation, in spite of good Mendelian ratios, prior to reading the de Vries paper.

729

Clearly de Vries was deeply disappointed at having been anticipated by Mendel, and this may be one reason why he did not pursue the more strictly genetic consequences of his findings but shifted instead to the evolutionary interpretation of progressive mutations. Speciation is where his major interest seems to have been all along. Evidently, de Vries thought that Mendelian inheritance was only one of several genetic mechanisms. How else can we explain his statement to Bateson (October 30, 1901), "It becomes more and more clear to me that Mendelism is an exception to the general rule of crossing." Hence, he more or less abandoned Mendelism to study other forms of inheritance which he considered to be of more importance for evolution.

For three reasons, de Vries will always be remembered as a great figure in the history of genetics: (1) because independently of Mendel he promoted the idea of dissecting differences among individuals into unit characters; (2) because he was the first to demonstrate the operation of Mendelian segregation in a wide variety of plant species; and (3) because he developed the concept of the mutability of genetic units. Thus, he was far more than a rediscoverer of Mendel. De Vries had, of course, a great advantage over Mendel. He was able to make use of the results of then recent cytological research when developing his theory. While Mendel wisely refrained from speculating on the nature of the "Elemente," the physical basis of his characters, de Vries related them to redefined Darwinian pangenes. With respect to inheritance, he synthesized Darwin and Mendel.

The case of Carl Correns (1864–1933),[5] the second rediscoverer of Mendelian inheritance, is more clear-cut. He states that the interpretation of Mendelian segregation came to him "like a flash" as he lay awake in bed toward morning (in October 1899).

He was busy with other researches, however, and did not read Mendel's memoir until a few weeks later (but referred to it in December 1899 in his xenia paper). Only when, on April 21, 1900, he received a reprint of de Vries's French Academy paper did he write up (in a single day) his results, which were reported in the April 27 session of the German Botanical Society and published about May 25. Correns, from the beginning, did not consider his part in the rediscovery very important, and includes a reference to Mendel ("Mendel's rule") in the title of his first communication. He realized that "the intellectual labor of finding out the laws anew for oneself was so lightened [through the research of the past 30 years, particularly the work of Weismann] that it stands far behind the work of Mendel." The only thing that might be suspect about Correns' independent rediscovery of Mendel is the fact that he was a student of Nägeli (whose niece he married) and might have known of Mendel's work all along. This possibility, however, is made implausible by the fact that it would have been most peculiar if Correns had not followed up this clue much sooner, if it had been available to him for twenty years.

The third person who is always listed as another independent rediscoverer of Mendel's rules is the Austrian plant breeder Erich Tschermak. As Stern (1966: xi) has shown, there is little justification in including Tschermak among the rediscoverers. He had indeed found Mendel's paper but had failed in his 1900 papers to understand the basic principles of Mendelian inheritance. Nevertheless, Tschermak had an important share in directing the attention of the plant breeders to the importance of Mendelian genetics.

Just why so many of the early Mendelians (Mendel, de Vries, Correns, Tschermak, Johannsen) were botanists has never been fully explained. Presumably there was a richer tradition of breeding varieties among horticultural and other cultivated plants, because plants are so much easier to cultivate and breed than animals. Perhaps there are also more discontinuous characters in leaves and flowers than one finds in domestic animals such as sheep, cattle, and pigs. Most characters studied by animal breeders were highly polygenic and not at all suitable for an elementary Mendelian analysis. Still, soon after 1900 Bateson started to work on the domestic fowl, Cuénot in France and (in 1902) Castle in the United States began to work on rodents, and in 1905 Castle introduced *Drosophila* as an experimental animal. Soon the work in animal genetics caught up with plant genetics and surpassed it when the schools of Morgan and Chetverikov got going. Already in 1914 A. Lang required 890 pages to report merely on the results of mammalian genetics obtained since 1900.

Plants (even higher plants) have a much richer diversity of genetic systems than animals. This can be very misleading for one who wants to establish universal "laws." Examples are the apomict systems of *Hieracium* which frustrated Mendel, the balanced heterozygous chromosome rings of *Oenothera* which led de Vries into an erroneous speciation theory, and the self-fertilizing near-homozygous beans (*Phaseolus*) which led Johannsen to belittle natural selection.[6] Cytoplasmic effects are apparently much more common in plants than in animals and have monopolized the attention of many plant geneticists (especially in Germany) without yielding (in the pre-molecular period) particularly interesting results. On the other hand, the plant kingdom has supplied not only the pea but also cereal species, particularly wheat, barley, maize, cotton (Gossypium), tobacco, and many other genetically highly informative species. No one has yet undertaken a comparative analysis of the positive (and negative) contributions made by the various species of animals and plants used in genetic research. Much work, it must be admitted, merely produced a confirmation of something already established by work on *Drosophila* or maize. Prior to the molecular period most genetic work was done either in botany or in zoology departments, and the interactions among plant and animal geneticists were not always as active as might have been desirable. After the 1930s lower plants (algae, fungi, yeasts) and prokaryotes (bacteria, viruses) became increasingly the favorite material of the geneticists. Realization of the pronounced differences in the genetic systems of eukaryotes and prokaryotes has revived an interest in the genetics of eukaryotes since the 1960s.

731

THE CLASSICAL PERIOD OF MENDELIAN GENETICS

The early history of genetics falls into two periods, the first from 1900 to about 1909, the second beginning with 1910. The earlier period, often designated as Mendelism, was preoccupied with evolutionary controversies and with doubts as to the universal validity of Mendelian inheritance. This period was dominated by de Vries, Bateson, and Johannsen, who have often been designated

as "the early Mendelians." The term "Mendelism" conveys different meanings to different people, depending on what aspect of
Mendelism one wants to emphasize. To the members of the genetics establishment it refers to the period in which particulate
inheritance was nailed down and the hardness of inheritance was
emphasized. To the evolutionists it means a period in which utterly erroneous ideas about evolution and speciation were pro-
mulgated by leading geneticists and during which mutation pressure was considered far more important than selection, ideas which
resulted in the alienation of the naturalists. The same term "Mendelism" was thus used sometimes with an approving and sometimes with an unfavorable connotation.

732

The second period, beginning in 1910 and dominated by the
Morgan school, was occupied much more intensively with purely
genetic problems such as the nature of the gene and the arrangement of the genes on the chromosome. The term "genetics," proposed by Bateson in 1906, was in due time accepted for this
broadened concept of the science dealing with inheritance.

It took 34 years for Mendel's publication to be rediscovered,
but the subsequent dissemination of Mendel's findings occurred
at an unprecedented rate. Both Correns and Tschermak learned
of de Vries' paper at the end of April 1900 and published their
own findings in May and June. In Great Britain William Bateson
reported on Mendel's experiments at the May 8 meeting of the
Royal Horticultural Society, and in France Cuénot also soon referred to Mendel's work.

As is true for most major scientific movements, subsequent
progress occurred at very different rates in different countries.
Great Britain, without any question, assumed leadership in Mendelian genetics, soon to be followed and eventually to be over-
taken by the United States (Castle,[7] East, Morgan, and others).
German genetics continuing the tradition of the 1880s, concentrated on developmental genetics and on unorthodox phenomena
(actual or seeming cytoplasmic inheritance, protozoan genetics, and
so forth). In France, after the promising beginnings made by
Cuénot,[8] not much happened until the 1930s. In Russia, as Gaissinovitch (1971: 98) has remarked, "Genetics began to develop as
a branch of science only in the Soviet period." In the nonwestern
world no science of genetics ever originated. Where genetics
flourished and in what direction it developed depended entirely
on the leading personalities in the field. Curiously, however, neither Correns nor de Vries played a major role in the subsequent

advances of Mendelian genetics. The major credit for this, at least in the early years, must be given to William Bateson (1861–1926),[9] who appreciated the importance of Mendel better than the so-called rediscoverers (Darden, 1977).

Bateson had been interested in discontinous variation (see Part II) since his stay in Professor W. K. Brooks's laboratory at Johns Hopkins University (1883, 1884) and had conducted breeding experiments since the 1880s, but intensively only since about 1897. On July 11, 1899, he presented a paper to the Royal Horticultural Society entitled "Hybridization and Cross Breeding as a Method of Scientific Investigation." It is evident from this lecture that at that time he had still not yet developed a theory of inheritance, in spite of many results now easily interpreted in Mendelian terms. The illumination did not occur to him until he read Mendel's original paper on May 8, 1900 (on the train from Cambridge to London). At once he became an enthusiastic Mendelian, had Mendel's paper translated, and published it with footnotes in the *Journal of the Royal Horticultural Society* (1900). Part of Bateson's enthusiasm was due to the fact that he saw in segregation a confirmation of his (erroneous) thesis (1894) that speciation is the result of discontinuous variation. De Vries had the same evolutionary theory and he also saw the discontinuity of Mendelian factors as important evidence for his saltational theory of speciation. Thus, paradoxically, much of the publicity and attention which Mendel got was for peripheral, if not the wrong, reasons. The opposition which the Bateson–de Vries thesis aroused is treated in Chapter 12 and I shall deal here only with Bateson's contributions to transmission genetics.

733

It is Bateson to whom we owe some of the most important technical terms in this field. He coined the term *genetics* for the new science (1906) and also (1901) the terms *allele* (orginally allelomorph), *heterozygote,* and *homozygote.* Availability of these semantically unambiguous terms greatly facilitated communication during this period. But Bateson and his collaborators were also responsible for important factual contributions to our understanding of inheritance. They were the first to discover certain deviations from the simple Mendelian situation (for example, polygeny and incomplete linkage). Through him, genetics gained an impetus in Britain that was entirely lacking in any other European country.

Bateson was a complex personality, pugnacious to the point of rudeness in his controversies, but at the same time completely

dedicated to research. He was a peculiar mixture of a revolutionary and a conservative who found it very difficult to accept new
ideas. In the first ten years after 1900 he was the major spark plug
of genetics; indeed, there is much justification in Castle's statement (1951) that Bateson "was the real founder of the science of
genetics." After 1910, however, his opposition to the chromosome
theory (see below) and his continuing advocacy of instantaneous
speciation seemed no longer constructive. In his role as a revolutionary he made the immortal remark (1908: 22): "Treasure your
exceptions; when there are none, the work gets so dull that no
one cares to carry it further. Keep them always uncovered and in
sight. Exceptions are like the rough rock work of a growing building which tells that there is more to come and shows where the
next construction is to be." In his own research he concentrated
very much on actual or seeming exceptions, and some of his important discoveries were the result of his following this motto.

734

Advances in Mendelian Genetics

The rate at which the new finds of genetics occurred after
1900 is almost without parallel in the history of science. Whether
we look at Lock's textbook of genetics (1906: particularly 163–
275), or Bateson's textbook (1909), we are surprised at the maturity which the understanding of Mendelian inheritance had
reached so soon after 1900. What were the reasons for this rapid
progress? One of them, of course, was that the beauty and simplicity of the new theory invited anyone to undertake genetic experiments to test its universality. Since the field was brand new,
almost anybody had a chance to make new discoveries. The Mendelian laws permitted predictions about modes of inheritance and
an immediate testing of these predictions. A second reason is more
doubtful: the magnificent achievements which cytological research
had made in the thirty-five years prior to 1900 had laid such a
sound foundation that it should have been possible to explain almost any purely genetic discovery in cytological, and more specifically in chromosomal, terms. Chromosomal cytology formed a
bridge to other areas of biology, and it was a bridge which had
been built before it could be used. But, curiously, even after it
could be used, it was almost completely ignored by the geneticists,
such as Bateson, Castle, and East, prior to Morgan.

The knowledge of the mechanism of inheritance was used to
shed new light on phenomena in various areas of biology, such as

evolutionary biology (see Chapters 12 and 13) or developmental physiology (a subject I will cover in a subsequent volume). In the following discussion the emphasis will be on aspects of transmission genetics.

Semidominance

Among the seven pairs of characters which Mendel had analyzed, he recognized only two variants of each pair: those that are dominant and those that are recessive. But this does not hold for all character pairs, as Mendel himself discovered. He remarked   *735* that flowering time "is almost exactly intermediate between that of" the parent plants. Correns (1900) found, likewise, that certain factors are not fully dominant but only "semidominant," thus producing an F_1 phenotype somewhat intermediate between that of the two parents. Two years later, Bateson found such semidominance when crossing white with black fowl. The F_1 was the blue Andalusian fowl.

This not only confirmed semidominance but also established the fact that Mendel's laws are as true for animals as they are for plants. At about the same time Cuénot demonstrated this on the basis of work on the coat color genes of the house mouse. Considering the fact that cells and nuclei of plants and animals show completely equivalent phenomena, this finding was perhaps not altogether unexpected. Nevertheless, the discovery that the Mendelian laws of heredity are valid in both kingdoms further contributed toward breaking down the ancient barrier between zoology and botany.

The Gene, the Unit of Inheritance

Prior to 1909 no generally accepted term was available for the genetic factor underlying a given visible character. Spencer, Haeckel, Darwin, de Vries, Weismann, and others who speculated about inheritance had postulated the existence of certain corpuscles with various qualities, but the names they had used were not widely adopted (see Chapter 16).

Mendel kept his speculation about the nature of the genetic material to a minimum, a wise decision on his part, considering the rudimentary understanding of nucleus and chromosomes in 1865. He referred in his experiments to traits ("Merkmale") and characters ("Charaktere"), essentially restricting himself to the phenotypic level, even though the symbols *A, Aa, a* used by him are generally considered to refer to the constitution of the genotype. He used the term "elements" ten times in his concluding remarks (1866: 41–42), several times very much as we would now use the work "gene," but he had no clear concept concerning the genetic material. Regardless of what was really in Mendel's mind, what he had described meant to the early Mendelians that which we would now call Mendelian inheritance.

The terms "phenotype" and "genotype" had not yet been coined in 1900, although Weismann made the implicit distinction between germ plasm and soma. For de Vries there was no real difference between the genetic material and the body (phenotype) since his pangens moved freely from the nucleus to the cytoplasm. A pangen corresponded for him to an elementary or unit character. He postulated the existence of a separate hereditary basis for each independently inheritable character. Sometimes de Vries referred to the genetic elements also as "factors," and Bateson, as well as the Morgan school, adopted this terminology in the beginning.

736

Like de Vries, Bateson also failed to make a clear distinction between an underlying genetic factor and the resulting phenotypic character. He referred to "unit characters" that "are alternative to each other in the constitution of the gametes" (1902). In order to be able to refer to such alternate conditions, like smooth versus wrinkled in the pea, Bateson introduced the term *allelomorph,* later shortened to *allele.* But again he failed to make a distinction between the somatic character and its determinant (gene) in the gamete. For a number of reasons, prior to about 1910 the silent assumption was made almost universally that there is a 1:1 relation between genetic factor (gene) and character. Hence, when one spoke of a unit character, it did not really matter whether one meant the underlying genetic basis or its phenotypic expression. It was, in part, this automatic assumption which led Castle to propose his contamination theory.

With the rapidly increasing genetic activity after 1900, the need arose for a technical term designating the material basis of an independently heritable character. The Danish geneticist W. L. Johannsen (1857–1927), realizing how similar the Mendelian factors behaved to de Vries's postulated pangens, proposed in 1909 to adopt a shortened version of pangen—*gene*—for the material basis of a hereditary character. Johannsen was a physicalist, and the last thing he wanted was to provide a definition of the term "gene" that was tainted by preformationist language. He chided those who had "a conception of the gene as a material, morphologically characterized structure which is very dangerous for the smooth advance of genetics; a conception which we must urgently warn against" (1909: 375). Consequently, instead of providing a definition for the gene, he merely said "the gene is thus to be used as a kind of accounting or calculating unit [*Rechnungseinheit*]. By no means have we the right to define the gene as a morphological structure in the sense of Darwin's gemmules or [Weismann's] biophores or determinants or other speculative morphological concepts of that kind. Nor have we any right to conceive that each special gene corresponds to a particular phenotypic unit-character or (as morphologists like to say) a 'trait' of the developed organism" (1909).

737

This definition reflects a conflict permeating biology at that period. The physicalists—and Johannsen, owing to his training, was strongly influenced by them—wanted to interpret everything in terms of forces. Embryologists coming from an epigenetic tradition were likewise quite unwilling to accept a corpuscular gene because it reminded them of preformation. Morgan's original reluctance to recognize genes, or at least corpuscular genes, was due to such reservations. Finally, there was still some influence of essentialism, which objected to any partitioning of the species essence. In 1917 Goldschmidt castigated the extreme caution of geneticists toward the gene: "We believe that this intellectual attitude toward the problem is the result of Johannsen's doctrine of agnosticism in regard to the nature of the gene, which resulted in a kind of mystic reverence, abhorring the idea of earthly attributes for a gene." In due time, of course, it was proven that the gene has precisely those (structural) characteristics which Johannsen had so carefully excluded from his definition. Indeed, from Morgan through Muller to Watson and Crick there was an ever closer approach toward a structural concept of the gene. Johannsen's term "gene" was soon universally adopted, since it filled a great need for a technical term designating the unit of inheritance. Yet, the absence of a definition was in part responsible for some of the controversies in the ensuing years. A further source of confusion was provided by the fact that, almost up to the present time, authors have been inconsistent in what they mean by gene. When talking about the white-eye gene of *Drosophila,* for example, some authors meant the white-eye allele, while others meant the locus at which the white-eye mutation had occurred, which is also the locus of all the alleles of white-eye.

The way from the coining of the word "gene" for the invisible, submicroscopic unit of inheritance to a full understanding of its nature was a long and tortuous one. Numerous geneticists, foremost among whom was H. J. Muller, devoted virtually their scientific career to this quest. In the end, as we shall see, it was found (in the 1950s) that the portion of the macromolecule which functions as the gene has indeed the structural complexity and specificity which Johannsen had rejected. How to get at the mystery of the gene was at first a very puzzling problem. Morgan and his associates quite rightly decided that to study changed genes, that is, "mutations," might be a hopeful entering wedge.

738

The Origin of New Variation (Mutation)

With the rediscovery of Mendel's law of segregation, the problem of the origin of genetic variation became acute. The existence of alleles demanded an explanation. Darwin had postulated a continuous replenishment of variation in order to have an abundant supply available on which natural selection could act. However, he was unable to account for its source. The time had now come to solve Darwin's puzzle, but at first the Mendelians made little progress in this. Indeed, they had to overcome formidable obstacles.

The main difficulty was that most students of variation still distinguished two kinds of variation. Darwin, for instance, recognized "many slight differences which may be called individual differences" (*Origin: 45*), later called individual variation, continuous variation, or fluctuating variation. His belief in the importance of this variation was one of the corner stones of his theory of evolution. However, Darwin also admitted that "some variations . . . have probably arisen suddenly, or by one step" (p. 30), and mentioned as examples "sports," like the turnspit dog and the ancon sheep. Bateson called these discontinuous variation. The belief in two kinds of variation had a long history which is intimately connected with Plato's concept of the *eidos* (essence). An essence is subject to accidental variation of minor amplitude, while any major deviation is possible only through the sudden origin of a new essence, that is, a new type. It was thought that the two kinds of variation had entirely different causations and that they played a very different role in evolution. This was the major bone of contention particularly in the battle between the biometricians and the Mendelians (see Chapter 12) but actually from the time of Lamarck to the evolutionary synthesis of the 1940s. De Vries's essay on variation (1909) illustrates the depth of the confusion (see also Mayr and Provine, 1980).

Individual or Continuous Variation

If one accepts the existence of soft inheritance, then one has no difficulty in accounting for individual variation. Any change of internal conditions or environmental influences (like nutrition or climate) could affect any character of an individual and change it. As Darwin explained it, "In the cases in which the organization [of the body] has been modified by changed conditions, the increased use or disuse of parts, or any other cause, the gemmules cast off from the modified units of the body will be themselves modified, and, when sufficiently multiplied, will be developed into new and changed structures" (1868, II: 397). Other believers in soft inheritance adopted similar explanations. Old characters would grade into new ones, yet the difference between them would be slight, manifesting itself as continuous variation. If any new genetic variation should originate by some unknown process, it would likewise be subject to soft inheritance and grade into previously existing variation. The essence, it was accepted, had the capacity to give rise to continuous individual variation. There was no major explanatory problem. The idea that the environment could affect genetic variability was widely held by animal and plant breeders (Pritchard, 1813; Roberts, 1929).

The situation changed fundamentally in 1883, when Weismann rejected the existence of soft inheritance. If "the conditions of life" cannot produce new variations and not even increase variability, what, then, can be the cause of individual variation? Neither Weismann nor de Vries had a sound theory for this, and the attention of the early Mendelians was so centered on discontinuous variation that they paid little attention, or none, to the problems of individual variation. How to reconcile discontinuous Mendelian factors with continuous variation puzzled them greatly.

It was not simply the absence of the right kind of information that delayed the solution of this problem but also the silent acceptance of a number of misconceptions. These included, in addition to the belief in two kinds of variation, the acceptance of soft inheritance (in spite of Weismann), of blending inheritance (in spite of Mendelism), of typological thinking, and a confounding of phenotype and genotype. A direct attack on the problem of the genetics of continuous variation and on the origin of its new components was not yet possible in the face of all these difficulties and

739

misconceptions. The solution actually came in through the back door via a study of discontinuous variation (even though it was believed to have nothing to do with continuous variation).

Discontinuous Variation

That an occasional individual may fall way outside the norm of variation of the population to which it belongs had already been known among the ancients. It was observed among wild animals, among domestic animals and cultivated plants, and even in man. Any variant that fell outside the normal variation of a population was a case of discontinuous variation. Albinos, individuals with six digits, or indeed any kind of freaks were described in the popular literature with fascination. In the fifteenth and early sixteenth centuries, when nature was credited with an enormous capacity for "generation," that is, for the origin of new things, monsters were described with loving detail, most of them real animals with birth defects (like two-headed calves), others purely mythical creatures, like chimaeras displaying a combination of human and animal body parts.[10]

In 1590 the apothecary Sprenger in Heidelberg discovered in his herb garden a celandine (*Chelidonium majus*) with an entirely different leaf-shape. He was able to propagate this plant and distribute its seeds widely. In due time specimens of it could be found in all major herbaria in Europe and descriptions of it in most seventeenth-century books of plants. The new variant was generally treated as a new species of *Chelidonium*. Three hundred and ten years later a similarly aberrant plant (in the genus *Oenothera*) inspired de Vries to propose a major new theory of evolution.

Conspicuously aberrant variants were found quite regularly among cultivated plants; indeed, they had given rise to many of the best-known horticultural varieties (particularly when affecting the color or shape of the flower). They were likewise discovered among domestic animals, such as hornless individuals in cattle, or sheep characterized by very short legs, a breed (ancon) quite popular for a while because these sheep were unable to jump fences or walls. In all these cases the breeders were able to develop pure lines by back-crossing to the parents and subsequent inbreeding, and reported what we would now call strictly Mendelian inheritance. There was no "blending," no gradual return to the parental type, in contrast to Kölreuter's findings with species hybrids. This fact, curiously, was completely ignored by Jenkin and Darwin in their famous controversy over blending inheritance (see Chapter 11).

By far the most celebrated case of an aberrant variant is that of the so-called *Peloria*. In 1741 a student at Uppsala brought a specimen of a plant to Linnaeus which at first sight seemed to be an ordinary butter-and-eggs plant (*Linaria*), being identical with it in its growth form, peculiar smell, characteristic color of the flower, calyx, fruit, and pollen.

Yet, while the common *Linaria* has the typical asymmetrical flower of a snapdragon, the *Peloria* had a radially symmetrical flower with five spurs. Linnaeus came to the conclusion that "this new plant propagates itself by its own seed, and is therefore a new species, not existing from the beginning of the world." More than that, according to the method of Linnaeus, *Peloria* was not merely a new species or genus but represented an entirely different class of flowers. This not only shook Linnaeus' concept of the constancy of species but seemed to refute even his axioms of classification (Larson, 1971: 99–104). At first Linnaeus thought hybridization was involved, but this idea soon had to be discarded. Ultimately, *Peloria* was shown not to be as constant as it had at first seemed, and Linnaeus finally decided to forget about this annoying "species," not even mentioning it in the *Species Plantarum* (1755).

More and more frequently such aberrant individuals or new varieties were found in the hundred years after Linnaeus, but they did not provide any new insight. Yet, a subtle change of emphasis during this period was perceptible. For Linnaeus and his contemporaries, such variants were discussed strictly in relation to the species concept. But as evolutionary thinking gradually emerged, varieties and the mode of their origin acquired a new significance. Unger's interest in this problem was, as we have seen, the stimulus for Mendel's experiments. After the publication of the *Origin*, variants were considered more and more frequently in relation to evolution.

The sudden appearance of seemingly new species was nothing but disturbing to the fundamentalists who believed in only a single episode of creation. By contrast, it was a comforting observation to those who were aware of the continuing extinction during geological time and who had to postulate new creations to fill the gaps. In the post-Darwinian period it was even more appealing to those evolutionists who were basically essentialists and could therefore envision speciation only as a process of sudden new origins (see Chapter 12).

Darwin's strong emphasis on the gradual nature of evolution—that is, the evolutionary importance of continuous variation—did not convince all of his contemporaries. Huxley, Kölliker, Galton, and others favored a saltational origin of new species and types by discontinuous variation. No one, however, was more convinced of the importance of discontinuous variation than Bateson (1894), who collected a formidable amount of material to prove his point (see Chapter 12).

De Vries and Mutation

742

It was not until after the rediscovery of Mendel's rules that these views on discontinuous variation matured into a major evolutionary theory, de Vries's *Die Mutations-theorie* (1901; 1903; see Chapter 12 for the role of this theory in evolutionary biology). When developing his new theory of inheritance, de Vries not only crossed varieties of cultivated plants but also studied variation in natural populations. In 1886 in a large population of the evening primrose *Oenothera lamarckiana* growing in an abandoned potato field in Holland, he found two plants which he considered to be sufficiently different from all the other individuals to be treated as a newly arisen species. When self-fertilized in de Vries's experimental gardens, they remained absolutely constant. Additional new types originated from individuals of *Oe. lamarckiana* which de Vries had transplanted from the old field to his gardens. In due time, in addition to many minor variants, more than twenty individuals were found by de Vries which he considered to be new species, and which indeed were constant when self-fertilized.

De Vries introduced the word *mutation* for the process by which these new "species" had originated. It may be useful to say a few words about this term, considering its great importance in the theory of inheritance. The term was used for any drastic change of form at least as far back as the middle of the seventeenth century (Mayr, 1963: 168). From the very beginning it was used both for discontinuous variation and for changes in fossils. In 1867 the term was formally introduced into paleontology by Waagen for the smallest distinguishable changes in a phyletic series. De Vries was well aware of this usage because he specifically refers (1901: 37) to Waagen. Like so many words in our language (such as "adaptation"), the word "mutation" has been used both for the process and for the product of the process. But there has been even further ambiguity. Sometimes the word was used to describe a change in the genotype, sometimes in the phenotype. To make matters even worse, mutation for de Vries was an evolutionary phenomenon, while in the subsequent history of genetics it became more and more an exclusively genetic phenomenon. This extensive confusion concerning the concept of mutation must be understood before one can appreciate the reasons for the extended controversy over the evolutionary role of mutations.

Although de Vries introduced the word "mutation" for the sudden production of new species, he knew of course nothing about the physical nature of these changes, and in actual practice, he used it as a term designating a sudden change of the phenotype. This was clearly established by subsequent students of *Oenothera,* who were able to demonstrate that nearly all of de Vries's so-called mutations were manifestations of chromosomal rearrangements (including polyploidy), very few of them being gene mutations in the now accepted sense (see below).

743

It required decades of genetic research before the term "mutation" was freed from the handicap imposed on it by its original ambiguity and by de Vries's assertion that mutation is a process of producing new species. De Vries clearly restricted the term to units of discontinuous variation: "The mutations . . . form a special division in the science of variability. They occur without transitions and are rare, while the ordinary variations are continuous and ever present . . . The contrast between these two major divisions, variability in the narrow sense and mutability, is at once evident if one assumes that the attributes of organisms are composed of definite units that are sharply distinct from each other. The occurrence of a new unit signifies a mutation; the new unit, however, is variable in its expression according to the same laws as the other previously existing elements of the species" (1901: iv–v).

Although de Vries was wrong in the evolutionary interpretation he gave to his mutations, he deserves credit for having emphasized, more than anyone before him, the actual origins of new genetic characters. Mendel and other students of inheritance had always dealt with the transmission of already existing factors and characters. De Vries forced attention on the problem of the origin of genetic novelties. Regardless of how much the meaning of the word "mutation" has changed since 1901, mutation from that date on remained an important problem of genetics.

De Vries describes how assiduously he had searched for the ideal plant that would clearly demonstrate instantaneous speciation through mutation. He studied more than a hundred species, but had to discard all but one because their variation did not live up to his expectations. He emphasized how exceptional *Oenothera* was, and yet apparently never realized how dangerous it was to base a fundamental new theory on phenomena observed in a single exceptional species.

Oenothera, as has since been established by the brilliant researches of Renner, Cleland, S. Emerson, and other geneticists (Cleland, 1972), has an extraordinary system of translocation chromosomes, permanently heterozygously balanced (owing to the lethality of the homozygotes). What de Vries had described as mutations actually consisted of segregation products of such chromosome rings. Nothing like it is found in other species of plants or animals (aside from a few rare, similarly balanced systems). De Vries's mutations were neither the source of normal variation, nor the normal process of species formation. Yet, his term "mutation" was retained in genetics because it was rescued by T. H. Morgan, even though he transferred it to a rather different genetic phenomenon.

THE EMERGENCE OF MODERN GENETICS

The year 1910 is almost as famous in the history of genetics as the year 1900; it was the year of Morgan's first *Drosophila* publication. The decade after the rediscovery of Mendel had been dominated by Bateson. Not only had he and his co-workers abundantly confirmed the Mendelian laws, but they had also found and explained a number of seeming exceptions, and Bateson had contributed importantly to the language of the field. It was also in this decade that the continuity and individuality of chromosomes was established by Boveri to the satisfaction of most.

One of those who felt altogether unconvinced by the Sutton-Boveri chromosome theory (see below) was the embryologist T. H. Morgan, E. B. Wilson's colleague at Columbia University in New York.[11] Even though Wilson and Morgan had the highest personal regard for each other and maintained close, friendly relations, at that time they thoroughly disagreed in the interpretation of the relation between chromosomes and inheritance. In 1908 Morgan started to conduct genetic experiments, at first with rats and mice. Perhaps his most fateful decision was to give up working with organisms like mammals, that have long generations, expensive maintenance, and susceptibility to disease. Two other American geneticists, W. E. Castle and Frank Lutz, had worked for years with the fruit fly *Drosophila melanogaster,* which produces a new generation every two to three weeks, can be maintained in discarded milk bottles, and is virtually immune to disease.[12] One further important attribute of *D. melanogaster* is that it has only four pairs of chromosomes, as against ±24 in most mammals. This made *Drosophila* specially suitable for studies in crossing over, such as were needed for the final substantiation of the chromosome theory.

Chromosomes and Mendelian Inheritance

After the mid-1890s a reaction set in against the speculative orgy of the Weismann years. In this new sober spirit the first accounts of the Mendelian laws by de Vries, Correns, and Bateson were rather descriptive, emphasizing ratios and the facts of segregation. Almost at once, however, a few students of inheritance, particularly those with a background in cytology, realized that one had to look for an explanation of the Mendelian phenomena, or, to be more specific, one had to search for a physical basis of Mendelian segregation. To these students it was obvious that there should be some connection between chromosomes and inheritance, a connection that was by no means acceptable to all.[13] In order to understand the opposition it is necessary to point out once more that the new science of genetics was born out of developmental biology. The original framework of the concepts of Weismann, Bateson, and Morgan was that of embryology. Although the battle between preformation and epigenesis had seemingly terminated a hundred years earlier with the decisive victory of epigenesis, embryologists continued to be overly-sensitive to any trace of preformational thinking. One has only to read some of Morgan's early (1903) discussions of Mendelism or Johannsen's discussions of the gene to get a feeling of their distaste for a corpuscular, hence for them preformationist, theory of Mendelian inheritance.

Authors who based their theories of inheritance on physical forces—for instance Bateson in his theory of dynamic vortices (Coleman, 1970)—saw a holistic, epigenetic unity in the genotype which seemed quite irreconcilable with a corpuscular theory. Such "dynamic" theories were held by certain geneticists long after the establishment of Mendelian genetics. R. Goldschmidt, for instance, as late as the 1950s, believed in "fields" of genetic forces and the possibility of systemic mutations of the entire genotype, another quite holistic concept. Johannsen's objection to defining the gene "as a morphological structure" seems to have had a similar background.

Their opponents opted for a morphological-corpuscular theory of inheritance, but were altogether uncertain how the genetic material was organized in the chromosomes. Much of the factual knowledge on which to base a chromosomal theory of inheritance was already available by the mid-1890s, but this did not lead to the elaboration of a viable theory. The reasons for this failure are manifold: (1) an aversion to a theory that might be branded as preformationist, (2) a failure to analyze phenomena of inheritance in terms of individual factors, (3) a peculiar emphasis in the period from 1885 to 1900 on the purely mechanical aspects of cell division, and (4) a predominant interest (particularly by Boveri) in purely developmental phenomena. Transmission genetics deals with populational phenomena that are quite inaccessible to the methods of functional analysis as practiced in cytology.

The developments after 1900 were influenced by a fortunate coincidence. The young American embryologist E. B. Wilson, during several stays in Europe, had developed an enthusiastic interest in cell biology, particularly under the influence of his friend Boveri. Even though at the time he himself had done only rather specialized original cytological research (on cell lineage), he composed a brilliant synthesis of the current understanding of the cell, and particularly of the chromosomes (1896; 2nd ed. in 1900), a work which more than anything else was instrumental in the subsequent synthesis of cytology and Mendelism. He greatly advanced the understanding of chromosomes in a series of eight classical studies (1905–1912), was the teacher and inspiration of all of T. H. Morgan's associates and, as a colleague and friend, had a great influence on Morgan. It is well justified to consider Wilson one of the fathers of the new science of genetics.[14]

Even though a number of authors in the 1890s expressed their conviction that the chromatin or nuclein of the chromosomes was the genetic material, this opinion alone was not sufficient for a substantial theory of inheritance. Thus, it was left to the decade after 1900 to establish, point by point, the relation between Mendelism and cytology. Speculation and assumption had to be replaced by solid evidence and ironclad proof.

To describe the steps by which this proof was assembled is difficult because the history of the chromosome theory grades into the history of the theory of the gene. Only by making some arbitrary cuts through a continuity is it possible to present the two histories separately. Yet, it is not only for didactic reasons but also for reasons of intellectual history that the two subjects are here dealt with separately: It would have been difficult, if not impossible, to develop a valid theory of the gene if there had not been the chromosome theory first.[15]

The rediscovery of Mendel's laws in 1900 brought about a drastic change in the situation. Not only did the almost feverish activity released by the rediscovery produce numerous new facts, but the cytological discoveries made in the 1880s and 90s suddenly acquired a new significance. The thought that the Mendelian laws are a logical consequence of the chromosomal organization of the genetic material occurred more or less independently to Montgomery (1901), Correns (1902), Sutton (1902), Wilson (1902), and Boveri (1902; 1904). Sutton and Boveri in particular presented a detailed substantiation of their conclusions. The conscious combining of cytological evidence and genetic argument by these authors resulted in the development of a new biological discipline, cytogenetics, in which Wilson and his students became the leaders. It is important to remember that Sturtevant, Bridges, and Muller had been Wilson's students before joining Morgan's research team.

The Sutton-Boveri Chromosome Theory

Nothing in the cytological advances made before and after 1900 was more important in the history of genetics than the demonstration of the individuality and continuity of chromosomes. Chromosomes are not visible between cell divisions; the resting nucleus merely shows slightly staining granules or a network of fine threads. The thesis that the chromosomes completely dissolve at the end of mitosis and are formed anew at the beginning of a new mitotic cycle seemed to be supported by microscopic observation. This explains why such experienced cytologists as Oskar Hertwig and R. Fick (1905; 1907) still maintained that thesis well into the Mendelian period. Indeed, the thesis that each chromosome maintains its individuality and integrity during the resting stage of the nucleus was based on inference; it could not be observed directly. Rabl (1885) was the first to formulate clearly the hypothesis of the individuality and continuity of each chromosome. He postulated that the threads of chromatin into which a given chromosome dissolves when the nucleus enters the resting stage consolidate again into the same chromosome at the beginning of the next mitotic cycle. This was strictly an inference from rather scanty data, mostly on stable chromosome numbers. Van Beneden (see Chapter 15) and Boveri subsequently claimed priority for the same inference. There is no doubt that Boveri, more than anyone else, supplied the decisive proof for the theory of chromosome individuality.[16] As early as 1891 he stated, "We may identify every chromatic element [chromosome] arising from a resting nucleus with a definitive element that entered into the formation of that nucleus." From this the remarkable conclusion follows that "in all cells derived in the regular course of division from the fertilized egg, one half of the chromosomes are of strictly paternal origin, the other half of maternal" (1891: 410).

748

Continuity through the resting stage of the nucleus and individuality of each chromosome seem to us today merely two sides of the same coin. It was not so in the 1890s. Weismann and others had suggested that each chromosome contained all the heritable properties of a species, that is, he denied the individuality of chromosomes in the Mendelian sense. If, however, a chromosome contains only part of the genetic endowment of an individual, each chromosome will have to be different from the others, that is, it will have to have individuality. In other words, if each chromosome is different from the others, it becomes necessary to prove both continuity and individuality.

The proof of continuity was supplied when it was shown by Montgomery (1901) and Sutton (1902) that some chromosomes are individually recognizable during mitosis and meiosis and that chromosomes with the same characteristics occur again and again at each cell division. More than that, they showed that always during the first prophase two similar chromosomes pair (synapsis) but separate from each other again during the reduction division (see below). This led to the conclusion that the chromosome set of a species consists of pairs of homologous chromosomes, one of which had been derived from the female gamete (the egg cell), the other from the male gamete (the spermatozoon), as had been observed by van Beneden in 1883. Evidently these chromosomes retain their identity from the time of fertilization (the formation of the zygote) through innumerable cell divisions up to the reduction division preceding the formation of the new gametes. Sutton ended his paper with the remarkable conclusion that "the association of paternal and maternal chromosomes in pairs and their subsequent separation during the reducing division . . . may constitute the physical bases of the Mendelian law of heredity" (1902). He expanded on this thought the next year (1903; see also McKusick, 1960).

These observations did not entirely eliminate the possibility that morphologically dissimilar chromosomes nevertheless also had similar genetic properties. This possibility was excluded by Boveri (1902; 1904) by an ingenious experiment. In a species of sea urchins with 36 chromosomes, Boveri was able, through proper manipulation (multiple fertilizations, and so forth), to produce embryos with highly variable chromosome numbers in the first four daughter cells. Yet, of all these embryos, only those that had 36 chromosomes in their daughter cells developed normally. Boveri concluded from this that each chromosome had a "different quality" and that the right combination of all of them had to be present in order to permit normal development.

749

It was now clearly established that the chromosomes follow the same rules as genetic characters, that is, they show segregation and independent assortment. Sutton and Boveri implictly or explicitly postulated that the genes are located on the chromosomes and that each chromosome has its particular set of genes. Clearly, especially as stated by Sutton (1903) and Boveri (1904), here was a well-rounded chromosomal theory of inheritance, deduced from the cytological evidence and from the independent assortment of Mendelian characters. It seemed to be able to explain all the facts of Mendelian inheritance.[17]

Curiously, the importance and universal application of the *Sutton-Boveri chromosomal theory of inheritance* (so called by Sutton's professor, Wilson, 1928) was not at all recognized at first. It was rejected not only by Bateson and Goldschmidt but also by other well-qualified biologists (such as E. S. Russell) as late as 1930. In part this was due to the fact that it was arrived at by inference based on observation. T. H. Morgan, for one, asserted he would not accept conclusions that were "not based on experiment," and similar statements were made by Johannsen. Actually, much of the Sutton-Boveri theory was based on experiment, an indication that there were deeper reasons for Morgan's resistance.

The evidence for the continuity of chromosomes through the resting stage was quite substantial by 1910; the evidence for their individuality rested mainly on Boveri's experiment. At first there was no definite evidence connecting a specific character trait with a definite chromosome. Sex determination was the character which first supplied such evidence. Eventually the most complete evidence came from linkage maps.

Sex Determination

What determines the sex of a child has been the subject of much speculation at least from the days of the Greeks.[18] We now know that all the early theories were wrong (Details can be found in Lesky, 1950, and Stubbe, 1965). Among the explanations were the position (or implantation) of the embryo in the left or right half of the uterus, the amount of sperm that had come from the left or the right testis, or the amount of semen, or the relative "heat" of the male or the female fluids, and so on. What all these theories had in common—and this is the decisive point—was that sex is not genetically determined but caused by purely environmental factors coincidental to the act of fertilization. Even after the discovery of the genetic basis of sex (after 1900), an environmental determination of sex was defended for several more decades by some leading embryologists and endocrinologists. And, as we shall see, there are indeed some organisms with nongenetic sex determination.

It did not escape some of the more perceptive Mendelians that the 1:1 sex ratio was the same as, to use Mendelian language, the ratio resulting from the cross of a heterozygote (Aa) and a homozygous recessive (aa). Mendel himself had already suggested this possibility to Nägeli on September 17, 1870. Others (Strasburger and Castle) made the same suggestion in the years after 1900, but Correns was the first to supply experimental proof by showing that half of the pollen of the dioecious plant *Bryonia* is male determining, the other half female determining, while all the eggs are identical with respect to sex determination. In this case the male is heterozygous, or to use Wilson's (1910) terminology, heterogametic, while the female is homogametic. Eventually, it was shown that in birds and lepidopterans females are heterogametic, while in mammals (including man) and dipterans (including *Drosophila*) the male sex is heterogametic. Could it be that sex is connected with a definite chromosome? Gradually the evidence accumulated that would substantiate this suggestion.

Sex Chromosomes

From the beginning of chromosome studies it was observed that not all chromosomes are necessarily identical in appearance.[19] In 1891 Henking observed during the meiosis of the insect *Pyrrhocoris* that half the spermatozoa received 11 chromosomes while the other half received not only these 11 chromosomes but an additional heavily staining body. Uncertain whether it was a chromosome or not, Henking designated this body as X. Nor did Henking associate this X body with one of the sexes.

During the next decade many additional cases were found of such extra chromosomes, or of the presence of one pair of chromosomes which differed in stainability, size, or other features from the remainder of the chromosome set. Since it was observed that half the spermatozoa would receive the X (accessory) chromosome, the other half not, McClung (1901) reasoned as follows: "We know that the only quality which separates the members of the species into two groups is that of sex. I therefore came to the conclusion that the accessory chromosome is the element which determines that the germ cells of the embryo shall continue their development past the slightly modified egg cell into the highly specialized spermatozoon," that is, that these somewhat unusual chromosomes are sex chromosomes, serving to determine sex. Some of the details of McClung's conclusions were wrong. The correct story of sex determination by sex chromosomes was soon worked out by Nettie Stevens (1905; see Brush, 1978) and E. B. Wilson (1905).

751

There are many patterns of sex determination, sometimes involving multiple sex chromosomes, and with either the male or in other cases the female being the heterozygous sex. All such detail can be found in every textbook of genetics or of cytology (see Wilson, 1925; White, 1973). The important point is that here was the demonstration that a phenotypic character, that of sex, is associated with a definite chromosome.

This was the first conclusive proof of such an association. Much of the genetic research of the ensuing years consisted in associating other characters either with the sex chromosomes or with other chromosomes, called autosomes. Leadership in this research, which solidified the chromosome theory of inheritance, was assumed by T. H. Morgan. The researches of his laboratory provided the final refutation of the theory of the genetic equivalence of all chromosomes. This theory had remained in vogue until after 1900, in spite of the discovery of species in which the chromosomes are of highly unequal size. The hold which this (to us) so improbable theory had on the biologists of the 1880s and 90s, was probably due to the fact that in some species all chromosomes indeed did look the same.

Now that the individuality of chromosomes had been conclusively established, and the association of at least one character, sex, with a definite chromosome had been discovered, genetics was ready to ask more precise questions about chromosomes and characters, or, to use Johannsen's more concrete terminology, about the relation between chromosomes and genes.[20] Does a chromosome as a whole control an entire set of characters, so to speak, as the control center of a developmental field, or are individual genes located at specific places on the chromosomes? And what is the mutual relation of different genes located on the same chromosome or on different chromosomes? These questions were answered within a remarkably short time (essentially between 1905 and 1915, indeed largely between 1910 and 1915) through brilliant genetic experiments constantly checked against the cytological evidence. The starting point, invariably, was some rather simple Mendelian phenomenon.

752

Morgan and the Fly Room

In 1909 Morgan began to breed *Drosophila.* He had been much impressed with de Vries' *Oenothera* mutations and apparently attempted to produce mutations in *Drosophila* by exposing his cultures to different chemicals, different temperatures, radium, and x-rays, but in this he had no success. However, in one of his pedigreed cultures a single white-eyed male appeared in a normal population of red-eyed flies.

This simple event, the occurrence of a single aberrant individual in a laboratory culture, started a veritable avalanche of investigations. There was, first of all, the question how this "white-eye" character had originated. Breeding the precious white-eyed male to its red-eyed sisters, Morgan found that although the F_1 progeny were red-eyed, white-eyed males reappeared in F_2, showing that the genetic factor for white-eyedness was recessive and that it must have originated by a sudden change of the red-eye gene. Morgan, who some years earlier had visited de Vries's laboratory in Holland, adopted de Vries' term "mutation" for the origin of a new allele. This transfer of the term was rather unfortunate in view of de Vries's evolutionary mutation theory and the chromosomal nature of the *Oenothera* mutations. Consequently, it resulted in considerable confusion during the ensuing twenty or thirty years (Allen, 1967; Mayr and Provine, 1980). Eventually, however, geneticists and evolutionists became conditioned to the new meaning of the term "mutation" given to it by Morgan.

There have been few investigators in the history of biology who have worked as closely with their collaborators as Morgan. It is therefore difficult to determine who should get the credit for some of the numerous findings of the Morgan laboratory. Indeed, some historians have tended to give nearly all the credit to his students and collaborators. This goes much too far. It must be remembered that in the two years following his first *Drosophila* paper of July 1910, Morgan published thirteen papers on the occurrence and behavior of some twenty sex-linked mutants in *Drosophila*. Very soon after "white-eye," two further sex-linked recessive mutants were found, "rudimentary wings" and "yellow body color." There is no doubt that much of the elucidation of the mechanism of Mendelian heredity was achieved by Morgan very early and was his own contribution. As Muller (1946) has characterized it: "However much the story of the formative period of the *Drosophila* work may be rewritten and reappraised in the future, there must remain agreement in regard to the fact that Morgan's evidence for crossing over and his suggestion that genes further apart cross over more frequently was a thunderclap, hardly second to the discovery of Mendelism." I want to emphasize here this singular contribution of Morgan's to the problem of linkage and crossing over, since in the ensuing analysis the emphasis is on the problems and not on the specific contribution made by each occupant of the fly room.

Morgan and his co-workers bred *Drosophila* flies in his "fly-room" at Columbia University by the tens and hundreds of thousands. Scrutinizing these flies carefully, he and his collaborators found a steady stream of new mutations. Soon (in the winter of 1910–11) Morgan took two Columbia undergraduates to work in his laboratory, Alfred H. Sturtevant and Calvin B. Bridges. Subsequently the group was joined by H. J. Muller, who also took his degree with Morgan. The splendid cooperative work of this group (Sturtevant, 1959; 1965a) is one of the sagas of biology: "There can have been few times and places in scientific laboratories with such an atmosphere of excitement and with such a record of sustained enthusiasm. This was due in large part to Morgan's own attitude, compounded of enthusiasm combined with a strong critical sense, generosity, open-mindedness, and a remarkable sense of humor."

Within a few years all major aspects of transmission genetics were elucidated by Morgan and his group. Where Bateson, de Vries, Correns, Castle, and the other early Mendelians had failed to find the right answers, in fact had failed even to ask the right questions, the Morgan group was brilliantly successful. One important reason for this was that Morgan, although by background an embryologist, deliberately concentrated on the problem of genetic transmission, pushing aside problems of gene physiology and ontogeny. Instead of speculating about laws of inheritance, he searched for facts and their simplest possible explanation. He was an empiricist through and through.

754

Alleles

Mendel had fully understood that phenotypic characters came in groups, specifically (in the characters chosen by him) in pairs. All the work done after 1900 confirmed that the material basis responsible for a phenotypic character could have alternate phenotypic manifestations or expressions. Translating this literally into Greek, these alternate determinants were "allelomorphs" (Bateson's term), or alleles. The discovery of the Mendelian inheritance of such alternate determinants of phenotypic characters cast an entirely new light on the problem of the causation of variation. It suggested that smooth versus wrinkled or yellow versus green in peas, or other analogous character pairs, might have a similar physical basis. Characters produced by different alleles ought to be just two versions of the same basic material (genotypic alternative).

In 1904 the French biologist L Cuénot discovered in the house mouse that there may be even more than two alleles for a set of traits; for instance, in the particular case of the mouse the coat color might be gray, yellow, or black. Bateson, Castle, Shull, Morgan, and other geneticists in due time found many other cases of such multiple allelism. The ABO blood group in man is a particularly familiar example. Sturtevant (1913) gave the first explanation of the phenomenon of multiple allelism, ascribing it to alternate states of the same gene (locus). This decisively refuted Bateson's presence-absence theory of gene action. In some special cases there may be more than fifty alleles of a single gene, as in blood group genes of cattle and certain compatibility genes in plants, and histocompatibility genes in vertebrates. True to Mendel's law, always only a single allele can be represented in a given gamete, but during fertilization it can combine with any one of a number of different alleles found in the gene pool of the population. Later in the history of genetics cases were found where genes behaved like alleles in certain crosses but did not do so in others (pseudallelism). The analysis of such cases (by Lewis and Green) led to a deeper understanding of nature of the gene (see below).

The researches of the Morgan group on the white-eye gene of *Drosophila* and on other *Drosophila* mutations clearly established that a gene could mutate to another allele, and this one again to a third and fourth one. Equally interesting was the discovery that these mutational steps are reversible and that a white-eyed fly could occasionally give rise to offspring with red eyes. The most important finding, perhaps, was that once a gene had given rise to a new allele, this new allele is perpetuated unchanged unless a new mutation occurs in one of its offspring. Genes are thus characterized by almost complete stability. The discovery of gene mutation, thus, was not a return to soft inheritance but, on the contrary, it confirmed the essential constancy of the genetic material. It was, so to speak, the final proof of hard inheritance, for the capacity for mutation allowed for evolutionary change in spite of the intrinsic constancy of the genetic substance.

755

The occurrence of mutations was soon confirmed in all sorts of other organisms, from man and other mammals down to the simplest animals, in all kinds of plants, and even in microorganisms. Indeed, from the 1920s to the 1950s the study of mutation seemed to be the most promising approach toward elucidating the nature of the genetic material. It was realized that the process of mutation posed formidable problems. Just exactly what is it that is happening to the gene during mutation? Also, is there any possibility of producing mutations under controlled conditions, that is, experimentally? De Vries had suggested as far back as 1904 "that the rays of Roentgen and Curie, which are able to penetrate into the interior of living cells, be used in an attempt to alter the hereditary particles in the germ cells" (in Blakeslee, 1936). From 1901 on attempts were made again and again to induce mutations through x-rays, radium, temperature shocks, or chemical agents. Owing to various technical deficiencies (heterogeneous material, small samples, and so forth), none of these numerous attempts yielded at first unambiguous results. It was not until H. J. Muller applied all of his perseverance and ingenuity to this problem that success was finally achieved in 1927.[21]

Independent Assortment of Characters vs. Linkage

One of Mendel's important findings was "that the behavior of each pair of differing traits in a hybrid association is independent of all other differences in the two parental plants" (1866: 22). This is now often referred to as the law of the independent assortment of characters. When Mendel crossed, for instance, a strain of peas with round yellow seeds (both characters being dominant) with another pure breeding strain with wrinkled green seeds (both characters recessive), he did not get a 3:1 ratio in the F_2 of round yellow seeds to wrinkled green seeds. Instead, he obtained in his particular experiment 556 seeds consisting of 315 round yellow, 101 wrinkled yellow, 108 round green, and 32 wrinkled green, approximately a 9:3:3:1 ratio. Thus, each individual pair of characters, round versus wrinkled and yellow versus green, had produced a 3:1 ratio (with round and yellow being dominant) but the two characters had segregated independently of each other. Mendel found the same to be true for the other five character pairs he studied, and for a while it was assumed that all characters obeyed this law of independent assortment.

This finding would not have been surprising if the nucleus were nothing but a bag full of pairs of gemmules, which would split up prior to gamete formation and be distributed independently. But since the nuclear material is organized into chromosomes, one should not expect more independent groups of characters than there are chromosomes, since chromosomes segregate as wholes during gamete formation. The fact that Mendel's seven characters had assorted independently was consistent with the fact, much later discovered, that *Pisum sativum* has only seven pairs of chromosomes (see below).

As more and more crosses were made during the period of feverish activity after the rediscovery of the Mendelian laws, exceptions to independent assortments were found (the first in *Matthiola* by Correns in 1900, others by Bateson's group) but, for reasons that will presently become apparent, they were not easy to interpret. Why there was no independent assortment of sex and eye color in the case of white-eyed flies was rather quickly established by Morgan after an initial incorrect hypothesis. When he interbred the F_1 flies (see above), red-eyed and white-eyed flies appeared in the F_2 at a ratio of 3:1, but all the white-eyed flies were males, while there were two females to one male among the red-eyed flies (see Fig. 2a). Some other crosses made by Morgan

756

gave what at first appeared to be even more unexpected results. For instance, when white-eyed females were crossed with normal red-eyed males, all of the female offspring were red-eyed and all of the male offspring white-eyed (see Fig. 2b). Obviously the sex gene and the eye-color gene did not assort independently.

Morgan concluded from these observations in 1910 that the factor for eye color (which had mutated from red to white) was coupled with the X factor which determined sex.[22] A year later (1911: 384) he specifically explained this coupling of characters in chromosomal terms: "Instead of random segregation in Mendel's sense we find 'association of factors' that are located near together in the chromosomes. Cytology furnishes the mechanism that the experimental evidence demands." Some other mutations, like yellow body color and miniature wings, were also found to be linked with sex, that is, to be located on the sex chromosome. Other linked groups of characters had nothing to do with sex and were apparently located on the other chromosomes of *Drosophila,* designated as *autosomes* (to distinguish them from the sex chromosomes).

757

De Vries, Correns, Boveri, and Sutton, in fact, had already predicted the occurrence of linkage on theoretical grounds. Their reasoning was based on the individuality of chromosomes and their continuity through the cell (mitotic) cycle.

Nondisjunction

Work by one of Morgan's collaborators, Calvin Bridges (1914), provided a still more convincing proof for the chromosome theory of inheritance. As we saw, when a white-eyed female *Drosophila* (carrying the recessive white allele on both X chromosomes) is crossed with a normal red-eyed male (carrying dominant red on its single X), it produces equal numbers of heterozygous red-eyed females and white-eyed males in the F_1. This is necessitated by the genetic constitution of the two parents. An abnormal strain of flies, however, turned up in Morgan's laboratory in which about 4.3 percent of the F_1 progeny in such a cross consisted of white-eyed females and red-eyed males. I shall not go into the details of the explanation, which can be found in any classical textbook of genetics. It was based on the prediction by Bridges that females in this stock had not only the two X chromosomes but also a male Y chromosome. Presumably the original XXY female had originated when an abnormal egg with two XX chromosomes (due to failure of reduction) had been fertilized by a Y sperm. During gamete formation of such an individual with three sex chromosomes (two

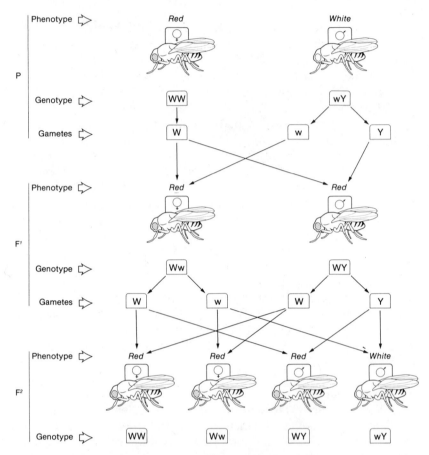

758

Figure 2a. F₁ sons of red-eyed mothers, having the X chromosome of the mother, are red-eyed. Fifty percent of the male F₂ of the F₁ females are again white-eyed, owing to normal Mendelian segregation.

X and one Y), either the two X chromosomes go to different gametes (eggs), resulting in X and XY eggs—this actually happening in 91.8 percent of the formed gametes—or else both X go to one egg and the Y to the other—in 8.2 percent of the cases. After fertilization with normal X or Y carrying spermatozoa, XXX and YY zygotes die, but a small percentage of exceptional red-eyed males (XY) and white-eyed females (XwXwY) are produced (see Fig. 3). Bridges' prediction was then confirmed by cytological examination, which indeed established the existence of XXY females and XYY males in this strain.

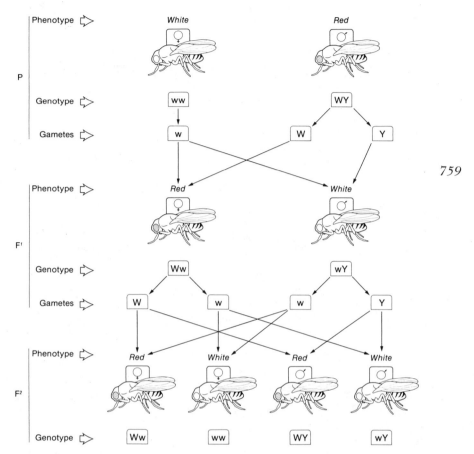

Phenotype ⯈

Genotype ⯈

Gametes ⯈

Phenotype ⯈

Genotype ⯈

Gametes ⯈

Phenotype ⯈

Genotype ⯈

759

Figure 2b. In the reciprocal cross, F_1 males receive the X chromosome of the mother. If the mother is homozygous recessive white-eyed, all her male offspring will be white-eyed, even though their father was red-eyed.

Previously (Wilson, 1909) and subsequently, other cases of nondisjunction had been found, including individuals with an extra autosome. In the human species, for instance, the presence of three number-21 chromosomes, referred to as trisomy of chromosome 21 due to nondisjunction, is the cause of Down's Syndrome (mongoloid idiocy). Individuals with an extra chromosome (trisomics) or with a missing autosome (monosomics) are viable in many plant species and have been used for interesting studies on the effects of different-balances of the same genes. In *Datura*, for instance, trisomy for any of the twelve pairs of chromosomes is

760

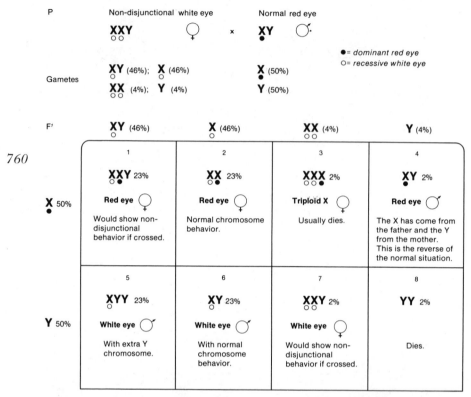

Figure 3. Bridges' demonstration that the sex-linked genes are located on the sex chromosome. It made use of a stock of Drosophila with a super-numerary X chromosome. This resulted in abnormal XY and XX gametes.

not only viable but characterized by a specific morphology. The same is true for monosomics of any of the 23 chromosome pairs of *Nicotiana*.

The importance of Bridges' work was that it supplied the first direct proof that sex-linked genes are carried by the X chromosome. His conclusion was confirmed again and again in the ensuing years. Thereafter, it became increasingly unreasonable to oppose the chromosome theory, even though a few authors, like Bateson and Goldschmidt, remained unconvinced and even Morgan retained some ambivalence.

Meiosis

Although after 1902 some biologists spoke freely of the chromosome theory of inheritance, there was still considerable uncertainty as to just what precisely this designation meant. Most of all it formalized Roux's suggestion of a linear arrangement of individually different genetic factors (we would now say genes) on the chromosomes. But this was not all. The cytologists had discovered in the 1870s to 1890s numerous chromosomal phenomena which surely had a bearing on inheritance. Such phenomena were intensely studied after 1900, and particularly by Morgan's group after 1910, and this greatly helped to broaden and strengthen the chromosome theory.

761

Let us begin with the behavior of chromosomes during gamete formation.[23] Egg nuclei and spermatozoa are "haploid," that is, they have half the number of chromosomes of (diploid) body cells. How is this halving of the chromosome set accomplished during gamete formation and how might this affect inheritance?

Reduction Division

Let us recall van Beneden's observation (1883) that during the fertilization of the egg of *Ascaris* the two chromosomes of the male nucleus join the two chromosomes of the nucleus of the egg cell, giving the new nucleus of the zygote four chromosomes. Every cell produced by the subsequent cleavage divisions of this fertilized egg cell has four chromosomes; it is "diploid," having twice the chromosome number of the gametes. If such a doubling of the chromosome number were to occur during each fertilization, each generation would have twice as many chromosomes as the parents had. Soon there would be thousands or millions of chromosomes in each cell. Obviously there had to be some process to compensate for the doubling of the amount of chromatin during fertilization, and Strasburger (1884: 133) and Weismann (1887) suggested that there had to be a "reduction division" prior to gamete formation. Boveri (1887–1888) concurred in Weismann's conclusion and Oskar Hertwig finally gave a full and completely correct description in 1890.

The cytologists were aware of the fact that during gamete formation in animals two consecutive cell divisions take place which differ strikingly from normal mitosis and for which eventually the term *meiosis* was introduced. Just exactly what it is that takes place during meiosis—or, to be more specific, how the reduction of the chromatic material is achieved—remained controversial for a long time. The only thing that was soon generally accepted was that the oocytes and gametocytes, that is, the cells which eventually give rise to the eggs and spermatozoa, have the same diploid chromosome number as ordinary body cells. Nevertheless, the gametes (eggs and spermatozoa) that result from the meiotic division have only half the chromosome number; they are haploid. After the nature of the meiotic divisions had been fully understood, it was evident that the maturation of the nucleus of the egg cells is completely analogous to that of the spermatocytes. But at first sight the processes in the two kinds of cells seem quite different and it is therefore advisable to describe them separately.

762

Let us remember that the chromosome set of each cell nucleus consists of a paired set of homologous chromosomes, one derived from the father, the other from the mother. During the first meiotic division, these homologous chromosomes attach themselves sidewise closely to each other, a pairing process designated as *synapsis*. Just what happens during this pairing process was not at all clear at first; indeed the contact is so close that microscopic analysis is unable to reconstruct exactly what happens during the prophase of the first meiotic division. It took almost another thirty years before genetic analysis essentially clarified the process (see below). This first nuclear division during meiosis is unique because the nucleus divides but the chromosomes do not. As a result, half the chromosomes go to one daughter nucleus, the other half to the other nucleus. In the case of *Ascaris*, the chromosomal number is hereby reduced from four to two. This first meiotic division is the *reduction division* postulated by Weismann and previously suggested by Galton.

In the second meiotic division the chromosomes split as in any mitotic division. In the egg cell both meiotic divisions take place near the periphery of the egg, and on both occasions one set of daughter chromosomes is extruded in a so-called *polar body*. The first polar body may or may not divide during the second meiotic division.

During the formation of the male gametes, the same two nuclear divisions take place, except no daughter nuclei are expelled as polar bodies. Instead, a set of four spermatozoa is formed, derived from the single nucleus in the spermatocyte mother cell entering the meiotic process, as shown by Hertwig (1890). Since there is no chromosome doubling during the second meiotic division, the four spermatozoa likewise have only half as many chromosomes as the "diploid" somatic cells. I have here presented the final interpretation of events occurring during the reduction division. The correct story was pieced together by findings and interpretations contributed by van Beneden, Hertwig, Weismann, and others. Hertwig and Weismann, at first, differed widely in their interpretation. The way to the gradual understanding of the problem has been excellently described by Churchill (1970). There were two major reasons for the original disagreement: Hertwig believed more or less in blending inheritance, the paternal and maternal chromosomes fusing during fertilization, and he thought that the chromosomes were dissolved after each cell divsion, to be reassembled from chromatic particles prior to the next cell division. The extrusion of polar bodies was for him a purely quantitative reduction of the chromatin mass. By contrast, Weismann believed in the continuing separateness of paternal and maternal chromosomes after fertilization, and he believed in the continuity of each chromosome through the entire mitotic and meiotic cycle and through the subsequent resting stage. In both questions Weismann's postulates turned out to be correct. The most important conclusion that emerged from the controversy was that the chromosomal content of the polar bodies extruded by the egg is exactly the same as that of the remaining egg nucleus, and that therefore the nuclear division (meiosis) during the formation of the female and male gametes are completely equivalent, even though they result in the formation of a single egg and three polar bodies in the one sex, and in four spermatozoa in the other. Although Hertwig and Weismann were overtly arguing over technical cytological detail, their positions were actually necessitated, as Churchill (1970) has shown very convincingly, by deep conceptual commitments. Hertwig represented the physicalist-physiological camp, Weismann the morphological-corpuscular-molecular school.

763

One can also express the cytological story of meiosis in genetic terms. Let us assume that during the formation of a new zygote as a result of the coming together of a paternal and a maternal chromosome set, a chromosome with allele A (derived from the father) paired with a chromosome with the allele a (derived from the mother), producing a zygote Aa. From the first cleavage division of the new zygote on these two homologous chromosomes will remain paired and all somatic cells in the developing organism will be heterozygous Aa. It is only in the second meiotic division during gamete formation (the reduction division) that the two homologous chromosomes separate, forming in equal number gametes with the gene *A* and gametes with the gene *a*. Which chromosome goes to which of the two reduced daughter cells is entirely a matter of chance (see below). Thus Mendelian segregation is perfectly explained by the observed phenomena of chromosomal behavior during fertilization and gamete formation (the Sutton-Boveri theory). Meiosis, as described up to this point, can account ·perfectly for linkage and segregation, but it is not the complete story. It is unable to explain the phenomena of incomplete linkage, referred to above. However, this can be accounted for by an additional process occurring during meiosis: crossing over. Meiosis in plants is chromosomally the same as in animals but it usually takes place at a different stage in the life cycle (prior to spore formation).

764

Crossing Over

Since any organism obviously has far more characteristics—as well as genetic determinants for these characters—than it has chromosomes, it was realized from the beginning (Correns, 1902; Sutton, 1903) that each chromosome must be the carrier of several if not many genes. This was soon confirmed by the work of Morgan's laboratory. The discovery of linkage groups, each associated with a definite chromosome, however, raised a new problem. If all genes on a chromosome were tightly linked together, an organism, for all practical purposes, would have only as many independent units of inheritance as it has chromosomes. This would pose an enormous constraint on recombination. Studying the F_2 generation of hybrids, de Vries (1903) concluded that the wealth of recombinations in an F_2 hybrid was far too great to be compatible with a theory of total linkage. He postulated therefore "an exchange of units" of the paired parental chromosomes during prophase I of meiosis. "How many and which [units would be exhanged] may then simply be left to chance" (1910: 243), provided that exchange is always strictly mutual. Such an exchange was also predicted by Boveri (1904: 118). Genetic analysis soon confirmed that the linkage of genes on the same chromosome is not complete. The first such observation was made by Bateson, Saunders, and Punnett (1905). In the F_2 of a cross of two varieties of sweet peas (*Lathyrus*) which differed in flower color and shape of the pollen grains, they got neither the expected 9:3:3:1 ratio nor a simple 3:1 but found 69.5 percent double dominants, 19.3 percent double recessives, and two classes of 5.6 percent heterozygotes. Evidently the genes for the two characters had neither assorted independently nor been completely linked (11 percent exceptions). Bateson proposed an *ad hoc* hypothesis to explain this, and since he did not believe in the chromosome theory, he did not consider crossing over.

It has frequently been remarked how curious it is that Mendel had not encountered linkage. The pea (*Pisum sativum*) has only seven pairs of chromosomes, and Mendel studied seven characters. Was it luck that they were not linked, thus sparing Mendel one additional complication? Presumably it was not. Mendel is known to have devoted several years to preliminary crosses before starting on his set of definitive experiments. It is quite likely that he had rejected characters (or at least one member of a character pair) which did not show independent assortment in the F_2. It is also possible that the seed producers from whom Mendel had acquired his material favored independently assorting characters. Finally, map distances of some of the genes are great enough to simulate independent assortment, even if some genes had been located on the same chromosome.[24]

765

Exceptions to complete linkage became a serious problem when the intensive analysis of the genetic constitution of the *Drosophila melanogaster* began in Morgan's laboratory. He and his associates found that the percentage of broken linkages ranged widely, sometimes being as low as 1 percent. How could this variability be explained?

Let us look more closely at one particular case. Let us take a set of three recessive genes—yellow body color (*y*), white eye color (*w*), and miniature wings (*m*) which are among the genes located in *Drosophila* on the X chromosome. If a male with these three genes is crossed with a normal female, one would expect the three recessive characters to turn up in the F_2 generation as a linked group. Actually the linkage for body color and eye color was broken in 1.3 percent of the flies, for eye color and wing size in 32.6 percent and for body color and wing size in 33.8 percent. How can these figures be explained?

The numerical values for these exceptions were too regular to be explained by such a haphazard process as a random exchange of units, as postulated by de Vries. However, cytological researches in the early 1900s permitted a different solution. The study of the details of meiosis had made enormous progress in the twenty years since the pioneer work of Boveri and Hertwig. No less than six different stages in the changes of the chromosomes (the chromatic material) were now distinguished during prophase I. During one of these stages the two paired chromosomes are still quite thin, but each has split into two chromatin threads (chromatids), the so-called four-strand stage. The two chromosomes form wavy loops and cross each other repeatedly.

The Belgian cytologist Janssens postulated (1909) that when the four chromatids were coiled around each other, one paternal and one maternal chromatid might break at a point where they crossed over another, and when the broken ends would be rejoining, it would be done by the paternal end healing to the maternal end, and vice versa. The other two chromatids would remain intact. Thus a "chiasma" would be formed (point where two chromosomes of a pair remain in contact during the later stages of prophase 1 of meiosis). The chiasma, in Janssen's views, is an indication of the crossing over of a paternal and a maternal chromatid. The final result would be a new chromosome consisting of pieces of the maternal and the paternal chromosome. The analysis of cases of incomplete linkage by the Morgan group was consistent with Janssen's theory.

The process of crossing over is so complex that it took some thirty years until it was finally decided which of various competing interpretations was the correct one (see Whitehouse, 1965, for a very readable presentation). However, it is now fully established that crossing over takes place at the four-strand stage and involves two of the four chromatids. Furthermore, it takes place at the very beginning of that stage (Grell, 1978).

Morgan and his collaborator A. H. Sturtevant (Lewis, 1961) figured that the amount of incomplete linkage due to crossing over represented the linear distance of factors along the chromosomes. The chance of a chromosomal break (and hence crossing over) between two genes would be the smaller the closer the location of two genes on a chromosome. With the help of this reasoning, Sturtevant (who was then only nineteen years old!) was able to calculate the position and sequence of genes on a chromosome and indeed to produce a chromosome map, the first one, for the X chromosome of *Drosophila melanogaster* (published in 1913). He thereby established that the genes then known for this chromosome are arranged along the chromosome in a linear sequence.

Among the early results there were some discrepancies. These were eliminated when Muller (1916) showed that there could be double crossovers on long chromosomes (simulating a nonoccurrence of crossing over for distant genes) and that the presence of a chiasma would interfere with the occurrence of further crossing over on nearby regions of the chromosome. Taking these two newly discovered phenomena (double crossover and interference) into account removed the discrepancies which had made some of Morgan's opponents skeptical of the validity of the crossing-over theory.

The chromosome theory of inheritance could now be supplemented by a theory of the gene (Morgan, 1926). By 1915 Morgan and his associates had studied more than a hundred mutant genes. These fell into four linkage groups corresponding most admirably to the four chromosomes of *Drosophila melanogaster*. The indirect proof for the chromosomal nature of linkage groups was thus complete. However, it was not until 1931 that Stern was able to use some abnormal chromosomes (a piece of the X chromosome attached to the small fourth chromosome) to provide cytological proof for the thesis of crossing over. A similar proof was provided in the same year by Creighton and McClintock (1931) for plants (maize). Indeed, maize turned out to be a remarkably favorable material for cytogenetic studies. Although it does not have giant chromosomes later so useful in *Drosophila* research, all ten chromosomes are morphologically distinct, and the occurrence of extra chromosomes is not infrequent. Barbara McClintock used these attributes of maize during thirty years of brilliant studies for an interpretation of gene action, the comprehensiveness of which was not generally realized until the molecular geneticists, years later, arrived at similar conclusions.

The story of crossing over as here presented is a considerable over-simplification, leaving out many complexities. For instance, the nature of the chiasmata (the bridges between chromosome segments that result from crossing over) has long been controversial. The number of chiasmata per chromosome arm is highly variable, and indeed there is no crossing over in some cases, for instance in the male *Drosophila*. There has been much argument as to the exact moment of the replication of chromosomes during the first meiotic division as well as the exact moment of chiasma formation (breakage and healing of chromatin strands), indeed even whether the presence of a chiasma always indicates genetic crossing over. Most importantly, the behavior of the different chromatid strands in a chromosome remained long controversial. As a result, the breakage-fusion theory of Janssens and Morgan as an explanation of the phenomena of crossing over was rejected

767

by certain authors. Belling, for instance, proposed the "copy-choice" theory, and Winkler the "gene-conversion" theory. Although in the end neither theory prevailed, both resulted in numerous experiments that led to a better understanding of crossing over and the nature of the gene. No comparative history of the three theories has yet been written. Textbooks of cytology and genetics must be consulted for full details on these technical matters (see also Grell, 1974). What is important is that all seeming exceptions could be explained ultimately in terms of the classical chromosome theory.

768

The reconstitution of chromosomes, effected by crossing over, is very important for the evolutionary process. It is an effective mechanism for the mixing of the paternal and maternal genes and, by producing new intrachromosomal combinations of genes, it provides a great abundance of new genotypes (far more than mutation) on which natural selection can act.

There is still another chromosomal process which facilitates recombination: the independent movement of maternal and paternal chromosomes during the reduction division in meiosis. Prior to 1902 there was a wide spread belief that the paternal and the maternal sets of chromosomes move as units. It was, for instance, believed by some authors that, during the maturation divisions of the egg cell, all the paternal chromosomes are eliminated in the polar bodies, to be replaced by a new set of father-derived chromosomes through fertilization. If this were true, no polar bodies would be produced during the maturation of the parthenogenetic eggs, but Boveri proved that polar body formation in parthenogenetic eggs does not differ in any way from that of sexual eggs. Furthermore, heterozygous females produce gametes with paternal genes. Finally, Carothers (1913) found that in species with unequal sized (heteromorphic) chromosome pairs, the larger chromosome moves randomly to either pole. This was conclusive proof that the maternal and paternal chromosome sets did not segregate as unitary packages. There is, however, a rare genetic condition ("meiotic drive") which prevents the random assignment of chromosomes to the polar bodies. This explains certain cases of the maintenance of otherwise deleterious genes in populations (see Part II).

Chromosomal Rearrangements

Chromosomes occasionally break completely during chiasma formation, and may then be reassembled in new ways, instead of simply healing together where the break had occurred. If there was a double break, the middle piece can be turned around, resulting in a chromosomal *inversion*. It is called a paracentric inversion if the centromere is not located on the inverted chromosome section, or a pericentric if it includes the centromere. A *translocation* occurs when a piece of one chromosome breaks off and attaches to (or is inserted into) another chromosome with which it is not homologous. Occasionally *unequal crossing over* occurs, resulting in two daughter chromosomes, one of which has a duplicated piece of chromosome, while the other has a deficiency. Two (acrocentric) chromosomes may fuse or a (metacentric) chromosome may undergo fission; such changes are called *Robertsonian rearrangements*. Finally, *polyploidy* refers to the presence of more than two of the basic number of chromosome sets. All these chromosomal changes are potentially of considerable evolutionary significance, but have not in any way weakened the chromosomal theory of inheritance. Chromosomal rearrangements that have a genetic effect are often called *chromosomal mutations*. I have avoided burdening the text with a history of the discovery of each one of these chromosomal mutations, because this would not contribute to the understanding of chromosomal evolution.[25]

769

Morgan and the Chromosome Theory

The claims made by some historians that Morgan and his group were the originators of the chromosome theory of inheritance is clearly not valid. The establishment of the individuality of the chromosomes (largely by Boveri) and Roux's convincing argument that the chromosomes must contain a linearly arranged set of qualitatively different genetic particles, combined with Mendel's findings of segregation, led in 1902–1904 inevitably, so to speak, to the Sutton-Boveri chromosome theory of inheritance. This theory was almost instantaneously accepted by most cytologists, since it was merely the capstone of an edifice erected by cytology during the preceding twenty years.

Considering how persuasive the theory was, the historian is somewhat puzzled why it encountered so much opposition, including that of some of the foremost geneticists like Bateson, Johannsen, and at first also Morgan. It is quite evident that a deep-seated conceptual disagreement between two major schools of biology was involved. Since the chromosome theory had been arrived at indirectly, by inference from disparate sets of facts, the opposition demanded proof, preferably experimental proof, which was eventually supplied by the Morgan group and others. But this was after 1910, following Morgan's conversion from an opponent to a supporter of the chromosome theory.[26]

Morgan had scathingly attacked the chromosome theory in a series of books and papers from 1903 to 1910 (Allen, 1966), basing his opposition on a number of arguments. First of all, the theory was mere "speculation," without empirical foundation. Nothing could be considered as science by Morgan that was not established by experiment. He had the utmost contempt for "philosophizing." More importantly, the claim that characters are controlled by particles and that these particles are located on individually different chromosomes was completely in conflict with his theory of biological phenomena (see below).

And yet in 1910, almost overnight, Morgan became one of the chief advocates of the chromosome theory and supplied some of the most decisive proofs in its favor. How can this dramatic conversion be explained? That it was sudden is documented by the dates of some of Morgan's pre- and post-conversion publications.

By an irony of fate, a 48-page paper in the *American Naturalist,* in which Morgan (1910a) severely attacked the chromosome theory, was published in August 1910 (although submitted in February), three weeks after Morgan's famous white-eye paper (1910b), submitted July 23 and published on July 27, which was instrumental in Morgan's abandoning of his opposition.

Morgan, who was forty-four years old in 1910 and well-known for his strong opinions, was able, in contrast to Bateson, to change his mind (up to a point!) when new experiments showed that his previous explanations were untenable. However, it is evident that Morgan's mind was strongly reenforced by his intellectual milieu. After all, his findings merely confirmed what his colleague and friend E. B. Wilson had urged on him for nearly a decade. Wilson's arguments were reenforced by Morgan's remarkable team of young collaborators. They were outstanding for diversity of talent and character, and for lacking Morgan's nineteenth-century biases. The principal characteristics of the members of the Morgan team have been well described by Jack Schultz (1967), himself a later member of the group: "Morgan's scepticism and Muller's system building . . . Sturtevant's extraordinary analytical powers, and the brilliance of Bridges' experimental talents." All the younger members of the team, most of them cooped up together in the small "fly room," worked on Morgan's reeducation. Just exactly who among the four members of the group contributed to what particular substantiation of the chromosome theory is impossible to reconstruct, and of little importance. Morgan's historiographic partisans are Sturtevant (1965a) and Allen (1967; 1978), while Muller is championed by Carlson (1966; 1974) and Roll-Hansen (1978b). Owing to their diversity, the inhabitants of the fly room supplemented each other most felicitously and as a team they practiced the hypothetico-deductive method quite admirably. Muller, Bridges, and Sturtevant presumably supplied after 1911 most of the hypotheses, and Morgan, forever, sternly insisted that they be tested thoroughly by experiment.

771

Even though Morgan himself had discovered (and correctly interpreted) crossing over and other essential evidence for the theory of the gene, there is much indirect evidence that he was a somewhat reluctant convert and occasionally tended to slip back into his pre-1910 thinking. As late as 1926 Morgan gave away his physicalist bias by claiming that students of heredity reach all their conclusions concerning genes "from numerical and quantitative data . . . The theory of the gene . . . derives the properties of the genes, so far as it assigns properties to them, from numerical data alone"—as if location on the chromosome was the only property genes have!

The consistency of the chromosome theory with the rapidly accumulating genetic data is presented already in 1915 with remarkable definiteness in *The Mechanism of Mendelian Heredity* by Morgan, Sturtevant, Muller, and Bridges. It is therefore somewhat puzzling why Bateson, Johannsen, and others continued in their opposition, and why, instead of ignoring them, Morgan's two closest associates, Sturtevant and Bridges, felt the need to substantiate the validity of the chromosome theory by ever new experiments. They delighted in finding seeming exceptions or discrepancies, in order to be able to prove that they nevertheless could be explained in terms of the chromosome theory. One wonders why they did not close the book and turn to entirely new problems, as did Muller. So far as I can judge, the highly ingenious and meticulously precise work in *Drosophila* genetics from 1915 to the 1930s did not result in any essential revision of the Sutton-Boveri theory. What it did do, however, is to prove the theory and to show its biological implications.

The answer to the question why the chromosome theory encountered so much resistance emerges from a study of the contemporary literature (Coleman, 1970; Roll-Hansen, 1978b). The chromosome theory was not merely one of the thousands of building stones in the edifice of biological knowledge; rather it was a test case for the validity of two basically different philosophies of biology—a confrontation of two *Weltanschauungen.* They were the same two schools that had differed on the nature of fertilization (contact vs. fusion) and in other nineteenth-century controversies, such as the origin of cell nuclei (see also Coleman, 1965; Churchill, 1971). It is difficult to describe the two opposing parties in terms that were not already obsolete in 1910. I can convey an impressionistic picture by saying that on one side were the physicalists-epigenesists-embryologists and on the other side the corpuscularists-preformationists-cytologists, but in doing so I am using designations that were inappropriate by 1910. For instance, labeling anybody a preformationist after 1800 is quite misleading. The physicalists, in principle, were extreme reductionists, but in this case they did not carry the analysis nearly as far as the corpuscularists. The physicalists were mechanists, but so were the corpuscularists. The physicalists always searched for movements and forces; they favored "dynamic" explanations; they attempted to quantify everything and express it in numerical values. The corpuscularists explained biological phenomena in terms of qualitatively different particles, in terms of structure, form, uniqueness, historical change, and populational aspects. Their "physical" explanations led them to invoke molecules (and thus chemistry) rather than forces (and thus physics).

One may argue how best to designate the two opposing camps, but there is little doubt as to the fundamental differences in their interpretation of the nature of organic matter. Bateson, Johannsen, and, at first, Morgan, were physicalists, and, if the chromosome theory of inheritance were correct, it could be interpreted as a refutation of their own conceptual framework. This was true in general as well as for particular aspects, as I shall now try to show.

The physicalists were horrified by the idea of having to recognize corpuscular genes. To them this was nothing less than reviving preformation in a modernized form. The argument of preformation vs. epigenesis had, of course, long been dead, if it is expressed in terms of an alternative between a homunculus and a *vis viva.* The idea of a homunculus was too absurd after the birth of embryology (ca. 1816–1828) to be considered any longer, but the epigenesists' belief in a generalized *vis viva* or general developmental force was equally untenable after biologists had become aware of the precision of inheritance. To Roux, Weismann, and Boveri it was completely obvious that the precision of inheritance required postulating an architecture of the germ plasm, that is, a structural complexity of the genetic material, later articulated in the Sutton-Boveri chromosome theory. The physicalists found it difficult to understand how one could hold such ideas without falling back into the naive preformationism of Bonnet.

An even sounder reason for the opposition was provided by embryology. Roux's brilliant 1883 theory of an equal division of the genetic material was soon seemingly refuted by Roux's own description of mosaic development and the results of cell lineage studies. One embryological discovery in the 1890s after the other seemed to be more easily explained by Weismann's theory of an unequal division of the germ plasm than by equal Mendelian division. The resolution of the seeming conflict between developmental phenomena and the Sutton-Boveri theory required many decades of analysis and reconceptualization.

Another reason for the opposition was the unrealistic simplicity of the first corpuscular genetic theory. One must remember that in the early 1900s no distinction had yet been made between genotype and phenotype. Even though the homunculus theory of preformation had been thoroughly discredited, it was replaced in the minds of certain embryologists and geneticists by a model in which each character of an organism was represented in the germ plasm by a specific genetic factor. The genotype was, so to speak, the phenotype in miniature, not as a homunculus but as a mosaic of hereditary particles (whether called gemmules, pangens, or what), each responsible for a definite component of the phenotype. This thinking is expressed in the "unit character" concept of the early Mendelians. De Vries (1889) had stated specifically that the pangens move from the nucleus into the cytoplasm, where they are responsible for development. The soma (the body) thus would consist of developed pangens. For the physicalists this was a morphological interpretation of inheritance, not different in principle from the old homunculus concept. Bateson and Johannsen quite specifically criticized what they considered this rather morphological interpretation of the chromosome theory.

The relation between transmission and development, which had been so puzzling to Weismann, Hertwig and the German embryologists, also played a role. Morgan and his group decided to deal with the two sets of problems separately, and to begin with

773

transmission genetics. Bateson and other opponents of the chromosome theory, continuing the Weismannian tradition, wanted a genetic theory that could explain transmission and development simultaneously. The theory of the presence of identical chromosomes (with linearly arranged corpuscular genes) in the most diverse tissues and organs of the body seemed to them incompatible with the observed phenomena of development.

As long as no distinction was made between genotype and phenotype, a corpuscularist was forced to think in terms of a kind of preformationism based on a one-to-one relationship between genetic factor and somatic character. It was stated by some adherents of the unit-character theory that there are as many genetic factors as an organism has characters. Weismann, with the consistency and logic that was so characteristic of him, postulated therefore that there must be different determinants for different characters at all developmental stages, for instance not only for each feature in the wing of the adult butterfly that can vary independently but likewise for each characteristic of the caterpillar. Since it was more or less taken for granted that the genetic material, through replication and growth, was directly converted into the phenotype, this was not merely a logical but one might say necessary conclusion. Consequently, when Castle discovered changes in the phenotype, which we now know to be due to modifying genes, he was forced to explain them in terms consistent with the one-gene-one-character hypothesis, inducing him to propose his contamination theory (see below).

The discovery of pleiotropy and polygeny (see below) eventually led to a rejection (or at least drastic modification) of the unit-character theory. This helped to narrow the gap between the two camps by freeing the adherents of the chromosome theory of the taint of crude preformationism. Yet, there is no doubt the controversy ended with a hands-down victory of the corpuscularists. Theirs was eventually called the molecular theory of inheritance. Carlson (1971) is quite right when he insists that conceptually Muller was a molecular biologist, but he was by no means the first one. A molecular basis of inheritance had been unequivocally postulated prior to Muller by Weismann, de Vries, and others, as far back as the 1880s.

It must be emphasized that this is a rather simplified presentation of the controversy and the positions in the two camps. Each participant, for instance Bateson, Johannsen, Weismann, Hertwig and Morgan, had his own peculiar mix of ideas, indeed sometimes rather illogical and contradictory mixes of ideas. Yet the chromosome theory was either consistent with their conception of living matter, or it was not. If it was not, they either had to try to refute it or give up long-held and cherished beliefs. No wonder Bateson and Johannsen were so recalcitrant.

Chromosome Research

Chromosome research continued to be exceedingly productive in the years after Boveri and E. B. Wilson. Cytogenetics, that is, the integration of chromosomal and genetic findings, made rapid advances owing to McClintock's analysis of the pachytene chromosomes of maize (1929), the rediscovery of the giant polytene chromosomes of dipterans by Heitz and Bauer (1933), the study of genetic systems of C. D. Darlington, the work of M. J. D. White, and the work of an ever increasing army of cytologists. A new era of active chromosome research began in the 1970s.[27]

775

The major advances in this field were made by the application of numerous new techniques. Chromosome numbers, for instance, can now be determined far more precisely than formerly by applying the squash technique, by tissue culture (producing enlarged cells), immersion in hypotonic solutions (also resulting in cell enlargement), and colchicin treatment (which inhibits spindle formation and contracts chromosomes). The newer techniques resulted, for instance, in the revision of the chromosome number of man from 48 to 46. In many researches, such as the localization of genes responsible for genetic diseases in man, the correct identification of individual chromosomes is highly important. Chromosomes are quite heterogeneous in composition and certain chemical treatments affect the various components differentially, resulting in the appearance of banding on the chromosome patterns. Depending on the technique used, one can recognize Q bands, G (for Giemsa) bands, R bands, T bands, and C bands (see Caspersson and Zech, 1972). A very different type of information is obtained by labeling chromosomes in live tissues by radioactive material (tritium).

Perhaps the most important finding of these researches is that the prokaryotes (bacteria and blue-green algae) possess the same genetic material (nucleic acid) as higher organisms, but that it is not organized into the kind of chromosomes found in higher organisms. However, precisely because the organization of the DNA (or RNA) in these microorganisms is so much simpler, they are specially suited for certain types of genetic analysis, particularly gene function and gene regulation. As a result, most molecular genetics until the early 1970s was based on research done with prokaryotes.

Although the organization of the DNA is now reasonably well understood for many prokaryotes, the eukaryote chromosome has been remarkably refractory to analysis (*Cold Spring Harbor Symposia,* 1978). It is known that the DNA is attached to (imbedded in?) a matrix of proteins (particularly histone), and there are strong indications that these proteins play a decisive role in gene activity. Yet, in spite of the mass of facts we have learned in recent years, it seems to me that we are still quite a distance from an internally consistent explanatory theory of structure and function of the eukaryote chromosome as a whole.[28] Thus, the acceptance of the chromosome theory of inheritance was by no means the end of the chromosome studies, but rather the entry into a new era of chromosome research.

776

18 ✍ Theories
of the gene

THE LAWS OF Mendelian genetics provided an excellent expla- nation of the phenomena of discontinuous variation. They were easy to apply whenever sharply defined characters were involved, such as green versus yellow or smooth versus wrinkled in peas. Literally hundreds of papers were published in the years after 1900 demonstrating the occurrence of Mendelian inheritance in many groups of animals or plants and establishing a Mendelian basis for any observable discontinuously varying character.

Nevertheless, for several decades there was widespread opposition to an acceptance of the universal applicability of Mendelian inheritance. It would be a mistake to ascribe this opposition to ignorance or conservatism, for such an interpretation is too simplistic. Actually, the opponents had what they considered perfectly valid arguments. Furthermore, to set the record straight, they did not deny the occurrence of some Mendelian inheritance; what they denied was that *all* inheritance is Mendelian. Since many of these opponents were first-rate biologists, it is important to ask what their reasons were.

Recent historians have tended to forget that most Darwinian zoologists and botanists at the turn of the century were interested in inheritance primarily for its connection with the species problem and the theory of evolution. Therefore, these Darwinians read only the writings of those two Mendelians who were most interested in evolution—de Vries and Bateson—and their theories forced the Darwinians inexorably into opposition. Both de Vries and Bateson preached that the discontinuity of inheritance proved the discontinuity of evolutionary origins. They were essentialists and saltationists (see Chapter 12) and neither placed much trust in natural selection. Thus, their views were totally at variance with those of the Darwinians, who saw evidence of gradual evolutionary change everywhere in nature. Owing to the claim of the Mendelians that there is a close correlation between the mode of genetic variation (that is, discontinuity) and the mode of evolution, and owing to their own conviction that evolution is gradual and continuous, the Darwinian naturalists were forced into the position of having to postulate some kind of non-Mendelian continuous form of inheritance to account for gradual evolution (Mayr and Provine, 1980).[1]

The greatest weakness of the Mendelian position, as the naturalists saw it, was that it left continuous variation unexplained. With nearly everyone at that period still accepting the duality of variation (continuous *and* discontinuous), Mendelism, so it was said, had no explanation for quantitative variation. Let us remember that Weismann, de Vries (1910: 73–74), and other authors of the 1880s and 90s had explained quantitative inheritance by unequal numbers of (identical) pangens or biophores contributed by the two parents. As stated by de Vries, "The pangens present may vary in their relative number, some may increase, others may decrease or disappear almost entirely . . . and finally the grouping of the individual pangens may possibly change. All of these processes will amply explain a strongly fluctuating [individual, continuous] variation" (1910: 74). This explanation became void when the Mendelian theory (one element only per contrasting character from either parent) was accepted. Continuous variation was now left without an explanation, and I can find no adequate replacement for the unequal distribution theory in de Vries's post-1900 writings.

The opponents of exclusively Mendelian inheritance asked: In the case of strictly quantitative characters, let us say size, does not the intermediacy of the offspring demonstrate the absence of discontinuous factors? Does this not indicate that there are two kinds of inheritance, Mendelian inheritance for discontinuous variation, and some other mode of inheritance for continuous variation? And isn't it much more important to explain the inheritance of continuous variation, because this is the variation on which Darwin's theory of gradual evolution was based? As a result of the absence of a theory of quantitative inheritance, a split developed among evolutionary biologists, the two opposing groups generally referred to as the Mendelians and the biometricians. This designation, however, is valid only for the period from 1900 and 1906, while the controversy had started in 1894 with the publication of Bateson's *Materials* and continued until the evolutionary synthesis in the 1930s and 40s. It caused a deep split in evolutionary biology, continuing for the first three decades of this century (Mayr and Provine, 1980). It was a clash between two philosophies, with the Mendelians favoring essentialistic thinking and the behavior of single units of inheritance, while the biometricians were interested in population phenomena and indulged in holistic interpretations. One might even go so far as saying that some of the polarities between the opponents go all the way back to the eighteenth century. Indeed, one of these ancient problems, that of blending inheritance, must be studied before we can proceed with the analysis of the post-1900 events.

Blending Inheritance

The naturalists and animal breeders, back into the eighteenth century, knew that "sports" (discontinuous variants), once they had arisen, might persist unchanged generation after generation. By contrast, when one crossed either different species or domestic and geographic races, they "blended." Darwin, for instance, uses the word "blending" almost invariably in connection with the crossing of species and races. So did Moritz Wagner and other naturalists writing about blending in the period after 1859. This term was based on the completely correct observation that there is very little detectable Mendelian segregation in the F_2 of most species crosses (see Kölreuter, Chapter 14). It must be emphasized that all these authors had phenotypes in mind, and since most differences between species are highly polygenic, the phenotypes of species and race crosses are usually rather intermediate, that is, they "blend." When originally coined, the term "blending" referred to the appearance of phenotypes.

Does this mean that these authors also believed in a blending or fusion of the genetic determinants of the observed phenotypic characters? Apparently they did so, but only in part. Darwin, for instance, makes many statements which apparently assume that the paternal and maternal gemmules may either fuse during fertilization or merely lie attached to each other, ready to separate again in later generations.[2] Darwin's great emphasis on the frequency of reversion completely refutes the notion that he had believed in universal fusion (blending). In the *Origin* (1859) he refers to reversion no less than eight times (pp. 13, 14, 25, 152, 160, 161, 163, and 473), and in the *Variation of Animals and Plants* (1868) he devoted an entire chapter (XIII) to it. In the second edition of the *Variation* (1893, II: 23) he states explicitly that it would "be more correct to say that the elements of both parent species exist in every hybrid in a double state, namely, blended together and completely separate." At other places he refers to "pure" and "hybridized" gemmules in the offspring of crosses. With particular approval Darwin refers to Naudin's views of the nonblending of parental characters in hybrids (see Chapter 14). Perhaps better than in any of his published writings, Darwin has stated his belief in particulate inheritance in 1856 in a letter to Huxley (*M.L.D.*, I: 103): "I have lately been inclined to speculate, very crudely and indistinctly, that propagation by true fertilisation will turn out to be a sort of mixture, and not true fusion, of two distinct individuals, or rather of innumerable individuals, as each parent has its parents and ancestors. I can understand on no other view the way in which crossed forms go back to so large an extent to ancestral forms."

780

Admittedly, in his subsequent writings Darwin never again stressed the particulate theory of inheritance quite so strongly as in this letter, but neither did he adopt a universal theory of blending, opposing claims notwithstanding. De Vries (1889) has stated quite correctly that Darwin's interpretation of inheritance is, on the whole, far more compatible with particulate than with blending inheritance. Darwin, although the author of a two-volume work on variation, was not primarily interested in the development of a genetic theory, and therefore he quoted reversion far more often as evidence for common descent than as evidence for a theory of inheritance. Support of common descent explains his intense interest in the occasional occurrence of zebra-like stripes on the legs and shoulders of the horse and ass.

Nägeli was one of the few biologists of the post-Darwinian period to have clearly adopted a theory of exclusively blending inheritance (perhaps also Oskar Hertwig), though a belief in blending inheritance was compatible with a hypothesis of gemmules, micelles, or other particles serving as the genetic material, provided maternal and paternal particles fuse during fertilization. All the others not only postulated particles as carriers of inheritance (some of which, of course, might fuse during fertilization), but postulated also that at least some of these particles were transmitted intact from generation to generation (see, for instance, Galton, 1876; and de Vries, 1889). The assertion, I believe first made by R. A. Fisher (1930), that Darwin and most pre-1900 students of variation had adopted a theory of exclusively blending inheritance is not supported by the evidence (see also Ghiselin, 1969; and Vorzimmer, 1970). How fully this was understood at that period is evident from a statement made by the American embryologist E. G. Conklin in 1898: "Many other phenomena, especially particulate inheritance, the independent variability of parts, and the hereditary transmission of latent and patent characters, can at present only be explained by referring them to ultra-microscopic units of structure" (from Carlson, 1966: 18). Considering the widespread acceptance prior to 1900 of the theory of particulate inheritance—that is, the theory that the genetic factors transmitted from the parents do not fuse after fertilization but retain their integrity throughout the entire life cycle—it is quite misleading to say that the most important effect of the rediscovery of Mendel's work in 1900 was the replacement of a universal belief in blending inheritance by one in particulate inheritance. Many authors, including Darwin, subscribed to a mixed theory. A continuing belief in blending inheritance, it would seem to me, played only a minor role in the resistance to Mendelism after 1900. What Fisher and others who accepted his interpretation forgot is that prior to about 1909 no distinction was made between genotype and phenotype, and that the term "blending" was traditionally used for an intermediate appearance of the phenotypes, particularly in species crosses. It did not necessarily indicate any commitment concerning the behavior of the genetic material.

781

It is thus necessary to clarify a second major confusion of the pre-Mendelian and early Mendelian period, the difference between phenotype and genotype.

The Difference between Phenotype and Genotype

The argument about blending has illustrated how important it is to make a distinction between the genotype (the genetic endowment of an individual) and the phenotype (the body into which this genotype has been transformed during development).

Almost the only nineteenth-century author who paid attention to this distinction was Galton. His new term "stirp" and redefined term "heredity" clearly referred to the genotype, and his terminology "nature vs. nurture" stressed the difference. Focus on this problem had been largely missing, not only in Darwin's writings but also in the post-Darwinian period. In 1900 when the science of genetics was born, this distinction had not yet been clearly made either terminologically or conceptually, except for Weismann's germ plasm and soma. The individual as a whole was for de Vries nothing but an enlarged version of the original set of pangens in the nucleus of the fertilized egg (zygote). This is why he never bothered to specify whether his term "mutation" referred to the phenotype or the underlying germ plasm.

But the animal breeders and those who raised cultivated plants knew all along that there is no such inexorable genetic determinism, as implied by de Vries's concept. There are many characters—let us say fruit size in tomatoes—that are affected both by genetic constitution and by environmental factors.

The first person who realized the need for a terminological distinction was the Danish geneticist Wilhelm Johannsen (1857–1927). Johannsen's background and education were quite unusual. He was largely self-taught, receiving most of his early education in pharmaceutical and chemical laboratories. When he finally turned to plant physiology he, like Galton, whom he greatly admired, stressed quantitative methods and statistical analysis. Also very much an essentialist, he was disturbed by the considerable size variation of beans remaining after a series of generations of self-fertilization, which presumably should yield beans that were genetically virtually identical and largely homozygous. To escape from this variation, he designated the statistical mean value of the sample as the "phenotype": "One might designate the statistically derived type . . . simply as phenotype . . . A given phenotype may be an expression for a biological unity, but by no means does it need to be. Most of the cases of phenotypes found in nature by statistical investigation are not!" (1909: 123). His terminology as well as his discussion showed clearly that Johannsen tried to reach the "pure essence," hence his pursuit of "pure lines." Subsequent workers found this typological definition of little use, and redefined the phenotype as the realized characteristics of an individual.[3] Although the terms are those of Johannsen, the modern usage of phenotype and genotype is actually closer to Weismann's somatoplasm and germ plasm.

After Johannsen had coined the word "gene" (see Chapter 17), he combined it with the root "type" to form the word *genotype,* as the counterpart to the term *phenotype.* "Genotype" refers to the genetic constitution of the zygote, formed by the union of two gametes: "This constitution we designate by the word genotype. The word is entirely independent of any hypothesis; it is *fact,* not hypothesis that different zygotes arising by fertilization can thereby have different qualities, that, even under quite similar conditions of life, phenotypically diverse individuals can develop" (1909: 165–170). Yet, on the whole, Johannsen tended to think, quite typologically, of the genotype of a population or species. Woltereck

782

(1909), at about the same time, adopted a different terminology to express the important insight that the same genotype can produce rather different phenotypes under different environmental conditions. What is inherited, said Woltereck, is merely a *norm of reaction*, a predisposition to react in a specific manner to any set of environmental conditions.

However, not until it was discovered (1944–1953) that the genotype consists of DNA and the body of proteins (and other organic molecules) was the fundamental difference between the genotype and the phenotype fully understood. In the early years of genetics considerable confusion continued, from which even Johannsen was not exempt. A failure to make a distinction between genotype and phenotype was at the bottom of many of the great controversies in the history of evolutionary biology, for instance those dealing with blending inheritance and with the nature of mutation. Indeed, a clear understanding of the difference between genetic endowment (genotype) and visible appearance (phenotype) was necessary for the final refutation of soft inheritance. It is no accident that Johannsen himself made a decisive contribution to this refutation, though he was helped by a fortunate choice of a test organism.[4]

What Johannsen did was to choose a self-compatible plant, the garden bean (*Phaseolus vulgaris*). Since plants of this species are normally self-fertilized, they are highly homozygous. As his basic breeding stock Johannsen chose nineteen plants, the product of several generations of self-fertilization. Within each of these "pure lines," he raised offspring from the largest and from the smallest beans. The variation within each group of descendants was virtually identical, regardless of the size of the parental bean. In other words, the genotypes of the large and the small beans within a pure line were the same, whereas the observed differences were phenotypic responses to varying environmental conditions. An important aspect of Johannsen's work was the precision with which he measured and weighed thousands of beans, as well as the careful statistical analysis of the findings. The inevitable conclusion was that differences in size due to differences in cultivation (fertilizer, light, water, and so on) cannot be transmitted to the next generation. There is no inheritance of acquired characters. Since the phenotype is the result of an interaction between genotype *and* the environment, it can not be considered an accurate representation of the genotype.

Johannsen's interesting pure-line experiments had a very ambiguous impact on biology. On one hand they helped to weaken the—at the time—still powerful and widespread belief in soft inheritance, but unfortunately the experiments were also cited by Johannsen himself and by others as evidence for the impotency of natural selection (see Chapter 12).

COMPETING THEORIES OF INHERITANCE

784 Now that we have clarified the problem of genotype vs. phenotype and of blending inheritance, we are ready to consider comprehensively the reasons for the opposition to the belief in the universal validity of Mendelian inheritance. The existence of competing theories of inheritance played a large role in this opposition. When the Mendelian laws were rediscovered in 1900, they were not able to occupy a vacant field. In fact several other theories—three major ones in particular—were already in existence which seemed to be able to explain gradual Darwinian evolution better than Mendelism.

Galton's Law of Ancestral Heredity

Darwin's cousin, Francis Galton, continued after 1875 to build on his early theory of inheritance (see Chapter 16). Virtually alone among early students of inheritance, he was interested in the populational aspects of genetic variation. In contrast to the hybridizers and Mendelians, he concentrated on quantitative characters, like height or skin color. He had observed that the mean value of such characters in a population remained on the whole the same from generation to generation. The tallest men had, on the average, shorter children than the mid-value between themselves and their spouses. Their offspring was *regressing* toward the population mean. The offspring of the shortest men, by contrast, was regressing upward toward the population mean. Galton's underlying reasoning very much appealed to common sense. He said every person receives about one half of his hereditary endowment from his father, the other half from his mother. Applying the same reasoning to the grandparental generation, a person receives about one fourth of his inheritance from each of the grandparents, one eighth from each great-grandparent, and so forth. The contribution of an ancestor, thus, would be halved in every generation. This was later called Galton's law of ancestral heredity.[5]

At first sight Galton's interpretation of inheritance seemed to

explain continuous variation far better than Mendelian segregation. Darwinians like Weldon and Pearson, who were committed
to Darwin's concept of gradual evolution, opted for Galton when
forced to choose between a theory of discontinuous and one of
continuous inheritance (even though Galton himself believed in
saltational evolution; see Chapter 12). The weaknesses of Galton's
law of ancestral heredity, even when modified by Pearson, were
many, one of them being that it was purely descriptive and did
not really provide any causal explanation at all, another that it did
not permit any predictions. Galton's worst mistake, however, was
that he transferred what was statistically true for the genotype as
a whole to the mode of inheritance of individual characters. Even
though Galton had adopted particles as the material basis of inheritance (see Chapter 16), he dealt with them in his reasoning as
if they were blending. The production of homozygous recessives
from heterozygous parents (in turn descended from heterozygous
grandparents) was completely inexplicable under Galton's law and
constituted its unequivocal refutation. Galton's law describes the
probable similarity of an individual to its ancestors reasonably well,
but is not applicable to individual genetic factors. However, it took
a considerable time for this to be fully understood, and Mendelism could not expect universal acceptance until all adherence to
Galton's law had been given up.

785

Even after Weldon had died in 1906 and Pearson (d. 1936)
and Galton (d. 1911) had turned to other fields, the problem of
inheritance of continuous variation remained controversial. To be
sure, in a prophetic paper the British mathematician Yule (1902:
234–235) had suggested that continuous variation might be due
to the coaction of multiple factors. This explanation was completely ignored by his contemporaries (see below.)

Contamination Theory

Efforts to explain continuous variation in a non-Mendelian
manner continued for many more years. William E. Castle, one
of the most ingenious experimenters in early genetics, observed
that albino guinea pigs derived from a cross with a black grandparent had more black pigmentation at the extremities, and occasionally elsewhere, than albinos from a pure strain of albinos.
He therefore developed the theory that in heterozygotes there
was some "contamination" of the white genetic factor by the black
one (and vice versa) during meiosis, so that the offspring showed
a slight degree of intermediacy. This was the last theory of "soft

inheritance" to be proposed by a respectable geneticist. Such a mutual influence of alternate characters would, of course, help considerably to explain continuous variation, and thus was a welcome theory for the Darwinians. Castle's contamination theory led to a controversy with Morgan and his students, particularly Muller.

When a crucial backcross experiment failed to confirm his predictions in 1919, Castle abandoned his theory. His thinking had been based on the unit-character concept of the early Mendelians, particularly Bateson, according to which each character was controlled by a single, specific genetic factor. If the character varied, as in Castle's crosses, this had to be due to a modification of the genetic factor. The multiple-factor theory (see below) led to the abandonment of the unit-character theory by showing that several, if not many, genes may affect (modify) a single character.

The Cytoplasmic Inheritance Theory

After Castle's gene contamination theory had been refuted, one last theory remained which attempted to explain continuous variation in a non-Mendelian manner. According to this theory, continuous variation was caused by a special "species substance," perhaps carried in the cytoplasm, and quite independent of the discontinuous Mendelian genes.

The idea that a uniform species substance was transmitted from generation to generation was only slowly displaced by the theory that inheritance is controlled by particulate genes located on chromosomes. Many observations made from the 1880s to the 1920s seemed to be better explained by postulating a relatively uniform, diffuse species-specific genetic substance, presumably residing in the cytoplasm and existing side by side with the chromosomal genes. The chromosomes, according to this idea, were the carriers of discontinuous characters, exemplified by de Vries's and Morgan's mutations, while continuous variation as well as that which was responsible for the "true nature" of species was carried in the cytoplasm. Such ideas were popular among embryologists. Observation and experiment had shown repeatedly that the cytoplasm of the mature ovum had a complex organization that seemed to be the primary control of early development. Recent work has fully confirmed these observations. This fact was responsible for Roux's shift from equal to qualitative cell division. It was only very much later that it was discovered that this organization of the cytoplasm is controlled by genes acting during the formation of the ovum while still in the ovary. At any rate, from Wilhelm His

786

(1874: 152) to Jacques Loeb, in 1916, many biologists openly expressed doubt whether the nucleus had anything to do with early development or the nature of species. Boveri, who himself had supplied the most decisive evidence in favor of the important role of the nucleus (see Chapter 17), continued to have reservations (1903, Roux's Archiv, 16: 356). Species characters, he suggested, can be partitioned into such that can be explained by chromosomal inheritance, but the inheritance of the characters that determine the assignment of a species to a higher taxon seemed to him to pose an open question. Many biologists in the pre-1930 period divided inheritance into that which is controlled by the nucleus and that which is controlled by cytoplasm. Even E. Baur (1929), the most consistent Darwinian among the continental geneticists, left open whether one could explain the characters of the higher taxa in the same manner as the species characters. There seemed to be nothing Mendelian in the variation of such characters.

787

The proponents of cytoplasmic inheritance had some seemingly valid arguments. The conspicuous effect of the egg cytoplasm in the early stages of embryogenesis was pointed out particularly by those, like Conklin and Guyer, who worked with species with highly unequal cleavage divisions. The naturalists remarked that the kind of mutations with which Morgan worked, like white eyes, yellow body color, missing bristles, crumpled wings, and so forth could be found not only in *Drosophila melanogaster* but also in other species of *Drosophila*, while—so they claimed—there was no evidence for a chromosomal inheritance for those subtle characters that distinguish species. The opponents of exclusive chromosomal inheritance could not conceive that the rich repertory of heritable characters could be confined to the minimal mass of the chromosomes. Winkler (1924) presents a good summary of the arguments in favor of cytoplasmic inheritance.

Botanists in particular discovered so many phenomena that seemed to demand the existence of cytoplasmic inheritance that Wettstein (1926) proposed to designate the genetic material located in the cytoplasm as *plasmon*, in contrast to the *genom* located in the nucleus. A considerable number of botanists, particularly German ones, discovered genetic effects of the cytoplasm, such as Correns (*Mirabilis* and other genera), Michaelis (*Epilobium*), Schwemmle (*Oenothera*), Oehlkers (*Streptocarpus*), Wettstein (mosses), and others.[6] In this milieu, Goldschmidt interpreted some of his findings in *Lymantria* also as due to cytoplasmic inheritance.

The emphasis on the cytoplasm in Germany was clearly a carry-over of the strong interest in developmental phenomena which had characterized German studies in inheritance during the 1880s and 90s. In retrospect it is clear that these studies of cytoplasmic phenomena were premature and that German genetics, in spite of the considerable number of workers involved, made a smaller contribution to our understanding of transmission genetics than did Bateson, Cuénot, Castle, or the Morgan school, who side-stepped the problem of cytoplasmic inheritance.

788

The belief in a major, general, independent contribution of the cytoplasm to inheritance was eventually refuted in a number of different ways (Wilson, 1925). There were first of all some theoretical considerations:

(1) The extraordinary precision which governs the division of the chromatic material of the nucleus is without parallel when it comes to the division of the cytoplasm.

(2) An essential identity of the paternal and maternal contribution to the genetic constitution of the offspring had been proven, for example, by reciprocal hybridization, in spite of the enormous inequalities in the amount of cytoplasm of male and female gametes in many species. This point was demonstrated with particular elegance by Boveri (1889), who was able to fertilize enucleated fragments of the large egg of one genus of sea urchins by the sperm of a different genus, with the developing embryo showing purely paternal characteristics, while genuine hybrid embryos were exactly intermediate between the two genera.

(3) The reduction division of the maturing female gametes (egg cells) affect only the chromatic material, not the cytoplasm. By contrast, the developing spermatozoa are stripped of most of the cytoplasm so that eventually an enormous inequality of maternal and paternal cytoplasm exists, yet there is complete equality in paternal and maternal genetic endowment.

More important than these theoretical considerations was the discovery of explanations that were able to account for the seeming exceptions. One such exception is known as *delayed Mendelian inheritance.*

When there is a large mass of egg cytoplasm, the first steps of development are sometimes controlled by factors in this cytoplasm which are of course the product of the maternal individual. For instance, the direction of the coiling of snails—either dextral (clockwise) or sinistral (counterclockwise—is laid down in the first cleavage division and is determined by the egg cytoplasm. Nevertheless, it was eventually shown that the direction of coiling is indeed controlled by a gene acting on the ovarian egg prior to fertilization, and dextral coiling is dominant over sinistral coiling, at least in the species of *Limnaea peregra* in which the classical work on this problem was conducted (Boycott and Diver, 1923). A sinistral female fertilized by a dextral male will produce sinistrally coiling offspring, but these in turn will all have dextral offspring, owing to the influence of the dominant dextral paternal gene in the formation of the egg cytoplasm. The textbooks of genetics report many cases of such delayed Mendelian inheritance, sometimes extending over several generations, which on first sight would seem to indicate the occurrence of cytoplasmic inheritance.

789

A second phenomenon cited as evidence for cytoplasmic inheritance is the inclusion in plant cells of chlorophyll grains and other so-called plastids and organelles which, to a lesser or greater extent, inherit their characteristics independently of the nucleus. Indeed, some of them have their own genetic material (DNA), a seeming heritage of their evolutionary origin. Variegation of leaves is one such maternally inherited plastid characteristic in certain species of plants. Organelles in animal cells, such as mitochondria, may likewise have their own DNA. Nevertheless, these phenomena do not fundamentally contradict the chromosomal theory of inheritance. The same is true for the interesting phenomena discovered by Sonneborn (1979) which indicate a good deal of autonomy of certain cytoplasmic structures in protozoans (ciliates).

A third group of phenomena that were at one time believed to document the existence of cytoplasmic inheritance is the infection of certain tissues by microorganisms which are handed on to the gametes during gamete formation. This includes such phenomena as the "petite colonie" phenomenon in yeast discovered by Ephrussi (1953), Sonneborn's kappa factor in *Paramecium* (Preer et al., 1974), a sex-ratio factor in *Drosophila*, the sterility factor in *Culex* (Laven), and others.

Thus, one after the other of the phenomena which at first seemed to indicate the occurrence of cytoplasmic inheritance were eventually shown to have a genic-chromosomal explanation. A final clarification of all aspects of possible cytoplasmic inheritance was made possible when cytoplasm could be dissected into its elements through electron microscopy and correlated chemical researches. This does not mean that the genetics of cytoplasm is now a closed chapter. The cytoplasm plays an important role in development and in the regulation of gene activity. Indeed, there are indications that the elaborate architecture of the cytoplasm plays a greater role than is now realized. It is also possible, if not probable—and Sonneborn's researches certainly strengthen this view—that this architecture of the cytoplasm is, in part, species-specific and is involved in many of the processes of the cell. The old belief that the cytoplasm is important in inheritance, thus, is not dead, although it has been enormously modified.

790

THE MENDELIAN EXPLANATION OF CONTINUOUS VARIATION

As one after the other of the non-Mendelian explanations of continuous variation were shown to be invalid, the conclusion became inevitable that continuous variation had to be explained in terms of discontinuous Mendelian genes. The solution was made possible when it was realized that a single aspect of the phenotype might be controlled by genes at several different loci. This, in fact, had already been worked out in full detail by Mendel (1866: 36) in explanation of the results of some of his species crosses (for example, *Phaseolus nanus* × *Ph. multiflorus*) and Gärtner's species crosses. Even Bateson recognized this as the potential resolution of the conflict: "If there were even so few, as, say, four or five pairs of possible allelomorphs, the various homo- and heterozygous combinations might, on seriation, give so near an approach to a continuous curve that the purity [that is, discontinuity] of the elements would be unsuspected, and their detection practically impossible" (1901: 234–235) for two, three, four, or many genes affecting single character, such as stature. He concluded, "Discontinuous variation must merge insensibly into continuous variation, simply owing to the compound nature of the majority of characters with which one deals." The conclusion that the inheritance of continuous variation can be explained in terms of the same discrete Mendelian factors as discontinuous variation was, however, slow to be accepted by the anti-Mendelians.

The first to demonstrate experimentally (1908–1911) that quantitative characters, resulting in continuous variation, can be inherited in a strictly Mendelian way was the Swedish plant breeder Nilsson-Ehle. In a cross between two varieties of wheat, one with red and the other with white seeds, he found only red-seeded plants in the F_1 and the F_2. When the F_2 plants were self-fertilized, a most peculiar segregation was observed in the F_3 (see genetics texts for details). His findings were consistent with the hypothesis that coloration was controlled by three separate genes that are inherited independently. It later turned out that Nilsson-Ehle had been lucky in studying this problem in wheat, since this cereal is a hexaploid, that is, a polyploid with three sets of chromosomes, each with a gene controlling red color. However, he later found other nonpolyploid cases where a single character was affected by two or three separate genes. East (1910) independently arrived at the same interpretation of continuous variation on the basis of work with maize, and Davenport (1910) through the study of skin color in man.[7] It is now known that the number of separate genes that may control a single character can be very large indeed. Mouse geneticists, for instance, have suggested that every coat color gene of the mouse simultaneously also has an effect on body size.

791

The remarkable effect of multifactorial inheritance is that it converts discontinuous variation of the genotype into continuous variation of the phenotype. In the case of Nilsson-Ehle's wheat, for instance, the more dominant genes for red were present in a plant, the deeper red was the color. In a population in which various individuals could range from being homozygous recessive for all red genes (thus having no red genes at all) to homozygous dominant for all three genes, there will be a continuous range of ever deeper red coloration. When components of nongenetic phenotypic variation are superimposed on this, a smooth curve of continuous variation results, even though the genetic basis of this variation consists of discrete, that is, discontinuous, Mendelian factors. The puzzle of the genetic basis of continuous variation was finally solved.

The expression of almost any gene, particularly those with quantitative effects, can be modified by other genes. The genes that modified the amount of pigmentation in Castle's hooded rats are a typical illustration. Modifying genes are particularly important in evolution since they readily respond to selection and since they provide populations with the needed flexibility to respond to sudden changes of the environment. The essence of multifactorial (polygenic) inheritance is that a single component of the .phenotype (a single character) can be controlled by several independent gene loci. Cases of multifactorial inheritance were discovered very early in the history of genetics, beginning with Mendel (one of his *Phaseolus* crosses). A celebrated case is that of the walnut comb in chickens, which Bateson and Punnett showed in 1905 to be due to the interaction of the pea comb and the rose comb loci; they also discovered a case of polygeny in the sweet pea. Nevertheless, there was a considerable reluctance among evolutionists to accept the multiple-factor hypothesis of continuous variation. To them it seemed a rather arbitrary *ad hoc* hypothesis to cover up a weakness of the Mendelian interpretation.

Even though multifactorial inheritance was discovered repeatedly from 1905 on, it seems to me that the Morgan school deserves the major credit for using it in order to refute the one-gene-one-character (that is, unit-character) theory of early Mendelism. Refutation of this theory permitted a much clearer separation of transmission genetics and physiological genetics. It removed some of the less acceptable, preformationist aspects of early Mendelian theory and required virtually no modification at all eventually to be translated into the language of molecular genetics ("genetic programs").

Multiple-factor inheritance, also referred to as *polygeny,* is not the only example of an interaction of different genes. In fact, the variety and magnitude of possible interaction among genes, and— as is now being realized—of different kinds of DNA, is becoming daily more apparent. Its importance was already appreciated by some of the early Mendelians, however. Bateson in particular was interested in *epistatic* (his term) interactions among different gene loci. To take a particularly simple case, an albino gene might suppress pigment production of several different pigment genes. As the Russian geneticist Chetverikov (1926) was the first to state clearly, all genes may contribute to the genetic milieu of other genes. This is of great importance in physiological and evolutionary genetics.

One particular kind of such interactions is *pleiotropy:* the phenomenon whereby a given gene may affect several characters, that is, different components of the phenotype. To know this is particularly important for the determination of the selective value of such genes. All the advances described in the preceding pages, including the discovery of polygeny and pleiotropy, made it ever more certain that all phenomena of inheritance could be interpreted in terms of discrete nuclear genes.

Genetics was now ready to analyze the continuous variation of the biometricians and show that it is consistent with Mendelian principles. Beginning with a highly original analysis by Fisher (1918) and subsequent analyses by Mather (1949) and by various animal breeders (Lerner, 1958), quantitative genetics made rapid progress from the 1940s on (Falconer, 1960; Thompson and Thoday, 1979; see also Part II).

The Demise of Soft Inheritance

The genes of Morgan and other geneticists of the second and third decade of the twentieth century clearly represented hard inheritance. The only way they could change was by mutation, converting by a single step one previously constant gene into another. One might think that the demonstration of this fact would have spelled the demise of any and all theories of soft inheritance, but this was not the case. In fact, soft inheritance died hard. There were many reasons for this. One was that the first proponents of hard inheritance, the early Mendelians (de Vries, Bateson, Johannsen), had quite unacceptable evolutionary ideas. Their opponents assumed, mistakenly, that to accept hard inheritance necessitated also the adoption of the clearly invalid evolutionary theories of the Mendelians. Furthermore, the genetic laws were worked out with the help of aberrant, if not clearly pathological, characters (albinism, extra digits, structural deficiencies, and so on). The naturalists thought that there was a continuing need for soft inheritance to explain gradual changes of evolutionarily important characters (against the mutationism of the Mendelians) and to account for adaptive geographic variation (climatic rules and so forth). The stronger the evidence became in favor of hard inheritance, the more efforts were made by the neo-Lamarckians to produce proof for an inheritance of acquired characters.

The weight of the accumulating evidence had become so convincing by the 1930s and 40s that even the last advocates among geneticists of the occurrence of some form of non-Mendelian inheritance were either converted or became silent. For another thirty years a belief in soft inheritance could occasionally be encountered among nongeneticists (Mayr and Provine, 1980), but as a viable scientific theory it was dead.

Perhaps one can attribute the demise of soft inheritance to three sets of factors. The first is that all attempts to find experimental evidence for the existence of soft inheritance were unsuccessful (see above). The second is that all studies of genes indicated their complete constancy (except for occasional mutation). The last is that all the phenomena that seemed to require the postulate of soft inheritance, such as continuous variation and the climatic rules, were eventually explained in terms of Mendelian genes and natural selection. Although by that time it was hardly necessary any longer, the molecular geneticists' demonstration in the 1950s that the pathway from nucleic acids to body proteins is a one-way street sounded the last death knell of soft inheritance.

Genetics had made giant strides in the fifty years after the rediscovery of Mendel's rules. Indeed in this period an understanding of almost all aspects of transmission genetics was achieved. It will help to follow the subsequent development of genetics to provide at this point a short summary of the findings made by about 1950.

(1) The genetic material is particulate, consisting of units, designated as genes, which have long-term stability ("hard inheritance").

(2) Definite characters are the product of determinants ("genes") located at well-defined loci on the chromosomes.

(3) Genes are "linked" on the chromosomes in a definite linear sequence, but the linkage can be broken by crossing over, which happens more frequently the farther apart the gene loci are on the chromosomes (except where reversed by double crossing over).

(4) In an individual of a sexually reproducing species, each gene is ordinarily represented twice, one of the two homologous units derived from the father, the other one from the mother (known as the principle of diploidy).

(5) A mutation is a discontinuous change of a gene.

(6) A strict separation must be made between genotype (the genetic material) and phenotype.

(7) Several genes may contribute to the expression of a single "character," that is, component of the phenotype (polygeny), and a single gene may affect several characters (pleiotropy).

Uncertainties about the Nature of the Gene

The basic features of Mendelian inheritance were reasonably well understood by about 1920, and genetics began to specialize. Population genetics emerged in the 1920s and flourished particularly from the 1930s to 1950s (see Chapter 13). Physiologists and embryologists began to realize that ultimately the phenomena studied by them had to be traced back to genes, and thus the study of gene function became an ever more important branch of genetics. But there was still a dark corner in transmission genetics. The questions had not yet been fully answered: What is the nature of the gene? How much "morphology" does it have? What kind of a molecule or aggregate of molecules is it? How large is a gene? How do different genes differ from each other chemically? Are all genes basically the same, or are there different kinds of genes? A multitude of similar questions concerning the exact nature of the material of inheritance remained unanswered, and attempts to answer them occupied the attention of a number of schools.

To report on the history of transmission genetics during the period from 1920 to 1960 is exceedingly difficult, because many of the problems studied at this period (such as variegation) are highly technical, and some of them are still not yet fully explained. Indeed, they will not be fully understood until the structure and function of the eukaryote chromosome has been clarified. Enormous efforts were made during that period to elucidate the nature of the gene, but what was found paled into insignificance or irrelevance when the actual structure of the DNA molecule was discovered in 1953. Nor was this period notable for the development of successful concepts. In fact, most of the new concepts advanced during this period, such as the genomere hypothesis and Goldschmidt's field theory of the gene, had to be abandoned, at least in the form in which they had been originally proposed. We still lack a truly critical history of the period, which would reconcile the findings and disputes of the period with the results of molecular genetics. Muller and Stadler, for instance, often arrived at different conclusions in the interpretation of their radiation experiments. Can the apparent conflict between some of their findings be resolved in the hindsight of the modern knowledge of the fine structure of the chromosome and its organization into enzyme and regulator genes? There is still much more historical analysis to be done, and the account which I present is tentative and may have to be revised considerably.

In order to facilitate the understanding of the doubts and controversies of the four decades from 1920 to 1960, I will begin by giving here a greatly simplified version of the classical gene-chromosome theory. The chromosome was likened to a string of beads, each bead representing a different gene. Each gene was assumed to be a discrete corpuscle, completely constant from generation to generation (except for exceedingly rare mutations), independent of the neighboring genes, and having no effect on them (except for rare cases of position effects). The gene was believed to have three capacities: (1) each gene controls (or affects) a char-

795

acter (the gene as unit of function), (2) each gene mutates independently of the others (the gene as unit of mutation), and (3) each gene can be separated from its nearest neighbor on the chromosome by the process of crossing over (the gene as unit of recombination). Mutation was believed to consist in a slight modification of the gene molecule, resulting in a new allele. Crossing over was considered a purely mechanical breaking of the string of beads, followed by a new fusion with the corresponding "piece of string" of a homologous chromosome.

796   The concept that genes are quite independent of their neighbors, and that their position on the chromosome is a purely fortuitous one, seemed strongly reinforced by the discovery of the Morgan school that adjacent genes on *Drosophila* chromosomes often controlled entirely unrelated characteristics, while the genes affecting a single feature, let us say the eye, were widely scattered over all the chromosomes. The proximity of genes was, so it was widely thought, simply the product of the history of former chromosome breakages. The fact that there are as many linkage groups as there are chromosomes was also consistent with the theory.

Furthermore, if the gene is a definite corpuscle, then it should be possible, with the help of various techniques, to calculate its approximate size, and use this as a second approach to estimate the number of genes that can be accommodated in the total chromatic thread of a nucleus. Muller (one is tempted to say "of course") was the first one (in Morgan, 1922) to make such calculations, which were refined in 1929. On the basis of a number of indicators (including frequency of mutation and certain aspects of crossing over) the best value for the total number of genes in *Drosophila melanogaster* was estimated by him in 1929 to be between 1,400 and 1,800. Various authors using the techniques of irradiation arrived later at estimates of 1,300 and 1,800 for the X chromosome alone, and thus a total of over 14,000 genes for all chromosomes.

Cytological observations seemed to give further support to the string-of-beads concept, and even to the possibility of counting the beads. The nuclear material during the leptotene stage of meiosis indeed often had the appearance of such a string, with definite "beads", referred to by the cytologists as *chromomeres*. It was postulated by some cytologists that each chromomere represented a different gene. Belling (1931) counted about 2,500 pairs of chromomeres in the nucleus of *Lilium*. Other cytogeneticists showed that some chromomeres contain several genes.

A spectacular advance in cytology seemed to confirm the chrommere theory. In 1933 Heitz and Bauer rediscovered the giant banded chromosomes in the salivary glands of *Drosophila,* and Painter and Koltsov suggested that the bands correspond to series of chromomeres in these highly polytene chromosomes and that the sequence of bands corresponds to the sequence of genes. Bridges (1938) counted at least 1,024 bands in the X chromosomes of the salivary gland of *D. melanogaster* and postulated a corresponding number of genes. By measuring the volume of the chromosomes, it was then possible to make estimates as to the size of the genes. However, these estimates were wrong by one or two orders of magnitude, and subsequent work on microbial genes showed that there is no fixed gene size; indeed, different genes of the same individual may vary through two orders of magnitude.

797

The discovery of the salivary chromosomes was of far greater importance for other problems of genetics than for the determination of the number and size of genes. A microscopic inspection of the salivary chromosomes often permitted a direct determination of the genotype without elaborate breeding tests. They revealed the presence of chromosomal mutations (rearrangements), inferred on the basis of genetic analysis. Inversions, deficiencies, duplications, and translocations in dipterans could now be studied easily. At the same time, the complexity of the banding pattern provided the first solid evidence for the complexity of the eukaryote chromosome and for the heterogeneity of the chromatic material.

Position Effect

At first all the known facts of inheritance seemed to be consistent with the string-of-beads model of genes and chromosomes, but eventually inconsistencies and contradictions were discovered.

The first serious discrepancy was provided by the discovery of the position effect by Sturtevant (1925).[8] There is a dominant gene on the X chromosome of *Drosophila melanogaster,* called "bar," which causes the eye to be narrow instead of round. This gene can mutate to more extreme narrowness (ultrabar) or revert to round-ness. Further analysis revealed two remarkable aspects of the case. First the bar phenotype is due not to a simple gene mutation but to a structural change of the chromosome. Study of the salivary gland chromosomes revealed that there are six bands at this locus

(S) in normal flies but in bar-eyed flies the six bands are duplicated (SS). Ultrabar flies had the same segment in triplicate (SSS). Normal, round-eyed flies that had arisen from bar by "mutation" had segment S only once. The structural change could be explained only by unequal (or "oblique") crossing over, as Sturtevant demonstrated by the behavior of mutated genes on either side of the bar locus. Detailed analysis of other genes in *Drosophila* and other organisms eventually showed that unequal crossing over is not rare at all—in other words, that the unit of recombination is not necessarily the gene. This was the first breach in the triple-capacity theory of the gene.

798

Perhaps even more startling was a second aspect of the bar gene. When two bar genes were adjacent to each other on the same chromosome they had a different effect on the number of ommatidia in the eye than when two genes were opposite each other on the two homologous chromosomes Sturtevant designated this as the *position effect*. The bar case, thus, proved that the function of a gene and thus its effect on the phenotype of an organism could be changed merely by altering the arrangement of the hereditary material on the chromosomes, without mutation and without any change in the quantity of the genetic material.

Pseudallelism

Another complication of the classical gene concept was introduced by the phenomenon called *pseudallelism*. The Morgan school in its early findings was particularly struck by the fact that adjacent genes usually seemed to have nothing to do with each other functionally. Genes affecting color of the eye, formation of the wing veins, formation of bristles, body color, and so forth might be adjacent to each other. "Genes" having very similar effects ordinarily were simply alleles of a single gene. If the gene is the unit of crossing over, no recombination among alleles should ever be expected. Indeed, early attempts (1913; 1916) by students of Morgan to discover crossing over of alleles at the white locus were unsuccessful, as later became evident, largely owing to small sample size. However, after Sturtevant (1925) had proposed the theory of unequal crossing over of the Bar duplication and Bridges (1936) had substantiated it on the basis of evidence provided by salivary chromosomes, it seemed time to try once more for recombination among seeming alleles. Oliver (1940) was the first to succeed, finding evidence for unequal crossing over of alleles at the lozenge locus of *Drosophila melanogaster*. Heterozygotes for two different alleles ($1z_g/1z_p$), bracketed by marker genes, reverted to wild-type with a frequency of about 0.2 percent. Recombination of the marker genes proved that crossing over among the "alleles" had occurred.

Genes that are so closely adjacent that crossing over can be recorded only in very large samples, and which therefore normally behave as if they were alleles, are designated *pseudoalleles* (Lewis, 1967). They share functional similarity with true alleles, and also share the capacity to produce a mutant phenotype when in transposition. They were found not only in *Drosophila* but also in maize and with particular frequency in certain microorganisms. Molecular genetics has shed much light on these questions, but a full understanding still eludes us, owing to the incompleteness of our understanding of gene regulation in eukaryotes.

799

But let us go back to the profound impact caused by the discovery of the position effect. In a review paper Dobzhansky drew these conclusions: "A chromosome is not merely a mechanical aggregation of genes, but a unit of a higher order . . . the properties of a chromosome are determined by those of the genes that are its structural units, and yet a chromosome is a harmonious system which reflects the history of the organism and is itself a determining factor of this history" (1936: 382).

Others were not satisfied with such a mild revision of the "bead concept" of the gene. Ever since the beginning of Mendelism there had been biologists (such as Riddle and Child) who cited seemingly weighty evidence against the corpuscular theory of the gene. The position effect was grist for their mill. Goldschmidt (1938; 1955) now became their most articulate spokesman. To replace the corpuscular theory of the gene, he proposed a "modern theory of the gene" (1955: 186). According to him there are no localized genes but instead "a definite molecular pattern in a definite section of a chromosome, and any change of pattern (position effect in the widest sense) changes the action of the chromosomal part and thus appears as a mutant." The chromosome as a whole is a molecular "field," and what used to be considered genes were discrete or even overlapping sections of this field; mutation, then, is a repatterning of the chromosomal field. This field theory was in conflict with too many facts of genetics to be accepted, but the mere fact that such an experienced geneticist as Goldschmidt could propose it seriously showed how insecure the theory of the gene still was. This is also reflected by a number of theoretical papers on the gene published from the 1930s to 1950s (Demerec, 1938; 1955; Muller, 1945; Stadler, 1954).

Unstable Genes

800

In some of his earliest genetic work, Hugo de Vries discovered in 1892 a variety of snapdragon (*Antirrhinum majus*) with red-striped flowers which in its offspring might produce a wide range of variegation, from small specks to narrow or broad streaks to large red sectors of the flower. Different flowers or flowers on different branches of the same plant might differ from each other in variegation. Since this first discovery, unstable genes have been found in many kinds of plants and animals and numerous explanations were proposed, such as shifts in dominance or the presence of "genomeres," subgenes of a major, highly complex, gene. This theory, owing to its extreme corpuscularity, was, so to speak, the polar opposite of the field theory. According to the genomere theory, certain (all?) genes were believed to be composed of different particles that were unequally distributed during mitotic divisions (shades of Weismann!). Correns, E. G. Anderson, Eyster, and Demerec favored, for a while, the genomere hypothesis, but the weight of the counterevidence made them abandon it by the beginning of the 1930s (Demerec, 1967; Carlson, 1966: 97–105). Demerec finally ascribed the instability to a "chemical instability of the genes," which explained nothing, of course, but simply transferred the bothersome phenomenon from the domain of the biologist to that of the chemist.

When, after a considerable lull, unstable genes again received attention, their behavior was ascribed to an interaction of gene loci or chromosomes. I am referring to the work of Barbara McClintock (1951), who showed that introduction of a structurally unstable chromosome g into certain genotypes of maize produced the "mutation" of many genes of chromosome g and of other chromosomes to unstable recessive forms. What was apparently involved was a reversible inhibition of the expression of these genes. Although the true significance of this "aberrant" finding (as it was then referred to) was not generally acknowledged until a dozen years later, when it was rediscovered in microbial genetics, here was clear evidence that a "mutation" at one locus could be simulated by regulatory activities at a different locus. In other words, the phenotypic expression of a gene may be changed by other genes, while the gene itself remains completely constant. No one knows how frequent such pseudomutations due to epistatic gene interactions are. In a period of over fifty years, numerous investigators had devoted a great deal of time and effort to the study of unstable genes in the belief that the explanation of the instability would shed important light on the nature of the gene. Alas, it turned out that the phenomenon is due not to some property of a single gene but to the functioning (interaction) of the entire system of genes.

The period from the 1930s to 1950s saw intense activity among the students of the gene, but it was also a period of considerable frustration. Microscopy did not succeed in providing a better picture of the gene than the purely genetic analysis. This was true even for the giant salivary chromosomes, which showed a bewildering variety of bands not closely correlated with any of the functions of the genes inferred to be located on or near these bands. Not being able to see genes directly, one could learn about them only by inference. And virtually the only way in which one could extract information about a gene was by studying its changes through mutation.

This was true in spite of the spectacular success of the study of the change in the chemistry of the gene products caused by mutation, particularly in microorganisms, beginning with the brilliant researches of Beadle and Tatum. But since this research quite deliberately confined itself to the study of the enzymes produced by genes, the light it could shed on the structure of the gene itself was limited.

Experimental Mutation and the Nature of the Gene

By 1920 it began to appear that not much more could be learned about the nature of genes simply by crossing experiments. Other ways would have to be found to produce entirely new information. Biochemistry and biophysics had prior to 1944 reached neither the conceptual maturity nor the technical proficiency to permit a solution of the gene problem via biochemistry. Under these circumstances it occurred to several investigators that the experimental production of mutations might be a way to shed light on the nature of genes. H. J. Muller was the first to realize that the haphazard way in which others had studied mutation, even experimental mutation, could never lead to unambiguous results. He therefore set out to meet certain essential conditions, particularly: (1) the genetic purity of the material to be tested,

(2) large numbers of individuals in the experimental and control samples to permit statistical tests of significance, and (3) the development of new methods, and particularly specially constructed strains (with appropriate lethals, markers, and cross-over suppressors) that would permit testing various hypotheses of gene structure. These special stocks of *Drosophila*, described in the textbooks of genetics, permitted Muller to calculate the actual frequency of newly occurring mutations. This was particularly important since many mutations are recessive and it is always difficult to determine the time when a recessive mutant has first occurred. Furthermore, many mutations are lethal in homozygous conditions, that is, when they occur on both homologous chromosomes. Homozygous lethals are, of course, lost, not showing in the offspring. Three steps were particularly important in Muller's technique: the placing of a marker gene on a chromosome for unambiguous identification, the installation of a crossing-over inhibiting mechanism on the chromosome, and the pairing of the marked chromosome with another chromosome suited to reveal any mutant change. When Muller had completed the development of these important stocks, he exposed some of his flies to various dosages of x-rays.

He used a stock of females which, when crossed with a male that had a lethal mutation on its X chromosome, would result in the death of all males in the F_2 generation. Thus, if one of the x-rayed males produced only daughters in the F_2 it indicated that a lethal mutation had been induced in his X chromosome.

When a normal untreated male is crossed to females of this stock, approximately only one cross in a thousand gave solely females in the F_2. This means that the chance of a lethal mutation occurring spontaneously at any one of the many loci of the normal X chromosome is one in 1,000 or 0.1%. This is the natural or spontaneous rate of mutation. When males were exposed to about 4,000 r-units of x-rays, females only appeared in the F_2 of about 100 in every 1,000 crosses. The rate of mutation in the x-rayed flies thus was 100 times as great as the spontaneous rate of mutation. Almost simultaneously with Muller, the plant geneticist L. J. Stadler (1896–1954) produced artificial mutations in barley and maize (1928).

Muller's findings, and particularly the elegant methods which he had developed, opened up an entirely new area of research. It was possible to place mutation research on a quantitative basis, for instance, to correlate rate of mutation with x-ray dosage. "The whole field of mutation research has been dominated by the ideas and experiments of Muller. He provided the conceptual framework, formulated the decisive questions, worked out ingenious experimental techniques, and at all stages guided the interpretation of the increasing mass of data into a coherent theory. Many of the ideas and suggestions put forward by him at a time when the means for testing them were not yet available have later been proved correct" (Auerbach, 1967).

Eventually it was shown that not only radiation but also some chemicals may have mutagenic effects. One of the first materials shown to be mutagenic was mustard gas. A British surgeon, Robson, made the astute observation that the burns caused by mustard gas were remarkably similar to x-ray burns. Accordingly he suggested to a geneticist (Auerbach) that she test mustard gas for mutagenicity, which indeed she found, as he had predicted (1941). Rapoport in the USSR found independently that formaldehyde is mutagenic. Since the 1940s the mutagenicity of numerous chemical compounds has been established (Auerbach, 1976). Each mutagen produces a wide range of mutations and no evidence was found for a specific action on a particular gene. However, the frequency of certain mutations caused by a chemical mutagen is often different from that caused by irradiation. A further particularly interesting finding is that some (many?) mutagens are also carcinogenic. This discovery led to the proposal of a rapid method for the screening of chemicals for possible carcinogenicity: exposing bacteria to the chemical and checking for an increase in rate of mutation.

What was far more important for Muller, however, was the thought that the artificial induction of mutation would shed light on the nature and structure of the gene. If a gene is a well-defined corpuscle of a definite size, bombarding it with ionizing radiation (electrons or short-wave rays) would produce "hits" on these corpuscles, and the resulting damage would show up as a mutation. This was the "hit theory" or target theory of mutation, articulated in more detail by the physicists K. G. Zimmer and M. Delbrück in a classical joint paper with Timofeeff-Ressovsky (1935).

The target theory did not lead to consistent results, however (Carlson, 1966: 158–165), and thus failed to produce a better understanding of the gene. Furthermore, it was found that even radiation of the substrate might increase mutation rate, and that many chemicals (mustard gas, phenol, and so on) are as mutagenic as irradiation. Anything that could interfere with the normal process of gene replication might result in a mutation. This led some authors to adopt as the definition of mutation "any error in gene replication" (recently shown not to apply to all cases of mutation).

The irradiation technique, however, encountered an even more fundamental difficulty. What was being irradiated were not genes in isolation but rather the chromosomes, that is, the genes and the matrix in which they are imbedded. Both the genes and the chromosomal matrix are vulnerable to x-ray damage, but the study of the resulting mutated phenotypes rarely revealed whether a gene mutation or a matrix (chromosome) mutation was involved. Cytological examination very often led to the discovery of small (often quite minute) deficiencies on the chromosomes and could thus be clearly identified as chromosomal mutations. The two most active leaders of x-ray mutation research, H. J. Muller (on *Drosophila*) and L. J. Stadler (on maize), held different views on the frequency of true gene mutations produced by x-ray treatment. Stadler accepted only those cases where the irradiation of the new mutant could produce a reversion to the pre-irradiation character. Such cases, at least in maize, were very rare. In all other cases Stadler suspected either the production of unstable genes or chromosome damage. As he said in his last publication (1954): "A mutant may meet every test of gene mutation, and yet, if it is not capable of reverse mutation, there is ground for the suspicion that it may be due to gene loss [chromosomal deletion], while, if it is capable of reverse mutation, there is ground for suspicion that it may be due to an expression effect [unstable gene]." Not every one, least of all Muller, took an equally pessimistic view of irradiation effects. Yet, even at best, there were severe limits to the information that could be extracted from the irradiation experiments.

Two facts were clearly established in this period: the first was that—in contrast to the initial impression—genes with similar function sometimes are closely adjacent on the chromosomes (gene complexes; Lewis, 1967), and the second, that genes must have considerable structural complexity ("morphology") to permit the partial independence of function, mutation, and recombination. This complexity had to be at the macromolecular level. More and more it became apparent to the geneticists that they were up against a wall which they could not scale with their genetic-cytological equipment.

One further observation made during the irradiation experi-

ments was disturbing. The earlier after the irradiation the mutation rate was determined, the higher the rate was found to be. It
seemed as if damaged chromosomes had the ability to "heal," at
least in part, or to restore sections that had been knocked out.
Subsequent research revealed indeed that there are some regular
repair mechanisms capable of repairing damaged genes and chromosomes (Hanawalt et al., 1978; Generoso et al., 1980). Observed
mutations, then, to put it bluntly, could be considered as errors
or failures of the repair genes.

As much as was learned in the 1920s, 30s, and 40s by the
dedicated work of the mutation workers, it added remarkably little to our understanding of the nature of the gene. It is quite true
what Demerec (1967), one of the most active participants in these
researches, said in retrospect, "Throughout the first half-century
of genetics, our concept of the physical structure of genes remained more or less static." No real progress was made until new
methods and different material were used.

Eukaryote chromosomes are so complex that even today we
lack an understanding of their organization and of the integration
of genes in these chromosomes (*Cold Spring Harbor Symposia,* 1978).
It has now become apparent that it was impossible in the first half
of the century to seek an access to the understanding of the gene
via the eukaryote chromosome. Real progress was not made until
the analysis was transferred from such eukaryotes as mice, *Drosophila,* and maize to the bacterium *Escherichia coli* and to viruses.
Since prokaryotes lack chromosomes and their genetic material is
organized much more simply, permitting an access to the DNA
unimpeded by the chromosomal matrix.

The most important lessons learned from the study of eukaryote chromosomes were of a negative kind. Unequal crossing over
showed that the functional gene is not necessarily the unit of recombination. Mutational analysis (particularly in microorganisms)
had shown that there might be several different mutational sites
within a single functional gene. And the position effect (cis-trans
differences) had shown that the gene is not necessarily the unit of
function. The original simple dogma that the gene is simultaneously the unit of recombination, of mutation, and of function thus
had to be abandoned. In view of these contradictions, Benzer
(1957) made the radical proposal to abandon the term "gene" altogether and to replace it by three different terms, *muton* for the
unit of mutation, *recon* for the unit of recombination (as determined by cross-over location), and *cistron* (from the cis-trans dif-

805

ference in position effects) for the unit of gene function. Among these three, the cistron comes closest to the traditional concept of the gene, since a gene is normally characterized by its effects. Ultimately, the term "gene" returned to universal usage, with the definition given by Benzer for the cistron. The terms "muton" and "recon" never came into general usage.

Different Gene Concepts

806   Precisely what concept of the gene various geneticists had in the period from 1900 to the 1950s is difficult to determine. This is true even if we limit our attention to those who accepted a corpuscular gene, disregarding those who had field theories or believed in a diffuse, continuous genetic substance. Since there is as yet no analysis by any historian available, I will try to offer a few preliminary comments. Matters are made more complex by the fact that several of the leading geneticists changed their views during their lifetimes. The four ways of looking at genes which I here describe by no means exhaust the possibilities.

Perhaps the oldest view was that the genes themselves are the building stones of the organism. Darwin's gemmule theory comes perhaps close to this view. It was somewhat modified by de Vries (1889) when he postulated that the pangens move from the nucleus into the cytoplasm of the cells that form the building stones of the tissues and organs of which an organism consist. Sometimes the silent assumption was made that genes consist of proteins.

Very widespread was a second view, going back in principle to Haberlandt (1887) and Weismann (1892), that the genes are enzymes, or act like enzymes, serving as catalysts for the chemical processes in the body. Since it was eventually found that enzymes are proteins, this would also imply that genes are proteins (Fruton, 1972). The discovery that the chromatic material consists of nucleoproteins, if not virtually pure nucleic acid, had remarkably little influence on the thinking of the enzyme school.

When the importance of nucleic acid was beginning to be appreciated, the gene was seen as a means of energy transfer. In 1947, three years after Avery and colleagues had demonstrated that DNA is the transforming agent, Muller advanced the idea that the chemical role of nucleic acid was its possible contribution to the energies of gene reactions. "It may be that nucleic acid in polymerized form provides a way of directing such a flow of energy into specific complex patterns for gene building or for gene relations upon the cell." As far as gene action was concerned, Muller concluded that "if the primary gene products are not like . . . the gene itself . . . then the gene must certainly act as an enzyme in producing them" (1973: 152; see Carlson, 1972). Still, Muller thought it was "too early to conclude that either the gene or its primary products do always, or usually, act as enzymes." Instead, Muller suggested a gene might "produce more molecules similar in composition (or complementary) to itself or to part of itself," and these gene products "might actually become used up in the reactions which they in turn take part in." Either of the two alternatives suggested by Muller had a strongly metabolic bias.

807

Finally, the gene was viewed by some as the conveyor of highly specific information. In a vague manner this idea had been around for a long time. It is such an obvious thought that some author or another must have articulated it specifically before 1953. Yet, I did not encounter such a hypothesis in a casual search of the literature. It requires, among other conceptual components, the acceptance of a complete separation of genotype and phenotype. The concept of the gene as a unit of information has, of course, now become the modern standard concept, after the structure of DNA and the role it plays in the production of proteins (transcription and translation) was discovered.

Each of these four concepts of the gene made certain assumptions concerning the chemical composition of genes and their function. How all-important the chemistry of the gene is in determining its nature was, however, not fully realized before about 1950.

19⌀ The chemical basis of inheritance

IN RETROSPECT it is evident that the methods available prior to the origin of molecular biology were altogether inadequate for a full understanding of the gene. During the period from 1910 to 1950, it became increasingly obvious that highly complex molecules formed the material basis of inheritance, and that the only way to achieve further progress was to learn more about the chemistry of the gene. To treat the molecular basis of inheritance either as a shapeless corpuscle or as a simple molecule was clearly inappropriate. The study of the gene was no longer a problem for the classical biologist; it had become a problem in the frontier region—and at first, the no-man's land—between biology, chemistry, and physics. And in the 1940s, when the problem was seriously attacked, it turned out that chemistry had already gone a long way toward the solution of the structure of the gene (Cairns, Stent, and Watson, 1966).

Let us remember that by the mid-1880s it was generally believed that the nucleus was the seat of inheritance (see Chapter 16), or, more narrowly, that the chromosomes—or, more specifically, the chromatin—were the genetic material. The term "chromatin" had been given by Flemming in 1879 to the stainable material in the nucleus. This at once raised the question as to the chemical nature of the chromatin: Was it a special substance, probably a protein, different from other substances or from the protein of the cytoplasm? Actually, an answer to this question had already been given ten years earlier (in 1869) by the Swiss physiologist and organic chemist Friedrich Miescher (1844–1895),[1] who had shown that chromatin is not a protein at all.

After Miescher had obtained his medical degree in 1868, he followed the advice of his uncle, the famous anatomist and histologist Wilhelm His, to take up histochemistry. As His said, "Since in my own histological research I had come to the conclusion again and again that the ultimate questions about the development of tissues can be solved only by way of chemistry, Miescher decided to obtain post-doctoral training in the laboratory of the famous organic chemist Hoppe-Seyler in Tübingen, arriving there Easter 1868."

Hoppe-Seyler suggested to Miescher that he study "the constitution of the lymphoid cells" because of their considerable medical importance. He used pus, which in those pre-antibiotic days was abundantly available in hospitals. Miescher, a most careful, conscientious, and competent worker, developed entirely new methods of purification and was soon able to separate the pus cells from all other components of the pus. He then tried to analyze the cytoplasm of the cells by separating it from the nuclei, and to determine its constituents. These endeavors were at first frustratingly unsuccessful. As the end-product of one of his extraction procedures, Miescher obtained a precipitate that did not have the properties of any of the known proteins. As his next step, Miescher washed whole pus cells in highly diluted hydrochloric acid, and all that was eventually left were nuclei. The unknown substance thus must have come from the nuclei. Since the research on the constitution of the cytoplasm had reached a dead end, Miescher decided to study instead the chemistry of the nuclei.

I have presented this sequence of events in some detail because the myth has developed that Miescher had started his research in order to solve the secret of inheritance. Nothing of the sort! Here was simply an organic chemist following the advice of his uncle to add to our knowledge of the chemistry of cells and tissues. What is impressive when one reads Miescher's work is his methodological inventiveness. Forever he applies new techniques, particularly new extraction and purification procedures, and he fully deserved, by his hard work and ingenuity, to become the discoverer of DNA. I think it is also correct to say that, prior to Miescher, biochemists worked with whole tissues, while Miescher isolated cells; or even parts of cells, such as nuclei. When Miescher analyzed the material he had obtained from the nuclei, he found that its outstanding characteristic was a rich content of phosphorus. Since this nuclear material was not one of the known organic substances, Miescher gave it a new name, *nuclein*.

Miescher arrived in Tübingen in the spring of 1868 and completed an account of his discovery in the late fall of 1869. The results were so novel, however, that Hoppe-Seyler did not at once publish the manuscript Miescher had submitted to him, but decided to check the results himself. Only after his own research,

809

and that of some of his other students, had confirmed everything Miescher had claimed was the nuclein manuscript published in the spring of 1871.

After he had returned to Basel in 1871, Miescher discovered that the sperm of the Rhine salmon provided a rich source of nuclein, for each spermatozoon is a single cell, and its head is essentially the nucleus. Miescher now had an almost unlimited supply of nuclein (as he jokingly said, the salmon testes supply "tons of nuclein") and he devoted the next years to its study. He found that it was associated with a protein which he named "protamine," and he was able to determine many of the chemical and physical properties of nuclein, including its empirical formula.

It is rather sad to have to say that after his first brilliant success, Miescher's subsequent research career was an anticlimax. This is the more surprising since he was an outstandingly gifted person. Perhaps it was due to the fact that, as the oldest of five brothers, he had all the characteristics of a firstborn. Thus he tended to ask conventional rather than revolutionary questions (Sulloway). Even though it soon became obvious that the nuclein was nothing other than the chromatin of the cytologists, Miescher never looked at it as a carrier of genetic information. Instead of asking genetic questions, he asked physiological or purely chemical questions, such as: "Where does the body get all the phosphorus to synthesize the large quantities of nuclein during sperm formation?" In 1872 he declared that he now wants to work on "the physiological aspects of nuclein, its distribution, chemical association, its appearance and disappearance in the body, its turnover."

Under the influence of Carl Ludwig, Julius Sachs, and Wilhelm His, Miescher had adopted the then fashionable physicalist and very mechanical way of looking at biological phenomena. This is well illustrated by his characterization of the process of fertilization, phrased in terms of the contact theory: "Suppose the nature of the egg cell, as compared to an ordinary cell, were to be determined by the circumstance that one link were missing in the series of factors which control the active organization? Because otherwise all the essential cell constituents are found in the egg. However, during the maturation of the egg, the protamin [protein in the nucleus] disintegrates under the formation of nitrogen (N) . . . and the otherwise perfect machine is brought to a complete standstill because one screw is still missing. The spermatozoon inserts again this screw in the right position, and thus restores the active organization. It does not require anything else. At the place where the chemical-physical quiescence was disturbed, the machine starts to work again, each cell produces protamin for its neighbors, and thus the motion spreads according to definite laws." Not a word is said about the combining of the genetic properties of the two parental gametes. How highly Miescher considered the purely mechanical aspects is illustrated by his concern for the question "in what direction and how deeply [do] the spermatozoa of different species penetrate into the soft protoplasmic mass of the egg?"

As if research on nuclein were comparatively unimportant, *811* Miescher turned to other studies and devoted himself in addition to his teaching activities, during a period of fourteen years (from 1874 to about 1887) to studies on the life history and metabolism of the salmon, the chemistry of the sperm tail, the detailed morphological structure of the sperm head, the chemistry of egg yolk, nutrition in Swiss federal institutions, and variation in human blood chemistry in relation to altitude. One has the impression that it was chance that determined his research objectives rather than considerations of their scientific importance. It was only later in life that Miescher returned to DNA research and, stimulated by Weismann's theories, began to ask the "right" kind of questions. It was too late, for unhappily he soon succumbed to tuberculosis at the age of fifty.

It is now known that the DNA is the chemical basis of the genetic program, and ever since the discovery of the structure of the DNA molecule by Watson and Crick in 1953, historians of science have taken a great interest in the history of DNA research. Some five or six books have been published on the subject as well as long chapters in various general histories of biochemistry.[2] My own treatment will touch only on the high spots and concentrate on the biological aspects of DNA research.

Miescher studied isolated nuclei, that is, nuclei that had been separated from the cytoplasm. This permitted him to test numerous chemical reagents for their reaction to nuclein. It seemed logical to apply the knowledge thus gained to whole cells. The cytologist Zacharias (1881) was the first to do so by observing under the microscope the reaction of cells to various reagents. Nuclei and chromosomes were found to be resistant to pepsin and dilute hydrochloric acid, to be soluble in alkali, and to swell in salt solution. All these were characteristics of Miescher's nuclein. Other cell elements, for instance the spindle fibers, did not show the nuclein reactions. This led Flemming (1882) to say "possibly chromatin is identical with nuclein, but, if not, it follows from Zacharias's work that one carries the other. The word chromatin may serve until its chemical nature is known, and meanwhile stand for that substance in the cell nucleus which is readily stained."

In the following years Hertwig, Strasburger, Kölliker, and Sachs all agreed that chromatin was to be considered identical with nuclein or at least virtually so. This was not just the private opinion of the German cytologists, for the Russion evolutionist Menzbir stated in 1891 (p. 217): "Thus, there is no doubt that chromatin alone is responsible for the hereditary transmission of characters from parents to children, and, in general, of characters of species from one generation to the next." Zacharias's evidence was also accepted by the chemists, for Kossel stated in 1893: "What histologists term chromatin is essentially a compound of nucleic acid with more or less albumin, to some extent it is perhaps pure nucleic acid."

In later years it was claimed that the nuclein of the early authors was a highly impure nucleoprotein, a mixture of DNA with a great deal of protein, hence irrelevant to the question whether these early authors should receive credit for having discovered that DNA is the genetic material. To be sure, the nuclein of Miescher and Kossel was not absolutely pure DNA, but it clearly was not badly contaminated by protein, as was later claimed. This is obvious from the empirical formulae given by Miescher and Kossel:

Miescher	$C_{29}H_{49}N_9O_{22}P_3$
Kossel	$C_{25}H_{36}N_9O_{20}P_3$
DNA (50%AT:50%GC)	$C_{29}H_{35}N_{11}O_{18}P_3$
	[Now known to be correct]

Miescher's sample may have been somewhat hydrated, but neither Miescher's nor Kossel's formula suggests the presence of protein. If protein had been present, the C and N values would have been much higher in relation to P_3 (as pointed out to me by W. McClure).

At the end of the century E. B. Wilson, in the second edition of his great classic *The Cell* (1900), stated, "Chromatin is probably identical with nuclein . . . that the nuclear substance, and especially the chromatin, is a leading factor in inheritance, is powerfully supported by the fact of maturation, fertilization, and cell division" (p. 332). However, in the end he expressed some doubts "whether the chromatin can actually be regarded as the idioplasm or physical basis of inheritance, as maintained by Hertwig and Strasburger" (p. 259).

Very soon after the discovery of nuclein the suggestion was made (Sachs, 1882: 718) that there ought to be a chemical difference between the nucleins of different species. As far back as 1871 it had been shown by Hoppe-Seyler that yeast had a nuclein, and by 1881 this had also been demonstrated for higher plants. In the 1880s when interest in phylogeny was at its peak, the nuclein of lower invertebrates was studied in the hope that a "primitive nuclein" would be discovered that would be much simpler than the nuclein of the salmon. It was a source of considerable disappointment when it was found that the nuclein of the sea urchin (*Arbacia*) was essentially the same as that of the salmon.

813

The Nature of Germ Plasm

Soon after it had been recognized that the chromatin consists (largely) of DNA and also that the chromatin is the germ plasm, the argument arose whether the essential nature of the chromatin was morphological or chemical. The biologists almost unanimously rejected a purely chemical explanation, saying that nuclein was chemically far too simple a substance to account for the incredibly complex architecture of the germ plasm. Boveri (1904) illustrated his viewpoint with an analogy. If a nucleus is compared to a watch, "the morphology of the nucleus deals with the entire machinery of the watch, while the chemistry of the nucleus at best can tell us about the metal of which the wheels have been made" (1904: 123). It was again the case of the blind men and the elephant, because the eventual solution was that the morphology of macromolecules (unknown in Boveri's days) explain the remarkable architecture of the germ plasm.

Among the early authors, de Vries, with his background both in botany and physical chemistry, had the soundest ideas. He emphasized that the germ plasm could not be a simple chemical: "The historically acquired characteristics require a molecular structure of such complexity, that present day chemistry is quite unable to provide an explanation" (1889: 31). Even before him Kölliker (1885: 41) had suggested "that the nuclei of identical chemical composition might be able to have different effects owing to the molecular structure of their effective substance (idioplasm)." Prophetic words!

By the end of the 1880s the cytologists had made all the contributions that they could make with their methods. They had shown as convincingly as is possible that the chromatin satisfied all the requirements of the genetic material and that the sperm head was virtually solid genetic substance. Just exactly what this substance was chemically did not particularly interest them, nor were they concerned with the size of the molecule and its structure. This was peculiar, because it should have been rather obvious that the role of DNA in inheritance could never be elucidated until its structure was understood. My search through the literature suggests that this question was never seriously asked, perhaps simply because methods to provide the information to answer this question were not yet available at that time.

At this point the problem was taken over by the chemists, and for more than half a century, the quest for the nature of DNA was a purely chemical affair. The first need was to confirm that nuclein was truly a substance altogether different from the proteins and that it had nothing to do with other phosphorus-rich substances in organisms (such as lecithin). Miescher still had confused ideas on these questions. To establish the unique characteristics of nuclein, it was necessary to develop methods of purifying nuclein and to make sure that the protein component was stripped off. Altmann (1889) was successful in this, and he named the protein-free portion of the nuclear material *nucleic acid*. How radically different the nucleic acids are from the proteins was far better realized by the chemists than by the biologists. As late as 1900, Wilson thought that pure nucleic acids grade into albumins through a series containing less and less phosphorus; "they vary in composition with varying physiological conditions" (p. 334).

As far as the study of the pure DNA was concerned, two ways were theoretically open to the investigators. They could either break up the DNA molecule and determine its components, or study the molecule as a whole, as was done after the 1920s when Staudinger had developed the principles of polymer chemistry. The second, however, was quite impossible within the conceptual framework of organic chemistry around the turn of the century when it was dominated by the concepts of colloid chemistry.

For the next fifty years the two great leaders in this research were A. Kossel and P. A. Levene. A description of the step-by-step elucidation of the chemical nature of the nucleic acid molecule is given by historians of biochemistry (Fruton, 1972; Portugal and Cohen, 1977). By 1910 it was generally agreed that the DNA molecule contained four bases: two purines (guanine and adenine) and two pyrimidines (cytosine and thymine), a phosphate, and a sugar. Yet it took another forty years before it was determined, in 1953, how these constituents are put together.

Kossel (1853–1927) started work on nuclein in Hoppe-Seyler's laboratory in 1879 and was able to demonstrate in that year the occurrence of a base, hypoxanthine, in the breakdown products of nuclein. Eventually he showed that hypoxanthine was derived from another base (adenine) and his work, in due time, led to the discovery or recognition of the other three bases.

In 1908 Phoebus Aaron Levene (1869–1940) entered the field of DNA research and soon became its leader. As early as 1893, Kossel had shown that a pentose sugar was part of the yeast nucleic acid, and in 1909 Levene and Jacobs identified it as a ribose. Other workers prepared their nucleic acid from the thymus gland of calves ("thymo-nucleic acid") and found a different sugar. Its identification turned out to be extremely difficult, but Levene and coworkers finally (1929) showed it to be 2-deoxyribose. For many years it was believed that ribose was the carbohydrate of plant nucleic acid and deoxyribose that of animal nucleic acid. But eventually ribonucleic acid (RNA) was found in the pancreas and in other animal cells and deoxyribonucleic acid (DNA) in the nuclei of plant cells. However, it was not until about the 1930s that it was fully understood that all animal and plant cells have both DNA and RNA. The cytochemists had only the vaguest idea of the function of nucleic acids in the cells, the ones most frequently supported were to serve as a pH buffer or to help with energy transfer.

Although much was learned about the chemical composition of DNA during the first three decades of the century, little progress was made in the understanding of the molecule as a whole and its biological action. Throughout this period the erroneous assumption was made that the four bases occurred in the nucleic acid in equal portions, a belief which served as the basis of the so-called tetranucleotide theory of the structure of DNA. This theory envisioned nucleic acid as a relatively small molecule with a molecular weight of about 1,500. It must be remembered that in order to get at the constituents of DNA, Kossel and Levene had to use the rather harsh analytical methods of organic chemistry. These, as we now know, destroy what is actually an extremely large molecule. At the time, however, the low molecular weights arrived at by various methods fitted quite well into the concepts of colloid chemistry prevalent at that time. New progress had to await the rise of polymer chemistry in the 1920s and 30s.[3]

THE SHIFTING FORTUNES OF THE NUCLEIC-ACID THEORY OF INHERITANCE

When the conviction spread that DNA was a rather small and simple molecule, the belief in its capacity to control development gradually lost persuasion. How could such a molecule be of importance in inheritance and control development from the fertilized zygote to the full-grown organism, considering the immense complexity of the developmental pathway? The large protein molecules with twenty different amino acids seemed, by contrast, to offer an absolutely unlimited number of permutations and combinations.

Not only chemical reasons made most biologists after 1900 abandon the idea that DNA is the genetic material. They were also disturbed particularly by the observation that during the mitotic cycle the chromosome material stained heavily only during the period when the chromatin was condensed into chromosomes. In the resting stage of the cell the chromosomes seemed to disintegrate into a poorly staining diffuse granular mass. (No DNA-specific stain was yet available at that time.) Boveri had suggested as early as 1888 that the chromatin is lost from the chromosomal framework during the resting stage and is reconstituted at the beginning of the mitotic cycle. This suggestion was eventually more and more frequently adopted, and by 1909 Strasburger opined that the chromatin "may form the nutritive material for the carriers of hereditary units . . . the chromatin itself cannot be the hereditary substance, as it afterwards leaves the chromosomes and the amount of it is subject to considerable variation in the nucleus, according to its stage of development" (p. 108). In 1920 Goldschmidt asserted emphatically, "If we consider the nuclein of the chromosomes as the genetic material, as is customary, then it is absolutely impossible to have a chemical idea of its diversified effects." Bateson (1916) stated in a similar vein: "The supposition that particles of chromatin, indistinguishable from each other and indeed homogeneous under any known test, can by their material nature confer all the properties of life surpasses the range of even the most convinced materialism."

816

Even after the highly specific sensitive Feulgen stain (see below) had been discovered in 1924, preparations were found (for example, sea-urchin oocytes) in which the nucleus seemed to contain no chromatin. By 1925 even E. B. Wilson had given up the idea that nuclein was the genetic material: "So far as the staining reactions show, it is not the basophilic component (nucleic acid) that persists, but the so-called achromatic or oxyphilic substance. The nucleic acid component comes and goes in different phases of cell activities."

The cause of the disenchantment was not only the destructiveness of the customary methods of organic chemistry and the absence of good methods to measure the amount of DNA in all stages of the mitotic cycle but also some rather obsolete ideas concerning the nature of chemical interactions. Strasburger (1910: 359), for instance, protested the conception "of true fertilization as a purely chemical process, therefore against any chemical theory of heredity . . . for me, the essence of fertilization lies in the union of organized elements." He was justified in 1910 to say so, for at that time rather simple-minded notions of chemical processes were still reigning, and the concept of complex, three-dimensional macromolecules had not yet been born.

The new concept of polymerized macromolecules had great attraction because it seemed to fulfill the old dream of so many mechanistic biologists that all biological material "ultimately consists of crystals." As soon as Staudinger's new theory of polymers became available Kol'tsov (1928; 1939) speculated freely on the crystalline nature of the material in the chromosomes. Sixteen years later Schrödinger (1944) proposed his theory of aperiodic crystals, avowedly influenced by a paper, the senior author of which, Timofeeff-Ressovsky, had been an associate of Kol'tsov.[4]

Since polymerized macromolecules are easily broken up into their components, their extraction requires far more delicate methods than had been practiced by Kossel and Levene. When such methods were employed, particularly the the Swedish school of Hammarsten, a product was obtained that was "snow white and of a peculiar consistency like gun cotton," quite different from the degraded products of the harsher extraction methods.

The study of such large molecules required entirely new methods, and when these (ultracentrifuging, filtration, light absorption, and so on) were applied in the 1930s and 40s by Caspersson and others, much to everyone's astonishment the DNA molecules with a molecular weight of half a million to one million, turned out to be larger by more than two orders of magnitude than the earlier estimates (1,500). Indeed they were considerably larger than protein molecules. These new findings completely removed one major objection to the theory of DNA as the carrier of the genetic information. What was needed next, and it was a considerably more difficult requirement, was to find a method by which to separate the DNA cleanly and completely from the protein and to demonstrate, by biological methods, that it was the DNA component that was responsible for hereditary transmission. This was achieved in 1944.

818

It was Oswald Theodore Avery (1877–1955) and his associates who supplied this proof when working on the transforming principle of the pneumonia bacterium.[5] It had long been known that pneumococci occur in several types, differing in their virulence. In 1928 the British bacteriologist F. Griffith (1877–1941) discovered that when he simultaneously injected mice with living type R (rough) avirulent pneumococci and with heat-killed type S (smooth) virulent cells, many of the animals died, and their blood contained living type S organisms. This discovery indicated that the living avirulent R bacteria had acquired something from the dead type S cells which led to the transformation of the harmless type R into virulent organisms of type S. As it was interpreted later, some genetic information had been transferred by the "transforming principle." After many years of experimentation, Avery, Macleod, and McCarthy (1944) succeeded in showing that the transforming principle in an aqueous cell-free solution was DNA. That this was indeed pure DNA and not an associated protein (as was claimed by some of Avery's opponents) was proven by a series of extremely sensitive tests (immune reactions and so on). The DNA solution did not show any reaction whatsoever with any tests for proteins. Furthermore, Avery and colleagues showed that no chemical mutagen was involved because the particular heritable modification was predictable. The autonomy of the material was further supported by its self-reproduction in transformed cells and, in certain later experiments, by linkage studies. Finally, when treated with a highly specific enzyme, desoxyribonuclease, a complete and irreversible inactivation of the transforming substance was brought about. The molecular weight was about 500,000 and the ultraviolet absorption properties were characteristic of nucleic acids.

Avery and his group remained extremely careful (perhaps too careful!) in the evaluation of their findings, but the evidence was so strong that they no longer needed to prove their point; the shoe was now on the other foot, and it was up to the opponents to refute Avery's claim.

The impact of Avery's findings was electrifying. I can confirm this on the basis of my own personal experience, having spent my summers during the second half of the 1940s at Cold Spring Harbor. My friends and I were all convinced that it was now conclusively demonstrated that DNA was the genetic material. Burnet, after visiting Avery's laboratory in 1943, wrote home to his wife: "Avery has just made an extremely exciting discovery which, put rather crudely, is nothing less than the isolation of a pure gene in the form of desoxyribonucleic acid" (Olby, 1974: 205). It was discussed at six major conferences in 1946 alone. Not everyone was converted, of course, and Muller (1947) expressed himself quite skeptically. Goldschmidt remained skeptical as late as 1955: "We conclude . . . that it cannot be stated as a dogma or as a proven fact that DNA is the genic material" (p. 56). Goldschmidt was 76 years old when he wrote this. The resistance was not restricted to the elder geneticists, however. Some biochemists, for example A. E. Mirsky, were even more skeptical.

The question asked by the skeptics was whether pure DNA is the transforming agent or a minute quantity of protein mixed with it, a possibility advanced by Mirsky and some other doubters. Among these, significantly, were most members of the "phage group," including Delbrück and Luria, neither of whom knew much biochemistry. Although they were fully aware of Avery's findings, they still accepted the tetranucleotide theory, so they could not believe that DNA could have the necessary complexity to be the genetic material. Their skepticism carried considerable weight since the phage group at that time was dominant in molecular biology. They finally were converted when members of their own group (Hershey and Chase) carried out an experiment with radioactive-labeled bacterial viruses (bacteriophages). The empty protein coats ("ghosts") (labeled with s^{35}) of phages could be largely separated experimentally from infected bacteria which had taken up little or no s^{35}, while the p^{32} (phosphorus) labeled content of these phages was found in the bacteria and not in the "ghosts." Although this was actually a less precise analysis than Avery's, it was accepted as conclusive by the phage group (Wyatt, 1974).

The publication of Avery's results released, as Erwin Chargaff later called it, a veritable "avalanche" of nucleic-acid research. Chargaff himself reported that he dropped everything he was doing at the time and immediately entered the nucleic-acid field (Chargaff, 1970). It must be remembered that only very few investigators were qualified to do so. The geneticists, in particular, no matter how keen they were on Avery's new discovery, had none of the necessary qualifications. An absence of research activity among them does not justify the claim that they, or at least the younger ones, were unaware of the importance of Avery's findings.

820   The two investigators who made the major contributions during the ensuing years were Chargaff and André Boivin. Chargaff (1950) showed that in any type of organism the ratio of adenine (purine) to thymine (pyrimidine) as well as that of guanine (purine) to cytosine (pyrimidine) is always close to 1, (that it is exactly 1 was apparently not at first realized by Chargaff, nor its molecular meaning), and that the ratio A + T to G + C differs from organism to organism. For instance, in his early work he found this ratio to be 1.85 in yeast and 0.42 in the tubercle bacillus. Chargaff's findings decisively refuted Levene's tetranucleotide hypothesis, according to which all bases occurred at equal frequency. The way was now open for new molecular theory of DNA. As it later turned out, the base pairing (purine with pyrimidine) discovered by Chargaff was one of the most important clues in the subsequent construction of the double helix.

Let us remember that there are two nucleic acids, deoxyribose nucleic acid (DNA) and ribose nucleic acid (RNA). After it had been demonstrated that they are not restricted to animals and plants, respectively, the question arose what role they play in the cell and, more specifically, where in the cell they are found. That DNA is characteristic of nuclei was known since Miescher's days, and there had long been indications that RNA is the typical nucleic acid of the cytoplasm, but whether diffused DNA also occurs in the cytoplasm and some RNA in the nucleus remained controversial. What was needed were techniques, applicable to intact cells, which would differentiate between DNA and RNA. In other words, further progress depended on technical breakthroughs. In 1923 the cytochemist R. Feulgen (1884–1955) introduced a staining method (an aldehyde reaction), later called Feulgen reaction, which was specific for DNA. This permitted the conclusive confirmation that DNA is restricted to the nucleus (except for the special DNA of some cellular organelles). It took a good many more years until a specific RNA reaction was discovered (Brachet, 1940; 1941; Caspersson, 1941).[6] This permitted the clear demonstration that RNA occurs in the nucleolus and in the cytoplasm.

The cytological research of the preceding generations permitted quantitative as well as qualitative predictions about nuclear DNA:

(1) Since chromatin is replicated and then divided equally at every cell division, all cells produced by mitosis should contain the identical amount of DNA.

(2) Gametes, owing to the reduction division, should have half the amount of DNA of somatic cells in diploid organisms.

(3) DNA should be an extremely stable compound, to judge from the relative rarity of mutations.

821

(4) Since during fertilization two rather different sets of DNA come together, they must have the capacity to work harmoniously together.

(5) Considering the tremendous genetic variation observed at every level, from local gene pool to highest taxa, DNA must be able to assume a very large number of possible configurations.

The new methods of determining the amount of DNA per cell soon permitted confirmation of the two quantitative predictions. Boivin and his collaborators, the Vendrelys (1948), by developing methods to determine the exact amount of DNA in a cell, were able to show that a diploid cell contains twice as much DNA as a haploid cell. Later it was found that polyploid cells have the expected multiples of haploid cells. All of these findings confirmed the association of DNA with the chromosomes. Further studies showed a very different behavior for DNA and RNA in cells with different metabolic activities. The amount of DNA in the cell nuclei was always constant, even, for example, in severely starved rats, while the amount of RNA declined rapidly in such individuals: "The invariability of the DNA appears as a natural consequence of the special function which is now attributed to it, that of being the depository of the hereditary characteristics of the species" (Mandel et al., 1948: 2020–2021).

THE DISCOVERY OF THE DOUBLE HELIX

Much was learned about DNA in these years, and the conclusions that were drawn were often remarkably prophetic. The metabolic inertia of DNA, for instance, seemed to confirm the widespread speculation of gene theoreticians that the gene functions as a "template": "The logical implication is that the gene need not 'do' anything [in the metabolism of the cell] but that it merely provides a blue print for synthesis" (Mazia, 1952: 115). The absolute quantitative constancy of the DNA agreed perfectly with this postulate.

In order to answer the question how the gene could serve as a template, it was necessary to learn more about the structure of the DNA molecule. This was appreciated by a number of investigators. From Levene on it had been realized that DNA must have a longitudinal structure consisting of a backbone of deoxyribose and phosphate to which somehow the bases are attached. What needed to be discovered was how the three kinds of molecules were attached to each other. Only then would it be possible to determine how DNA could carry out its genetic function. Three laboratories, in particular, were in hot pursuit of this goal and had, when they started to work on this project, perhaps an equal chance of success. One was that of Linus Pauling at the California Institute of Technology (Pasadena), who had discovered the alpha helix structure of proteins and had made such important contributions to our understanding of the forces that bond molecules together.

The second group, Maurice Wilkins and his associates, worked at Kings College, London. Their special competence was x-ray crystallography, and Rosalind Franklin in this group succeeded in taking some excellent pictures of the x-ray diffraction pattern of DNA. From her work, and from other findings, arose several questions: Is the backbone of the DNA molecule straight or is it twisted into a helix? Furthermore, is there only a single helix or two, or three? Finally, how are the purine-pyrimidine bases attached to the backbone? Are the bases attached on the outside like the bristles of a bottle brush? Or, if there is a double or triple helix, would the bristles be on the inside, and how are these bases attached to each other? These and many other questions had been raised by Pauling and the Kings College group, but had not yet been settled when a third group of investigators started DNA work at Cambridge: James D. Watson and Francis Crick.

The details of the forward steps as well as the wrong guesses and multiple frustrations of the three groups of investigators need not be told again, since this has already been done so often and so well (Olby, 1974; Judson, 1979). What is important to mention is that one of the investigators, James D. Watson, more than any of the others, realized the decisive importance of the DNA molecule in biology, and it was this understanding which urged him relentlessly to push this work toward a successful conclusion, in spite of his rather modest technical qualifications for this task.

Wilkins as late as 1950 had wondered "what nucleic acid is in cells for."

Watson (b. 1928) had done his Ph.D. research at Blooming-ton, Indiana, under S. E. Luria. There and at Cold Spring Harbor he had learned of the importance of DNA, and when some other research plans of his could not be activated for technical reasons, he decided to go to England and take up DNA research. At the Cavendish laboratory at Cambridge he encountered a kindred soul in Francis Crick (b. 1916). Crick, equally brilliant as Watson, had some of the technical know-how which Watson lacked but was, at least in the beginning, not nearly as compulsive about the impor-tance of DNA as Watson. Both of them were highly deficient of certain kinds of knowledge, but by talking to many people, visit-ing relevant laboratories, and endlessly experimenting with var-ious models, they eventually came up with the right solution in February and March of 1953. Cut-out models of the various com-ponent molecules were of great help to them in arriving at the three-dimensional structure of DNA.

823

The most crucial "bit of information" was Chargaff's (1950) discovery of the 1:1 ratio of purines and pyrimidines (AT and GC). Although this discovery had been available for two years, it had been more or less ignored by all three groups of DNA inves-tigators. When Watson and Crick finally realized the importance of this numerical relationship, it took them only another three weeks of fiddling with their cut-out models to hit on the correct structure.

The final result, with which every high school student is now familiar, is that DNA is a double helix, and that the two strands are connected, like the steps of a circular staircase, by a sequence of base pairs (one purine and one pyrimidine). It is the sequence of the four possible base pairs (AT, TA, CG, GC) which, as was shortly afterwards discovered, provides the genetic information. This information serves as the blueprint in the assembling of polypeptides and proteins and thus controls cellular differentia-tion. Watson and Crick's double helix fitted all the facts so per-fectly that it was accepted by everyone almost at once, including the two most actively competing laboratories, those of Pauling and Wilkins. This dispelled all the remaining doubts whether or not DNA was truly the genetic material.

As Roux had conceived it in 1883, the basic process of trans-mission genetics is the division of a nucleus into "two identical halves." This phrasing misplaces the emphasis. The decisive event is actually the doubling of the genetic material, followed by its segregation into two daughter cells. The most crucial event in cell division, thus, is the exact replication of the DNA. Just how this could be done was a complete mystery until the double helix was discovered. This was at once clear to Watson and Crick, as they said (rather coyly) in their original paper (1953a: 737): "It has not escaped our notice that the specific pairing we have postulated immediately suggests a possible copying mechanism for the genetic material." As they outline in a subsequent publication, an untwisting of the helix together with the breaking of the bonds between the purine and pyrimidine bases produces two templates that serve as the replicating mechanism of the DNA.

824

Understanding the double helix and its function has had a profound impact not only on genetics but also on embryology, physiology, evolutionary theory, and even philosophy (Delbrück, 1971). The genotype and phenotype problem could now be stated in definitive terms and the last nail could be hammered into the coffin of the theory of the inheritance of acquired characters. Even though a suspicion had been repeatedly expressed as early as the 1880s and 90s that the genetic material might be different from the building material of the body, and even though the terms "genotype" and "phenotype" had been introduced in 1908, it was not fully realized until 1944 how fundamentally different the genotype is from the phenotype. Only from 1953 on was it understood that the DNA of the genotype does not itself enter into the developmental pathway but simply serves as a set of instructions. The breakthrough of molecular biology in the 1950s coincided with the birth of information science and some of the key words of that field, like program and code, became available for molecular genetics.

The coded "genetic program," modified from generation to generation and incorporating historical information, became a familiar and powerful concept. A history of the antecedents of this concept has not yet been written. Hering (1870) and Semon's (1904) concept of the *mneme*, although originally introduced to bolster the idea of an inheritance of acquired characters, is definitely in this tradition. Even closer is His's comparison (1901) of the activity of the germ plasm to the production of messages, the consequences of which can be far more complex than the simple message. Nevertheless, the concept of the genetic program as an unmoved mover (Delbrück, 1971) was so novel that nobody had come even close to it prior to the 1940s.

There has hardly been a more decisive breakthrough in the whole history of biology than the discovery of the double helix. I agree with Beadle's (1969: 2) judgment: "I have said many times that I regard the working out of the detailed structure of DNA one of the great achievements of biology in the twentieth century, comparable in importance to the achievements of Darwin and Mendel in the nineteenth century. I say this because the Watson-Crick structure immediately suggested how it replicates or copies itself with each cell generation, how it is used in development and function, and how it undergoes the mutational changes that are the basis of organic evolution."

825

The understanding of the double helix opened up an immense new field of exciting research and it is no exaggeration to say that as a result molecular biology completely dominated biology for the next fifteen years.[7] The long search for the true nature of inheritance had ended. The open questions became increasingly physiological, dealing with the function of genes and their role in ontogeny and neurophysiology. However, the story of transmission genetics was completed.

None of the findings of transmission genetics (summarized in Chapter 17) was modified by the findings of molecular biology in any major way. It is worth remembering that the fine structure analysis of the gene (Benzer's recognition of subunits) was achieved by the methods of classical genetics, not by biochemical methods. The claim, sometimes made, that, owing to the new approaches of molecular biology, transmission genetics had been "reduced to" molecular genetics is not at all substantiated (Hull, 1974). That genes are molecules had been postulated by biologists as far back as the 1880s, a hypothesis subscribed to by the majority of the leading Mendelians. However, prior to 1944 this was only a hypothesis. It is the unquestioned achievement of molecular biology to have provided the chemical explanation for the phenomena of transmission genetics. The structure of DNA (double helix) (1) explains the nature of the linear sequence of genes, (2) reveals the mechanism for the exact replication of genes, (3) explains in chemical terms the nature of mutations, (4) and shows why mutation, recombination, and function are separable phenomena at the molecular level.

Far greater has been the impact of molecular biology on our understanding of gene function, where an entirely new frontier has been opened up. The classification of genes into various categories, such as structural, regulatory, and repetitive DNA, is still in a preliminary stage. The role of nucleosomes and the various proteins in the eukaryote chromosomes is understood only in a rudimentary way. The role of introns, transposons, and supposedly "silent" DNA is mysterious. New phenomena are discovered almost every month that pose new puzzles. What little we do understand seems to indicate that all these phenomena are involved with regulating gene function. Molecular genetics is still very much of an unfinished story.

GENETICS IN MODERN THOUGHT

Few other branches of biology have had as profound an impact on human thinking and human affairs as genetics. This is far too large a subject to be dealt with adequately in a few pages, and all I can do is to call attention to the diverse applications of genetic thought.

That certain human diseases might have a genetic causation had long been known, since they often run in families. Hemophilia, so widespread among male descendants of Queen Victoria, is perhaps the best-known example. Polydactyly was described by Maupertuis and Réaumur in the eighteenth century. By now many hundreds of genetic diseases of man are known, and in many cases it is now established on what chromosome the mutant gene is located (McKusick, 1973).[8]

Three aspects of human genetics are of particular interest. The first is that some of the genetic diseases of man represent failures of metabolism. The English physician Garrod, as early as 1902, had suggested that the disease alkaptonuria was caused by a block in a metabolic reaction sequence and that the block was due to the congenital deficiency of a specific enzyme (see also Garrod, 1909). Although largely ignored when first published, Garrod's theory, when rediscovered by Beadle and Tatum, played an important role in the development of physiological genetics.

A second important aspect of human genetics is that it has forced geneticists to study those phenotypic conditions that have a somewhat unorthodox mode of inheritance. Thus it is now rather evident that the gene or set of genes responsible for schizophrenia has low "penetrance." This means that a person may not show the manifestations of this disease even though he or she has the requisite genetic condition. Genes with low penetrance are quite widespread in *Drosophila* (as shown by the work of Timofeeff-Ressovsky and of Goldschmidt) but, for obvious reasons, their study is avoided by geneticists. There are other genes that are variable in the intensity of their expression (for instance, apparently the genes for diabetes), and the study of such genes has likewise expanded our understanding of the modes of inheritance.

Perhaps the most far-reaching impact that genetic thinking has had on modern man is to raise the possibility that almost all human characteristics may have a partial genetic basis. This claim is made not only for physical but also for mental or behavioral attributes. The relative contribution of the genetic constitution on nonphysical human characteristics, particularly intelligence, is one of the most controversial current biological and social subjects.

827

Finally, it might be pointed out how important genetics has become in animal and plant breeding. Milk production in dairy cattle and egg production in chickens are two examples of the magnificent achievements of animal geneticists. The breeding of crop plants for disease resistance and the development of hybrid corn and short-stem cereals are other illustrations. Even if the so-called green revolution was not quite as successful as predicted, it has helped nonetheless to increase, sometimes dramatically, the productivity of many crop plants. Primitive man in thousands of years made less progress in his endeavors to improve crop plants than modern genetics was able to accomplish in a ten-year period.

Anyone who studies a modern textbook of genetics is overwhelmed by the wealth of facts and interpretations. As far as a nonspecialist is concerned, even the most elementary text contains not just "all you want to know about heredity" but actually far more than you want to know. The situation is aggravated by the fact that modern genetics has more or less separated into three or four largely independent fields: transmission or classical genetics, evolutionary or population genetics, molecular genetics, and physiological or developmental genetics.

This creates a formidable difficulty for the historian of ideas who wants to summarize in few sentences the major concepts that have emerged from the mass of researches conducted and published from 1865 to 1980. My own attempt to do so is admittedly provisional:

(1) The most spectacular and—until the 1940s—totally unexpected finding is that the genetic material, now known to consist of DNA, does not itself participate in building the body of a new individual but merely serves as a blueprint, as a set of instructions, designated as the "genetic program."

(2) The code, with the help of which the program is translated into the individual organisms, is identical throughout the

living world, from the lowest microorganism to the highest plants and animals.

(3) The genetic program (genome) in all sexually reproducing diploid organisms is double, consisting of one set of instructions received from the father and another from the mother. The two programs are normally strictly homologous, and act together as a unit.

(4) The program consists of DNA molecules, associated in the eukaryotes with certain proteins (like histones) whose detailed function is still uncertain but which apparently assist in regulating the activity of different loci in different cells.

(5) The pathway from the DNA of the genome to the proteins of the cytoplasm (transcription and translation) is strictly a one-way track. The proteins of the body cannot induce any changes in the DNA. An inheritance of acquired characters is thus a chemical impossibility.

(6) The genetic material (DNA) is completely constant ("hard") from generation to generation, except for a very occasional (about one in 100,000) "mutation" (that is, an error in replication).

(7) Every individual in sexually reproducing organisms is genetically unique, because several different alleles may be represented at tens of thousands of loci in a given population or species.

(8) This enormous store of genetic variation supplies almost unlimited material for natural selection.

20 ✐ Epilogue: Toward a science of science

INCREASINGLY often one reads references to a "science of science." What is meant by this designation? It relates to an evolving discipline that would combine the sociology of science, the history of science, the philosophy of science, and the psychology of scientists with whatever generalizations one can make about the activities of scientists and about the development and methodology of science. It would include generalizations on the intellectual growth and style of work of the great leaders of science and, for that matter, also of the great army of other scientists who make contributions to the gradual progress of our knowledge and understanding.

Philosophers and sociologists of science have asked, and to some extent answered, numerous questions. For instance, what generalizations can one make concerning the origin of new research traditions, their flowering, decline, and replacement? Is it true that there are scientific revolutions and, if so, is their life history consistent with Thomas Kuhn's description? What factors in the milieu of science and scientists are most important for the occurrence of revolutionary or at least innovative periods? What is the relative proportion of scientific advances made possible by new technologies, new observations, or new kinds of experiments versus those made possible by new conceptualizations? Furthermore, is it legitimate to recognize such a division, or is the carrying out of new experiments and the collecting of new observations merely the testing of new hypotheses and theories?

No theory of science so far proposed has been accepted universally. Logical positivism has had a well-developed theory of science, dealing both with discovery and explanation. But since the extensive criticism of recent decades indicates clearly that this theory is badly in need of revision, if it is valid at all, I shall not try to present it here. Numerous endeavors of replacement have been made (those of Popper, Feyerabend, Lakatos, Laudan, and others) but we still seem to be far from a synthesis.

The observations and generalizations of the sociologists of science (such as Merton) seem, on the whole, less vulnerable; indeed, so far as they go, they describe the situation quite well. However, their work deals with rather specific problems, such as multiple independent discoveries or the role of priority in the reward system of scientists. At the present, no sociologist could or would claim that we have a well-rounded sociology of science. What we have so far are "contributions toward a sociology of science."

830

All the writings dealing with the science of science published in the past were heavily skewed in favor of the physical sciences. The following notes and comments may serve to bring the biological sciences more definitely into this field. Unfortunately, I have been unable to compose a well-rounded science of biological science, and my contribution is what Schopenhauer would have called *parerga kai paralipomena*. I hope that it will stimulate others to do better.

SCIENTISTS AND THE SCIENTIFIC MILIEU

The growth of science is the growth of the ideas of scientists. Each new or modified idea was born in the brain of an individual scientist. This is fully acknowledged by historians and is even reflected in scientific language when we refer to Mendel's laws, to Darwinism, or to Einstein's relativity theory. For the sake of simplicity, the thought of the great intellectual leaders and conceptual innovators is often presented in histories of science as if it were of monolithic unity and constancy. One is referred to Lamarck 1809 or Darwin 1859, as if the better-known accounts of the development of Darwin's ideas, the doubts, hesitations, inconsistencies, contradictions, and frequent changes of mind were quietly ignored, and the development of his ideas was presented as a logical chain of inferences and conclusions. How misleading this was became evident when historians began to study Darwin's works and correspondence critically, and particularly when they analyzed Darwin's notebooks and unpublished manuscripts (1975; 1980). Limoges (1970), Gruber (1974), Kohn (1975; 1980), Herbert (1977), Schweber (1977), and Ospovat (1979) show how misleading the traditional representation of the birth of Darwin's theories was. His ideas on speciation, for instance, underwent a dramatic change in the 1850s (Sulloway, 1979), and he allowed more for soft inheritance in the 1870s than in the 1850s.

The thought of many of the greatest scientists has been characterized by a long process of maturation and quite often by complete reversals. Linnaeus, for instance, who at first had proclaimed so emphatically the constancy and permanence of species, developed later in life a theory of an origin of species by hybridization. Lamarck believed in the constancy of species up to age 55, then adopted evolution but moved more and more from a straight line in the next fifteen or twenty years to a tree-like concept of evolution. Rensch, Sumner, and Mayr were neo-Lamarckians in their younger years but fully adopted selectionism later on. In fact, some of the greatest scientists have modified their ideas most frequently and most profoundly. One can never understand the impact of a thinker throughout his lifetime if one does not understand the permutations of his thought. The same is true for many philosophers. The Kants of the *Cosmogony* (1755), of the *Critique of Pure Reason* (1781), and of the *Critique of Judgment* (1790) were three rather different thinkers. Scientists who never changed their major ideas from the beginning to the end of their career are probably a small minority. No one, as far as I know, has yet made a special study of the drastic changes in the thinking of great scientists (some of them veritable conversions). There are many unanswered questions in this area. Are there special ages at which such changes occur with particular frequency? What brings these changes about? Is it true that some scientists "backslide" in their later years?

All interpretations made by a scientist are hypotheses, and all hypotheses are tentative. They must forever be tested and they must be revised if found to be unsatisfactory. Hence, a change of mind in a scientist, and particularly in a great scientist, is not only not a sign of weakness but rather evidence for continuing attention to the respective problem and an ability to test the hypothesis again and again.

There are great differences in the personalities of different authors, and this dominates the style of their research. Ostwald (1909) classified scientists into romantics and classics. The romantic is bubbling over with ideas that have to be dealt with quickly to make room for the next one. Some of these ideas are superbly innovative; others are invalid if not silly. The romantic usually does not hesitate to abandon his less successful ideas. The classic, by contrast, concentrates on the perfection of something that already exists. He tends to work over a subject exhaustively. He also tends to defend the status quo. Sulloway (1982) has shown that statistically there is a drastic difference in the style of firstborns and later-borns. Firstborns tend to be conservatives, often corresponding rather well to Ostwald's designation of classics. During scientific revolutions they tend to defend the exisiting paradigm. Later-borns, by contrast, tend to be revolutionaries who propose unorthodox theories.

Hardly anyone has described the workings of the mind of a successful scientist better than Darwin. He stated repeatedly that he could not make any observations without "speculating," as he called it. Everything he saw raised questions in his mind. Another characteristic of successful scientists is flexibility—a willingness to abandon a theory or assumption when the evidence indicates that it is not valid. Several of the architects of the evolutionary synthesis of the 1930s abandoned previously held beliefs when they were shown to be erroneous. A third generalization that can be made about almost all great scientists is that they have a considerable breadth of interest. They are able to make use of concepts, facts, and ideas of adjacent fields in the elaboration of theories in their own fields. They make good use of analogies and favor comparative studies.

Research Strategies

Medawar (1967) very perceptively has emphasized how important it is for a scientist to have a feasible research program. For instance, all geneticists from Nägeli and Weismann to Bateson failed to develop successful theories of heredity because they attempted to explain simultaneously inheritance (transmission of the genetic material from generation to generation) and development. The wish to do so was not surprising, since nearly all of them had approached genetic problems from the field of embryology. It was Morgan's genius to put aside all developmental-physiological questions (even though he himself had also come from embryology) and concentrate strictly on the problems of transmission. His pioneering discoveries from 1910 to 1915 were entirely due to this wise restriction. The developmental problems which his findings (and those of his collaborators) raised were simply set aside. This was very fortunate, because some of these problems—for instance why genes in cis-position may have a different effect from those in trans-position (position effect)—are still not yet fully solved, more than fifty years after their discovery.

There are many potential reasons why a problem may not yet

be ready for solution. The technical tools for their analysis may not yet have been forged, and certain concepts, particularly when requiring the help of neighboring fields, may not yet have been developed. In such cases the as-yet-insoluble problems must be treated as "black boxes," to be opened when their time arrives, as Darwin did in regard to the source of unlimited variation in nature.

In referring to Weismann I have already mentioned a second poor research strategy, a failure to partition a problem into its components. The study of heredity, for example, could make no progress until the problems of transmission and development were properly separated. This need to separate the components of a complex problem is true for both science and philosophy; the failure to separate the concept of teleology into its four components (see Chapter 2) and the failure to distinguish between species category and taxon (Chapter 4) are but two examples.

Another research strategy that has been, in the long run, even more damaging for the progress of science is the ever-repeated confirmation of findings that no longer require confirmation. The great comparative anatomists of the nineteenth century—Haeckel, Gegenbaur, and Huxley—used comparative anatomy with splendid success to confirm Darwin's theory of common descent. Yet comparative anatomists continued to consider the establishment of homologies and the search for common ancestors as their exclusive object long after any opposition to the theory of common descent had disappeared (Coleman, 1980). Only the Russian school of Severtsov was somewhat outside of this tradition, and so was Böker, who asked different questions but ones which unfortunately were marred by his Lamarckian philosophy. It took almost a hundred years after Darwin before the introduction of new *Fragestellungen* revitalized comparative anatomy (through the work of D. D. Davis, W. Bock, and others). When one reads the publications of the Morgan school after 1920, one has the feeling that its members also suffered from this deficiency. The main emphasis of the research continued to be, in the face of opposition from Bateson, to prove the correctness of the chromosome theory of inheritance, even though by this time this theory had been established beyond any reasonable doubt. The significant advances in genetics during the 1930s and 40s, as a result, were made by representatives of other schools of genetics.

Another poor research strategy is to limit oneself to the mere piling up of facts and descriptions without using them to develop

833

new generalizations or concepts. The detractors of taxonomy have not without reason ridiculed those taxonomists who seem to have no other research objective than to describe ever more new species, as if this activity were the alpha and omega of taxonomic science. As necessary as undoubtedly the inventory taking of the diversity of the organic world is, any inquisitive systematist wants to go beyond this Linnaean stage. The same criticism applies to certain practitioners in almost any branch of biology. The censusing of "quadrats" in early ecology was another such purely descriptive enterprise.

834

There is a law of diminishing returns in research as in so many other aspects of human activity, and it is the ability of the perceptive scientist to recognize when this point has been reached. Curiously, there are some investigators who repeatedly were near a major discovery but perversely dropped their line of research to take up an entirely new problem. The reason generally seems to have been that they had failed to ask the appropriate meaningful questions and had therefore thought that the line of research in which they had been engaged had been exhausted. This is one more piece of evidence for the importance of asking meaningful questions.

The Power of Ideologies

It has often been observed that different scientists may draw entirely different, sometimes diametrically opposed, conclusions from the same facts. How can this be? Evidently such a divergence of interpretation is the result of a drastic difference in the ideologies (*Weltanschauungen*) of the respective scientists. For example, two mid-nineteenth-century scientists might fully agree on how admirably insects are adapted for the visiting flowers, and flowers, in turn, for being pollinated by insects. Yet, a pre-Darwinian natural theologian would consider these facts as striking evidence for the wisdom of the creator, while a Darwinian would consider the same facts as excellent confirmation for the power of natural selection. Whether an author subscribes to essentialism or to population thinking, whether he believes in reductionism or emergentism, and whether or not he clearly understands the difference between proximate and ultimate causations—all such basic differences in ideology will determine what biological theories are acceptable to him. For this reason, the change and replacement of individual scientific theories is far less important in the history of science than the waxing and waning of the major ideologies that may influence the thinking of scientists.

The study of the basic philosophies or ideologies of scientists is very difficult because they are rarely articulated. They largely consist of silent assumptions that are taken so completely for granted that they are never mentioned. The historian of biology encounters some of his greatest difficulties when trying to ferret out such silent assumptions; and anyone who attempts to question these "eternal truths" encounters formidable resistance. In biology, for hundreds of years, a belief in the inheritance of acquired characters, a belief in irresistible progress and in a *scala naturae,* a belief in a fundamental difference between organic beings and the inanimate world, and a belief in an essentialistic structure of the world of phenomena are only a few of the silent assumptions that influenced the progress of science. Basic ideological polarities were involved in all of the great controversies in the history of biology, indicated by such alternatives as quantity vs. quality, reduction vs. emergence, essentialism vs. population thinking, monism vs. dualism, discontinuity vs. continuity, mechanism vs. vitalism, mechanism vs. teleology, statism vs. evolutionism, and others discussed in Chapter 2. Lyell's resistance to evolutionism was due not only to his natural theology but also to his essentialism, which simply did not allow for a variation of species "beyond the limits of their type." Coleman (1970) has shown to what large extent Bateson's resistance to the chromosome theory of inheritance was based on ideological reasons. One can go so far as to claim that the resistance of a scientist to a new theory almost invariably is based on ideological reasons rather than on logical reasons or objections to the evidence on which the theory is based. For an excellent discussion of the causes for resistance to new ideas, see Barber (1961).

835

Mutually Incompatible Components

When one closely studies the thinking of any innovative scientist, one finds almost without exception that it contains mutually contradictory components. This is perhaps least surprising in the case of Lamarck, who was 55 years old when he underwent his drastic change from believing in a constant world to believing in an evolving world. He superimposed his newer ideas on the traditional thinking of the eighteenth century, and numerous flagrant contradictions were the understandable result.

When one analyzes the set of beliefs of an author of bygone days, one must carefully avoid judging inconsistencies on the basis of modern insights. Probably no scientist has ever escaped internal inconsistencies in his conceptual framework. Lyell preached uniformitarianism, but it struck even some of his contemporaries as inconsistent how close to a nonuniformitarian theory was his explanation of the origin of new species. Darwin applied population thinking while explaining adaptation by natural selection, but he employed dangerously typological language in some of his discussions of speciation. None of the Darwinians stressed natural selection more forcefully than A. R. Wallace, and yet he was unable to apply it to man. Darwin and many other pre-1900 geneticists frequently stressed the integrity of the genetic particles (as demonstrated by reversion and other phenomena), but all of them also allowed for a certain amount of fusion (later called blending) of equivalent particles. Historians of science, in my opinion, have not paid nearly enough attention to such contradictions and conceptual incompatibilities. Too often the thought of a scientist is presented as a well-rounded, harmonious system, while in reality it usually consisted of bits and pieces that were constantly being revised, but revised piecemeal so that certain components no longer harmonized with others. It would be a fascinating task to examine the thought of the leaders of biological research for such contradictions.

Premature or Unfashionable?

Major scientific discoveries are often largely or completely neglected by their contemporaries. Many instances of this are quoted in the literature, the best-known example perhaps being Mendel's rules, published in 1866 and neglected until 1900. Avery's demonstration that the transforming agent of pneumococcus is nucleic acid is often cited as another instance. This discovery was published in 1944, but until 1953 it did not receive nearly as much attention as such a spectacular discovery deserves. My own discovery of the importance of peripherally isolated populations for speciation and macroevolution, published in 1954, was hardly referred to until the 1970s but has now become so fashionable that it was more often referred to in a recent textbook of macroevolution (Stanley, 1979) than the work of any paleontologist.

It has been claimed that such neglect occurs because the discoveries are "premature." Stent (1972) has given the definition that "a discovery is premature if its implications cannot be connected by a series of simple logical steps to canonical, or generally accepted knowledge." Actually, it would seem dubious to designate a discovery premature if it has been made by someone, particularly if the discoverer had been deliberately looking for some such solution, as was the case with Mendel. My own analysis of the situation, without wanting to go into all the details, is that a discovery is likely to be ignored if it is made in a field that is not fashionable at the time, that is, if it lies outside the dominant research interests of the period. In the case of Mendel, most hybridizers were interested in the "species substance," and an analysis of individual characters was beyond the boundaries of their problem. The embryologists who did most of the speculating in genetics during that period were only (or at least primarily) interested in the developmental aspects of inheritance. Segregation and ratios were irrelevant as far as they were concerned.

In the case of Avery's discovery, to take up the second example cited in the literature, my eyewitness experience leads me to believe that its importance or at least its implications were fully realized by many geneticists, and it was through them that Watson was indoctrinated in the importance of the problem. However, the analysis of the structure of the DNA (and thus its suitability as an information receiving and conveying molecule) was outside the area of competence of these biologists. The problem had to be taken up by chemists, and this was indeed done by Chargaff and others. In that instance there was certainly no case of prematurity, except in the sense that most chemists and biophysicists who worked on DNA were not nearly as aware of the importance of this molecule as were the biologists. Finally, to take up the third example, the importance of peripherally isolated populations was ignored by nearly all geneticists because it did not fall within their research competence. An ideal situation such as the frequent establishment of peripherally isolated populations of *Drosophila* in the Hawaiian Islands had to arise before a geneticist (Carson) took up this problem. Peripherally isolated species were also ignored by the paleontologists because paleontologists prior to 1972 were virtually limiting themselves to "vertical" thinking. It was not a coincidence that one of the two paleontologists (S. J. Gould) who applied the concept to paleontology had been associated with me in the preceding years in the teaching of an advanced course in evolutionary biology.

I conclude from this that prematurity is perhaps not the best word for this phenomenon. It is simply that there is only limited contact among the workers of different research areas and most of them will not relate discoveries in adjacent fields to the problems of their own field. Most scientists are truly interested only in researches that have a bearing on their own work and that are accessible to their research techniques and tools.

Form of Publication

It has often been remarked that Mendel's work would not have been ignored for 34 years if it had been published in one of the better known and more prestigious botanical journals instead of in the proceedings of a local natural-history society. It is indeed true that the particular outlet through which a scientific discovery or new generalization is published is of considerable import and deserves to be emphasized far more than has been done in the past. Castle and Weinberg published their findings, now designated as the Hardy-Weinberg Law, in relatively obscure places and thus their priority was long overlooked, while Hardy published his in *Science,* where everybody encountered it quickly.

I remember several illustrations of the importance of the place of publication from my own work. In the early 1930s it was quite generally accepted that the sexual dimorphism in plumage color in birds was due to the suppression in females of the neutral (male) plumage by female hormone. In 1933 I discovered that in several species of birds there was drastic geographic variation in the nature and degree of sexual dimorphism on the islands of Indo-Australia. In one species (*Petroica multicolor*) there was standard sexual dimorphism on some islands exactly as in Australia, where the species had originated. On other islands, however, the males were hen-feathered, both sexes thus having the cryptic coloration of the females, while on still other islands the females were cock-feathered, both sexes having the brilliant black, white, and red coloration normal for adult males. Since it was highly unlikely that there was any geographic variation of the sex hormones in this species, I concluded that the sexual dimorphism was directly controlled by the potential of the feather germs. I published this (to me, as a young man, epoch-making) discovery in the publications of the American Museum of Natural History (1933; 1934), where, of course, no endocrinologist or developmental physiologist ever read it and thus it was completely ignored.

Up to the middle of the nineteenth century almost the only places of publication for a biologist were journals of academies and various scientific and natural-history societies, most of which were distributed by exchange. Except for those of the Paris Academy, the Linnean Society of London, and the Zoological Society of London, most society journals were little read, at least internationally. The situation improved as more and more specialized journals were founded, many specialized branches of biology having an almost meteoric rise as soon as such a specialized journal became available.

The publication of books, experience shows, is, or at least was during past generations, of great importance for the prestige of a scientist. In the early editions of *American Men of Science,* the most prominent scientists were designated by an asterisk, and it was generally known that this would happen as soon as a scientist had published a book. However, the publication of books also has its drawback. Somehow, it is assumed that books summarize the state of the art in a certain field or concerning a certain problem. If an author develops original ideas in a book that, in other parts, is a summary of the literature, it is very likely that the new ideas will be overlooked. One must advise young authors therefore to publish novel ideas separately in journal articles, where there is much less danger that their thoughts will be ignored.

One can make one other generalization. It is unwise for an author to combine highly heterogeneous material in a single publication. The title of such a publication in most cases refers only to one of the topics, and the others are likely to be ignored. This has been characteristically true for the taxonomic literature. If someone publishes interesting and new ideas on the species concept, on speciation, or on the theory of biogeography in a taxonomic monograph entitled *A Revision of the Family XX of Beetles (or Fishes),* he should not be surprised when no one pays any attention to his ideas. Now that there are technical journals available for almost any subdivision or discipline of biology, it is easier for an author to direct his contributions to the most appropriate journals and have them read by his peers.

THE MATURATION OF THEORIES AND CONCEPTS

The backbone of science is the system of generalizations, theories, and concepts which form the explanatory framework of the observed phenomena. It has been the major objective of the philosophy of science to determine how theories are formed and tested; how one distinguishes between hypotheses, laws, and theories; what differences there are between the logic of discovery and that of explanation; and how all related problems are to be dealt with. I will not attempt to go over this ground again but will instead discuss some particular factors that have played a role, for better or worse, in the development of scientific theories and concepts.

Constructive Contributions toward Improved Concepts

840

A new idea is rarely fully developed when it first occurs. Darwin added a great deal to his concept of natural selection after it had first occurred to him in September 1838. Indeed, when one reads an author's first expression of an idea, one is usually surprised at how vague it is. Also it may be intermingled with extraneous or even contradictory elements.

Concepts and theories are usually part of the total research tradition of a specific branch of science, and it is in some respects more instructive to study the factors which contribute to (or impede) the maturation of such a scientific discipline than to try to do this for a particular concept. Let me now discuss some of these factors, not necessarily in any order of importance.

The Elimination of Invalid Theories or Concepts

Nothing strengthened the theory of natural selection as much as the refutation, one by one, of all the competing theories, such as saltationism, orthogenesis, an inheritance of acquired characters, and so forth. Another example can be found in the maturation of the modern concept of inheritance. About a dozen previously held concepts from the Greeks to 1900 had to be refuted in order to make room for the current concepts of transmission genetics (see Chapter 16).

The Elimination of Inconsistencies and Contradictions

Inconsistencies and what strikes us as internal contradictions are often not at all apparent during less mature stages of theories. When a thinker subscribes simultaneously to seemingly incompatible concepts, he acts as if the different concepts were localized in different brain compartments without channels of intercommunication. For instance, most of the believers in soft inheritance in the eighteenth and nineteenth centuries were essentialists and ought to have believed in unchangeable essences. The early Mendelians, to give another example, ascribed evolutionary change to random mutation, ignoring the fact that such a random process could never lead to the remarkable adaptations in the living world.

Some early evolutionists, for instance Asa Gray, devoutly believed in a personal God and yet accepted natural selection and other aspects of Darwinism which other of their contemporaries considered quite incompatible with creationism. Severe dilemmas arise whenever scientific facts or theories come into conflict with the basic philosophy or ideology of a scientist. Usually in such cases it is easier to live with a contradiction than to give up either science or one's adopted ideology. However, where the contradictions merely affect competing theories, either one or the other will eventually be shown to be invalid and this will produce a clear scientific advance.

841

Input from Other Fields

Many major advances in the maturation of concepts and theories have been due to an input of ideas or techniques from other fields. Such input may come from other branches of biology, as when genetics received ideas from applied animal and plant breeding, cytology, systematics, or from the physical sciences (particularly chemistry), or mathematics. Well-developed theories or models are often available in one science that are equally applicable and most productive when transferred to another science.

The Elimination of Semantic Confusions

Technical terms, when clearly defined and well understood, are a great help in the advance of scientific understanding. Contrariwise, when a term is inadvertently transferred to a different concept (as was done by T. H. Morgan with the term "mutation") or when the same term is used for different concepts, considerable confusion will ensue until the ambiguity is cleared up. The introduction of new technical terms often helps to disentangle this kind of confusion. Examples are the term "taxon" (for which previously the term "category" had also been used), the term "subspecies" (for which taxonomists had previously used the term "variety" which was also being used for individual variants), or the term "isolating mechanisms" (for which previously no term had existed). Examples could be quoted from any branch of biology where the introduction of new terms resulted in clarification in a previously confused area. The evolutionary synthesis of the 1930s and 40s was greatly facilitated by the introduction of the terms "polytypic" by Huxley and Mayr, "sympatric" and "allopatric" by Mayr, "gene pool" by the Russian school, "genetic drift" by Sewall Wright, and other terms such as "founder principle" and "genetic homeostasis." When properly defined and clearly delimited against other phenomena with which they had been previously confounded, such terms helped to eliminate controversies.

When a term is transferred from concept to concept no matter how much the underlying conceptualization is changing, misunderstandings inevitably result. And yet, in most cases the retention of the technical term has been preferable to a continuous introduction of new terms whenever there was a slight or gradual change of the underlying concept. For example the term "gene", when proposed by Johannsen, was specifically given for a "nonmaterial" entity, an "accounting unit." In the Morgan school the term was soon applied to a definite, distinctly material locus on the chromosome and in molecular genetics to a certain set of base pairs, likewise a strictly material entity. One could go on and on giving examples of this kind.

Metaphors play an important role in the history of science. There are felicitous metaphors and those that are unfortunate. Darwin's term "natural selection" is on the borderline of the two categories and was strenuously resisted by the majority of his contemporaries. They tended to personify whatever did the selecting and to insist that there was no real difference between selection by Nature and creation by God. When, at the urging of his friends, Darwin adopted Spencer's term "survival of the fittest," he jumped from the frying pan into the fire, because this new metaphor suggested circular reasoning. The term "genetic drift," introduced by Sewall Wright to designate the stochastic processes of changes in allele frequencies in small populations, was misinterpreted by certain authors as a steady one-directional drift. A study of the introduction and subsequent fate of metaphors in biology would be an interesting task for a historian.

The Eclectic Fusion of Two Competing Theories

Biological theories are usually rather complex. It rarely happens that one theory has an unchallenged monopoly. More often, two or more theories are competing with each other, and the controversy as to which is correct may continue over decades if not centuries. The ultimate solution is rarely the complete victory of one of the alternate theories but more often a synthesis of the best elements of several theories. For instance, the modern eclectic theory of recapitulation combines the valid components of two previously feuding theories, that of the *Naturphilosophen* and that of K. E. von Baer, with the Darwinian theory of common descent:

ontogeny recapitulates—with minor or major deviations—the ontogenetic (but *not* the adult) stages of the ancestors.

The controversy over the nature of the genetic material, which raged from approximately 1880 until well into the twentieth century, is another example. The physicalists believed it to be either a physical force or something "purely chemical," while the embryologists and naturalists were so impressed by the incredible specificity and precision of inheritance that, from Darwin and Weismann on, they postulated a well-structured—or, as their opponents called it, a "morphological"—basis of inheritance. Macromolecules, of course, were unknown during the greater part of the duration of this controversy. When the final answer was given in 1953, it turned out that the genetic material was *both* chemical *and* highly structured. The resolution of the controversy was a synthesis of the opposing viewpoints.

843

Impediments to the Maturation of Theories and Concepts

Historians of science have described numerous factors which had the effect of delaying the maturation of a research program or of preventing the adoption of a correct theory, but there are two factors that are sometimes not given sufficient weight.

Failure to Consider the Alternatives

I have shown above how it is sometimes possible to synthesize two competing theories by a process of eclectic fusion. Unfortunately, this is not what usually happens. When a scientific theory is wrong in part, what usually occurs is not that it is improved by a replacement of the wrong components, but rather that a counter theory is proposed which serves as a sort of antithesis, as if the original theory had been completely wrong. And yet this counter theory will be wrong in certain respects that were correct in the original theory. For instance, when embryological researches made it evident that preformation (in the sense of incapsulation) was wrong, it was not replaced by a modified theory of preformation (genetic program) but rather by a theory of pure epigenesis. To take another example, the theory of recapitulation of the adult stages of the ancestors was opposed by a theory of embryogenesis that denied any effect of ancestry but ascribed the similarities of ontogenetic stages to a purely fortuitous parallel progression from less to more specialized. Finally, neo-Lamarckian theories of evolution, which relied on the influence of the environment, were opposed by mutational theories, in which evolutionary change was ascribed entirely to "mutation pressure" (repeated mutation in the same direction) and in which any role of the environment, even as an agent of natural selection, was excluded.

As a result, the history of science is characterized by wide swings of the pendulum. Whenever an entirely new theory, or even more so when an entirely new research tradition, is introduced, certain truths that had previously been accepted are abandoned. In some instances this seems unnecessary. In other instances the "antithesis stage" seems to be necessary before a balanced synthesis can be achieved. For instance, theories of sympatric speciation were proposed so frequently and so uncritically between 1859 and 1940 that it was perhaps necessary to assert the prevalence of geographic speciation with an almost intolerant emphasis in order to enforce a more critical approach to the problem of sympatric speciation.

The swinging of the pendulum may result in the complete giving-up of a research tradition. The introduction of physicalism into physiology by Carl Ludwig and J. Müller's students resulted in the abandonment of a very promising beginning of ecological physiology (for example, Bergmann's work) and in fact of all why-questions in physiology. Even though this led to a brilliant flowering of a physiology of proximate causes, it took almost a hundred years before a new start in ecological physiology was made, which concentrates on the adaptational nature of physiological processes.

Many of the long-standing controversies in science were caused by the failure of opponents to see that the two opposing viewpoints did not exhaust the number of possible explanatory choices. One wonders whether the old axiom of logical division — *Tertium non dat*— is not the subconsious norm for this attitude. The explanation of organic diversity for Louis Agassiz was that it is either due to the plan of the creator or that it is the accidental by-product of the blind play of physical forces (Mayr, 1959e). Darwin's explanation (natural selection) was so far ouside Agassiz's alternate explanatory models that it was not touched by anything he had argued. Agassiz's argument, of course, was only a version of the old alternative "chance vs. necessity." Even Monod (1970) failed to see that the process of natural selection offers an option that bypasses the unpalatable choice between chance and necessity. The classical positions in the preformation–epigenesis battle (Roe, 1981) or in the von Baer–Haeckel formulations of the recapitulation theory are other examples. It would be interesting to tabulate in how many of the major controversies in the history of biology the posing of such incomplete alternatives was involved. The frequency of such incomplete alternatives should warn the participants in any controversy to study carefully whether there is not a third option that would avoid the seeming stalemate of the controversy.

A second kind of false alternative involves cases where an "either/or" question is asked when in reality the two so-called alternatives are actually merely two sides of the same coin. An example is the claim (White, 1978) that speciation is very often chromosomal rather than geographical. White is of course right in saying that chromosomal rearrangements are often vitally involved in speciation, but this does not in the least necessitate abandoning the process of geographic speciation. Quite the contrary, such chromosomal rearrangements are most easily established in peripherally isolated founder populations, that is, in geographic isolation. The two-sides-of-the-same-coin principle was ignored in the recommendation recently made by a population biologist to ignore species since they are merely arbitrary inventions of the taxonomist, and to study instead populations. This author overlooked the fact that a population has relations to two other kinds of populations: those which do not share the same space but which do share the same isolating mechanisms (that is, conspecific populations) and those which share the same space but which are reproductively isolated, (that is, different species).

False alternatives have been at the bottom of nearly all major controversies in the history of evolutionary biology: isolation or natural selection (M. Wagner), mutation or natural selection (de Vries, Bateson, Morgan), gradual evolution or discontinuous inheritance (Mendelians vs. biometricians), importance of the environment or natural selection (neo-Lamarckians and their opponents), behavior or mutation (preadaptationists), to mention just a few. The two-sides-of-the-same-coin principle must be kept in mind in every biological problem, for each phenomenon in biology has both proximate and evolutionary causations. Neither *Entwicklungsmechanik*, which deals with proximate factors, nor comparative (phylogenetic) embryology, which deals with evolutionary causations, can tell the complete story. Sexual dimorphism (hormonal vs. selectionist interpretation) is another example of such causal duality, and so are all seasonal phenomena, like bird migration (Mayr, 1961). The two types of explanation are not, as was erroneously believed by some authors, alternate solutions to the problem; rather, they both must be explored before we can achieve a full explanation of the phenomenon.

Erroneous Search for Laws

This constitutes a second impediment to the maturation of theories and concepts. As far as the application of laws in biology is concerned, it must be remembered that the physical sciences provided all norms in science for some four hundred years. It is only since 1859 that the biological sciences have begun to emancipate themselves from the dominance of the physical sciences. Whatever regularities or generalizations a biologist found prior to that date (and to a large extent even afterwards) he felt obliged to explain in terms of the language and conceptual framework of the physical sciences.

I have pointed out in this work many instances where physicalism has had a deleterious effect on developments in biology. For instance, in the physical sciences when a law is valid for one particular set of phenomena, it is usually equally valid for all similar sets, unless the very fact of the failure of the applicability of the law to a set of phenomena shows that they are *not* similar phenomena. This consideration has proven of considerable heuristic value in the physical sciences. In biology, where so many unique phenomena are encountered and where virtually all so-called laws have exceptions, the belief in the universality of laws has led to numerous invalid generalizations and to controversy. Again and again when observations made on one species or on one higher taxon were extended by generalization to all other taxa, it was found that the generalization did not hold.

Quinarianism is one of the many ill-founded endeavors to make biology "scientific" by making it quantitative or by making it obey definite "laws." It seemed most unscientific that taxa should be of unequal size; hence attempts were made to squeeze all organisms into groups of a fixed number, usually five. Having such a numerical classification made systematics, for the adherents of quinarianism, as scientific as Galileo and Newton had made physics.

Another example is Schwann's endeavors to explain the origin of cells as analogous to the origin of crystals. "The principal result of my investigation is that a uniform developmental principle controls the individual elementary units of all organisms, analogous to the finding that crystals are formed by the same laws in spite of the diversity of their forms" (1839: iv).

When Edgar Anderson discovered in the 1930s and 1940s how widespread clandestine hybridization is in plants, he, Epling, Stebbins, and other botanists were convinced that the reason why zoologists had not discovered an equally high frequency of hybridization in animals was that they had not properly looked for it. Considerable efforts were made in the ensuing twenty-five years to discover it in animals, but the results were, on the whole, negative. Higher animals simply are different genetic systems from plants. The same is true for the occurrence of polyploidy. About 50 percent of the flowering plants are polyploids, and some leading cytogeneticists of the 1920s, 30s, and 40s were convinced that "therefore" polyploidy would be equally common in animals. Actually, except for some groups that have abandoned sexual reproduction, polyploidy is very rare in the animal kingdom, and differences in chromosome number that were at one time interpreted as being due to polyploidy have in most cases a different interpretation (White, 1973; 1978).

847

To give another example, certain groups of animals, like freshwater fishes, have very poor dispersal facilities. Normally they can spread from one area of distribution to another one only if the respective land masses are in physical contact with each other. Certain biogeographers who specialize in the distribution of freshwater fishes or of other poor colonizers have therefore jumped to the conclusion that the distribution of all animal groups reflects the former history of land masses. In reality, most species in many groups of organisms can disperse across rather wide water gaps; it would lead to wrong conclusions to use the distribution pattern of such easily dispersing groups as the basis for a reconstruction of former land connections.

Uniqueness is characteristic for most complex systems. Physical scientists, of course, also encounter uniqueness. During the recent space explorations the findings on each of the explored planets showed that their atmosphere and surface geology was unique. This does not mean that generalizations are not possible in sciences rich in unique phenomena; it simply means that they have to be formulated in probabilistic terms and it also means that such probabilistic generalizations (whatever one may want to call them) are of far greater importance in the daily life of a scientist than so-called universal laws.

The Heuristic Value of Erroneous Theories
It is curious how often erroneous theories have had a beneficial effect for particular branches of science. Such theories often

stimulate a search for supporting facts that would be ignored by opposing theories but that would be most useful as support for a different explanatory scheme. For instance, Geoffroyism—the belief in a direct impact of the environment—stimulated a very active search for correlations between the environment and adaptive features. Eventually this extensive literature became powerful support for the theory of natural selection. Selectionists did not need to discover such correlations because they had already been collected and carefully tabulated by the neo-Lamarckians.

848

Vitalism had perhaps a more beneficial effect on the development of physiology in the eighteenth and early nineteenth centuries than did mechanism. The vitalist Bichat had a greater impact on the subsequent researches of Magendie and Bernard than mechanists like LaMettrie and Holbach. Natural theology produced a splendid collection of observations upon all sorts of adaptations in nature. This material could be incorporated in toto into evolutionary biology as soon as "design" was replaced by natural selection. The behavioral observations of natural theologians like Reimarus and Kirby formed the most valuable basis of the subsequent study of animal behavior.

This suggests that if a research tradition is able to gather a massive amount of facts that seemingly support it, something is amiss in the opposing theories. It also confirms the old observation that facts, as long as they are correct, never lose their value, while hypotheses and theories can stimulate research regardless of whether they are valid or not.

THE SCIENCES AND THE EXTERNAL MILIEU

The emphasis in the preceding sections has been on developments within science. The sociologists of science, however, have correctly stressed the fact that science is not going on in a vacuum but reflects inevitably the general *Zeitgeist* of the period. I have attempted to describe this in some detail in Chapter 3, and have also referred to the controversy between externalists and internalists in Chapter 1. In this chapter I shall attempt to discuss a few rather specific problems.

In such a major research area as biology there is nearly always a dominant field that sets the fashion for a given period, like systematics in the time of Linnaeus, physiology from the 1830s to the 1850s, evolution and phylogeny in the 1860s and 70s, genetics in the first two decades of the twentieth century—but eventually sharing the limelight with *Entwicklungsmechanik*—molecular biology from the 1950s on, and perhaps now ecology. These periods are not strictly consecutive, since the waxing and waning of each of these periods is sufficiently stretched out so that usually two or more coexist simultaneously. Superimposed on these developments within biological disciplines are some broader influences that affect all branches of biology simultaneously. Romanticism and *Naturphilosophie* in Germany from the 1780s to the 1830s was one such influence; natural theology in England in the first half of the nineteenth century is another example; physicalistic reductionism in much of the twentieth century is still another one. Each of these overarching ideologies had a beneficial effect on some parts of biology but an inhibiting if not distinctly deleterious effect on others. The only broad generalization I am willing to make at this time is that each of these broad influences benefited either functional biology or evolutionary biology but affected the alternate branch of biology adversely. Only within recent decades has it been realized how drastically different the conceptual backgrounds of these two major divisions of biology are.

849

Each dominant research tradition favors certain explanatory models, and there is a great danger that such explanations will be applied to situations where they are entirely inappropriate. When "movements and forces" was fashionable as an explanation in the physical sciences, physiological processes in organism were explained by the "movements of molecules." When Newton unified terrestrial and cosmic mechanics through the introduction of the force of gravity, a "force of life" seemed at once to explain all phenomena of living organisms. Since classes of inanimate objects usually consist of identical members, that is, are of a homogeneous composition, the geneticist Johannsen, who had had much of his training in physiological and physical chemistry, attempted to "purify" genetically heterogeneous populations by isolating "pure lines." Numerous similar instances could be recorded, where the adoption of fashionable concepts or techniques failed to produce meaningful results.

Window Dressing or Genuine Influence?

As Merton (1973) has correctly pointed out, scientists crave recognition. They are afraid that their work will be unnoticed if it is stated in unfashionable language or imagery. Whenever possible they will cite some distinguished scientist or philosopher in support of their conclusions. This has been naively interpreted by some historians of science as proof of a direct influence of the cited authorities on the thought of the respective scientist; and yet a close study of the writings of these scientists often reveals that they arrived at their conclusions quite independently and had attached the "label of approval" of the distinguished author only during the write-up of their work.

When Locke was at the height of his fame, scientists claimed that they had arrived at their results through Lockean empiricism, even though they had not changed their approach in the slightest after reading Locke. In recent years, when Karl Popper was the great fashion among scientists, opposing schools of taxonomists outclaimed each other as being the true followers of Popper. At the time when Darwin did his work, induction (or what was believed to be induction) had great prestige and accordingly Darwin proclaimed solemnly that he was following "the true Baconian method" when in fact his hypothetico-deductive approach was anything but inductionism. After Dobzhansky had called attention to the mathematical population analyses of Fisher, Haldane, and Wright, most self-respecting evolutionists listed the papers of these three workers in their bibliographies but later admitted that they either had not read them at all in the original or only to a very small extent. In the Renaissance, when the method of logical (dichotomous) division was at the height of its influence, all botanists proudly proclaimed that they were following the Aristotelian method of classification, even though Aristotle himself had explicitly pointed out that dichotomy was not the way to construct biological classifications, and even though it has now become quite evident that these botanists themselves arrived at their grouping by inspection rather than by logical division. I am pointing this out as a warning to those who are trying to reconstruct influences. The mere fact that an author cites a certain work or says he is following the principles of this or that philosopher or scientist does not necessarily mean that this cited work really had a decisive influence on his thinking.

As long as mathematics, physics, and chemistry enjoyed high prestige throughout the eighteenth and nineteenth centuries, it was a sound strategy for a scientist to use the appropriate labels in order to give visibility to his work. Window dressing for these reasons is, therefore, encountered particularly often when an author includes mathematics in a paper even though it does not add anything to his previously obtained results. A well-known taxonomist asked his mathematician wife to add an appendix to all of his taxonomic papers which contained elaborate statistics of his measurements even though he virtually never made use of these statistics in his taxonomic conclusions.

851

Conversely, several instances are known in the history of biology, perhaps many more if one would really look for them, where a law, principle, or generalization was ignored when first stated because it was phrased in words rather than in the form of a mathematical equation. When finally stated mathematically, it was hailed and generally accepted. For instance, Castle (1903) showed that the genotypic composition of a population remains constant when selection ceases, but this was ignored until Hardy and Weinberg provided a mathematical formulation. In 1939 I showed that the bird fauna of an island in the Pacific is the result of a balance between colonization and extinction, and analyzed this principle in detail with respect to New Caledonia. Again, this was ignored for twenty-five years, until MacArthur and Wilson phrased it in mathematical terms in their theory of island biogeography (1967).

Traditionally there has been a tendency among scholars to refer to the approach of their opponents in terms that are meant to be unflattering if not derogatory: "My work is dynamic, yours is static; mine is analytical, yours purely descriptive; my explanation is strictly mechanistic (that is, it explains everything in chemico-physical terms), your explanation is holistic (it leaves much that is unexplained)." Curiously, his opponent might make some of the same claims. To be able to explain everything in terms of Newtonian movements and forces was the ideal during much of the nineteenth century to such an extent that the "right words" were used even when there was no trace of an actual Newtonian analysis. Nägeli's mechanical/physiological theory of inheritance (1884) is an apt illustration. All that Nägeli was actually able to present was pure speculation (and all that was new in it turned out to be wrong!) and yet Nägeli boasted of having proposed a strictly "mechanistic" theory. "Mechanistic" meant scientific. This must be mentioned here because a historian, looking at such statements from the outside, might fail to realize that such claims were purely pyschological weapons. Lowering your opponent raises your own status. This is why the physicist Rutherford referred to biology as "postage stamp collecting."

The Sources of Influence

852   It is a well-known phenomenon that authors may ignore available facts and ideas for years if not decades, until they can be used at a favorable moment in the construction of a new theory or concept. For example, the exponential growth of populations in the absence of opposing factors should have been well known to Darwin since his student days in Cambridge. At that time he carefully read Paley who wrote brilliantly about "superfecundity." Many other authors consulted by Darwin in the ensuing ten years stressed the same principle, and yet it was not until September 28, 1838, that Darwin connected this with the widely held concept of struggle for existence and made it the basis for his theory of natural selection.

Nothing is truer than Pasteur's famous statement that only "the prepared mind" makes discoveries. But little thought has been given up to now to the process by which the mind is prepared. A mere knowledge of certain facts is not enough, nor the presence of certain concepts and ideas, if they are hidden away in a different brain compartment. An astonishingly high proportion of major new concepts and theories is based on components that had long before been available but which no one had been able to tie together properly. This must be remembered in any search for external influences in the development of scientific ideas. Ideas derived from sociology, economics, anthropology, and ethics may be stored away in memory centers that have no open channels to evolutionary biology, ecology, or ethology.

When Darwin developed the concept of character divergence, for instance, he claimed to have been decisively influenced by Milne-Edwards' concept of functional division of labor, whereby the division of labor among organs in the body, was parallel to the division of labor in manufacturing and social economy. Schweber (1977) wondered why Darwin did not credit this thought to the British authors who, from Adam Smith on, never failed to emphasize the importance of a division of labor and of competition, along with related subjects. Without question Darwin was well acquainted with these ideas, having read most of the relevant literature. However, he stored this knowledge in a compartment of his brain which he never tapped when speculating on evolutionary divergence. It was not until Milne-Edwards made the connection that Darwin saw what should have been obvious to him for the preceding fifteen years.

The whole problem of the relation of different bodies of knowledge to one another is in need of more study. The majority (virtually all!) of paleontologists between 1859 and Simpson explained macroevolutionary phenomena with the help of either saltations or orthogenetic tendencies (or both). When the genetic evidence made it virtually inescapable that neither of these two explanatory schemes could be valid, Simpson demonstrated that the macroevolutionary phenomena were perfectly consistent with the Darwinian theory. He did not "prove" this, because how are you to prove this? Yet, from this point on, it was up to the opponents of Darwinism to refute Simpson's thesis.

The same was true in my own case. I showed that the phenomena of speciation, of the biology of species, of adaptive geographic variation, of the formation of higher taxa, and so on, were entirely consistent with a Darwinian explanation, and I showed, furthermore, that the discordant explanations advanced by the Mendelians were not consistent with the evidence of the systematist. It is neither possible to derive the phenomena at the level of population and species from those at the level of the gene nor vice versa. But it can be shown that they are consistent. The reductionists postulated that the phenomena at one level are the inexorable consequences of the phenomena at a different level, but this is *not* the case.

The refutation of a portion of a theory or research tradition does not necessarily affect its principal thesis. For instance, Darwin accepted a certain amount of soft inheritance in his theory, yet, the subsequent demonstration that there is no soft inheritance did not weaken the theory of natural selection. If anything, it strengthened it. In any mixed or complex theory the different components may show a considerable amount of independence of one another.

853

THE ROLE OF TECHNICAL ADVANCES IN SCIENTIFIC RESEARCH

Historians of science have consistently stressed the importance of improvements in technology in the history of science. This is abundantly documented in all scientific disciplines. I have called attention repeatedly to the importance of the invention of the microscope; and the entire history of cytology is a history of the effect of technical improvements. It begins with the invention of the microscope by Jansson and Jansson (about 1590), the achromatic lens (1823), oil immersion lens (1878), apochromatic objective (1886), phase-contrast microscope (1934), and electron microscope (1938). Correspondingly, there were steady improvements in the making of microscopic preparations (microtomes, squash techniques), in fixing, and in staining. The true understanding of the cytoplasm had to wait for the invention of the electron microscope. The importance of new instrumentation and new techniques is even greater in molecular biology. Here virtually every new insight was achieved by a new technique.

Another important aspect of the technology of biology is the use of different kinds of experimental organisms. Mendel showed that the pea was ideal for the study of the units of inheritance, while both he and de Vries discovered that other plants (*Hieracium* and *Oenothera*) had complex properties that gave misleading results. Most species of animals and plants are suitable for selection experiments but not the self-fertilizing and nearly homozygous garden bean chosen by Johannsen for his selection experiments. C. W. Woodworth, a student of W. E. Castle, called to his teacher's attention "that the rapid-breeding pomace fly *Drosophila* had distinct advantages in breeding experiments over the laboratory mammals which Castle was then using" (Davenport, 1941). From Castle's lab the use of *Drosophila* spread to the laboratories of Lutz and Morgan.

The history of genetics supplies many instances of a fortunate or unfortunate choice of experimental animals and plants. Beadle and Tatum's *Neurospora* and the subsequent use of bacteria (*Escherichia coli*) and of various viruses were fortunate choices. Nägeli's unfortunate choice of the apomictic *Hieracium,* on the other hand, led him to question Mendel's laws, De Vries's *Oenothera* led him to postulate speciation by single mutations, and Johannsen's garden bean led him to deny the importance of natural selection. It is particularly important for workers who enter biology from the physical sciences, where most generalizations have universal validity, to realize that all organisms have unique properties and that one cannot automatically transfer the findings made in one organism to others; they should also realize that certain species are much more favorable for certain investigations than other species. Organisms are complex biological systems, each of which has unique characteristics. When one looks at the literature of behavioral biology prior to the 1940s, one discovers that a major part of it is devoted to studies on "The rat," comments and discussions implying that what was found for the rat would be equally true for any animal (Beach, 1950). In later studies on primates, the experimental animal was often simply referred to as "The monkey," as if all monkeys had the same characteristics. In physiological and embryological studies of birds, reference was usually made either to "The chicken" or "The pigeon," as if this covered the total diversity of the 9,000 species of birds.

855

Much of the progress of cytology made in the 1870s and 1880s was due to the discovery of ever new types of cytological material, each of which had certain advantages over the others. Van Beneden's discovery of *Ascaris* and Boveri's discovery of certain sea urchins permitted conclusions that could not be arrived at with any other material then known

Not only the choice of the right technique and of an appropriate biological material is crucial for scientific advance in biology, but more broadly the choice of method in general. No one questions that the appropriate technique for the study of functional phenomena is the experiment; but it must be emphasized that the causal explanation of historical (evolutionary) phenomena ordinarily must rely on inferences from observations. The blindness of many experimentalists to the findings of the naturalists was caused to a large extent by their stubborn insistence that only the experiment can give answers to scientific questions. That a historical development like speciation, or, more broadly, all of evolution, can be interpreted only by inference based on an appropriate seriation of observations was not only not recognized by the experimentalists in the first third of the twentieth century but indeed vehemently rejected. Even today some authors consider the experiment the exclusive method of science. Quite recently one of them remarked that "an experimental approach to the origin of species is curiously absent from the works of Darwin." It was this attitude which made Bateson so blind to the findings of the taxonomists that, as late as 1922, he completely ignored their conclusive results. When part of a historical narrative consists of functional processes, they can be tested by experiment. But the historical sequence as such, usually involving populations or other complex systems, can be reconstructed only on the basis of inferences derived from observations. It was his obsession with the exclusive value of the experiment which misled de Vries into believing that mutations explained the origin of species. It would be interesting to go through the history of science and see how often a misplaced insistence on experiment has caused research to move into unsuitable directions.

PROGRESS IN SCIENCE

Considering the seemingly ever increasing number of unsolved scientific problems, the doubt is occasionally expressed whether science is really making any progress. It is not easy to define scientific progress. It is characterized by an improved understanding of previously puzzling phenomena, by the removal of contradictions, by the opening of black boxes, by the possibility of making better probabilistic predictions, and by the establishment of causal connections between previously unconnected phenomena. In spite of the difficulty of definition, a working scientist is rarely in doubt as to whether or not a new discovery, theory, or concept contributes to the progress of science. However, it has been pointed out (Kuhn, 1962; Feyerabend, 1975) that science is often quite irrational in its methods and that progress in one direction may be attended by losses in other directions.

As I have emphasized in Chapter 1, it seems to me that progress in the biological sciences is characterized not so much by individual discoveries, no matter how important, or by the proposal of new theories, but rather by the gradual but decisive development of new concepts and the abandonment of those that had previously been dominant. In most cases the development of major new concepts has not been due to individual discoveries but rather to a novel integration of previously established facts. Darwin's theory of descent with modification through natural selection is a good illustration of this principle. Other such major concepts, largely based on previously known facts, are that of biological species, the genetic program, genetic recombination, accelerated speciation in peripherally isolated populations, the cell theory, and even the gene.

The most drastic changes in the conceptual framework of a science are usually designated as scientific revolutions, a subject about which much has been written in the last twenty years. As I have pointed out previously (Mayr, 1972b), the Darwinian revolution, like nearly all major biological controversies, was protracted over far more years than is usually credited to a scientific revolution. I cannot think of a single case in biology where there was a drastic replacement of paradigms between two periods of "normal science." On the other hand, there is no doubt that certain discoveries, new concepts, reformulations of old concepts, and new techniques have had profound impact on the subsequent development of biology. I need only to mention the publication of the *Origin,* the rediscovery of Mendel, the evolutionary synthesis, and the discovery of the structure of DNA. Although the concept of scientific revolutions does not adequately reflect what happens during the growth of a science, it would be equally unrealistic to believe in a steady and even rate of progress in science.

857

Perhaps the skepticism about the overall progress of science is greater in the physical sciences, where the spectacular achievements from the seventeenth to the end of the nineteenth centuries were followed by developments like the principle of complementarity, the uncertainty principle, the puzzles of the elementary particles, relativity, and others that have introduced previously unsuspected uncertainties. It would seem that progress in the biological sciences has been far steadier and more clearly visible: the displacement of a static by an evolving world, common descent, natural selection, particulate inheritance, the role of hormones and enzymes, the population concept, the biological species, the control of development by a genetic blue print, various components of ethological theory, and important contributions to the understanding of the function of all organs of the body—to mention only a small fraction of the untold number of concrete advances in our understanding. Even though great mysteries remain, particularly with respect to the functioning of complex systems, no one can doubt the enormous progress biological science has made and continues to make.

Yet, when it comes to developing a truly comprehensive science of science, it can be done only by comparing the generalizations derived from the physical sciences with those of the biological and social sciences, and by attempting to integrate all three branches. I rather suspect that the raw material for such comparisons and for an integration is already available and that it is only necessary that someone adopts this as the objective of his research.

Notes
References
Glossary
Index

Notes

1. How To Write a History of Biology

1. This is well illustrated by the two most extensive and best known histories of biology, those of Radl (1907–08) and Nordenskiöld (1928). Both of them were written not only more than fifty years ago but by authors each of whom had a very definite viewpoint. Radl, for instance was something of a romantic and very much impressed by the importance of Paracelsus, Schelling, and Hegel. Nordenskiöld's history of biology, even though it is still authoritative, has many weaknesses. In particular, he wrote at a period when Darwinism had reached its nadir of prestige, at least on the European continent. Owing to his anti-Darwinian bias, his presentation of the history of evolutionary biology is virtually useless. By contrast his treatments of the history of anatomy, embryology, and physiology are still quite sound. Both authors stress the lexicographic and biographical aspects, and proceed essentially in chronological order. Nordenskiöld recognized that the history of biology should be "a segment of general cultural history" and stated that it had been his endeavor to concentrate "in his treatment on theoretical principles and generalizations . . . apparent in biological research." In his actual presentation of the material he did not follow his own eminently sound guidelines nearly as much as is demanded by modern scientific historiography.

2. The Place of Biology in the Sciences

1. The relations between science and religion are, however, far more complex than can be presented here. See also Hooykaas (1972), Maritain (1942), Simpson (1949), Merton (1938), Dillenberger (1960), and Moore (1979).

2. The history of the hypothetico-deductive method has not yet been written. Its beginnings go back to Descartes's deductive method. Locke, Hume, and other philosophers have used it casually. Whewell was a forceful advocate of this method. In addition to Darwin, various evolutionists and other scientists employed it in the nineteenth century. Weismann (1892: 303) described it excellently; in fact it was his principal method throughout his career. Hempel, Popper, and other philosophers have sponsored it in recent decades. See also Hull (1973) and Ruse (1975b; 1979a).

3. One well-known author, for instance, claimed that in biology "the experimental has replaced the encyclopedic method" in the nineteenth (sic!) century. Actually, Gesner (1551) and Aldrovandi (1599)

were the last encyclopedists, and to denigrate authors like Ray, Tournefort, Buffon, Adanson, and Linnaeus as encyclopedists displays shocking ignorance. When not calling the work of naturalists encyclopedic, experimentalists have designated their work as "purely descriptive."

4. Not infrequently in the history of science and philosophy a committed physicalist changed his mind at some period in his life and began to recognize the autonomy and methodological independence of biology. Cassirer has well presented this in his *The Problem of Knowledge* (1950: 118–216). No one illustrates this better than Kant. In his *Metaphysical Elements of Natural Science* (1786) Kant still declared that the scientific and the mathematical views of nature were one, and that in any particular theory there was only as much real science as there was mathematics. By the time he published *The Critique of Judgment* only a few years later (1790), he confessed that the problems of biology, in particular those of diversity and adaptation, could not be solved with the methodology and limited conceptual framework of the physical sciences. That Kant himself failed rather miserably to develop a philosophy of biology is irrelevant in this connection. What is important is his clear recognition that the Newtonian model of movements and forces, and the reductionist approach to organisms, simply gave no answers to the most important questions of biology. Even Leibniz, who among the pre-Kantian philosophers seemed to have the most sympathy with biology, demanded that all phenomena of nature be explained according to the same laws, that is, mathematically and mechanically.

5. Glossaries such as the ones in my *Animal Species and Evolution* (1963) or *Principles of Systematic Zoology* (1969) are veritable lists of the concepts important in the respective areas of biology.

6. Not every conceptualization stressing the individual leads to "population thinking." The nominalists, for instance, stressed that only individuals exist, bracketed together into classes by names. Yet, when one reads their writings, one discovers that they deal with individuals in a strictly philosophical (logical) sense, not stressing their unique biological individuality. Lamarck, for instance, when he denied the existence of species and said "there are only individuals," treated these individuals as identical since, at any one locality, they were exposed to identical "circumstances." The Scottish philosophers of the eighteenth century who also spoke a good deal of individuals stressed the political aspects, that is, the individual versus the society or the state, again ignoring the characteristic of biological uniqueness.

The role of nominalism in the rise of population thinking is still unclear. Nominalism had a long history. The sophist Antisthenes already promoted nominalist ideas. The grammarian Martianus Capella (ca. A.D. 400) defined the genus entirely nominalistically as "the gathering of many species under one name." His writings were popular throughout the Middle Ages and undoubtedly influenced the establishment of nominalism. Of more immediate concern is that the writings of Locke are a curious mixture of essentialism and nominalism. Empiricism certainly was favorable to population thinking, but, so far as I know, no one has yet traced the connection.

7. Interestingly, two thousand years later, K. E. von Baer applied the very same reasoning. Because the development of the chick embryo from the fertilization of the egg on was so clearly goal-directed, therefore the universe as a whole and all historical processes in it, including organic evolution, had to have some unknown factor which "determines" development in a goal-directed manner.

8. The possibility exists that a historical narrative could be dissected into numerous discrete steps, each of them affected by a constellation of conditions and relevant laws, providing high predictability for each of these steps. Yet, such an analysis would not be practical owing to the large number of steps and of factors influencing the course of events. On prediction see also Suppes (1971), Williams (1973a), Good (1973), Fergusson (1976), and *Amer. Nat.* 111: 386–394.

9. As Mandelbaum (1971: 380) has pointed out, the concept of emergence has a long history. It was held by materialists like Marx and Engels, by positivists such as Comte, by the nondualists Alexander and Sellars, and by the dualists Lovejoy and Broad. G. H. Lewes in his *Problems of Life and Mind* (especially Vol. II, 1874–75) developed the first full-fledged philosophy based on the principle of emergence. Emergence was the most characteristic aspect of the philosophy of Claude Bernard. For Engels the emergence of new and irreducible properties in nature was taken to be a manifestation of the fundamental dialectical self-transformation of matter, and an acceptance of such properties did not therefore collide with materialism (Mandelbaum, 1971: 28).

Modern physics has more and more accepted emergentism. From a mere knowledge of atomic structure, said Weisskopf (1977: 406), "theorists [unacquainted with the real world] would never predict the existence of liquids." (Niels Bohr told me the same in 1953.) Weisskopf continues, "A knowledge of the basic laws is insufficient for a real understanding how the 'parts' are related to the 'whole' at each step of the set of hierarchies" (p. 410).

10. Nothing brings out more dramatically the fundamental change in the philosophy of biology in recent decades than a comparison of these recent books with the older literature, represented by H. Driesch, H. Bergson, A. N. Whitehead, A. Arber, J. S. Haldane, R. S. Lillie, J. von Uexküll, W. E. Agar, but also the writings of L. J. Henderson, J. H. Woodger, L. L. Whyte, G. Sommerhoff, J. D. Bernal, and E. S. Russell.

3. The Changing Intellectual Milieu of Biology

1. Almost every paragraph in this overview is a condensation of a far more complex story than is apparent from my account. The full story will be told in Chapters 4–19 and, as far as functional biology is concerned, in a later volume. The task I have set myself is formidable. I hope I have succeeded reasonably well in capturing the changing mood of biology as a whole, as well as the ups and downs of various biolog-

863

ical disciplines. Where I know that I have failed is in giving an adequate picture of the total intellectual-cultural-social background of each period, and how, and to what extent, it affected the development of thought in various areas of biology. I lack the expertise that a social or intellectual historian would bring to this task. Perhaps no single person can provide a balanced picture since a scientist and a social historian, when attempting to analyze the causation of scientific developments, inevitably differ widely in assigning weights to internal factors and to those of the general contemporary milieu.

2. For other treatments of the changing milieu of biology see the various histories of biology quoted above. Also Smith (1976), Hall (1969), Leclercq (1959), Taton (1958; 1964), Pledge (1939), Allen (1975), Coleman (1971), Dawes (1952).

3. For the rise of Ionian philosophy on the coast of Asia Minor, see Sarton (1952, Vol. I). For further details on the effect of the Greek philosophers on the history of biology, see Chapters 7–19. For further reading see Adkins (1970), Kirk and Raven (1971), Freeman (1946), Guthrie (1965), de Santillana (1961).

4. The terms "chance" and "necessity" have repeatedly changed their meaning in the history of philosophy. I am using them in their customary contemporary meaning.

5. I am using the terms "essentialism" and "essentialistic" consistent with Popper's definition (see Chapter 2).

6. There is an immense literature on Aristotle, much of it, of course, dealing with problems of little interest to a biologist. For an overall evaluation I found best Düring (1966). Special topics are treated by Balme, 1970 (Aristotle's zoology), 1980 (Aristotle's nonessentialism), and Gotthelf, 1976. There are excellent introductory discussions in some of the English editions of Aristotle, particularly in the Loeb Library. See also the books of Randall (1960) and Grene (1963).

Part I. The Diversity of Life

1. To present an account of the intellectual history of systematics is particularly difficult because there is no real precedent. How little most historians of science understand the conceptual development of systematics is evident from almost any of the better known histories of science or of specific periods. The history of systematics is dealt with in most histories of botany (Sachs, Jessen, Green, Mägdefrau) but with major emphasis on biography and classificatory detail. There are some excellent special treatments, as of the herbalists, Ray, Linnaeus, Buffon, Cuvier, Darwin, and so on, but no one so far has presented "a view from the distance." To extract the essential history of the changing concepts from the overwhelming mass of detail has not been easy. I hope my treatment will not be found too unbalanced.

2. For literature on folk taxonomy see also Conklin (1962); Berlin, Breedlove, and Raven (1974); Majnep and Bulmer (1977), and Gould (1979).

3. Unfortunately there is no good history of the major museums or plant

and animal collections in the world. There are histories of individual museums, for example, Anon. (Murray) 1904–1912, or Lingner (1970). See also Stresemann (1975: 367–373) for a history of the bird collections in American museums. For plant collections see Lanjouw and Stafleu (1956). A history of the Museum d'Histoire Naturelle (Paris) by Limoges is in preparation.

4. Leidy (1853), Küchenmeister (1857), Leuckart (1886), Hoeppli (1959), Foster (1965).

5. The best information—biography centered—on the history of protozoology can be found in Corliss (1978: pt. I; 1979: pt. II, with an excellent bibliography). See also Cole (1926) and historical information in the textbooks of Manwell (1961) and Kudo (1966). For further details consult histories of botany, such as Sachs (1882), Ballauff (1954), Jessen (1864), Mägdefrau (1973), Green (1909).

6. For a history of paleontology consult: Geikie (1897), Zittel (1899), Gillispie (1951), Haber (1959), Hölder (1960), Edwards (1967), Schneer (1969), Rudwick (1972). See also Guyénot (1941: 337–358); Ley (1968: 191–221).

7. For Louis Agassiz the natural system of organisms was "a system devised by the Supreme Intelligence and manifested in these objects" (1857: 4). He summarized his views in the *Essay on Classification:* "All organized beings exhibit in themselves all those categories of structure and of existence upon which a natural system may be founded, in such a manner that, in tracing it, the human mind is only translating into human language the Divine thoughts expressed in nature in living realities" (p. 136).

8. The literature on the history of classifications is endless. As far back as 1763 Adanson gave a history of plant classifications. Hyman (1940) gives a short history of each of the phyla and classes of Invertebrates treated by her. More recently a number of specialized treatments of individual taxa have been published. As examples may be mentioned Winsor (1976b) on the Radiata and Corliss (1978–79) on that of the ciliates. Indispensable as a guide are the great zoological handbooks, such as the German one by Kükenthal (1923) and the French one by Grassé (1948ff., *Traité de zoologie*).

9. A large number of authors have participated in bringing about this intellectual advance. This includes Beckner (1959), Cain (1958; 1959a; 1959b), Hennig (1950; 1966), Huxley (1940; 1958), Hull (1970), Inger (1958), Mayr (1942; 1963; 1969), Michener (1957), Remane (1952), Rensch (1934; 1947), and Simpson (1945; 1961).

To a surprising extent the current controversies in systematics are the same as those of the post-Darwinian period; indeed, many of them go back to Linnaeus, to the Renaissance botanists, or even to Aristotle. Chronology under these circumstances is less important than a clear and rather detailed analysis of the problems. A problem-oriented treatment would seem, under the circumstances, the most appropriate for a history of ideas. I have chosen this method of presentation, even though I risk being accused of having written a textbook of the principles of systematics rather than a chronological history.

4. Macrotaxonomy, the Science of Classifying

1. For the best available analyses of Aristotle's method of classification, see Lloyd (1961) and Peck (1965). Also Balme (1962) and Stresemann (1975: 3–6).

2. The higher taxa recognized by Aristotle were the following: Among the blooded animals (vertebrates) he recognized six rather unequal groups: (1) hairy viviparous quadrupeds (mammals); (2) birds; (3) Cetacea; (4) fish; (5) serpents; (6) oviparous blooded quadrupeds (most reptiles and amphibians).

   Among the bloodless animals he recognizes four groups: (1) Malakia (cephalopods and soft-shelled crustaceans); (2) Crustacea; (3) Testacea (most mollusks, echinoderms, ascidians, and other hard-shelled marine animals); and (4) insects. This classification (particularly of the blooded animals) is nowhere clearly tabulated by Aristotle. Rather it is extracted from his writings by later compilers, although *Historia animalium* 4.523a–b has a rather straightforward listing of the bloodless animals.

3. For other treatments of Aristotle see Chapters 3 and 7.

4. What the ancients knew about the animals and plants of foreign countries has been reported by Sarton (1927–1948). For a readable account of this, including the "travelers tales" retold by Herodotus and a description of the credulity of the early compilers, see Ley (1968).

5. For an excellent discussion of the methods of classification among the herbalists, and a critique of the treatment by Sachs (1890), see Larson (1971). For further literature on herbals see Meyer (1854–1857), Fischer (1929), Arber (1938), and Thorndike (1945).

6. The works of Brunfels, Bock, and Fuchs greatly stimulated the description of plants. Here I can merely mention authors like Mattioli (1500–1577), Lobel (1538–1616), Cordus (1515–1544), and Lécluse (Clusius) (1526–1609). Not only did they introduce many new species into the literature, but several of them, particularly Lobel, adopted Bock's method of placing together plants that "resemble one another." The origin (or rediscovery) of "upward classification," later emphasized so strongly by Adanson, clearly originated in the writings of Bock, Lobel, and other botanists of the sixteenth century. None of them applied it as consistently as Bauhin. (See Mägdefrau, 1973: 21–31; Zimmermann, 1953: 114–120; Larson, 1971: 6–20.)

7. To gain a better understanding of scholastic logic, as applicable to downward classification, see Maritain (1946), Cain (1958: 1959a), Stearn (1959), and Stafleu (1971). For a history of the development of diagnoses see Hoppe (1978).

8. "Weighting" is the term used for the evaluation of taxonomic characters. It is based on the assumption that a reliance on certain characters would lead to better, more "natural," classifications than a use of other characters. How the weight of a character is to be determined has remained controversial. Traditionally certain characters have been considered to be more helpful (to have more "weight") than others (Mayr, 1969: 217–228).

9. The best analysis of Cesalpino's method is by Bremekamp (1953b). See also Sachs (1890), Stafleu (1969), Larson (1971), Sloan (1972), and Sprague (1950: 7–12).

10. For Ray see Sachs (1890: 69–74), Raven (1950), Sprague (1950: 19–20), Crowther (1960), Larson (1971: 37–44), Sloan (1972), Mägdefrau (1973: 43–46).

11. For Turner see Raven (1947: 38–137), Stresemann (1975). For Rondelet, see Guyénot (1941). For Belon, see Guyénot (1941: 44–47, 139, 210), Stresemann (1975: 16–18, 24–26, 45). For Gesner, see Stresemann (1975: 18–21, 25), Larson (1971: 15–18), Ley (1929), Fischer (1967). For Aldrovandi, see Stresemann (1975: 19, 21, 27, 41), Guyénot (1941: 52, 139).

12. Already Linnaeus' father is said to have adopted the name Linnaeus, and nearly all of Carl's writings were published under this name. Late in life (1762), when ennobled, he took the name Carl von Linné.

13. The following titles permit a good entry into the concepts and practices of Linnaeus. Among these Hagberg, Larson, and Stafleu are most comprehensive. My own treatment is not biographical. Daudin (1926); Hagberg (1939); Svenson (1945); Stearn (1962); Cain (1958); Hofsten (1958); Larson (1971); Stafleu (1971); Stearn (1971). Certain aspects of the thinking of Linnaeus are treated elsewhere in this volume, for instance his evaluation of taxonomic characters, his concept of a natural system, his species concept, and binomial nomenclature.

14. For an excellent analysis of Linnaeus's philosophy of nature see Hofsten (1958). Also Stearn (1962), Egerton (1973) and Limoges (in Linné, 1972).

15. For the kind of logic that was particularly important for Linnaeus see: Joseph (1916) or Maritain (1946). Linnaeus's logic is also discussed by Larson (1971), Stafleu (1971), and particularly Cain (1958).

16. For Linnaeus's concept of the genus see particularly Cain (1958), Stearn (1962), Larson (1971: 73–93), and Stafleu (1971: 68–76, 91–103).

17. For an analysis of the history of the concepts of description and diagnosis from Aristotle to Darwin, see Hoppe (1978).

18. Sir Joseph Banks was able to persuade the Admiralty to permit him and his naturalist, the Swede Solander (a pupil of Linnaeus), to accompany Cook on the first of his great voyages to the South Seas (1768–1771). This set an example later followed by other voyagers, particularly Cook's voyage on the *Resolution* (1772–1775) from which the Forsters brought back their famous collections. It was their example which inspired Alexander von Humboldt (as he himself emphasizes) and, in turn, Darwin on the *Beagle*. Equally important was that Banks gave social and scientific prestige to natural history. He served for many years as the President of the Royal Society and actively supported natural-history collections and museums. It was he who persuaded young J. E. Smith to buy the Linnaean collections which, five years later (1789), led to the founding of the Linnean Society of London.

867

868

19. Buffon's early concept of a natural classification is described in the "Premier Discours" (*Oeuvr. Phil.*: 13a–b) as follows: "It seems to me that the only way to make an informative and natural classification is to place together things that are similar and to separate those that differ from each other [upward classification!]. If the individuals show a perfect resemblance to each other, or show differences which can only be perceived with difficulty these individuals will be of the same species [continues to describe how to recognize different species of the same genus and of different genera] . . . This is the methodical order that must be followed in the arrangement of natural productions. Understand, however, that the resemblances and differences are to be taken not simply from one part, but from the whole ensemble [of characters], and will take into account the form, the size, the exterior aspect, the number and position of the different parts and the very substance of the thing itself."

20. For more information on Cuvier see other chapters. See also Coleman (1964), Winsor (1976b: 7–27), and more specifically Cain (1959a: 186–204). Most of Cuvier's new classes of invertebrates had already been known to Linnaeus as orders within his class Vermes. Through his dissections Cuvier discovered how fundamentally different they were from each other.

21. A fuller treatment of the *scala naturae* is given in other chapters. See also Lovejoy (1936) and Ritterbush (1964: 122–141).

22. Abraham Trembley (1700–1784), although best known for his work on regeneration, contributed much to our understanding of freshwater invertebrates by a number of studies made from 1741 to 1746. He was the first to realize the animal character of the zoophytes, based on his work on hydra and the bryozoan *Lophopus*. Furthermore, he made important contributions to our knowledge of the ciliate protozoans and the rotifers (Baker, 1952: 102–129).

23. For an excellent discussion of quinary and other numerological systems in avian classification see Stresemann (1975: 170–191); for quinarianism in invertebrate zoology, see also Winsor (1976b).

24. For de Candolle see Nordenskiöld (1928: 436–438), Cassirer (1950: 135–136), Cain (1959a: 7–12), and Mägdefrau (1973: 64–66).

25. The term "taxon" was first proposed by Meyer-Abich (1926: 126–137) and taken up by the Dutch botanist Lam in the late 1940s. After being officially adopted by the International Botanical Congress in 1950, and made the name of a new journal (*Taxon*) devoted to systematics (particularly that of plants), the term became current in the 1950s. As far as animals are concerned, it is not yet used in Simpson's *Classification of Mammals* (1945, but written earlier) but is used in Mayr, Linsley, and Usinger (1953). Yet, the term "category" is still occasionally used in the current literature when reference is made to taxa.

5. Grouping According to Common Ancestry

1. For literature on Darwin's taxonomy see Simpson (1959; 1961), Cain (1959a: 207–216), Ghiselin (1969: 78–88).

2. In addition to his Cirripedia monographs and the *Origin*, Darwin has

presented his views on classification in various works (Ghiselin, 1969), particularly in the *Descent of Man* (1875, 2nd. ed.: 146–165), and in his orchid book, *The Various Contrivances by Which Orchids Are Fertilized by Insects* (1862).

3. An excellent history of phylogenetic research from Haeckel to the end of the century is Alfred Kühn (1950). It deals particularly with the attempts of F. Müller, Dohrn, Claus, and Hatschek to clarify the phylogeny of the arthropods, to discover the position of the Pycnogonida and of *Limulus,* and to determine the mutual relationship of crustaceans, arachnids, and insects. The numerous speculations of this period concerning the derivation of the chordates and the evolution of the classes of vertebrates are described and criticized very competently. There is also an excellent bibliography.

869

4. Some insight into these controversies can be gained by studying the following papers and the literature cited by these authors: Dougherty (1963), Clark (1964), Ulrich (1972), Siewing (1976), Salvini-Plawen and Splechtna (1979).

5. The claims of numerical phenetics evoked a large number of rebuttals. The following literature (and the included bibliographies) provides a good overview of the more important objections raised by critics. Simpson (1964a), Mayr (1965a; 1969: 203–232; 1981b), Hull (1970), Johnson (1968).

6. Autapomorph characters are derived (apomorph) characters that have evolved in only one of two sister groups, while synapomorph characters are derived characters shared by both sister groups.

7. The following are some recent critiques of cladistics: Ashlock (1974), Mayr (1976: 433–476), Michener (1977), Szalay (1977), Hull (1979).

8. Treatments of evolutionary taxonomy: Simpson (1945; 1975: 3–19), Bock (1973; 1977), Michener (1977), Ashlock (1979), Mayr (1981b).

9. I will only refer to a few early publications. Now new reviews, symposium volumes, and textbooks are published annually. They are far too numerous and constantly changing to be mentioned.

10. A few titles (with the included bibliographies) may give access to the vast literature on the epistemology of classification: Beckner (1959), Hull (1965), Bock (1977), Hull (1978). Furthermore, much of the literature quoted under n. 5 and n. 7 as well as the standard philosophical literature, from Mill and Jevons, to Gasking, Hempel, and Popper, is helpful.

11. The classification of most organisms has now matured to such a point that a major reorganization is rarely necessary. In the phyla of invertebrates a complete reorganization of the Turbellaria became necessary in the second third of this century when it was realized that the branching of the intestines (Triclada, Polyclada) was a purely adaptive feature and the structure of the reproductive system and of the pharynx were chosen instead as the major diagnostic features. It was, however, possible to salvage in the new classification large pieces of the old one. The classification of the prokaryotes (Fox et al., 1980) is another case where the use of a new character (ribosomal RNA) has

resulted in a fundamentally different classification, at least in certain groups.

6. Microtaxonomy, the Science of Species

1. The number of species definitions in the biological literature, reflecting different species concepts, is virtually unlimited. A historial survey of such definitions was given in Mayr (1957). This includes an extensive bibliography, with references to previous summaries, such as those of Bachmann (1905), Besnard (1864), Du Rietz (1930), Plate (1914), Spring (1838), and Uhlmann (1923). For detailed discussions of the species problem see Mayr (1963: 12–30, 334–359, 400–423); Simpson (1961: 147–180); Poulton (1908: 46–94).

   For a further discussion on the relation of species taxon to species category see Mayr (1969: 23–53, 181–197).

   For other recent papers on the species problem see Dobzhansky (1935); Camp and Gilly (1943); Hull (1965); Beaudry (1960); Heslop-Harrison (1963); Vent (1974); Wiley (1978); Slobodchikoff (1976).

2. "Phenon" (pl. phena) is a convenient term for the different forms or phenotypes that may be encountered within a single population—the "varieties" of much of the older literature. This includes the sexes (when there is sexual dimorphism), age stages, seasonal variants, and individual variants (morphs, and so on). For a more detailed treatment of phena see Mayr (1969: 5, 144–162).

3. The term "taxon" was first proposed by Meyer-Abich (1926), adopted by some botanists in the 1940s, incorporated in 1950 in the International Code of Botanical Nomenclature, and used in zoology in 1953 by Mayr, Linsley, and Usinger. See Chapter 5 for a further discussion.

4. For the earlier history of the species concept see Balme (1962), Crombie (1961: 150–151), Zirkle (1959), Mayr (1968), Sloan (1978).

5. In 1798 Cuvier defined species as follows (*Tabl. élém.:* 11): "The totality ('collection') of all organized bodies, born from each other, or from common parents, and of all those which resemble them as much as they resemble each other, is called species." The scholastic heritage in his thinking is revealed by the additional statement: "Organized bodies which . . . do not seem to differ from a species except by accidental causes . . . qualify as varieties of that species."

6. For the species concept of Linnaeus and its changes throughout his life, see Ramsbottom (1938), Cain (1958), Larson (1971: 99–111), and Stafleu (1971: 134). Other authors who have discussed Linnaeus's species concept are Greene (1912), Daudin (1926), Svenson (1945; 1953), Zimmermann (1953), Bremekamp (1953c), and von Hofsten (1958), as listed in Stafleu's bibliography.

7. In addition to Kölreuter, there were numerous other species hybridizers (for example Gärtner and Naudin) who thought they could determine the true nature of species "by experiment," that is, by crossing species. Darwin was much interested in the subject. The aim of these hybridizers was not to discover the laws of inheritance, as has been erroneously assumed by certain historians of genetics.

8. For Buffon's species concept see Lovejoy (1959b: 84–113), Roger (1963: 567–577), Stafleu (1971: 302–310), Farber (1972), Sloan (1978: 531–539), and Larson (1979).

9. For Lamarck's later species concept see *Nouv. Dict., nouv. éd*, X (1817): 441–451, and Szyfman (1977).

10. The difficulties which botanists seem to have with the biological species concept are reflected in the publications of Davis and Heywood (1963), Raven (1977), and Cronquist (1978), and in the papers cited by these authors.

11. For a more complete listing of Darwin's views on species, as given in his notebooks, see Kottler (1978: 278–280) and Sulloway (1979). His notebooks on transmutation of species (NBT) are referred to as B, C, D, and E (de Beer, 1960–61; de Beer et al., 1967). *871*

12. He finally adopted the biological species concept in 1975.

13. The gradual clarification of the biological species concept is well reflected by Mayr's series of papers on this subject: Mayr, 1940; 1942; 1946; 1948; 1955; 1957; 1963; 1969a; 1969b; several of them are reprinted in Mayr, 1976.

14. For discussion of isolating mechanisms see Mayr (1963), chapters 5 and 6; Blair (1961); and various recent review papers.

15. Darwin says about species borders: "The range of the inhabitants of any country by no means exclusively depends on insensibly changing physical conditions, but in large part on the presence of other species, on which it depends, or by which it is destroyed, or with which it comes into competition . . . the range of any one species, depending as it does on the range of others, will tend to be sharply defined" (*Origin:* 175).

16. Pioneers in plant taxonomy were Kerner (1869), Wettstein (1898), Cajander (1921). See also Langlet (1971).

17. The finest work in biosystematics was perhaps that of the Carnegie Institution group of Clausen, Keck, and Hiesey (Clausen, 1951). See also Stebbins (1950) and Grant (1971).

18. The discrimination between sibling species is of great importance in applied biology. The malaria mosquito *Anopheles maculipennis* complex is a classical illustration; others are the discovery of a sibling species (*M. subarvalis*) confused until the 1960s with the meadow mouse *Microtus arvalis*, notorious for its enormous population outbreaks from western Europe to the Pacific coast of Siberia; or the much studied polychaete *Capitella capitata*, an excellent indicator of marine pollution, who was shown only recently by enzyme analysis to consist of six sibling species (Grassle, 1976).

19. The same point has been made, quite emphatically, by Wagner in a symposium on biosystematics (1970).

20. Part of this literature is reviewed in Richardson (1968). It is quite deficient with respect to the German literature (Pallas, Esper, Gloger, early writings of Moritz Wagner, and the straight taxonomic literature). Stresemann (1975) fills many of these gaps and has a superior grasp of the problems involved. See also Mayr (1963: 334–339).

21. For more details on the history of the subspecies concept see Mayr (1963: 346–351) and Stresemann (1975). For the rich but now obsolete terminology of infraspecific variants, see Plate (1914: 124–143), and Rensch (1929).

22. Wiley (1978) has attempted to combine the criteria of the biological species definition with those of the evolutionary species, but his definition refers both to taxa and to categorial ranking and retains some of the weaknesses of the evolutionary species concept.

Part II. Evolution

1. There are numerous histories of evolutionary science in English, French, German, Russian, and other languages. Some of the earlier ones are too uncritical to be still useful, like Osborn (1894) or Perrier (1896). The best introduction to the coming of evolutionary thought is Toulmin and Goodfield (1965). The best sourcebook, although concentrating on phylogeny, is Zimmermann (1953). Informative and very readable is Greene (1959). See also Ostoya (1951).

7. Origins without Evolution

1. See Reiser (1958: 38–47), Hull (1965), Mayr (1959a), Popper (1945).

2. For literature on Aristotle, see Chapters 3 and 4. See also Balme (1970).

3. Nominalism was originated by Roscellinus and Abelard, promoted by Duns Scotus and Roger Bacon, and brought to its height by Occam. Has clearly influenced inductive philosophy in England (for example, that of Locke).

4. Descartes was somewhat ambivalent and said on another occasion that one could also assume that God could have been satisfied merely to create the laws of nature, and that this would have resulted in the development of the world as we now see it. For Descartes's attitude on evolution see Zimmermann (1953: 161–166).

5. The concept of heterogony—the conversion of one species into another one—was promoted particularly by Theophrastus (*Enquiry into Plants*, bk. 2). Virgil described in the *Georgics* how wheat and barley mutated into wild oats. The history of the concept is well discussed by Zirkle (1959). See also Chapter 6.

6. The volumes of Gillispie (1951), Schneer (1969), and Rudwick (1972) provide an excellent introduction to the early history of geology and its literature. See also Albritton (1980) and Blei (1981).

7. Werner was apparently the first to establish sedimentary strata. Being basically a mineralogist, he tried (unsuccessfully) to determine the age of strata by mineralogical criteria. Unfortunately, he made several major errors, such as including granite and other igneous rocks among sedimentary rocks; nevertheless his role as a pioneer in geology is now more and more recognized.

8. Blumenbach (1790: 18) made fun of those who wanted to derive the entire living fauna from the passengers on Noah's ark by saying, "I find it quite incomprehensible how the sloth could have made the

pilgrimage from Mt. Ararat to South America since it requires an hour to crawl 6 feet."

9. For the history of paleontology consult Geikie (1897), von Zittel (1899), Zimmermann (1953: 187–195), Hölder (1960), Scherz (1971), Rudwick (1972), and Blei (1981).

10. See Cuvier (1812), Coleman (1964), Rudwick (1972). See also Hofsten (1936).

11. For van Leeuwenhoek see Chapter 4.

12. For a broad general survey see Bowler (1974b). For general works on the Enlightenment, consult Cassirer (1951), Hazard (1954), Gay (1966), and Hampson (1968).

13. See also Nisbet (1979), Bury (1920), Leibniz (ca. 1712).

14. For Bonnet see Lovejoy (1936: 283–286), Zimmermann (1953: 210–219), Savioz (1948).

15. For Maupertuis see Brunet (1929), Glass (1959), Roger (1963), Jacob (1970).

16. For Buffon see Wilkie (1959), Lovejoy (1959b), Roger (1962; 1963), Piveteau (1964), Hanks (1966), Bowler (1973), and Farber (1975).

17. See also Chapters 4 and 6 for other aspects of Buffon's thought.

18. His statement that "les animaux ne sont à beaucoup d'égards que des productions de la terre" ("animals are in most respects nothing but the productions of the earth") largely eliminated the historical moment from zoogeography. Yet this view was adopted by Alexander von Humboldt, by other contemporary biogeographers, and by Herder. This is why Darwin was so surprised when finding that the tropical and temperate zone faunas of South America were more similar to each other than to those of the corresponding zones of South America and Africa.

19. See Diderot (1749; 1769 [1966]), Vartanian (1953), Mayer (1959), and Crocker (1959).

20. Useful as a first introduction to the Reimarus literature, particularly his deism is Reimarus (1973). Also Stresemann (1975).

21. See Berlin (1974), Toulmin and Goodfield (1965: 135–140), Zimmermann (1953: 238–245), Lovejoy (1959c: 207–221), Cassirer (1950: 217–225).

22. See King-Hele (1963), Ritterbush (1964: 159–175), Harrison (1972: 247–264).

23. Zimmermann (1953: 195–210) points out correctly how familiar Linnaeus was with many of the facts later used by Darwin as evidence for evolution. This includes the occurrence of a striking mutation (like *Peloria*) remaining constant in the ensuing generations and yet crossable with the parent "species" (*Linaria*). He was aware of the rich fossil fauna of some parts of his country but was unable to give it an evolutionary interpretation. Although fully aware of the distinctness of the faunas and floras of the different continents, Linnaeus was prevented from drawing the kind of conclusions Darwin later did by the literal interpretation of the naming of all creatures by Adam and

of the emptying of Noah's ark at a single locality. However, he admitted that local floras in North America, Africa, and other parts of the world contain so many similar species "that their orginal origin from a single species appears possible" (see particularly his *Fundamentum Fructificationis*, 1762; also *Amoenitates Academicae* 6: 279–304). Linnaeus's treatment of the systematics of man and the anthropoid apes lent itself readily to an evolutionary interpretation. Finally, scattered through his writings are indications that he had come around to accept a great age for the world (Nathorst, 1908). See also Hofsten (1958).

8. Evolution before Darwin

1. There is a rich Lamarck literature, competently listed and discussed by Burkhardt (1977), by far the best and most comprehensive modern treatment of Lamarck. For some additional aspects of Lamarck's thought see also Hodge (1971a; the best statement of the philosophical basis of Lamarck's thought, although underestimating the impact of his zoological researches on Lamarck) and Mayr (1972a). Three older treatments are still of interest: Packard (1901), Landrieu (1909), and Daudin (1926). See also Schiller (1971), Kühner (1913), Tschulok (1937). For complete bibliographies of Lamarck's publications see Landrieu (1909) and Burkhardt (1977). There are English translations of two of his works, *The Zoological Philosophy* (1914) to which all page numbers refer, and the *Hydrogeology* (1964). For a bibliography of the German literature on Lamarck, see Zirnstein (1979); also Kohlbrugge, 1914.

2. Schiller (1971: 87–103) denies that Lamarck's "série des corps organisés" has anything to do with the *scala naturae*, but I am not persuaded by his arguments. I readily admit that the discovery (by comparative anatomy) of discrete anatomical types, each with a unique organization and constellation of organs, had converted the continuous chain of Bonnet into a more or less discontinuous one, as Daubenton pointed out emphatically (Burkhardt, 1977: 124). Yet Lamarck never fails to stress the continuity (no matter how tenuous) of the types ("masses") of organisms.

3. See also Chapter 4. Most earlier histories of evolution were accounts of Darwin's putative forerunners. Examples of this literature are Osborn (1894), Perrier (1896), and Glass et al. (1959). See also Kohlbrugge (1915). Of particular interest (although woefully incomplete so far as non-British authors are concerned) is Darwin's own "Historical Sketch," inserted in the *Origin of Species* from the third edition on (1861).

4. I. Geoffroy Saint-Hilaire (1847), Russell (1916), Schuster (1930), Cahn (1962), Bourdier (1969), Rudwick (1972).

5. Cuvier (1812; 1813), Russell (1916), Daudin (1926), Coleman (1964), Bourdier (1969), Rudwick (1972: 101–163). My own account heavily leans on that of Coleman.

6. This did not faze Lamarck in the slightest. Since he had estimated the age of the continents in his *Hydrogéologie* to be many millions of

years, he considered 3,000 years to be a negligible span of time, quite irrelevant for the problem of the constancy of species, particularly in view of the constancy of conditions in the Nile Valley.

7. I have perhaps exaggerated the difference between the theology of the average physicist and that of the average naturalist. Actually there were quite a few physicists who are also best classified as "natural theologians." Indeed, some of them (including Newton) went so far as to postulate that the Lord intervened in the running of the world whenever he was not satisfied with the working of his laws. Still, a fundamental difference between the attitude of physicists and naturalists cannot be doubted.

8. The literature on natural theology is vast. See for instance Gillispie (1951), Hooykaas (1959), McPherson (1972).

9. In addition to Hooykaas (1959), Rudwick (1972), Wilson (1972), and Bowler (1976), see Cannon (1961), Simpson (1970), and Rudwick (1971). These publications contain references to numerous other publications on the subject.

10. The invalidity of Huxley's thesis was first pointed out by Hooykaas (1959 and earlier) and confirmed by Cannon (1960) and by virtually all subsequent writers on the subject (including Rudwick, Bartholomew, and Ospovat).

11. For Lyell's creationism see K. Lyell (1881), Lovejoy (1959a: 356–414), Cannon (1961), Wilson (1970), Mayr (1972: 981–989), Bartholomew (1973), Ruse (1975), Ospovat (1977).

12. Millhauser (1959), Lovejoy (1959a), Hodge (1972), Gillispie (1951), Ruse (1979a: 98–116).

13. For a discussion of this question see Hofstadter (1959), Medawar (1969: 45–67), Freeman (1974), Nichols (1974).

14. Summarized in Weismann (1893; 1895), and Freeman (1974).

15. For von Baer's evolutionism see also Holmes (1947), Oppenheimer (1959: 292–322), and Raikov (1968).

16. Potonié (1890), Kohlbrugge (1915), Uhlmann (1923), Schindewolf (1941), Temkin (1959: 323–355). In a letter to Cuvier on March 9, 1801, Kielmeyer expresses his belief that many recent species are modified descendants of fossil species (Kohlbrugge, 1912: 291–295). In 1850, in the second edition of *Die Pflanze* Schleiden accepted evolution, so did H. P. D. Reichenbach in 1854.

9. Charles Darwin

1. With the possible exception of Freud, there is no other scientist about whom so much has been written and continues to be written as about Darwin. Each year scores of new articles and books appear. Two recent bibliographies are those of Loewenberg (1965) and Greene (1975). One of the best recent books is that of Ruse (1979a), which is a reliable guide into the Darwin literature. It is the first comprehensive treatment of Darwin that has made use of the immense handwritten material (in Cambridge University Library) that has become available since 1959. Equally sound is the somewhat more spe-

cialized *Darwin on Man* by Gruber (1974). The older treatments of Irvine (1955) and Eiseley (1958) are dated, and some others, particularly that of Himmelfarb (1959; see review by Anthony West in *The New Yorker* of Oct. 10, 1959, pp. 176–189; also *Scientific American,* 1959) are too biased to be useful.

876

Darwin's previously unpublished manuscript of his "species book" was finally published in 1975 under the title *Natural Selection,* owing to the dedicated efforts of R. C. Stauffer. An unexpurgated version of Darwin's autobiography was recently published (Darwin, 1958). The first edition of the *Origin* is now available in an inexpensive facsimile edition (Darwin, 1964). Darwin's *Notebooks on Transmutation* were published by de Beer (1960–67), who also wrote a readable Darwin biography (1963). Transcriptions of other notebooks were published by Barrett (1974) and by Herbert (1980). The best treatments of Darwin's method are by Ghiselin (1969), Gruber (1974), Ruse (1979 and earlier) and Hodge (forthcoming.) The intellectual milieu in 1859 is best revealed by the study of the critical reviews of the *Origin* (Hull, 1973).

2. Darwin's own account (1839; 1845, 2nd ed.) of his observations on the voyage of the *Beagle* is one of the most delightful and exciting travel books in existence. Every observation he makes stimulates him to ask challenging questions. Even today the *Journal of Researches* is fascinating reading. See also Darwin (1958: 71–82), Moorhead (1969), Keynes (1978).

3. For the development of Darwin as a naturalist and the high level of his professionalism on his return from the *Beagle,* see Herbert (1974–77).

4. Francis Darwin (1887), usually cited as *L.L.D.*; Darwin and Seward (1903), usually cited *M.L.D.* An edition of the complete Darwin correspondence, to comprise some ten volumes, is now in preparation (Smith and Burkhardt, eds.).

5. His autobiography is our most direct source concerning Darwin's religious beliefs (pp. 85–95), but it is as unreliable on this subject as in many other respects. Darwin wrote it for the benefit of his family, including his deeply religious wife, Emma. It is not surprising, therefore, that Darwin was very guarded in how he reported his loss of faith. The statement that he might still have been a theist when he wrote the *Origin* seems rather unbelievable. The best modern account of Darwin's religious views is that of Gillespie (1979), although I would ascribe Darwin's choice of words to caution and consideration for friends and relatives to a greater extent than does Gillespie. See also Gruber (1974) and Ospovat (1980).

6. For those of Lyell's views on species, speciation, and evolution, so far as they have a bearing on the development of Darwin's thought, see Lyell (1881, I: 467–469), Cannon (1961), Coleman (1962), Rudwick (1970), Mayr (1972b), Bartholomew (1973), Wilson (1972), Bowler (1976), Ospovat (1977), Hodge (1982?).

British historians have tended to credit Lyell with the introduction of ecological causal factors into the discussion of evolution. This may be correct as far as the British literature is concerned. But ecological questions played a large role not only in the writings of Buffon and Linnaeus, but also in those of many other continental authors such as Pallas, Blumenbach, the Forsters, E. A. W. Zimmermann, Willdenow, v. Humboldt, de Candolle, von Buch and others. There is a considerable need for a comparative study of their writings, although an excellent beginning has been made by Hofsten (1916) as far as the discontinuity problem is concerned.

7. For an analysis of the changes in Darwin's thought see Gruber (1974), Herbert (1974), Kohn (1981), Sulloway (forthcoming), and Hodge (forthcoming).

8. My account on Darwin and the Galapagos owes a great deal to F. Sulloway's original researches embodied in an unpublished manuscript (1970) "Geographic isolation in Darwin's thinking: a developmental study of the growth of an idea"; Sulloway (1979); and Sulloway (ms.), "Darwin's Genius," devoted to the development of Darwin's ideas.

9. For an analysis of Darwin's views on speciation, see Mayr (1959b), Herbert (1974), Kottler (1978), and particularly Sulloway (1979), where the changes in Darwin's thinking are well documented. For a further discussion of Darwin's difficulties with the role of isolation in speciation see Vorzimmer (1970: 159–185).

10. Darwin prepared three early drafts of his views. The dating of the presumed first of these is uncertain, perhaps 1839 (Vorzimmer, 1975) perhaps later. In June 1842 he wrote down a sketch of 35 handwritten pages and in the summer of 1844 an essay of 189 pages (copied on 231 pages). Sketch and essay were published by F. Darwin in 1909 and republished again by de Beer (Darwin and Wallace, 1958).

11. Extensive lists of Wallace's publications are provided by Marchant (1916) and McKinney (1972). The latter provides an excellent presentation of Wallace's thought and a historical reconstruction of how he arrived at his ideas.

12. Wallace spent every penny to build up his personal library and read all available books in the public library. He reminds one of Harry Truman, who claimed that he had read all the 3,000 volumes of the local libraries by the time he was 13 years old.

10. Darwin's Evidence for Evolution and Common Descent

1. The history of geology and paleontology in relation to evolution has been splendidly covered in several recent books, for instance Rudwick (1972) and Bowler (1976). Since my own treatment of this exciting period is regrettably short, I refer to these excellent volumes, but see also Gillispie (1951), Schneer (1969) and literature cited in Chapter 7, n. 9. See Hull (1973) for the conversion of Pictet to evolutionism.

2. In the *Origin* Darwin explained numerous phenomena as consistent with the theory of common descent but as inexplicable or capricious under the assumption of special creation; see for example pp. 4, 55,

59, 95, 129, 133, 139, 152, 155, 159, 167, 188, 194, 203, 352, 355, 372, 390, 394, 396, 398, 406, 411–458, 469, 473, 478, 482, 488. See also Gillespie (1979).

3. For a more detailed presentation of Darwin's theory of classification see Chapter 4.

4. The fine study of M. Winsor (1976) sheds much light on the crucial period from 1800 to the 1850s, but it should be extended to the work of far more French, German, and British taxonomists, particularly entomologists and vertebrate zoologists.

5. There is a superb history of early biogeography by Hofsten (1916). It deals particularly with the pre-Darwinian period and with the explanation of distributional discontinuity.

6. For the development of de Candolle's thought see Asa Gray (1863) and Hofsten (1916).

7. For discussions of Darwin's biogeography, see particularly Darlington (1959), Egerton (1968), and Ghiselin (1969).

8. The post-Darwinian history of biogeography builds on Darwin to such an extent by further developing his pioneering ideas that it is best treated as a straight continuation of Darwinian biogeography. Indeed, the concepts of the leading current biogeographers are remarkably similar to those of Darwin (except, of course, he did not yet know of plate tectonics). There are a number of modern texts of biogeography. The standard work, but pre-plate tectonics, Darlington (1957). The following recent titles open access to a great deal of biogeographic literature: Schmidt (1955), Elton (1958), de Lattin (1967), Udvardy (1969), Carlquist (1974), Pielou (1979), Müller (1980). See also n. 11 below.

9. In the 1930s and 40s Mayr showed in a series of analyses the invalidity of previous claims that the distribution of birds in the Indo-Australian archipelago had taken place via land bridges. Rather, the distribution patterns are entirely consistent with the hypothesis of transoceanic dispersal. This was documented in papers published 1931 to 1965 in which a number of methodological innovations were introduced. References to these publications, and a reprinting of most of them, can be found in Mayr (1976). Among these the two papers most important for the methodology of landbridge analysis are Mayr (1941; 1944a). A good entry into Simpson's massive contributions to biogeography is provided by two of his summarizing publications (Simpson, 1965; 1980).

10. In the period from the 1890s to the 1940s biogeography was completely dominated by land-bridge building. This is reflected in the major books of the period: Arldt (1907), Scharff (1907; 1912), Gadow (1913); and among botanists, Skottsberg (1956). Even such a perceptive author as Rensch (1936) accepts without question the necessity of explaining all distribution patterns in the Malay archipelago in terms of former land bridges.

11. For various views on plant dispersal, in addition to Carlquist, see

Guppy (1906), Ridley (1930), Gulick (1932), Skottsberg (1956), Gressitt (1963), Baker and Stebbins (1965).

12. Takeuchi, Uyeda, and Kanamori (1970), Hallam (1973), Wilson
(1976), Uyeda (1978).

13. Croizat (1958; 1964), Croizat, Nelson, Rosen (1974), and Nelson and
Rosen (1980).

14. Cody and Diamond (1975), MacArthur and Wilson (1967).

15. There is probably no other branch of biology for which we have as
superb a history as for morphology: Russell (1916) is unsurpassed to
this day, remarkable for its fresh analysis of the primary sources. See
also Cole (1944), Simpson (1959), and Gould (1970). Also *Amer. Zool.*
15:294–481. Anatomy is well covered in Nordenskiöld (1928); see
also Singer (1926).

16. There is an enormous, largely rather uncritical Goethe literature. See
Russell (1916), Troll (1926), Bräuning-Oktavio (1956), Wells (1967),
Nisbet (1972), Eyde (1975).

17. Among twentieth-century morphologists who promoted idealistic
morphology were Naef (1919; 1931), Meyer (1926), Lubosch (1931),
Kälin (1941), and Zangerl (1948); among botanists, Troll.

18. For Cuvier on morphology, see Russell (1916), Daudin (1926), and
Coleman (1964).

19. The best treatment of Geoffroy still is that by Russell (1916: 52–78).
A somewhat different interpretation, using unpublished sources, is
given by Bourdier (1969). See also Cahn (1962), Buffetaut (1979).

20. This debate between Geoffroy and Cuvier is often reported as a debate between an evolutionist and an antievolutionist, which is quite
misleading (Lubosch, 1918: 357; Piveteau, 1950; Coleman, 1964;
Bourdier, 1969). The problem of evolution was involved only peripherally. It was simply a debate about different interpretations of
animal structure and about the effects of function on structure. It
was, if we want to say so, primarily a debate on the methodology of
comparative anatomy, and secondarily on the philosophy of nature.

21. See Owen (1848), Russell (1916: 102–112), Ruse (1979a: 116–127,
133–137, 227–228), MacLeod (1965).

22. For discussions of the type of evidence useful to substantiate or refute the inference of homology, see Remane (1952), Simpson (1961),
and Bock (1979).

23. Nothing illuminates the difficulties of assigning morphology to a particular area of biology better than the lack of communication among
different of its schools. There were the phylogenetic morphologists,
like Gegenbaur, Haeckel, and Huxley (up to Remane and Romer);
there was a strong remnant of idealistic morphology (Naef, Kälin, Lubosch), and there were the evolutionary morphologists (Böker, D.
Davis, Bock, von Wahlert). There were also some national schools,
none as influential as that of Severtsov and Schmalhausen in the USSR
(see Adams, 1980b). This is not even mentioning the various schools
dealing with proximate causations such as His and the *Entwicklungs-*

mechaniker from Roux to Harrison and Spemann, or "idealistic" functionalists, like D'Arcy Thompson (Davis, 1955).

24. The seminal literature in this field are two publications by D. Davis (1960; 1964); other contributions are by Bock (1959), Bock and von Wahlert (1965), and Frazetta (1975).

25. Much of the history is excellently discussed in Russell (1916). The most up-to-date treatment of the evolutionary aspects of embryology is by Gould (1977). See also von Baer (1828), de Beer (1940; 1951), Lovejoy (1959a), Oppenheimer (1959), Coleman (1973), Ospovat (1976). For Meckel, Serrès, Agassiz, Haeckel, and so on, see Gould (1977).

26. The fascinating history of the discovery of the chorda dorsalis in various groups of chordates is well presented by Arzt (1955).

27. Most older and some recent text books of evolution have concentrated on listing proofs for evolution, for instance Plate (1925) and Moody (1962). More recent text books concentrate more on problems of causation and on evolutionary mechanisms.

11. The Causation of Evolution: Natural Selection

1. Darwin was an omnivorous reader, not only of geology and of natural-history subjects, but also of philosophy and general intellectual literature. It is evident that this reading must have contributed to the shaping of his ideas; therefore, quite rightly, several recent authors have analyzed Darwin's reading: Gruber (1974), Schweber (1977), Ruse (1979), Manier (1978), and Kohn (1981). Darwin's journals indicate how much he read each week, in addition to all his other activities, and it is evident that some books he merely glanced through, searching for specific information. Schweber thinks that reading Brewster's review of Comte and the writings of Quetelet were specially important; Ruse considers the reading of Herschel and Whewell decisive. All this gave Darwin a "prepared mind," but the excitement reflected by his notebook entry of September 28, 1838, as well as the definite statement in his autobiography (p. 120) indicate to me that the reading of the Malthus sentence had a decisive impact on Darwin's thinking.

2. There is a vast literature on the nature of the Darwinian revolution and on the origin and growth of Darwin's theory of natural selection. The most important recent publications are: Limoges (1970), Herbert (1971; 1974; 1977), Greene (1971), Mayr (1972b; 1977a), Gruber and Barrett (1974), Schweber (1977), Ruse (1979a), Ospovat (1979), Kohn (1981), Hodge (forthcoming). Even though there is far-reaching agreement among these authors, there are still some unresolved issues, most of them matters of emphasis. Among these are: (1) Did Darwin's interpretation of the struggle for existence change slowly from 1836 to September 1838 from the benign concept of natural theology to the Malthusian fierce struggle, or did this happen all at once on September 28, 1838? (2) Did Darwin's stress on the genetic uniqueness of the individual come exclusively from the experiences of the breeders and taxonomists, or also from Scottish philosophers?

(3) More generally, to what extent did Darwin actually get new ideas from philosophers and to what extent did he simply cloth his ideas in the language of contemporary fashionable philosophy (particularly Herschel and Whewell) in order to make them more respectable?

3. From his own observations in South America and his reading Darwin had learned that natural populations of species may undergo enormous fluctuations. A drought in the Pampas had killed millions of cattle, for instance. Such observations brought home to Darwin the insight of the superior fitness of the few survivors, and reenforced Lyell's emphasis of competition in the struggle for existence. For numerous references concerning variable population size, competition, and survival in Darwin's notebooks and letters, see Egerton (1968); also much material in Stauffer (1975). The best statement of the historical changes in the balance of nature concept is by Egerton (1973). See also Stauffer (1960) on Linnaeus's influence on Darwin.

881

4. Several authors, more or less independently, recognized the importance of Darwin's shift from competition among species to competition among individuals: Vorzimmer (1969), Herbert (1971), Ghiselin (1969; 1971–72). The lesson that Darwin's struggle for existence is a *reproductive* competition among individuals was, alas, not understood at all by most of those who subsequently operated most freely with the struggle for existence, particularly racists and adherents of so-called social Darwinism (or more correctly, social Spencerism). This is true, for instance, for most authors cited by Greene (1977), but occasionally even Darwin expressed himself ambiguously (Greene 1981).

5. For instance, Young (1969; 1971), but see Freeman (1974).

6. Even more inconceivably, as shown below, there were still two others advancing a theory of natural selection prior (1818, 1831) to Darwin and Wallace. They all, however, had more in common than meets the eye! All were British, at least three of them had read Malthus, and all were exposed to the same zeitgeist. Yet, it remains a puzzle why so few others were affected by this zeitgeist.

7. Unfortunately, all of Wallace's own accounts as to how he came to think of natural selection were written forty to fifty years later, when metaphors like survival of the fittest had been imprinted in everybody's mind. We can therefore not say with complete certainty how the various pieces of Wallace's explanatory model came together in 1858. See also Smith (1972).

8. See for instance Darlington (1959). Darwin himself reported on his real or putative forerunners in a historical sketch included in the *Origin* beginning with the third edition. Most of the cases listed by Zirkle (1941) as earlier proposals of natural selection, however, fall under the category of "elimination" (see above). For excerpts of the writings of Wells, Matthew, Blyth and Chambers see McKinney (1971).

9. Lyell, accused by "German critics" (apparently he referred to Bronn) of "the direct and miraculous intervention of the First Cause, as often

882

as a new species is introduced, and hence [that] I have overthrown my own [uniformatarian] doctrine of populations, carried on by a regular system of secondary causes," insisted that he agreed with Herschel that such a creation of new species would have to "be carried on through the intervention of intermediate causes." But when he describes under what circumstances this would take place, he describes the crassest miraculous special creation, as he himself later called it, a "perpetual intervention hypothesis." It would give us an idea "so far as regards the attributes of the Presiding Mind." It would permit us to imagine "the circumstances that must be contemplated and foreknown, before it can be decided what powers and qualities a new species must have in order to enable it to endure for a given time, and to play its part in due relation to all other beings destined, to coexist with it, before it dies out. It might be necessary, perhaps, to be able to know the number by which each species would be represented in a given region 10,000 years hence, etc. etc." (Lyell, 1881: 467–469). Everything about each new species is "preordained," "predetermined," "appointed." No secondary or intermediate causes are conceivable that can do any of that. Lyell's beliefs, his own disclaimers notwithstanding, amounted to rank special creationism (see also Cannon, 1961; Mayr, 1972b).

10. This ground has been well covered by numerous books, referred to in any history of evolutionism.

11. There is still considerable uncertainty as to the nature of scientific revolutions. They all have one thing in common, which is that after the revolution things are not quite what they were before. In most other respects each revolution is unique. Greene (1971) and Mayr (1972b) have shown how little the Darwinian revolution conforms to Kuhn's description of scientific revolutions. For instance, there was no crisis situation in the 1850s, there was no replacement of one paradigm by another, and some two hundred years passed by between the beginning of the Darwinian revolution (Buffon, 1749) and its end (evolutionary synthesis in 1947), even though there was a climax at the midpoint (1859).

12. The writings of Poulton (1896), Kellogg (1907), Delage and Goldsmith (1912), Plate (1913), and Tschulok (1922) provide good introductions into the controversies over natural selection.

13. For more detailed discussions of the changes wrought in metaphysical concepts by the theory of natural selection, see also Passmore (1959), Bowler (1977b), Ruse (1979a), and Gillespie (1979).

14. For a full discussion of the problem of teleology see Chapter 2.

15. For summaries on the genetics of mimicry, with reference to much additional literature, see Ford (1964: 201–246) and Turner (1971; 1977). Much of the genetic analysis was made by P. M. Sheppard and C. A. Clarke. For further literature on mimicry see Wickler (1968) and Blaisdell (1976).

16. The term "neo-Lamarckism" was proposed by A. S. Packard in 1884. See also Cope (1887; 1896), Kellogg (1907), Pfeifer (1965), Boesiger (1980), Bowler (1977a), Dexter (1979), and Burkhardt (1980).

17. Lack of space made me resist the temptation to provide a far more extensive treatment of the fascinating history of the theory (or better, "theories") of orthogenesis. For further reading see Nägeli (1865; 1884), von Baer (1876), Eimer (1888), Kellogg (1907), Ospovat (1978), and Bowler (1979).

18. I believe C. O. Whitman (1919, I: 9–11) was the first to recognize clearly the importance of developmental and variational constraints. When published, this work did not fit at all into the atomistic ("bean-bag") thinking of the Mendelians and was ignored (see also Mayr, 1963; Bowler, 1979: 68).

19. For a further discussion of evolutionary progress see: Ayala (1974), Huxley (1942), Goudge (1961), Mandelbaum (1971), Simpson (1974), and Thoday (1975).

20. For neo-Darwinian interpretations of evolutionary trends, see Franz (1920; 1935), Huxley (1942), Simpson (1949, Rensch (1960), and Stebbins (1969; 1974).

12. The Diversity and Synthesis of Evolutionary Thought

1. An excellent account of the reception of Darwinism in Germany is that of Montgomery (1974); see also Mullen (1964), Querner (1975), Stresemann (1975), Gregory (1977), Mayr and Provine (1980).

2. Poulton (1896), Vorzimmer (1970), Hodge (1974).

3. Haller (1963), Pfeifer (1974), Cravens (1978). For the impact of Darwinism on the intellectual life of America see Wilson (1967) and Russett (1976).

4. Mention must be made here, for the sake of completeness, of a movement in popular thinking and in social theory usually referred to as social Darwinism. Actually it was Herbert Spencer who was the intellectual father of this concept and it would be better to call it social Spencerism. It praised struggle for existence, unmerciful competition, and social bias under the excuse that this is what Darwin had taught. Unfortunately the historiography of this subject is as biased as was the movement itself. Since social Darwinism is not part of the history of ideas in biology, it would be out of place to discuss it in detail. I must refer to the relevant literature instead. For social Darwinism, particularly in America, see Hofstadter (1944), Freeman (1974), Greene (1977; 1981), Bannister (1979).

5. Farley (1974), R. E. Stebbins (1974), Conry (1974), Limoges (1976), Durand (1978), and Boesiger (1980).

6. For Weismann's most important evolutionary writings, see Weismann (1883; 1886; 1892; 1896) and Gaupp (1917). Weismann had already postulated in an earlier publication (1868: 27) that the genetic constitution of an organism must exercise a constraining influence upon its capacity for variation.

7. Darlington's original account (1932) was still very typological. It was almost exclusively concerned with the mechanisms, virtually never asking evolutionary ("why?") questions.

8. See also Allen (1979), where the differences between the two camps

883

are well described. Allen, however, tends to be a little vague on the relations between mutationism and particulate (Mendelian) inheritance. More authors than he believes were opposed to mutationism but accepted Mendelian inheritance for discontinuous characters (Mayr and Provine, 1980).

9. Bateson never wavered in his insistence that all evolutionarily significant changes are due to the origin of major discontinuities. He restated it in 1913, in his Melbourne address of 1914, and in his Toronto address of 1922. For instance, "Modern research lends not the smallest encouragement or sanction to the view that gradual evolution occurs by the transformation of masses of individuals, though that fancy has fixed itself on popular imagination" (1914: 18). No one disagreed with him more vigorously than Poulton (1908a).

10. Allen (1969), Bowler (1978). My own interpretation with its stress on de Vries's typological thinking rather than his revulsion by social Darwinism is closer to Allen than to Bowler. For *Oenothera* genetics see Cleland (1972).

11. Several readers in evolutionary genetics are now available: Peters (1959), Spiess (1962), Jameson (1977).

12. Provine (1971), Cock (1973), Norton (1973), de Marraise (1974).

13. Gloger (1833). In spite of his interest in variation, Gloger was not at all an evolutionist. He considered geographic varieties to be of the same nature as age and sex differences. And since the latter do not lead to the formation of new species, "climatic species cannot originate, only varieties" (p. 106). By reversing climatic conditions, climatic races would "within a few years" return to the ancestral condition (p. 107).

14. For literature on the role of isolation see Mayr (1942; 1955; 1963), Lesch (1975), Stresemann (1975), and Sulloway (1979), which give references to the original literature of Wagner, Romanes, Gulick, Wallace, Seebohm, K. Jordan, D. S. Jordan, J. Grinnell, and other participants in the controversy.

13. Post-Synthesis Developments

1. To be mentioned among recent textbooks are Nei (1975), Grant (1977), Dobzhansky, Ayala, Stebbins, and Valentine (1977), and Futuyma (1979).

2. It is most interesting that from the beginning of his genetics work in Russia (1922) until 1936 Dobzhansky worked on gene action problems (pleiotropy, position effect, causation of sterility, and so on), even after he had started to read Sewall Wright's papers and started to work on *Drosophila pseudoobscura*. His rediscovery of population evolution (of which he was fully aware during his coccinellid researches in Russia) was due to his *Drosophila pseudoobscura* collections, his literature researches during the writing of *Genetics and the Origin of Species,* and his conversations with naturalists while staying in the east in 1936.

3. The method of starch gel electrophoresis was developed by Oliver Smithies (1955). C. L. Markert combined it with histochemical staining techniques to resolve and identify isozymic forms of enzymes. For a review of the field see Markert (1975).

4. For a more detailed discussion of Darwin's thoughts on spontaneous generation and the origin of life see Gruber (1974: 151–156).

5. There is an enormous literature on the origin of life, new volumes or journal articles being published annually. A very good introduction (but not the last word in everything) is Miller and Orgel (1974). See also Eigen and Schuster (1977–78) and the English translation of Oparin's work (1938). To this day none of the various theories of the origin of life has been generally adopted. Some of them are discussed in Fox and Dose (1977) and Dickerson (1978). See also Monod (1974a). For spontaneous generation see Farley (1977).

6. There is a large literature on non-Darwinian evolution. The case for neutrality is made by Kimura and Ohta (1971) and Nei (1975). The case for selectionism by Dobzhansky et al. (1977), Clarke (1976), and in various contributions in Ayala (1976) and Salzano (1975). See also Lewontin (1974).

7. The question to what extent sexual dimorphism is the result of sexual selection and to what extent the result of natural selection has been debated since Darwin's day. Darwin and Wallace strongly disagreed on this, as discussed by Kellogg (1907: 106–128), Mayr (1972c), Blaisdell (1976), and Kottler (1980). See also Turner (1978), West-Eberhard (1979), Hamilton and Barth (1962), and Endler (1978). The amount and the kind of sexual dimorphism is the end product of a compromise among many partially opposing selection pressures. Each case is potentially different from every other case.

8. The classical concept of sexual selection was limited to a narrowly defined mating success. Recent studies show that any behavioral trait may be selected that leads to improved survival or reproductive success. This includes parent–offspring interactions, sibling rivalry, and all sorts of interactions of related and unrelated conspecific individuals, whether selfish or altruistic. For an introduction to this vast subject see Alexander (1979) and Trivers (ms.). For the evolutionary significance of social competition see West-Eberhard (1979).

9. Hamilton (1964a,b), Williams (1975), Wilson (1975), Ghiselin (1974a), Smith (1978), Caplan (1978a), Ruse (1979b), Dawkins (1976), Gregory, Silvers, and Dutch (1978), Barash (1979), Alexander (1979), Barlow and Silverberg (1980), Blum and Blum (1979).

10. For recent literature on sympatric speciation see also Futuyma and Mayer (1980), Paterson (1981), and Mayr (1982).

11. Templeton assumed that his modified interpretation of genetic revolutions would require the introduction of a new term ("genetic transilience"). However, this change of interpretation is far less than between the species of Linnaeus, the gene of Johannsen, the mutation of de Vries, and the current concepts designated by these terms. We would drown in terminology if a new term were introduced every

885

time a scientific concept was modified. Furthermore, Galton coined the term "transilience" for a major saltation in a single individual.

12. The seminal contributions to the new thinking were made by Walter Bock. Of particular importance are his publications on function and role (Bock and von Wahlert, 1965), on preadaptation and multiple pathways (1959), and on microevolutionary sequences (1970).

13. See Cowan (1977, I: 133–208), Haller (1963), Ludmerer (1972), Bajema (1977), Pickens (1968), and Searle (1976).

Part III. Variation and Its Inheritance

1. There is a formidable amount of literature on the history of genetics. Comprehensive histories are: Barthelmess (1952), Brink (1967), Dunn (1951; 1965a), Stubbe (1965), Sturtevant (1965a). Anyone reading these works will notice how much I owe to these authors.

Two textbooks present the discoveries of genetics excellently in a historical framework. They are particularly suitable for a historian who wants to learn more about genetics: Moore (1963), Whitehouse (1965).

The field is also extremely well supplied by sourcebooks, only some of which can be listed here: Krizenecky (1965), Moore (1972), Peters (1959), Stern and Sherwood (1966), Voeller (1968), Spiess (1962), Levine (1971), Taylor (1965).

14. Early Theories and Breeding Experiments

1. For reviews of the ideas of the ancients on inheritance and generation see His (1870), Zirkle (1935; 1936; 1946; 1951), Balss (1936), and Lesky (1950). Also Hall (1969, I: 13–163).

2. For further details on the species hybridizers and plant breeders, see Roberts (1929), Olby (1966), Stubbe (1973). Both Roberts and Olby have excellent accounts of the work of Kölreuter.

3. See Roberts (1929: 129–136), Dunn (1965a: 30); Olby (1966: 62–65, 167–170).

4. He and others, including Darwin, got 3:1 ratios but did not recognize them as such. See Dunn (1965b: 31), Roberts (1929: 276, 283), and Zirkle (1951).

15. The Germ Cells, Vehicles of Heredity

1. The history of the study of cells has been presented repeatedly so well that only an outline will be presented here. Anyone wanting a more detailed treatment should consult the following publications: Baker published a particularly valuable series of studies (including bibliography) in the history of cytology (1948–1955). See also: Coleman (1965), Hughes (1959), Klein (1936), Maulitz (1971), Moore (1963), Pickstone (1973; very detailed), Studnicka (1931), Wilson (1896; the great classic!). These works contain references to the classical literature, such as the writings of Brown (1833), Schleiden (1838), Schwann (1839), Virchow (1858), and other works published from 1800 to 1900 and cited on the following pages.

For a history of improvements in the microscope and microscopic techniques, see Hughes (1959).

2. Baker (1948–1955), Berg (1942), Jacob (1973), Lindeboom (1970), Wilson (1944).

3. It is doubtful that there is any connection between this concept of crystallization and the widespread eighteenth-century concept of organic crystallization. See Coleman (1964: 161–162) and Maulitz (1971).

4. For further details on sex and fertilization see Hughes (1959: 29–76), Barthelmess (1952: 97–121), Olby (1966: 86–100), Coleman (1965), Stubbe (1965: 194–207). So far as I know the literature, the best account of the discovery of sexuality in plants is still that of Sachs (1875): Camerarius, Kölreuter, Sprengel, their forerunners and opponents, as well as the discovery of sexuality and fertilization in the cryptogams, are treated on pp. 359–444.

887

5. Ghiselin (1974a), G. W. Williams (1975), White (1978: 696–758), Maynard Smith (1976). See also Stebbins (1950) and Grant (1971) for plants.

6. Hughes (1959: 62–67). The secondary literature on the history of cytology unfortunately suffers from national bias which must be watched for. Wilson (1896) is perhaps the most impartial reporter. See also Barthelmess (1952) and Klein (1936).

7. The various schemes have been reviewed in detail by Strasburger (1884), Hertwig (1884), de Vries (1889), Weismann (1892), Delage (1895), Wilson (1896), and more recently by Baker (1948–1955), Barthelmess (1952), Coleman (1965), Dunn (1965a: 33–49), and Geison (1969).

8. Among terms for such particles may be mentioned: physiological units (Spencer, 1864), gemmules (Darwin, 1868), plastidules (Ellsberg, 1874; Haeckel, 1876), micellae (Nägeli, 1884), idioblasts (Hertwig, 1884), pangens (de Vries, 1889), biophores (Weismann, 1892), and plastosomes (Wiesner, 1892). See also Hall (1969 II: 304–354).

9. For more details on the chromosomal aspects of inheritance see Coleman (1965: 145–154), Wilson (1896: 182), Voeller (1968), Barthelmess (1952: 103–219), Hughes (1959: 55–73), Moore (1972: 19–47). Voeller (1968), an excellent reader, contains more or less lengthy excerpts from the writings of Kölreuter, Oskar Hertwig, Fol, Strasburger, Weismann, Flemming, Roux, Van Beneden, Montgomery, McClung, Boveri, Sutton, Wilson, Stevens, Mendel, Morgan, Sturtevant, and others.

10. For detailed descriptions of cell division (mitosis) see any modern textbook of biology or cytology. For the historical aspects see Wilson (1896; 1925), Hughes (1959: 55–73) and Coleman (1965: 129–133).

11. Coleman (1965: 145–154) gives a superb review of these developments. See also Barthelmess (1952: 112–113), Voeller (1968: 21–39), and Wilson (1896: 182).

12. I have been unable to allocate precise credit for the first clear recognition that there is a separate genetic substance and that it is confined to the nucleus. Haeckel (1866), Galton (1876), Weismann (1883 and later papers), Nägeli (1884), Hertwig (1884), and Strasburger (1884) have all contributed to this insight.

13. Unfortunately only few of the brilliant contributions of Theodor Boveri (1862–1915) can be mentioned (see also Chapter 16). His life and researches are excellently recounted by Baltzer (1962).

16. The Nature of Inheritance

1. For a somewhat different interpretation see Bowler (1974a).

2. A typical example is the French horticulturist Verlot (1865), whose conclusions are well described by Roberts (1929: 136–143). Similar views were held by Romanes (1896: 267–268).

3. For a particularly perceptive analysis see Geison (1969), though he occasionally confounds soft and blending inheritance. See also Zirkle (1946), Olby (1966), Ghiselin (1969: 181–186; 1975), Vorzimmer (1970). For the early history of the theory of pangenesis see Lesky (1950).

4. See particularly Olby (1966: 70–79), Pearson (1914–1930), Galton (1872; 1876), Cowan (1972).

5. Gaupp (1917; full biography of Weismann), Schleip (1934; analysis of his scientific contribution), Churchill (1968; development of Weismann's thought).

6. After Weismann had published his theory of the continuity of the germ plasm, it was called to his attention that similar theories had been published previously. These Weismann discusses in his *Keimplasma* (1892). On p. 260 he mentions Owen (1849), Galton (1872; 1876), Jäger (1878), and Nussbaum (1880) as alleged forerunners. There is little doubt that Weismann was unaware of these earlier authors and developed his ideas independently. As a matter of fact, nobody paid any attention to these ideas until Weismann's essay on the continuity of the germ plasm was published.

7. Occasionally a genotype abandons sexuality, but asexuality is not very widespread in the animal kingdom. Evidently, sexuality must have a selective advantage, though its evolutionary role remains a controversial issue (see Chapter 13).

8. De Vries (1889), Heimans (1962), Darden (1976).

9. Mendel (1866). See Stern and Sherwood (1966; includes Mendel's letters to Nägeli), Krizenecky (1965; Mendel's classic paper, in German, and a collection of 27 original papers published during the rediscovery era), Iltis (1932), Olby (1966), Gustafson (1969). All page references to Mendel refer to the German original. For Unger's influence on Mendel, see Olby (1971). The continuing results of the Mendel research are published in the *Folia Mendeliana*, Brno.

10. Mendel's understanding of such errors of sampling (which he presumably owed to his Vienna physics professors) was of vital importance. In his smaller crosses Mendel encountered deviations from expected 3:1, ranging from 32:1 to 14:15. Such deviations induced Nägeli and Weldon (and presumably other opponents of Mendelian inheritance) to reject Mendel's interpretation, because they did not understand the nature of statistical fluctuations (errors of sampling).

11. Also, Mendel did not exclude the possibility that in the case of spe-

cies hybrids "der hybride Embryo aus gleichartigen Zellen gebildet wird, in welchen die Differenzen gänzlich and bleibend vermischt sind" ("the hybrid embryo is formed by 'gleichartige' cells in which the differences are mixed completely and permanently"; 1866: 41).

17. The Flowering of Mendelian Genetics

1. I can give only a rather streamlined account of the history of genetics after 1900. Barthelmess (1952), Dunn (1965a), Sturtevant (1965a), Carlson (1966), and much of the specialized literature have shown that progress in the understanding of genes, their mutation, and inheritance was less direct than here presented. Unorthodox theories, particularly some proposed by Bateson, Castle, and Goldschmidt, will not be discussed for lack of space, even though these theories are of considerable interest because they illustrate the particular conceptual framework which induced these workers to adopt different interpretations from the Morgan group. For further details, one must consult the mentioned histories of genetics.

   889

   Once the field of genetics had been established, its literature grew at an exponential rate and still continues to do so. For literature on early Mendelism see also Brink (1967), Dunn (1951), Krizenecky (1965), and Olby (1966). (See also review in *Isis,* 59: 233–224; and essay review by Mayr, 1973).

2. For further details on the rediscovery of Mendel, see *Genetics,* 35 (1950), suppl. to no. 5, pt. 2: 1–47. Also Krizenecky (1965), Olby (1966), Roberts (1929), Stern and Sherwood (1966), Stubbe (1965), Sturtevant (1965b), and Dunn (1966).

3. See Heimans (1962; 1978), Darden (1976), and Zirkle (1968).

4. See Cleland (1972), also Heimans (1978), Olby (1966), Zirkle (1968).

5. See Correns (1924), Stein (1950), Wettstein (1939), and the *Dictionary of Scientific Biography,* III: 421–423.

6. Darlington (1939). For genetic systems in plants, consult Stebbins (1950) and Grant (1964).

7. W. E. Castle, a professor of zoology at Harvard, established genetics in America well before T. H. Morgan. Castle and his students, including Sewall Wright, C. C. Little, and L. C. Dunn, were particularly active in mammalian genetics. See Dunn (1965b), Provine (1971), Carlson (1966: 23–38; a rather biased account), and Castle (1951). Although of roughly the same age (b. 1867) as Morgan (b.1866), Castle for a long time had only two graduate students in the seminal period of 1900–1910, because Mark, the professor of zoology at Harvard, had the privilege of taking care of the Ph.D. candidates. For a review of Castle's school, see Russell (1954).

8. Cuénot (1902; 1928); Limoges (1976).

9. See W. Bateson (1928), B. Bateson (1928), Coleman (1970), and Darden (1977).

10. See Ley (1968), Jacob (1973), Bateson (1894), Stubbe (1965), and Larson (1971: 99–104).

11. See Allen (1978), Muller (1946), and Sturtevant (1959).

890

12. See Allen (1975), Davenport (1941), Castle (1951), and Sturtevant (1959).

13. See Hughes (1959: 77–111), Wilson (1925), and Morgan (1903). For collections of classical papers on chromosomes and genetics, see footnote 20, below.

14. See Baxter (1976), Muller (1943; 1966), Wilson (1896), and Roll-Hansen (1978b).

15. For the difference between the chromosome theory and the theory of the gene, see also Darden (1980) and Darden and Maull (1977).

16. Baltzer (1962) and Gilbert (1978).

17. See McKusick (1960). Others besides Sutton and Boveri, arrived at essentially the same conclusions in the years 1902 to 1904: Correns (1902), de Vries (1903; 1910), and Cannon (1902). See also Wilson (1925), Baltzer (1962), and Moore (1972).

18. See Allen (1966), Zirkle (1946a), and Harris and Edwards (1970).

19. For a modern review of sex chromosomes in animals see White (1973: 573–695).

20. There are a number of valuable collections of papers on chromosomes, for example, Voeller (1968) and Phillips and Burnham (1977).

21. See Muller (1973), Pontecorvo (1968), and Carlson (1972; 1966).

22. Doncaster and Raynor had described the first case of sex-linked inheritance (in the moth *Abraxas*) already in 1906.

23. The processes that take place during the two cell divisions which precede gamete formation are very complex, and a few controversial aspects still remain. I cannot provide a detailed analysis of the cytological processes themselves (see cytology texts) or of the tortuous history of the gradual working out of the correct story. For the early history (up to about 1890) see Churchill (1970). For later developments see Whitehouse (1965) and Grell (1978) (on the very earliness of the occurrence of crossing over during meiosis).

24. Even though Mendel did not encounter any linkage, his seven character pairs did not have to be determined by genes on seven different chromosomes. It is now probable that only four or five chromosomes were involved. Map distances were big enough to allow as much recombination through crossing over or by chromosomal movements to conceal "synteny" (that is, location on same chromosome). See Nowitski and Blixt (1978).

25. See White (1973), Grant (1964), and Stebbins (1971).

26. For more details see Coleman (1970), Roll-Hansen (1978b; a good analysis, particularly of Johannsen's views, although I consider Roll-Hansen's use of the terms "reductionist" and "holist" rather misleading), also Dunn (1965a), Carlson (1966), and Allen (1978).

27. There is a good summary of classical cytogenetics by Swanson (1957). See also Grant (1964) and much recent symposium and periodical literature.

28. The latest summary of research on the fine structure of eukaryote

chromosomes can be found in the Cold Spring Harbor Symposium on chromatin (1978).

18. Theories of the Gene

1. For literature on this controversy see Provine (1971), Froggatt and Nevin (1971), Norton (1973; 1975), Cock (1973), Provine (1979), Mayr and Provine (1980), de Marraise (1974), and Yule (1902).

2. For a discussion of Darwin's concept of blending, see Ghiselin (1969: 161–164, 173–180), Olby (1966: 55–70), Vorzimmer (1970: 28–38, 97–126), Kottler (1978: 288–291). See also Cowan (1972: 391–394) on the history of the concept of reversion.

3. For a detailed analysis see Churchill (1974). See also Roll-Hansen (1978a: 202–206).

4. For further analysis see Churchill (1974: 5–30) and Whitehouse (1965: 23–25, 32–33).

5. Galton repeatedly changed the formulation of this law, and it was further modified by Karl Pearson, who had enthusiastically adopted most of Galton's ideas. For a detailed account of the complicated story of Galton's law, see Provine (1971: 19–35, 179–187), Swinburne (1965), Cowan (1972), Pearson (1914–1930), Froggatt and Nevin (1971). Galton firmly established the concept of "heredity" as meaning those traits of an individual that are due to inheritance from his ancestors and not due to an adaptive response to the environment. Heredity is the "nature" portion of the nature vs. nurture polarity. Owing to his population thinking, Galton was able to develop two major new concepts of statistics, regression and correlation. It is curious that nevertheless Galton failed to understand natural selection.

6. Winkler (1924), Wettstein (1926), and Correns and Wettstein (1937) present excellent surveys of the evidence for cytoplasmic inheritance. This includes genetic properties of cytoplasmic organelles (plastids and so on), but it also discusses phenomena that may be due to regulatory genes or other kinds of genetic determination that have not yet been reduced to simple Mendelian factors. These phenomena are of particular importance in physiological genetics. As late as 1926 Johannsen was of the opinion that the study of the four hundred known mutations of *Drosophila melanogaster* had not touched the central core of its genotype. In retrospect it appears that he was not altogether wrong. Sumner in his pre-1927 writings strongly endorsed this same idea.

   For modern reviews of cytoplasmic inheritance in plants, see Caspari (1948), Dunn (1951: 291–314), Michaelis (1954), Hagemann (1964), Sager (1972), Grant (1975: chap. 12), and Grun (1976).

7. The work of MacDowell (1914) contributed to the acceptance of multifactorial inheritance.

8. For a review of the position effect, see Sturtevant (1965b).

19. The Chemical Basis of Inheritance

1. See Miescher (1897), Fruton (1972: 180–261), Portugal and Cohen (1977), and Olby (1968).

891

2. See for example Hess (1970; contains many important references), Watson (1968, Olby (1974), Sayre (1975), and Judson (1979).

3. See Fruton (1972), Olby (1974, cf. *Science*, 187, 1975: 827–830), and Portugal and Cohen (1977).

4. See Koltzoff (1928), Kol'tsov (1939), Timofeeff-Ressovsky, Zimmer, and Delbrück (1935), Schrödinger (1944), and Olby (1971).

5. Hotchkiss (1965; 1966), Olby (1974), Dubos (1976), Cairns, Stent, and Watson (1966).

6. Brachet had the bright idea to search for stains that would clearly distinguish between the two nucleic acids. Going back to the inexhaustible literature of the German stain technologists, he found in a paper by Unna what he was looking for. By systematically applying these stains to all sorts of tissues that were actively producing proteins, he came to the conclusion that RNA was involved in protein synthesis. For a good summary of the developments in the 1930s and 40s, see Brachet (1957).

7. At Harvard University it was in the year of unrest (1968) that student interest rather suddenly began to veer away from molecular biology, toward ecology, behavior, and evolution. The visible evidence for this change was a petition of two-thirds of the biology concentrators for more faculty appointments in the nonmolecular branches of biology and for a reduction in the nonbiological portion of concentration requirements. Although molecular biology continued to flourish, its monopoly was broken.

8. See Stern (1968: 1–26), McKusick (1975), and McKusick and Ruddle (1977).

References

Abel, Othenio. 1924. *Lehrbuch der Paläozoologie*. 2nd ed. Jena: Gustav Fischer.

Adams, Mark. 1968. The founding of population genetics: contributions of the Chetverikov School, 1924–1934. *J. Hist. Biol.* 1: 23–29.

——— 1970. Towards a synthesis: populations concepts in Russian evolutionary thought, 1925–1935. *J. Hist. Biol.* 3: 107–129.

——— 1980a. Sergei Chetverikov, the Kol'tsov Institute and the evolutionary synthesis, in Mayr and Provine (1980), pp. 242–278.

——— 1980b. Russian morphology and the evolutionary synthesis, in Mayr and Provine (1980), pp. 193–225.

Adanson, M. 1772. Examen de la question, si les espèces changent parmi les plantes. *Mém. Acad. Sci.* Paris 1772: 31–48.

Adkins, A. W. H. 1970. *From the Many to the One*. London: Constable.

Agassiz, Louis. 1857. Essay on classification, in *Contributions to the Natural History of the United States*, vol. 1. Boston: Little, Brown & Co. (Reprinted 1962, ed. Edward Lurie. Cambridge: Harvard University Press.)

Albritton, C. C., Jr. 1980. *Changing Conceptions of the Earth's Antiquity after the Sixteenth Century*. San Francisco: Freeman, Cooper.

Alcock, John. 1979. *Animal Behavior*. Sunderland, Mass.: Sinauer Associates.

Alexander, R. M. 1968. *Animal Mechanics*. Seattle: University of Washington Press.

Alexander, Richard D. 1979. *Darwinism and Human Affairs*. Seattle: University of Washington Press.

Allen, D. E. 1976. *The Naturalist in Britain*. London: Allen Lane.

Allen, Garland E. 1966. Thomas Hunt Morgan and the problem of sex determination, 1903–1910. *Proc. Amer. Phil. Soc.* 110: 48–57.

——— 1968. Thomas Hunt Morgan and the problem of natural selection. *J. Hist. Biol.* 1: 113–139.

——— 1969. Hugo de Vries and the reception of the mutation theory. *J. Hist. Biol.* 2: 55–87.

——— 1975a. The introduction of *Drosophila* into the study of heredity and evolution: 1900–1910. *Isis* 66: 322–333.

——— 1975b. *Life Science in the Twentieth Century*. New York: John Wiley & Sons.

——— 1978. *Thomas Hunt Morgan: The Man and His Science*. Princeton: Princeton University Press.

—————— 1979. Naturalists and experimentalists: the genotype and the phenotype. *Stud. Hist. Biol.* 3: 179–209.

Allen, J. E. 1877. The influence of physical conditions in the genesis of species. *Radical Rev.* 1: 108–140.

Allen, John M. 1963. *The Nature of Biological Diversity.* New York: McGraw-Hill.

Alvarez, Luis. 1980. Asteroid theory of extinctions strengthened. *Science* 210: 514.

Alvarez, L. W., W. Alvarez, F. Asaro, and H. V. Michel. 1980. Extraterrestrial cause for the Cretaceous-Tertiary extinction. *Science* 208: 1095–1108.

Anderson, E. 1949. *Introgressive Hybridization.* New York: John Wiley & Sons.

Anderson, P. W. 1972. More is different. *Science* 177: 393–396.

Arber, Agnes. 1938. *Herbals: Their Origin and Evolution. A Chapter in the History of Botany: 1470–1670.* Cambridge: Cambridge University Press.

Arldt, Theodor. 1907. *Die Entwicklung der Kontinente und ihrer Lebewelt: Ein Beitrag zur vergleichenden Erdgeschichte.* Leipzig: Engelmann.

Artz, Th. 1955. Die Erforschungsgeschichte der Chorda dorsalis und die Entstehung des Chordaten-Begriffes im 19. Jahrhundert. *Nova Acta Leopoldina,* N.S. 17: 361–409.

Ashlock, Peter D. 1974. The uses of cladistics. *Ann. Rev. Ecol. Syst.* 5: 81–99.

—————— 1979. An evolutionary systematist's view of classification. *Syst. Zool.* 25: 441–450.

—————— 1982 (in press). Empty internodes and hidden paraphyly.

Auerbach, Ch. 1967. The chemical production of mutations. *Science* 158: 1141–47.

—————— 1976. *Mutation Research: Problems, Results, and Perspectives.* London: Chapman & Hall.

Ayala, F. J. 1968. Biology as an autonomous science. *Amer. Sci.* 56: 207–221.

—————— 1972. Mendelism versus Darwinism. [A review of Provine (1971)]. *Nature* 239: 235.

—————— 1974a. Biological evolution: natural selection or random walk? *Amer. Sci.* 62: 692–701.

—————— 1974b. The concept of biological progress, in Ayala and Dobzhansky (1974), pp. 339–355.

—————— 1975a. Genetic differentiation during the speciation process. *Evol. Biol.* 8: 1–78.

—————— 1975b. Scientific hypotheses, natural selection, and the neutrality theory of protein evolution, in Salzano (1975), pp. 19–42.

—————— (ed.). 1976. *Molecular Evolution.* Sunderland, Mass.: Sinauer.

—————— and Th. Dobzhansky (eds.). 1974a. *Studies in the Philosophy of Biology: Reduction and Related Problems.* Berkeley and Los Angeles: University of California Press.

—————— et al. 1974b. Genetic variation in natural populations of five *Drosophila* species and the hypothesis of selective neutrality of protein polymorphism. *Genetics* 77: 343–384.

Baer, K. E. v. 1828. *Entwicklungsgeschichte der Thiere: Beobachtung und Reflexion.* Königsberg: Bornträger.

—————— 1876. *Studien aus der Geschichte der Naturwissenschaften.* St. Petersburg: H. Schmitzdorf.

Bailey, C. 1928. *The Greek Atomists and Epicurus: A Study.* Oxford: Oxford University Press.

Bajema, Carl Jay. 1977. *Eugenics: Then and Now.* Stroudsburg, Pa.: Dowden, Hutchinson and Ross.

Baker, H.G., and G. Ledyard Stebbins (eds.). 1965. *The Genetics of Colonizing Species.* New York: Academic Press.

Baker, J. N. L. 1931. *History of Geographical Discovery and Exploration.* London: Hasrap.

Baker, John R. 1938. The evolution of breeding seasons, pp. 161–177, in de Beer (1938).

—————— 1948–1955. The cell theory: a restatement, history, and critique. *Quart. J. Microscopical Science* 89: 103–123; 90: 87–108; 93: 157–190; 96: 449.

—————— 1952. *Abraham Trembley, Scientist and Philosopher.* London: Edward Arnold.

Baker, V. R. 1978. The Spokane flood controversy and the Martian outflow channels. *Science* 202: 1249–56.

Baldwin, James Mark. 1909. *Darwin and the Humanities.* Baltimore: Review Publishing.

Ball, Ian R. 1977. On the phylogenetic classification of aquatic planarians, in T. G. Karling and M. Meinander (eds.), The Alex. Luther Centennial Symposium on Turbellaria. *Acta Zoologica Fennica* 154: 21–35.

Ballauff, Theodor. 1954. *Die Wissenschaft vom Leben: Eine Geschichte der Biologie.* München: Karl Alber.

Balme, D. M. 1962a. Aristotle's use of differentiae in zoology, in *Aristote et les problèmes de méthode.* Louvain, p. 205.

—————— 1962b. Genos and Eidos in Aristotle's biology. *Classical Quarterly,* N.S. 12: 81–98.

—————— 1965. Aristotle's use of the teleological explanation. Inaugural Lecture (Queen Mary College), pp. 1–27.

—————— 1970. Aristotle and the beginnings of zoology. *J. Soc. Biblphy. Nat. Hist.* 5: 272–285.

—————— 1980. Aristotle's biology was not essentialist. *Archiv für Geschichte der Philosophie* 62(1): 1–12.

Baltzer, Fritz. 1962. *Theodor Boveri: Leben and Werk.* Stuttgart: Wissenschaftliche Verlagsgesellschaft.

Bannister, R. C. 1979. *Social Darwinism: Science and Myth in Anglo-American Social Thought.* Philadelphia: Temple University Press.

Barash, D. 1979. *The Whisperings Within.* New York: Harper & Row.

895

Barber, B. 1961. Resistance of scientists to scientific discovery. *Science* 134: 596–602.

Barlow, G. W., and J. Silverberg (eds.). 1980. *Sociobiology: Beyond Nature/Nurture?* Boulder: Westview Press.

Barlow, Nora (ed.). 1945. *Charles Darwin and the Voyage of the Beagle.* London: Pilot Press. (See also Darwin, C., [1958], [1963].)

Baron, W. 1963 Die Anschauungen Johann Friedrich Blumenbachs über die Geschichtlichkeit der Natur. *Sudhoff's Archiv* 47: 19–26.

Barrett, P. H. 1974. Darwin's early and unpublished notebooks, in Gruber (1974), pp. 259–425.

Barthélemy-Madaule, M. 1979. *Lamarck ou le mythe du précurseur.* Paris: Seuil.

Barthelmess, A. 1952. *Vererbungswissenschaft.* Freiburg: Karl Alber.

Bartholomew, M. 1973. Lyell and evolution: an account of Lyell's response to the project of an evolutionary ancestry for man. *Brit. J. Hist. Sci.* 6: 261–303.

Bateman, A. J. 1948. Intra-sexual selection in *Drosophila. Heredity* 2: 349–368.

Bates, Henry Walter. 1862. Contributions to an insect fauna of the Amazon Valley. *Trans. Linn. Soc. London* 23: 495–566.

———— 1863. *The Naturalist on the River Amazon.* London.

Bateson, B. (ed.). 1928. *William Bateson, Naturalist.* London: Cambridge University Press.

Bateson, William. 1894. *Materials for the Study of Variation.* London: Macmillan.

———— 1908. *The Methods and Scope of Genetics.* Cambridge: Cambridge Univeristy Press.

———— 1909. *Mendel's Principles of Heredity.* London: Cambridge University Press.

———— 1913. *Problems of Genetics.* New Haven: Yale University Press.

———— 1914. Presidential address. *Brit. Ass. Adv. Sci.,* Sydney, Australia, pt. 2.

———— 1916. [Review of Morgan et al. (1915).] *Science* 44: 536–543.

———— 1922. Evolutionary faith and modern doubts. *Science* 55: 55–61.

———— 1928. *Scientific Papers,* ed. R. C. Punnett. 2 vols. Cambridge: Cambridge University Press.

Bather, F. A. 1927. Biological classification: past and future. *Proc. Geol. Soc.* 83: 62–104.

Baur, E. 1911. *Einführung in die experimentelle Vererbungslehre.* 1st ed. Berlin: Borntraeger, pp. 258–268 (3rd ed., 1929).

Baxter, Alice L. 1976. Edmund B. Wilson as a preformationist: some reasons for his acceptance of the chromosome theory. *J. Hist. Biol.* 9: 29–57.

Beach, F. A. 1950. The snark was a boojum. *Amer. Psychol.* 5: 115–124.

Beaudry, J. R. 1960. Symposium on the species problem and taxonomy.

Rev. Canad. Biol. 19: 215–325.

Beck, B. B. 1980. *Animal Tool Behavior: The Use and Manufacture of Tools by Animals.* New York: Garland STPM Press.

Beckner, M. 1959. *The Biological Way of Thought.* New York: Columbia University Press.

———— 1974. Reduction, hierarchies, and organicism, in Ayala and Dobzhansky (1974), pp. 163–177.

Beddall, B. G. 1957. Historical notes on avian classification. *Systematic Zool.* 6: 129–136.

Bell, P. R. (ed.). 1959. *Darwin's Biological Work.* Cambridge: Cambridge University Press.

Belon, P. 1555. *L'histoire de la nature des oyseaux.* Paris.

Benzer, S. 1957. The elementary units of heredity, in McElroy and Glass (1957), pp. 70–93.

Berg, A. 1942. Die Lehre von der Faser als Form- und Funktionselement des Organismus. *Virchow Archiv für pathologische Anatomie und Physiologie* 309: 394ff.

Berg, L. 1926. *Nomogenesis.* London: Constable.

Bergmann, C. 1847. Über die Verhältnisse der Wärmeökonomie der Thiere zu ihrer Grösse. *Göttinger Studien* 1: 595–708.

Bergmann, G. 1957. *Philosophy of Science.* Madison: University of Wisconsin Press.

Berlin, B., D. E. Breedlove, and P. H. Raven. 1974. *Principles of Tzeltal Plant Classification.* New York and London: Academic Press.

Berlin, I. 1960. The philosophical ideas of Giambattista Vico, in Harold Action et al., *Art and Ideas in Eighteenth Century Italy.* Lectures given at the Italian Institute, 1957–1958. Roma: Edizioni di storia e letteratura.

———— 1976. *Vico and Herder: Two studies in the history of ideas.* London: Hogarth Press.

Bertalanffy, L. v. 1932. *Theoretische Biologie.* 2 vols. Berlin: Borntraeger.

———— 1949. Das biologische Weltbild. *Die Stellung des Lebens in Natur und Wissenschaft,* Vol. 1. Bern: Francke. (Eng. trans. 1952. *Problems of Life.* New York: John Wiley & Sons.)

Bessey, C. E. 1908. The taxonomic aspect of the species. *Amer. Nat.* 42: 218–224.

Blair, W. Frank (ed.). 1961. *Vertebrate Speciation.* Austin: University of Texas Press.

Blaisdell, M. B. 1976. *Darwinism and Its Data: The adaptive Coloration of Animals.* Ph.D. diss., Harvard.

Blakeslee, A. F. 1936. Twenty-five years of genetics (1910–1935). *Brooklyn Bot. Gard. Memoirs* 4: 29–40.

Blandino, Giovanni. 1969. *Theories on the Nature of Life.* New York: Philosophical Library.

Blei, W. 1981. Erkenntwiswege zur Erd- und Lebensgeschichte. Wissenschaftliche Taschenbücher, No. 219. Berlin: Akademie-Verlag.

Blum, M. S., and N. A. Blum (eds.). 1979. *Sexual Selection and Reproductive Competition in Insects.* New York: Academic Press.

Blumenbach, J. F. 1790. *Beyträge zur Naturgeschichte.* Göttingen.

Blunt, W. 1971. *Compleat Naturalist.* New York: Viking Press.

Bock, Walter. 1959. Preadaptation and multiple evolutionary pathways. *Evolution* 13: 194–211.

———— 1970. Microevolutionary sequences as a fundamental concept in macroevolutionary models. *Evolution* 24: 704–722.

———— 1973. Philosophical foundations of classical evolutionary classification. *Syst. Zool.* 22: 375–392.

———— 1977. Foundations and Methods of Evolutionary classification, in Major Patterns in Vertebrate Evolution, *NATO Advanced Study Institute,* Ser. A, 14: 851–895.

———— 1979. The synthetic explanation of macroevolutionary change: a reductionist approach. *Bull. Carnegie Mus. Nat. Hist.* 13: 20–69.

———— and W. de W. Miller. 1959. The scansorial foot of the woodpeckers. *Amer. Mus. Novit.* 1931: 1–45.

———— and G. von Wahlert. 1965. Adaptation and the form-function complex. *Evolution* 19: 269–299.

Boesiger, E. 1980. The state of evolutionary biology in France at the time of the synthesis, in Mayr and Provine (1980), pp. 309–321.

Boivin, A., R. Vendrely, and C. Vendrely. 1948. L'acide desozyribonucléique du noyau cellulaire, etc. *C.r. hebd. Séanc. Acad. Sci.,* Paris 226: 1061–63.

Böker, Hans. 1935, 1937. *Einführung in die vergleichende biologische Anatomie der Wirbeltiere.* 2 vols. Jena: Gustav Fischer.

Bolk, L. 1915. Überlagerung, Verschiebung, und Neigung des Foramen magnum am Schädel der Primaten. 3. *Morph. Anthrop.* 7: 611–692.

Bonde, Niels. 1974. Review of interrelationships of fishes. *Syst. Zool.* 23: 562–569.

———— 1975. Origin of "higher groups": viewpoints of phylogenetic systematics, in *Problemes actuels de paléontologie (Evolution des vertèbres).* Colloques Internationaux du Centre National de la Recherche Scientifique.

Bondi, Hermann. 1977. The lure of completeness, in R. Duncan and M. Weston-Smith (eds.), *Encyclopedia of Ignorance,* Vol. 1. Oxford: Pergamon, pp. 5–8.

Bonner, J. T. 1980. *The Evolution of Culture in Animals.* Princeton: Princeton University Press.

Bonnet, C. 1769. La palingénésis philosophique. 2 vols. Geneva: C. Philibert and B. Chirol.

Born, M. 1949. *Natural Philosophy of Cause and Chance.* London: Oxford University Press.

Boucot, A. J. 1978. Community evolution and rates of cladogenesis. *Evol. Biol.* 11: 454–655.

Bourdier, F. 1969. Geoffroy Saint-Hilaire versus Cuvier: the campaign for paleontological evolution (1825–1838), in Schneer (1969), pp. 36–61.

898

Boveri, Th. 1903. Über den Einflus der Samenzelle auf die Larvencharaktere der Echiniden. *Roux's Arch.* 16: 356.

Bowler, Peter J. 1973. Buffon and Bonnet: theories of generation and the problem of species. *J. Hist. Biol.* 6: 259–281.

—— 1974a. Darwin's concept of variation. *J. Hist. Med. Allied Sci.* 29: 196–212.

—— 1974b. Evolutionism in the Enlightenment. *Hist. Sci.* 12: 159–183.

—— 1975. The changing meaning of "evolution." *J. Hist. Ideas* 36: 95–114.

—— 1976. *Fossils and Progress.* New York: Science History Publications.

—— 1977a. Edward Drinker Cope and the changing structure of evolutionary theory. *Isis* 68: 249–265.

—— 1977b. Darwinism and the argument from design: suggestions for a reevaluation. *J. Hist. Biol.* 10: 29–43.

—— 1978. Hugo de Vries and Thomas Hunt Morgan: the mutation theory and the spirit of Darwinism. *Ann. Sci.* 35: 55–73.

—— 1979. Theodor Eimer and orthogenesis: evolution by "definitely directed variation." *J. Hist. Med. Allied Sci.* 34: 40–73.

Boycott, A. E., and C. Diver. 1923. On the inheritance of sinistrality in *Limnaea peregra. Proc. Roy. Soc.* London, Ser. B, 95: 207–213.

Boycott, A. E. et al. 1930. The inheritance of sinistrality in *Limnaea peregra. Phil. Trans. Roy. Soc. London,* Ser. B, vol. 219: 51–131.

Boyle, Robert. 1674. *About the Excellency and Grounds of a Mechanical Hypothesis, Some Considerations Occasionally Proposed to a Friend.* London: Printed by T. N. for Henry Herrington.

Brachet, Jean. 1957. *Biochemical Cytology.* New York: Academic Press.

Brandon, R. N. 1978. Adaptation and evolutionary theory. *Stud. Hist. Phil. Sci.* 9: 181–206.

Bräuning-Oktavio, H. 1956. Vom Zwischenkieferknochen zur Idee des Typus: Goethe als Naturforscher in den Jahren 1780–1786. *Nova Acta Leopoldina* N.S. 126: 1–144.

Bremekamp, C. E. B. 1953a. Linné's significance for the development of phytography. *Taxon* 2: 57–67.

—— 1953b. A re-examination of Cesalpino's classification. *Acta. Bot. Neerl.* 1: 580–593.

—— 1953c. Linné's views on the hierarchy of the taxonomic groups. *Acta. Bot. Neerl.* 2: 242–253.

Bridges, C. B. 1916. Non-disjunction as proof of the chromosome theory of heredity. *Genetics* 1: 1–52, 107–163.

Brillouin, L. 1962. *Science and Information Theory.* 2nd ed. New York: Academic Press.

Brink, R. A. (ed.). 1967. *Heritage from Mendel.* Madison: University of Wisconsin Press.

Bronowski, J., and B. Mazlish. 1960. *The Western Intellectual Tradition.* New York: Harper & Row.

Brunet, P. 1929. *Maupertuis.* Paris: A. Blanchard.

Brush, S. G. 1978. Nettie M. Stevens and the determination of sex determination by chromosomes. *Isis* 69: 163–172.

Buch, Leopold von. 1825. *Physicalische Beschreibung der Canarischen Inseln.* Berlin: Kgl. Akad. Wiss., pp. 132–133.

Buchdahl, G. 1973. Leading principles and induction: the methodology of Matthias Schleiden, in Giere and Westfall (1973), pp. 23–52.

Buddenbrock, W. 1930. *Biologische Grundprobleme und ihre Meister.* Berlin: Borntraeger.

Buffetaut, E. 1979. The evolution of the crocodilians. *Sci. Amer.* 241 (Oct.): 130–144.

Buffon, Georges Louis. 1749–1804. *Histoire naturelle, générale et particulière.* 44 vols. Paris: Imprimerie Royale, puis Plassan.

——— 1779. Les époques de la nature. (Critical edition by J. Roger. 1962. *Mém. Mus. Nat. d'Hist. Nat.,* Ser. C, 10: i–clii, 1–343.)

——— 1954. *Oeuvres philosophiques,* ed. J. Piveteau. Paris: Presses Universitaires de France.

Bumpus, H. C. 1896. The variations and mutations of the introduced sparrow, *Passer domesticus,* in *Biol. Lectures, Marine Biol. Lab., Wood's Hole* (1896–1897), pp. 1–15.

——— 1899. The elimination of the unfit as illustrated by the introduced sparrow *Passer domesticus,* in *Biol. Lectures, Marine Biol. Lab., Wood's Hole* (summer session 1897 and 1898).

Burchfield, Joe D. 1975. *Lord Kelvin and the Age of the Earth.* New York: Science History Publications.

Burkhardt, R. W., Jr. 1977. *The Spirit of System: Lamarck and Evolutionary Biology.* Cambridge: Cambridge University Press.

——— 1980. Lamarckism in Britain and the United States, in Mayr and Provine (1980), pp. 343–352.

Burtt, B. L. 1966. Adanson and modern taxonomy. *Notes from the Royal Botanic Garden Edinburgh* 26: 427–431.

Bury, J. B. 1920. *The Idea of Progress, An Inquiry into Its Growth and Origin.* (Rpt. 1955. New York: Dover.)

Bush, G. L. 1975. Modes of animal speciation. *Ann. Rev. Ecol. Syst.* 6: 339–364.

Butterfield, Herbert. 1931. *The Whig Interpretation of History.* London: Bell. (Rpt. 1965. New York: Norton Library.)

——— 1957. *The Origins of Modern Science, 1300–1800.* London: Bell. (Rpt. 1965. New York: Free Press.)

Cahn, Th. 1962. *La vie et l'oeuvre d'Etienne Geoffroy Saint-Hilaire.* Paris: Presses Universitaires de France.

Cain, Arthur J. 1954. *Animal Species and Their Evolution.* London: Hutchinson's University Library.

——— 1958. Logic and memory in Linnaeus's system of taxonomy. *Proc. Linn. Soc. London* 169: 144–163.

900

―――― 1959a. Deductive and inductive methods in post-Linnaean taxonomy. *Proc. Linn. Soc. London* 170: 185–217.

―――― 1959b. Taxonomic concepts. *Ibis* 101: 302–318.

―――― and G. A. Harrison. 1958. An analysis of the taxonomist's judgment of affinity. *Proc. Zool. Soc. London* 131: 85–98.

―――― 1960. Phyletic weighting. *Proc. Zool. Soc. London* 135: 1–31.

Cairns, J., G. S. Stent, and J. D. Watson. 1966. *Phage and the Origins of Molecular Biology.* (Delbrück Festschrift.) Cold Spring Harbor Lab. of Quantitative Biology.

Cajander, A. K. 1921. Einige Reflexionen über die Entstehung der Arten insbesondere innerhalb der Gruppe der Holzgewächse. *Acta Forest. Fenn.* 21: 1–21.

Cameron, H. C. 1952. *Sir Joseph Banks: The Autocrat of the Philosophers.* London: Batchworth Press.

Camp, W. H., and C. L. Gilly. 1943. The structure and origin of species. *Brittonia* 4: 323–385.

Campbell, B. (ed.). 1972. *Sexual Selection and the Descent of Man, 1871–1971.* Chicago: Aldine.

―――― 1974. *Human Evolution: An Introduction to Man's Adaptations.* Chicago: Aldine.

Candolle, A. de. 1855. *Géographie botanique raisonné*, vols. 1–2. Paris: Bibliothèque Universelle.

―――― 1862. *Etude sur l'espèce, á l'occasion d'une révision de la famille des cupulifères.* Paris: Bibliothèque Universelle.

Cannon, W. A. 1902. A cytological basis for the Mendelian laws. *Bull. Torrey Bot. Club* 29: 657; 30: 133–172, 519–543.

Cannon, Walter F. 1960. The Uniformitarian-Catastrophist debate. *Isis* 51: 38–55.

―――― 1961. The impact of uniformitarianism: two letters from John Herschel to Charles Lyell, 1836–37. *Proc. Amer. Phil. Soc.* 105: 301–314.

―――― (Susan Faye). 1978. *Science in Culture: The Early Victorian Period.* New York: Dawson and Science History Publications.

Caplan, Arthur L. 1976. Ethics, evolution, and the milk of human kindness. *Hastings Center Report* (April 1976). Rpt. in A. L. Caplan (ed.), 1978a, *The Sociobiology Debate.* New York: Harper & Row.

Caplan, A. L. 1977. Tautology, circularity, and biological theory. *Am. Nat.* 111: 390–393.

―――― 1978b. Testability, disreputability, and the structure of the modern synthetic theory of evolution. *Erkenntnis* 13: 261–278. (See also *Amer. Nat.* 101: 390–393.)

Carlquist, Sherwin. 1965. *Island Life: A Natural History of the Islands of the World.* Garden City: Natural History Press.

―――― 1974. *Island Biology.* New York: Columbia University Press.

―――― 1975. *Ecological Strategies of Xylem.* Berkeley and Los Angeles: University of California Press.

901

Carlson, E. A. 1966. *The Gene: A Critical History.* Philadelphia: Saunders.

———— 1972. H. J. Muller (1890–1967). *Genetics* 70: 1–30.

———— 1974. The *Drosophila* group: the transition from the Mendelian unit to the individual gene. *J. Hist. Biol.* 7: 31–48.

Carr, E. H. 1961. *What Is History?* London: Macmillan. (Rpt. 1964. Harmondsworth: Penguin Books.)

Carson, Hampton L. 1975. The genetics of speciation at the diploid level. *Amer. Nat.* 109: 83–92.

———— 1976. The unit of genetic change in adaptation and speciation. *Annals of the Missouri Botanical Garden* 63: 210–223.

————, D. E. Hardy, H. T. Spieth, and W. S. Stone. 1970. The evolutionary biology of the Hawaiian *Drosophilidae,* in *Essays in Evolution and Genetics in Honor of Theodosius Dobzhansky.* New York: Appleton-Century-Crofts, pp. 437–543.

———— and K. Y. Kaneshiro. 1976. *Drosophila* of Hawaii: Systematics and ecological genetics. *Ann. Rev. Ecol. Syst.* 7: 311–345.

Caspari, E. 1948. Cytoplasmic inheritance. *Adv. Genet.* 2: 1–66.

Caspersson, T., and L. Zech (eds.). 1972. *Chromosome Identification.* Stockholm: Nobel Foundation; New York: Academic Press.

Cassirer, Ernst. 1950. *The Problem of Knowledge: Philosophy, Science, and History since Hegel.* New Haven and London: Yale University Press.

———— 1951. *The Philosophy of the Enlightenment,* trans. Fritz C. A. Koelln and James P. Pettegrove. Princeton: Princeton University Press.

Castle, W. E. 1903. The laws of heredity of Galton and Mendel, and some laws governing race improvement by selection. *Proc. Amer. Acad. Arts Sci.* 39: 233–242.

———— 1951. The beginnings of Mendelism in America, in Dunn (1951), pp. 59–76.

Caullery, M. 1931. *L'évolution.* Paris: Payot.

Causey, Robert. 1977. *The Unity of Science.* Dordrecht: D. Reidel.

Chamberlin, T. C. 1890. The method of multiple working hypotheses. *Science* 15: 92. (See also *Science* 148: 754–759.)

Chargaff, E. 1950. Chemical specificity of nucleic acids and mechanism of their enzymatic degradation. *Experientia* 6: 201–209.

———— 1971. Preface to a grammar of biology. *Science* 172: 637–642.

Chatton, H. 1925. *Pansporella perplexa. Ann. sci. nat. Zool.* 8: 5.

Chetverikov, S. S. 1926. On certain aspects of the evolutionary process from the standpoint of modern genetics. *J. Explt. Biol.* (Russian) A2: 3–54. (English trans. [1961]. *Proc. Amer. Phil. Soc.* 105: 167–195.)

———— 1927. Über die genetische Beschaffenheit wilder Populationen. *Verhandlungen d. V Internat. Kongres. f. Vererbungswissenschaft. Berlin.* 2: 1499–1500. Leipzig: Borntraeger.

Churchill, F. B. 1968. August Weismann and a break from tradition. *J. Hist. Biol.* 1: 91–112.

———— 1970. Hertwig, Weismann, and the meaning of reduction division, circa 1890. *Isis* 61: 429–457.

———— 1974. William Johannsen and the genotype concept. *J. Hist. Biol.* 7: 5–30.

———— 1976. Rudolf Virchow and the pathologist's criteria for the inheritance of acquired characteristics. *J. Hist. Med.* 31: 117–148.

———— 1979. Sex and the single organism: biological theories of sexuality in mid-nineteenth century. *Stud. Hist. Biol.* 3: 139–177.

Clark, R. B. 1964. *Dynamics in Metazoan Evolution: The Origin of the Coelum and Segments.* Oxford: Oxford University Press.

Clarke, B. 1962. Balanced polymorphism and the diversity of sympatric species, in *Taxonomy and Geography.* Systematics Association Publ., no. 4, pp. 47–70.

———— 1975a. The contribution of ecological genetics to evolutionary theory: detecting the direct effects of natural selection on particular polymorphic loci. *Genetics Suppl.* 79: 101–113.

Clausen, J. 1951. *Stages in the Evolution of Plant Species.* Ithaca, N.Y.: Cornell University Press.

Cleland, R. E. 1972. *Oenothera: Cytogenetics and Evolution.* New York: Academic Press.

Clifford, H. T., and W. Stephenson. 1975. *An Introduction to Numerical Classification.* New York: Academic Press.

Cloyd, E. L. 1972. *James Burnett, Lord Monboddo.* London: Clarendon Press, Oxford University Press.

Cock, A. G. 1973. William Bateson, Mendelism, and biometry. *J. Hist. Biol.* 6: 1–36.

———— 1977. Bernard's symposium—the species concept in 1900. *Biol. J. Linn. Soc.* 9: 1–30.

Cody, Martin L., and Jared M. Diamond. 1975. *Ecology and Evolution of Communities.* Cambridge: Harvard University Press, Belknap Press.

Cohen, I. B. 1982 (in press). Three notes on the reception of Darwin's ideas on natural selection. *J. Hist. Biol.*

Cold Spring Harbor Symposia on Quantitative Biology. 1978. Vol. 42: Chromatin (2 vols.). Cold Spring Harbor Laboratory.

Cole, F. J. 1926. *The History of Protozoology.* London: University of London Press.

———— 1944. *A History of Comparative Anatomy, From Aristotle to the Eighteenth Century.* London: Macmillan.

Coleman, W. 1962. Lyell and the "reality" of species. *Isis* 53: 325–338.

———— 1964. *Georges Cuvier, Zoologist.* Cambridge: Harvard University Press.

———— 1965. Cell, nucleus, and inheritance: an historical study. *Proc. Amer. Phil. Soc.* 109: 124–158.

———— 1970. Bateson and chromosomes: conservative thought in science. *Centaurus* 15: 228–314.

———— 1971. *Biology in the Nineteenth Century: Problems of Form, Function, and Transformation.* New York: John Wiley & Sons.

903

——— 1973. Limits of the recapitulation theory: Carl Friedrich Kielmeyer's critique of the presumed parallelism of earth history, ontogeny, and the present order of organisms. *Isis* 64: 341–350.

——— 1976. Morphology between type concept and descent theory. *J. Hist. Med. Allied Sci.* 31: 149–175.

——— 1980. Morphology in the evolutionary synthesis, in Mayr and Provine (1980), pp. 174–180.

——— (ed.). 1967. *The Interpretation of Animal Form.* New York and London: Johnson Reprint Corp.

Collingwood, R. G. 1939. *An Autobiography.* Oxford: Oxford University Press.

Colp, R. 1977. *To Be an Invalid.* Chicago: University of Chicago Press.

Coluzzi, M., A. Sabatini, V. Petraca, and M. A. Di Deco. 1977. Behavioral differences between mosquitoes with different inversion karyotypes in polymorphic populations of the *Anopheles gambiae* complex. *Nature* 266: 832–833.

Conklin, H. C. 1962. Lexicographical treatment of folk taxonomies. *Int. J. Amer. Linguistics* 28: 119–141.

Conry, Y. 1974. *L'introduction du darwinisme en France au XIXe siècle.* Paris: Vrin.

——— 1980. L'ideé d'une 'marche de la nature' dans la biologie pré-Darwinienne au XIXe siècle. *Rev. Hist. Sci.* 33: 97–149.

Cope, Edward Drinker. 1887. *The Origin of the Fittest.*

——— 1896. *The Primary Factors of Organic Evolution.* Chicago.

Corliss, J. O. 1978–1979. A salute to fifty-four great microscopists of the past: a pictorial footnote to the history of protozoology. Pt. 1 (1978) *Trans. Amer. Micr. Soc.* 97: 419–458; Pt. 2 (1979) *Trans. Amer. Micr. Soc.* 98: 26–58.

Correns, C. 1900. Mendel's Regel über das Verhalten der Nachkommenschaft der Rassenbastarde. *Ber. Dtsch. Bot. Ges.* 18: 158–168.

——— 1902. Scheinbare Ausnahme von der Mendelschen Spaltungsregel für Bastarde. *Ber. Dtsch. Bot. Ges.* 20: 159–172.

——— 1905. Gregor Mendel's Briefe an Carl Nägeli, 1866–1873. *Abh. Math.-Phys. Kl. K. Sächs. Ges. Wiss.* 29: 189–265. (English trans. in Stern and Sherwood [1966]; also *Genetics* 35, no. 5, pt. 2: 1–29.)

——— 1924. *Gesammelte Abhandlungen zur Vererbungswissenschaft, 1899–1924,* ed. F. v. Wettstein. Berlin: J. Springer.

———, and F. v. Wettstein. 1937. Nicht mendelnde Vererbung. *Handb. Vererb. Wiss.* Berlin II H, pp. 1–158.

Cowan, Ruth Schwartz. 1972. Francis Galton's contributions to genetics. *J. Hist. Biol.* 5: 389–412.

——— 1977. Nature and nurture: the interplay of biology and politics in the work of Francis Galton. *Stud. Hist. Biol.* 1: 133–208.

Cravens, H. 1978. *The Triumph of Evolution: American Scientists and the Heredity-Environment Controversy, 1900–1941.* Philadelphia: University of Pennsylvania Press.

Creighton, H. B., and B. McClintock. 1931. A correlation of cytological and genetical crossing over in *Zea mays*. *Proc. Nat. Acad. Sci.* 17: 492–497.

Crick, F. 1966. *Of Molecules and Men*. Seattle: University of Washington Press.

Croce, B. 1913. *The Philosophy of Giambattista Vico*, trans. R. G. Collingwood. London: Howard Latimer.

Crocker, L. G. 1959. Diderot and eighteenth century French transformism, in Glass, Temkin, and Strauss (1959), pp. 114–143.

Croizat, L. 1958. *Panbiogeography*. Caracas: published by the author.

—— 1964. *Space, Time, Form: The Biological Synthesis*. Caracas: published by the author.

——, G. Nelson, and D. E. Rosen. 1974. Centers of origin and related concepts. *Syst. Zool.* 23: 265–287.

Crombie, A. C. 1950. The notion of species in Medieval philosophy and science. *Actes VI Cong. Int. d'Hist. Sci.* 1: 261–269.

—— 1952. *Augustine to Galileo*, vols. 1–2. London: Heinemann. (Rpt. 1961. Cambridge: Harvard University Press.)

—— (ed.). 1965. *Scientific Change: Historical Studies in the Intellectual, Social, and Technical Conditions for Scientific Discovery and Technical Invention, from Antiquity to the Present*. London: Heinemann.

Cronquist, A. 1978. Once again, what is a species?, in *Biosystematics in Agriculture*. Montclair, N.J.: Allanheld, Osmun, and Co.; New York: John Wiley & Sons, pp. 3–20.

Crow, J. F., and M. Kimura. 1970. *An Introduction to Population Genetics Theory*. New York: Harper & Row.

Crowther, J. G. 1960. *Founders of British Science*. London: Cresset.

Cuellar, O. 1977. Animal parthenogenesis. *Science* 197: 837–843.

Cuénot, L. 1902. La loi de Mendel et l'herédité de la pigmentation chez les souris. *Compt. rend. Acad. Sci.* 134: 779–791.

—— 1928. Génétique des Souris. *Bibl. genetica* 4: 179–242.

—— 1951. *L'évolution biologique: Les faits, Les incertitudes*. Paris: Masson.

Cuvier, G. 1812. *Recherches sur les ossemens fossiles des quadrupèdes*, etc. 4 vols. Paris: Déterville.

—— 1813. *Essay on the theory of the earth*. Edinburgh.

—— and C. Duméril. 1829. Rapport sur un mémoire de M. Roulin . . .*Ann. des sci. natur.* 17: 107–112.

Darden, L. 1976. Reasoning in scientific change: Charles Darwin, Hugo de Vries, and the discovery of segregation. *Stud. Hist. Phil. Sci.* 7: 127–169.

—— 1977. William Bateson and the promise of Mendelism. *J. Hist. Biol.* 10: 87–106.

—— 1980. Theory construction in genetics, in Nickles, T. (ed.), *Scientific Discovery: Case Studies*. New York: D. Reidel, pp. 151–170.

—— and Nancy Maull. 1977. Interfield theories.*Phil. Sci.* 44: 43–64.

905

Darlington, C. D. 1932. *Recent Advances in Cytology*. Philadelphia: P. Blakiston's Sons.

———— 1939. *The Evolution of Genetic Systems*. Cambridge: Cambridge University Press.

———— 1959. *Darwin's Place in History*. Oxford: Blackwell.

Darlington, P. J. 1957. *Zoogeography*. New York: John Wiley & Sons.

———— 1959. Darwin and zoogeography. *Proc. Amer. Phil. Soc.* 103: 307–319.

Darwin, C. 1844. Essay. (First published in F. Darwin [1909].)

———— 1859 (24 Nov.). *On the Origin of Species by Means of Natural Selection or the Preservation of Favored Races in the Struggle for Life*. London: Murray.

———— 1861. An historical sketch of the progress of opinion on the origin of species. (In the later editions of the *Origin of Species*.)

———— 1862. *The Various Contrivances by Which Orchids Are Fertilized by Insects*. London: Murray.

———— 1868. *The Variation of Animals and Plants under Domestication*, vols. 1–2. London: Murray.

———— 1871. *The Descent of Man*. London: Murray.

———— 1872. *The Expression of the Emotions in Man and Animals*. London: Murray.

———— 1933. *Charles Darwin's Diary of the Voyage of H.M.S. "Beagle,"* ed. N. Barlow. Cambridge: Cambridge University Press.

———— 1958. *The Autobiography of Charles Darwin*, ed. Nora Barlow. London: Collins.

———— 1960–1967. Darwin's notebooks on transmutation of species, ed. G. de Beer. *Bull. Brit. Mus.* (*Nat. Hist.*) 2: 27–200; 3: 129–176. (Notebooks B, C, D, E.)

———— 1963. Darwin's ornithological notes, ed. G. de Beer. *Bull. Brit. Mus.* (*Nat. Hist.*) Histor. Ser. 2: 201–278.

———— 1964. *On the Origin of Species*. (Facsimile of first edition, ed. Ernst Mayr.) Cambridge: Harvard University Press.

———— 1967. Darwin's notebooks. Pages excised by Darwin, ed. G. de Beer, M. J. Rowlands, and B. M. Skramovsky. *Bull. Brit. Mus.* (*Nat. Hist.*) 3: 129–176.

———— 1975. *Natural Selection*, ed. R. C. Stauffer. Cambridge: Cambridge University Press.

———— 1980. The Red Notebooks of Charles Darwin, ed. Sandra Herbert. *Bull. Brit. Mus.* (*Nat. Hist.*) Histor. Ser. 7: 1–168.

———— and A. R. Wallace. 1958. *Evolution by Natural Selection*, ed. G. de Beer. Cambridge: Cambridge University Press.

Darwin, Erasmus. 1796. *Zoonomia*. London: J. Johnson.

Darwin, F. 1887. *The Life and Letters of Charles Darwin*, vols. 1–3. London: Murray. (Reprinted 1969. New York: Johnson Reprint Corp.)

———— 1909. *The Foundations of the Origin of Species, by Charles Darwin*. Cambridge: Cambridge University Press.

—— and A. C. Seward. 1903. *More Letters of Charles Darwin.* 2 vols. London: Murray.

Daudin, Henri. 1926. *De Linné à Jussieu. Méthodes de la classification et idée de série en botanique et en zoologie (1740–1790).* Paris: Felix Alcan.

—— 1926. *Cuvier et Lamarck: Les classes zoologiques et l'idée de série animale (1790–1830).* 2 vols. Paris: Felix Alcan.

Davenport, Charles B. 1941. The early history of research with *Drosophila. Science* 93: 305–306.

Davidson, Eric H., and Roy J. Britten. 1979. Regulation of gene expression: possible role of repetitive sequences. *Science* 204: 1052–59.

Davis, D. Dwight. 1955. (Comparative) Anatomy, in K. P. Schmidt (ed.), pp. 618–622.

—— 1960. The proper goal of comparative anatomy, *Proc. Cent. Bicent. Cong. Biol.,* Singapore, December 2–9, 1958, ed. R. D. Purchon. Singapore: University of Malaya Press, pp. 44–50.

—— 1964. The giant panda: a morphological study of evolutionary mechanisms. *Fieldiana: Zool. Memoirs* Chicago 3.

Davis, P. H., and V. H. Heywood. 1963. *Principles of Angiosperm Taxonomy.* Edinburgh and London: Oliver & Boyd.

Dawes, Ben. 1952. *A Hundred Years of Biology.* London: Gerald Duckworth.

Dawkins, R. 1976. *The Selfish Gene.* Oxford: Oxford University Press.

de Beer, G. R. (ed.). 1938. *Evolution: Essays on Aspects of Evolutionary Biology.* Oxford: Clarendon Press.

—— 1940. *Embryos and Ancestors.* Oxford: Clarendon Press. (2nd ed. 1951.)

—— 1954. *Archaeopteryx Lithographica.* London: British Museum (Nat. Hist.). (See also C. Darwin [1960, 1967].)

Delage, Yves, and M. Goldsmith. 1909. *Les théories de l'évolution.* Paris: Flammarion. (English trans., B. W. Heubsch. 1912. *The Theories of Evolution.* New York.)

Delbrück, Max. 1949. A physicist looks at biology. *Trans. Conn. Acad. Arts Sci.* 38: 173–190.

—— 1971. Aristotle-totle-totle, in J. Monod and E. Borek (eds.), *Of Microbes and Life.* New York: Columbia University Press, pp. 50–55.

Demerec, M. 1938. Eighteen years of research on the gene. Carnegie Institute Washington, Publ. no. 501, pp. 295–314.

—— 1955. What is a gene—twenty years later. *Amer. Nat.* 89: 5–20.

—— 1967. Properties of genes, in Brink (1967), pp. 49–61.

Derham, W. 1713. *Physico-Theology, or, Demonstration of the Being and Attributes of God from His Works of Creation.* London.

De Vries, H. 1889. *Intracelluläre Pangenesis.* Jena: Gustav Fischer. (English trans. 1910. Chicago: Open Court Publishing Co.)

—— 1901–1903. *Die Mutationstheorie. Versuche und Beobachtungen über die Entstehung der Arten im Pflanzenreich,* vol. 1. *Die Entstehung der Arten durch*

907

Mutation; vol. 2 *Elementare Bastardlehre.* Leipzig: Veit. (English trans. J. B. Farmer, and A. D. Darbishire. 1909–1910. Chicago: Open Court Publishing Co.)

——— 1906. *Species and Varieties: Their Origin by Mutation.* 2nd ed. Chicago: Open Court Publishing Co.

——— 1909. Variation, in Seward (1909), pp. 66–84.

Dewey, John. 1909. The influence of Darwinism on philosophy. *Pop. Sci. Month.* 75: 90–98. (Reprinted in Loewenberg, B. J. 1957. *Darwinism: Reaction or Reform.* New York: Rinehart & Co.)

De Witt, N. W. 1965. *Epicurus and His Philosophy.* Minneapolis: University of Minnesota Press.

Dexter, R. W. 1979. The impact of evolutionary theories on the Salem group of Agassiz zoologists (Morse, Hyatt, Packard, Putnam). *Essex Institute Historical Collections* 115(3): 144–171.

Diamond, J. 1966. Zoological classification system of a primitive people. *Science* 151: 1102–4.

Dickerson, R. E. 1978. Chemical evolution and the origin of life. *Sci. Amer.* 239 (3): 70–86.

Diderot, D. 1749. *Lettre sur les aveugles.*

——— 1754. *Pensées sur l'interpretation de la nature.*

——— 1769. *Le rêve de d'Alembert.* (English trans. L. W. Tancock. 1966. *The Dream of D'Alembert.* Middlesex: Harmondsworth.)

Dijksterhuis, E. J. 1961. *The Mechanization of the World Picture,* trans. C. Dikshoorn. Oxford: Clarendon Press.

Dillenberger, J. 1960. *Protestant Thought and Natural Sciences.* Garden City, N.Y.: Doubleday.

Dirac, P. A. M. 1977. Heisenberg's influence on physics. *Commentarii* 3 (14): 1–15.

Dobell, C. 1960. *Antony van Leeuwenhoek and His "Little Animals."* New York: Dover.

Dobzhansky, Th. 1935. A critique of the species concept in biology. *Phil. Sci.* 2: 344–355.

——— 1936. Position effects of genes. *Biol. Rev.* 11: 364–384.

——— 1937. *Genetics and the Origin of Species.* New York: Columbia University Press.

——— 1951. *Genetics and the Origin of Species.* 3rd ed. New York: Columbia University Press.

——— 1970. *Genetics of the Evolutionary Process.* New York: Columbia University Press.

——— 1972. [Review of Provine (1971)]. *Perspec. Biol. Med.* 1972: 645–646.

———, and C. Epling. 1944. Contributions to the genetics, taxonomy, and ecology of *Drosophila pseudoobscura* and its relatives. Carnegie Inst. Wash., Publ. no. 554, pp. 1–183.

———, F. J. Ayala, G. L. Stebbins, and J. W. Valentine. 1977. *Evolution.* San Francisco: W. H. Freeman.

Dohrn, Anton. 1875. *Der Ursprung der Wirbelthiere und das Princip des Functionswechsels*. Leipzig: Engelmann.

Donahue, J. 1978. Review of *Heracletean Fire: Sketches from a Life before Nature*, by Erwin Chargaff. *Nature* 276: 133.

Doncaster, L., and G. H. Raynor. 1906. Breeding experiments with Lepidoptera. *Proc. Zool. Soc. London* 1: 125–133.

Dougherty, E. C. (ed.) 1963. *The Lower Metazoa: Comparative Biology and Phylogeny*. Berkeley and Los Angeles: University of California Press.

Doyle, J. A. 1978. Origin of angiosperms. *Ann. Rev. Ecol. Syst.* 9: 365–392.

Driesch, H. 1899, 1901, 1909. *Philosophie des Organischen*. Leipzig: Quelle und Meyer.

—— 1905. *Der Vitalismus als Geschichte und als Lehre*. Leipzig: J. A. Barth.

—— 1951. *Lebenserinnerungen*. München and Basel: Ernst Reinhard.

Dubos, René. 1965. *Man Adapting*. New Haven: Yale University Press.

—— 1976. *The Professor, the Institute, and DNA*. New York: Rockefeller University Press.

Dunn, E. R. 1922. A suggestion to zoogeographers. *Science* 56: 336–338.

Dunn, L. C. 1965a. *A Short History of Genetics*. New York: McGraw-Hill.

—— 1965b. William Ernest Castle (1867–1962). *Biogr. Mem. Nat. Acad. Sci.* 38: 31–80.

—— 1966. Xenia and the origin of genetics. *Proc. Amer. Phil. Soc.* 117: 105–111.

—— 1973. Wilhelm Ludwig Johannsen (1857–1927), in *Dict. Sci. Bio.* 7: 113–115.

—— (ed.) 1951. *Genetics in the Twentieth Century*. New York: Macmillan.

Durand, J. A. 1978. L'idée d'évolution dans l'oeuvre d'Albert Gaudry (1827–1890). D.Sc. diss., University of Paris VI.

Du Rietz, G. E. 1930. The fundamental units of biological taxonomy. *Svensk. Bot. Tidskrift* 24: 333–428.

Düring, I. 1966. *Aristoteles, Darstellung und Interpretation seines Denkens*. Heidelberg: Carl Winter.

Edwards, W. N. 1967. *The Early History of Paleontology*. London: British Museum (Natural History).

Egerton, F. N. 1968. Studies of animal populations from Lamarck to Darwin. *J. Hist. Biol.* 1: 225–259.

—— 1973. Changing concepts of the balance of nature. *Quart. Rev. Biol.* 48: 322–350.

—— 1975. Aristotle's population biology. *Arethusa* 8: 307–330.

Ehrendorfer, F. (ed.) 1970. Biosystematics at the crossroads. *Taxon* 19: 137–304.

Ehrlich, P. R., and P. H. Raven. 1965. Butterflies and plants: a study in co-evolution. *Evolution* 18: 586–608.

—— 1969. Differentiation of populations. *Science* 165: 1228–32.

Eichler, Wolfdietrich. 1968. Kritische Einwände gegen die Hennigsche kladistische Systematik. *Biol. Rdsch.* 16: 175–185.

Eigen, M. 1971. The hypercycle. *Naturwiss.* 58: 465, 519.

────── and P. Schuster. 1977–1978. The hypercycle. *Naturwiss.* 64: 541–565; 65: 7–41, 341–369.

────── and Ruthild Winkler. 1975. *Das Spiel, Naturgesetze steuern den Zufall.* München: R. Piper & Co.

Eimer, Th. 1888. *Die Entstehung der Arten auf Grund von Vererbung erworbener Eigenschaften* I. Jena: Gustav Fischer.

Eiseley, L. C. 1958. *Darwin's Century.* New York: Doubleday.

────── 1959. Charles Darwin, Edward Blyth, and the theory of natural selection. *Proc. Amer. Phil. Soc.* 103: 94–158.

Eldredge, Niles. 1977. Trilobites and evolutionary patterns, in Hallam (1977), pp. 305–332.

────── and S. J. Gould. 1972. Punctuated equilibria: an alternative to phyletic gradualism, in Schopf and Thomas (1972), pp. 82–115.

Ellegard, A. 1957. The Darwinian theory and nineteenth century philosophies of science. *J. Hist. Ideas* 18: 362–393.

────── 1958. *Darwin and the General Reader.* Göteborg: Göteborg Universitets Arsskrift.

Elsberg, L. 1876. On the plastidule hypothesis. *Proc. Seventy-fifth Meeting Amer. Assoc. Adv. Sci.* 1876: 178–187.

Elton, Charles S. 1958. *The Ecology of Invasions by Animals and Plants.* London: Methuen.

Endler, John A. 1977. *Geographic Variation, Speciation, and Clines.* Princeton: Princeton University Press.

────── 1978. A predator's view of animal color patterns. *Evol. Biol.* 11: 319–364.

Engler, A. 1899. *Die Entwicklung der Pflanzengeographie in den letzten hundert Jahren. In Wissensch. Beiträge zum Gedächtnis der hundertsten Wiederkehr des Antritts von A. von Humboldt's Reise nach Amerika.* Berlin: Gesellschaft für Erdkunde.

────── 1914. Pflanzengeographie, in *Kultur der Gegenwart,* pt. 2, Sec. 4, 4: 187–263.

Eyde, Richard H. 1975. The foliar theory of the flower. *Amer. Sci.* 63: 430–437.

Eyles, J. M. 1969. William Smith: some aspects of his life and work, in Schneer (1969), pp. 142–158.

Falconer, D. S. 1960. *Introduction to Quantitative Genetics.* Edinburgh and London: Oliver & Boyd.

Farber, Paul L. 1972. Buffon and the concept of species. *J. Hist. Biol.* 5: 259–284.

────── 1975. Buffon and Daubenton: divergent traditions within the Histoire Naturelle. *Isis* 66: 63–74.

———— 1977. The development of taxidermy and the history of ornithology. *Isis* 68: 550–566.

Farley, J. 1974. The initial reactions of French biologists to Darwin's *Origin of Species. J. Hist. Biol.* 7: 275–300.

———— 1977. *The Spontaneous Generation Controversy from Descartes to Oparin.* Baltimore and London: Johns Hopkins University Press.

Ferguson, A. 1976. Can evolutionary theory predict? *Amer. Nat.* 110: 1101–4.

Ferris, G. F. 1928. *The Principles of Systematic Entomology.* Stanford, Calif.: Stanford University Press.

Feyerabend, P. 1975. *Against Method.* London. NLB.   *911*

Fichman, M. 1977. Wallace: zoogeography and the problem of land bridges. *J. Hist. Biol.* 10: 45–63.

Fischer, Hermann. 1929. *Mittelalterliche Pflanzenkunde.* Hildesheim: Georg Oìms.

———— 1967. *Conrad Gessner, Universalgelehrter, Naturforscher, Arzt.* Zürich: Orell Füssli.

Fisher, R. A. 1918. The correlations between relatives on the supposition of Mendelian inheritance. *Trans. Roy. Soc.* Edinburgh 52: 399–433.

———— 1922. On the dominance ratio. *Proc. Roy. Soc.* Edinburgh 42: 321–341.

———— 1930. *The Genetical Theory of Natural Selection.* Oxford: Clarendon Press. (Rev. ed. 1958. New York: Dover).

———— 1959. Natural selection from the genetical standpoint. *Aust. J. Sci.* 22: 16–17.

Fleischmann, Albert. 1901. *Die Descendenztheorie.* Leipzig: Arthur Georgi.

Florkin, M. 1972ff. *A History of Biochemistry.* 4 vols. New York: Elsevier.

Forbes, E. 1846. On the connection between the distribution of the existing fauna and flora of the British Isles and the geological changes which have affected their area, especially during the epoch of the Northern Drift. *Mem. Geological Survey, Great Britain,* vol. 1. London.

Ford, E. B. 1964. *Ecological Genetics.* London: Methuen.

Forster, J. R. 1778. *Observations Made during a Voyage round the World.* London.

Foster, W. D. 1965. *A History of Parasitology.* Edinburgh and London: Livingstone.

Fothergill, P. G. 1952. *Historical Aspects of Organic Evolution.* London: Hollis & Carter.

Foucault, M. 1966. *Les mots et les choses: Une archéologie des sciences humaines.* Paris: Gallimard. (English trans. 1971. *The Order of Things.* New York.)

Fox, G. E. et al. 1980. The phylogeny of prokaryotes. *Science* 209: 457–463.

Fox, S. W., and K. Dose. 1979. *Molecular Evolution and the Origin of Life.* Rev. ed. New York: Dekker.

Franz, Victor. 1920. *Die Vervollkommnung in der lebenden Natur: Eine Studie über ein Naturgesetz.* Jena: Gustav Fischer.

——— 1935. *Der biologische Fortschritt.* Jena: Gustav Fischer.

Frazer, J. G. 1909. Some primitive theories of the origin of man, in Seward (1909), pp. 152–170.

Frazzetta, T. H. 1975. *Complex Adaptations in Evolving Populations.* Sunderland, Mass.: Sinauer Associates.

Freeman, D. 1974. The evolutionary theories of Charles Darwin and Herbert Spencer. *Curr. Anthrop.* 15: 211–237.

Freeman, K. 1946. *The Presocratic Philosophers.* Oxford: Blackwell.

Froggatt, P., and N. C. Nevin. 1971. The "law of ancestral heredity" and the Mendelian-ancestrian controversy in England, 1889–1906. *J. Med Gen.* 8: 1–36.

——— 1972. Galton's Law of Ancestral Heredity: its influence on the early development of human genetics. *Hist. Sci.* 10: 1–27.

Fruton, J. S. 1972. *Molecules and Life.* New York: Wiley-Interscience.

Futuyma, Douglas J. 1979. *Evolutionary Biology.* Sunderland, Mass.: Sinauer Associates.

——— and Gregory C. Mayer. 1980. Non-allopatric speciation in animals. *Syst. Zool.* 29 (3): 254–271.

Gadow, H. 1913. *The Wanderings of Animals.* New York: G. P. Putnam's Sons.

Gaissinovitch, A. E. 1973. Problems of variation and heredity in Russian biology in the late nineteenth century. *J. Hist. Biol.* 6: 97–123.

Galton, Francis. 1872. On blood relationships. *Proc. Roy. Soc.* 20: 394–402.

——— 1876. A theory of heredity. *J. Anthrop. Inst. Great Britain and Ireland* 5: 329–348.

——— 1894. Discontinuity in evolution. *Mind* N.S. 3: 362–372.

Gardiner, P. 1952. *The Nature of Historical Explanation.* London: Oxford University Press.

Gardner, Eldon J. 1960. *History of Biology.* Minneapolis, Minn.: Burgess Publishing Co.

Garrod, A. E. 1909. *Inborn Errors of Metabolism.* London: Frowde & Hodder.

Gaupp, Ernst. 1917. *August Weismann: Sein Leben und sein Werk.* Jena: Gustav Fischer.

Gay, Peter. 1966. *The Enlightenment: An Interpretation.* New York: Alfred Knopf.

Gegenbaur, Carl. 1859, 1870. *Grundzüge der Vergleichenden Anatomie.* 1st and 2nd eds. Leipzig.

Geikie, A. 1897. *The Founders of Geology.* London.

Geison, G. L. 1969. Darwin and heredity: the evolution of his hypothesis of pangenesis. *J. Hist. Med. Allied Sci.* 24: 375–411.

Generoso, W. M. et al. (eds.). 1980. *DNA Repair and Mutagenesis in Eukaryotes*. New York: Plenum Books.

Geoffroy, Isidore St. Hilaire. 1830. *Principes de philosophie zoologique discutés en Mars 1830*. Paris: Pichon.

———— 1847. *Vie, travaux, et doctrine scientifique d'Etienne Saint-Hilaire par son fils*. Paris: Bertrand.

Ghiselin, Michael T. 1969. *The Triumph of the Darwinian Method*. Berkeley and Los Angeles: University of California Press.

———— 1971–1972. The individual in the Darwinian revolution. *New Lit. Hist.* 3: 123.

———— 1972. [Review of Provine (1971)]. *Science* 175: 507.

———— 1974. *The Economy of Nature and the Evolution of Sex*. Berkeley and Los Angeles: University of California Press.

———— 1974. A radical solution to the species problem. *Syst. Zool.* 23: 536–544.

———— 1975. The rational of pangenesis. *Genetics* 79: 47–57.

———— and L. Jaffe. 1973. Phylogenetic classification in Darwin's monograph on the subclass *Cirripedia*. *Syst. Zool.* 22: 132–140.

Giere, R. N., and R. S. Westfall (eds). 1973. *Foundations of Scientific Method: The Nineteeth Century*. Bloomington and London: Indiana University Press.

Gilbert, S. F. 1978. The embryological origins of the gene theory. *J. Hist. Biol.* 11: 307–351.

Gillespie, N. C. 1979. *Charles Darwin and the Problem of Creation*. Chicago and London: University of Chicago Press.

Gillispie, C. C. 1951. *Genesis and Geology: The Impact of Scientific Discoveries upon Religious Beliefs in the Decades before Darwin*. New York: Harper & Brothers.

Gilmour, J. S. L. 1940. Taxonomy and philosophy, Huxley, J. S. (1940), pp. 461–474.

———— 1951. The development of taxonomic theory since 1851. *Nature* 168: 400–402.

Gilson, E. 1960. *The Christian Philosophy of St. Augustine*. New York: Random House.

Gingerich, Philip D. 1976. Paleontology and phylogeny: patterns of evolution at the species level in early tertiary mammals. *Amer. J. Science* 276: 1–28.

———— 1977. Patterns of evolution in the mammalian fossil record, in Hallam (1977), pp. 469–500.

———— 1979. Paleontology, phylogeny, and classification: an example from the mammalian record. *Syst. Zool.* 28: 451–464.

Glacken, Clarence J. 1967. *Traces on the Rhodian Shore. Nature and Culture in Western Thought from Ancient Times to the End of the Eighteenth Century*. Berkeley and Los Angeles: University of California Press.

Glass, B. 1959a. The germination of the idea of biological species, in Glass, Temkin, and Strauss (1959), pp. 30–50.

913

———— 1959b. Maupertuis, pioneer of genetics and evolution, in Glass, Temkin, and Strauss (1959), pp. 51–83.

Glass, Bentley, O. Temkin, and W. L. Strauss, Jr. (eds.). 1959. *Forerunners of Darwin, 1745–1859.* Baltimore: Johns Hopkins University Press.

Glick, T. F. (ed.) 1974. *The Comparative Reception of Darwinism.* Austin and London: University of Texas Press.

Gloger, C. L. 1827. Etwas über die der Aufstellung neuer Vogelarten durch Hn. Brehm zum Grunde liegende Ansicht. *Isis von Oken* 20, col. 688.

———— 1833. *Das Abändern der Vögel durch Einfluss des Klimas.* Breslau: August Schulz.

Goethe, W. 1795. *Erster Entwurf einer allgemeinen Einleitung in die vergleichende Anatomie.* (Published 1820. *Morphologie* 1 [2].)

———— 1807. *Bildung und Umbildung organischer Naturen.*

Goldschmidt, R. 1938. *Physiological Genetics.* New York: McGraw-Hill.

———— 1940. *The Material Basis of Evolution.* New Haven: Yale University Press.

———— 1955. *Theoretical Genetics.* Berkeley and Los Angeles: University of California Press.

Goodfield, June. 1974. Changing strategies: a comparison of reductionist attitudes in biological and medical research in the nineteenth and twentieth centuries, in Ayala and Dobzhansky (1974), pp. 65–86.

Gotthelf, Allan. 1976. Aristotle's conception of final causality. *Rev. Metaphysics* 30: 226–254.

Goudge, T. A. 1961. *The Ascent of Life.* Toronto: University of Toronto Press.

———— 1965. Another look at emergent evolutionism. *Dialogue* 4: 273–285.

Gould, S. J. 1970. Evolutionary paleontology and the science of form. *Earth-Science Reviews* 6: 77–119. (See also *Amer. Zool.* 15: 294–481.)

———— 1977a. *Ontogeny and Phylogeny.* Cambridge: Harvard University Press, Belknap Press.

———— 1977b. The return of hopeful monsters. *Nat. Hist.* 86 (June–July): 22–30.

———— 1979. A quahog is a quahog. *Nat. Hist.* 88 (7) (August–September): 18–26.

Grant, Verne. 1957. The plant species in theory and practice, in Mayr (1957), pp. 39–80.

———— 1964. *The Architecture of the Germ Plasm.* New York: John Wiley & Sons.

———— 1971. *Plant Speciation.* New York: Columbia University Press.

———— 1975. *Genetics of Flowering Plants.* New York: Columbia University Press.

———— 1977. *Organismic Evolution.* San Francisco: W. H. Freeman.

Grassé, P. P. 1977a. *Evolution of Living Organisms.* New York: Academic Press. (Trans. of *L'evolution du vivant.* 1973. Paris: Albin Michel.)

914

———— 1977b. Les gènes en surimpression: une priorité. *C. R. Acad. Sci., Paris* Ser. D, 284: 141–142.

Grassle, J. P., and J. F. Grassle. 1976. Sibling species in the marine pollution indicator *Capitella* (Polychaeta). *Science* 192: 567–569.

Gray, Asa. 1876. *Darwiniana.* New York: D. Appleton. (Rpt. 1963. Cambridge: Harvard University Press.)

Gray, James. 1953. *How Animals Move: The Royal Institution Christmas Lectures 1951.* Cambridge: Cambridge University Press.

Green, J. R. 1909. A history of botany 1860–1900, being a continuation of Sachs "History of Botany," 1530–1860. (Facsimile reprint 1967. New York: Russell & Russell.)

Greene, E. L. 1912. *Carolus Linnaeus.* Philadelphia: Christopher Sower.

Greene, John C. 1959. *The Death of Adam: Evolution and Its Impact on Western Thought.* Iowa City: University of Iowa Press.

———— 1967. Review of M. Foucault's *Les mot et les choses. Social Science Information* 6 (4): 131–138.

———— 1971. The Kuhnian paradigm and the Darwinian revolution, in Roller (1971), pp. 3–25.

———— 1975. Reflections on the progress of Darwin studies. *J. Hist. Biol.* 8: 243–273.

———— 1977. Darwin as a social evolutionist. *J. Hist. Biol.* 10: 1–27.

———— 1981. *Science, Ideology, and World View.* Berkeley and Los Angeles: University of California Press.

Gregory, F. 1977. *Scientific Materialism in Nineteenth Century Germany.* Dordrecht and Boston: D. Reidel.

Gregory, M. S., A. Silvers, and D. Dutch (eds.). 1978. *Sociobiology and Human Nature.* San Francisco: Jossey-Bass.

Gregory, W. K. 1914. Convergence and allied phenomena in the mammalia. *Trans. Brit. Assoc. Adv. Sci.* Report on the 83rd meeting. Birmingham, 1913.

———— 1936. On the meaning and limits of irreversibility of evolution. *Amer. Nat.* 70: 517–528.

Grell, K. G. 1972. *Trichoplax:* Eibildung, Furchung. *Z. Morphol. Tiere* 73: 297–314.

Grell, R. F. 1978. Time of recombination in the *Drosophila melanogaster* oocyte: evidence from a temperature-sensitive recombination-deficient mutant. *Proc. Nat. Acad. Sci.* 75: 3351–54.

———— (ed.) 1974. *Mechanisms in Recombination.* New York: Plenum Press.

Grene, M. 1959. Two evolutionary theories. *Brit. J. Phil. Sci.* 9: 110–127, 185–193.

———— 1963. *A Portrait of Aristotle.* London: Faber.

Gressitt, J. L. 1956. Some distribution patterns of Pacific Island faunae. *Syst. Zool.* 5: 11–47.

Gruber, H. E. 1974. *Darwin on Man.* New York: Dutton.

Grun, P. 1976. *Cytoplasmic Genetics and Evolution.* New York: Columbia University Press.

Gulick, A. 1932. Biological peculiarities of oceanic islands. *Quart. Rev. Biol.* 7: 405–427.

Gulick, J. T. 1872. Diversity of evolution under one set of external conditions. *J. Linn. Soc. of London, Zoology* 11: 496–505.

——— 1888. Divergent evolution through cumulative segregation. *J. Linn. Soc.* 20: 189–274, 312–380.

Guppy, H. B. 1906. Observations of a naturalist in the Pacific between 1891 and 1899. Vol. 2 (*Plant Dispersal*). London: Macmillan.

Gustafson, A. 1969. The life of Gregor Johann Mendel—tragic or not? *Hereditas* 62: 239–258.

Guthrie, W. K. C. 1965. *A History of Greek Philosophy*. Cambridge: Cambridge University Press.

Guyénot, E. 1941. *Les sciences de la vie aux XVIIe et XVIIIe siècles: L'idée d'évolution*. Paris: Albin Michel.

Haber, F. C. 1959. Fossils and the idea of a process of time in natural history, in Glass, Temkin, and Strauss (1959), pp. 222–261.

Haeckel, E. 1866. *Generelle Morphologie der Organismen: Allgemeine Grundzüge der organischen Formen-Wissenschaft, mechanisch begründet durch die von Charles Darwin reformirte Descendenz-Theorie*. 2 vols. Berlin: Georg Reimer.

——— 1868. *Natürliche Schöpfungsgeschichte*. Berlin: Georg Reimer.

——— 1875. *Ziel und Wege der heutigen Entwicklungsgeschichte*. Jena: Hermann Duffl.

——— 1876. *Die Perigenesis der Plastidule oder die Wellenzeugung der Lebensteilchen*. Berlin.

Haffer, J. 1974. *Avian Speciation in Tropical South America*. Cambridge, Mass.: Nuttall Ornithological Club Publ. no. 14.

Hagberg, K. 1939. *Carl Linnaeus*. Stockholm. (English ed. 1952. London: Jonathan Cape.)

Hagemann, R. 1964. *Plasmatische Vererbung*. Jena: Gustav Fischer.

Haldane, J. B. S. 1924–1932. A mathematical theory of natural and artificial selection. 9 parts. *Trans. Proc. Camb. Phil. Soc.*

——— 1929. The origin of life. *Rationalist Ann.*, p. 3.

——— 1932. *The Causes of Evolution*. New York: Longmans, Green.

——— 1937. The effect of variation on fitness. *Amer. Nat.* 71: 337–349.

——— 1949. Human evolution: past and future, in Jepsen, Mayr, and Simpson (1949), pp. 405–418.

——— 1957. The cost of natural selection. *J. Genetics* 55: 511–524.

Hall, A. R. 1954. *The Scientific Revolution, 1500–1800: The Formation of the Modern Scientific Attitude*. London: Longmans, Green.

Hall, T. S. 1969. *Ideas of Life and Matter*. 2 vols. Chicago: University of Chicago Press.

Hallam, A. 1973. *A Revolution in the Earth Sciences: From Continental Drift to Plate Tectonics*. Oxford: Clarendon Press.

——— (ed.). 1977. *Patterns of Evolution as Illustrated by the Fossil Record.* Amsterdam and New York: Elsevier.

Haller, M. H. 1963. *Eugenics.* New Brunswick: Rutgers University Press.

Halvorson, Harlyn O., and K. E. Van Holde (eds.). 1980. *The Origins of Life and Evolution.* New York: Alan R. Liss.

Hamburger, Viktor. 1980. Evolutionary theory in Germany: a comment, in Mayr and Provine (1980), pp. 303–308.

Hamilton, T. H., and R. H. Barth. 1962. The biological significance of season change in male plumage appearance in some new world migratory bird species. *Amer. Nat.* 96: 129–144.

Hamilton, W. D. 1964. The genetical evolution of social behavior, pts. 1–2. *J. Theoret. Biol.* 7: 1–52.

Hampson, Norman. 1968. *The Enlightenment.* Middlesex: Harmondsworth.

Hanawalt, P. C., E. C. Friedberg and C. F. Fox (eds.). 1978. *DNA Repair Mechanisms.* New York: Academic Press.

Hanks, L. 1966. *Buffon avant l'histoire naturelle.* Paris: Presses Universitaires de France.

Hardy, G. H. 1908. Mendelian proportions in a mixed population. *Science* N.S. 28: 49–50.

Harper, J. L. 1977. *Population Biology of Plants.* New York: Academic Press.

Harris, G. W., and R. G. Edwards (eds.). 1970. A discussion on the determination of sex. *Phil. Trans. Roy. Soc.* B259, no. 828: 1–206.

Harris, H. 1966. Enzyme polymorphisms in man. *Proc. Royal Soc.* B164: 298–316.

Harrison, J. 1972. Erasmus Darwin's views on evolution. *J. Hist. Ideas* 32: 247–264.

Hartmann, Max. 1924. *Allgemeine Biologie: Eine Einführung in die Lehre vom Leben.* Jena: Gustav Fischer. (3rd ed. 1947.)

——— 1941. *Zur Lehre vom Eidos bei Platon und Aristoteles.*

Hawkes, J. G. (ed.). 1968. *Chemotaxonomy and Serotaxonomy.* London and New York: Academic Press.

Hazard, Paul. 1957. *European Thought in the Eighteenth Century: From Montesquieu to Lessing,* trans. J. Lewis May. New Haven: Yale University Press.

Heberer, G. 1943a. *Evolution der Organismen.* Jena: Gustav Fischer.

——— 1943b. Das Typenproblem in der Stammesgeschichte, in Heberer (1943a), pp. 545–585.

Hecht, M. K., and W. C. Steere (eds.). 1970. *Essays in Evolution and Genetics.* New York: Appleton-Century-Crofts.

Hedberg, O. (ed.). 1958. Systematics of today. *Uppsala Univ. Årsskrift* 1958: 6.

Hegnauer, R. 1962–1966. *Chemotaxonomie der Pflanzen.* Basel and Stuttgart: Birkhauser.

Heimans, J. 1962. Hugo de Vries and the gene concept. *Amer. Nat.* 96: 93–102.

917

—— 1978. Hugo de Vries and the gene theory. *Proc. XV Int. Cong. Hist. Sci., Edinburgh*, pp. 469–480.

Heincke, Fr. 1898. *Naturgeschichte des Herings.* I. *Die Lokalformen u. die Wanderungen.* Berlin: O. Salle (Abh. d. deutsch. Seefischerei - Ver. II).

Heitz, E., and H. Bauer. 1933. Beweise für die Chromosomennatur der Kernschleifen in den Knäuelkernen von Bibio hortulanus. *Z. Zellforsch.* 17: 67–82.

Hempel, C. G. 1965. *Aspects of Scientific Explanation.* New York: Free Press.

——, and P. Oppenheim. 1948. Studies in the logic of explanation. *Phil. Sci.* 15: 135–175.

Henderson, L. J. 1913. *The Fitness of the Environment.* New York: Macmillan. (Rpt. 1958. Boston: Beacon Press.)

Hennig, Willi. 1950. *Grundzüge einer Theorie der Phylogenetischen Systematik.* Berlin: Deutscher Zentralverlag.

—— 1965. Phylogenetic Systematics. *Ann. Rev. Entomol.* 10: 97–116.

—— 1966. *Phylogenetic Systematics,* trans. D. D. Davis and R. Zangerl. Urbana: University of Illinois Press.

—— 1974. Kritische Bemerkungen zur Frage "Cladistic analysis or cladistic classification?" *Z. zool. Syst. Evolut.-forsch.* 12: 279–294.

Hensen, V. 1887. Über die Bestimmung des Planktons oder des im Meere treibenden Materials an Pflanzen und Thieren. *Ber. Komm. wiss. Unters. deutschen Meere, Kl.* 5: 1–107.

Herbert, Sandra. 1971. Darwin, Malthus, and selection. *J. Hist. Biol.* 4: 209–217.

—— 1974. The place of man in the development of Darwin's Theory of Transmutation, pt. 1. *J. Hist. Biol.* 7: 217–258.

—— 1977. The place of man in the development of Darwin's Theory of Transmutation, pt. 2. *J. Hist. Biol.* 10: 243–273.

Herbert, W. 1837. *Amaryllidaceae.* London: James Ridgway.

Hertwig, O. 1884. Das Problem der Befruchtung und der Isotropie des Eies, eine Theorie der Vererbung. *Jena. Z. Naturwiss.* 18: 21–23.

—— 1921. Zur Abwehr des ethischen, des sozialen, und des politischen Darwinismus. 2nd ed. Jena: Gustav Fischer.

Hertwig, R. 1927. *Abstammungslehre und neuere Biologie.* Jena: Gustav Fischer.

Heslop-Harrison, J. W. 1963. Species concepts, in T. Swain (ed.), *Chemical Plant Taxonomy.* New York: Academic Press, pp. 17–40.

Hess, Eugene L. 1970. Origins of molecular biology. *Science* 168: 664–669.

Hesse, R., W. C. Allee, and Karl P. Schmidt. 1951. *Ecological Animal Geography.* New York: John Wiley & Sons.

Hilts, Victor L. 1973. Statistics and social science, in Giere and Westfall (1973), pp. 206–233.

Himmelfarb, Gertrude. 1959. *Darwin and the Darwinian Revolution.* Garden City, N.Y.: Doubleday.

His, W. 1871. Die Theorien der geschlechtlichen Zeugung. *Arch. f. Anthropol.* 4: 197–220, 317–332. (See also 5 [1872]: 69–111.)

——— 1874. *Unsere Körperform und das physiologische Problem ihrer Entstehung.* Leipzig: Vogel.

——— 1901. Das Princip der organbildenden Keimbezirke und die Verwandtschaften der Gewebe: Historisch kritische Bemerkungen. *Arch. Anat. Phys. Anat. Abh.* 1901: 307–337.

Hodge, M. J. S. 1971a. Lamarck's science of living bodies. *Brit. Jour. Hist. Sci.* 5: 323–352.

——— 1971b. Species in Lamarck, Lamarck (1971), pp. 31–46.

——— 1972. The universal gestation of nature: Chambers' *Vestiges* and *Explanations. J. Hist. Biol.* 5: 127–151.

——— 1974. Darwinism in England, in Glick (1974), pp. 3–31, 32–80.

——— 1977. The structure and strategy of Darwin's "long argument." *Brit. J. Hist. Sci.* 10: 237–246.

——— 1981 (in press). Darwin and natural selection: his methods and his methodology.

——— 1981 (in press). (Darwin on extinction.)

Hoeppli, R. 1959. *Parasites and Parasitic Infections in Early Medicine and Science.* Singapore: University of Malaya Press.

Hoffmann, H. 1881. Rückblicke auf meine Variations-Versuche von 1855–1880. *Bot. Zeitung* 39: 345–425.

Hofmeister, W. 1851. *Vergleichende Untersuchungen der Keimung, Entfaltung und Fruchtbildung höherer Kryptogamen,* etc. Leipzig.

Hofstadter, Richard. 1944. *Social Darwinism in American Thought.* Philadelphia: University of Pennsylvania Press.

Hofsten, N. v. 1916. Zur älteren Geschichte des Diskontinuitätsproblems in der Biogeographie. *Zool. Annal.* 7: 197–353.

——— 1936. From Cuvier to Darwin. *Isis* 24: 361–366.

——— 1936b. Ideas of creation and spontaneous generation prior to Darwin. *Isis* 25: 80–94.

——— 1958. Linnaeus's conception of nature. *Kungl. Vetensk.-Soc. Arsbok* (1957), pp. 65–105.

——— 1963. A system of "double entries" in the zoological classification of Linnaeus. *Zool. Bidr. Uppsala* 35: 603–631.

Hölder, H. 1960. *Geologie und Paläontologie in Texten und ihre Geschichte.* Freiburg and München: Karl Alber.

Holmes, F. L. 1977. Conceptual history. *Stud. Hist. Biol.* 1: 209–218.

Holmes, S. J. 1947. K. E. von Baer's perplexities over evolution. *Isis* 37: 7–14.

Hook, S. (ed.). 1963. *Philosophy and History.* New York: New York University Press.

Hooker, J. D. 1853. *The Botany of the Antarctic Voyage II. Flora Novae-Zelandiae.* London.

919

—— (ed. Leonard Huxley). 1918. *Life and Letters of Sir Joseph Dalton Hooker.* 2 vols. London: Murray.

Hooykaas, R. 1952. The species concept in eighteenth century mineralogy. *Arch. int. hist. sci.* 5: 18–19, 45–55.

—— 1959. *Natural Law and Divine Miracle.* Leiden: Brill.

—— 1972. *Religion and the Rise of Modern Science.* Edinburgh and London: Scottish Academic Press.

Hoppe, B. 1978. Der Ursprung der Diagnosen in der botanischen und zoologischen Systematik. *Sudhoff's Archiv* 62: 105–130.

Hotchkiss, R. D. 1965. Oswald T. Avery. *Genetics* 51: 1–10.

—— 1966. Gene, transforming principle, and DNA, in Cairns, Stent, and Watson (1966), pp. 180–200.

Hubby, J. L., and R. Lewontin. 1966. The number of alleles at different loci in *Drosophila pseudoobscura. Genetics* 54: 577–594.

Hughes, A. 1959. *A History of Cytology.* London and New York: Abelard-Schuman.

Hull, David L. 1964. Consistency and monophyly. *Syst. Zool.* 13: 1–11.

—— 1965. The effect of essentialism on taxonomy—two thousand years of stasis. *Brit. J. Phil. Sci.* 15: 314–366; 16: 1–18.

—— 1967. Certainty and circularity in evolutionary taxonomy. *Evolution* 21: 174–189.

—— 1970. Contemporary systematic philosophies. *Ann. Rev. Ecol. Syst.* 1: 19–54.

—— 1973. *Darwin and His Critics.* Cambridge: Harvard University Press.

—— 1974. *Philosophy of Biological Science.* Englewood Cliffs, N.J.: Prentice-Hall.

—— 1975. Are species really individuals? *Syst. Zool.* 25: 174–191.

—— 1978a. A matter of individuality. *Phil. Sci.* 45: 335–360.

—— 1978b. The principles of biological classification: the use and abuse of philosophy. *Philosophy of Science Association* 2.

—— 1979. The limits of cladism. *Syst. Zool.* 28: 416–440.

—— 1981 (in press). Central subjects and historical narratives.

Humboldt, A. v. 1795. Die Lebenskraft oder der rhodische Genius: Schiller's Horen, Jahrg. 1795, St. 5, 90–96. (Rpt. in *Ansichten der Natur.* 1849. 3rd ed. 297–314.

—— 1805. *Essai sur la géographie des plantes.* Paris.

Hume, David. 1738. *Treatise of Human Nature: An Attempt to Introduce the Experimental Method of Reasoning into Moral Subjects.*

—— 1779. *Dialogues concerning Natural Religion.*

Huxley, J. S. (ed). 1940. *The New Systematics.* Oxford: Clarendon Press.

—— 1942. *Evolution, The Modern Synthesis.* London: Allen & Unwin.

—— 1958. Evolutionary processes and taxonomy with special reference to grades. *Uppsala Univ. Arsskr.* (1958), pp. 21–39.

Huxley, T. H. 1860. The Origin of Species. *West. Rev.* 17: 541–570.

—— 1863. *Evidence as to Man's Place in Nature.* New York.

Hyman, Libbie M. 1940–1959. *The Invertebrates.* 5 vols. New York: McGraw-Hill.

Iltis, H. 1932. *Life of Mendel.* London: Allen & Unwin.

Inger, R. F. 1958. Comments on the definition of genera. *Evolution* 12: 370–384.

Irvine, W. 1955. *Apes, Angels and Victorians.* New York and London: McGraw-Hill.

Jacob, Francois. 1970. *La logique du vivant: Une histoire de l'hérédité.* Paris: Gallimard.

—— 1973. *The Logic of Life.* New York: Pantheon Books.

—— (1977). Review of *La logique du vivant,* by Frederic L. Holmes, in *Studies in History of Biology* 1: 209–218.

—— 1977. Evolution and tinkering. *Science* 196: 1161–66.

Jaenike, J. 1981. Criteria for ascertaining the existence of host races. *Amer. Nat.* 117: 830–834.

Jäger, G. 1878. *Lehrbuch der allgemeinen Zoologie,* vol. 2. Leipzig.

Jameson, D. L. (ed.). 1977. *Genetics of Speciation.* Stroudsburg, Penna.: Dowden, Hutchinson & Ross.

Jardine, N., and R. Sibson. 1971. *Mathematical Taxonomy.* London: John Wiley & Sons.

Jenkin, Fleeming. 1867. The Origin of Species. *The North British Review* 46: 277–318. (Rpt. in Hull [1973]: pp. 302–344.)

Jepsen, G., E. Mayr, and G. G. Simpson (eds.). 1949. *Genetics, Paleontology, and Evolution.* Princeton: Princeton University Press.

Jessen, Karl F. W. 1864. *Botanik der Gegenwart und Vorzeit.* Leipzig: F. U. Brockhaus. (Rpt. 1948. Waltham, Mass.: Chronica Botanica Co.)

Jevons, W. S. 1877. *Principles of Science.* London: Macmillan.

Johannsen, W. 1903. *Über Erblichkeit in Populationen und in reinen Linien.* Jena: Gustav Fischer.

—— 1909. *Elemente der Exakten Erblichkeitslehre.* Jena: Gustav Fischer.

—— 1915. Experimentelle Grundlagen der Deszendenzlehre: Variabilität, Vererbung, Kreuzung, Mutation in C. Chun and W. Johannsen (eds.), *Die Kultur der Gegenwart.* III. 4. Leipzig and Berlin: B. G. Teubner, pp. 597–660.

Johnson, L. A. S. 1968. Rainbow's end: the quest for an optimal taxonomy. *Proc. Linn. Soc. New South Wales* 93: 8–45. (Rpt. 1970. *Syst. Zool.* 19: 203–239.)

Jordan, K. 1896. On mechanical selection and other problems. *Novit. Zool.* 3: 426–525.

—— 1905. Der Gegensatz zwischen geographischer und nichtgeographischer Variation. *Z. wiss. Zool.* 83: 151–210.

Joseph, H. W. B. 1916. *An Introduction to Logic.* Oxford: Oxford University Press.

Judson, H. F. 1979. *The Eighth Day of Creation*. New York: Simon and Schuster.

Kälin, J. A. 1941. *Ganzheitliche Morphologie und Homologie.* Freiburg (Schweiz) and Leipzig: Universitätsbuchhandlung.

Kammerer, Paul. 1924. *The Inheritance of Acquired Characteristics.* New York: Boni & Liveright.

Kant, I. 1790. *Kritik der Urteilskraft.* (English trans. J. H. Bernard. 1914. *Critique of Judgment.* London.)

Kearney, Hugh F. 1964. Galileo and the mathematical universe, in *Origins of the Scientific Revolution. Problems and Perspectives in History Series.* London: Longmans, Green, pp. 125–126.

Keast, J. A. 1961. Bird speciation on the Australian continent. *Bull. MCZ* 123: 306–495.

Kellogg, Vernon L. 1907. *Darwinism Today.* New York: Henry Holt.

Kerkut, G. A. 1960. *Implications of Evolution.* Oxford: Pergamon Press.

Kerner, A. von Marilaun. 1866. *Gute und schlechte Arten.* Innsbruck.

——— 1869. *Die Abhängigkeit der Pflanzengestalt von Klima und Boden.* Innsbruck.

Keynes, R. D. (ed.). 1979. *The Beagle Record.* Cambridge: Cambridge University Press.

Kim, K. C., and H. W. Ludwig. 1978. Phylogenetic relationships of parasitic *Psocodea* and taxonomic position of the *Anoplura. Ann. Ent. Soc. Amer.* 71: 910–922.

Kimura, M. 1960. Optimum mutation rate and degree of dominance as determined by the principle of minimum genetic load. *J. Genetics* 57: 21–34.

——— and T. Ohta. 1971. *Theoretical Aspects of Population Genetics.* Princeton: Princeton University Press.

King, J. L., and T. H. Jukes. 1969. Non-Darwinian evolution. *Science* 164: 788–798.

King-Hele, D. 1963. *Erasmus Darwin.* London: Macmillan.

Kirk, G. S., and J. E. Raven. 1971. *The Presocratic Philosophers.* Cambridge: Cambridge University Press.

Klein, Marc. 1936. *Histoire des origines de la théorie cellulaire.* Paris. (Vol. 3 of *Exposés d'histoire et philosophie des sciences*).

Kohlbrugge, J. H. F. 1912. G. Cuvier und K. F. Kielmeyer. *Biol. Centralbl.* 32: 291–295.

——— 1914. J. B. de Lamarck und der Einfluss seiner Deszendenztheorie von 1809 bis 1859. *Z. Morph. Anthrop.* 18: 191–206.

——— 1915. War Darwin ein originelles Genie? *Biol. Centralbl.* 35: 93–111.

Kohn, David. 1980. Theories to work by: rejected theories, reproduction, and Darwin's path to natural selection. *Studies Hist. Biol.* 4: 67–170.

Kölliker, A. 1899. *Erinnerungen aus meinem Leben.* Leipzig: Engelmann.

Kol'tsov, N. K. 1928. Physikalisch-chemische Grundlage der Morphologie. *Biol. Zentralbl.* 48: 345–369.

——— 1939. Les molécules hereditaires. *Act. sci. industr.* 776: 1–60. Paris: Hermann.

Korschelt, E. 1922. Lebensdauer, Altern, und Tod. 2nd ed. Jena: Gustav Fischer.

Korschinsky, S. 1899. Heterogenesis und Evolution. *Naturw. Wochenschr.* 14: 273–278.

——— 1901. Heterogenesis und evolution. *Flora, Erg.bd.* 89: 240–368.

Kottler, Malcolm J. 1978. Charles Darwin's biological species concept and theory of geographic speciation: the transmutation notebooks. *Ann. Sci.* 35: 275–297.

——— 1979. Hugo de Vries and the rediscovery of Mendel's laws. *Ann. Sci.* 36: 517–538.

——— 1980. Darwin, Wallace, and the origin of sexual dimorphism. *Proc. Amer. Phil. Soc.* 124: 203–226.

Koyré, Alexander. 1965. Commentaries, in Crombie (1965), p. 856.

Krizenecky, J. 1965. *Fundamenta Genetica.* Folia Mendeliana, vol. 6. Prague: Czechoslovakian Academy of Science.

Küchenmeister, F. 1855. *Die in und an dem Körper des lebenden Menschen vorkommenden Parasiten.* 2 vols. Leipzig: B. G. Teubner.

——— 1857. The animal and vegetable parasites of the human body.

Kudo, Richard R. 1966. *Protozoology.* 5th ed. Springfield, Ill.: Thomas.

Kühn, Alfred. 1950. Anton Dohrn und die Zoologie seiner Zeit. *Pubb. Stag. Zool. Napoli,* supp. 1950.

Kuhn, T. 1962. *The Structure of Scientific Revolutions.* Chicago: University of Chicago Press.

——— 1971. The relations between history and history of science. *Daedalus* (Spring 1971): 271–304.

Kühner, F. 1913. *Lamarck, Die Lehre vom Leben.* Jena: Diederichs.

Kükenthal, Willy. 1923ff. *Handbuch der Zoologie.* Berlin and Leipzig: Walter de Gruyter.

Lacépède. 1800. Discours sur la durée des espèces, in *Hist. Nat. des Poissons.* Paris, 2: xxiii–lxiv.

Lack, David. 1944. Ecological aspects of species formation in passerine birds. *Ibis* 86: 260–286.

——— 1945. The Galapagos finches: a study in variation. *Occas. Pap. Calif. Acad. Sci.,* no. 21.

——— 1947. *Darwin's Finches.* Cambridge: Cambridge University Press.

——— 1949. The significance of ecological isolation, in Jepsen, Mayr, and Simpson (1949), pp. 299–308.

——— 1954. *The Natural Regulation of Animal Numbers.* Oxford: Clarendon Press.

——— 1966. *Population Studies of Birds.* Oxford: Clarendon Press.

————— 1968. *Ecological Adaptations for Breeding in Birds.* London: Methuen.

Lakatos, I. 1970. Falsification and the methodology of scientific research programmes, in I. Lakatos and A. Musgrave (eds.), *Criticism and the Growth of Knowledge.* Cambridge: Cambridge University Press.

Lamarck, Jean-Baptiste. 1801. *Système des animaux sans vertèbres . . . précédé du discours d'ouverture du cours de zoologie, donné dans le Muséum National d'Histoire Naturelle, l'an VIII de la République.* Paris.

————— 1802a. *Hydrogéologie.* Paris: L'Auteur, etc. (English trans. A. V. Carozzi. 1964. Urbana: University of Illinois Press.)

————— 1802b. *Recherches sur l'organisation des corps vivants.* Paris.

————— 1809. *Philosophie zoologique, ou exposition des considérations relatives à l'histoire naturelle des animaux.* Paris. (English trans. Hugh Elliot. *The Zoological Philosophy.* 1914. London: Macmillan.)

————— 1815–1822. *Histoire naturelle des animaux sans vertèbres.* 7 vols. Paris.

————— 1907. Discours d'ouverture (an VII, an X, an XI, et 1806), ed. A. Giard. *Bull. Sci. de la France et de la Belgique,* Paris.

————— 1944. La biologie: texte inédite, ed. P. Grassé. *Rev. Scient.* 82: 267–276.

————— 1971. *Colloque International "Lamarck,"* ed. J. Schiller. Paris: A. Blanchard.

Landrieu, M. 1909. *Lamarck, le fondateur du transformisme.* Paris: Société Zoologique de France.

Langlet, Olof. 1971. Two hundred years of genecology. *Taxon* 20: 653–722.

Lanjouw, J., and F. A. Stafleu. 1956. *Index Herbariorum.* 3rd ed. Utrecht: International Association for Plant Taxonomy.

Lankester, Ray. 1909. *Treatise on Zoology.* London: Adams and Charles Black.

Larson, James L. 1971. *Reason and Experience: The Representation of Natural Order in the Work of Carl von Linné.* Berkeley and Los Angeles: University of California Press.

————— 1979. Linné's French critics. *Svensk. Linné. Arss.,* 1978, Uppsala, pp. 67–79.

Lattin, G. de 1967. *Grundriss der Zoogeographie.* Jena: Gustav Fischer.

Laudan, Larry. 1968. Theories of scientific method from Plato to Mach. *Hist. Sci.* 7: 1–63.

————— 1977. *Progress and Its Problems: Toward a Theory of Scientific Growth.* Berkeley and Los Angeles: University of California Press.

Leclercq, Jean, and Pierre Dagnelle. 1966. *Perspectives de la Zoologie Européenne.* I. *Histoire. Problèmes contemporains.* Gembloux: J. Duculot.

Leibniz, G. W. (ca. 1712). Monadology, in Loemker, L. E. (ed.), *G. W. Leibniz: Philosophical Papers.* 2nd ed. 1969. Dordrecht: Reidel.

Leicester, H. M. 1974. *Development of Biochemical Concepts from Ancient to Modern Times.* Cambridge: Harvard University Press.

Leidy, J. 1853. *A Flora and Fauna within Living Animals.* Smithsonian Contributions to Knowledge, no. 44. Washington D.C.

Leone, C. A. (ed.). 1964. *Taxonomic Biochemistry and Serology.* New York: Ronald Press.

Lerner, I. M. 1954. *Genetic Homeostasis.* Edinburgh: Oliver & Boyd.

―――― 1958. *The Genetic Basis of Selection.* New York: John Wiley & Sons.

―――― 1972. Noise of conflict. [A review of Provine (1971).] *Mendel Newsletter* 8 (October 1972).

Lerner, M.-P. 1969. *Recherches sur la notion de finalité chez Aristote.* Paris: Presses universitaires de France.

Lesch, J. E. 1975. The role of isolation in evolution: George J. Romanes and John T. Gulick. *Isis* 66: 483–503.

Lesky, Erna. 1950. Die Zeugungs-und Vererbungslehren der Antike und ihr Nachwirken. *Abh. Akad. Wiss., Mainz, Geistes-u. Sozial. Kl.,* no. 19.

Leuckart, Rudolf. 1848. *Über die Morphologie und die Verwandtschaftsverhältnisse der wirbellosen Thiere: Ein Beitrag zur Characteristik und Classification der thierischen Formen.* Braunschweig: Vieweg.

―――― 1879–1886. *Die menschlichen Parasiten.* 2nd ed. Leipzig: C. F. Winter.

―――― 1886. *The Parasites of Man.* Edinburgh: Pentland.

Levine, L. (ed.). 1971 *Papers on Genetics.* St. Louis: C. V. Mosby.

Lewes, G. H. 1874–1875. *Problems of Life and Mind.* 2 vols. London: Longmans, Green.

Lewis, E. B. 1967. Genes and gene complexes, in Brink (1967), pp. 17–47.

Lewis, H. 1962. Catastrophic selection as a factor in speciation. *Evolution* 16: 257–271.

―――― 1966. Speciation in flowering plants. *Science* 152: 167–172.

Lewontin, R. C. 1969. The bases of conflict in biological explanation. *J. Hist. Biol.* 2: 35–45.

―――― 1974. *The Genetic Basis of Evolutionary Change.* New York and London: Columbia University Press.

―――― , J. A. Moore, W. B. Provine, and Bruce Wallace. 1981. *Dobzhansky's Genetics of Natural Populations* I–XLIII. New York: Columbia University Press.

Ley, W. 1929. *Konrad Gesner: Leben und Werk.* München: Münchener Beiträge zur Geschichte der Naturwissenschaften, Heft. 15/16.

―――― 1968. *Dawn of Zoology.* Englewood Cliffs, N.J.: Prentice-Hall.

Liebig, Justus von. 1842. Die organische Chemie in ihrer Anwendung auf Physiologie und Pathologie.

―――― 1863. *Über Francis Bacon von Verulam und seine Methode der Naturforschung.* München.

Lima-De-Faria, A. 1975. Where is molecular biology going? *Hereditas* 81: 113–118.

925

Limoges, Camille. 1970. *La sélection naturelle*. Paris: Presses Universitaires de France.

────── 1976. Natural selection, phagocytosis, and preadaptation: Lucien Cuénot, 1886–1901. *J. Hist. Med. Allied Sci.* 31: 176–214.

Lindeboom, G. A. 1970. Boerhaave's concept of the basic structure of the body. *Clio Medica* 5: 203–208.

Lindroth, C. H. 1973. Systematics specializes between Fabricius and Darwin: 1800–1859, in Smith, R. F., T. E. Mittler, and C. N. Smith, *History of Entomology*. Palo Alto: Annual Reviews, pp. 119–154.

Lingner, E. (ed.) 1970. Museum für Naturkunde an der Humboldt-Universität zu Berlin—200 Jahre. *Wiss. Zeitschr. Humboldt-Universität. Math.-Nat. R.* 19: 123–315.

Linnaeus, C. 1739. Rön om wäxters plantering, grundat nā naturen. Svensk. Wetensk. Acad. Handl. 1

────── 1749–1769. *Amoenitates academicae*. 7 vols. Stockholm. (Partial trans. Linnaeus, c. 1781. *Select Dissertations from the Amoenitates*. London: Robinson; Ann Arbor: University Microfilms, 1974.)

────── 1753. *Species plantarum*. Halmiae.

────── 1758. *Systema naturae*. 10th ed. Stockholm.

Linné, C. 1972. *L'équilibre de la nature*. Introduction by C. Limoges. Paris: Vrin.

Lloyd, G. E. R. 1961. The development of Aristotle's theory of the classification of animals. *Phronesis* 6: 59–81.

Lock, R. H. 1906. *Recent Progress in the Study of Variation, Heredity, and Evolution*. London: Murray.

Loewenberg, B. J. 1965. Darwin and Darwin studies. *Hist. Sci.* 4: 15–54.

Lorenz, Konrad. 1973a. *Die Rückseite des Spiegels*. München and Zürich: R. Piper & Co.

────── 1973b. The fashionable fallacy of dispensing with description. *Naturwiss.* 60: 1–9.

Lovejoy, A. O. 1936. *The Great Chain of Being*. Cambridge: Harvard University Press.

────── 1959a. The argument for organic evolution before the *Origin of Species, 1830–1858*, in Glass, Temkin, and Strauss (1959), pp. 356–414.

────── 1959b. Buffon and the problem of species, in Glass, Temkin, and Strauss (1959), pp. 84–113.

────── 1959c. Herder: progressionism without transformism, in Glass, Temkin, and Strauss (1959), pp. 207–221.

────── 1959d. Kant and evolution, in Glass, Temkin, and Strauss (1959), pp. 173–206.

Lubosch, W. 1918. Der Akademiestreit zwischen Geoffroy St. Hilaire und Cuvier im Jahre 1830 und seine leitenden Gedanken. *Biol. Zentralbl.* 38: 357–384, 397–456.

────── 1931. Geschichte der vergleichenden Anatomie, in Bolk, W. et al. (eds.) *Handbuch der vergl. Anatomie Wirbeltiere*. Berlin: Urban und Schwarzenberg.

926

Ludmerer, K. 1972. *Genetics and American Society: A Historical Appraisal.* Baltimore: Johns Hopkins University Press.

Ludwig, W. 1940. Selektion und Stammesentwicklung. *Naturwiss.* 28: 689–705.

———— 1941. Zur evolutorischen Erklärung der Höhlentiermerkmale durch Allelelimination. *Biol. Zentralbl.* 62: 447–455.

———— 1943. Die Selektionstheorie, in Heberer (1943a).

Lyell, Charles. 1830–1833. *Principles of Geology, being an Attempt to Explain the Former Changes of the Earth's Surface, by Reference to Causes Now in Operation.* 3 vols. London. (Facsimile ed. 1970. Germany: J. Cramer.)

Lyell, K. 1881. *Life, Letters, and Journals of Sir Charles Lyell.* London: Murray.

Lyon, John. 1976. The "Initial Discourse" to Buffon's Histoire Naturelle. (The first complete English translation). *J. Hist. Biol.* 9: 133–181.

MacArthur, R., and E. O. Wilson. 1967. *The Theory of Island Biogeography.* Princeton: Princeton University Press.

MacDowell, E. C. 1914. Multiple factors in Mendelian inheritance. *J. Exp. Zool.* 16: 177–194.

MacLeod, Roy M. 1965. Evolutionism and Richard Owen, 1830–1868: an episode in Darwin's century. *Isis* 56: 259–280.

Mägdefrau, K. 1973. *Geschichte der Botanik.* Stuttgart: Gustav Fischer.

Magnus, R. 1906. *Goethe als Naturforscher.* Leipzig: J. A. Barth.

Maier, A. 1938. *Die Mechanisierung des Weltbildes.* Forschungen zur Geschichte der Philosophie und der Pädagogik, Heft 18. Leipzig.

Maillet, Benoit de. 1748. *Telliamed: Conversations between an Indian Philosopher and a French Missionary.* (English trans. A. V. Carozzi. 1968. Urbana: University of Illinois Press.)

Mainardi, Danilo. 1980. L'evoluzione del comportamento. No. 51. Rome: Accademia Nazionale dei Lincei.

Mainx, F. 1955. Foundations of biology. *Int. Encycl. Unif. Sci.* 1 (9): 1–86.

Majnep, I. S., and R. Bulmer. 1977. *Birds of My Kalam Country.* Auckland: Auckland University Press.

Malthus, T. R. 1798. *An Essay on the Principle of Population, as It Affects the Future Improvement of Society.* London: J. Johnson. (Darwin actually read the 6th ed. [1826]. London: Murray.)

Mandel, P., L. Mandel, and M. Jacob. 1948. Sur le comportement comparé . . . des deux ácides nucléiques des tissus animaux. *C. r. hebd. Séanc. Acad. Sci.,* Paris 226: 2019–21.

Mandelbaum, M. 1965. *History and Theory,* Supp. 5.

———— 1971. *History, Man, and Reason.* Baltimore: Johns Hopkins University Press.

Manier, E. 1978. *The Young Darwin and His Cultural Circle.* Dordrecht and Boston: D. Reidel.

927

Manwell, Reginald C. 1961. *Introduction to Protozoology.* New York: St. Martin's Press.

Marchant, J. (ed.) 1916. *Alfred Russel Wallace: Letters and Reminiscences.* New York and London: Harper.

Margulis, Lynn. 1970. *Origin of Eukaryotic Cells: Evidence and Research Implications for a Theory of the Origin and Evolution of Microbial, Plant, and Animal Cells on the Precambrian Earth.* New Haven: Yale University Press.

—————— 1981. *Symbiosis in Cell Evolution: Life and Its Environment on the Early Earth.* San Francisco: W. H. Freeman.

Maritain, J. 1942. Science and wisdom, in Ausben, R. N. (ed.), *Science and Man.*

—————— 1946. *Formal Logic.* New York: Sheed & Ward.

Markert, C. L. 1975. Biology of isozymes, in Markert, C. L. (ed.), *Isozymes I Molecular Structure.* New York: Academic Press, pp. 1–9.

Marraise, Robert de. 1974. The double-edged effect of Sir Francis Galton: a search for the motives of the biometrician-Mendelian debate. *J. Hist. Biol.* 7: 141–174.

Mather, K. 1943. Polygenic inheritance and natural selection. *Biol. Rev.* 18: 32–64.

—————— 1973. *Genetical Structure of Populations.* London: Chapman & Hall.

Matthew, P. 1831. *On Naval Timber and Arboriculture.* London: Longman. (Reprinted in *Gardener's Chronicle* (7 April 1860), also McKinney [1971] pp. 29–40.)

Matthew, W. D. 1915. Climate and evolution. *Ann. New York Acad. Sci.* 24: 171–318.

Maulitz, R. D. 1971. Schwann's way: cells and crystals. *J. Hist. Med. Allied Sci.* 26: 422–437.

Mayer, J. 1959. *Diderot homme de science.* Rennes: Imprimerie bretonne.

Maynard Smith, J. 1972. On evolution, in *The Status of Neo-Darwinism.* Edinburgh: Edinburgh University Press, pp. 82–91. (Originally in Waddington (ed.), *Towards a Theoretical Biology,* 2: 82–89.)

—————— 1978. *The Evolution of Sex.* Cambridge: Cambridge University Press.

Mayr, E. 1933. Notes on the variation of immature and adult plumages in birds and a physiological explanation of abnormal plumages. *Amer. Mus. Novit.* no. 666, pp. 1–10.

—————— 1934. Notes on the genus Petroica. *Amer. Mus. Novit.* no. 714, pp. 1–19.

—————— 1940. Speciation phenomena in birds. *Amer. Nat.* 74: 249–278.

—————— 1941. The origin and the history of the bird fauna of Polynesia. *Proc. 6th Pacific Sci. Cong.* 4: 197–216.

—————— 1942. *Systematics and the Origin of Species.* New York: Columbia University Press.

—————— 1944a. The birds of Timor and Sumba. *Bull. Amer. Mus. Nat. Hist.* 82: 127–194.

—————— 1944b. Wallace's line in the light of recent zoogeographic studies. *Quart. Rev. Biol.* 29: 1–14.

——— 1945. Symposium on age of the distribution pattern of gene arrangements in *Drosophila pseudoobscura:* introduction and some evidence in favor of a recent date. *Lloydia* 8: 69–83.

——— 1946a. History of the North American bird fauna. *Wils. Bull.* 58: 3–41. (See also Mayr [1976], pp. 566–588.)

——— 1946b. The naturalist in Leidy's time and today. *Proc. Acad. Nat. Sci. Phil.* 98: 271–276.

——— 1948. The bearing of the new systematics on genetical problems: the nature of species, in *Advances in Genetics,* vol. 2. New York: Academic Press, pp. 209–237.

——— 1954. Change of genetic environment and evolution, in Huxley, J., A. C. Hardy, and E. B. Ford (eds.). 1954. *Evolution as a Process.* London: Allen & Unwin, pp. 157–180. (See also Mayr [1976], pp. 188–210.)

——— 1955. Karl Jordan's contribution to current concepts in systematics and evolution. *Trans. Roy. Entomol. Soc. London* 107: 45–66.

——— 1957. Species concepts and definitions, in *The Species Problem.* Amer. Assoc. Adv. Sci., Publ. no. 50. Washington D.C., pp. 1–22.

——— 1959a. Darwin and the evolutionary theory in biology, in *Evolution and Anthropology: A Centennial Approach.* Washington D.C.: Anthropological Society of America. (See also Mayr [1976], pp. 26–29.)

——— 1959b. Isolation as an evolutionary factor. *Proc. Amer. Phil. Soc.* 103: 221–230. (See also Mayr [1976], pp. 129–134.)

——— 1959c. Trends in avian systematics. *Ibis* 101: 293–302.

——— 1959d. Where are we? *Cold Spring Harbor Symposia Quant. Biol.* 24: 1–14. (See also Mayr [1976], pp. 307–328.)

——— 1959e. Agassiz, Darwin, and evolution. *Harvard Library Bulletin* 13: 165–194.

——— 1960. The emergence of evolutionary novelties, in Tax (1960), pp. 349–380. (See also Mayr [1976], pp. 88–113).

——— 1961. Cause and effect in biology. *Science* 134: 1501–6.

——— 1963. *Animal Species and Evolution.* Cambridge: Harvard University Press.

——— 1964. The new systematics, in Leone (1964), pp. 13–32.

——— 1965a. Numerical phenetics and taxonomic theory. *Syst. Zool.* 14: 73–97.

——— 1965b. What is a fauna? *Zool. Jb. Syst.* 92: 473–486. (See also Mayr [1976], pp. 552–564.)

——— 1968. Illiger and the biological species concept. *J. Hist. Biol.* 1: 163–178.

——— 1969. *Principles of Systematic Zoology.* New York: McGraw-Hill.

——— 1970. *Populations, Species, and Evolution.* Cambridge: Harvard University Press.

——— 1972a. Lamarck revisited. *J. Hist. Biol.* 5: 55–94. (See also Mayr [1976], pp. 222–250.)

929

—— 1972b. The nature of the Darwinian revolution. *Science* 176: 981–989. (See also Mayr [1976], pp. 277–296.)

—— 1972c. Sexual selection and natural selection, in B. Campbell (1972), pp. 87–104.

—— 1973. The recent historiography of genetics. (Essay review). *J. Hist. Biol.* 6: 125–154. (See also Mayr [1976], pp. 329–353.)

—— 1974a. Behavior programs and evolutionary strategies. *Amer. Sci.* 62: 650–659. (See also Mayr [1976], pp. 694–711.)

—— 1974b. The challenge of diversity. *Taxon* 23: 3–9.

—— 1974c. Cladistic analysis or cladistic classification? *J. zool. Syst. Evol. forsch.* 12: 94–128. (See also Mayr [1976], pp. 433–476.)

—— 1974d. Teleological and teleonomic: a new analysis. *Boston Stud. Philos. Sci.* 14: 91–117. (See also Mayr [1976], pp. 383–404.)

—— 1976. *Evolution and the Diversity of Life.* Cambridge: Harvard University Press.

—— 1977a. Darwin and natural selection. *Amer. Sci.* 65: 321–327.

—— 1977b. The study of evolution, historically viewed, in C. E. Goulden (ed.), *The Changing Scenes in Natural Sciences, 1776–1976.* Philadelphia: Academy of Natural Sciences, Special Pub. No. 12, pp. 39–58.

—— 1981a. *La biologie de l'évolution.* Paris: Hermann et Co.

—— 1981b. Biological classification: Toward a synthesis of opposing methodologies. *Science* 214: 510–516.

——, E. G. Linsley, and R. L. Usinger. 1953. *Methods and Principles of Systematic Zoology.* New York: McGraw-Hill.

—— and Lester L. Short. 1970. *Species Taxa of North American Birds.* Cambridge, Mass.: Nuttall Ornithological Club, Pub. No. 9.

—— and William Provine (eds.). 1980. *The Evolutionary Synthesis.* Cambridge: Harvard University Press.

Mazia, D. 1952. Physiology of the cell nucleus, in E. S. G. Barron (ed.), *Modern Trends in Physiology and Biochemistry.* New York: Academic Press, pp. 77–122.

McClintock, B. 1929. A cytological and genetical study of triploid maize. *Genetics* 14: 180–222.

—— 1951. Chromosome organization and genic expression. *Cold Spring Harbor Symposia Quant. Biol.* 16: 13–47.

McElroy, W. D., and B. Glass (eds.). 1957. *A Symposium on the Chemical Basis of Heredity.* Baltimore: Johns Hopkins University Press.

McKinney, H. L. 1971. *Lamarck to Darwin: Contributions to Evolutionary Biology 1809–1859.* Lawrence, Kans.: Colorado Press.

—— 1972. *Wallace and Natural Selection.* New Haven and London: Yale University Press.

McKusick, V. A. 1960. Walter S. Sutton and the physical basis of Mendelism. *Bull. Hist. Med.* 34: 487–497.

—— 1975. *Mendelian Inheritance in Man.* 4th ed. Baltimore: Johns Hopkins University Press.

—— and F. H. Ruddle. 1977. The status of the gene map of the human chromosomes. *Science* 196: 390–405.

McPherson, T. 1972. *The Argument from Design*. London: Macmillan.

Meckel, J. F. 1821. *System der vergleichenden Anatomie*. 7 vols. Halle: Rengersche Buchhandlung.

Medawar, P. B. 1967. *The Art of the Soluble*. London: Methuen.

Mendel, Johann (Gregor). 1866. Versuche über Pflanzen-hybriden. *Verh. Natur. Vereins Brünn* 4 (1865): 3–57.

—— 1905. [Letters to Nägeli.] *Abh. der Math.-Phys. Kl. K. Sächs. Ges. Wiss.* 29: 189–265.

—— 1966. [Letters to Nägeli] in Stern and Sherwood (1966), pp. 56–102.

Mendelsohn, Everett. 1964. The emergence of science as a profession in Nineteenth-century Europe. Chapter 1 in *The Management of Scientists* (1964. Boston: Beacon Press, pp. 3–47).

Menzbier, M. A. 1893. Experience of the theory of heredity. *Russkaya Mysl*, no. 10, pp. 214–215.

Merton, Robert K. 1938a. Science, technology, and society in seventeenth century England. *Osiris* 4: 360–632.

—— 1938b. Motive forces in the new science, in *Science, Technology, and Society in Seventeenth Century England*. Bruges: Saint Catherine Press, Ch. 5.

—— 1961. Singletons and multiples in scientific discovery. *Proc. Amer. Phil. Soc.* 105: 470–486.

—— 1973. *The Sociology of Science: Theoretical and Empirical Investigations*. Chicago: University of Chicago Press.

Merz, John T. 1896–1914. *A History of European Thought in the Nineteenth Century*. 4 vols. London.

Meyer, A. 1926. *Logik der Morphologie*. Berlin: J. Springer.

Meyer, E. H. F. 1854–1857. *Geschichte der Botanik*. Königsberg.

Michaelis, P. 1954. Cytoplasmic inheritance in Epilobium and its theoretical significance. *Adv. Genet.* 6: 287–401.

Michener, C. D. 1977. Discordant evolution and the classification of allodapine bees. *Syst. Zool.* 26: 32–56; 27: 112–118.

—— and R. R. Sokal. 1957. A quantitative approach to a problem in classification. *Evolution* 11: 130–162.

Miescher, F. 1897. *Die histochemischen und physiologischen Arbeiten*, ed. W. His. 2 vols. Leipzig: Vogel.

Mikulinsky, S. R. 1978. Internalism-externalism controversy as a phony problem, in *Proc. XV Int. Cong. Hist. Sci., Edinburgh*, pp. 88–101.

Miller, S. J. 1953. A production of amino acids under possible primitive earth conditions. *Science* 117: 528.

Miller, S. L., and L. E. Orgel. 1974. *The Origin of Life on Earth*. Englewood Cliffs, N.J.: Prentice-Hall.

Millhauser, M. 1959. *Just before Darwin: Robert Chambers and "Vestiges."* Middletown, Conn.: Wesleyan University Press.

931

Mills, S. K., and J. H. Beatty. 1979. The propensity interpretation of fitness. *Phil. Sci.* 46: 263–286.

Mirsky, A. E., and H. Ris. 1951. The deoxyribonucleic acid content of animal cells and its evolutionary significance. *J. Genet. Physiol.* 34: 451–462.

Möbius, K. 1877. *Die Auster und die Austernwirtschaft.* Berlin.

Monod, J. 1970. *Le Hasard et la necessité.* Paris: Seuil. (English ed. 1971. *Chance and Necessity.* New York: A. A. Knopf.)

—— 1974a. *Problems of Scientific Revolutions.* (Herbert Spencer lectures, 1973.) Oxford: Clarendon Press.

—— 1974b. Préface in E. Mayr, *Populations, espèces, et évolution.* Paris: Hermann, pp. xv–xxii.

Montgomery, William M. 1974. Germany, in Glick (1974), pp. 81–116.

Moody, P. A. 1962. *Introduction to Evolution.* New York: Harper.

Moore, J. A. 1963. *Heredity and Development.* New York: Oxford University Press.

—— 1972. *Readings in Heredity and Development.* New York: Oxford University Press.

—— 1979. Creationism in California. *Daedalus* 103: 173–189.

Moore, James R. 1979. *The Post-Darwinian Controversies: A Study of the Protestant Struggle to Come to Terms with Darwin in Great Britain and America, 1870–1900.* Cambridge: Cambridge University Press.

Moorhead, A. 1969. *Darwin and the Beagle.* New York: Harper & Row.

Moorhead, P. S., and M. M. Kaplan (eds.). 1967. *Mathematical Challenges to the Neo-Darwinian Interpretation of Evolution.* Wistar Institute Symposium Monograph, no. 5. Philadelphia: Wistar Institute Press.

Morgan, Conway Lloyd. 1923. *Emergent Evolution.* London: Williams & Norgate.

—— 1933. *The Emergence of Novelty.* London: Williams & Norgate.

Morgan, T. H. 1903. *Evolution and Adaptation.* New York: Macmillan.

—— 1910a. Chromosomes and heredity. *Amer. Nat.* 44: 449–496.

—— 1910b. Sex limited inheritance in *Drosophila. Science* 32: 120–122.

—— 1910c. Chance or purpose in the origin and evolution of adaptation. *Science* 31: 201–210.

—— 1916. *A Critique of the Theory of Evolution.* Princeton: Princeton University Press. (Reissued 1925. *Evolution and Genetics.*)

—— 1926. *The Theory of the Gene.* New Haven: Yale University Press.

—— 1932. *The Scientific Basis of Evolution.* New York: W. W. Norton.

——, A. H. Sturtevant, H. J. Muller, and C. B. Bridges. 1915. *The Mechanism of Mendelian Heredity.* New York: Henry Holt.

Mudford, P. G. 1968. William Lawrence and the natural history of man. *J. Hist. Ideas* 29: 430–436.

Mullen, P. C. 1964. *The Preconditions and Reception of Darwinian Biology in Germany, 1800–1870.* Ph.D. diss., Berkeley. Unpublished.

Müller, Fritz. 1864. Für Darwin, in Moller, A. (ed.). *Fritz Müller, Werke, Briefe, und Leben.* Jena: Gustav Fischer, pp. 200–263.

———— 1879. Ituna and Thyridia: a remarkable case of mimicry in butterflies. *Trans. Entomol. Soc. London,* p. xx. (English trans. of *Kosmos* 26: 497.)

Muller, H. J. 1939. Reversibility in evolution considered from the standpoint of genetics. *Biol. Reviews* 14: 261–280.

———— 1943. Edmund B. Wilson—an appreciation. *Amer. Nat.* 77: 5–37, 142–172.

———— 1945. The gene. *Proc. Roy. Soc. Biol.* 134: 1–37.

———— 1946. Thomas Hunt Morgan (1866–1945). *Science* 103: 550–551.

———— 1950. Our load of mutations. *J. Human Genetics* 2: 111–176.

———— 1966. [Introduction to reprint of Wilson (1925).]

———— 1973. *The Modern Concept of Nature,* ed. E. A. Carlson. Albany: State University of New York Press.

Müller, Paul. 1980. *Biogeographie.* Stuttgart: Eugen Ulmer.

Munson, R. 1971a. Biological adaptation. *Phil. Sci.* 38: 200–215.

———— 1971b. *Man and Nature: Philosophical Issues in Biology.* New York: Delta.

Anon. (Murray). 1904–1912. *The History of the Collections Contained in the Natural History Department of the British Museum.* 2 vols. London.

Naef, Adolf. 1919. *Idealistische Morphologie und Phylogenetik.* Jena: Gustav Fischer.

———— 1931. Die Gestalt als Begriff und Idee, in Bolk, W. et al. (eds.), *Handbuch der vergl. Anatomie Wirbeltiere.* Berlin: Urban und Schwarzenberg, pp. 77–118.

Nagel, Ernest. 1961. *The Structure of Science: Problems in the Logic of Scientific Explanation.* New York: Harcourt, Brace & World.

Nägeli, C. 1865. *Entstehung und Begriff der Naturhistorischen Art.* München: K. Bayr. Akademie.

———— 1884. *Mechanisch-physiologische Theorie der Abstammungslehre.* Leipzig: Oldenbourg.

Nathorst, A. G. 1908. Carl von Linné as a geologist. *Ann. Report Smiths. Inst.,* pp. 711–743.

Nei, Masatoshi. 1975. *Molecular Population Genetics and Evolution.* Amsterdam and New York: North-Holland Publishing Co.

Nelson, G., and D. E. Rosen (eds.). 1980. *Vicariance Biogeography: A Critique.* New York: Columbia University Press.

Nevo, Eviatar. 1978. Genetic variation in natural populations: patterns and theory. *Theoret. Pop. Biol.* 13: 121–177.

Newton, Alfred. 1888. Early days of Darwinism. *Macmillan's Magazine* 57: 241.

Nichols, Ch. 1974. Darwinism and the social sciences. *Phil. Soc. Sci.* 4: 255–277.

933

Niggli, P. 1949. *Probleme der Naturwissenschaften erläutert am Begriff der Mineralart.* Basel: Birkhäuser.

Nisbet, H. B. 1970. *Herder and the Philosophy and History of Science.* Cambridge: Cambridge University Press.

—— 1972. *Goethe and the Scientific Tradition.* London: University of London (Institute of Germanic Studies).

Nisbet, Robert A. 1969. *Social Change and History: Aspects of the Western Theory of Development.* New York: Oxford University Press.

Noll, Alfred (ed.). 1971. *Die Lebenskraft in den Schriften der Vitalisten und ihrer Gegner.* Leipzig: NDC.

Nordenskiöld, E. 1928. *The History of Biology.* New York: A. A. Knopf. (Translated from the Swedish edition, 1920–1924.)

Norton, B. J. 1973. The biometric defense of Darwinism. *J. Hist. Biol.* 6: 283–316.

—— 1975. Biology and philosophy: the methodological foundations of biometry. *J. Hist. Biol.* 8: 89–93.

Norton, H. T. J. 1915. Table, in Punnett, R. C., *Mimicry in Butterflies.* Cambridge: Cambridge University Press.

Novitski, E., and S. Blixt. 1978. Mendel, linkage, and synteny. *Bioscience* 28: 34–35.

Nussbaum, M. 1880. Zur Differenzirung des Geschlechts im Thierreich. *Archiv f. Mikroskopische Anatomie* 18: 1–121.

Nuttall, G. H. F. 1904. *Blood Immunity and Blood Relationships.* London: Cambridge University Press.

O'Donald, Peter. 1980. *Genetic Models of Sexual Selection.* New York: Cambridge University Press.

Ohno, S. 1970. *Evolution by Gene Duplication.* Berlin: J. Springer.

Olby, R. C. 1966. *The Origins of Mendelism.* London: Constable.

—— 1968. Miescher's study of the nucleus. *Actes XII Congr. Int. Hist. Sci.,* pp. 135–138.

—— 1971. Schrödinger's problem: what is life? *J. Hist. Biol.* 4: 119–148.

—— 1971. [Unger's influence on Mendel]. *Proc. Gregor Mendel Coll.,* Brno, p. 99–103.

—— 1974. *The Path to the Double Helix.* London: Macmillan.

—— 1979. Mendel no Mendelian? *Hist. Sci.* 17: 53–72.

Oparin, A. I. 1938. *The Origin of Life.* New York: Macmillan.

Oppenheimer, J. 1959. An embryological enigma in the *Origin of Species,* in Glass, Temkin, and Strauss (1959), pp. 292–322.

Orel, Vítezslav. 1971. A reconstruction of Mendel's experiments and an attempt at an explanation of Mendel's way of presentation. *Folia Mendeliana* 6: 45.

Orgel, L. E., and F. H. C. Crick. 1980. Selfish DNA: the ultimate parasite. *Nature* 284: 604–607.

934

Osborn, F. 1968. *The Future of Human Heredity.* New York: Weybright and Talley, p. 81.

Osborn, H. F. 1894. *From the Greeks to Darwin.* New York: Columbia University Press.

Ospovat, Dov. 1976. The influence of Karl Ernst von Baer's Embryology, 1828–1859: a reappraisal in light of Richard Owen's and William B. Carpenter's palaeontological application of von Baer's Law. *J. Hist. Biol.* 9: 1–28.

——— 1977. Lyell's theory of climate. *J. Hist. Biol.* 10: 317–339.

——— 1978. Perfect adaptation and teleological explanation: approaches to the problem of the history of life in the mid-nineteenth century. *Stud. Hist. Biol.* 2: 33–56.

——— 1979. Darwin after Malthus. *J. Hist. Biol.* 12: 211–230.

——— 1980. God and natural selection: the Darwinian idea of design. *J. Hist. Biol.* 13: 169–194.

Ostoya, R. 1951. *Les théories d'évolution.* Paris: Payot.

Ostwald, W. 1909. *Grosse Männer.* Leipzig: Akademische Verlagsgesellschaft.

Owen, Richard. 1848. *Report on the Archetype and Homologies of the Vertebrate Skeleton.* London: Voorst.

——— 1849. *On the Nature of Limbs.* London.

Packard, A. S. 1901. *Lamarck, The Founder of Evolution: His Life and His Work.* London: Longmans, Green.

Paley, William. 1802. *Natural Theology: Or, Evidences of the Existence and Attributes of the Deity, Collected from the Appearances of Nature.* London: R. Fauldner.

Pallas, Simon Peter. 1811. *Zoographia Rosso-Asiatica.* St. Petersburg.

Pantin, C. F. A. 1968. *The Relations between the Sciences.* Cambridge: Cambridge University Press.

Passmore, J. A. 1959. Darwin and the climate of opinion. *Aust. J. Sci.* 22: 14–15.

——— 1965. Comments on historical assumptions of the history of science, in Crombie (1965), pp. 853–861.

Pattee, Howard H. 1970. Can life explain quantum mechanics?, in Bastin, T. (ed.), *Quantum Theory and Beyond.* Cambridge: Cambridge University Press, p. 307.

——— (ed.). 1973. *Hierarchy Theory: The Challenge of Complex Systems.* New York: George Braziller.

Pauly, A. 1905. *Darwinismus und Lamarckismus.* München: E. Reinhardt.

Pearson, K. 1914–1930. *The Life, Letters, and Labours of Francis Galton.* 4 vols. Cambridge: Cambridge University Press.

Peck, A. L. 1965. *Introduction to Aristotle's Historia Animalium.* Loeb edition. Cambridge: Harvard University Press.

Perrier, Edmond. 1888. *Le transformisme.* Paris: Baillière.

935

—— 1896. *La philosophie zoologique avant-Darwin* 3rd ed. Paris: Felix Alcan.

Peters, J. A. (ed.). 1959. *Classical Papers in Genetics*. Englewood Cliffs, N.J.: Prentice-Hall.

Peters, R. H. 1976. Tautology in evolution and ecology. *Amer. Nat.* 110: 1–12.

Petit, C., and L. Ehrman. 1969. Sexual selection in *Drosophila*. *Evol. Biol.* 3: 177–223.

Pfeifer, E. J. 1965. The genesis of American neo-Lamarckism. *Isis* 56: 156–157.

—— 1974. United States, in Glick (1974), pp. 168–226.

Phillips, R. L., and C. R. Burnham (eds.). 1977. *Cytogenetics*. Benchmark Papers in Genetics, no. 6. Stroudsberg, Pa.: Dowden, Hutchinson & Ross.

Pickens, D. P. 1968. *Eugenics and the Progressives*. Nashville: Vanderbilt University Press.

Pickstone, J. V. 1973. Globules and coagula: concepts of tissue formation in the early nineteenth century. *J. Hist. Med. Allied Sci.* 28: 336–356.

Pielou, E. C. 1979. *Biogeography*. New York: John Wiley & Sons.

Pirie, N. W. 1969. Gardyloo. *The Listener* 82: 331.

Pittendrigh, Colin S. 1958. Adaptation, natural selection, and behavior, in Roe and Simpson (1958), pp. 390–416.

Piveteau, J. 1950. Le débat entre Cuvier et Geoffroy Saint-Hilaire sur l'unité de plan et de composition. *Rev. d'hist. sci.* 3: 343–363.

—— (ed.). 1954. *Oeuvres philosophiques de Buffon*. Paris: Presses Universitaires de France.

Plate, Ludwig. 1903. *Über die Bedeutung des Darwinschen Selectionsprinzip und Probleme der Artbildung*. Leipzig: Engelmann. (Rev. eds. 1908 and 1913.)

—— 1914. Prinzipien der Systematik mit besonderer Berücksichtigung des Systems der Tiere. *Kultur der Gegenwart* 3 (4): 119–159.

—— 1925. *Die Abstammungslehre*. Jena: Gustav Fischer.

Platt, J. R. 1964. Strong inference. *Science* 146: 347–353.

Pledge, H. T. 1939. *Science since 1500*. London: H.M. Stationery Office.

Pohlenz, M. 1948. *Die Stoa: Geschichte einer geistigen Bewegung*. Göttingen: Vandenhoeck und Ruprecht.

Pontecorvo, G. 1968. Hermann Joseph Muller, 1890–1967. *Biogr. Mem. Fellows Roy. Soc.* 14: 349–389.

Popper, Karl. 1945. *The Open Society and Its Enemies*. I. *The Spell of Plato*. London: Routledge & Kegan Paul.

—— 1972. *Objective Knowledge*. Cambridge: Cambridge University Press.

—— 1974. *Unended Quest: An Intellectual Autobiography*. La Salle, Ill.: Open Court Publishing Co.

Portugal, F. H., and J. S. Cohen. 1977. *A Century of DNA*. Cambridge: M.I.T. Press.

Potonié, H. 1890. Aufzählung von Gelehrten, die in der Zeit von Lamarck bis Darwin sich im Sinne der Deszendenz-Theorie geäussert haben. *Naturwiss. Wochenschr.* 5: 441–445.

Poulton, E. B. 1896. *Charles Darwin and the Theory of Natural Selection*. London: Cassel and Co.

———— 1903. What is a species? *Proc. Ent. Soc. London*, pp. lxxvi–cxvi. (Rpt. with revisions in Poulton, 1908a.)

———— 1908a. *Essays on Evolution*. Oxford: Clarendon Press.

———— 1908b. Thomas Henry Huxley and the theory of natural selection, in Poulton, 1908a, pp. 193–219.

Powell, J. R., and C. E. Taylor. 1979. Natural selection in ecologically diverse environments. *Amer. Sci.* 67: 590–596.

Preer, J. R., L. B. Preer, and A. Jurand. 1974. Kappa and other endosymbionts in *Paramecium aurelia*. *Bact. Revs.* 38: 113–163.

Provine, William B. 1971. *The Origins of Theoretical Population Genetics*. Chicago and London: University of Chicago Press.

———— 1979. Francis B. Sumner and the evolutionary synthesis. *Stud. Hist. Biol.* 3: 211–240.

————, P. Froggart, and N. C. Nevin. 1971. The "law of ancestral heredity" and the Mendelian-biometrician controversy in England, 1889–1900. *J. Med. Gen.* 8: 1–36.

Pynchon, T. 1973. *Gravity's Rainbow*. New York: Viking.

Querner, H. 1975. Darwin's Deszendenz- und Selektionslehre auf den deutschen Naturforscher-Versammlungen. *Acta Hist. Leopoldina*, no. 9. Leipzig: J. A. Barth.

Radinsky, Leonard. 1978. Do albumin clocks run on time? *Science* 200: 1182–83.

Radl, E. 1907–1908. *Geschichte der biologischen Theorien in der Neuzeit*. 2 vols. Leipzig: Engelmann (Rev. ed. 1913.)

Raikov, B. E. 1968. Karl Ernst von Baer, 1792–1876: Sein Leben und sein Werk. *Acta Hist. Leopoldina*, no. 5. Leipzig: J. A. Barth.

Ramsbottom, J. 1938. Linnaeus and the species concept. *Proc. Linn. Soc. London* 165: 164–166.

Randall, J. H. 1960. *Aristotle*. New York: Columbia University Press.

Rathke, H. 1825. Kiemen bey Säugthieren. *Isis*, pp. 747–749.

Raup, D. M. 1972. Approaches to morphologic analysis, in Schopf and Thomas (1972), pp. 28–45.

Raven, Charles E. 1947. *English Naturalists from Neckam to Ray*. London: Cambridge University Press.

———— 1950. *John Ray, Naturalist: His Life and Works*. 2nd ed. Cambridge: Cambridge University Press.

Raven, P. 1976. Systematics and plant population biology. *Syst. Bot.* 1: 284–316.

———— 1977. The systematics and evolution of higher plants, in *The*

Changing Scenes in Natural Sciences, 1776–1976. Academy of Natural Sciences, special publication no. 12, pp. 59–83.

Raven, Peter H., Brent Berlin, and Dennis E. Breedlove. 1971. The origins of taxonomy. *Science* 174: 1210–13.

Ravin, A. W. 1977. The gene as catalyst; the gene as organism. *Stud. Hist. Biol.* 1: 1–45.

Regenbogen, Otto. 1931. Eine Forschungsmethode antiker Naturwissenschaft. *Quellen Studien zur Geschichte der Mathematik,* Abt. B: Studien. I. Berlin: Springer, pp. 131–182.

Reichert, C. 1837. Über die Visceralbogen der Wirbeltiere im allgemeinen und deren Metamorphosen bei den Vögeln und Säugethieren. *Müller's Archiv f. Anat. Physiol. Wiss. Med.,* pp. 120–122.

Reif, W. E. 1975. Lenkende and limitierende Faktoren in der Evolution. *Acta Biotheor.* 24: 136–162.

Reimarus, H. S. 1973. *Hermann Samuel Reimarus (1695–1768), ein bekannter Unbekannter der Aufklärung in Hamburg: Veröffentlichung der Joachim Jungius Gesellschaft.* Göttingen: Vandenhoeck und Ruprecht.

Reiser, O. L. 1968. The concept of evolution in philosophy, in Buchsbaum, R. (ed.), *A Book That Shook the World.* Pittsburgh, Penna.: Pittsburgh University Press, pp. 38–47.

Remane, Adolf. 1952. *Die Grundlagen des Natürlichen Systems, der vergleichenden Anatomie und der Phylogenetik.* Leipzig: Akademische Verlagsgesellschaft.

Rensch, Bernhard. 1929. *Das Prinzip geographischer Rassenkreise und das Problem der Artbildung.* Berlin: Borntraeger.

——— 1933. Zoologische Systematik und Artbildungsproblem. *Verh. Dtsch. Zool. Ges.* 1933: 19–83.

——— 1934. *Kurze Anweisung für zoologische-systematische Studien.* Leipzig: Akademische Verlagsgesellschaft.

——— 1936. *Die Geschichte des Sundabogens.* Berlin: Borntraeger.

——— 1947. *Neuere Probleme der Abstammungslehre.* Stuttgart: Enke.

——— 1948. Organproportionen und Körpergrösse bei Vögeln und Säugetieren. *Zool. Jahrb. (Physiol.)* 61: 337–450.

——— 1960. The laws of evolution, in Tax (1960), 1: 95–116.

——— 1968. *Biophilosophie auf erkenntnistheoretischer Grundlage.* (Panpsychistischer Identismus). Stuttgart: Gustav Fischer.

——— 1971. *Biophilosophy.* New York: Columbia University Press.

Rhoades, M. M. 1954. Lewis J. Stadler, geneticist. *Science* 120: 553–554.

——— 1957. Lewis John Stadler, 1896–1954, in *Biogr. Mem. Nat. Acad. Sci.,* 30: 329–347.

Ricklefs, R. E. 1978. *Ecology.* Portland, Ore.: Chiron Press.

Riddle, Oscar. 1954. *The Unleashing of Evolutionary Thought.* New York: Vantage Press.

Ridley, H. N. 1930. *The Dispersal of Plants throughout the World.* Kent: Ashford.

938

Ritter, W. E. 1919. *The Unity of the Organism, or, the Organismal Conception of Life*, vols. 1–2. Boston: Gorham Press.

Ritterbush, Philip C. 1964. *Overtures to Biology: The Speculations of Eighteenth Century Naturalists*. New Haven and London: Yale University Press.

Robbins, H. 1974. The statistical mode of thought, in J. Neyman (ed.), *The Heritage of Copernicus: Theories "More Pleasing to the Mind."* Cambridge: M.I.T. Press, pp. 417–432.

Roberts, H. F. 1929. *Plant Hybridization before Mendel*. Princeton: Princeton University Press.

Robson, G. C., and O. W. Richards. 1936. *The Variation of Animals in Nature*. London: Longmans, Green.

Roe, A., and G. G. Simpson (eds.). 1958. *Behavior and Evolution*. New Haven: Yale University Press.

Roe, Shirley A. 1981. *Matter, Life, and Generation*. Cambridge: Cambridge University Press.

Roger, Jacques (ed.). 1962. Buffon. Les époques de la nature. *Mém. Mus. Nat. d'Hist. Nat.* N.S. Sér. C, vol. 10. Paris: Editions du Muséum.

——— 1963. *Les sciences de la vie dans la pensée Francaise du XVIIe et XVIIIe siècle*. Paris: Armand.

Roller, D. H. D. (ed.). 1971. *Perspectives in the History of Science and Technology*. Norman: University of Oklahoma Press.

Roll-Hansen, Nils. 1978a. The genotype theory of Wilhelm Johannsen and its relation to plant breeding and the study of evolution. *Centaurus* 22: 201–235.

——— 1978b. *Drosophila* genetics: a reductionist research program. *J. Hist. Biol.* 11: 159–210.

Romanes, George John. 1892–1897. *Darwin, and after Darwin: An Exposition of the Darwinian Theory and a Discussion of Post-Darwinian Questions*, vols. 1–3. Chicago: Open Court Publishing Co.

——— 1893. *An Examination of Weismannism*. London: Longmans, Green.

——— 1896. *Life and Letters*. London: Longmans, Green.

Rothschild, W. L., and K. Jordan. 1903. Lepidoptera collected by Oscar Neumann. *Novit. Zool.* 10: 492.

Rousseau, G. 1969. Lamarck et Darwin. *Bull. Mus. Hist. Nat.* 41: 1029–41.

Roux, W. 1883. *Über die Bedeutung der Kerntheilungsfiguren*. Leipzig: Engelmann. (Also in Roux, W. 1895. *Ges. Abh.* 2: 125–143.)

Rudwick, M. J. S. 1970. The strategy of Lyell's *Principles of Geology*. *Isis* 61: 5–33.

——— 1971. Uniformity and progression: reflections on the structure of geological theory in the age of Lyell, in Roller (1971), pp. 209–227.

——— 1972. *The Meaning of Fossils*. London: Macdonald.

——— 1977. Historical analogies in the geological work of Charles Lyell. *Janus* 64: 89–107.

Ruse, M. 1973. *The Philosophy of Biology*. London: Hutchinson.

—— 1975a. Charles Darwin and artificial selection. *J. Hist. Ideas* 36: 339–350.

—— 1975b. Darwin's debt to philosophy: an examination of the influence of the philosophical ideas of John F. W. Herschel and William Whewell on the development of Charles Darwin's theory of evolution. *Stud. Hist. Phil. Sci.* 6: 159–181.

—— 1975c. The relationship between science and religion in Britain, 1830–1870. *Church Hist.* 44: 505–522.

—— 1979a. *The Darwinian Revolution*. Chicago: University of Chicago Press.

—— 1979b. *Sociobiology: Sense or Nonsense*. Boston: D. Reidel.

Russell, C. A. (ed.). 1973. *Science and Religious Belief: A Selection of Recent Historical Studies*. London: University of London Press.

Russell, E. S. 1916. *Form and Function: A Contribution to the History of Animal Morphology*. London: Murray.

—— 1930. *The Interpretation of Development and Heredity*. Oxford: Clarendon Press.

Russell, Elizabeth S. 1954. One man's influence: a tribute to William Ernest Castle. *J. Hered.* 45: 210–213.

Russett, Cynthia E. 1976. *Darwin in America: The Intellectual Response, 1865–1912*. San Francisco: W. H. Freeman.

Ryan, Michael J. 1980. Female mate choice in a neotropical frog. *Science* 209: 523–525.

Sachs, Julius. 1882. Vorlesungen über Pflanzenphysiologie. Leipzig.

—— 1890. *History of Botany (1530–1860)*, trans. Henry E. F. Garnsey. Oxford: Clarendon Press. (Originally published in German, 1875.)

Sager, R. 1972. *Cytoplasmic Genes and Organelles*. New York: Academic Press.

Sageret, Augustin. 1826. Considerations sur la productions des hybrides des variantes et des variété en general, et sur celles des Cucurbitacées en particulier. *Annales des Sci. Nat. Ser.* 1, 8: 294–314.

Salvini-Plawen, L. v., and Ernst Mayr. 1977. On the evolution of photoreceptors and eyes. *Evol. Biol.* 10: 207–263.

Salvini-Plawen, L. v., and H. Splechtna. 1979. Zur Homologie der Keimblätter. *Z. f. zool. Syst. Evol-forsch.* 17: 10–30.

Salzano, F. M. (ed.). 1975. *The Role of Natural Selection in Human Evolution*. Amsterdam: North-Holland Publishing Co.

Santillana, G. de. 1961. *The Origins of Scientific Thought*. New York: Mentor Books.

Sarton, George. 1927–1948. *Introduction to the History of Science*. 3 vols. Baltimore: Williams and Wilkins.

Savioz, R. 1948. *La philosophie de Charles Bonnet de Genève*. Paris: Vrin.

Sayre, A. 1975. *Rosalind Franklin and DNA*. New York: W. W. Norton.

Schaffner, K. F. 1969a. Theories and explanations in biology. *J. Hist. Biol.* 2: 19–33.

—— 1969b. The Watson-Crick model and reductionism. *Brit. J. Phil. Sci.* 20: 325–348.

Scharff, R. F. 1907. *European Animals: Their Geological History and Geographical Distribution.* New York: Dutton.

—— 1912. *Distribution and Origin of Life in America.* New York: Macmillan.

Schaxel, J. 1919. *Die Grundzüge der Theorienbildung in der Biologie.* Jena: Gustav Fischer.

Scherz, Gustav. 1971. *Dissertations on Steno as a Geologist.* Odense: University Press.

Schiemann, E. 1935. Erwin Baur. *Ber. Deutsch. Bot. Ges.* 52: 51–114.

Schierbeck, A. 1967. *Jan Swammerdam: His Life and Works.* Amsterdam: Swets and Zeitlinger.

Schindewolf, O. H. 1936. *Paläontologie, Entwicklungslehre, und Genetik.* Berlin: Borntraeger.

—— 1941. Einige vergessene deutsche Vertreter des Abstammungsgedankens aus dem Anfange des 19. Jahrhunderts. *Paläont. Zeitschr.* 22: 139–168.

—— 1950. *Grundfragen der Paläontologie.* Stuttgart: Schweizerbart.

—— 1969. Über den "Typus" in der morphologischen und phylogenetischen Biologie. *Abh. Akad. Wiss. u. Lit., Mainz, Math.-Nat. Kl.,* no. 4, pp. 58–131.

—— 1972. Phylogenie und Anthropologie aus paläontologischer Sicht, in Gadamer, H. G., and Paul Vogler, *Neue Anthropologie,* München: Deutscher Taschenbuch Verlag. I: 247.

Schlegel, H. 1844. *Kritische Übersicht der europäischen Vögel.* Leiden: Arnz und Comp.

Schleiden, M. J. 1842. *Grundzüge der wissenschaftlichen Botanik.* Leipzig.

—— 1863. *Das Alter des Menschengeschlechts, die Entstehung der Arten und die Stellung des Menschen in der Natur. Drei Vorträge für gebildete Laien.* Leipzig: Engelmann.

Schleip, W. 1934. [Analysis of Weismann's contribution.] *Naturwiss.* 22: 33–41.

Schmalhausen, I. I. 1949. *Factors of Evolution: The Theory of Stabilizing Selection.* Philadelphia: Blakiston.

Schmidt, J. 1918. Racial studies in fishes. I. Statistical investigations with Zoarces viviparus L. *J. Genet.* 7: 105–118.

Schmidt, Karl P. 1955a. Herpetology, in *A Century of Progress in the Natural Sciences 1853–1953.* San Francisco: California Academy of Sciences, pp. 591–627.

—— 1955b. Animal geography, in *A Century of Progress in the Natural Sciences, 1853–1953.* San Francisco: California Academy of Sciences, pp. 767–794.

Schmidt-Nielsen, K. 1979. *Animal Physiology: Adaptation and Environment.* 2nd ed. Cambridge: Cambridge University Press.

Schmitt, C. B. 1976. Science in the Italian universities in the sixteenth

941

and early seventeenth centuries, in Crosland, M. (ed.), *The Emergence of Science in Western Europe.* New York: Science History Publications, pp. 35–56.

Schneer, C. J. (ed.). 1969. *Toward a History of Geology.* Cambridge: M.I.T. Press.

Schopf, J. W. 1978. The evolution of the earliest cells. *Sci. Amer.* 239: 110–138.

Schopf, T. J. M., and J. M. Thomas (eds.). 1972. *Models in Paleobiology.* San Francisco: Freeman, Cooper.

Schrödinger, E. 1944. *What Is Life?* Cambridge: Cambridge University Press.

Schumacher, I. 1975. Die Entwicklungstheorie des Heidelberger Paläontologen und Zoologen Heinrich Georg Bronn (1800–1862). Ph.D. diss. Heidelberg. (Unpublished.)

Schultz, Jack. 1967. Innovators and controversies. [Review of Carlson. (1966).] *Science* 157: 296–301.

Schuster, A. 1911. *The Progress of Physics.* Cambridge: Cambridge University Press.

Schuster, J. 1930. Die Anfänge der wissenschaftlichen Erforschung der Geschichte des Lebens durch Cuvier und Geoffroy St. Hilaire. *Arch. Gesch. Math. Naturw. u. Technik,* 12, pt. 3. Leipzig: Vogel.

Schwann, Th. 1839. *Mikroskopische Untersuchungen über die Übereinstimmung in der Struktur und dem Wachstum der Tiere und Pflanzen.* Berlin.

Schwartz, J. S. 1974. Charles Darwin's debt to Malthus and Edward Blyth. *J. Hist. Biol.* 7: 301–318.

Schweber, S. 1977. The origin of the *Origin* revisited. *J. Hist. Biol.* 10: 229–316.

Sclater, P. L. 1858. On the general geographical distribution of the members of the class Aves. *J. Proc. Linn. Soc. London (Zoology)* 2: 130–145.

Scott, W. B. 1894. On variations and mutations. *Amer. J. Sci.* (3) 48: 355–374.

Scriven, M. 1959. Explanation and prediction in evolutionary theory. *Science* 130: 477–482.

Searle, G. R. 1976. *Eugenics and Politics in Britain.* Leyden: Noordhoff.

Sepkoski, J. J. 1979. A kinetic model of Phanerozoic taxonomic diversity. II. *Paleobiol.* 5: 222–251.

Serrès, E. 1860. Principes d'embroyologenie, de zoogenie, et de teratogenie. *Mém. Acad. Sci.* 25: 1–943.

Severtzoff, A. N. 1931. *Morphologische Gesetzmässigkeiten der Evolution.* Jena: Gustav Fischer.

Seward, A. C. (ed.). 1909. *Darwin and Modern Science.* Cambridge: Cambridge University Press.

Sherrington, C. S. 1906. *The Integrative Action of the Nervous System.* New Haven: Yale University Press.

Siewing, R. 1976. Probleme und neuere Erkenntnisse in der Grossystematik der Wirbellosen. *Verh. Dtsch. Zool. Ges.* 1976: 59–83.

Simon, H. A. 1962. The architecture of complexity. *Proc. Amer. Phil. Soc.* 106: 467–482.

Simon, Michael A. 1971. *The Matter of Life: Philosophical Problems of Biology.* New Haven: Yale University Press.

Simpson, George Gaylord. 1940. Mammals and land bridges. *J. Wash. Acad. Sci.* 30: 137–163.

―――― 1943. Turtles and the origin of the fauna of Latin America. *Amer. J. Sci.* 241: 413–429.

―――― 1944. *Tempo and Mode in Evolution.* New York: Columbia University Press.

―――― 1945. The principles of classification and a classification of mammals. *Bull. Amer. Mus. Nat. Hist.* 85: 1–350.

―――― 1947. Holarctic mammalian faunas and continental relationships during the Cenozoic. *Bull. Geol. Soc. Amer.* 58: 613–688.

―――― 1949. *The Meaning of Evolution.* New Haven: Yale University Press.

―――― 1953. *The Major Features of Evolution.* New York: Columbia University Press.

―――― 1959. Anatomy and morphology: classification and evolution, 1859 and 1959. *Proc. Amer. Phil. Soc.* 103: 286–306.

―――― 1961a. *Principles of Animal Taxonomy.* New York: Columbia University Press.

―――― 1961b. Lamarck, Darwin, and Butler. *Amer. Scholar* 30: 238–249.

―――― 1963. Biology and the nature of science. *Science* 139: 81–88.

―――― 1964a. Numerical taxonomy and biological classification. *Science* 144: 312–313.

―――― 1964b. *This View of Life.* New York: Harcourt, Brace & World.

―――― 1965. *The Geography of Evolution: Collected Esssays.* Philadelphia: Chilton Books.

―――― 1970. Uniformitarianism: an inquiry into principle, theory, and method in geohistory and biohistory, in Hecht and Steere (1970), pp. 43–96.

―――― 1974. The concept of progress in organic evolution. *Social Research,* pp. 28–51.

―――― 1975. Recent advances in methods of phylogenetic inference, in W. P. Luckett and F. S. Szalay (eds.), *Phylogeny of the Primates.* New York: Plenum Press, pp. 3–19.

―――― 1980. *Splendid Isolation.* New Haven: Yale University Press.

―――――, A. Roe, and R. C. Lewontin. 1960. *Quantitative Zoology.* Rev. ed. New York: Harcourt, Brace, & World.

Singer, C. 1926. *The Evolution of Anatomy.* New York: A. A. Knopf.

Skottsberg, C. J. F. 1956. *The Natural History of Juan Fernandez and Easter Island.* I. *Geography, Geology, Origin of Island Life.* Uppsala: Almquist & Wiksell, pp. 193–438.

Sloan, P. R. 1972. John Locke, John Ray, and the problem of the natural system. *J. Hist. Biol.* 5: 1–53.

943

———— 1978. The impact of Buffon's taxonomic philosophy in German biology: the establishment of the biological species concept. *Proc. XV Int. Cong. Hist. Sci., Edinburgh,* pp. 531–539.

Slobodchikoff, C. N. (ed.). 1976. *Concepts of Species (A Reader).* Benchmark Papers in Systematics and Evolutionary Biology, no. 3.

Smart, J. J. C. 1963. *Philosophy and Scientific Realism.* London: Routledge & Kegan Paul.

———— 1968. *Between Science and Philosophy.* New York: Random House.

Smit, P. 1974. *History of the Life Sciences: An Annotated Bibliography.* New York: Hafner Press.

Smith, C. U. M. 1976. *The Problem of Life.* London: Macmillan.

Smith, Roger. 1972. Alfred Russel Wallace: philosophy of nature and man. *Brit. J. Hist. Sci.* 6 (22): 177–199.

Smith, William. 1815. *A Memoir to the Map and Delineation of the Strata of England and Wales with Part of Scotland.* London.

Smithies, O. 1955. Zone electrophoresis in starch gels: group variation in the serum proteins of normal human adults. *Biochem. J.* 61: 629–641.

Smuts, J. C. 1926. *Holism and Evolution.* London: Macmillan.

Sneath, P. H. A. 1957. The application of computers to taxonomy. *J. Gen. Microbiol.* 17: 201–226.

———— 1962. The construction of taxonomic groups, in *Microbial Classification, Symposia Society General Microbiol.,* no. 12. Cambridge: Cambridge University Press, pp. 289–332.

Snow, C. P. 1959. *The Two Cultures and the Scientific Revolution.* New York: Cambridge University Press.

Sokal, Robert R. 1977. Clustering and classification: background and current directions, in *Classification and Clustering.* New York: Academic Press.

————, and P. H. A. Sneath. 1963. *Principles of Numerical Taxonomy.* London: W. H. Freeman. (Rev. ed. 1973.)

————, and T. J. Crovello. 1970. The biological species concept: a critical evaluation. *Amer. Nat.* 104: 127–153.

Solbrig, Otto (ed.). 1979. *Fifty Years of Plant Taxonomy.* New York: Columbia University Press.

————, Subodh Jain, George B. Johnson, and Peter H. Raven. 1979. *Topics in Plant Population Biology.* New York: Columbia University Press.

Sonneborn, T. M. 1957. Breeding systems, reproductive methods, and species problems in protozoa, in Mayr (1957), pp. 155–324.

———— 1975. The *Paramecium aurelia* complex of fourteen sibling species. *Trans. Amer. Micr. Soc.* 94: 155–178.

———— and M. V. Schneller. 1979. A genetic system for alternative stable characteristics in genomically identical homozygous clones. *Developmental Genetics* 1: 21–46.

Spencer, Herbert. 1862–1896. *A System of Synthetic Philosophy.* 10 vols. I. *First Principles.* 1862. London: Williams & Norgate.

—— 1868. *Essays: Scientific, Political, and Speculative.* London.

Spiess, H. (ed.). 1962. *Papers on Animal Population Genetics.* Boston: Little, Brown & Co.

Sprague, T. A. 1950. The evolution of botanical taxonomy from Theophrastus to Linnaeus, in *Lectures on the Development of Taxonomy.* London: Linnean Society, pp. 1–23.

Stadler, L. J. 1954. The gene. *Science* 120: 811–819.

Stafleu, F. A. 1963. Adanson and his "Familles des plantes," in Adanson, Carnegie Institute of Technology Monograph Series, no. 1. Pittsburgh, Penna.: Hunt Botanical Library, 1: 123–264.

—— 1969. Biosystematic Pathways Anno 1969. *Taxon* 18: 485–500.

—— 1971. *Linnaeus and the Linnaeans.* Utrecht: International Association for Plant Taxonomy.

Stanier, R. Y., and C. B. van Niel. 1942. The concept of a bacterium. *Arch. Microbiol.* 42: 17–35.

——, M. Doudoroff, and E. A. Adelberg. 1970. *The Microbial World* 3rd ed. Englewood Cliffs, N.J.: Prentice-Hall.

Stanley, S. M. 1979. *Macroevolution: Pattern and Process.* San Francisco: W. H. Freeman.

Stannard, J. 1979. Natural history, in Lindberg, D. C. (ed.), *Science in the Middle Ages.* Chicago: University of Chicago Press.

Starck, D. 1962. *Der heutige Stand des Fetalisationsproblems.* Hamburg-Berlin: Parey.

Staudinger, H. 1920. Über Polymerisation. *Ber. Dtsch. Chem. Ges.* 53: 1073–85.

Stauffer, R. S. 1960. Ecology in the long manuscript version of Darwin's *Origin of Species* and Linnaeus' *Oeconomy of Nature. Proc. Amer. Phil. Soc.* 104: 235–241. (See also Darwin, C. [1976].)

Stearn, W. T. 1959. The background of Linnaeus's contributions to the nomenclature and methods of systematic biology. *Syst. Zool.* 8: 4–22.

—— 1962. *Three Prefaces on Linnaeus and Robert Brown.* Weinheim: J. Cramer.

—— 1971. Linnaean classification, nomenclature, and method, in Blunt, W. (ed.), *The Compleat Naturalist.* London, pp. 242–252.

Stebbins, G. Ledyard. 1950. *Variation and Evolution in Plants.* New York: Columbia University Press.

—— 1969. *The Basis of Progressive Evolution.* Chapel Hill: University of North Carolina Press.

—— 1971. *Chromosomal Evolution in Higher Plants.* Reading, Mass.: Addison-Wesley Publishing Co.

—— 1974. *Flowering Plants: Evolution above the Species Level.* Cambridge: Harvard University Press.

—— 1977. In defense of evolution: tautology or theory? *Am. Nat.* 111: 386–394.

—— 1979. Fifty years of plant evolution, in Solbrig (1979), pp. 18–41.

Stebbins, R. E. 1974. France, in Glick (1974), pp. 117–167.

945

Stein, E. 1950. Dem Gedächtniss von Carl Erich Correns. *Naturwiss.* 37: 457–463.

Stent, G. 1972. Prematurity and uniqueness in scientific discovery. *Sci. Amer.* 227: 84–93.

Stern, C. 1968. Mendel and human genetics, in *Genetic Mosaics.* Cambridge: Harvard University Press, pp. 1–26.

―――― and Eva R. Sherwood (eds.). 1966. *The Origin of Genetics: A Mendel Source Book.* San Francisco: W. H. Freeman.

――――, and E. Stern. 1978. A note on the "three discoveries" of Mendelism. *Folia Mendeliana* 13: 237–240.

Storer, N. W. 1973. Introduction to Merton (1973). Chicago: University of Chicago Press.

Strasburger, E. 1884. *Neue Untersuchungen über den Befruchtungsvorgang bei den Phanerogamen, als Grundlage für eine Theorie der Zeugung.* Jena: Gustav Fischer.

―――― 1909. The minute structure of cells in relation to heredity, in Seward (1909), pp. 102–111.

Stresemann, E. 1919. Über die europäischen Baumläufer. *Verh. Orn. Ges. Bayern* 14: 39–74.

―――― 1975. *Ornithology: From Aristotle to the Present.* Cambridge: Harvard University Press.

Strickland, H. E. 1840. Observations upon the affinities and analogies of organized beings. *Ann. Mag.Nat. Hist.* 4: 219–226.

―――― 1846. On the structural relations of organized beings. *Phil. Mag.* 28: 354–364.

Stubbe, H. 1965. *Kurze Geschichte der Genetik bis zur Wiederentdeckung der Vererbungsregeln Gregor Mendels.* 2nd ed. Jena: Gustav Fischer. (English trans. 1973. Cambridge: M.I.T. Press.)

―――― 1966. *Genetik und Zytologie von Antirrhinum.* Jena: Gustav Fischer.

Studnicka, F. K. 1931. Aus der Vorgeschichte der Zellentheorie. *Anat. Anz.* 73: 390–416.

Sturtevant, A. H. 1913. The linear arrangement of six sex-linked factors in *Drosophila,* as shown by their mode of association. *J. Exp. Zool.* 14: 43–59.

―――― 1939. On the subdivision of the genus *Drosophila. Proc. Nat. Acad. Sci.* 25: 137–141.

―――― 1942. *The Classification of the Genus Drosophila.* University of Texas Publ. no. 4213.

―――― 1944. Book review: *Drosophila pseudoobscura. Ecology* 25: 476.

―――― 1959. Thomas Hunt Morgan, 1866–1945. *Biogr. Mem. Nat. Acad. Sci.* 33: 283–325.

―――― 1961. *Genetics and Evolution: Selected Papers of A. H. Sturtevant* (ed. E. B. Lewis). San Francisco: W. H. Freeman.

―――― 1965a. The early Mendelians. *Proc. Amer. Phil. Soc.* 109: 199–204.

―――― 1965b. *A History of Genetics.* New York: McGraw-Hill.

Sulloway, Frank J. 1970. Geographic isolation in Darwin's thinking: a developmental study of the growth of an idea. (Unpublished paper, Harvard.)

———— 1979. Geographic isolation in Darwin's thinking: the vicissitudes of a crucial idea. *Stud. Hist. Biol.* 3: 23–65.

———— (1982). The *Beagle's* collection of Darwin's finches (Geospizae). *Bull. Brit. Mus. (Nat. Hist.) Zool. Ser.*

———— (In prep.). Darwin's genius.

———— (ms). *The roots of intellectual rebellion: family constellations and scientific revolutions.*

Sumner, F. 1932. Genetic, distributional, and evolutionary studies of the subspecies of deer-mice *(Peromyscus). Bibl. genetica.* 9: 1–106.

Suppes, P. 1969. *Studies in the Methodology and Foundations of Science.* Dordrecht: D. Reidel.

Svenson, H. K. 1945. On the descriptive method of Linnaeus. *Rhodora* 47: 273–302, 363–388.

———— 1953. Linnaeus and the species problem. *Taxon* 2: 55–58.

Swanson, Carl P. 1957. *Cytology and Cytogenetics.* Englewood Cliffs, N.J.: Prentice-Hall.

Swinburne, R. G. 1965. Galton's Law—formulation and development. *Ann. Sci.* 21: 15–31.

Szalay, F. S. 1977. Ancestors, descendants, sister groups, and testing of phylogenetic hypotheses. *Syst. Zool.* 26: 12–18.

Szyfman, L. 1977. Lamarck's later species concept. *Bull. Biol. France* 777: 209–229.

Takeuchi, H., S. Uyeda, and H. Kanamori. 1970. *Debate about the Earth.* San Francisco: W. H. Freeman.

Taton, René. 1958. *La science moderne (de 1450 a 1800).* Paris: Presses Universitaires de France.

———— 1964. *The Beginnings of Modern Science, from 1450 to 1800.* London: Thames & Hudson.

Taylor, J. H. (ed.). 1965. *Selected Papers on Molecular Genetics.* New York: Academic Press.

Tax, Sol (ed.). 1960. *Evolution after Darwin.* I. *The Evolution of Life: Its Origin, History, and Future.* Chicago: University of Chicago Press.

Temkin, O. 1959. The idea of descent in post-Romantic German biology: 1848–1858, in Glass, Temkin, and Strauss (1959), pp. 323–355.

———— 1963. Basic science, medicine, and the romantic era. *Bull. Hist. Med.* 37: 97–129.

Templeton, Alan R. 1980. The theory of speciation via the founder principle. *Genetics* 94: 1011–38.

Thoday, J. M. 1966. Mendel's work as an introduction to genetics. *Advancement Sci.* 23: 120–134.

———— 1975. Non-Darwinian "evolution" and biological progress. *Nature* 255: 675–677.

947

Thompson, d'Arcy. 1917. *On Growth and Form.* Cambridge: Cambridge University Press.

Thompson, J. N., Jr., and J. M. Thoday. 1979. *Quantitative Genetic Variation.* New York: Academic Press.

Thorndike, L. 1945. *The Herbal of Rufinus.* Chicago: University of Chicago Press.

―――― 1958–1960. *A History of Magic and Experimental Science.* 8 vols. New York: Columbia University Press.

Thorne, R. F. 1973. The "Amentiferac" or Hamamelidae as an artificial group. *Brittonia* 25: 395–405.

Thorpe, W. H. 1930. Biological races in insects and allied groups. *Biol. Rev.* 5: 177.

―――― 1940. Ecology and the future of systematics, in Huxley, J. S. (1940), pp. 341–364.

Throckmorton, L. H. 1968. Concordance and discordance of taxonomic characters in *Drosophila* classification. *Syst. Zool.* 17: 355–387.

―――― 1978. Molecular phylogenies, in Romberger, J. A. , R. H. Foote, L. Knutson, and P. L. Lentz (eds.), *Beltsville Symposia in Agricultural Research 2. Biosystematics in Agriculture.* Montclair, N.J.: Allanheld, Osmun and Co.; New York: John Wiley & Sons.

Timofeeff-Ressovsky, H. H., and N. W. Timofeeff-Ressovsky. 1927. Genetische Analyse einer freilebenden Drosophila melanogaster Population. *Roux Arch. Entw. Mech.* 109: 70.

Timofeeff-Ressovsky, N. W. 1925. Studies on the phenotypic manifestation of hereditary factors. *Zurn. Eksp. Biol.* 1.

―――― , K. G. Zimmer, and Max Delbrück. 1935. Über die Natur der Genmutation und der Genstruktur. *Nachr. Ges. Wiss. Göttingen, Math.-Phys. Kl.* 6: 189–245.

Toulmin, S. 1972. *Human Understanding.* Princeton: Princeton University Press.

―――― , and J. Goodfield. 1965. *The Discovery of Time.* New York: Harper & Row.

Tristram, H. B. 1859. On the ornithology of Northern Africa (Sahara). *Ibis* (October): 429–433. (See also I. B. Cohen [in press, 1982].)

Trivers, Robert. 1972. Parental investment and sexual selection, in Campbell (1972), pp. 136–179.

Troeltsch, E. 1922. *Der Historismus und seine Probleme.* Tübingen: Mohr.

Troll, W. (ed.). 1926. *Goethe's Morphologische Schriften.* Jena: Eugen Diederichs.

Tschulok, S. 1922. *Deszendenzlehre (Entwicklungslehre): Ein Lehrbuch auf historischkritischer Grundlage.* Jena: Gustav Fischer.

―――― 1937. *Lamarck, eine kritischhistorische Studie.* Zürich and Leipzig: Max Nichan.

Turesson, G. 1922. The genotypic response of the plant species to the habitat. *Hereditas* 3: 211–350.

Turner, John R. G. 1971. Studies of Müllerian mimicry and its evolution

in burnet moths and heliconid butterflies, in Creed, R. (ed.), *Ecological Genetics and Evolution*. Oxford: Blackwell, pp. 224–260.

——— 1977. Butterfly mimicry: the genetical evolution of adaptation. *Evol. Biol.* 10: 163–206.

——— 1978. Why male butterflies are non-mimetic: natural selection, sexual selection, group selection, modification, and sieving. *Biol. J. Linn. Soc.* 10: 385–432.

———, M. S. Johnson, and W. F. Eames. 1979. Contrasted modes of evolution in the same genome: allozymes and adaptive change in *Heliconius*. *Proc. Nat. Acad. Sci.* 76: 1924–28.

Tuxen, S. L. 1973. Entomology systematizes and describes: 1700–1815, in Smith, R. F., T. E. Mittler, and C. N. Smith, *History of Entomology*. Palo Alto: Annual Reviews, pp. 95–118.

Udvardy, M. D. F. 1969. *Dynamic Zoogeography*. New York: Van Nostrand Reinhold.

Uhlmann, E. 1923. Entwicklungsgedanke und Artbegriff. *Jena. Z. Naturwiss.* 59: 1–116.

Ulrich, W. 1972. Die Geschichte des Archicoelomatenbegriffs und die Archicoelomatennatur der Pogonophoren. *Z. f. zool. Syst. Evol.-forsch.* 10: 301–320.

Unger, F. 1852. *Versuch einer Geschichte der Pflanzenwelt*. Wien: Braumüller.

Ungerer, E. 1966. *Die Wissenschaft vom Leben*, vol. 3. Freiburg and München: Karl Alber.

Uyeda, S. 1978. *The New Crew of the Earth*. San Francisco: W. H. Freeman.

Vandel, A. 1968. *La genèse du vivant*. Paris: Masson.

Van Valen, L. 1973. A new evolutionary law. *Evol. Theory* 1: 1–30.

——— 1976. Ecological species, multispecies, and oaks. *Taxon* 25: 233–239.

Vartanian, Aram. 1950. Trembley's polyp, La Mettrie, and eighteenth-century French materialism. *J. Hist. Ideas* 11: 259–286.

——— 1953. *Diderot and Descartes*. Princeton: Princeton University Press.

Vendrely, R., and C. Vendrely, 1949. La teneur de noyau cellulaire en ácide désoyribonucleique à travers les organes, les individus et les espèces animales. *Experientia* 5: 327–329.

Vent, W. (ed.). 1974. *Widerspiegelung der Binnenstruktur und Dynamik der Art in der Botanik*. Berlin: Akademie-Verlag.

Verlot, B. 1864. Mémoire sur la production et la fixation des variétés dans les plantes d'ornement. *J. soc. hort. Paris* 10.

Vico, Giambattista. 1725. *Scienzia Nuova* (English trans. T. G. Bergin and M. H. Fish. 1968. *The New Science of Giambattista Vico*. Ithaca: Cornell University Press.)

Virchow, R. 1858. *Die Cellularpathologie in ihrer Begründung auf physiologische und pathologische Gewebelehre*. (English trans. 1971. *Cellular Pathology*. New York: Dover.)

949

Voeller, Bruce (ed.). 1968. *The Chromosome Theory of Inheritance: Classic Papers in Development and Heredity.* New York: Appleton-Century-Crofts.

Vorzimmer, P. 1963. Charles Darwin and blending inheritance. *Isis* 54: 371-390.

———— 1965. Darwin's ecology and its influence upon his theory. *Isis* 56: 148–156.

———— 1969. Darwin, Malthus, and the theory of natural selection. *J. Hist. Ideas* 30: 527–542.

———— 1970. *Charles Darwin, The Years of Controversy: The Origin of Species and Its Critics, 1859–1882.* Philadelphia: Temple University Press.

———— 1975. An early Darwin manuscript: the "outline and draft of 1839." *J. Hist. Biol.* 8: 191–217.

Voss, E. G. 1952. The history of keys and phylogenetic trees in systematic biology. *J. Sci. Labs. Denison University* 43: 1–25.

Waddington, C. H. 1957. *The Strategy of the Genes.* London: Allen & Unwin.

———— (ed.). 1968–1972. *Towards a Theoretical Biology.* 4 vols. Edinburgh: Edinburgh University Press.

Wagner, Moritz. 1841. *Reisen in der Regentschaft Algier in den Jahren 1836, 1837, and 1838.* Leipzig: Leopold Voss.

———— 1889. *Die Entstehung der Arten durch räumliche Sonderung.* Basel: Benno Schwalbe.

Wagner, W. H. 1970. Biosystematics and evolutionary noise. *Taxon* 19: 146–151.

Wallace, A. R. 1853. *A Narrative of Travels on the Amazon and Rio Negro.* London: Reeve & Co.

———— 1855. On the law which has regulated the introduction of new species. *The Annals and Magazine of Natural History* ser. 2, 16: 184–196. (Rpt. in McKinney [1971].)

———— 1858. On the tendency of varieties to depart indefinitely from the original type. *J. Proc. Linn. Soc. (Zoology)* 3: 53–62.

———— 1866. On the phenomena of variation and geographical distribution as illustrated by the *Papilionidae* of the Malayan region. *Trans. Linn. Soc. London* 25: 1–72.

———— 1870. *Contributions to the Theory of Natural Selection.* London: Macmillan.

———— 1876. *The Geographical Distribution of Animals.* 2 vols. London: Macmillan.

———— 1880. *Island Life.* London: Macmillan.

———— 1891. *Natural Selection and Tropical Nature.* London: Macmillan.

———— 1905. *My Life: A Record of Events and Opinions.* 2 vols. New York: Dodd, Mead & Co.

Walsh, B. D. 1864. On phytophagic varieties and phytophagic species, with remarks on the unity of coloration in insects. *Proc. Ent. Soc. Philadelphia* 5: 194–215.

Warburton, F. E. 1967. The purposes of classification. *Syst. Zool.* 26: 241–245.

Warming, J. E. B. 1896. *Lehrbuch der ökologischen Pflanzengeographie*. Berlin.

Watson, J. D. 1968. *The Double Helix*. New York: Atheneum.

Watson, J. D. , and F. H. C. Crick. 1953a. Molecular structure of nucleic acids: a structure for deoxyribose nucleic acid. *Nature* 171: 737–738.

—— 1953b. Genetical implications of the structure of deoxyribonucleic acid. *Nature* 171: 964–967.

Wegener, A. 1915. *Die Entstehung der Kontinente und Ozeane*. Braunschweig: Vieweg.

Weidenreich, F. 1929. Vererbungsexperiment und vergleichende Morphologie. *Paläont. Zeitschr.* 11: 275–286.

Weiling, F. 1966. Hat J. G. Mendel bei seinen Versuchen "zu genau" gearbeitet? *Der Züchter* 36, H.8.

Weinstein, A. 1962. The reception of Mendel's paper by his contemporaries. *Proc. 10th Cong. Hist. Sci.* 17: 492–497.

Weismann, August. 1868. *Über die Berechtigung der Darwinschen Theorie*. Leipzig: Engelmann.

—— 1872. *Über den Einfluss der Isolirung auf die Artbildung*. Leipzig: Engelmann.

—— 1883. *Über die Vererbung*. Jena: Gustav Fischer.

—— 1885. *Die Kontinuität des Keimplasmas als Grundlage einer Theorie der Vererbung*. Jena: Gustav Fischer.

—— 1886. *Die Bedeutung der sexuellen Fortpflanzung für die Selektionstheorie*. Jena: Gustav Fischer.

—— 1889. *Essays upon Heredity*. Oxford: Clarendon Press. (2nd ed. 1892.)

—— 1892. *Das Keimplasma: Eine Theorie der Vererbung*. Jena: Gustav Fischer. (English ed., 1893.)

—— 1893. *Die Allmacht der Naturzüchtung*. Jena: Gustav Fischer.

—— 1895. *Neue Gedanken zur Vererbungsfrage*. Jena: Gustav Fischer.

—— 1896. *Über Germinal Selektion: Eine Quelle bestimmt gerichteter Variation*. Jena: Gustav Fischer.

—— 1899. Thatsachen und Auslegungen in Bezug auf Regeneration. *Anat. Anz.* 15: 1–31.

—— 1909. The selection theory, in Seward, (1909) pp. 18–65.

Weisskopf, V. 1977. The frontiers and limits of science. *Amer. Sci.* 65: 405–411.

Wells, George A. 1967. Goethe and evolution. *J. Hist. Ideas* 28: 537–550.

Wells, K. D. 1971. Sir William Lawrence (1783–1867): a study of pre-Darwinian ideas on heredity and variation. *J. Hist. Biol.* 4: 319–361.

—— 1974. The historical context of natural selection: the case of Patrick Matthew. *J. Hist. Biol.* 6: 225–258.

Wells, William Charles. 1818. *An Account of a Female of the White Race of*

951

Mankind, Part of Whose Skin Resembles That of a Negro, with Some Observations on the Cause of the Differences in Colour and Form between the White and Negro Races of Man. In Appendix to Two Essays: One upon a Single Vision with Two Eyes, The Other on Dew. London: Archibald Constable.

Werner, B. 1975. Bau und Lebensgeschichte des Polypen von *Tripedalia cytophora* (Cubozoa, class. nov., Carybdeidae) und seine Bedeutung für die Evolution der *Cnidaria. Helgol. wiss. Meeresunters.* 27: 461–504.

West-Eberhard, Mary Jane. 1979. Sexual selection, social competition, and evolution. *Proc. Amer. Phil. Soc.* 123: 222–235.

Wettstein, F. v. 1926. Über plasmatische Vererbung. *Nachr. Ges. Wiss. Göttingen Math.-Phys. Kl.* 250–281.

—— 1940. Experimentelle Untersuchungen zum Artbildungs-problem. II. Zur Frage der Polyploidie als Artbildungs-faktor. *Ber. Dtsch. Bot. Ges.* 58: 374–388.

Wettstein, R. 1898. *Grundzüge der geographisch-morphologischen Methode in der Pflanzensystematik.* Jena: Gustav Fischer.

White, Leslie A. 1959. *The Evolution of Culture.* New York: McGraw-Hill.

White, Lynn. 1967. The historical roots of the ecological crisis. *Science* 255: 1203–17.

White, M. 1963. The logic of historical narration, in Hook (1963), p. 4.

White, M. J. D. 1973. *Animal Cytology and Evolution* 3rd ed. London: Cambridge University Press.

—— 1978. *Modes of Speciation.* San Francisco: W. H. Freeman.

Whitehouse, H. L. K. 1965. *Towards an Understanding of the Mechanism of Heredity.* London: Edward Arnold.

Whitman, Charles Otis. 1919. *Posthumous Works of Charles Otis Whitman* (ed. Harvey A. Carr). I. *Orthogenetic Evolution in Pigeons.* Washington, D.C.: Carnegie Institution.

Whyte, L. L., A. Wilson, and D. Wilson. (eds.). 1969. *Hierarchical Structures.* New York: Elsevier.

Wickler, W. 1968. *Mimicry in Plants and Animals.* London: Weidenfeld & Nicolson.

Wiesner, J. 1892. *Die Elementarstructur und das Wachstum der lebenden Substanz.* Wien.

Wiley, E. O. 1978. The evolutionary species concept reconsidered. *Syst. Zool.* 27: 17–26.

Wilkie, J. S. 1959. Buffon, Lamarck, and Darwin: the originality of Darwin's theory of evolution, in Bell (1959), pp. 262–307.

Williams, G. C. 1966. *Adaptation and Natural Selection.* Princeton: Princeton University Press.

—— 1975. *Sex and Evolution.* Princeton: Princeton University Press.

Williams, Mary B. 1973a. Falsifiable predictions of evolutionary theory. *Phil. Sci.* 40: 518–537.

—— 1973b. The logical status of the theory of natural selection and other evolutionary controversies, in Bunge, M. (ed.), *The Methodological Unity of Science.* Dordrecht: D. Reidel, pp. 84–102.

Willis, J. C. 1922. *Age and Area.* Cambridge: Cambridge University Press.

———— 1940. *The Course of Evolution by Differentiation or Divergent Mutation Rather than by Selection.* Cambridge: Cambridge University Press.

Wilson, A. C., S. S. Carlson, and T. J. White. 1977. Biochemical evolution. *Ann. Rev. Biochem.* 46: 573–639.

Wilson, A. C., V. M. Sarich, and L. R. Maxon. 1974. The importance of gene arrangement in evolution: evidence from studies on rates of chromosomal, protein, and anatomical evolution. *Proc. Nat. Acad. Sci.* 71: 3028–30.

Wilson, E. B. 1896. *The Cell in Development and Inheritance.* New York: Macmillan.

———— 1925. *The Cell in Development and Heredity.* 3rd ed. New York: Macmillan.

Wilson, E. O. 1975. *Sociobiology.* Cambridge: Harvard University Press.

Wilson, J. T. (ed.). 1976. *Continents Adrift and Continents Aground.* San Francisco: W. H. Freeman.

Wilson, J. W. 1944. Cellular tissue and the dawn of the cell theory. *Isis* 35: 168–173.

Wilson, Leonard G. 1972. *Charles Lyell, The Years to 1841: The Revolution in Geology.* New Haven: Yale University Press.

———— 1980. Geology on the eve of Charles Lyell's first visit to America, 1841. *Proc. Amer. Phil. Soc.* 124: 168–202.

———— (ed.). 1970. *Sir Charles Lyell's Scientific Journals on the Species Question.* New Haven: Yale University Press.

Wilson, R. J. 1967. *Darwinism and the American Intellectual: A Book of Readings.* Homewood, Ill.: Dorsey Press.

Wimsatt, W. C. 1972. Teleology and the logical structure of function statements. *Stud. Hist. Phil. Sci.* 3: 1–80.

Winkler, H. 1924. Über die Rolle von Kern und Protoplasma bei der Vererbung. *Z. ind. Abst. Vererb.* 33: 238–253.

Winsor, Mary P. 1969. Barnacle larvae in the nineteenth century: a case study in taxonomic theory. *J. Hist. Med. Allied Sci.* 24: 294–309.

———— 1976a. The development of Linnaean insect classification. *Taxon* 25: 57–67.

———— 1976b. *Starfish, Jellyfish, and the Order of Life.* New Haven: Yale University Press.

Woese, Carl R., and George E. Fox. 1977. Phylogenetic structure of the prokaryotic domain: the primary kingdoms. *Proc. Nat. Acad. Sci.* 74: 5088–90.

Wollaston, T. V. 1860. Review of the *Origin of Species. Ann. Mag. Nat. Hist.* 5: 132-143.

Wood, R. J. 1973. Robert Bakewell (1725–1795), pioneer animal breeder and his influence on Charles Darwin. *Casopis Morav. Musea* 58: 231–242 (*Folia Mendel.* 8).

Woodger, J. H. 1929. *Biological Principles: A Critical Study.* London: Routledge & Kegan Paul.

953

Wright, Sewall. 1930. The genetical theory of natural selection: a review. *J. Hered.* 21: 349–356.

—— 1931a. Evolution in Mendelian populations. *Genetics* 16: 97–159.

—— 1931b. Statistical theory of evolution. *Amer. Stat. J.* March supp., pp. 201–208.

—— 1932. The roles of mutation, inbreeding, crossbreeding, and selection in evolution. *Proc. 6th Int. Cong. Genetics, Ithaca* 1: 356–366.

—— 1967. Comments on the preliminary working papers of Eden and Waddington, in Moorhead and Kaplan (1967), pp. 117–120.

—— 1978. *Evolution and the Genetics of Populations.* IV. *Variability within and among Natural Populations.* Chicago and London: University of Chicago Press.

—— and Th. Dobzhansky. 1946. Genetics of natural populations. XII. *Genetics* 31: 125–156.

Wyatt, H. V. 1974. How history has blended. *Nature* 249: 803–805.

Wynne-Edwards, V. C. 1962. *Animal Dispersion in Relation to Social Behaviour.* Edinburgh and London: Oliver & Boyd.

Young, R. M. 1969. Malthus and the evolutionists: the common context of biological and social theory. *Past and Present* 43: 109–141.

—— 1971. Darwin's metaphor: does nature select? *Monist* 55: 442–503.

Yule, G. U. 1902. Mendel's laws and their probable relations to intraracial heredity. *New Phytologist* 1: 194–238.

Zangerl, Rainer. 1948a. The methods of comparative anatomy and its contribution to the study of evolution. *Evolution* 2: 351–374.

—— 1948b. The vertebrate fauna of the Selma formation of Alabama. II. The Pleurodiran turtles. *Field. Geol. Mem.* 3: 23–56.

Zimmermann, E. A. W. 1778–1783. *Geographische Geschichte des Menschen und der allgemein verbreiteten Tiere,* vols. 1–3. Leipzig.

Zimmermann, W. 1935. Rassen- und Artbildung bei Wildpflanzen. *Forsch. u. Fortschr.* 11: 272–274.

—— 1938. *Vererbung "Erworbener Eigenschaften" und Auslese.* Jena: Gustav Fischer.

—— 1953. *Evolution: Die Geschichte ihrer Probleme und Erkenntnisse.* Freiburg: Karl Alber.

Zirkle, C. 1935. *The Beginnings of Plant Hybridization.* Philadelphia: University of Pennsylvania Press.

—— 1941. Natural selection before the "Origin of Species." *Proc. Amer. Phil. Soc.* 84: 71–123.

—— 1946. The discovery of sex-influenced, sex-limited, and sex-linked heredity, in *Studies and Essays . . . in Honor of George Sarton,* pp. 169–194.

—— 1946. The early history of the idea of the inheritance of acquired characters and of pangenesis. *Trans. Amer. Phil. Soc.* N.S. 35: 91–151.

—— 1951a. Gregor Mendel and his precursors. *Isis* 42: 97–104.

—— 1951b. The knowledge of heredity before 1900, in Dunn (1951), pp. 35–57.

—— 1959. Species before Darwin. *Proc. Amer. Phil. Soc.* 103: 636–644.

—— 1968. The role of Liberty Hyde Bailey and Hugo de Vries in the rediscovery of Mendelism. *J. Hist. Biol.* 1: 205–218.

Zirnstein, G. 1979. Die Hauptaspekte von Lamarck's Evolutionshypothese und die Biologie von 1859. *Biol. Rdsch.* 17: 345–366.

Zittel, K. A. v. 1899. *Geschichte der Geologie und Paläontologie bis Ende des 19. Jahrhunderts.* München und Leipzig: Oldenbourg.

Zuckerkandl, E. 1975. The appearance of new structures and functions in proteins during evolution. *J. Mol. Evol.* 7: 1–57.

—— and L. Pauling. 1962. In Kasha, M., and B. Pullman (eds.), *Horizons in Biochemistry.* New York: Academic Press, pp. 189–225.

Glossary

For a more detailed glossary of terms relating to systematics, see Mayr, 1969; for evolutionary biology, see Mayr, 1970. Biological terms defined in the text are omitted from the Glossary (see index).

Allele. One of several alternate forms of a gene occupying the same chromosomal locus.

Allotetraploid. An individual or species resulting from the doubling of the chromosomes of a species hybrid.

Angiosperms. Flowering plants.

Apomixis. In plants, asexual reproduction, corresponding to parthenogenesis in animals.

Asexual reproduction. Any form of propagation not resulting from zygote formation (the fusion of two gametes).

Autosome(s). Any chromosome that is not a sex chromosome.

Biota. Fauna and flora of a region.

Chiasma. Place where, during meiosis, two homologous chromosomes establish close contact and where usually an exchange of homologous parts between nonsister chromatids takes place.

Chromatids. Two longitudinal units of a chromosome resulting from a split in early prophase, becoming daughter chromosomes later in mitosis.

Chromatin. The stainable material in the nucleus, now known to consist of DNA.

Chromosome. Discrete longitudinal bodies in the nucleus into which the genetic material is organized.

Cistron. The gene of function; the functional unit of inheritance.

Coelom. Body cavities bordered by mesoderm.

Crossing over. The reciprocal exchange of homologous parts between nonsister chromatids.

Cryptogams. Plants that are not seed bearing plants, as ferns, mosses, fungi.

Cytoplasm. Part of cell outside the nucleus.

Dendrogram. A tree-like diagram of relationship.

Diakinesis. Stage in meiosis, at end of prophase, during which chromosomes are strongly condensed and chiasmata are particularly well visible.

Dominant. Allele which in a heterozygote determines the phenotype.

Ecotype. A local population of plants selected for the edaphic and biotic conditions of its habitat, and expressing this in its phenotype.

Endosperm. Tissue nourishing the embryo in seed plants.

Epigenesis. Origin during ontogeny of structures from undifferentiated material.

Epistasis. Interaction between nonallelic genes.

Fauna. The animal life of a region.

Gamete. Reproductive cell, i.e., egg in female and spermatozoon in male.

Gametophyte. Haploid phase of life cycle of plants.

Gemmules. Hypothetical invisibly small carriers of genetic attributes.

Genotype. The total genetic constitution of an organism.

Herbal. An illustrated book in which plants, particularly of medicinal use, are named and described.

Heterozygote. Individual having different alleles at the same locus in the two homologous chromosomes.

Homozygote. Individual having the same allele at the corresponding loci of two homologous chromosomes.

Idioplasm. Nägeli's term for the genetic material.

Infusorians. Obsolete term for small aquatic organisms (mostly protozoans, rotifers, one-celled algae); primarily used for protozoans.

Inheritance, blending. The complete fusing of the paternal and maternal genetic materials.

Inheritance, multifactorial. Control of a character by several genes (polygeny).

Inheritance, particulate. The nonfusion of the parental genetic material during zygote formation.

Isolating Mechanism. Biological properties of individuals that prevent the interbreeding of sympatric populations.

Linkage. The association of certain genes owing to their location on the same chromosome.

Macrogenesis. Evolution by discontinuous change; saltational evolution.

Mastodon. An extinct relative of the elephant.

Meiosis. The two successive divisions of the nucleus preceding the formation of gametes.

Mesozoic. The geological era that lasted from about 225 million years to 65 million years; the age of reptiles.

Mitosis. The division of the nucleus.

Monophyletic. Of a taxon, the members of which are descendants of the nearest common ancestor.

Mutation. Discontinuous change in chromosomal DNA, ordinarily an error in DNA replication.

Neo-Darwinism (Romanes, 1896). Darwin's theory of evolution, but rejecting any inheritance of acquired characters.

Niche. The multidimensional resource space of a species; its ecological requirements.

Nondisjunction. Failure of the two homologous chromosomes of a pair to go to opposite poles at the first meiotic division; as a result one daughter cell has both chromosomes and the other neither.

Nuclein. The name given by Miescher to the phosphorus-rich compound in the nuclei.

Orthogenesis. Hypothesis that a rectilinear trend in evolution is due to a built-in finalistic principle.

Pachytene. State during the prophase of meiosis during which the homologous chromosomes are completely paired.

Pangenesis. Hypothesis that all parts of the body contribute genetic material to the reproductive organs, and particularly to the gametes.

Parapatric. Referring to two species having contiguous geographic ranges but no (or only minimal) interbreeding in the zone of contact.

Parthenogenesis. Development of an egg without fertilization.

Phage. Bacterial virus.

Phenotype. The totality of characteristics of an individual.

Plankton. Small organisms (animals and plants) that float in the water; particularly algae and crustaceans.

Plate tectonics. Geological theory according to which the earth's crust consists of moving continental plates.

Pleiotropic. A gene that affects several characteristics of the phenotype.

Polygeny. The determination of a phenotypic character by several genes.

Polyploid. Having more than two sets of haploid chromosomes.

Position effect. A change in the phenotypic effect of a gene owing to a change of its position on the chromosome.

Preformation. The theory that all the structures of an organism are present in one of the gametes.

Proboscideans. Relatives of the elephant, including the extinct mammoth and mastodon.

Prokaryotes. Primitive organisms (bacteria and bluegreen algae) without a nucleus and with its nucleic acid organized into a single string.

Pure line. A genetically uniform (i.e., homozygous) population.

Recessive. Allele which in a heterozygote is not expressed in the phenotype.

Reduction division. One of the two meiotic divisions, usually the first one, in which the number of chromosomes is halved.

Semidominance. Intermediacy of the phenotype of the heterozygote between the phenotypes of the two homozygotes.

Sibling species. Reproductively isolated but morphologically identical or nearly identical species.

Soft inheritance. Inheritance during which the genetic material is not constant from generation to generation but may be modified by the effects of the environment, by use or disuse, or other factors.

Spontaneous generation. The spontaneous origin of life from inanimate matter.

Sporophyte. Diploid phase in life cycle of plants.

959

Sympatric. Coexisting at the same locality.

Synapsis. The pairing of homologous chromosomes during the first division of meiosis.

Taxon. A taxonomic group, at any categorical level.

Terebratulas. An extinct group of brachiopods (invertebrates).

Tertiary. The most recent of the major geological eras, extending from about 65 million years ago to the Recent.

Weighting. Assigning a value to a taxonomic character.

Xenia. Effect of pollen on the characters of the endosperm.

Zygote. The cell produced by the union of two gametes and their nuclei.

Index

971

973